Mechanics of Accuracy in Engineering Design of Machines and Robots

Volume I: Nominal Functioning and Geometric Accuracy

Vladimir T. PORTMAN

Ben-Gurion University of the Negev, Beer Sheva, Israel

ASME PRESS

ASME Press books are available at special quantity discounts to use as premiums or for use in corporate training programs. For more information, contact Special Sales at CustomerCare@asme.org

Library of Congress Cataloging-in-Publication Data

Names: Portman, V. T. (Vladimir Teodorovich) author.
Title: Mechanics of accuracy in engineering design of machines and robots / Vladimir T. Portman.
Description: New York : ASME Press, 2018- | Includes bibliographical references and index. Contents: Volume I. Nominal functioning and geometric accuracy.
Identifiers: LCCN 2017049008 | ISBN 9780791861615
Subjects: LCSH: Machinery--Design and construction. | Robots--Design and construction.
Classification: LCC TJ230 .P625 2018 | DDC 620/.0042--dc23
LC record available at https://lccn.loc.gov/2017049008

Series Editor's Preface

Dear Reader,

The ASME Press Robotics Engineering Book Series continues to present research and engineering accomplishments in the key areas of robotics and mechanisms. This time, I am delighted to introduce a new book by Professor Vladimir T. Portman - *Mechanics of Accuracy in Engineering Design of Machines and Robots*.

Professor Portman has been a longtime collaborator of ASME Press. He is a member of the Advisory Group at the ASME Robotics Engineering Book Series. His first book on *Accuracy of Machine Tools* was published by ASME Press in 1988.

This new monograph offers a comprehensive study on accuracy as an essential problem in engineering design of machines and robots. The book presents a detailed consideration and mathematical analysis of the whole range of technical issues related to accuracy, including accuracy models and methods of assessment, form-shaping functions of various manipulators, kinematic accuracy and error analysis, and many others. The subject-matter of the book is a necessary component of mechanical engineering education.

I believe this work by Professor Portman has laid a technical foundation for advanced engineering and manufacturing of future robotic systems and machines. The book can be equally useful for researchers and graduate students, university professors teaching engineering and technological courses, and engineers and technologists in industry.

The Robotics Engineering Book Series is committed to both assisting eminent scholars in publishing their findings and introducing outstanding new researchers to the world of books in Robotics. We look forward to close collaboration with researchers and engineers around the world.

Vladimir V. Vantsevich, ScD, PhD, ASME Fellow
Founding Editor, ASME Press Robotics Engineering Book Series

November 29, 2017

ASME Press Robotics Engineering Book Series

Advisory Group Members

Contents

Foreword

Accuracy is one of the fundamental characteristics and one of the most important indexes of the quality of machines and robots. It significantly defines and, in turn, depends on their structure and applications. Accuracy provision, maintenance, and enhancement are permanently hot problems in modern manufacturing and manufacturing science. In the very beginning, the accuracy was considered as a purely technological problem. Both academic and engineering approaches lay with development and standardization of the metrological requirements, interchangeability, and technical measurements. In the design-related disciplines, accuracy analysis was reduced to the tolerances and fit theory [1].

As the machine structures and machine kinematics have become more and more complicated, both from structural and informativeness viewpoints, the accurate manufacture of machine parts and their dimensional chains is not sufficient for providing machine accuracy, and there is a need to consider a machine as a whole subject in its own right [2, 3]. Error propagation has to be considered from the active element AE (cutting tool, measuring probe, working instrument fitted to the end-effector, etc.) throughout all machine structure up to elements to be processed – target element (TE) – taking into account not only geometry but also kinematics errors, static and dynamic deformations, and so on. This approach is of fundamental importance for industrial robots (IR), machine tools, and coordinate measuring machines (CMM), whereat the multi-coordinate structure with complicated kinematics is accompanied by heavy demands on accuracy and productivity.

General features set off machine tools, IRs, and CMMs in a special group – the precision working machines (PWM). Interrelationships between the manipulating system and the TE and AE of the PWMs are introduced by means of an integrated object called a *form-shaping system* (FSS). The form-shaping process is dedicated to producing and/or reproducing the form and shape[1] of the TE using two types of input information: set of kinematic motions and geometry of AE. The FSS of the PWM is featured by combination of high requirement for accuracy and productivity and complicated trajectories of the AE relative to TE. Hence, the PWM simultaneously provides both the general *form* of the TE and its accurate *shape*. For example, in the machine tools, the process results in predetermined transformation of the form and shape of the blank or semi-finished item into the designed product fixed by the drawing.

The features of the machine tool FSS are methodologically formalized by means of a mathematical model called the form-shaping function (FSF) derived in the book *Accuracy of Machine Tools* [4]. Here, the FSF-based approach is extended to all PWM and applied not only to serial-kinematics machines, but also to machines with parallel and hybrid kinematics. Mathematically, the FSF presents a product of the manipulation matrix, modelling a set of the system movements, by vector of the system's AE associated with setup-related

---

[1]According to *Longman Dictionary of Scientific Usage* [Goldman, A., Paynem E.M.F., Longman Group, Ltd, Harlow, 1979, p. 13], "the form is the general overall character of an object." On the other hand, "the form is generalization of shape, e.g., each horse has its individual shape, but there is a large number of possible shapes that have the form of a horse. A knowledge of the form of a horse allows us to distinguish it from the other animals, and a knowledge of the shape of a horse allows us to distinguish it from the other horses".

constraints among system parameters. In this connection, the FSF models the machine hardware and bridges the real data processing done at the hardware level with virtual procedures carried out by the computerized control systems.

The effective tool for machine motion modeling is the matrix technique for multistage coordinate transformations from the AE to TE, and vice versa. The technique allows unification of a large body of data relating to errors of different physical natures aimed at their summation and further comparison with allowable values of the performance indexes formulated at all stages of machine design. As a native widening, the same matrix technique is applied to studying infinitesimal displacements, and the fact of their smallness allows significant simplification of the PWM accuracy-related problems [4–6].

Despite the fact that the matrix technique was successfully long used for a wide spectrum of industrial problems in the theory of gearing [7], mechanism and machine theory [8–10], machine tools and manufacturing technology [4–5], and robotics [11–14], in the last few decades great progress has been made by commercialization of powerful computer instruments – general software systems for mathematical and other applications – carrying out both traditional calculations and non-trivial intellectual procedures such as symbolic computations, graphical presentation of results, etc. In this book, the Mathematica® software system [15] is systematically applied to overcome mathematical and technical difficulties in the considered problems.

In terms of Mathematica, the synthesis of the FSF is reduced to ordered multiplication of a set of matrices selected (with possible repetitions) from six elementary matrices modelling elementary motions of rigid bodies. As applied to the FSF of real machines and robots, the set is combined according to coordinate codes of the machine fixing the number, nomenclature, and order of kinematic motions. For accuracy-related problems, a standard matrix of geometric and kinematic errors is added to six abovementioned matrices. The final step of the machine functioning – a setup for a specified operation – is formalized through simultaneous equations of different complexity connecting system parameters. This approach is successfully applied to the synthesis of both linear and non-linear models for serial-, parallel-, and hybrid-kinematics machines.

The book comprises the Introduction and four parts including 19 chapters with consecutive numbering, and three appendices. Each part combines a theoretical chapter with applications to machines and robots with different kinematic types. The formalized consideration is accompanied with application examples, which, as a rule, are explained with numerical solutions using realistic initial values.

In the *Introduction* (Chapters 1 and 2), the general approach to accuracy problems in manufacturing, especially to the PWM (IRs, machine tools, CMMs, etc.) accuracy is described, and basic definitions in the field are entered.

In *Part I* "Model of Nominal Functioning" (Chapters 3–7), geometric and kinematic problems relating to nominal, i.e., without external and internal disturbances, performance of the PWM are considered. After general FSS and FSF definitions, they are applied to the PWM of different kinematics types. The consideration is finished by setup problems, in which the theoretical approach is illustrated by various examples: industrial robots for welding and painting, lathes, milling machines, grinding machines, gear- and thread-cutting machines, CMMs, etc.

Part II "Geometric and Kinematic Accuracy" (Chapters 8–12) describes geometric problems and mathematical models related to actual (with disturbances) performance of PWMs with serial- and parallel-type kinematics. In Chapters 8–11, the linearized mathematical model – a budget of geometric and kinematic errors – is formulated using variation of nominal FSF and the standard form of error matrices. The developed approach is illustrated by some practical examples: error budgets of robots, machine tools, and CMMs. In Chapter 12, non-linear problems are considered by definition of two FSFs describing a pair of geometrically-closed surfaces, and some real-life examples are solved.

Part III "Stiffness-Related Accuracy Problems" (Chapters 13–16) consists of formulation of the static problems and synthesis of the stiffness-compliance matrices. The synthesis uses the Jacobian matrix obtained through transformation of the error budget expression from *Part II*. To overcome the difficulties caused by heterogeneous units of measurement of the stiffness matrix entries, a convenient performance index – the collinear stiffness value (CSV) – is applied. The CSV is used for natural stiffness evaluation in both regular and singular configurations, and the minimum CSV is applied to construct a machine workspace satisfying preliminary given stiffness limits. In particular, the CSV allows computation of a virtual protective barrier, keeping the machine from approaching the vicinity of singular configurations of the parallel-kinematics mechanisms. As examples, the CSV-based local and global stiffness features of the machines are simulated, investigated, and visualized. An approach to synthesis of the stiffness matrices for the closed chains and net of links is considered.

Part IV "Deterministic Metrology of Machines and Robots" consists of material relating to assessments and enhancement of the PWM accuracy. An approach of engineering metrology developed in the ISO and ASME standards based on coordinate methods is applied to enhance the PWM accuracy.

Appendices consist of the basic information about application of the Mathematica® software system for solution of typical analytical and numerical problems on accuracy of the PWM.

Generally, the book is built using the form-shaping function as a basic model allowing a unified approach for synthesis of the calculating model of the precision machine mechanics.

The book is issued in two volumes. Volume 1 includes Introduction, Parts I and II (Chapters 1–12), and hence, embraces the general accuracy definitions, nominal machine functioning models, and geometrical accuracy problems

The book is intended for machine and robot designers and researchers, university graduate and senior undergraduate students, and may be also useful for instructors in the mechanical engineering field for teaching processes.

References

[1] Shigley, J.E., Mischke, C.R., Budynas, R.G, 2003, Mechanical Engineering Design, 7th ed., McGraw-Hill.

[2] Hocken, R., "Machine Tool Accuracy", Chairman, Report of Working Group 1 of the Machine Tool Task Force, Vol. 5, UCRL-52960-S, 1980.

[3] Slocum, A., 1992, Precision Machine Design, Prentice-Hall, Englewood Cliffs, NY.
[4] Reshetov, D. N., Portman, V. T., 1988,"Accuracy of Machine Tools", ASME Press, New York, NY, 304 pp.
[5] Faux, I.D., Pratt, M.J., 1979, Computational Geometry for Design and Manufacture, Ellis Harwood, Chichester, UK.
[6] Mooring, B.W., Roth, Z.S., Driels, M.R., 1991, Fundamentals of Manipulator Calibration, John Wiley, New York, 330 pp.
[7] Litvin, F.L., 1968, Theory of Gearing, 2nd ed., Nauka, Moscow (*in Russian*).
[8] Sheth, P.N. and Uicker, J.J., IMP (Integrated Mechanism Program), A Computer Aided Design Analysis System for Mechanisms and Linkages, ASME Journal of Engineering for Industry, May 1972, pp. 454-464.
[9] Hunt, K.H. , 1978, Kinematic Geometry of Mechanisms, Clarendon Press, Oxford.
[10] Dimentberg, F.M., 1978, Theory of Screws and Its Applications, Nauka, Moscow (*in Russian*).
[11] Paul, P.R., 1981, Robot Manipulators: Mathematics, Programming, and Control, MIT Press.
[12] Tsai, L-W., 1999, Robot Analysis, John Wiley & Sons, Inc., New York, NY.
[13] Merlet, J.-P., 2006, Parallel Robots, 2nd ed., Kluwer Academic Publishers, Dordrecht, Netherlands.
[14] Ceccarelli, M., 2004, Fundamentals of Mechanics of Robotic Manipulation, Kluwer Academic Publishers, Dordrecht, Netherlands.
[15] Wolfram, S., 2003, The Mathematica Book, 5th ed., Wolfram Media, 1488 pp.

Acknowledgement

This work started in the Experimental and Research Institute for Metal Cutting Machine Tools (ENIMS), Moscow, USSR, one of the most qualified scientific institutes in the field of manufacturing technology in the former USSR, and continued in Ben-Gurion University of the Negev (BGU), Beer-Sheva, Israel. I have had the good fortune to work in both these organizations.

Prof. Dmitry Nikolaevich Reshetov (1908-2000), advisor of my Ph.D. thesis and co-author of our book *Accuracy of Machine Tools* was one of the persons establishing the field of my scientific interests. Prof. Yugene Rivin (1932-2011), Wayne State University, Detroit, Mi, USA, and Prof. Roland Weill (1923-2013), Technion - Israel Instritute of Technology, Haifa, Israel, were permanent sparring partners in discussing our common scientific problems.

Two conferences organized by Prof. Ichiro Inasaki (KEIO University, Japan, now Shubu University) in collaboration with Prof. Takeshi Kishinami, dedicated to this book's concepts, helped me to clarify the book's objectives. I am grateful to Prof. Alexander Slocum for his valuable comments relating to my papers and discussions about this book subject-matter.

My co-authors Victoria G. Shuster, M.Sc. in Mathematics and M.Sc. in Engineering science, Prof. Ben-Zion Sandler (Ben Gurion University of the Negev, Beer Sheva, Israil), Dr. Emmanuil Kushnir, and my former students Dr. Yanina Rubenchick, Dr. Yair Shneor, and Dr. Vladimir Chapsky significantly contributed to the material of this book.

I express appreciation to copyeditor Edna Oxman for styling the manuscript.

Mrs. Mary Grace Stefanchik and Prof. Vladimir Vantsevich significantly helped me with preparation of the book publication.

And the last but not the least, my infinite gratitude for my wife Alla and family for their permanent help and support.

1. Problem of accuracy

1.1 Definitions and qualitative description

The word "accuracy" has a wide spectrum of everyday applications. "Merriam-Webster's Online" dictionary defines the *accuracy* as:

(1) freedom from mistake or error, i.e., correctness;

(2a) conformity to truth or to a standard or model, i.e., exactness;

(2b) degree of conformity of a measure to a standard or a true value; this meaning is close to *precision*.

In turn, the precision is defined as the degree of refinement with which an operation is performed or a measurement stated.

To illustrate a principle of assessments of accuracy and precision, consider, as an example, some targets for sport shooting shown in Figure 1.1. Four sportsmen shot five times. Targets 1 and 2 each have three hits in the bull's-eye and two hits in the nearest ring. Thus, sportsmen's shots in these targets are more accurate than those of the shooters, whose results are fixed in targets 3 and 4. Additionally, it is not quite correct to apply the terms "accuracy" and "precision" to target 4. The sportsman, who shot in this target, missed his aim three times, and the rest are not so good, either. Hence, it would be more proper to establish that in this case there is no reason to talk about the *accuracy* of the shooting. In other words, comparison from the accuracy viewpoint, i.e., estimation in terms of "more or less" accurate, may make sense when the results do *not deviate too much* from the target.

Now let us return to targets 1 and 2. Obviously, the closer the holes are to the accepted value, the more accurate the results are. Both shots are of identical accuracy from the viewpoint of closeness to the target center since their bullet-holes result in the same scores. However, the results of the first shooter are *more precise* since his shots are grouped more tightly together. In the other words, a set of shots is considered as more precise if the set occupies a smaller area within the target. Thus,

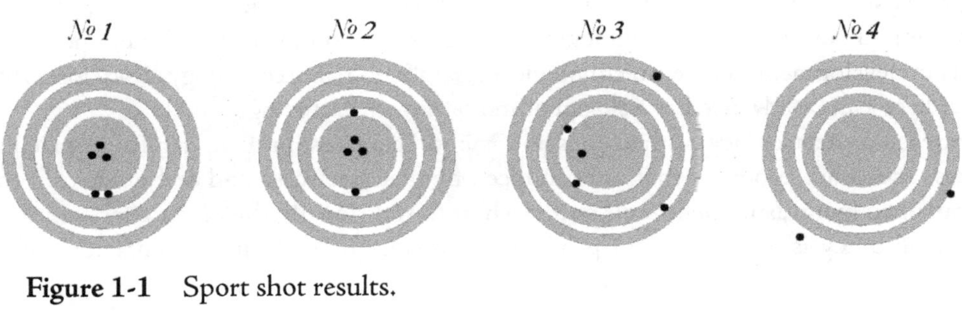

Figure 1-1 Sport shot results.

- The accuracy is estimated through closeness of the bullet-holes to the target; the closer the bullet-holes are to the target, the more accurate is the shooting;
- The precision is estimated through the tightness of the bullet-holes to each other: the smaller area the bullet-holes occupy, the more precise is the shooting;
- The results as in target 3 are less accurate and less precise than those in targets 1 and 2;
- The results shown in target 4 are neither accurate nor precise. In other words, there is no point in application of these terms, and we can recommend shot 4 to upgrade his shooting skill.

In manufacturing technology, the accuracy problem is of paramount importance, since accuracy is one of the most important indices of the product quality. Note that in manufacturing measurements the word "precision" is equivalent to "repeatability". The provision and enhancement of accuracy requirements for produced parts are established and simultaneously "hot" problems in modern manufacturing.

Accuracy-related problems have been intensively studied for more than 250 years as applied to machine tools and measuring devices, for less than a century as applied to coordinate measuring machines, and about 60 years as applied to robotics. During the second half of the last century, important problems such as reliability, flexibility, computerization, and "just-in-time" have been in the forefront, but concurrent with them the accuracy requirements are permanently increased.

The accuracy parameters significantly affect all criteria of machine performance including quick-acting, energy efficiency, metal consumption, reliability, durability, and ecology conditions. There is direct correspondence between the accuracy of production and measuring machines and robots, on the one hand, and the accuracy of the machined parts and technological and human-type operations, on the other. Hence, the accuracy is a fundamental characteristic of mechanical systems of many machines and robots defining their structure and applications and, in turn, depending on their structure and applications.

The modern development of technology is associated with continuously increasing accuracy requirements, including maintenance under load and time. New technological solutions must primarily satisfy the accuracy criteria. In robotics, the recent extensive investigations of parallel kinematic machines derive from the fact that they are more accurate and stiffer than articulated robots. In machine tool technology, multi-step successful development of numerical control and flexible manufacturing allow transfer of progressive methods of the accuracy provision from highly skilled persons into the memory of computerized control systems. The development of three-dimensional metrology allowed important detailing of requirements for accuracy of machines and machined parts with widespread applications of comprehensive requirements for geometrical parameters – accuracy of form, position, and orientation. Transfer of the technological processes into submicron and nanometer areas results in development of a new technological discipline –nanotechnology. Therefore, "Accuracy at first" can be advanced as a slogan of modern manufacturing.

Increasing the accuracy of machines and robots increases, in turn, quality of machines, devices, and other products that are produced by machine tools and technological robots accompanied by measurement tools and machines. Additionally, the direct effect of increasing the accuracy is an increase in productivity of machines during manufacture of new machines.

Accuracy affects the operations of machines with respect to the following criteria:

- dynamic loading, vibrations, and noise associated with manufacturing errors – disbalance and runout of the components, kinematic errors, clearances in mating parts, surface waviness;
- strength with respect to dynamic loads, distribution of loads in statically indeterminate systems, use of multiple-contact designs, conditions of engagement, and stress concentration;
- friction, heating, and wear losses associated with dynamic loads, harmful loads due to redundant links in the mechanisms and pressure concentration, non-optimum clearances and states of the surfaces;
- reliability associated with all performance criteria, their mutual influence, and also with initial accuracy margin;
- quick-acting associated with dynamic loads and heating;
- material consumption in view of speed limitation and additional harmful loads.

For increasing the accuracy, the multi-element design is commonly used: wave and multiline gearing, multi-wedged and toothed belt drives, flexible couplings, and splined, flat-toothed, comb-shaped, herringbone joints. Of important significance for production machinery are the requirements for the quality of the products under specified productivity and minimum waste; in particular, for metal-working machines – the requirements for the accuracy of the manufacturing products (workpieces); for printing presses – the quality of printing, in particular, color printing; for textile machinery – the quality of the fabrics; for energy-converting machines – thermal efficiency and speed; for transport machines – thermal efficiency and fuel economy, attainable speed, low specific metal content, reliability, handling, and comfort; for hydraulic machines and hydraulic equipment – efficiency and absence of or minimum leakage; for chemical machines and machines of the food industry – accuracy of metering, weighing, and so on; for testing machines – accuracy of measurements.

In the following devices, the accuracy is particularly important: astronomical devices due to the large distances to the measured; optical and optoelectronic techniques that determine the success of measurements in microscopes and photo and filming techniques; time devices and other physical devices determining, to a considerable extent, the successes in physics and chemistry; medical devices including robotics-based medical devices that allow one observation, records, and other important service functions during the most complex processes of the human body.

The accuracy of measuring devices must be very high, since the errors of measurement must be far less than the errors of the measured quantities (not more than 1/10 for products of high accuracy). The accuracy of components and mechanisms with respect to their satisfactory operation must ensure the exact location and motion of the elements, limitation of both the dynamic loads and concentration of loads and stresses, optimum clearances and tensions in the joints, and others.

The requirements for roughness and accuracy of surfaces of the components are established by taking into account their function and required surface properties to which pertain: exact basing; physical and chemical properties (e.g., corrosion resistance), reflectivity, thermo-electrical conductivity, adhesion ability; frictional properties; strength and rigidity (resistance to volumetric and contact fatigue); bearing pressure; dynamic and static sealing capability; and so on. These functions determine, to a great extent, the requirements for surface texture. Of significant influence on performance are waviness, surface roughness, profile and lay.

Errors of machines and robots can be divided into:

- Instrumental errors; their own structural errors not linked with external actions: trajectory errors caused by programming features, manufacturing errors and systematic errors caused by the substituted (non-ideal) operation principle of the machine and technological process (for example, layout errors caused by changing the complex profile of the product during machining into a simplified one);
- Errors from external actions without irreversible changes in the system (elastic and temperature strains);
- Errors from irreversible changes of machine parameters (wear, aging, distortions).

The majority of precision technological machines and their components become out of order not as a result of aging or failures and not as a result of general obsolescence, but because of loss of accuracy as a result of wear, press-downs, and distortions leading, finally, to considerable reduction of the main output characteristics. In some cases, in order to prolong the service life of machines, they are fabricated with a so-called accuracy margin that takes into account the inevitable loss of accuracy during exploitation and the rise of the requirements for accuracy.

1.2 Accuracy development: milestones

The historical process of accuracy increase took place rapidly. Accuracy of calculations, measurements, and production of parts and machines had increased throughout the history of science and technology. It is generally agreed that the accuracy of the obtained scientific results was and remains an exact measure of the scientific success or failure. What is more, the progress of exact sciences and manufacturing technology may be represented in terms of increasing accuracy of scientific results. The impossibility to produce manufacturing components with the necessary accuracy delayed the appearance and development of the first machines. At the very beginning, machine components were matched to each other, i.e., the fundamental principle of modern manufacturing – interchangeability of the parts – was unavailable.

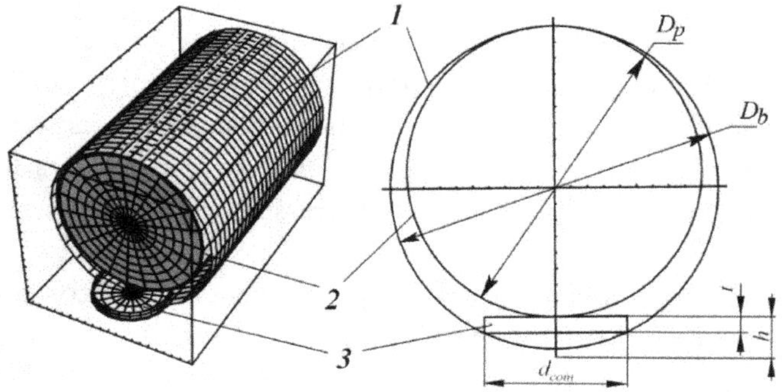

Figure 1-2 A possible scheme of measurement by Watt: measurement of clearance using a six-pence coin.

By 1776, according to James Watt, it was possible to obtain an accuracy of boring a cylinder of 72-inch (1828.8 mm) length "being not further from absolute truth than the thickness of a thin sixpence in the worst part" of the boring cylinder [1]. Hence, Watt used the sixpence coin as a measuring tool (in modern terms, feeler gauge) under a default assumption that the dimensions of the coin are more or less fixed. An unconfirmed scheme of measurements of the clearance is shown in Figure 1.2, where 1, 2, 3 are the cylinder, piston, and coin, respectively.

A evaluated value is the difference of the bore diameter D_b and piston diameter D_p defines clearance $s = D_b - D_p$. The coin is operating as a GO-NO-GO gauge, whose measurement result corresponds to clearance s as follows:

(1) The coin is a NO-GO gauge if the clearance is less than the coin thickness, $s \leq t$.

(2) The coin is a GO gauge if the coin completely penetrates the clearance, and a result of measurement h depends on coin sizes as follows:

$$h = t + \frac{1}{2}(D_b - \sqrt{D_b^2 - d_c^2}) \tag{1.1}$$

where t_c is the coin thickness, and d_c is the coin diameter.

The nominal diameter of the cylindrical connection in Newcomen's and Watt's steam engines was 21-24 inches (533-600 mm). For D_b = 560 mm, t = 1 mm, and d_c = 19 mm, we have $h = 1 + (1/2)\ (560 - \sqrt{560^2 - 19^2}) = 1.16121$ mm. Hence, Watt could evaluate whether the clearance lies in the range $1.0 < h < 1.16$ mm.

Now there is an opportunity to demonstrate a very important principle, widely used in accuracy and precision numerical operations. Since the coin diameter d_c = 19 mm is far less than the bore and piston diameters, $d_c << D_b$ and then $d_c/D_b << 1$, Eq. (1.1) can be significantly simplified by means of *linearization* of the sub-duplicate value:

$$D_b - \sqrt{D_b^2 - d_c^2} = D_b - D_b\sqrt{1 - \left(\frac{d_c}{D_b}\right)^2} \approx D_b - D_b\left(1 - \frac{d_c^2}{2D_b^2}\right) = \frac{d_c^2}{2D_b} \tag{1.2}$$

Substitution of Eq. (1.2) into Eq. (1.1) yields:

$$h \approx t + \frac{d_c^2}{4D_b} = 1.161\,\text{mm} \tag{1.3}$$

The relative deviation of results obtained through Eqs. (1.1) and (1.3) is less than 0.03%, far less than the actual errors in the considered method of measurement. Obviously, in this case, the linearization is not so necessary since the original expression, Eq. (1.1), is quite simple. However, this approach is widely distributed in the general theory of accuracy since the errors of different physical natures are, as a rule, some orders less than the nominal sizes that made it possible to neglect the highest-degree error-related terms (the terms of more than the first order). For example, in Eq. (1.2) the binomial formula where the terms of more than the first order are ignored is applied,

$$\sqrt{1 - a} \approx 1 - a/2 \tag{1.4}$$

This is the reason that the linearization will be in considerable use in the following material. In particular, it allows formulating infinitesimal matrices significantly facilitating computation of geometric and kinematic errors of complicated mechanical systems.

Two very simple yet very important and widespread quantitative assessments are as follows. If N measurements of the evaluated value are performed, and results of measurements $\delta_1, \delta_2, \ldots, \delta_N$ present a sample of deviations of the measured values from its common nominal value, the average value δ and its standard deviation σ are defined:

$$\bar{\delta} = \frac{\delta_1 + \delta_1 + \ldots + \delta_N}{N} \tag{1.5}$$

$$\sigma = \sqrt{\frac{(\delta_1 - \bar{\delta})^2 + (\delta_2 - \bar{\delta})^2 + \ldots + (\delta_N - \bar{\delta})^2}{N}} \tag{1.6}$$

The standard deviation σ characterizes the *precision* of the results of measurements, while evaluation of accuracy is needed in both $\bar{\delta}$ and σ values. If a priori the results of measurements are known to have the normal (Gaussian) distribution and number N is as long as is reasonably large, the maximal and minimal deviations can be obtained using the "3-sigma rule",

$$\delta_{max} = \bar{\delta} + 3\sigma \text{ and } \delta_{min} = \bar{\delta} - 3\sigma \tag{1.7}$$

Additionally, if δ_{max} and δ_{min} are available deviations, their difference presents the tolerance of the measured value,

$$T = \delta_{max} - \delta_{max} \tag{1.8}$$

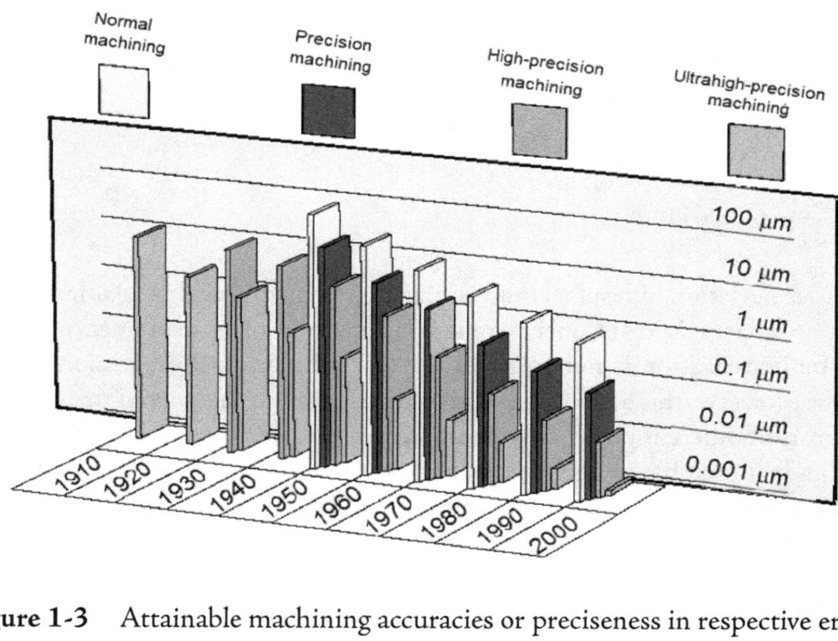

Figure 1-3 Attainable machining accuracies or preciseness in respective eras [2].

The rate of increase of attainable accuracy of manufacturing is particularly high at the present time. It rises by factor of approximately 10 every 20 years: the relative error of manufacturing in 1940 amounted to 10^{-4}, in 1960 to 10^{-5}, and in 1980 to 10^{-6}. Taniguchi [2] gives a compact presentation of the accuracy progress in modern manufacturing from normal machining to nano-technological requirements, i.e., up to 1 nm = 10^{-9} m (Fig. 1.3).

The progress concerns the development of two interrelated groups of manufacturing equipment: processing equipment (machine tools) and measuring instrument (inspection equipment). The way from regular lathes and milling machines to atom, molecule, and ion beam machining characterizes the progress of the former group (Table 1.1). Accordingly, the way from Vernier calipers and micrometers up to the laser-based devices, tunnel microscope, and other super-high accurate measuring instruments describes the development of the latter group. The importance of the machine tools and measuring technique for accuracy development and control must be emphasized.

Table 1-1 Equipment related to accuracy range [2].

Accuracy range	Processing equipment Machine tools	Measuring instrument Inspection equipment
> 0.1 mm = 10^2 μm	Lathes Milling machines	Vernier calipers
0.01 ÷ 0.1 mm = 10 ÷ 10^2 μm	Precision lathers, Grinding machines, Lapping machines, Honing machines	Mechanical comparators, Micrometers, Dial indicators
0.001 ÷ 0.01 mm = 1 ÷ 10 μm	(Curve generators), Precision jig borers, Jig grinding machines, Super finishing machines.	(Electronic or pneumatic micrometers) Optical comparators
10^{-1} ÷ 1 μm	Precision grinding machines, Precision lapping machines (optical lens grinding machines), Precision diamond lathes.	Optical or magnetic scales, Electrical comparators, Electronics comparators (non- contact)
10^{-2} ÷ 10^{-1} μm = 101 ÷ 10^2 nm	(D.U.V. mask aligner), (Optical lens finish grinding machines), Precision diamond grinding machines, Ultra precision grinding machines.	Laser measuring instrument, Optical fibers, Talysurf, Talyrond.
10^{-3} ÷ 10^{-2} μm = 1 ÷ 10 nm	(Diffraction grating ruling engine), (Electron beam lithographer), Super high precision grinding machines, Super high precision grinding machines, Super high precision lapping and polishing machines.	High precision laser measuring instrument (Doppler, multi-reflection) Talysteps
0.3 nm ÷ 1 nm	Atom, molecule or ion machining, atom or molecule deposition, (Substance synthesizing)	Scanning electron microscope, Transmission electron microscope, Electron diffraction equipment, Ion analyzer, X-ray micro analyzer, Auger analyzer, ESCAR.

1.3 Accuracy of a part

To provide their designed functions, the part-associated deviations must be fall within the preliminary specified ranges. A corresponding set of requirements is given by Geometrical Product Specifications (GPS), whose main concepts are:

- Dimensional tolerances, including linear and angular tolerances of size and special cases such tolerances of pitch, distances, etc.;
- Geometrical tolerances, including tolerances of form, orientation, location, and run-out;
- Tolerances of surface texture, including roughness and waviness.

Let us briefly consider these groups of the GPS concepts.

1.3.1 Dimensional (parametric) tolerancing

In the target-related example in Figure 1.1, both accuracy and precision were previously qualitatively described in comparative terms – closer, tighter, etc. However, some elements of quantitative approach were also used: the target is divided into some zones and the number of bullet-holes lying in one or another zone is calculated. In official events, each target zone corresponds to some points – maximum number of points corresponds to the central circle and minimum relates to the periphery ring. The total number of points, i.e., the shooting accuracy, defines the sportsman's point position. If there are two or more top-scorers, i.e., the two have the same shooting accuracy, the precision factors are applied: the relative position of the bullet-holes (precision) is taken into account.

In science and manufacturing, the definitions of accuracy and precision are imperatively accompanied with numerical evaluations using numerous accuracy-related parameters. In other words, the accuracy and precision are quantified and this gives a possibility to say that one result is more or less accurate than another is. A basic parameter of quantitative estimation presents the (statistical) deviation Δ of actual (measured, observed) size D_a from the nominal one D_{nom}:

$$\Delta = D_a - D_{nom} \tag{1.9}$$

Relative deviation δ presents the dimensionless form of statistical deviation:

$$\delta = \Delta / D_{nom} = (D_a - D_{nom}) / D_{nom} \tag{1.10}$$

Physical nature of size D in Eqs. (1.9) and (1.10) can be different: diameters, lengths, distances, angles, temperature, pressure, etc. To satisfy application conditions, deviation Δ have to lie within a specified interval, between two limit deviations Δ_{min} and Δ_{max}:

$$\Delta_{min} \leq \Delta \leq \Delta_{max} \tag{1.11}$$

Hence,

$$D_{min} \leq D_a \leq D_{max}$$

$$D_{min} = D_{nom} + \Delta_{min} \text{ and } D_{max} = D_{nom} + \Delta_{max}$$

Inequalities (1.11) allow numerical evaluation of the accuracy problem. The quantification of the accuracy problems is the subject of the *dimensional tolerancing and metrology*. The difference of the allowable interval boundaries presents the tolerance

$$T=\Delta_{max}-\Delta_{min}=D_{max}-D_{min} \tag{1.12}$$

The tolerance system is one of the first and most important systems of national and international standards. It has established an idea of degrees of accuracy caused by principles of interchangeability and mechanization in regular manufacturing for two centuries. The tolerance values for dimensional parameters intensively developed in the mid-1900s are given by a system of international standards that defines dependence of tolerance on nominal value and requirements for accuracy. In the medium range of sizes, 1–500 mm, the numerical values of tolerances for some tolerance grades IT are formulated as follows:

$$\mathrm{IT} = a\,i \tag{1.13}$$

where IT is the tolerance (International Tolerance) in μm; a is the tolerance grade factor defining the quality level (the factor is dimensionless and size-independent); and i is the tolerance unit (in μm), which is assumed to be proportional to the cubic root of nominal size D (in mm!) with a small linear correction,

$$i=0.45\sqrt[3]{D}+0.001D \tag{1.14}$$

The nominal sizes up to 500 mm (so called medium dimension) is divided by 16 intervals: above 0 up to (including) 3 mm; above 3 up to 6 mm,..., above 315 up to 400 mm; and above 400 up to 500 mm. The second term in Eq. (1.14) is assumed to take into account the error of measurements. In the specified range of sizes between D_{min} and D_{max}, the nominal size D figuring in Eq. (1.14) is defined as a geometrical mean of the range boundaries:

$$D=\sqrt{D_{min}D_{max}} \tag{1.15}$$

There are 19 tolerance grades: IT01, IT0, IT1, IT2,...,IT16, and IT17. In Table 1.2, factor a is listed for the tolerance grades from IT5 (International Tolerance No. 5) to IT17.

Note that the values of coefficient a in Eq. (1.14) for grades from IT6 and coarser approximately present the geometric progression with factor 1.6. The tolerances for grades more accurate than IT5 are defined as follows: IT01 = 0.3 + 0.008 D; IT0 = 0.5 + 0.012 D; IT1 = 0.8 + 0.020 D; $\mathrm{IT3}=\sqrt{\mathrm{IT1}\times\mathrm{IT5}}$; $\mathrm{IT2}=\sqrt{\mathrm{IT1}\times\mathrm{IT3}}$; and $\mathrm{IT4}=\sqrt{\mathrm{IT3}\times\mathrm{IT5}}$, where D is nominal size in mm; and IT is tolerance in μm. For example, the tolerance value for tolerance grade IT7 (a = 16) within the range above 80 up to 120 mm (i.e., $D=\sqrt{80\times120}=97.980$ mm) is $\mathrm{IT7}=a\,i=16\,(0.45\sqrt[3]{97.980}+0.001\times97.980)=34.8\,\mu\mathrm{m}\approx35\,\mu\mathrm{m}$, and the last number is accepted as tolerance IT7 = 0.035 mm for sizes $80 < D \le 120$ mm.

Table 1-2 Tolerance factors for different tolerance grades.

IT5	IT6	IT7	IT8	IT9	IT10	IT11	IT12	IT13	IT14	IT15	IT16	IT17
a = 7	10	16	25	40	64	100	160	250	400	640	1000	1600

In Table 1.3 the tolerance values for the range up to 500 mm and tolerance grades IT1 ÷ IT16 are shown. The tolerances for large sizes – more than 500 up to 3150 mm – are calculated by equation IT = $a\ I$, where the tolerance unit I in μm is proportional to the nominal dimension D in mm, instead of the cube root of D in Eq. (1.14), with addition of a constant term $I = 0.004\ D + 2.1$.

To illustrate the concepts of accuracy and precision, a statistical approach to accuracy evaluation adopted in mass-production is considered. As is known, mass- and serial-production present the widespread method for producing goods in large quantities at low cost per unit.

Consider as an example a preliminary turning of a hole of diameter 21.8±0.1 mm (Fig. 1.4a). According to drawing requirements: (a) the nominal hole diameter of the machined hole is D_{nom} = 21.8 mm; (b) the (current technological) tolerance for the machined diameter is 0.2 mm; and (c) the machined diameters must lie in the range between D_{max} = 21.9 and D_{min} = 21.7 mm. To test whether the requirement is satisfied in the current machining process, the parts produced in an automated transfer line are measured immediately after machining the hole of diameter 21.8 mm. Results of measurements of 40 parts are listed in Table 1.4, and a histogram is shown in Figure 1.4b. Note that accuracy of the measurements is 0.005 mm. As is known, a ratio of error of measurements to error of the measured value must be less than 1/10. In this case, the requirement is satisfied since the ratio is 0.0050/0.1 = 1/20 < 1/10.

The set of measurements is considered as a statistical sample. The most important numerical assessments of the sample are the mean (average) value D_m and standard deviation σ_D, and its evaluation s_D:

$$D_m = \frac{1}{N} \sum_{i=1}^{N} D_i \tag{1.16}$$

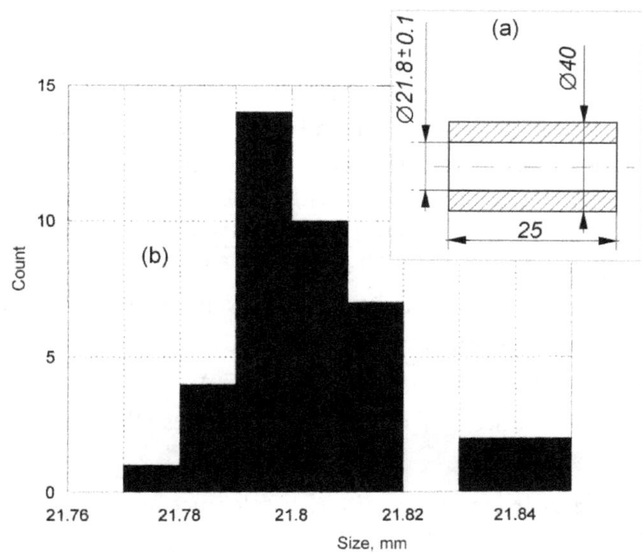

Figure 1-4 (a) The machined part; (b) the histogram of the measured sizes.

Table 1-3 IT system for 0 500 mm.

Size range from.... ....incl	*High precision and ultra- preision machining*					*Semi-finished machinig*					*coarse machiniwng, pressing, sting casting*					
	1	2	3	4	5	6	7	8	9	10	11	12	13	14	15	16
	$\pm\frac{IT1}{2}$	$\pm\frac{IT2}{2}$	$\pm\frac{IT3}{2}$	$\pm\frac{IT4}{2}$	$\pm\frac{IT5}{2}$	$\pm\frac{IT6}{2}$	$\pm\frac{IT7}{2}$	$\pm\frac{IT8}{2}$	$\pm\frac{IT9}{2}$	$\pm\frac{IT10}{2}$	$\pm\frac{IT11}{2}$	$\pm\frac{IT12}{2}$	$\pm\frac{IT13}{2}$	$\pm\frac{IT14}{2}$	$\pm\frac{IT15}{2}$	$\pm\frac{IT16}{2}$
0....3	0.0004	0.0006	0.001	0.0015	0.002	0.003	0.005	0.007	0.0125	0.020	0.030	0.050	0.070	0.125	0.200	0.300
3....6	0.0005	0.0007	0.0012	0.002	0.0025	0.004	0.006	0.009	0.015	0.024	0.0375	0.060	0.090	0.150	0.240	0.375
6...10	0.0005	0.0007	0.0012	0.002	0.003	0.0045	0.0075	0.011	0.018	0.029	0.045	0.075	0.110	0.180	0.290	0.450
10...18	0.0006	0.001	0.0015	0.0025	0.004	0.0055	0.009	0.0135	0.0215	0.035	0.055	0.090	0.135	0.215	0.0350	0.550
18...30	0.0007	0.0012	0.002	0.003	0.0045	0.0065	0.0105	0.0165	0.026	0.042	0.065	0.105	0.165	0.260	0.420	0.650
30...50	0.0007	0.0012	0.002	0.0035	0.0055	0.008	0.0125	0.0195	0.031	0.050	0.080	0.125	0.195	0.310	0500	0.800
50...80	0.001	0.0015	0.0025	0.004	0.0065	0.0095	0.015	0.023	0.037	0.060	0.095	0.150	0.230	0.370	0.600	0.950
80...120	0.0012	0.002	0.003	0.005	0.0075	0.011	0.0175	0.027	0.0435	0.070	0.110	0.175	0.270	0.435	0.700	1.100
120...180	0.0017	0.0025	0.004	0.006	0.009	0.0125	0.020	0.0315	0.050	0.080	0.125	0.200	0.315	0.500	0.800	1.250
180..250	0.0022	0.0035	0.005	0.007	0.010	0.0145	0.023	0.036	0.0575	0.0925	0.145	0.230	0.360	0.575	0.925	1.450
250..315	0.003	0.004	0.006	0.008	0.0115	0.016	0.026	0.0405	0.065	0.105	0.160	0.260	0.405	0.650	1.050	1.600
315..400	0.0035	0.0045	0.0065	0.009	0.0125	0.018	0.0285	0.0445	0.070	0.115	0.180	0.285	0.445	0.700	1.150	1.800
400..500	0.004	0.005	0.0075	0.01	0.0135	0.020	0.0315	0.0485	0.0775	0.125	0.200	0.315	0.485	0.775	1.250	2.000

Table 1-4 Actual diameters D_i (mm) of measured boring.

i	D_i	i	D_i	i	D_i	i	D_i	i	D_i
1	21.820	9	21.800	17	21.770	25	21.780	33	21.800
2	21.830	10	21.820	18	21.820	26	21.780	34	21.810
3	21.820	11	21.810	19	21.810	27	21.810	35	21.810
4	21.780	12	21.805	20	21.810	28	21.790	36	21.790
5	21.800	13	21.800	21	21.785	29	21.810	37	21.800
6	21.820	14	21.840	22	21.810	30	21.840	38	21.800
7	21.790	15	21.790	23	21.800	31	21.805	39	21.800
8	21.780	16	21.830	24	21.820	32	21.790	40	21.820

$$\sigma_D = \sqrt{\frac{\sum_{i=1}^{N}(D_i - D_m)^2}{N}} \text{ and } s_D = \sqrt{\frac{\sum_{i=1}^{N}(D_i - D_m)^2}{N-1}} \tag{1.17}$$

where N is the sample volume (in this case, $N = 40$).

Calculations by formulae (1.16) and (1.17) yield $D_m = 21.805$ mm and $s_D = 0.016736$ mm. The mean deviation is $\Delta_m = D_m - D_{nom} = 21.805 - 21.800 = 0.005$ mm. According to the abovementioned definitions, the mean deviation Δ_m characterizes (and evaluates) the sample *accuracy*, while the standard deviation s_D describing dispersion of the sample elements around the mean value characterizes the sample *precision*.

Mean value D_m and standard deviation s_D depend only on the sample components and are independent of the distribution law (in the case shown in Table 1.4, the subject is the distribution law of the measured diameters). If the additional information about the distribution low is known, standard deviation allows prediction of the dispersion zone with preliminary fixed reliability of the results. For example, if the hypothesis of normal distribution is accepted, the limit deviation with reliability 99.73% is defined using "3-sigma rule". In the considered example, accurate values of the mean value and standard deviation are unknown since they can be estimated from measurements of a sample of a very big volume. Note that there are two reasons for application of normal- based estimations: (a) the distribution law is really close to the normal one; and (b) the sample volume $N = 40$ is sufficiently not small.

The approximate limit interval is defined for the sample in Table 1.4 as follows:

$$B_{min} = D_m - 3\,s_D = 21.805 - 3 \times 0.016736 = 21.7548 \text{ mm}$$

$$B_{max} = D_m + 3\,s_D = 21.805 + 3 \times 0.016736 = 21.85521 \text{ mm} \tag{1.18}$$

Since both tolerance limits fall within the region defined through Eqs. (1.18) and (1.19), $D_{max} = 21.9 < B_{max} = 21.855$ and $D_{min} = 21.7$ mm $> B_{min} = 21.755$, the machining process can be currently continued without additional corrections and fittings. A graphical presentation of the results of measurements is shown in Figure 1.4b. Note that the actual fiducial probability of the region defined through Eqs. (1.18) is not as great as 99.73%, but in engineering practice the described approach to evaluation of current machining accuracy is successfully applied and the standard deviation is regularly defined through the tolerance T:

$$\sigma = T/6 \tag{1.19}$$

1.3.2 Geometrical tolerancing

According to ISO 1101 [3], geometrical tolerances include form, orientation, location, and run-out tolerances (Table 1.5). Tolerances of form include straightness, flatness, roundness, cylindricity, profile of any line and profile of any surface [4]. An example is shown in the design drawing (Fig. 1.5): cylindricity of the cylindrical bar must be no more than 0.05 mm. The tolerance is shown in a rectangular frame divided into two compartments: the left one is a symbol of the toleranced characteristic of form (in this case, the cylindricity symbol – a circle between two parallel lines) and the right one is the tolerance value in mm.

To examine whether the form tolerance is satisfied, the toleranced feature (a line or a surface) is measured on the CMM, and then a minimum zone (MinZ) for the examined feature is built. The MinZ for cylindricity presents a pair of coaxial cylinders (Fig. 1.5) satisfying the following requirements: (a) all points of the measured cylinder lie between two coaxial cylinders; and (b) the difference of the cylinders' radii $t = R_{max} - R_{min}$ must be minimal. For the part in Figure 1.5, the difference must be less than the tolerance, $t \leq 0.05$ mm.

Tolerances of orientation include parallelism, perpendicularity, angularity, profile of line, and profile of any surface. The orientation tolerances are fixed with respect to a datum or (a set of) datums. A datum can appear as an axis, a plane, a straight line, etc. An example is shown in the design drawing (Fig. 1.6a): perpendicularity of the plane with respect to datum A is 0.1 mm. The tolerance is shown in a rectangular frame divided into more than three compartments. The left-hand compartment consists of a symbol of the orientation tolerance (in this case, it is the perpendicularity symbol), the intermediate compartment displays the tolerance value $t = 0.1$ mm, and the datum symbol A is in the right-hand compartments. An explanation of datum A as the lower plane of the part is given in the drawing (Fig. 1.6a). In this case, all the points of the toleranced surface S must lie between planes $B1$ and $B2$ perpendicular to datum plane A (Fig. 1.6b). The two planes must be spaced at $b \leq t = 0.1$ mm. The space between the boundaries $B1$ and $B2$ presents the minimum zone for the considered tolerance.

The limit deviations of the form and position of cylindrical surfaces (roundness, cylindricity, radial runout) and the limit deviations from alignment are proportional to the cubic root of the size as it takes place in Eq. (1.14) for dimensional tolerances. The limit deviation of form and location of plane surfaces (flatness, straightness, parallelism, perpendicularity, axial runout) are proportional to the square root of the size.

1.3.3 Waviness and roughness of the machined surface

Between two boundaries of the minimum zone defining the geometrical tolerance, the profile of the controlled line or surface varies in magnitude. The size and form deviations present a

Table 1-5 Geometrical Tolerances.

Designation		Definition	
⏤	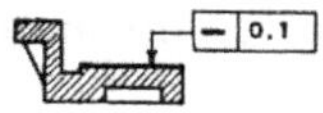	Straightness	Form
⏥	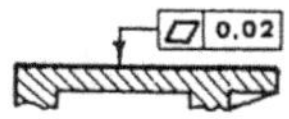	Flatness	
○	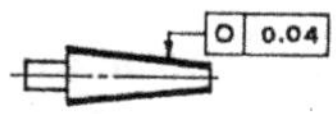	Circularity	
⌭	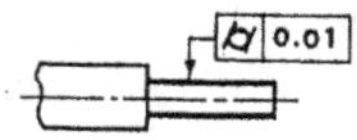	Cylindricity	
⌒	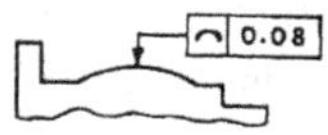	Profile of any line	
⌓	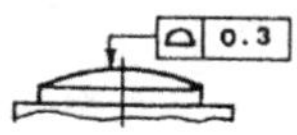	Profile of any surface	
//	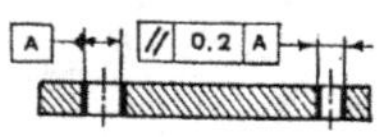	Parallelism	Orientation
⊥	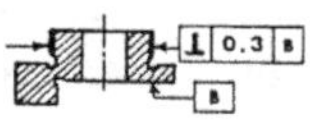	Perpendicularity	
∠	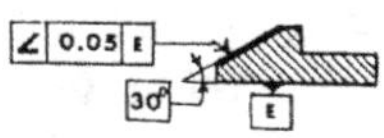	Angularity	
⌖	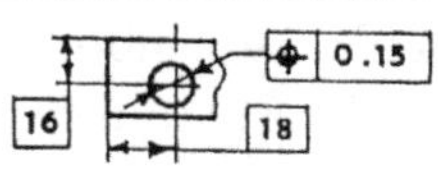	Position	Location
◎	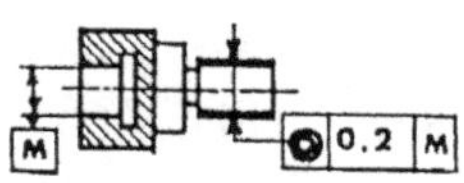	Concentricity and coaxiality	
⌯	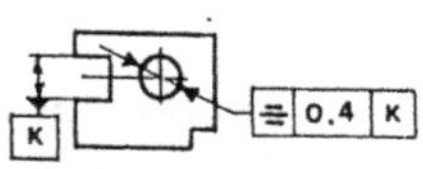	Symmetry	
↗	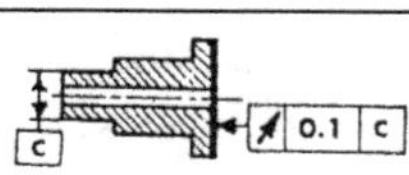	Circular run-out	Run-out
⌰	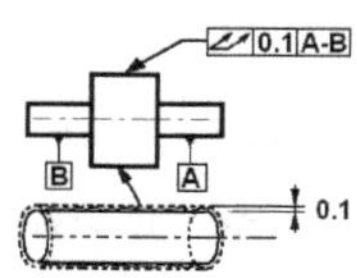	Total run-out	

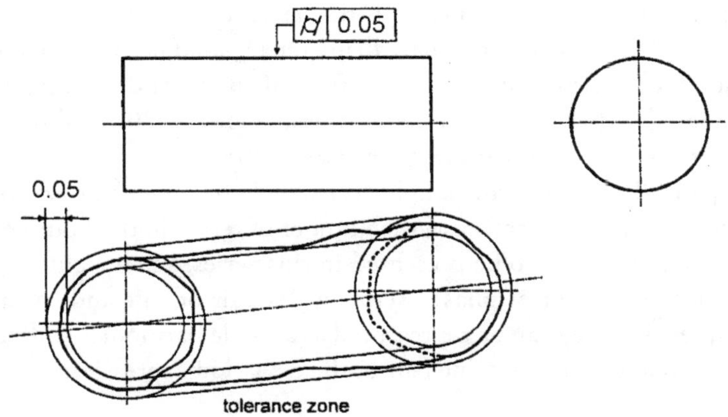

Figure 1-5 Cindricity evaluation.

large scale deviation, i.e., the deviation regarding the whole feature. They ignore the small-scale deviations of the actual profile of the part from the nominal form. Important additional results concerned with the part accuracy are obtained by spectral analysis of the profile. The analysis defines a spectrum of the form deviations, and surface waviness and roughness. Interpretation of the profile irregularities on length L depends on pitch p between two neighbouring profile peaks: an interval $p = L/2 \div L/20$ corresponds to the form deviations, $p = L/7 \div L/50$ is typical for waviness, and deviations with $p < L/40$ define the profile (or surface) roughness. Note that the frequency boundaries between form and waviness as well as between waviness and roughness are diffuse and they can have similar values.

A circular profile with different deviations from the nominal form is shown in Figure 1.7. In Figure 1.7a, the profile has the triangular form. Its form deviation – roundness – is specified by difference of internal and external radii of the ring-type minimum zone. The same profile with waviness of frequency 20 is shown in Figure 1.7b (the pitch of the deviations is $\pi/10$). In Figure 1.7c, the high-frequency deviations – roughness with frequency 100 – are superimposed on the original profile.

In this case, the deviation Δ of the actual profile from the nominal one can be expressed as follows:

$$\Delta=\Delta_D/2+e\cos\varphi+b\cos 3\varphi+w\cos 20\varphi+h\cos 100\varphi \tag{1.20}$$

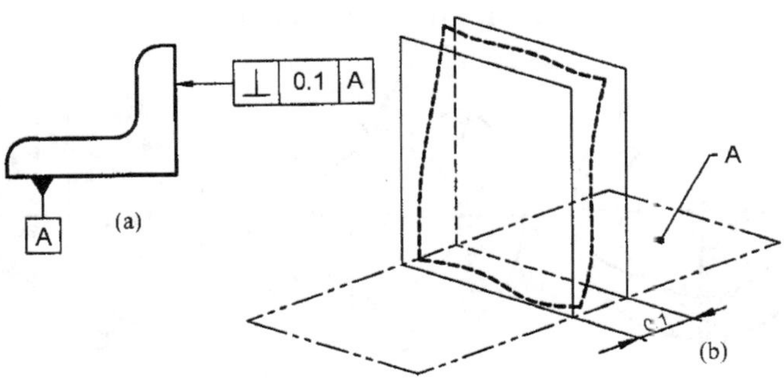

Figure 1-6 Perpendicularity evaluation.

where φ is the polar angle of the profile; Δ_D is the deviation of the diameter, i.e., the difference $\Delta_D = D_{lms} - D$ of the least-mean-square diameter D_{lms} and nominal diameter D; e is the position deviation (eccentricity); b is the magnitude of the form deviation; w is the magnitude of the waviness (in this case, the term with frequency $k = 20$); and h is the roughness magnitude (in this case, the term with frequency $k = 100$).

Note that Eq. (1.20) presents the simplified example of the profile where all the terms are only cosine terms. In the general case, the number of terms in the Fourier series is much more; and, the sine terms have to be presented in this series, i.e., in contrast to Eq. (1.20), different harmonics can be out of phase. Actually, if a pair of sine and cosine components has identical frequencies, they can be represented as a single harmonics shifted in phase. For example, there is a following sine-cosine pair in the deviation expression

$$\Delta_i = a_i \cos k_i\varphi + b_i \sin k_i\varphi, \text{ with } (a_i)^2 + (b_i)^2 \neq 0 \tag{1.21}$$

This pair is identical to the general sinusoidal function

$$\Delta_i = A_i \sin(k_i\varphi + \lambda_i) \tag{1.22}$$

with

$$A_i = \sqrt{a_i^2 + b_i^2} \text{ and } \lambda_i = \arctan(b_i / a_i)$$

Usually, the low- and high-frequency components are separately measured. The separation is carried out by application of different filtration methods: physical filtration by means of different radii of the sensitive contacting probe as well as by analog and software filtration. For example, measurements on coordinate measuring machines allow effective evaluation of accuracy of size, form, location, and orientation as well as waviness.

The surface roughness presenting the profile deviation component with maximal frequency is separately measured at profile segments of limited measurement length, which is chosen in accordance with actual heights of the profile peaks (Fig. 1.8). The results of measurements can be expressed through the Fourier series consisting of N harmonics:

$$\Delta = \sum_{i=1}^{N} A_i \sin(k_i\varphi + \lambda_i)$$

where designations are the same as in Eqs. (1.21) and (1.22).

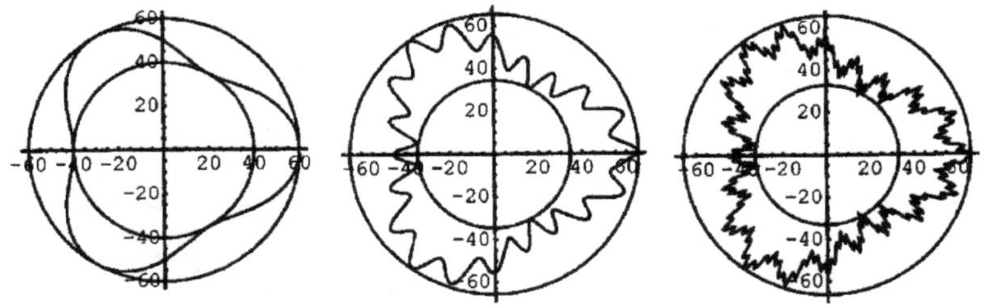

Figure 1-7 Circle form deviations of different frequencies.

To provide a wide spectrum of accuracy requirements that can be imposed on manufactured parts, technological machines and robots, in turn, have to satisfy a set of specific accuracy requirements. Historically, the requirements of this type were directly applied to measuring devices and then extended to machine tools, whose geometrical accuracy, stiffness, and accuracy of the test workpieces are established by national standards and technical specifications for machines of different types.

1.3.4 Abbe errors

Development of methods of engineering metrology and measuring devices allows transfer from trial-and-error fitting to interchangeability of matched parts in assembly. In this frame, the following foundations of mechanical accuracy can be emphasized [5, 6]: (a) development of standards of lengths; (b) development of standards of geometry; (c) methods of dividing the circle; (d) methods of roundness measurements; and (e) development of coordinate-measurement machines.

One of the important methods of accuracy optimization as applied to working machines is minimization of the Abbe offset, originally formulated for evaluation of parallax-caused deviations in optical devices for mechanical measurements. In Figure 1.9, the Abbe error is explained. Point *A* of a moving body in its nominal position is located at non-small offset *H* from the line of measurement (in this case, the *X*-axis). The point *A* is bound to be displaced from the nominal position to another nominal position A_1. However, as a result of guideways errors the body is rotated at small angle α, and the point is displaced from the position A_1 to actual position A_2. The Abbe offset can be defined as the distance *H* between the axis of the displacement measuring system of a machine and the axis of the measured displacement. A position error resulting from the angle α and offset *H* has two radically different values depending on the direction of measurement:

$$\delta_x = H \sin\alpha \approx H\alpha$$

$$\delta_z = H - H\cos\alpha = H(1-\cos\alpha) \approx H\alpha^2/2$$

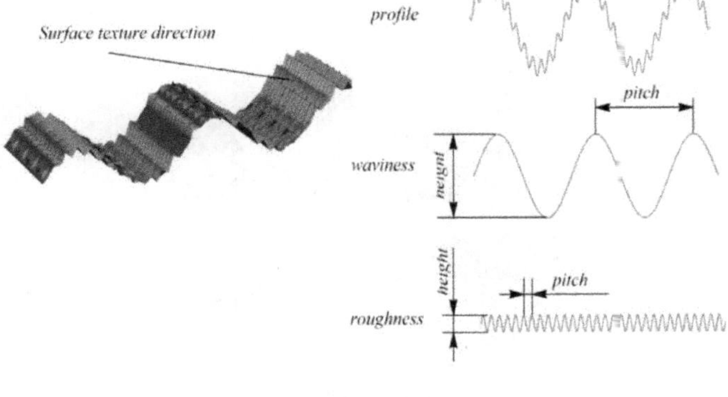

Figure 1-8 Frequency comparison of the form deviation, waviness and surface roughness.

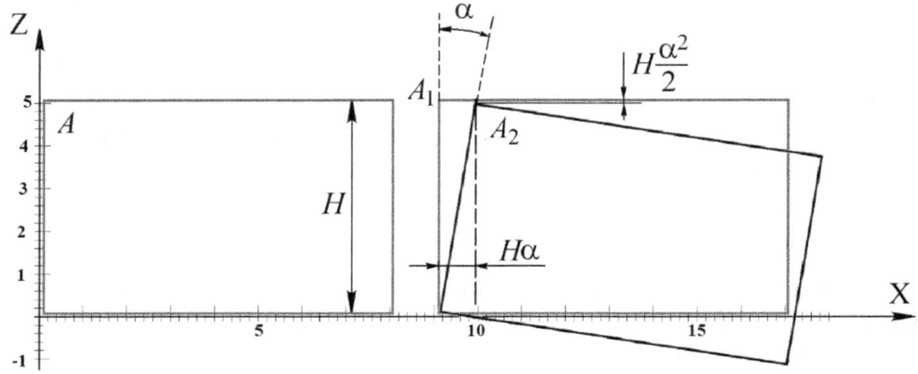

Figure 1-9 The Abbe error explanation.

where δ_x and δ_z are the errors measured along the line of measurement (X-axis) and normally to it (Z-axis), respectively.

In the case shown in Figure 1.9, the collinear deviation δ_x (the sinus-type deviation in Abbe terms) presents the Abbe error, while another error component δ_z can be (and is usually) ignored since it is one order of magnitude smaller than δ_x. It is important that in real structures of machines and robots, the offset distances, as a rule, take place, and there must be an approach to decrease them whenever possible.

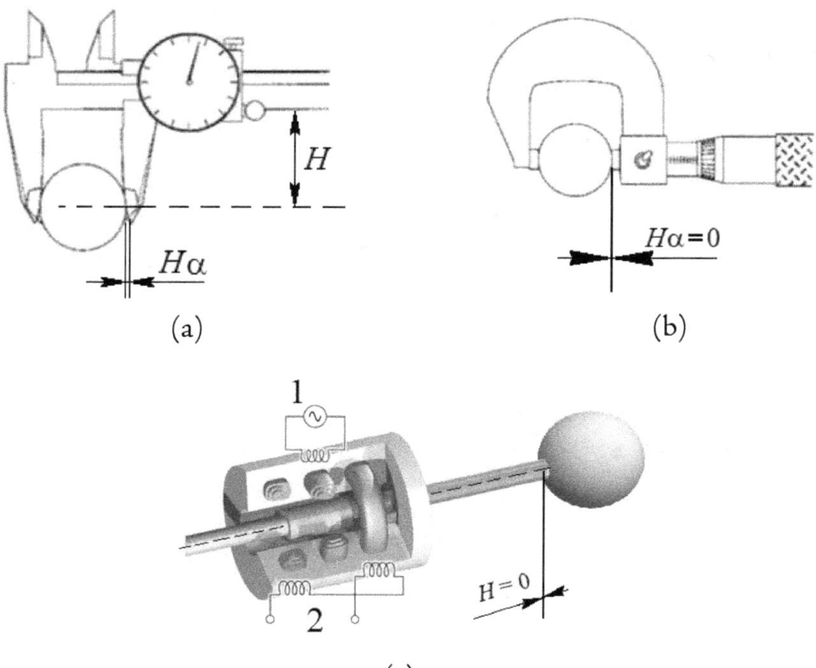

Figure 1-10 Comparison of Abbe offsets in measurement instruments: (a) Vernier caliper; (b) micrometer; (c) linear variable differential transformer (LVDT).

The Abbe error is decreased as this distance is decreased, and it dies away in the optimal structure when the offset equals zero ($H = 0$), and the measuring reference (measuring line) is placed coaxially with the axis along which the measured feature is oriented or reference point of motion is displaced.

In Figure 1.10a and b, two most-used workshop-level measuring instruments are considered from the Abbe error viewpoint. When the same part is measured by a caliper, there is a distance H between the measured diameter and the measuring scales and the Abbe errors can occur. Contrary to that, when a circular part is measured by the micrometer, the measured diameter lies coaxially with the measuring reference – an axis of precise measuring screw. Thus, the Abbe offset is zero and the measurement is free of the Abbe errors. The same is valid for the linear variable differential transformer (LVDT) (Fig. 1.10c) – the measurement is free of the Abbe errors since the measuring tip is coaxial with the moving core centrally located with respect to primary coil 1 and secondary coils 2.

1.3.5 Standardization

The concerned contributor into accuracy progress in manufacturing is the national and international standardization covering the most important manufacturing problems. The first such subject was the standard of meter as a common length unit. Prior to 1960, the meter was defined as the distance between two lines marked on the platinum-iridium bar – the international prototype of the meter made in 1889. The structure and conditions of measurement were under standard requirements. The original artifact of the meter is still kept at the International Bureau of Weights and Measures (BIPM, Sèvres, France) under the conditions specified in 1889. To reduce the uncertainty, with which the meter may be realized, the artifact-based definition was replaced in 1960 using a definition based on a wavelength of krypton-86 radiation. From 1983, the meter was defined as the length of the path traveled by light during a time interval of 1/299792458th of a second. This definition fixes the speed of light in vacuum at exactly 299,792,458 m/sec.

Another subject of international standardization that is crucial in relating to accuracy problems was dimensional tolerancing in the drafting standards established in 1935. The further development of this group of standards resulted from widespread use of the coordinate measuring machine and computer-aided control systems at the workshop level, allowing an immediate numerical presentation of the measured parts. Then the presentation is transformed into a set of accuracy characteristics regardless of their number and complexity. The standard approach to mathematical presentation of dimensioning and tolerancing is defined in standards ISO 1101, ASME Y14.5M: 1994 for dimensioning and tolerancing [7] and ASME Y14.5.1M: 1994 [8]. The latter standard formulates these parameters in terms of vector algebra, i.e., the mathematical technique, typical for an academic approach, becomes the regular manufacturing tool for solution of accuracy problems.

The establishment of the optimum accuracy level is a technical and economic optimization problem. These dependencies under about the same manufacturing technology are schematically shown in Figure 1.11: as tolerance T decreases (accuracy is enhanced), the consumptions into manufacturing increase (line 1), while those under operating conditions are decreased (line 2). The sum of these consumptions (line 3) has the minimum associated

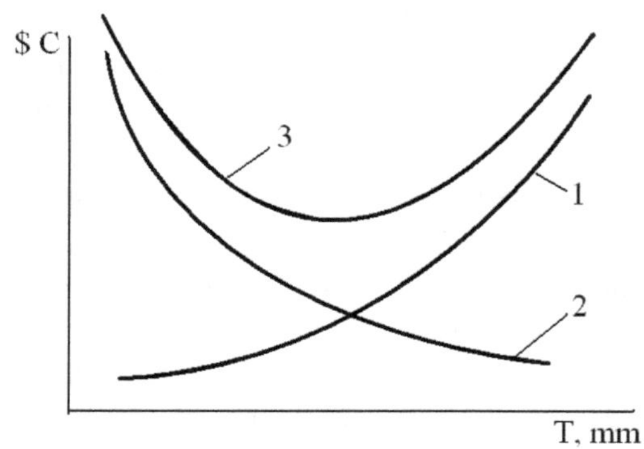

Figure 1-11 Consumption *vs.* tolerance value.

with the optimal level T_{opt}. Note that this minimum is usually weak, and hence, there are some available technological decisions.

Approximate evaluation of line 2 in Figure 1.11 is given in Figure 1.12 for typical turning operations [9]. If, for example, the cost of the finished turn operation applied to steel parts is assumed as 100%, the costs of the semi-finished turn and rough turn are about 60% and 25%, respectively, while the costs of the grinding and honing operations are about 240% and 390%, respectively.

However, if there is a need in mass-product manufacturing, the accuracy requirements are usually satisfied by task-oriented technological processes, whose accuracy cannot be described by abovementioned considerations. Glowing examples of this are ultra-precision design of mechanical elements in modern computers, audio-, video-technique, and so on, whose accuracy cannot be satisfied by regular enhancement of traditional manufacturing technology.

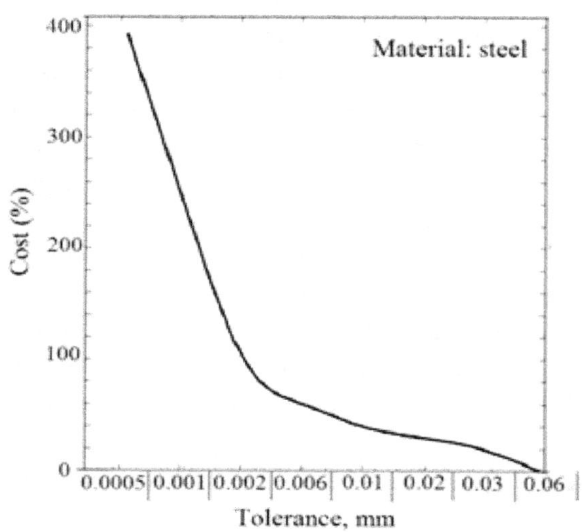

Figure 1-12 Approximate machining cost for typical operations [9].

There are rare cases of the existence of optimum values of accuracy in purely technical formulations. For example, there is an optimum level of surface roughness for rubbing pairs in conditions of boundary friction. Optimum arrangements of tolerance fields are also used; in particular, the allowed cambers of deviation from straightness of the guideways of the machine tool reduce the effect of elastic strains and wear. The allowed deviations at the angles of profiles of the threads of screws and nuts increase the pliability of the screws and decrease the non-uniformity of load distribution over the thread helixes. A certain dissipation of the natural frequencies of vibrations of some machine components, in particular, turbo-machinery valves, is favorable from the standpoint of dynamic stability.

1.4 Effect of manufacturing accuracy on operation of machine components

1.4.1 Accuracy of fixed joints

The torque transferred through a *fixed conical joint* (Fig. 1.13a) significantly depends on the difference of angles of internal and external associated cones as well as on their form deviations [10]. The transferred friction moment T vs. difference $\delta\alpha = \alpha_{ex} - \alpha_{in}$, where α_{ex} and α_{in} are the actual angles of external and internal cones, is shown in Figure 1.13d. As evident from the graph, any non-zero difference $\delta\alpha$results in reduction of the moment T. However, in the case, in which the external cone angle is more than the internal one (Fig. 1.13b), the transferred moment is reduced less than in the case $\alpha_{ex} > \alpha_{in}$ (Fig. 1.13c). For example, with $\alpha_{ex} - \alpha_{in} = +10'$, the moment is 20% more than that in the case $\alpha_{ex} - \alpha_{in} = -10'$. Actually, in the case $\alpha_{ex} < \alpha_{in}$ (Fig. 1.13b), the contact in the conical joint is concentrated in the vicinity of the maximal cone diameter, and this results in better contact than in the case $\alpha_{ex} > \alpha_{in}$, and vice versa.

In *interference fit joints,* there is a linear correlation between the contact pressure p created between mating cylinders *1* and *2* (Fig. 1.14a) and total interference value N. If the inner cylinder *1* is solid and the mating parts have identical Young's modulus E, the correlation becomes

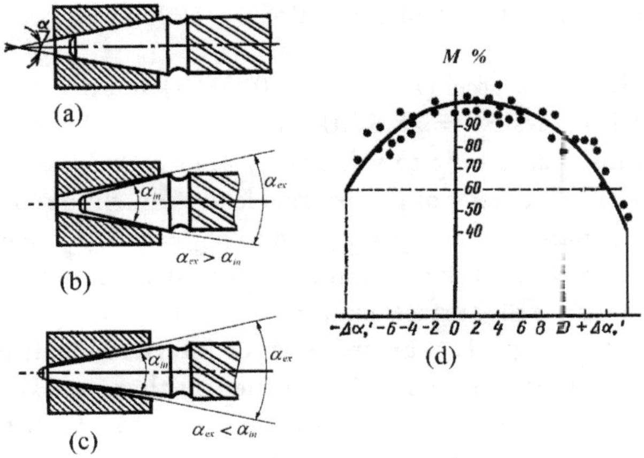

Figure 1-13 Contact conditions in conical joint.

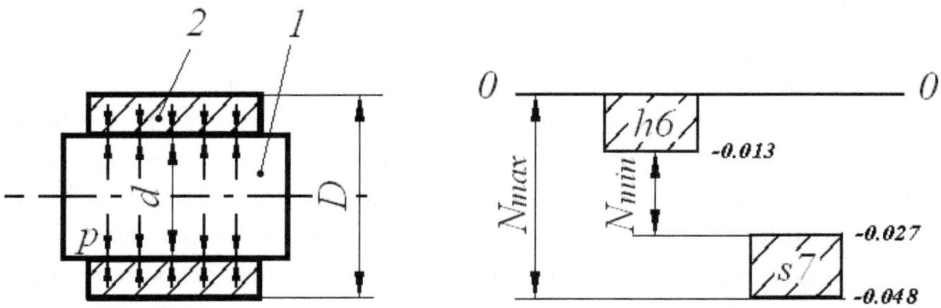

Figure 1-14 Interference fit joint.

$$p = \frac{EN_{eff}}{d}\left(1 - \frac{d^2}{d_{ex}^2}\right) \tag{1.23}$$

with

$$N_{eff} = \Delta_{sh} - \Delta_{hole} - 1.2(R_{z1} + R_{z2}),$$

where d is the nominal diameter of the joint; and d_{ex} is the external diameter of the part *2*; N_{eff} is the total interference value; Δ_{sh} and Δ_{hole} are the deviations of the mating shaft (part *1*) and hole (part *2*), respectively (taking into account the deviation sign); R_{z1} and R_{z2} are the ten-point heads – roughness of parts *1* and *2*, respectively.

As an example, a scheme of tolerance intervals of one of the standard interference fits Ø20S7/h6 is shown in Figure 1.14b. The tolerance sizes are: hole $\varnothing 20^{-0.027}_{-0.048}$ and shaft $\varnothing 20_{-0.013}$. Corresponding surface roughness is R_{z1} = 3.2 μm and R_{z2}= 6.3 μm. Hence, minimal and maximal effective interference values are:

$$(N_{eff})_{min} = -0.013 - (-0.027) - 1.2\,(0.0032 + 0.0063) = 0.0045 \text{ mm}$$

$$(N_{eff})_{max} = -0.013 - (-0.048) - 1.2\,(0.0032 + 0.0063) = 0.0255 \text{ mm}$$

If the material of the pair is steel ($E = 2.1 \times 10^5$ MPa) and d_{ex} = 32 mm, the internal pressure p, Eq. (1.23), becomes p_{min} = 28.8 MPa and p_{max} = 163.2 MPa. As is known [11], interference fits are chosen according to value $(N_{eff})_{min}$, i.e., the pressure p_{min} = 28.8 MPa must assure guarantee 100% transfer of the external load through the considered fit. Hence, a necessity to obtain the pressure value $p < p_{min}$ up to $p_{max} = 5.67p_{min}$ significantly increases the weight of the structure. One of the natural ways to reduce this occurrence is to choose a more accurate fit and enhance the surface quality.

A large spread of fit strength is characteristic for interference fit joints upon reaching high mean strength. This is associated with the fact that the values of the interference in the joints are commensurable with the tolerances for the fitting diameter, and the coefficient of friction is dependent on many factors representing random quantities with large natural spread (ingress of lubricating material, fouling, surface roughness, and so on). The coefficient of variation v of fit strength ($v = S/T_{mean}$, where S and T_{mean}

are the standard deviation and the mean value of the strength) can be evaluated as follows:

$$v=\sqrt{v_N^2+v_f^2}$$

with

$$v_N=(1/3)(N_{max}-N_{mix})/(N_{max}+N_{mix}) \text{ and } v_f=(1/3)(f_{max}-f_{mix})/(f_{max}+f_{mix})$$

where N is the interference value; f is the friction coefficient; and v_N and v_f are the coefficient of the deviation of the interference and friction coefficient, respectively; the evaluation is based on application of the Gaussian distribution and probability 0.9973.

Thus, the boundary probabilistic values of the transmitted moments are expressed through mean value T_{mean} and the coefficient of variation as follows:

$$T_{max,min}=T_{mean}(1\pm 3v)$$

In Figure 1.15, performance indexes of the compressor vs. piston-to-cylinder clearance are shown [10]. All indexes are adversely affected with use in consequence of wear-caused clearance increasing from 0.1 mm (initial clearance value) up to 1.1 mm. Relative values of the indexes of mass capacity and indicated power are degraded (curves 1 and 2, respectively), while the oil egress is increased (curve 3).

Misalignment of the shafts considerably affects the operation of couplers, shafts, and supports. In the regular case, the relative deviation of the axes of two shafts connected through toothed gearing is limited by 0.001 rad.

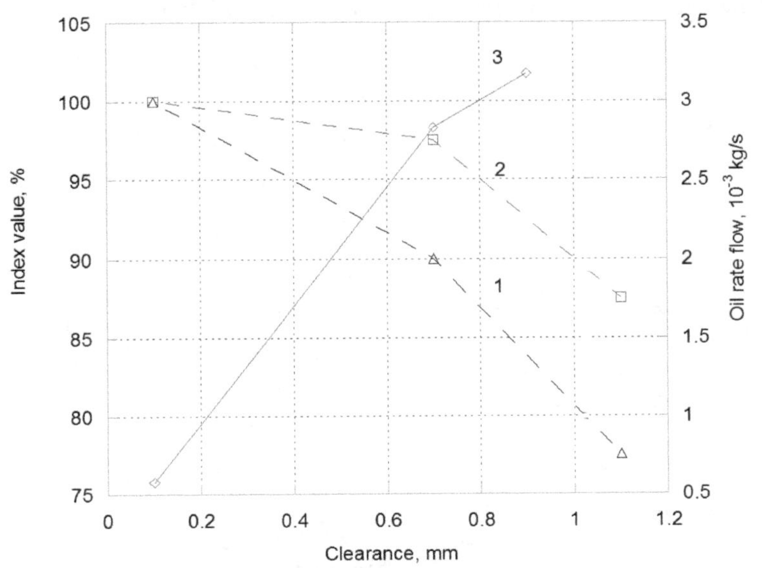

Figure 1-15 Compressor performance indexes vs. piston-to-cylinder clearance [10].

1.4.2 Load-carrying capacity of toothed gears

The effect of manufacturing accuracy on the load-carrying capacity of gears is felt through the participation of one or several pairs of teeth in the transfer of load, through the value of the internal dynamic load and load concentration along the tooth length.

The bending load-carrying capacity of spur wheels is usually calculated from the condition of the action of the total load on the tip of the teeth under one-pair engagement. However, if the greatest probabilistic difference f_{pb} between the circular pitch of one toothed wheel and the base pitch of the other gear is less than the elastic deformations of the teeth, then two pairs of teeth are working. This occurs in drives of the sixth degree of accuracy and higher (with respect to smoothness), and also in stressed drives of the seventh degree of accuracy.

Then, from the condition of mutuality of motion translations, the applied theoretical bending force to the tip of one pair of meshed teeth and, consequently, the bending stress are reduced as a result of multiplication by the coefficient $K_\alpha \leq 1$. In particular, for a straight spur gear $K_\alpha = 0.42 + 5.5f_{pb}/w$, where $f_{pb} = 0.8\sqrt{f_{pb1}^2 + f_{pb2}^2}$ is the highest probability difference of deviations of the main pitches of gear p_{b1} and wheel p_{b2}, μm; and w is the specific peripheral force, N/mm.

For drives of the seventh degree of accuracy, the limit deviations of the main pitches of geared wheels are shown in Fig. 1.16. For transmissions of the nth degree of accuracy, the deviations for $n = 7$ are multiplied by coefficient k:

n	3	4	5	6	7	8
k	0.18	0.28	0.44	0.70	1.0	1.4

The effects of manufacturing errors are manifested more in non-hardened gear wheels than in hardened gears for which elastic deformations are higher.

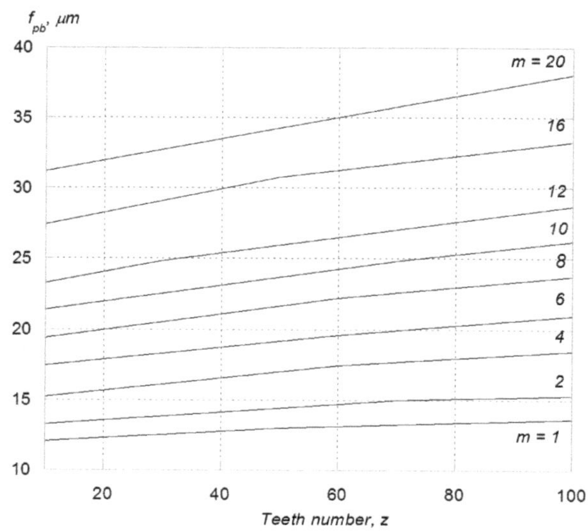

Figure 1-16 Deviations of the main pitches of geared wheels (the 7th degree of accuracy).

The internal dynamics of straight gear drives is mainly determined by the difference of the main pitches of the gear wheels in operation and is characterized by the dynamic factor

$$K_v = 1 + \frac{w_d}{w}, \text{ with } w_d = v\delta(n-2)\sqrt{a/u}$$

where w_d is the specific dynamic load; v is the peripheral speed, m/s; $u = Z_2/Z_1 \geq 1$ is the gear ratio; a is the distance between axes, mm; $(n - 2)$ is a factor characterizing the effect of the calculated error of the main pitches; δ is a coefficient of proportionality also characterizing the effect of the dynamic load on the strength.

Under bending, irrespective of surface hardness, δ is equal to: 0.16 for straight teeth, 0.11 for straight teeth with head modification, and 0.06 for oblique teeth. The same values are roughly applicable for contact strength of hardened teeth. However, if the hardness of at least one of the wheel pair $HB \leq 350$, then, in view of running-in, this coefficient is reduced by about 50 percent. The above formula is applicable in the sub-resonance region, i.e., as a rough guide, for straight tooth drives when $vz_1/100 \leq 10$ and for oblique tooth drives when $vz_1/100 \leq 14$.

The load dynamic factor K_v is usually assumed to be 1.3–1.4 at total useful load, and in machines with a strongly variable regime at small loads (for example, in machine-tool gear boxes at the higher levels of rotational speeds) the load dynamic factor could be larger than 2.

Load concentrations along the tooth length from manufacturing errors are mainly affected by the deviations of direction of tooth lines, deviations from the alignment of the recesses, and radial play of external and internal bearing rings.

The coefficient of load concentration for hardened (inadequately run-in) gears and for unhardened gears prior to running-in is $K_1 \approx 1 + 0.4bc\gamma/w$, where b is the width of the rim behind the residue of the roundings of the end, mm; c is the specific stiffness of the teeth, N/mm^2; γ is the angle of skewness from elastic deformations and manufacturing errors, rad.

The coefficient K_1 directly characterizes the reduction of the load-carrying capacity of the drives by contact strength. The coefficient $K_F = \zeta K_1$ of non-uniformity of distribution of the bending stresses in the teeth over the width of the rim is somewhat less than K_1, where ζ, is a coefficient taking into account redistribution of the bending stresses in the teeth compared with the distribution of the specific load, $\zeta = 0.7, 0.8$, and 0.9, for $b/m = 4, 10$, and 20, respectively (here m is the module, mm).

The results of experiments directly aimed at studying the effect of manufacturing accuracy of gears on their load-carrying capacity are available. Upon consistent variation of accuracy by various parameters, a decrease of the load-carrying capacity is observed from the simultaneous effect of the errors on the joint operation of the teeth, dynamic loads, and other factors.

Table 1-6 Load-carrying capacity of the working surfaces vs. pitch error.

Data	F/b (force/width ratio) for pitch error	
	$fp_b < 2\ \mu m$	$fp_b = 15–16\ \mu m$
Spur gears $z_1 = z_2 = 25$	$F/b = 190$ N/mm	$F/b = 135$ N/mm
Helical gears $z_1 = 15, z_2 = 24, \beta_0 = 39°$	$F/b = 275$ N/mm	$F/b = 125$ N/mm

Under testing, the bending load-carrying capacity F (normal to the measured surface of the tooth) as a function of pitch error f_{pb} of the gears (m_n = 2 mm; z_1/z_2 = 40/65; b = 18 mm; β_0 = 32°; n_1 = 3000 min^{-1}; material – steel 16MnCr5; heat-treatment: cementation and hardening) varied as follows:

f_{pb}, μm	5	17	35	60
F, N	12,500	10,000	8000	6500

The load-carrying capacity (force/width ratio F/b) of the working surfaces vs. pitch error f_{pb} for gears m_n = 5 mm; b = 10 mm (material – steel 45; lubrication by oil 6.5 VU at 50°C) varied as follows:

It follows from these data that the working surfaces of helical tooth gears have higher load-carrying capacity only for low pitch errors. The load-carrying capacity of working surfaces as a function of profile errors in tests of analogous drives for a peripheral speed of 18 m/s, when the errors increase two times (from 8-10 to 18-20 μm), decreased in straight tooth gears by 30% and in oblique tooth gears by 40%. Roughness of the working surfaces of the teeth reduces their actual load-carrying surface and gives rise to stress peaks on profile peaks. For example, when the mean deviation of the profile R_a decreases from 1.25–0.63 to 0.63–0.32 and to 0.32–0.16 μm, the design load-carrying capacity of the working surfaces increases by 10% and 20%, respectively.

1.4.3 Bearing units

The accuracy of manufacturing of components mating with *rolling bearings* significantly influences the work of the bearings in view of the skewness of the rings, form distortion of roller tracks and shaft play, and harmful loads on the separators. The allowable angular skewness of bearing rings from the flexure of the shafts and manufacturing errors is:

Table 1-7 The allowable angular skewness of bearing rings.

Radial single-row ball bearings with normal clearance	8′
Radial thrust ball bearing with angle of contact α:	
when α = 12°	6′
α = 25°	5′
α = 36°	4′
Spherical roller bearings, double row	1.5–4°
Radial roller bearings with short rollers:	
Without cage 4′ with cage	4′
Conical bearings:	10′
with rollers without cage	2′
with rollers with cage	6′
with cage on the outer ring	14′
Spherical roller, double row	0.5–2.5°

The allowable manufacturing misalignment of the rings is estimated as 1/4 of the total. The allowable deviations of the form of mounting surfaces under the rings of the bearings are estimated for the bearings of accuracy classes 0 and 6 as 1/2 of the tolerance for the diameter of the mounting surface, and for bearings of accuracy classes 5, 4, 3, and 2 as 1/4 of the above tolerance.

The load-carrying capacity of *friction sliding bearings* with liquid lubrication under the condition of constant temperature (oil cooling) is inversely proportional to the square of the value of the clearance, which is of the same order as the tolerances for the diameters of the journals and sleeve.

Misalignment of shaft and bearing axes is the most serious cause of reducing the lifetime of satisfactorily lubricated bearings and decreasing their load-carrying capacity, which is usually high under exact parallelism of the axes.

Due to local manufacturing errors in bearings, particularly in bearings of heavy machinery, hot zones are formed from which bearing breakdown commences.

1.4.4 Threaded joints

Deviations of *thread* pitch under a normal manufacturing method do not noticeably affect the static strength of the joints, but can reduce the durability limit by up to 20%. Moderate design pitch deviations (negative for a screw or positive for a nut), compensating for the elongation of the screw under load, can also theoretically increase the durability limit.

Deviations of the angle of thread profile, which are not identical for the screw and nut, make the conditions of contact worse in the thread over the profile, but reduce the stiffness of the helixes, which leads to some equalization of load distribution between them. The durability limit in this case increases and, sometimes, the difference between the angles of the profiles of the screw and the nut within 5° is artificially created as a means for increasing fatigue resistance. There is no need for rigidly limiting the deviations of the angles of the profile.

Deviations with respect to thread diameter from the standpoint of static compressive and shearing strengths of the helixes are characterized by the height of helix overlap. For a normal nut height and normal manufacturing accuracy, the strength of the helixes is assured.

The effect of closeness of thread contact on fatigue is expressed by a curve having an extremum. The greatest durability is observed in the region of zero and small tensions. Under large tensions (around 60 μm) a sharp decrease in durability is observed. A decrease of durability by 20-30% is also observed with large clearances. The effect of manufacturing accuracy on the static strength of regular threaded joints can be disregarded.

The perpendicularity limit of the supporting plane of nuts relative to the thread axis amount to 48′ for precise class; 1°11′ for medium class, and 1°37′ for rough class. Perpendicularity limit of the supporting plane of the screw head of a diameter up to 30 mm are 48′ for precise and medium classes and 1°37′ for rough class; deviations of screw heads of a diameter greater than 30 mm are half as large.

In the balance of skewness components, the skewness of the supporting plane of the nuts (heads) dominates. Among them, the parallelism of the tightened surfaces amount about 35-50% of the above skewness, parallelism of the thread and body, the skewness of the thread axis relative to the supporting surface of the part and the parallelism of the planes of the washers from the supporting plane of the nut contribute 35–45%, 45–50%, and 25–35%, respectively.

The total skewness in the joints is approximately determined by the geometrical addition:

$$\alpha \approx \sqrt{\sum \alpha_i^2 \cos\theta_i + \sum \alpha_i^2 \sin\theta_i}$$

where α_i is the skewness of the *i*th element of the joint at angle θ_i.

The actual (99.7%) interval of the spread of skewness, which can be regarded as the design interval, turned out to be 1.8–2 times smaller than the interval computed by algebraic summing according to the minimum-maximum method.

According to experimental data, the total manufacturing and operating skewness (resulting from deformations) of the supporting surfaces of up to $\alpha = 4°$ does not significantly affect the static strength of steel and titanium bolts. Some experiments show that significant reduction of static strength is absent at angles of bolt skewness of up to 8° with the exception of high-strength bolts, which are extremely sensitive to skewness.

Durability of threaded joints decreases sharply due to the skewness α of the supporting surfaces, and is greater the stronger the material of the bolts. For example, the durability of bolts with $\sigma_b = 1080$ MPa at $\alpha = 30'$ and 2° decreases by 12% and 50%, respectively; durability of bolts from steel with $\sigma_b = 1120$ MPa at $\alpha = 30'$ decreases by 19%, and at $\alpha = 1°$ by 50%, and when $\sigma_b = 1560$ MPa – by 48% and 79%, respectively, and for bolts from titanium alloys-by 65% and 71%, respectively. The limit stress amplitudes for incomplete fit of the nut face (head) of the screw decrease by 50–60%, even for screws of 4.6 and 6.8 strength classes.

The effect of skewness of threaded joints on their performance depends, to a large extent, on the flexibility of the screw characterized by the ratio of its length l to the diameter d. Therefore, the allowable misalignment must be established as a function of l/d and not of d. In particular, for screws with small $l/d = 2 \div 4$ the allowable skewness must be reduced by one half compared with normal ratios of l/d of precise class of joints.

1.5 Error interrelationships in machine structures

1.5.1 Error accumulation in serial-kinematics machines

Industrial robots (IR) came into existence in modern concerns only after the mid-fifties of the past century. The modern era in robotics can be reckoned from 1961 when General Motors installed industrial robot Unimate on its production line in New Jersey. Since mechanics of robots and systems of trajectory control have much in common with those used in machine tools and coordinate-measuring machines, it is worth comparing the formation and enhancement of their accuracy with data illustrated in Figure 1.3 and Table 1.1. The IR can be divided in robots with serial- and parallel- and hybrid-kinematics.

As typical examples of *serial-kinematics* robots, the 4-DOF SCARA robot and 6-DOF Stanford arm [12] are shown in Figure 1.17. The former has three revolute joints 1, 2, 3, 4 with parallel vertically oriented axis and one prismatic joint (along the last rotation axis). The latter robot has one fixed link (column 1) and some movable links connected by revolute

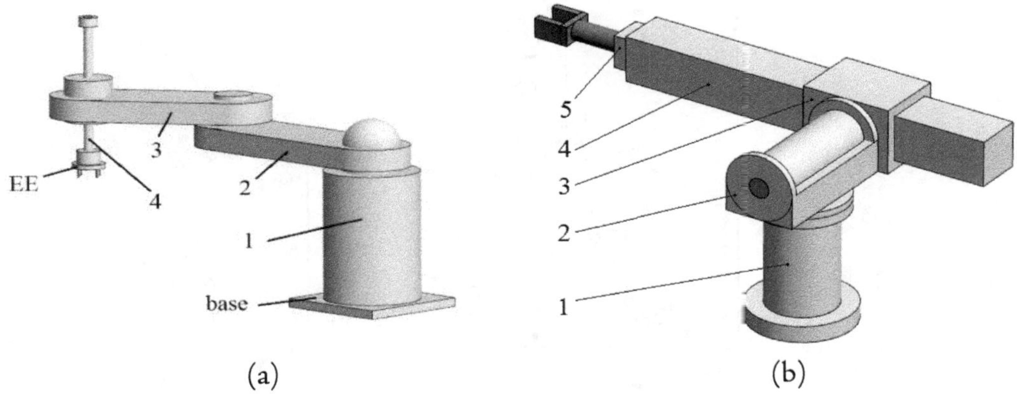

Figure 1-17 The serial-type robot manipulators: (a) the 4-DOF SCARA robot; (b) the 6-DOF Stanford hand.

or prismatic joints. For example, link 2 of the Stanford arm rotates around the vertical axis of the column; link 3 rotates around the horizontal axis intersecting the vertical axis orthogonal to it; an axis of link 4 intersects the previous one perpendicularly; links 3 and 4 compose a sliding pair; link 4 carries out a translational motion relative to link 3; and link 5 rotates about the axis, which is in line with the axis of translation.

Each pair of links forms a 1-DOF kinematic pair, and thus, the manipulator consists of six kinematic pairs, whose motions are driven and controlled by six drives. As is clear from the structure of the serial-kinematics robot, the errors of all links and joints of the serial-kinematics robot caused by its geometry, kinematics, statics and dynamics are accumulated along the way from the base (column) up to the EE. Moreover, small errors of the geometric and kinematic parameters can lead to non-small deviations of the EE from the nominal pose when the EE is away from the column axis.

For a robot of regular dimensions, deviations of the EE from a preliminary given linear position can be of the order of 10–20 mm and more, and orientation errors of the order of some degrees. Hence, the robot has poor *absolute accuracy* of position and orientation. To improve this, the robot calibration procedures are applied [13], carried out in two stages. First, the systematic errors along a required trajectory are measured, and second, these errors regardless of their physical nature are compensated by means of the computerized control system of the robot. The procedure improves the accuracy by at least a factor of 10 and it may reach up to 0.5÷1 mm.

The *repeatability* of the robots (the deviation from the average displacement after repeated positioning) is usually far better than their accuracy. It is significant that the repeatability values (not accuracy) are usually given in the manufacturers notice. Depending on the type, size, and load the repeatability is from ±0.02 up to ±0.05 mm for 4-DOF SCARA-type robots (Fig. 1.17a) and from ±0.1 up to ±0.25 (±0.5) for 6-DOF spherical robots (Fig. 1.17b). Thus, the accuracy and precision characteristics of the serial robots are significantly worse (one order or more of magnitude) than those of machine tools.

As a simple example of accumulation of the links errors, the planar articulated robot is considered (Fig. 1.18). The robot must repeatedly supply a technological tool 5 attached to

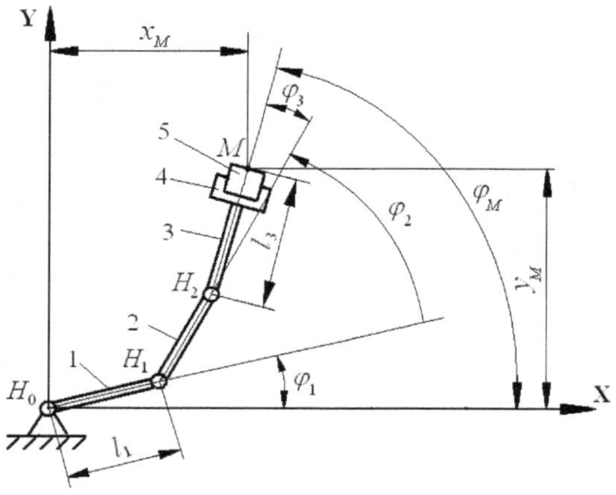

Figure 1-18 The planar RRR robot.

the end-effector 4 to a given position in the horizontal plane. Links 1, 2, and 3 are moved relative to a fixed base. The movable links are connected in series through three revolute joints H_1, H_2, and H_3. Controlled motions of the robot are three rotations given by angles φ_1, φ_2, and φ_3 providing programmed position and orientation of the end-effector. The robot accuracy is estimated by ability of the tool 5 to achieve a specified pose, i.e., tool position coordinate x_M, y_M and tool orientation angle $\varphi = \varphi_1 + \varphi_2 + \varphi_3$.

The deviations Δx_M and Δy_M of the target coordinates x_M and y_M as well as the deviation of orientation from their nominal values caused by:

- Angular deviations $\Delta_{\varphi 1}$, $\Delta_{\varphi 2}$, and $\Delta_{\varphi 3}$ of the angles φ_1, φ_2, φ_3; deviation $\Delta_{\varphi i}$ ($i = 1, 2, 3$) presents sum of the kinematic error and drive-compliance-induced angular error
- Geometrical errors Δ_{l1}, Δ_{l2}, Δ_{l3} of the link lengths l_1, l_2, l_3; and,
- Radial run-out Δ_{H1}, Δ_{H2}, Δ_{H3} of the bearings H_1, H_2, H_3.

As a result, the horizontal and vertical deviations Δx_M and Δy_M of point M and angular deviation $\Delta\varphi$ of tool orientation become:

$$\Delta x_M = (\Delta_{l1} + \Delta_{H1} + l_1\Delta_{\varphi 1}) \cos \varphi_1 + (\Delta_{l2} + \Delta_{H2} + l_2\Delta_{\varphi 2}) \cos(\varphi_1 + \varphi_2) + (\Delta_{l3} + \Delta_{H3} + l_3\Delta_{\varphi 3}) \cos (\varphi_1 + \varphi_2 + \varphi_3) \quad (1.24)$$

$$\Delta y_M = (\Delta_{l1} + \Delta_{H1} + l_1\Delta_{\varphi 1}) \sin \varphi_1 + (\Delta_{l2} + \Delta_{H2} + l_2\Delta_{\varphi 2}) \sin (\varphi_1 + \varphi_2) + (\Delta_{l3} + \Delta_{H3} + l_3\Delta_{\varphi 3}) \sin (\varphi_1 + \varphi_2 + \varphi_3) \quad (1.25)$$

$$\Delta \varphi = \Delta_{\varphi 1} + \Delta_{\varphi 2} + \Delta_{\varphi 3} \quad (1.26)$$

As is clear from Eqs. (1.24–26), geometrical errors (here are manufacturing and assembly errors) and errors of motions (here are angular kinematic errors and run-out) are summarized and accumulated. The accumulation presents a principal disadvantage of the articulated robots, significantly reducing their accuracy. In turn, this results in high magnification of linear errors when the end-effector is located at large distances from its initial position. Other significant contributors are the low angular stiffness of the robot actuators and accumulation of the position errors of the intermediate members at the end-effector.

To enhance the robot accuracy, calibration procedures are developed [13]. The procedures present the software-based compensation of individual deviations of the robot elements and units by means of correction of robot motions based on results of accuracy tests. In the case study (robot Puma 560), the calibration results in enhancement of the position error of the end-effector by at least a factor of 10 – from 0.400 to 0.040 inch.

Since robotics tends to human-intellect-related terms, this approach is referred to as the "*teaching method*" using so-called "*training*" operations. The goal of these procedures is elimination of systematic errors of robot member motions. Approximately, the articulated robot regular accuracy has position accuracy of about 1 mm along each coordinate axis.

Currently, the articulated robots are produced with a relatively high level of accuracy. To avoid the possible error sources, the elements and units of the robot hardware have the stiffer structure and tighter tolerances, and their systematic errors are preliminary compensated using software-aided procedures. For example, 6-axis (Fig. 1.19a) and 5-axis (Fig. 1.19b) Mitsubishi Industrial Robots MELFA RV-EN with maximum reach radius 500-650 mm and payload 2–4 kgf have position accuracy of 0.1 mm along each of the coordinate axis and repeatability of ±0.04 mm [14]. Mechanical issues of the high accuracy of the robot result from very compact robot origin posture, rigid two-sided symmetrical structure of the revolute axes, the ability to hold the full or partial symmetry in many configurations, etc.

Obviously, the error accumulation is also featured for machine tools and CMMs since the vast majority of these machines have serial kinematics, and many attempts at development of high-accuracy multi-axis machines using parallel kinematics systems have not been so successful. Geometrical and kinematics errors are accumulated when the active element (the cutting tool in machine tools and the measuring probe in CMM) moves from one point of the target element (e.g., workpiece or measurand) to another. In particular, translational motions are accompanied with the Abbe error since the working surface of the machine table has structural offset from the guideways. For example, point A in the workspace of the horizontal-boring machine (Fig. 1.20a) is offset by distances shown through vectors A_x, A_y, and A_z from the X-, Y-, and Z-guideways, respectively. Projections of these vectors on coordinate axes represent nine Abbe errors for point A.

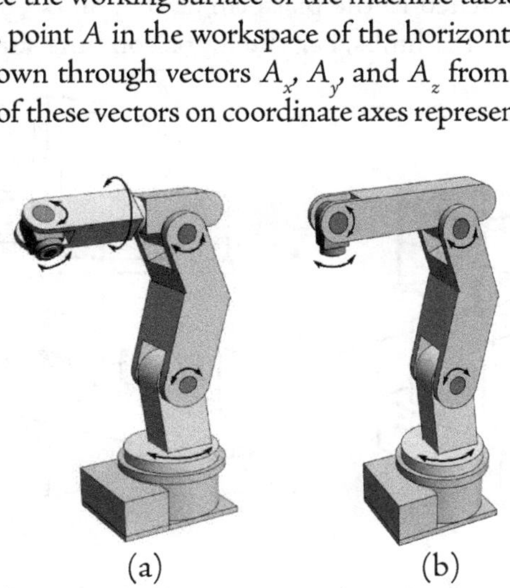

(a) (b)

Figure 1-19 Mitsubishi Industrial Robots MELFA RV-EN with 5 and 6 axes [14].

However, there is a significant difference from the viewpoints of accumulation of geometric and static deformations. Actually, the robot base is located on the end of the kinematic chain of the robot links, whereas the bed of machine tools and CMM, as a rule, presents an intermediate link of kinematic chain, i.e., the chain of machine tools and CMM is bifurcated. Hence, the base of the serial-kinematic robot accumulates deformations of all robot links and joints, while the bed of machine tool or CMM accumulates deformations of only a part of links and joints. Since the bed usually presents the stiffest element of mechanical system, the bifurcation increases machine stiffness and accuracy.

For example, the base of the vertical-milling machine combines two interconnected branches (Fig. 1.20b): the column presents the base for the spindle unit and spindle head carrying out the *Z*-translation and spindle rotation, and the bed bears the saddle and the table worked out the *X*- and *Y*-translations. Obviously, this structure decreases displacements of the cutting tool relative to the workpiece caused by static deformations within the guideways and the other joints.

1.5.2 Error averaging in parallel kinematics machines

New possibilities in seemingly well-studied design areas of machine kinematics have opened through development of the parallel-kinematics machines (PKM) used over a wide range of load and accuracy requirements. Pioneering publications about real-life applications of parallel-kinematics (PK) mechanisms date back about 60 years [15, 16], i.e., approximately at the same time as serial robots. However, their intensive investigations and applications started in the late 1980s as the necessary computerized control technology became available.

As typical examples of the PK robot, the Gough-Stewart platform [15] and Eclipse mechanism [17] are shown in Figure 1.21. In the Gough-Stewart platform (Fig. 1.21a), the moving platform *Pl* is connected with a fixed base by six extensible limbs ("legs" is also in use) 1, 2,...,6 of controlled-variable lengths. Varying the length is actively performed through prismatic joints connecting the upper and lower parts of the limb. The lower ends of the limbs are attached to the base at six points $M_1, M_2,\dots,M_6$ by universal (Hook's, Cardan) joints. The upper ends are attached to the platform at six points $Q_1, Q_2,\dots,Q_6$ by spherical joints. The drives are built into the limbs, i.e., the limbs combine supporting and kinematic

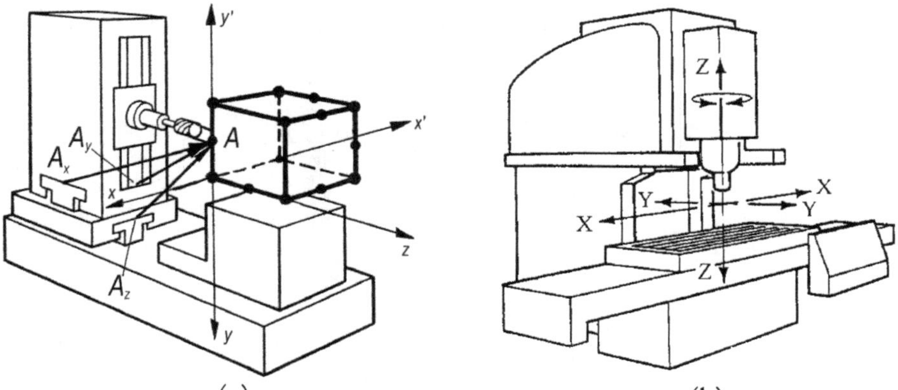

(a) (b)

Figure 1-20 Machine tool structures: (a) the workspace of the horizontal-boring machine; (b) layout of the vertical-milling machine.

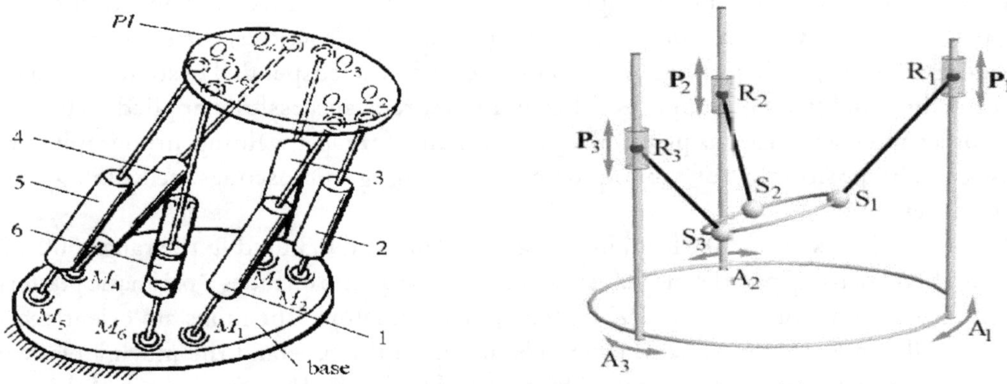

(a) (b)

Figure 1-21 The 6-DOF parallel mechanisms: (a) the Gough-Stewart platform (hexapod); (b) the Eclipse mechanism (tripod).

functions for moving platform *Pl* with respect to its base. Thus, the platform is connected with the base by six independent identical kinematic chains.

The specific structure and exterior of the 6-DOF platform mechanisms with six limbs have resulted in the entomology-related name *hexapod*. Correspondingly, the name *tripod* is applied to mechanisms with *three* limbs. The Eclipse mechanism [17] (Fig. 1.21b) has three vertical columns that can slide along the horizontal-base-located circular guideways A_1, A_2, A_3 and three sliders P_1, P_2, P_3 translate along the columns. One end of the limb is connected to the mobile platform through a spherical joint; the opposite end is connected through a revolute joint to the slider, which translates along the fixed guideways on the columns.

In both mechanisms, each pair of the limbs in common with the fixed base and the platform forms a closed kinematic loop. Thus, forces acting on the EE – mobile platform – are absorbed by some supports (limbs), and, conceptually, they have high potential stiffness and accuracy in comparison with those of the serial-kinematics robots. The same is also true for dynamics features due to low moving mass of the moving platform compared with a large chain of links of the serial robots.

To provide the high accuracy of the PK robot, a calibration-compensation procedure is necessary. However, the procedure has some special features due to parallel kinematics since any positioning operation requires joint action of all active axes. Application of parallel kinematics systems in multi-axis machine tools shows that the former are quite competitive with the latter. The accuracy of the 5-axis and 3-axis PK machine tools is 10÷30 and 10÷15 μm, respectively [18-20], i.e., one-two orders of magnitude better than that of serial-kinematics robots. Similar correspondence takes place for stiffness values, but a portion of kinematically-admissible space has to be eliminated from the workspace. In the parallel- and hybrid-kinematics machine tools, the stiffness values in the directions orthogonal to spindle axis are in the range $(1.5 \div 3.0) \times 10^7$ N/m.

From the accuracy viewpoint, a specific group of parallel-kinematics robots forms robots for small and very small displacements – *MEMS and nano-robots*.

The best accuracy during multi-DOF motion in the sub-millimeter range with sub-micron resolution is reached by means of parallel mechanisms with so-called orthogonal architecture. The principle scheme is shown in Figure 1.22, where the moving platform is shown as a cube. In contrast to classical structures, the six supports are oriented in pairs parallel to the coordinate axes, i.e., the suspended system consists of three orthogonal pairs of supports. This architecture is successfully applied to 6-DOF ultraprecision stages used as positioning instruments for laser welding, lithography processes, probe positioning of atomic force and scanning microscopes, cell biology, and medical research.

For example, resolution 0.01 μm in the range 20 μm and 0.3 μrad in the range 600-700 μrad is obtained using parallel mechanisms [22]. The structure includes a prismatic platform based on six limbs with build-in PZT piezoelectric actuators. The limbs are orientated in pairs parallel to Cartesian axes. They are connected to the base and the moving platform through flexure hinges without external friction and backlash. The mechanisms of this type provide stiffness values up to $(0.2 \div 2) \times 10^7$ N/m.

Parallel kinematics machines have higher accuracy and stiffness in greater part of their workspace compared with articulated robots. As an example, let us consider the position and orientation errors of the moving platform of the Gough-Stewart platform caused by deviations $\Delta_1, \Delta_2, \ldots, \Delta_6$ of the limb lengths from the programmed values. These deviations result from kinematic errors of the preprogrammed limb elongations and static errors caused by compliance of three elements: robot drive system, and spherical and universal joints. In the central location of the horizontally-oriented platform, nominal lengths of all six limbs coincide and present function of height position z, i.e., $l_1 = \ldots = l_6 = l_z$. The position error Δ_z associated with the central point of the platform along the vertical direction is:

$$\Delta_z = \frac{l(z)}{z}\frac{(\Delta_1 + \Delta_2 + \Delta_3 + \Delta_4 + \Delta_5 + \Delta_6)}{6} \tag{1.27}$$

In particular, if the limbs are of identical accuracy and stiffness, i.e., $\Delta_1 = \ldots = \Delta_6 = \Delta$, Eq. (1.27) yields $\Delta_z = (l_z/z)\,\Delta$. Hence, error Δ_z of the platform position results from averaging the errors of six limbs including geometrical and kinematic errors, and compliance-induced

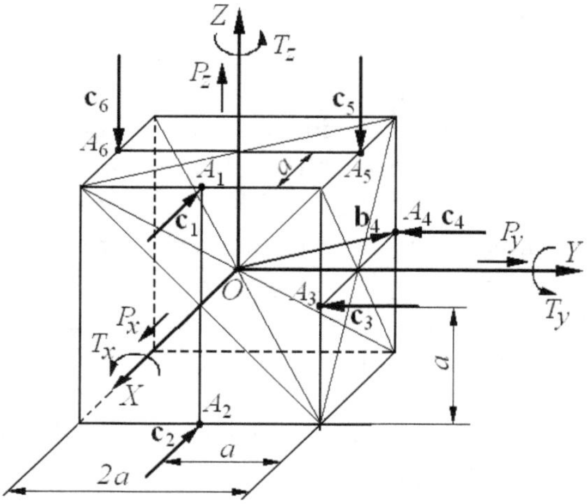

Figure 1-22. Isotropic suspension system (2-2-2 version).

deformation errors, in contrast to articulated robots where the errors are accumulated. This is the main reason for the higher accuracy and stiffness of the parallel robot in comparison with the articulated one. Note that the complete averaging takes place only for vertical motion of the mobile platform in its central location and horizontal orientation, and only for the error measured along the vertical axis. In the general pose of the platform and along an arbitrary direction of measurement, averaging of six errors also takes place but not in so simple form as in the symmetric case described by Eq. (1.27). Moreover, the parallel-kinematics mechanisms usually have singular configurations wherein the compliance-caused deformations (static errors) not only are average, but can drastically increase.

1.6 Conclusion

The general approach to accuracy problems is considered. The problem is considered as applied to the multifunctional and reprogrammed precision working machines such as machine tools, industrial robots and coordinate measuring machines. These machines share the common properties since they generate complex predesigned accurate (or relatively accurate) trajectories by means of a series of relatively simple motions of their units using different working tools of relatively simple form.

References

[1] Dickenson, H.W., 1935, James Watt: Craftsman and Engineer, Cambridge University Press, Cambridge, UK.

[2] Taniguchi, N., 1992, Future-Trends of Nanotechnology, International Journal of the Japan Society for Precision Engineering, Vol. 26, 1, pp. 1-7.

[3] ISO 1101, 2004, Technical drawings; Geometrical tolerancing; Tolerances of form, orientation location and run-out; Generalities, definitions, symbols, indication on drawings.

[4] Humienny, Z., (ed.), Osanna, P.H., Tamre, M., Weckenmann, A., Blunt, L., Jacubiec., W., 2001, Geometrical Product Specifications: Course for Technical Universities, Warsaw University of Technology Printing House, Warsaw.

[5] Moore, W.R., 1970, Foundations of Mechanical Accuracy, The Moore Special Tool Co., Bridgeport, CT, USA.

[6] Hocken, R., 1980, Machine Tool Accuracy, in Technology of Machine Tools, Vol. 5, Lawrence Livermore Lab., University of California, UCRL-52960-5, Livermore, CA.

[7] ASME Y14.5M-1994, Dimensioning and Tolerancing, An ASME National Standard, ASME, New York, NY, ASME14.5-2009.

[8] ASME Y14.5.1M-1994, Mathematical Definition of Dimensioning and Tolerancing Principles, An ASME National Standard, ASME, New York, NY.

[9] Sutton, G.P., in Hocken (ed.), 1980, in Technology of Machine Tools, Vol. 5, Lawrence Livermore Lab., University of California, UCRL-52960-5, Livermore, CA, pp. 9.3-1.

[10] Yakushev, A.I., Vorontsov, L.N., and Fedotov, N.M., 1986, Interchangeability, Standardization, and Technical. Measurements, Mashinostroenie, Moscow, (in Russian).

[11] Shigley, J., Mischke. C.R., 1989, Mechanical Engineering Design, 5th ed., McGraw-Hill Book Company, New York.

[12] Tsai, L.-W., Robot Analysis: The Mechanics of Serial and Parallel Manipulators, 1999, John Wiley, New York, NY.

[13] Mooring, B.W., Roth, Z. S., Driels, M.S., 1991, Fundamentals of Manipulator Calibration, John Wiley, New York, NY.

[14] Mitsubishi, Industrial Robot - MELFA RV-EN, Specification Manual, 1998.
[15] Stewart, D., 1965, A platform with six degrees of freedom, Proceedings 1965/66 of the Institute of Mechanical Engineers, Part1, 180 (15), 371-386.
[16] Gaugh, V.E., et al, 1962, Universal tire test machine, Proc. 9th Int. Automobile technical congress FISITA, ImechE, London, Vol. 117, pp. 117-137.
[17] Kim, J. W., Park, F.C., Lee, J.M., 1999, New parallel mechanism machine tool capable of five-face machining, CIRP Annals - Manufacturing Technology, 48/1, pp. 337-340.
[18] Schwenke, H., Knapp, W., Haitjema, H., Weckenmann, A., Schmitt, R., Delbressine, F., 2008, Geometric error measurement and compensation of machines - An update, *CIRP Annals - Manufacturing Technology*, 57 (2), pp. 660-675.
[19] Brecher, C., Flore, J., Wenzel, C., 2013, Systematic analysis and identification of 5-axis machining accuracy, Proceedings of the 13th International Conference of the European Society for Precision Engineering and Nanotechnology, EUSPEN 2013 Vol. 2, 2013, pp. 331-334.
[20] Slocum, A., 1992, Precision Machine Design, Prentice Hall, Englewood Cliffs, NJ.
[21] Tomita, Y., Sato, F., Ito, K., Koyanagawa, T., 1992, Decoupling Method of Ultraprecision Stage using Parallel Linkage Mechanism, Int. J. Soc. Prec. Eng., Vol. 26, No. 1, pp. 47-53.

2. Accuracy-associated models

2.1 Models of machine structure

2.1.1 Working machine and working process

A working machine (WM) presents equipment (assemblage of parts) transmitting energy, materials, and information by means of *controlled* mechanical motions with the purpose to displace, replace and simplify physical and/or intellectual works of manual and non-manual workers as well as increase productivity and quality of the manufactured products. Typical representatives of the WMs are industrial robots (IR), machine tools (MT), and coordinate measuring machines (CMM). As modern machinery is developed, automation and computerization level of the WMs is increased to improve such important performance indexes as productivity, accuracy, flexibility, universality, effectiveness, etc. As a rule, it adds complexity to the WM kinematics, drives, and control system: the number of operation-required motions increases and their interrelationship becomes more involved.

Two parallel processes are simultaneously acting in the WM (Fig. 2.1). The power process transforms the input power from a supply of electrical, chemical, beam, mechanical or other input energy into the *mechanical energy of movable units* by means of drives and actuators accompanied if necessary by power transmissions. This energy provides manipulation and transportation functions as applied to objects to be processed. The spectrum of the working processes is very wide: machining, welding, painting, deposition, measuring, assembling, packaging, stockholding and so on.

In parallel with the power process, the information process forms, transforms, and fixes the target information from an active element (tool) to a processed (target) element providing its form and shape.

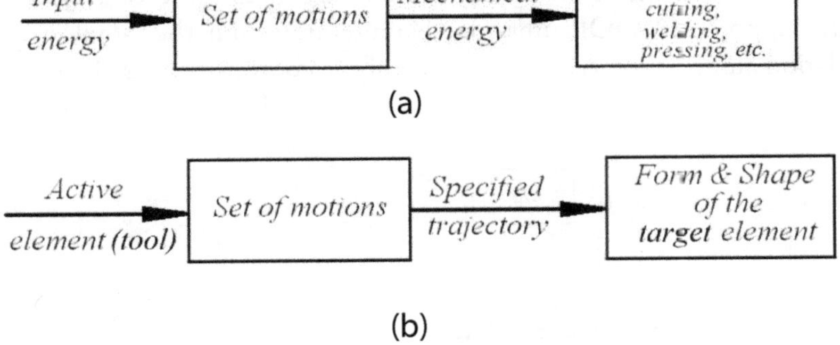

Figure 2-1 Machine working processes: (a) energy transformation; (b) flow of information.

From the informatics viewpoint, any WM includes three basic systems: mechanical, actuating, and control (Fig. 2.2). The direct information way is as follows: the control system (*CS*) obtains the preprogrammed information from an external source and transfers it through the drives-and-actuators system (*AS*) and power transmissions to the mechanical system (*MS*) whose movable units carry out prescribed motions. There is also the feedback information trace: the *CS* takes up the signals about a current situation within *AS* and *MS* and accordingly corrects the outputs.

Distinctive features of the WMs result from the necessity to combine complicated functional assignment with predetermined (usually, high or stringent) quality requirements among them the accuracy one of the most important features. To satisfy these requirements, the mechanical system must have precise and stiff structure designed for multi-DOF motion. A permanent tendency to meet these requirements replace mechanical elements composing the machine structures for electrical, electronic and other non-mechanical elements.

The native means for this tendency is computerization of the manufacturing processes and equipment through built-in NC and CNC systems. For example, the kinematic chain between the spindle and main leadscrew in the engine lathe is very cumbersome and, in particular, includes a change gear train, which requires the hand operations for changing the toothed gears, a place for gear storage, and so on. Contrary to that, the CNC turning center uses a separate power motor for the main (cutting) motion and servo or stopped motors for feeds, while the kinematic connection of their motions is worked out through regular software.

The same tendency can be demonstrated in the development of translational feed drives (Fig. 2.3). Widespread feed drives (Fig. 2.3a) obtain the rotary motion from an electric engine *M* and transform it through belt transmission *1*, gearbox 2, and ball screw-nut pair *3-4* into translation of the working table WT. The next generation (Fig. 2.3b) has a high-torque engine *HTM* directly attached to lead screw-nut pair. Some modern machines have a linear engine 5 (Fig. 2.3c), i.e., no rotational elements, the low-inertial structure with minimum mechanical elements, and highest dynamic properties.

Note that the group of machines considered in this book—machine tools, industrial robots and CMM—become, from the viewpoint of their structure, the machines best suited for computerized motion control since they realize target complicated trajectories using a set of serially supported 1-DOF units. Each unit carries out the simplest elementary motion—translation along or rotation around one of the coordinate axes. This is the reason

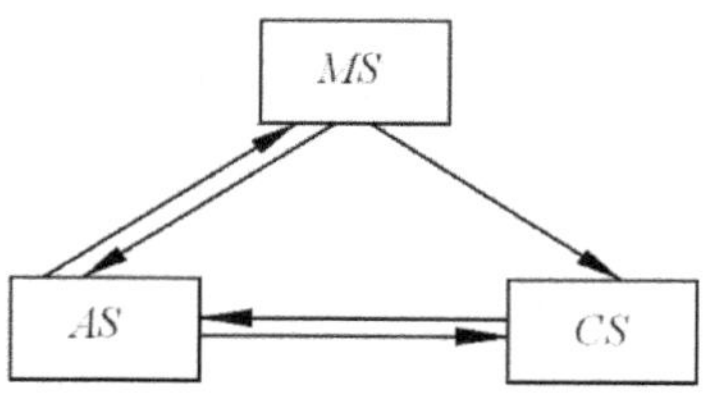

Figure 2-2 Information model of a working machine: *MS* – mechanical system; *AS* – actuating system (drives); *CS* – control system.

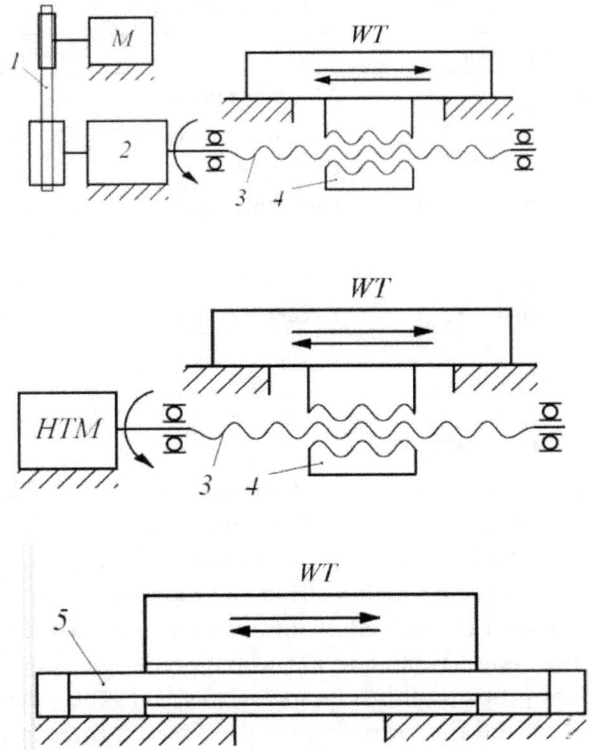

(a)

(b)

(c)

Figure 2-3 Translational feed drives development: (a) an electric engine with-power transmission; (b) a high-torque engine directly attached to lead screw-nut pair; (c) a linear engine.

why in modern structures of these machines, the applied numerical control systems (CNC) are similar. It is fair to say that application of these systems began from machine tools and only after this the CNC became a regular part of CMMs and IRs. General features set off machine tools, IRs, and CMMs in a special group – the precision WM (PWM).

A target-oriented part of the PWM mechanical system is a manipulation system presenting an ordered aggregate of machine links, the relative positions and mutual movements of which provide execution of the design-specified operations (Table 2.1). The goal is to obtain designed (prescribed, preprogrammed) trajectories of active elements (AE) specified for the PWM relative to target elements (TE) subjected to a working process. The manipulation system presents a chain of mechanical links composing a base and some (no fewer than one) movable links. In the stationary IR, the base link is located on one of the ends of the manipulating system [e.g., 1-5], while in machine tools and CMMs, the stationary link, as a rule, presents an intermediate link of the manipulation system.

The AE and TE are changeable parts usually attached to the extreme (ending and initial, respectively) links of the manipulation system. Being carried out the working process, the AE acts on a local area of the TE at the moment and extends the action over all the predesigned area of the TE owing to the manipulation process.

Table 2-1 General terms relating to precision working machines.

Concept	Machine tool	CMM	Industrial Robot
Active element (tool)	Cutting tool	Measuring probe	Attached to EE[(a)]
Target (passive) element	Workpiece	Part to be measured	Processed element
Main functional system	Machining system	Coordinate system	Manipulating system
Power process	Cutting	Manipulation	Assembly, welding, etc.
Informatics function	Form-shaping	Accuracy evaluation	Function dependent

[(a)]EE – end-effector

2.1.2 Form-shaping system: description

Interrelationships between the manipulating system and the TE and AE of the PWMs (machine tools, IRs, and CNNs) are introduced by means of an integrated object called a *form-shaping system (FSS)*. The FSS of the WM is referred to as an aggregate of joints connecting mechanical links, whose mutual position and movement ensure the travel trajectory of the AE relative to the TE specified by information and working processes [6].

The FSS consists of at least one AE and one TE, whose predesigned and controlled relative motions are worked out by means of the manipulation system and interrelated through the working process. The FSS and the form-shaping processes as their target function are the central concerns of this book. The form-shaping process is dedicated to producing and/or reproducing the form and shape of the TE using two type of input information: a set of kinematic motions and the geometry of active elements.

The FSSs of PWMs feature the combination of high requirement for accuracy and productivity and complicated trajectories of the AE relative to TE. Hence, the PWMs simultaneously provide both the general *form* of the TE and its accuracy features (*shape*). For example, in the machine tools, the process results in predetermined transformation of the form and shape of the blank or semi-finished item into the designed product fixed by the drawing.

Main examples of the AEs, TEs, and *manipulation systems* of industrial robots, machine tools and CMMs are shown in Table 2.1. In industrial robots, the AE are varied technological tools attached to the robot end-effector: an electrode for arc welding, an air spray gun for painting and cleaning, fitting tools for assembly, etc. In machine tools, the AEs present a wide spectrum of cutting tools and ground wheels. In coordinate measuring machines (CMM) and measuring robots, the AEs present a measuring probe, measuring heads, and so on.

Some examples of basic elements of form-shaping systems are listed in Table 2.2. For example, the AEs of industrial robots are electrodes for plasma arc welding, spray guns for painting, different grippers for assembling; the AEs of the CMMs and measuring robots are measuring probes, the AEs of machine tools are different cutting tools for turning, milling, drilling, grinding, etc. The TEs of the industrial robots are the parts to be welded, painted, assembled, measured, and so on. The TEs of the machine tools present billets, blanks, workpieces, and in-process and machined parts.

Table 2-2 FSS basic elements.

Working machine	Active element	Working process	Target element
Welding robot	Electrode(s)	Welding	Weld
Assembly robot	Workpart gripper	Assembly	Conjugate part
Painting robot	Sprayer	Painting	Painted surface
Plasma cutting robot	Plasma torch	Plasma cutting	Flat sheets of steel
Measuring robot	Measuring head	Measurement	Measured part
Lathe	Cutting tool	Turning	Workpiece
Milling machine	Milling cutter	Milling	Workpiece
Broaching machine	Broach	Broaching	Workpiece
Drilling-boring machine	Drill, boring tools	Drilling-boring	Hole; bore
Grinding machine	Grinding wheel	Grinding	Surface to be ground
Gear hobbing machine	Hob	Hobbing	Toothed gear
CMM	Measuring probe	Measurement	Measured part

The working processes can be accompanied by presence or absence of a direct mechanical contact between the AE and TE. The former takes place during machining processes when mechanical cutting tools are applied. In this case, the FSS becomes a mechanically-closed loop. The latter is typical for processes using an active medium such as welding arc, laser beam, electrical discharge, high-pressure water, etc. In this case, the FSS can be considered as an open kinematic chain.

Depending on the process type, identical subjects can be produced with different combinations of the input kinematic and geometric parameters. For example, the molding of a sculpture surface uses a single kinematic motion and an active element (die mold) of the complicated form presenting a nearly ideal mirror image of the target surface, while the machining of the same surface has to be executed by application of five motions and one milling cutter or grinding wheel of a simple (for example, spherical) form.

The list of form-shaping processes includes a wide spectrum of technological processes such as machining (material removing), molding, welding, assembling, measuring, and so on. In this book, only processes requiring application of machines with complicated kinematics are considered. Obviously, mechanical systems of the machines simultaneously perform both power and information processes.

The appropriate *target elements* are the object to be processed by the robot (e.g., fixing part of the assembled pair when an active element of assemblage is located and manipulated by the end-effector), the part to be machined on the machine tool, the part to be measured on the CMM, etc.

The *manipulation system* provides the relative motion of the AE attached to one ending link of the kinematic chain with respect to the TE attached to another ending link. The link of the FSS is an aggregate of the machine elements and units, whose shift relative to each other is nominally absent during the functioning of the machine, i.e., the link includes the elements that permanently remain relatively stationary during the machine operation.

For example, the spindle of a lathe together with the chuck and clamped part to be machined are regarded as a link. However, if necessary, the three elements (the spindle, chuck, and workpiece) can be considered as a chain of links unmoved relative to each other.

The FSS comprises the bed and movable units of the machine tool together with the final drive-transmission elements. The initial and ending links of the FSS are the workpiece and cutting tool, respectively. In some cases, the FSSs can be reconfigurable, i.e., their composition can partially change due to change of some units as well as the workpieces and cutting tools associated with the FSS.

The consideration in this book starts from the FSS of serial-type robots and machine tools, with further extension to machines with parallel and hybrid kinematics. The interrelation of the FSS with the other system is depicted in Figure 2.4. The FSS takes up the motion from the actuators and sends the feedback signals indicating the positions of different links through sensors of the control system. During the process of adjustment, the composition of the FSS can partially change due to the change of the some elements including the *AE* and *TE*.

The simplest, and simultaneously, the most widespread form of the FSS is the chain of links represented by the chain-type graph shown in Figure 2.5a. It comprises one *AE* (tool) processing one *TE*. In this case, the intermediate links of the chain include the supports for two adjacent links, while the ending links are connected by the supports with only one neighboring link. In Figure 2.5a, the *AE* and workpiece *TE* are shown as an input and an output signal, respectively, while all the links of the FSS are considered as a unique vertex a_0 symbolizing the multistage transformation law resulting from the sequence of link motions.

To enhance machine productivity, the multiple-tool and multiple-positioning schemes are widely used, especially in equipment for high-volume production These FSSs are described by tree-graphs (Fig. 2.5b, c, d), whose terminals (external vertexes) correspond to the *AEs* and *TEs*. The multiple-tool FSS (Fig. 2.5b) consists of several tools

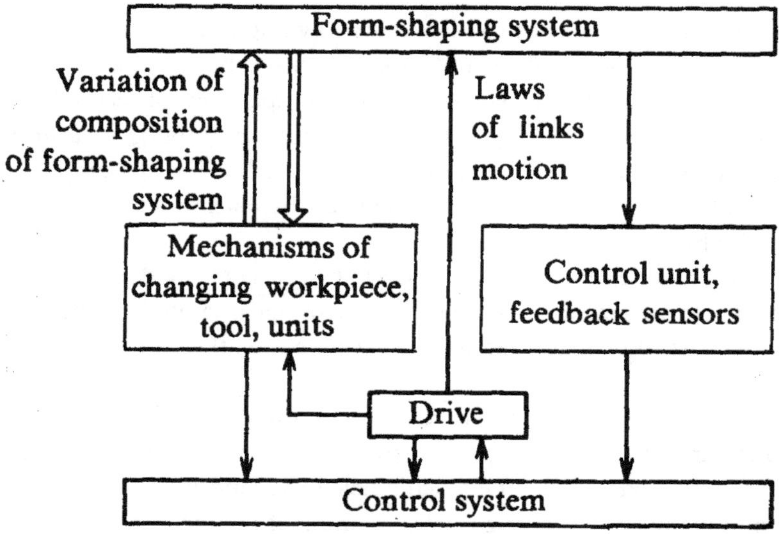

Figure 2-4 Interrelation of form-shaping system with other machine systems.

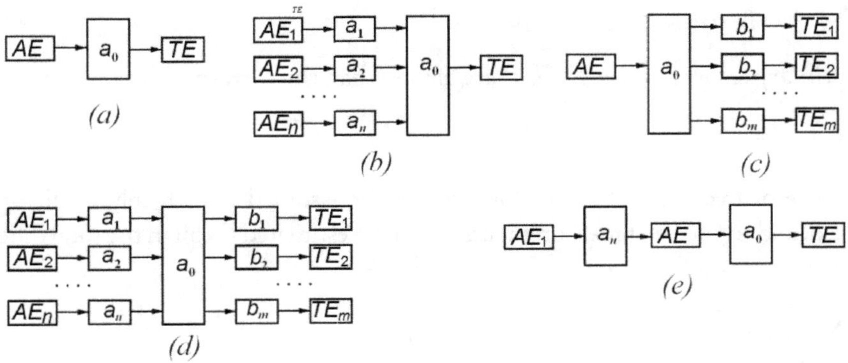

Figure 2-5 FSS structures: (*a*) chain structure; (*b*) multiple-tool FSS; (*c*) multiple-positioning FSS; (*d*) parallel-action structure; (*e*) extended FSS; AE_1, AE_2,..., AE_n– active elements; TE_1, TE_2, ..., TE_m– target elements; a_0 – manipulation system; a_1, a_2,..., a_n– tool associated elements; b_1, b_2, ..., b_m– *TE* associated elements.

AE_1, AE_2, ..., *AEn* processing a single *TE* (e.g., drilling machines with multiple spindle holders conventionally applied in the automotive industry). The multiple-positioning FSS (Fig. 2.5c) has one *AE* and machines several workpieces *TEs* simultaneously. As examples, one can cite face-grinding machines for working rollers, rings, and so on. Figure 2.5d shows the layout of a machine for parallel machining by several tools of several workpieces. The turning multiple-spindle machines have such an FSS.

In Figure 2.5e, a graph of an extended (two-stage) FSS chain is shown. The cutting tool is considered as an output link of the first stage of the FSS. In turn, the cutting tool presents the input link of the second stage. Such situation takes place in the FSS of grinding machines. The grinding wheel has to be preliminarily dressed with aid of a diamond or a carbide-tipped tool and only thereafter the wheel can serve as the tool for machining the piece to be ground.

The main mathematical model of the FSS synthesizing information relating to a manipulation system, a technological tool and the object to be processed, is the form-shaping function (FSF) formulated in Chapter 3.

2.2 Accuracy Calculation

2.2.1 Differential method of accuracy evaluation

Historically, the problem of calculating the accuracy of complicated systems arose on the basis of the problem of estimating the accuracy of indirect measurements and dimensional chains when the estimated quantity x can be presented as a function of a set of other quantities $q_1, \ldots, q_n$:

$$x = f(q_1, \ldots, q_n) \tag{2.1}$$

If quantities $q_1, \ldots, q_n$ are known with the error of measurements or manufacturing (primary errors) $\Delta q_1, \ldots, \Delta q_n$, correspondingly, the error of measurements δx of quantity x is estimated as

$$\delta x = f(q_1 + \Delta q_1, \ldots, q_n + \Delta q_n) - f(q_1, \ldots, q_n) \quad (2.2)$$

If the primary errors Δq_i ($i = 1, \ldots, n$) are far less than the corresponding nominal values, i.e.,

$$\Delta q_i << q_i, \quad (2.3)$$

two- and more-degree terms of the Taylor series can be assumed as negligibly small, Eq. (2.2) can be linearized, and the error of measurements δx can be estimated through the total differential dx of function $f(q_1, \ldots, q_n)$, Eq. (2.1):

$$dx = \sum_{i=1}^{n} \frac{\partial f}{\partial q_i} \Delta q_i \quad (2.4)$$

where the partial derivative $\frac{\partial f}{\partial q_i}$ ($i = 1, \ldots, n$) is the transfer factors (or the transfer functions if it depends on one or more arguments) associated with the input errors Δq_i, and Eq. (2.4) presents the rule of error propagation from the input errors to the output one.

Equation (2.4) defines the *differential method* of accuracy calculation, which is based on interpretation of position errors as a total differential of the position function, Eq. (2.4), and successfully applied to the calculations of the *accuracy of mechanisms* [7]. A basic characteristic of the mechanism kinematics is the *position function* expressing the displacement of the output link through the given time-history displacement(s)of the input link(s). In the case of the mechanism with one degree of freedom (1-DOF), there is a single independent variable, and the *nominal* position function is expressed in the form of Eq. (2.1) as follows:

$$x(\theta) = f(a_1, \ldots, a_n) \quad (2.5)$$

where θ is the independent variable; and $a_1, \ldots, a_n$ are the dimensional parameters of the mechanism.

If only the proper deviations of the mechanism components without errors of initial data are taken into account, Eq. (2.4) becomes:

$$df(\theta) = \frac{\partial f(\theta)}{\partial \theta} \Delta\theta + \sum_{i=1}^{n} \frac{\partial f(\theta)}{\partial a_i} \Delta a_i \quad (2.6)$$

where $\partial f(\theta)/\partial\theta$ presents the transfer factors of *kinematic error* $\Delta\theta$; and $\delta f(\theta)/\delta a$ ($i = 1, \ldots, n$) is the θ-dependent transfer functions for geometric *errors* Δa_i *associated* with parameter a_i occurring during the manufacture of the links of the mechanism and its assembly.

Let us consider two relatively simple examples, which demonstrate some problems relating to application of the differential method.

Example 2.1: The accuracy of an in-line crank-slider mechanism (Fig. 2.6). A position function of this 1-DOF mechanical linkage expresses a current position of the slider as a function of one kinematic and two dimensional parameters:

$$x = x(\theta, R, L) = R + L - R\cos\theta - \sqrt{L^2 - R^2\sin^2\theta} \quad (2.7)$$

where x is the distance from the extreme right position of point C (the center of the rod-piston hinge), $0 \le x \le 2R$; θ is the angle of rotation of crank 4, $0 \le \theta \le 2\pi$; $R = |AD|$ is the radius of the crank; and $L = |DC|$ is the length of the connecting rod 3.

The total differential of the right-hand side of position function, Eq. (2.7), presents a sum of the position and kinematic errors:

$$dx = \frac{\partial x}{\partial \theta}\Delta\theta + \frac{\partial x}{\partial R}\Delta R + \frac{\partial x}{\partial L}\Delta L = \left(R\sin\theta + \frac{R^2\sin^2\theta}{2\sqrt{L^2 - R^2\sin^2\theta}} \right)\Delta\theta + \left(1 - \cos\theta + \frac{R\sin^2\theta}{\sqrt{L^2 - R^2\sin^2\theta}}\right)\Delta R + \left(1 - \frac{L}{\sqrt{L^2 - R^2\sin^2\theta}}\right)\Delta L \qquad (2.8)$$

Obviously, the differential of position function takes into account deviations of all parameters included in position function, and for mechanisms they present the important or sometimes main part of the total error. However, in mechanical systems, particularly in those that must meet high-accuracy requirements, there are some error sources that the total differential, Eq. (2.4), cannot take into account since the corresponding components are *outside the position function*, Eq. (2.1). In particular, run-outs of three revolute joints in the crank-slider mechanism shown in Figure 2.6 result in additional deviations of the displacement x, but they are absent in position function, Eq. (2.7), and hence, not taken into consideration by the total differential, Eq. 2.8. Obviously, in the regular accuracy requirements, these deviations can be ignored, but for high-accuracy systems, they have to be taken into account using additional calculating schemes.

Example 2.2: Accuracy calculation of a cylinder machined on a lathe using longitudinal turning (Fig. 2.7). In the simplest case, the position function of the lathe describes a current position of the single-point cutting tool. The position vector of the machined cylinder of radius R, whose axis oriented along the Z-axis, has the form:

$$r = \begin{bmatrix} R\cos\varphi \\ R\sin\varphi \\ z \end{bmatrix} \qquad (2.9)$$

where φ and z is the polar angle and the longitudinal coordinate of a current cylinder point.

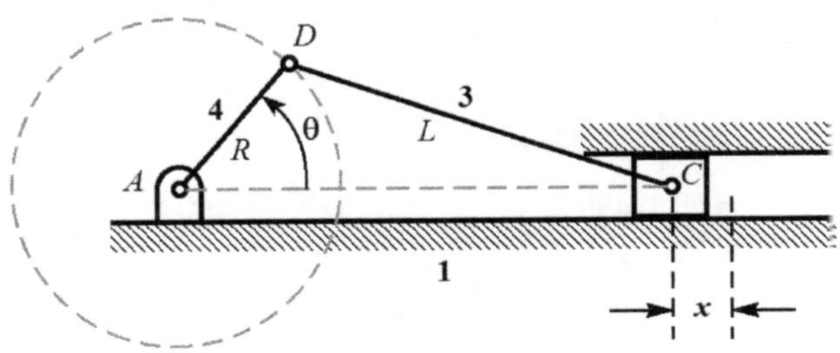

Figure 2-6 The in-line crank-slider mechanism.

The kinematic parameters φ and z are formed by rotational and translational movements of the single active point of the cutting tool, while the radius R is given by fitting position of the cross slide. The total differential of the position vector (2.9) combine two kinematics and one dimensional error:

$$d\mathbf{r} = \begin{bmatrix} -R\sin\varphi \\ R\cos\varphi \\ 0 \end{bmatrix} d\varphi + \begin{bmatrix} 0 \\ 0 \\ 1 \end{bmatrix} dz + \begin{bmatrix} \cos\varphi \\ \sin\varphi \\ 0 \end{bmatrix} dR = \begin{bmatrix} -(R\sin\varphi)d\varphi + (\cos\varphi)dR \\ (R\cos\varphi)d\varphi + (\sin\varphi)dR \\ dz \end{bmatrix} \tag{2.10}$$

Dimensional and geometrical errors of the machined surface are defined through the normal error Δn—projection of the vector $d\mathbf{r}$ on the normal $\mathbf{n}$ at the surface (Fig. 2.7):

$$\mathbf{n} = [\cos\varphi, \sin\varphi, 0]^T \tag{2.11}$$

$$\Delta_n = d\mathbf{r} \cdot \mathbf{n} = \begin{bmatrix} -(R\sin\varphi)d\varphi + (\cos\varphi)dR \\ (R\cos\varphi)d\varphi + (\sin\varphi)dR \\ dz \end{bmatrix} \cdot \begin{bmatrix} \cos\varphi \\ \sin\varphi \\ 0 \end{bmatrix} =$$

$$(-R\cos\varphi\sin\varphi + R\cos\varphi\sin\varphi)d\varphi + (\cos^2\varphi + \sin^2\varphi)dR = dR \tag{2.12}$$

Thus, the differential method results in a single component of the normal error – the error of the setup for radius. However, this result does not include such important issues of the dimensional errors as deviations of movements of two units taking part in the cylinder turning: spindle run-out, parallelism of the longitudinal movement to spindle axis and straightness of the longitudinal movements. The reason is that corresponding parameters are ignored by the position function, and thus cannot appear in its total differential.

The examples have been clearly demonstrated that the differential method does not take into account some important types of errors of mechanical structures and needs for extension.

2.2.2 Variational method of accuracy evaluation

A convenient method for significant extension of counted errors presents consideration of the machine mechanical system as a set of ordered interrelated links having six position and orientation errors each. This approach is called the *variational method* of accuracy evaluation since it uses a variation of the position function instead of its differentiation. It should be particularly emphasized that the set of errors obtained by differential method become a part of the general set resulted from the variational method.

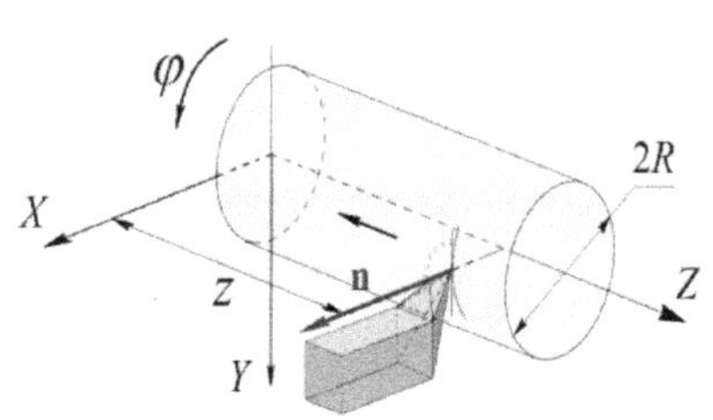

Figure 2-7 Turning the cylinder on the lathe.

Let us return to Eq. (2.2). Two terms in its right-hand side present the same function with different values of the arguments. Thus, differential dx takes into account only deviations of the mechanism parameters. However, eccentricities, clearances, misalignments, and other deviations from nominal size and form result in a small violation of the function $f(q_1, \ldots, q_n)$ in itself, i.e., during the operations and measurements the realized function is

$$x_1 = f_1(q_1, \ldots, q_n)$$

and, in addition, the difference

$$\delta f = f_1(q_1, \ldots, q_n) - f(q_1, \ldots, q_n) \tag{2.13}$$

is small in magnitude.

A small function δf of variables $q_1, \ldots, q_n$, Eq. (2.9), is called the variation of function $f(q_1, \ldots, q_n)$. On the whole, the total error Δx of the mechanism presents the total variation Δf of function $f(q_1, \ldots, q_n)$:

$$\Delta x = \delta x + \delta f, \tag{2.14}$$

where δx, Eq. (2.2), evaluates the parametric errors caused by deviations of the input parameters $q_1, \ldots, q_n$; and δf, Eq. (2.13), presents isoparametric variation of function $x = f(q_1, \ldots, q_n)$ resulting from small deviation of function at constant values of input variables $q_1, \ldots, qn$.

Figure 2.8 illustrates the concepts of function variations for the case of a single argument. The discussion above indicates the bounds of the differential method and demonstrates why the variational method its presents further natural development. The variational method takes into account both groups of errors entering into Eq. (2.14) and will be described with applications as the total variation of the form-shaping function (FSF) [8, 9](see Part II of this book).

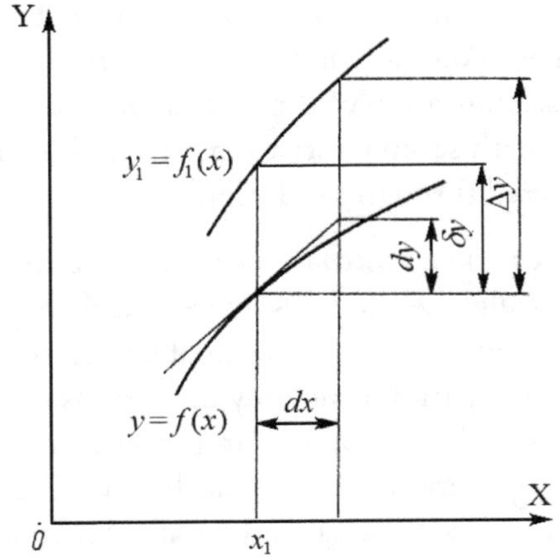

Figure 2-8 Comparison of total differential dy with variations of function $y = f(x)$: δy is the isochronic variation and Δy is the total variation.

Note that the variational method:

(a) Simultaneously takes into account both errors entering into the total differential of the position function and additional errors resulting from the other DOF of the system links, i.e., it presents a necessary extension of the differential method needed for high-accuracy requirements.

(b) It is specifically suitable for such universal, multifunctional, reconfigurable and reset machines as the PWMs, which are characterized by many output performance indexes.

2.3 Book Structure

The book consists of comprehensive descriptions, mathematical formulations, and practical applications of accuracy problems as applied to machines with complicated kinematics and strong accuracy requirements such as machine tools, coordinate measuring machines, industrial robots combined into a group named precision working machines (PWM). The general principle "the higher part accuracy the better functionality" presents only necessary but not sufficient requirements to provide operation of working machines with predesigned requirements for accuracy of machine functions. For example, with the same manufacturing accuracy of robot parts, the robot accuracy is differs depending on whether the robot has parallel-, serial, or hybrid-type kinematics. In this connection, the problems are considered in connection with machine architecture, kinematics, and structure taking into account geometric and kinematic errors, and deformations of the structural elements. The mathematical consideration is accompanied by real-life examples.

Complicated kinematics of machines and robots caused by a necessity to carry out the multi-DOF motions structurally realized by chain of spatial motions of machine units requires cumbersome coordinate transformations between the unit-associated coordinate frames. The conventionally adopted matrix technique uses to advantage for this goal. Generally, the problems considered in the book indiscriminately require permanent application of matrix mathematical technique and, in this connection, can be effectively solved with and only with regular applications of software systems. In this book, the accuracy- and stiffness-related matrix algebra operations are formulated, solved, computed, and demonstrated using the Mathematica® software system. In Table 2.1, general themes considered in the book and their mechanical, mathematical, and computational models are declared.

The book combines two volumes. Volume I includes the Introduction, Parts I and II, and three Appendices. In Volume I, application of the matrix technique is limited, wherever possible, to the simplest operations with seven reference matrices allowing analysis and synthesis of a rigid body system connected through rigid supports.

Part 1 describes the model of the *nominal mechanical* system and its functioning without any external distortion. Mathematically, the model is expressed through the form-shaping function (FSF) containing basic information about nominal structure and kinematics. The FSF allows formulation of PWM kinematics using only operations on six matrices modelling six elementary nominal motions — three rotations and three translations. The standard form for a wide spectrum of actual applications, such as machine tools including gearing and threading machines, CMMs, and robots with serial and parallel kinematics, is described. In what follows, the FSF serves also as a basic formal tool for synthesis of the models of *actual functioning*, which differs from the nominal one by the fact that manufacturing errors and load-induced deformations are taken into account.

Table 2-3 Models considered in this book.

Content	Volume I		Volume II	
	Part 1	Part 2	Part 3	Part 4
Theme	Nominal functioning	Geometrical and kinematic accuracy	Stiffness-related problems	Task-oriented metrology
Mechanical system	Kinematics of rigid bodies	Rigid bodies on rigid supports	Rigid/elastic bodies on elastic supports	Rigid bodies under tolerancing
Mathematical model	Form-shaping function (FSF)	FSF variation	Kinematics-dependent stiffness/compliance	Dimensional and geometric tolerances
Computational technique	Motion matrices	Infinitesimal matrices	Stiffness/compliance matrices, FEM	Linear programming

Part 2 describes the models relating to geometrical and kinematic accuracy of the machines and robots. Mathematically, the models represent a variation of the FSF allowing formulation of the linear error balance in the cases, in which the errors are significantly smaller than nominal sizes, and non-linear components in the other cases. An infinitesimal error matrix consisting of six position and orientation errors is added to the six abovementioned matrices of elementary nominal motions to describe an actual movement of the system, with geometrical and kinematic errors. For both nominal and actual functioning of the mechanical system, the 4×4 homogeneous transformation matrices (HTM) are used.

The specific mathematical features of the HTM are as follows: the matrices of motions constitute a *group with multiplication* operation and the error matrices constitute a *group with addition* operation. From the application viewpoint, it is very important that the matrices, operations, and results of the operations have the direct engineering interpretation: among matrices, there are matrices of translation, matrices of rotation, matrices of motion, and error matrices; the multiplication is interpreted as coordinate transformation of two neighboring links of a kinematic chain;and the addition is interpreted as accumulation of different kinematic and geometric errors.

Appendices included in Volume I consists of descriptions of the Mathematica® commands associated with basic matrix technique and their applications for solution of engineering problems. Examples demonstrate calculations of the geometrical accuracy-related performance indexes as applied to the real IRs, machine tools and CMMs.

Volume II includes Parts III and IV, and Appendices. The FSF-based matrix technique is augmented by the use of the Jacobian, stiffness and compliance matrices.

Part 3 of the bookis dedicated to the statics and stiffness analysis of machines and robots. The approach uses the traditional mathematical technique of the structural mechanics modified for the mechanical systems of the machines, wherein the external load depends on a set of kinematic parameters. In particular, important problems concerning limitations of the working volume of the modern machines, caused by possible singular configurations with zero stiffness or zero compliance values, are analyzed. To solve these problems, an effective computational approach, protecting against reducing the stiffness value beyond the lowest preliminary given level, is developed. Elastic properties of the mechanical system stiffness/compliance values of links and joints resulting in deformations of elastic elements and supports under external loads are considered.

The stiffness/compliance features of a separate link are described through 6×6 matrices, and for the systems with n moving links, the features are evaluated through the $6n \times 6n$ matrices, which can be represented as a partitioning matrix combining the 6×6 joint-associated submatrices. The corresponding procedures are well known from structural mechanics theory and widely used in civil engineering.

To evaluate the stiffness characteristics under the general type of load (both external force and torque), the collinear stiffness value (CSV) is proposed. This approach allows limitation of the machine workspace satisfying stiffness value requirements, and building the virtual protective barrier to keep from approaching both singular configurations and resonance frequencies.

In **Part IV**, the 3-D metrological problems, concerning accuracy and precision evaluation from the viewpoint of satisfying the parametric (dimensional) and geometric tolerances requirements, are considered. Dimensional chains presenting a part of these problems, closest to the drawing-related design stage, are used to formalize some spatial tolerances. With development of CMM-based systems and built-in machine dimensional devices, the problem of computer-aided to lerancing becomes an emerging one. In particular, a point cloud of results of measurements represented through $3N$ numbers (N is the number of points of measurement) needs to be reduced to a set of regular engineering performance indexes. This problem established by ISO and ASME standards is formulated and solved using the substitute geometry based approach and linear-programming optimization procedures in close connections with results developed in the previous parts of the book.

Appendices included in Volume II consists of examples of calculations of the accuracy-related stiffness-compliance problems as applied to the PWMs.

References

[1] Asfahl, C. R., 1992, Robots and Manufacturing Automation, 2nd ed., John Wiley and Sons, New York, NY.

[2] Paul, R. P., 1981, Robot Manipulators: Mathematics, Programming, and Control, MIT Press, Cambridge, MA.

[3] Tsai, L.-W., 1999, Robot Analysis, John Wiley & Sons, New York, NY.

[4] Craig, J. J., 2005, Introduction to Robotics: Mechanics and Control, Prentice Hall, Upper Saddle River, NJ.

[5] Ceccarelli, M., 2004, Fundamentals of mechanics of robotic manipulation, Kluwers Academic Publishers, Dordrecht, The Netherlands.

[6] Reshetov, D. N., and Portman, V. T., 1988, Accuracy of Machine Tools, ASME Press, New York, NY.

[7] Bruevich, N. G., 1946, Accuracy of Mechanisms, Moscow: GITTL (*in Russian*).

[8] Portman V. T., 1980, Error Summation in the Analytical Calculation of Accuracy, Machines & Tooling, 1, pp. 7-10.

[9] Portman, V. T., 1981, Universal Method for Calculating the Accuracy of Mechanical Devices, Soviet Engineering Research, 1 (7), 11-15.

Part I: Model of nominal functioning

3 Form-shaping function: mathematical formulation

3.1 Geometric and kinematic subjects

3.1.1 Rigid body

3.1.1.1 *Coordinate frame and points*

The mathematical model of the rigid body presents the Cartesian coordinate frame $S \equiv OXYZ$ associated with the body (Fig. 3.1). The origin O and the X-, Y-, and Z-axis are usually associated with the mechanical and manufacturing bases of the link, such as guideways surfaces, axes, accurate planes and cylindrical surfaces, local supports, etc. The point K of the rigid body (Fig. 3.1a) is defined through the Cartesian coordinated x_K, y_K, and z_K presenting elements of a position vector:

$$\mathbf{r}_K = \begin{bmatrix} x_K \\ y_K \\ z_K \end{bmatrix} \tag{3.1}$$

Let K_1 and K_2 be two body points defined by position vectors $\mathbf{r}_1$ and $\mathbf{r}_2$ as well as their difference – vector $\mathbf{r}_{12}$ (Fig. 3.1b),

$$\mathbf{r}_1 = \begin{bmatrix} x_1 \\ y_1 \\ z_1 \end{bmatrix} \text{ and } \mathbf{r}_2 = \begin{bmatrix} x_2 \\ y_2 \\ z_2 \end{bmatrix} \tag{3.2}$$

$$\mathbf{r}_{12} = \mathbf{r}_2 - \mathbf{r}_1 \tag{3.3}$$

The difference can be represented in coordinate notation

$$\mathbf{r}_{12} = \begin{bmatrix} x_{12} \\ y_{12} \\ z_{12} \end{bmatrix} = \begin{bmatrix} x_2 - x_1 \\ y_2 - y_1 \\ z_2 - z_1 \end{bmatrix} \tag{3.4}$$

The distance d between two points defined by vectors $\mathbf{r}_1$ and $\mathbf{r}_2$ is defined as the Euclidean norm $||\mathbf{r}_{12}||$ of vector $\mathbf{r}_{12}$,

$$d = ||\mathbf{r}_{12}|| = \sqrt{x_{12}^2 + y_{12}^2 + z_{12}^2} \tag{3.5}$$

The fundamental feature of the rigid body: the distance between two arbitrary points of the rigid body independent of the body position, orientation, and motion, d = Const.

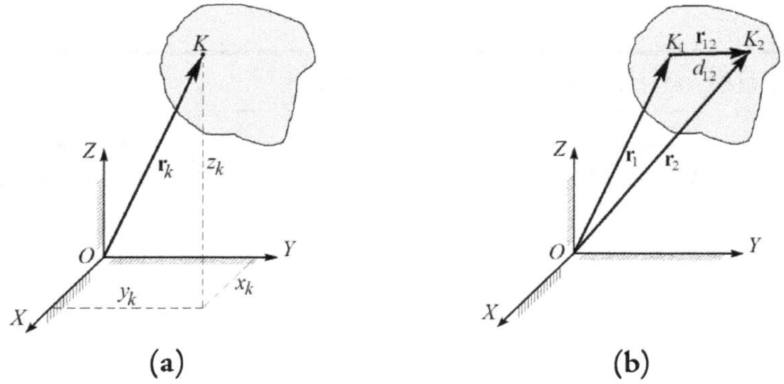

Figure 3-1 Rigid body: (a) coordinate frame; (b) interrelationships of body points.

3.1.1.2 Body finite spatial displacement

To describe the body motion in three-dimensional space, a pair of the Cartesian coordinate frames is entered (Fig. 3.2): the fixed frame $S = OXYZ$ associated with the base (fixed body) and the moving frame $S_p = O_pX_pY_pZ_p$ associated with the moving rigid body (Fig. 3.2a). An arbitrary point of the rigid body (point Q) is shown in Figure 3.2b. Since two different coordinate systems are considered, point Q can be differently defined in the entered frames:

$$\mathbf{r}_Q = \begin{bmatrix} x_Q \\ y_Q \\ z_Q \end{bmatrix} \text{ and } \boldsymbol{\rho}_Q = \begin{bmatrix} \rho_x \\ \rho_y \\ \rho_z \end{bmatrix} \tag{3.6}$$

where $\mathbf{r}_Q = \overline{OQ} = [x_Q, y_Q, z_Q]^T$ is the position vector of point Q as referred to the fixed frame S *depending on* the current position and orientation of the moving body; and $\boldsymbol{\rho} = \overline{O_pQ} = [\rho_x, \rho_y, \rho_z]^T$ is the position vector of point Q as referred to the coordinate frame S_p associated with the moving body, *independent of* the current position and orientation of the moving body.

As is clear from Figure 3.2, the translation of the moving body is described by transfer of the origin O into point O_p, while the three axes of frame $O_pX'Y'Z'$ remain parallel to the corresponding axes of the base-associated frame. Then a rotation around point O_p results in the required orientation of the axes X_p, Y_p, and Z_p. The position vectors $\mathbf{r}_Q$ and ρ of the same point are interconnected as follows:

$$\mathbf{r}_Q = \mathbf{r}_p + \mathbf{A}_{rot}\boldsymbol{\rho} \tag{3.7}$$

where $r_Q = \overline{OQ_p}$ is the position vector of origin O_p; and $\mathbf{A}_{rot}$ is the 3×3 rotation matrix combining three unit vectors **a**, **b**, **c**, which define orientation of axes X_p, Y_p, and Z_p with respect to the axes X, Y, and Z, respectively,

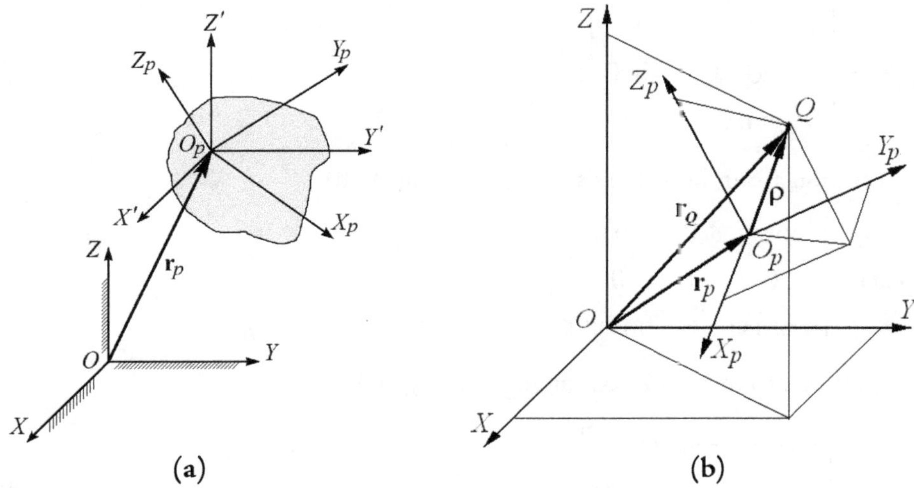

Figure 3-2 Displacement of the rigid body: (a) fixed and moving coordinate frames in 3-D space; (b) position vectors of point Q in two coordinate frames.

$$\mathbf{r}_p = \overline{OO_p} = \begin{bmatrix} p_x \\ p_y \\ p_z \end{bmatrix} \tag{3.8}$$

$$\mathbf{A}_{rot} = [\mathbf{a}\,\mathbf{b}\,\mathbf{c}] \tag{3.9}$$

$$\mathbf{a} = \begin{bmatrix} a_x \\ a_y \\ a_z \end{bmatrix}; \mathbf{b} = \begin{bmatrix} b_x \\ b_y \\ b_z \end{bmatrix}; \mathbf{c} = \begin{bmatrix} c_x \\ c_y \\ c_z \end{bmatrix} \tag{3.10}$$

Components of the unit vectors **a**, **b**, and **c** are their directional cosines, and the vectors are the orthogonal to each other,

$$\|\mathbf{a}\| = \sqrt{a_x^2 + a_y^2 + a_z^2} = 1; \|\mathbf{b}\| = \sqrt{b_x^2 + b_y^2 + b_z^2} = 1; \|\mathbf{c}\| = \sqrt{c_x^2 + c_y^2 + c_z^2} = 1 \tag{3.11}$$

$$\mathbf{a} \cdot \mathbf{b} = \mathbf{a} \cdot \mathbf{c} = \mathbf{b} \cdot \mathbf{c} = 0$$

Hence, the rotation matrix $\mathbf{A}_{rot}$, Eq. (3.9) is orthogonal [1], and therefore, its determinant equals one, $\det[\mathbf{A}_{rot}] = 1$, and inverse and transpose matrices are identical, $\mathbf{A}_{rot}\,(\mathbf{A}_{rot})^{T} = 1$. Any rotation of the moving body can be defined by three successive rotations around the coordinate axes at three Euler angles θ, ψ, and φ. Each of three rotations can be described by one of three 3 × 3 rotation matrices:

(a) rotation around the X-axis at angle θ (Fig. 3.3a)

$$\mathbf{A}_x(\theta) = \begin{bmatrix} 1 & 0 & 0 \\ 0 & \cos\theta & -\sin\theta \\ 0 & \sin\theta & \cos\theta \end{bmatrix}; \tag{3.13}$$

(b) rotation around the Y-axis at angle ψ (Fig. 3.3b)

$$\mathbf{A}_y(\psi) = \begin{bmatrix} \cos\psi & 0 & \sin\psi \\ 0 & 1 & 0 \\ -\sin\psi & 0 & \cos\psi \end{bmatrix}; \tag{3.14}$$

(c) rotation around the Z-axis at angle φ (Fig. 3.3c)

$$\mathbf{A}_y(\varphi) = \begin{bmatrix} \cos\varphi & -\sin\varphi & 0 \\ \sin\varphi & \cos\varphi & 0 \\ 0 & 0 & 1 \end{bmatrix} \tag{3.15}$$

A general case of rotation in 3-D space presents three successive rotations around X-, Y-, Z-axes at Euler angles θ, ψ, φ, respectively, shown in Figure 3.4. Since the final orientation of the body depends on a sequence of rotations, the sequence must be preliminarily fixed. In what follows, the next sequence will be applied (Fig. 3.4):

(1) Rotation around the X-axis of frame $OXYZ$ at angle θ described through Eq. (3.13); the moving frame is $OX_1Y_1Z_1$, the X- and X_1-axes are identical;
(2) Rotation around the Y_1-axis of frame $OX_1Y_1Z_1$ at angle ψ described through Eq. (3.14); the moving frame is $OX_2Y_2Z_2$, the Y_1- and Y_2-axes are identical (Fig. 3.4b); and,
(3) Rotation around the new Z_2-axis of frame $OX_2Y_2Z_2$ at angle φ described through Eq. (3.15); the moving frame is $OX_3Y_3Z_3$, the Z_2- and Z_3-axes are identical (Fig. 3.4c).

The general 3 × 3 rotation matrix $\mathbf{A}_{rot} = \mathbf{A}_{rot}(\theta, \psi, \varphi)$ presents the ordered product of three rotation matrices modeling rotations around the three coordinate axes, Eqs. (3.13–15),

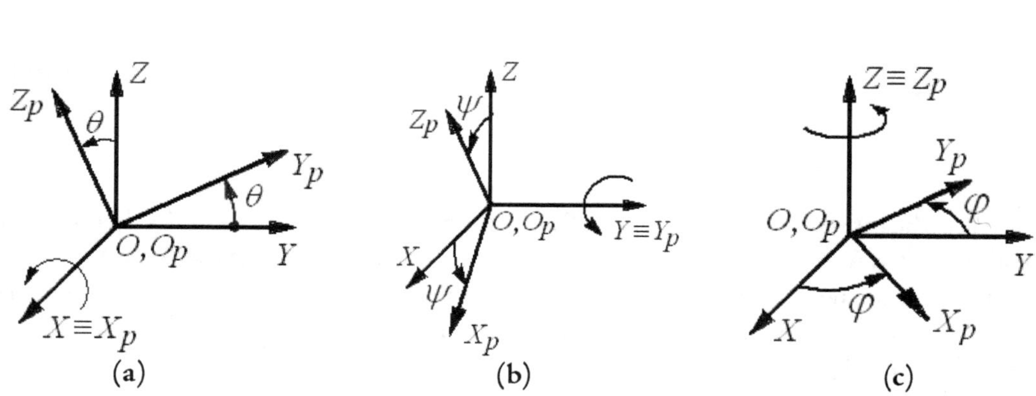

Figure 3-3 Body rotations around coordinate axes: (a) the X-axis; (b) the Y-axis; (c) the Z-axis.

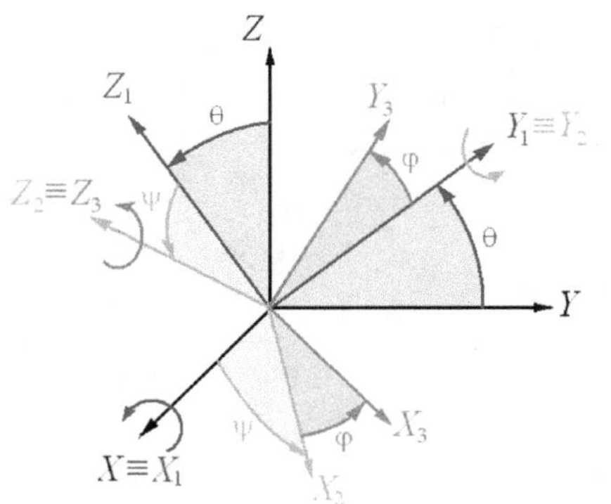

Figure 3-4 The Euler angles: θ of rotation around the X-axis, ψ of rotation around the Y_1-axis, and φ of rotation around the Z_2-axis.

$$\mathbf{A}_{rot} = \begin{bmatrix} 1 & 0 & 0 \\ 0 & \cos\theta & -\sin\theta \\ 0 & \sin\theta & \cos\theta \end{bmatrix} \begin{bmatrix} \cos\psi & 0 & \sin\psi \\ 0 & 1 & 0 \\ -\sin\psi & 0 & \cos\psi \end{bmatrix} \begin{bmatrix} \cos\varphi & -\sin\varphi & 0 \\ \sin\varphi & \cos\varphi & 0 \\ 0 & 0 & 1 \end{bmatrix} =$$

$$\begin{bmatrix} \cos\psi\cos\varphi & -\cos\psi\sin\varphi & \sin\psi \\ \sin\theta\sin\psi\cos\varphi+\cos\theta\sin\varphi & -\sin\theta\sin\psi\sin\varphi+\cos\theta\cos\varphi & -\sin\theta\cos\psi \\ -\cos\theta\sin\psi\cos\varphi+\sin\theta\sin\varphi & \cos\theta\sin\psi\sin\varphi+\sin\theta\cos\varphi & \cos\theta\cos\psi \end{bmatrix} \quad (3.16)$$

Note that Eq. (3.16) is not the only presentation of the general rotation: this is one of 24 versions of the possible combinations and sequences of the Euler angles forming different expressions for the rotation matrix.

Numerical Example 3.1. **Given:** The moving body (frame $O_3X_3Y_3Z_3$) is rotated around frame $OXYZ$ at the Euler angles $\theta = \psi = \varphi = \pi/2$. **Required:** Calculation of the rotation matrix $\mathbf{A}_{rot}$, Eq. (3.16), and illustration of the sequence of the successive rotations.

Solution:

Calculation by Eq. (3.16) yields:

$$\mathbf{A}_{rot} = \begin{bmatrix} 1 & 0 & 0 \\ 0 & 0 & -1 \\ 0 & 1 & 0 \end{bmatrix} \begin{bmatrix} 0 & 0 & 1 \\ 0 & 1 & 0 \\ -1 & 0 & 0 \end{bmatrix} \begin{bmatrix} 0 & -1 & 0 \\ 1 & 0 & 0 \\ 0 & 0 & 1 \end{bmatrix} = \begin{bmatrix} 0 & 0 & 1 \\ 0 & -1 & 0 \\ 1 & 0 & 0 \end{bmatrix}, \quad (3.17)$$

i.e., the X- and Z-axes coincide with the Z_3- and X_3-axes, respectively; and the Y- and Y_3-axes are oriented along the same line, but opposite in direction.

In Figure 3.5, three steps of rotations are shown: (a) initial orientation; and (b), (c), and (d) orientations after successive rotations at angles $\theta = \pi/2$, $\psi = \pi/2$, and $\varphi = \pi/2$, respectively. The obtained picture illustrates the geometric meaning of matrix $\mathbf{A}_{rot}$, Eq. (3.17):

- The first column shows that the X_3-axis is identical to the Z-axis;
- The second column shows that the Y_3-axis is opposite to the Y-axis; and,
- The third column shows that the Z_3-axis identical to the X-axis.

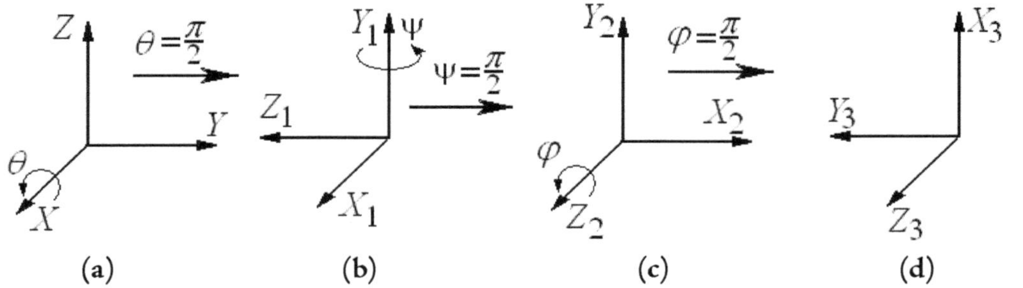

Figure 3-5 The three-stage transformation from the XYZ- to $X_3Y_3Z_3$-frame.

Six simplest motions of the frame S_i relative to S_{i-1} – three translations along the coordinate axes and three rotations around the axes – are symbolized through six natural numbers (Table 3.1). Any position, orientation, and motion of the body with respect to the given base can be described through a set of the subsequent simplest motions listed in Table 3.1. This allows description of the relative position and orientation of the body by six parameters: three coordinates of a body point and three angles of a body rotation.

Note once more that the final position of the body after some successive motions depends on the order, in which the motions are carried out. In Figure 3.6, two motions – rotation and translation – result in different final positions of the transferred body since the motions are carried out in different orders. This demonstrates the non-commutativity of the successive motions.

3.1.2 Lines in 3-D space

3.1.2.1 Straight line

Let $\mathbf{r}_{12}$ be defined through Eq. (3.3) and $||\mathbf{r}_{12}|| \neq 0$. Two points defining the vector $\mathbf{r}_{12}$ simultaneously define a straight line passing through these points. Unit vector **e** along the line and its moment **n** about the origin of the coordinate frame are as follows:

$$\mathbf{e} = \begin{bmatrix} e_x \\ e_y \\ e_z \end{bmatrix} = \mathbf{r}_{12} / \| \mathbf{r}_{12} \| \tag{3.18}$$

Table 3-1 Axis-associated motions.

Motion	Symbol
Translation along the axis	
X	1
Y	2
Z	3
Rotation around the axis:	
X	4
Y	5
Z	6

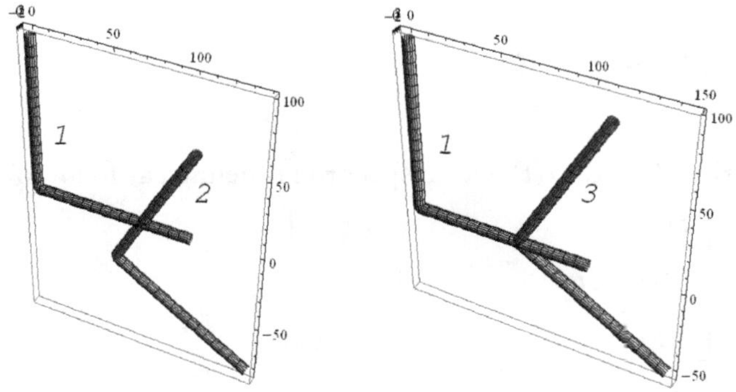

Figure 3-6 Non-commutativity of successive motions: 1 – body in the initial position; 2 – body after rotation and translation; 3 – body after translation and rotation.

$$\mathbf{n} = \mathbf{r}_1 \times \mathbf{e} = \mathbf{r}_2 \times \mathbf{e} \tag{3.19}$$

where $\mathbf{r}_1$ and $\mathbf{r}_2$ are the position vectors of the point K_1 and K_2 (Fig. 3.1.b)

The vectors **e** and **n** completely determine both orientation and position of the line in space. Six components of the vectors **e** and **n** present *Plücker coordinates* of the line, which can be assembled into the six-order vector

$$\mathbf{u} = [\mathbf{e}^{\mathrm{T}}, \mathbf{n}^{\mathrm{T}}]^{\mathrm{T}} = [e_x, e_y, e_z, n_x, n_y, n_z]^{\mathrm{T}} \tag{3.20}$$

Note that only four of six Plücker coordinates are independent since they are related to each other by two equations:

$$\|\mathbf{e}\| = \sqrt{e_x^2 + e_y^2 + e_z^2} = 1 \text{ and } \mathbf{e} \cdot \mathbf{n} = 0 \tag{3.21}$$

The line passing through the point $\mathbf{r}_1$ in the direction of the vector **e** is parametrically represented as a function of a single parameter *t*:

$$\mathbf{r} = \mathbf{r}_1 + t\,\mathbf{e}\,,\; t_{\min} \le t \le t_{\max} \tag{3.22}$$

with

$$\mathbf{r} = [x, y, z]^{\mathrm{T}};\; \mathbf{r}_1 = [x_1, y_1, z_1]^{\mathrm{T}};\; \mathbf{e} = [e_x, e_y, e_z]^{\mathrm{T}}$$

Numerical Example 3.2. The line missing through the point $\mathbf{r}_1 = [2, 2, 5]^{\mathrm{T}}$ in the direction of the vector $\mathbf{e} = [0.5, \sqrt{3}/2,\; 0]^{\mathrm{T}}$ is defined through equation

$\mathbf{r} = [x, y, z]^{\mathrm{T}} = [2 + 0.5t,\; 2 + t\sqrt{3}/2,\; 5]^{\mathrm{T}}$.

In Figure 3.7a, the line is shown for $t_{\min} = 0$ and $t_{\max} = 10$. ■

3.1.2.2 Curves in 3-D space: Parametric representation

A curve associated with the rigid body is defined through three functions $x(t)$, $y(t)$, and $z(t)$ of a single real parameter *t*,

$$\mathbf{r}(t)=\begin{bmatrix} x(t) \\ y(t) \\ z(t) \end{bmatrix}, t_{min} \leq t < t_{max} \tag{3.23}$$

The first derivative of vector $\mathbf{r}(t)$ with respect to t is defined as follows:

$$d\mathbf{r} / dt \equiv \dot{\mathbf{r}}(t) = \begin{bmatrix} \dot{x}(t) \\ \dot{y}(t) \\ \dot{z}(t) \end{bmatrix} \neq 0 \tag{3.24}$$

The arc length s is defined through the differential

$$ds = \sqrt{[\dot{x}(t)]^2 + [\dot{y}(t)]^2 + [\dot{z}(t)]^2}\, dt = \sqrt{\dot{\mathbf{r}}(t) \cdot \dot{\mathbf{r}}(t)}\, dt \tag{3.25}$$

$$s \equiv \int_0^t ds = \int_0^t \sqrt{\dot{\mathbf{r}} \cdot \dot{\mathbf{r}}}\, dt \tag{3.26}$$

Equation (3.26) allows definition of the curve and derivative equations through parameter s:

$$\mathbf{r}(s) = \begin{bmatrix} x(s) \\ y(s) \\ z(s) \end{bmatrix}, \; s_{min} \leq s < s_{max} \tag{3.27}$$

$$\mathbf{r}'(s) = \frac{\partial \mathrm{r}(s)}{\partial s} = \frac{\partial \mathrm{r}(t)}{\partial t} \Big/ \frac{\partial s}{\partial t} \tag{3.28}$$

Three *unit* vectors **t**, **n**, and **b** are directed along the tangent, the principal normal, and the binormal, respectively,

$$\mathbf{t} = \mathbf{r}' \tag{3.29}$$

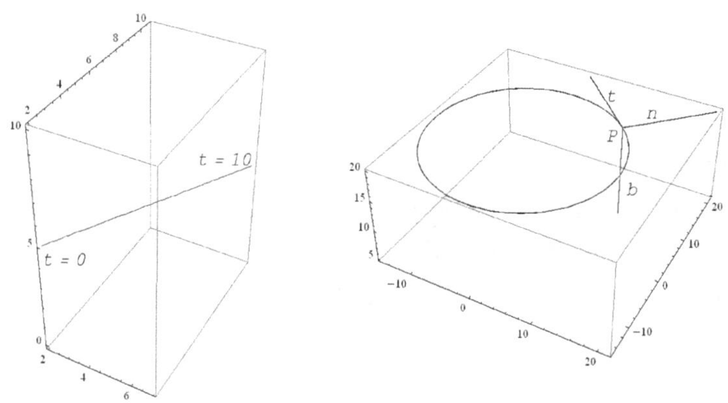

Figure 3-7 The lines in 3-D space: (a) the straight line for *Numerical Example 3.2*; (b) the circle for *Numerical Example 3.3*.

$$\mathbf{n} = \mathbf{r}''/|\mathbf{r}''| \tag{3.30}$$

$$\mathbf{b} = \mathbf{t} \times \mathbf{n} \tag{3.31}$$

where $|\mathbf{r}''|$ is the magnitude of vector $\mathbf{r}''$, $|\mathbf{r}''| = (\mathbf{r}'' \cdot \mathbf{r}'')^{1/2}$.

Note that $\mathbf{r}(s)$, Eq. (3.23), can have no derivatives at some points: singular points such as an acnode (isolated point), a crunode (double point), a cuspidal point, etc. At the regular point of the curve, the curvature K, curvature radius R, and torsion T are

$$K = |\mathbf{r}''|;\ R_K = 1/K = 1/|\mathbf{r}''|;\ \text{and}\ T = (R_K)^2\,(\mathbf{r}'\mathbf{r}''\mathbf{r}''') \tag{3.32}$$

Equations of the tangent, principal normal, and binormal forming the moving trihedron at the point $\mathbf{r}_1 = [x_1, y_1, z_1]^T$ of the curve have the following forms

$$\mathbf{r}_t = \mathbf{r}_1 + u\mathbf{t};\ \mathbf{r}_n = \mathbf{r}_1 + u\mathbf{n};\ \text{and}\ \mathbf{r}_b = \mathbf{r}_1 + u\mathbf{b} \tag{3.33}$$

3.1.2.3 *Circle*

A vector equation of the circle lying in the horizontal plane at the height h is

$$\mathbf{r}(t) = \begin{bmatrix} R\cos t + a \\ R\sin t + b \\ h \end{bmatrix},\ 0 \le t < 2\pi \tag{3.34}$$

where t is the polar angle; R is the circle radius; a and b are the coordinate of the center.

Calculations according Eqs. (3.24–27) yield

$$\dot{\mathbf{r}}(t) = R\begin{bmatrix} -\sin t \\ \cos t \\ 0 \end{bmatrix} \ne 0 \tag{3.35}$$

$$ds = \sqrt{(-R\sin t)^2 + (R\cos t)^2}\,dt = R\,dt \tag{3.36}$$

$$s = \int_0^t \sqrt{\dot{\mathbf{r}} \cdot \dot{\mathbf{r}}}\,dt = \int_0^t \sqrt{R\begin{bmatrix} -\sin t \\ \cos t \\ 0 \end{bmatrix} \cdot R\begin{bmatrix} -\sin t \\ \cos t \\ 0 \end{bmatrix}}\,dt = Rt \tag{3.37}$$

Hence,

$$t = s/R\ \text{and}\ \mathbf{r}(s) = \begin{bmatrix} R\cos(s/R) + a \\ R\sin(s/R) + b \\ h \end{bmatrix} \tag{3.38}$$

Trihedron orts, Eqs. (3.29–31), are

$$\mathbf{t} = \mathbf{r}' = \frac{\partial \mathbf{r}(s)}{\partial s} = \begin{bmatrix} -\sin(s/R) \\ \cos(s/R) \\ 0 \end{bmatrix} = \begin{bmatrix} -\sin t \\ \cos t \\ 0 \end{bmatrix} \tag{3.39}$$

$$\mathbf{r}'' = \frac{\partial \mathbf{r}'(s)}{\partial s} = \frac{1}{R}\begin{bmatrix} \cos(s/R) \\ \sin(s/R) \\ 0 \end{bmatrix} \text{and } |\mathbf{r}''| = 1/R \tag{3.40}$$

$$\mathbf{n} = \mathbf{r}''/|\mathbf{r}''| = \begin{bmatrix} \cos(s/R) \\ \sin(s/R) \\ 0 \end{bmatrix} = \begin{bmatrix} \cos t \\ \sin t \\ 0 \end{bmatrix} \tag{3.41}$$

$$\mathbf{b} = \mathbf{t} \times \mathbf{n} = \begin{bmatrix} -\sin t \\ \cos t \\ 0 \end{bmatrix} \times \begin{bmatrix} \cos t \\ \sin t \\ 0 \end{bmatrix} = \begin{bmatrix} 0 \\ 0 \\ -1 \end{bmatrix} \tag{3.42}$$

The curvature K, its radius R_K, and the torsion T, Eqs. (3.32), are

$$K = |\mathbf{r}''| = 1/R \text{ and } R_K = R$$

$$T = (R_K)^2\,(\mathbf{r}'\mathbf{r}''\mathbf{r}''') = R^2\left(\begin{bmatrix} -\sin(s/R) \\ \cos(s/R) \\ 0 \end{bmatrix}, \frac{1}{R}\begin{bmatrix} \cos(s/R) \\ \sin(s/R) \\ 0 \end{bmatrix}, \frac{1}{R^2}\begin{bmatrix} -\sin(s/R) \\ \cos(s/R) \\ 0 \end{bmatrix}\right) = 0$$

Thus, the radius of curvature of the circle is identical to the circle radius and the torsion equals zero.

Numerical Example 3.3.

Given: The circle of radius 15 mm, whose center is at the point $x = y = 0$, is located on the horizontal plane at a height of 20 mm. **Required:** Construction of the circle and the moving trihedron at the point $\mathbf{r}_1 = \mathbf{r}(t = \pi/4)$.

Solution:

(1) Equation (3.35) becomes

$$\mathbf{r}(t) = \begin{bmatrix} R\cos t + a \\ R\sin t + b \\ h \end{bmatrix} = \begin{bmatrix} 15\cos t \\ 15\sin t \\ 20 \end{bmatrix}, 0 \le t < 2\pi; \mathbf{r}_1 = \mathbf{r}_{t=\pi/4} = \begin{bmatrix} 15\cos(\pi/4) \\ 15\sin(\pi/4) \\ 20 \end{bmatrix}$$

(2) The position vector of the point associated with $t = \pi/4$ is

$$\mathbf{r}_1 = \mathbf{r}_{t=\pi/4} = \begin{bmatrix} 15\cos(\pi/4) \\ 15\sin(\pi/4) \\ 20 \end{bmatrix} = \begin{bmatrix} 15/\sqrt{2} \\ 15/\sqrt{2} \\ 20 \end{bmatrix}$$

(3) Orts of the moving trihedron at the point $\mathbf{r}_1$, Eqs. (3.29–31), become

$$\mathbf{t}_1 = \begin{bmatrix} -\sin t \\ \cos t \\ 0 \end{bmatrix} = \begin{bmatrix} -1/\sqrt{2} \\ 1/\sqrt{2} \\ 0 \end{bmatrix}; \mathbf{n}_1 = \begin{bmatrix} \cos t \\ \sin t \\ 0 \end{bmatrix} = \begin{bmatrix} 1/\sqrt{2} \\ 1/\sqrt{2} \\ 0 \end{bmatrix}; \mathbf{b}_1 = \begin{bmatrix} 0 \\ 0 \\ -1 \end{bmatrix}$$

(4) Equations (3.33) of the tangent, principal normal, and binormal forming the moving trihedron become

$$\mathbf{r}_t = \mathbf{r}_1 + u\mathbf{t}_1 = \begin{bmatrix} 15/\sqrt{2} - (1/\sqrt{2})u \\ 15/\sqrt{2} + (1/\sqrt{2})u \\ 20 \end{bmatrix}; \mathbf{r}_n = \mathbf{r}_1 + u\mathbf{n}_1 = \begin{bmatrix} 15/\sqrt{2} + (1/\sqrt{2})u \\ 15/\sqrt{2} + (1/\sqrt{2})u \\ 20 \end{bmatrix}; \mathbf{r}_b = \mathbf{r}_1 + u\mathbf{b}_1 = \begin{bmatrix} 15/\sqrt{2} \\ 15/\sqrt{2} \\ 20-u \end{bmatrix}$$

In Figure 3.7b, the circle is built as well as the tangent, principal normal, and binormal (with $0 \le u \le 15$) are shown.

3.1.2.4 *Helical line*

Vector equation of a circular helix coiled around the cylinder of radius a:

$$\mathbf{r}(t) = \begin{bmatrix} a\cos t \\ a\sin t \\ bt \end{bmatrix} \tag{3.43}$$

where t is the current angle; b is screw parameter expressed through the screw lead h,

$$b = \frac{h}{2\pi} \tag{3.44}$$

$$s = \int_0^t \sqrt{\dot{\mathbf{r}} \cdot \dot{\mathbf{r}}}\,dt = \int_0^t \sqrt{\begin{bmatrix} -a\sin t \\ a\cos t \\ b \end{bmatrix} \cdot \begin{bmatrix} -a\sin t \\ a\cos t \\ b \end{bmatrix}}\,dt = t\sqrt{a^2 + b^2} \tag{3.45}$$

$$t = s/\sqrt{a^2+b^2} \text{ and } \mathbf{r}(s) = \begin{bmatrix} a\cos\dfrac{s}{\sqrt{a^2+b^2}} \\ a\sin\dfrac{s}{\sqrt{a^2+b^2}} \\ bs/\sqrt{a^2+b^2} \end{bmatrix} \tag{3.46}$$

The first and second derivatives and the trihedron orts, Eqs. (3.29-31),

$$\mathbf{t} = \mathbf{r}' = \frac{\partial \mathrm{r}(s)}{\partial s} = \frac{\partial \mathrm{r}(t)}{\partial t} / \frac{\partial t}{\partial s} = \frac{1}{\sqrt{a^2+b^2}} \begin{bmatrix} -a\sin t \\ a\cos t \\ b \end{bmatrix} \tag{3.47}$$

$$\mathbf{r}'' = \frac{\partial \mathrm{r}'(s)}{\partial s} = \frac{-a}{a^2+b^2} \begin{bmatrix} \cos t \\ \sin t \\ 0 \end{bmatrix} \text{ and } |\mathbf{r}''| = \frac{a}{a^2+b^2} \tag{3.48}$$

$$\mathbf{n} = \mathbf{r}''/|\mathbf{r}''| = \begin{bmatrix} -\cos t \\ -\sin t \\ 0 \end{bmatrix} \tag{3.49}$$

$$\mathbf{b} = \mathbf{t} \times \mathbf{n} = \frac{1}{\sqrt{a^2+b^2}} \begin{bmatrix} -a\sin t \\ a\cos t \\ b \end{bmatrix} \times \begin{bmatrix} -\cos t \\ -\sin t \\ 0 \end{bmatrix} = \frac{1}{\sqrt{a^2+b^2}} \begin{bmatrix} b\sin t \\ -b\cos t \\ a \end{bmatrix} \tag{3.50}$$

The curvature, its radius, and the torsion, Eqs. (3.32),

$$K = \mathbf{r}'' = \frac{a}{a^2+b^2} \text{ and } R_K = \frac{a^2+b^2}{a} \tag{3.51}$$

$T = (\mathrm{R}_K)^2(\mathbf{r}'\mathbf{r}''\mathbf{r}''') =$

$$\frac{(a^2+b^2)^2}{a^2} \left(\frac{1}{\sqrt{a^2+b^2}} \begin{bmatrix} -a\sin t \\ a\cos t \\ b \end{bmatrix}, \frac{-a}{a^2+b^2} \begin{bmatrix} \cos t \\ \sin t \\ 0 \end{bmatrix}, \frac{a}{(a^2+b^2)^{3/2}} \begin{bmatrix} \sin t \\ -\cos t \\ 0 \end{bmatrix} \right) = \frac{b}{a^2+b^2} \tag{3.52}$$

Numerical Example 3.4. **Given:** The screw line diameter is d = 25 mm; the lead h = 10 mm. **Required:** Construction of the line for $-\pi/2 \le t \le 7\pi$ and the moving trihedron at the point $\mathbf{r}_1 = \mathbf{r}(t = 7\pi)$.

Solution:

(1) According Eq. (3.44), parameter $b = 10/(2\pi)$. Position vector, Eq. (3.43), is

$$\mathbf{r}(t) = \begin{bmatrix} 12.5\cos t \\ 12.5\sin t \\ 5t/\pi \end{bmatrix}$$

(2) The position vector of the point associated with $t = \pi/4$ is

$$\mathbf{r}_1 = \mathbf{r}_{t=7\pi} = \begin{bmatrix} 12.5\cos t \\ 12.5\sin t \\ 5t/\pi \end{bmatrix} = \begin{bmatrix} -12.5 \\ 0 \\ 35 \end{bmatrix}$$

(3) Equations (3.30-33) of the tangent, principal normal, and binormal forming the moving trihedron become:

$$\mathbf{r}_t = \mathbf{r}_1 + u\mathbf{t}_1 = \begin{bmatrix} -12.5 \\ 0 \\ 35 \end{bmatrix} + \frac{u}{\sqrt{a^2+b^2}} \begin{bmatrix} -a\sin t \\ a\cos t \\ b \end{bmatrix} = \begin{bmatrix} -12.5 \\ -0.9920u \\ 35+0.12630u \end{bmatrix}$$

$$\mathbf{r}_n = \mathbf{r}_1 + u\mathbf{n}_1 = \begin{bmatrix} -12.5 \\ 0 \\ 35 \end{bmatrix} + \begin{bmatrix} -\cos t \\ -\sin t \\ 0 \end{bmatrix} u = \begin{bmatrix} -12.5-u \\ 0 \\ 35 \end{bmatrix}$$

$$\mathbf{r}_b = \mathbf{r}_1 + u\mathbf{b}_1 = \begin{bmatrix} -12.5 \\ 0 \\ 35 \end{bmatrix} + \frac{u}{\sqrt{a^2+b^2}} \begin{bmatrix} b\sin t \\ -b\cos t \\ a \end{bmatrix} = \begin{bmatrix} -12.5 \\ 0.12630u \\ 35+0.9920u \end{bmatrix}$$

In Figure 3.8a, the circular helix according the *Numerical Example 3.4* ($d = 25$ and $h = 10$ mm) as well as tangent **t**, normal **n**, and binormal **b** at point $t = 7\pi$ are shown.

As an expansion, the screw line with a variable lead is shown in Figure 3.8b. Its equation is

$$\mathbf{r}(t) = \begin{bmatrix} a\cos t \\ a\sin t \\ b(1+q\frac{t}{2\pi})t \end{bmatrix} \tag{3.53}$$

Each subsequent lead of the screw is increased by q percent relative to the previous one. For example, the screw shown in Figure 3.8b has $q = 0.2$, i.e., its lead h has a progressive increase of 20% from initial 10 mm, $h = 10(1 + 0.2\,k)$ mm, with $k = 0, 1, 2, 3$.

In Figure 3.8c, the conical helix of the variable diameter is shown,

$$\mathbf{r}(t) = \begin{bmatrix} a(1-q\dfrac{t}{2\pi})\cos t \\ a(1-q\dfrac{t}{2\pi})\sin t \\ bt \end{bmatrix} \tag{3.54}$$

The basic diameter is $d = 25$ mm, and then $d = 25(1 - 0.2\,k)$ mm, where $k = 0, 1, 2, 3$.

3.1.3 Surfaces in 3-D space

3.1.3.1 *General formulation*

A surface associated with the rigid body is defined through functions $x(u, v)$, $y(u, v)$, and $z(u, v)$ of two real parameters u and v presenting curvilinear (or Gaussian) coordinates of the surface,

$$\mathbf{r}(u,v) = \begin{bmatrix} x(u,v) \\ y(u,v) \\ z(u,v) \end{bmatrix}, u_{min} \le u \le u_{max} \text{ and } v_{min} \le v \le v_{max} \tag{3.55}$$

$$\mathbf{N} = \mathbf{r}_u \times \mathbf{r}_v \neq 0 \tag{3.56}$$

where u_{min} and u_{max}, v_{min} and v_{max} are the bounders of an open interval, whereat the surface is regular; $\mathbf{N}$ is the normal of the surface; and $\mathbf{r}_u$ and $\mathbf{r}_v$ are the partial derivatives of the vector function, Eq. (3.55), with respect to parameters u and v, respectively,

$$\mathbf{r}_u = \frac{\partial \mathbf{r}(u,v)}{\partial u} = \begin{bmatrix} x_u \\ y_u \\ z_u \end{bmatrix} \text{ and } \mathbf{r}_v = \frac{\partial \mathbf{r}(u,v)}{\partial v} = \begin{bmatrix} x_v \\ y_v \\ z_v \end{bmatrix} \tag{3.57}$$

with

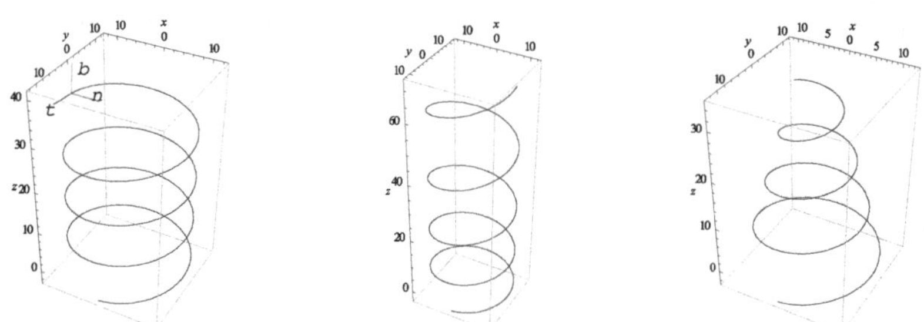

Figure 3-8 Screw lines in space: (a) circular helix of 25 mm diameter and 10 mm lead, **t**, **n**, and **b** are the tangent, normal, and binormal at point $t = 7\pi$; (b) circular helix of the variable lead; (c) conical helix of the variable diameter.

$$x_u \equiv \frac{\partial x}{\partial u}; y_u \equiv \frac{\partial y}{\partial u}; z_u \equiv \frac{\partial z}{\partial u}; x_v \equiv \frac{\partial x}{\partial v}; y_v \equiv \frac{\partial y}{\partial v}; \text{ and } z_v \equiv \frac{\partial z}{\partial v} \tag{3.58}$$

The surface points, at which the partial derivatives exist but the product $\mathbf{r}_u \times \mathbf{r}_v$ tends to zero, correspond to special points (ribs, vertices, and so on). A line on the surface given by conditions u = Const is called the coordinate v-line. Correspondingly, v = Const defines the coordinate u-line. The derivatives $\mathbf{r}_u$ and $\mathbf{r}_v$ are directed along the tangents to the coordinate lines passing through point $\mathbf{r}(u, v)$. A set of u- and v-lines presents the coordinates system on the surface.

Equation (3.56) presents the requirement for the existence and inequality to zero of the normal to the surface at each point $\mathbf{r}(u, v)$ in the range of the values u and v under consideration. The unit vector $\mathbf{n}$ of the normal $\mathbf{N}$, Eq. (3.56), is of particular importance,

$$\mathbf{n} = \frac{\mathbf{N}}{N} = \pm \frac{\mathbf{r}_u \times \mathbf{r}_v}{\|\mathbf{r}_u \times \mathbf{r}_v\|} = \pm \frac{\mathbf{r}_u \times \mathbf{r}_v}{\sqrt{\begin{vmatrix} y_u & z_u \\ y_v & z_v \end{vmatrix}^2 + \begin{vmatrix} z_u & x_u \\ z_v & x_v \end{vmatrix}^2 + \begin{vmatrix} x_u & y_u \\ x_v & y_v \end{vmatrix}^2}} \tag{3.59}$$

where $N = \|\mathbf{r}_u \times \mathbf{r}_v\|$ is the magnitude of normal vector $\mathbf{N}$.

The plus or minus sign in Eq. (3.59) defines the internal or external direction of the normal with respect to a rigid body, Eq. (3.55). Note that the equation of the surface in space can be also specified in the implicit form

$$f(x, y, z) = 0 \tag{3.60}$$

If $\partial f/\partial z \neq 0$, this equation can be transformed into the explicit form $z = z(x, y)$ and then represented in the form (3.55) as follows:

$$\mathbf{r}(x,y) = \begin{bmatrix} x \\ y \\ z(x,y) \end{bmatrix}, \; x_{min} \le x \le x_{max} \text{ and } y_{min} \le y \le y_{max} \tag{3.61}$$

3.1.3.2 *Plane*

Parametric representation of a plane has the form

$$\mathbf{r}(u,v) = \mathbf{r}_0 + u\mathbf{a} + v\mathbf{b} = \begin{bmatrix} x_0 + a_x u + b_x v \\ y_0 + a_y u + b_y v \\ z_0 + a_z u + b_z v \end{bmatrix}, \; u_{min} \le u \le u_{max} \text{ and } v_{min} \le v \le v_{max} \tag{3.62}$$

where u and v are curvilinear coordinates (straight lines) of the plane; $\mathbf{r}_0$ is the position vector of the point; $\mathbf{a}$ and $\mathbf{b}$ are the vectors parallel to the plane, $\mathbf{r}_0 = [x_0, y_0, z_0]^T$; $\mathbf{a} = [a_x, a_y, a_z]^T$; and $\mathbf{b}_2 = [b_x, b_y, b_z]^T$.

The normal $\mathbf{N}$, Eq. (3.56), and unit vector of the normal $\mathbf{n}$, Eq. (3.59), are

$$\mathbf{N} = \mathbf{r}_u \times \mathbf{r}_v = \mathbf{a} \times \mathbf{b} \tag{3.63}$$

$$\mathbf{n} = \pm \frac{\mathbf{N}}{|\mathbf{r}_u \times \mathbf{r}_v|} = \pm \frac{1}{\sqrt{d_1^2 + d_2^2 + d_3^2}} \begin{bmatrix} N_x \\ N_y \\ N_z \end{bmatrix} \tag{3.64}$$

where

$$\begin{aligned} d_1 &= -a_z b_y + a_y b_z \\ d_2 &= a_z b_x - a_x b_z \\ d_2 &= -a_y b_x + a_x b_y \end{aligned} \tag{3.65}$$

The implicit presentation, Eq. (3.60), of the plane in the space is

$$Ax + By + Cz + D = 0 \tag{3.66}$$

For example, if $C \neq 0$, the parametric representation has the form

$$\mathbf{r}(x,y) = \begin{bmatrix} x \\ y \\ z(x,y) \end{bmatrix} = \begin{bmatrix} x \\ y \\ -(A/C)x - (B/C)y - D/C \end{bmatrix}, \; x_{min} \le x \le x_{max}\,;\; y_{\min} \le y \le y_{\max} \tag{3.67}$$

Numerical Example 3.5.

Given: The parametric equation, Eq. (3.67), of the plane:

$$\mathbf{r}(x,y) = \begin{bmatrix} x \\ y \\ 3x \end{bmatrix} \quad 0 \le x \le 2; 0 \le y \le 5$$

Required: Construction of the plane and its normal at the point P_1: $x_1 = 2$ and $y_1 = 2.5$.
Solution: The normal and its unit vector are calculated through Eqs. (3.1.3.9-10):

$$\mathbf{N} = \mathbf{r}_x \times \mathbf{r}_v = \begin{bmatrix} 1 \\ 0 \\ 3 \end{bmatrix} \times \begin{bmatrix} 0 \\ 1 \\ 0 \end{bmatrix} = \begin{bmatrix} -3 \\ 0 \\ 1 \end{bmatrix}$$

$$\mathbf{n} = \pm\frac{\mathbf{N}}{|\mathbf{r}_x \times \mathbf{r}_y|} = \pm\frac{1}{|\sqrt{(-3)^2+0^2+1^2}}\begin{bmatrix}-3\\0\\1\end{bmatrix} = \pm\begin{bmatrix}-3/\sqrt{10}\\0\\1/\sqrt{10}\end{bmatrix}$$

Point P_1 given by parameters $x_1 = 2$ and $y_1 = 2.5$ is

$$\mathbf{r}_1 = \begin{bmatrix}x_1\\y_1\\3x_1\end{bmatrix} = \begin{bmatrix}2\\2.5\\6\end{bmatrix}$$

The equation of the normal at point P_1:

$$\mathbf{r}_n = \mathbf{r}_1 + \mathbf{n}u = \begin{bmatrix}2\\2.5\\6\end{bmatrix} \pm \begin{bmatrix}-3/\sqrt{10}\\0\\1/\sqrt{10}\end{bmatrix}u = \begin{bmatrix}2\\2.5\\6\end{bmatrix} \pm \begin{bmatrix}-3/\sqrt{10}\\0\\1/\sqrt{10}\end{bmatrix}u = \begin{bmatrix}2 \mp 3u/\sqrt{10}\\2.5\\6 \pm u/\sqrt{10}\end{bmatrix}$$

where the sign is chosen depending on the direction of the normal.

The plane and normal are shown in Figure 3.9a. The curvilinear coordinates of the plane are two sets of the orthogonal straight lines.

3.1.3.3 *Cylinders*

Right circular cylinder and its unit vector **n** of the normal have the parametric equations

$$\mathbf{r}(\varphi, z) = \begin{bmatrix}R\cos\varphi\\R\sin\varphi\\z\end{bmatrix} \quad 0 < \varphi < 2\pi;\ z_{min} < z < z_{max} \tag{3.68}$$

$$\mathbf{N} = \frac{\partial \mathrm{r}}{\partial z} \times \frac{\partial \mathrm{r}}{\partial \varphi} = \begin{bmatrix}R\cos\varphi\\R\sin\varphi\\0\end{bmatrix} = R\begin{bmatrix}\cos\varphi\\\sin\varphi\\0\end{bmatrix}$$

$$\mathbf{n} = \frac{\mathbf{N}}{|\mathbf{N}|} = \pm\frac{\mathbf{N}}{R} = \pm\begin{bmatrix}\cos\varphi\\\sin\varphi\\0\end{bmatrix} \tag{3.69}$$

where R is the circle radius; z is the longitudinal coordinate; and φ is the polar angle.

The cylinder, Eq. (3.68), its normal **n** and two tangents $\mathbf{t}_1$ and $\mathbf{t}_2$ at a cylinder point are shown in Figure 3.9b. The curvilinear coordinates system is formed by the cross-section circles $z = z_i$ and the straight lines parallel to the circle axis in the axial sections, $\varphi = \varphi_i$.

Involute cylinder. The surface of the tooth flank with involute tooth system (Fig. 3.10a) presents an involute cylinder. Its equation $\mathbf{r}(\varphi, z)$, normal **N**, magnitude N, and unit vector of the normal **n** are as follows:

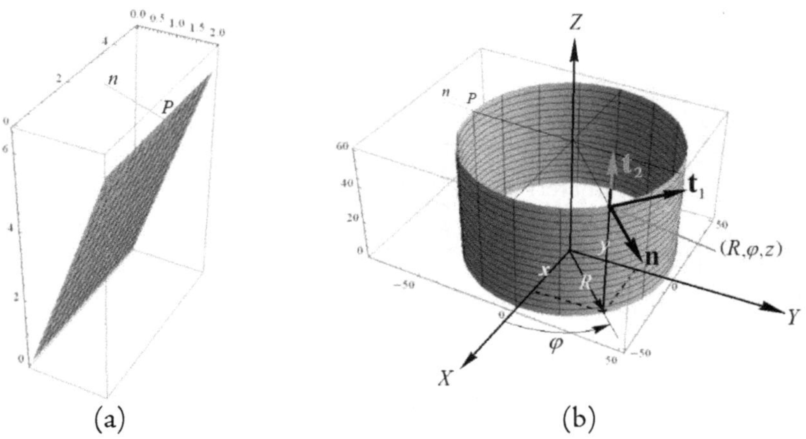

Figure 3-9 Surfaces in 3-D space: (a) a plane; (b) a cylinder.

$$\mathbf{r}(\varphi, z) = \begin{bmatrix} R_b(\sin\varphi - \varphi\cos\varphi) \\ R_b(\cos\varphi + \varphi\sin\varphi) \\ z \end{bmatrix}, 0 < \varphi < \varphi_{max} \text{ and } z_{min} < z < z_{max} \tag{3.70}$$

$$\mathbf{N} = \frac{\partial \mathbf{r}}{\partial z} \times \frac{\partial \mathbf{r}}{\partial \varphi} = \begin{bmatrix} R_b\varphi\cos\varphi \\ -R_b\varphi\sin\varphi \\ 0 \end{bmatrix} = R_b\varphi \begin{bmatrix} \cos\varphi \\ -\sin\varphi \\ 0 \end{bmatrix} \tag{3.71}$$

$$\mathbf{n} = \frac{\mathbf{N}}{\|\mathbf{N}\|} = \pm\frac{\mathbf{N}}{R_b\varphi} = \pm\begin{bmatrix} \cos\varphi \\ -\sin\varphi \\ 0 \end{bmatrix} \tag{3.72}$$

where R_b is the radius of the base circle; z is the longitudinal coordinate of the tooth profile, and φ is the angular parameter presenting the unrolling angle during involute generation.

The normal **N** is shown to exist in all the surface points with the exception of the boundary points for $\varphi = 0$, where the involute has the common point with the base circle. The curvilinear coordinates are the set of the intersected involutes and straight lines (Fig. 3.10a).

Generally, the involute equation has the form

$$\mathbf{r}(\varphi, z) = \begin{bmatrix} R_b\sin(\varphi - \Phi) - R_b\varphi\cos(\varphi - \Phi) \\ R_b\cos(\varphi - \Phi) + R_b\varphi\sin(\varphi - \Phi) \\ z \end{bmatrix}, 0 < \varphi < \varphi_{max} \text{ and } z_{min} < z < z_{max} \tag{3.73}$$

where Φ is the angular pitch – the angle between two neighboring involutes.

In Fig. 3.10b, involutes 1 and 2 associated with the common base surface are shown, where the angle between them is the angular pitch (in this case, $\pi/4$).

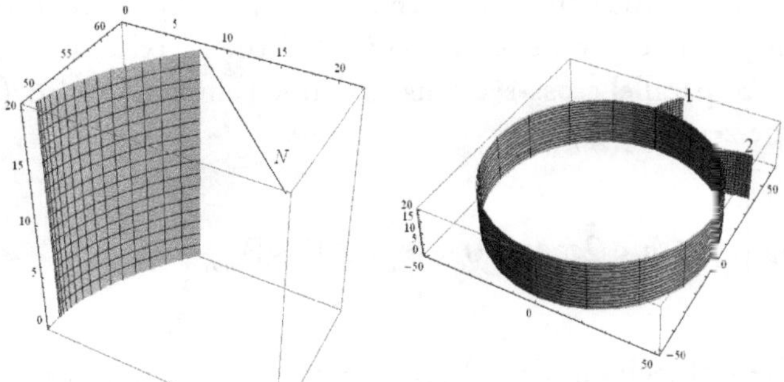

(a) (b)

Figure 3-10 (a) The involute cylinder N is the normal at a point; (b) the base circle and two associated involute cylinders.

3.1.3.4 Sphere

Spherical coordinates (Fig. 3.11) are used to express position vectors of a sphere and its normal $\mathbf{n}(\varphi, \theta)$:

$$\mathbf{r}(\varphi,\theta) = \begin{bmatrix} R\sin\theta\cos\varphi \\ R\sin\theta\sin\varphi \\ R\cos\theta \end{bmatrix}, 0 \le \varphi < 2\pi \text{ and } 0 \le \theta \le \pi \tag{3.74}$$

$$\mathbf{n}(\varphi,\theta) = \begin{bmatrix} \sin\theta\cos\varphi \\ \sin\theta\sin\varphi \\ \cos\theta \end{bmatrix} \tag{3.75}$$

where R is the sphere radius (Fig. 3.12a); φ is the azimuthal angle, $0 \le \varphi < 2\pi$, counted from the X-axis; and θ are the zenith angle counted from the Z-axis, $0 \le \theta < \pi$.

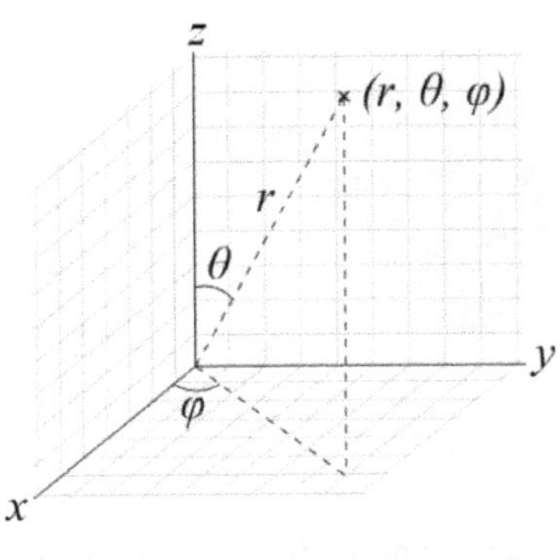

Figure 3-11 Spherical coordinates of the point.

Curvilinear coordinates of the sphere are two sets of circles lying in cross sections passing through (meridians) and perpendicular to (latitudes) the Z-axis. The spherical segment (Fig. 3.12b) between two parallel cross-sections, is defined similarly to Eq. (3.75), but with other boundaries of the zenith angle θ,

$$\mathbf{r}=\begin{bmatrix} R\sin\theta\cos\varphi \\ R\sin\theta\sin\varphi \\ R\cos\theta \end{bmatrix}, 0\leq\varphi<2\pi \text{ and } 0<\theta_{min}\leq\theta\leq\theta_{max}<\pi \tag{3.76}$$

Height h of the spherical segment (Fig. 3.12b) between two parallel cross-sections is calculated as follows:

$$h = R\,(\cos\theta_{min} - \cos\theta_{max}) \tag{3.77}$$

Numerical Example 3.6. **Given:** $\theta_{min} = \pi/12$ and $\theta_{max} = \pi/2$, i.e., the low boundary is the great circle (Fig. 3.12c), then $h = R\,(\cos\pi/12 - \cos\pi/2) = 0.9659\,R$.

3.2 Form-shaping system

3.2.1 Kinematics and structure

3.2.1.1 Kinematic chain

The form-shaping system (FSS) composes the manipulation system together with an active element (*AE*), or tool, and target element (*TE*). The FSS of the machine or robot is referred to as an aggregate of mechanical elements, the mutual positions and movements of which ensure the specified trajectory of the *AE* relative to *TE* (e.g., the cutting tool relative to the machined workpiece). The FSS comprises the base and movable units of the manipulation system together with the final links of the drives. The FSS fulfills the target function of the machine and is regarded as the most vital structural complex determining the designed purpose.

The FSS with one *AE* and one *TE* forms a *chain of links*, wherein the links are placed in series one after another. The links are associated with coordinate frames S_0, S_1, S_2,...,S_n,

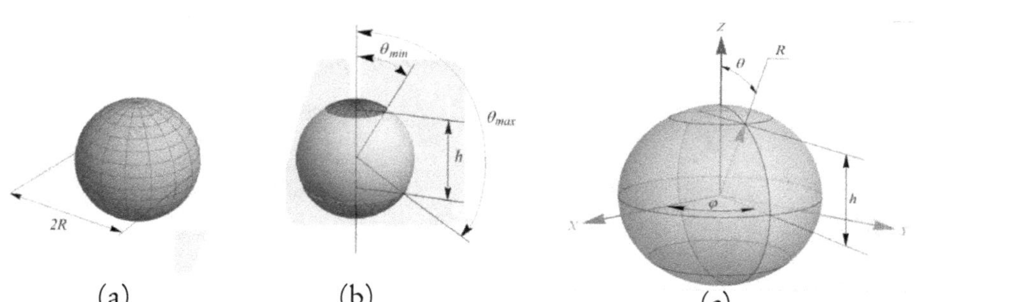

Figure 3-12 (a) Latitude and longitude (meridians and parallels); (b) the spherical zone and the height of the spherical segment; (c) an illustration for *Numerical Example 3.6.*

where n is the number of movable links (frames). The *AE* and *TE* are rigidly connected with the last link (frame S_n) and initial link (frame S_0), respectively. The minimal chain includes only two links – either of them is stationary and the other is movable. The proposed enumeration is convenient if the link motions are considered relative to the *TE*, i.e., the *TE* is considered as a stationary link while the others including the absolutely-fixed link carry out the motions relative to the *TE*.

The *i*th link has l_i degrees of freedom DOF ($l_i = 0 \div 6$) in its motion with respect to the previous link

$$l_i = \begin{cases} 0, \text{ if the } i\text{th link is stationary relative to } (i\text{-}1)\text{th link} \\ 1, \text{ if the } i\text{th link has 1-DOF in motion relative to the } (i\text{-}1)\text{th one} \\ \cdots\cdots \\ 6, \text{ if the } i\text{th link has 6-DOF in motion relative to the } (i\text{-}1)\text{th one} \end{cases}, i = 1,\ldots,n \quad (3.78)$$

A link can represent a chain of parts rigidly clamped together. For example, the lathe-spindle-related set comprising the spindle, chuck and workpiece presents a unique link. The links are alternated by the DOF number under two types:

- *Geometric* link with $l_i = 0$. A pair of neighboring geometrical links of a dimensional chain composes relatively unmoved links, which either are offset by a *constant distance* from one another or are turned through a *constant angle* by one another;
- *Kinematic* link with $l_i = 1 \div 6$ that perform either linear or angular time-dependent displacements. If $l_i = 1$, the pair of the $(i - 1)$th and ith links is connected by a 1-DOF kinematic pair. If $l_i = 2 \div 6$, the pair of the $(i - 1)$th and ith links presents the parallel kinematics mechanism.

A sum of the link DOF is mobility number l in the motion of the *AE* relative to the *TE*:

$$l = \sum_{i=1}^{n} l_i, \text{ with } 0 \le l_i \le 6 \text{ and } l \ge 1 \quad (3.79)$$

A convenient form for description of the FSS is the "link-motion" chain-type graph (Fig. 3.13). The ith graph vertex ($i = 0, 1,\ldots,n$) models the ith FSS link and its coordinate frame S_i. The graph edge connecting the ith and $(i - 1)$th vertices represents the l_i DOF, Eq. (3.78), in motion of the ith link relative to the $(i - 1)$th one.

The FSS combines l *kinematic* and $(n - l)$ *geometrical* links. Depending on the DOF numbers of the *kinematic links*, there are three types of the FSS kinematics:

- *Serial kinematics* FSS combines only a 1-DOF *kinematic* link, $l \le n$. The serial FSS are inherent in machine tools and coordinate measuring machines (CMM) of "the traditional structures", articulated robot-manipulators with and without translational links, and so on. As a rule, these machines have *orthogonal kinematics*, i.e., their axes of rotations and directions of translations are orthogonal or parallel to each other.
- *Parallel kinematics* FSS combines only two links ($n = 1$) – basic link S_0 and a single movable link S_1 (moving platform), and the FSS mobility number is

$$l = 2 \div 6 \quad (3.80)$$

- *Hybrid kinematics* FSS combines both 1-DOF and multi-DOF kinematic links.

3.2.1.2 Coordinate code and motion identifiers

Some important features of the FSS can be described using simple numerical codes. Formulation of the *integer codes* of the FSS allows a compact representation of the machine kinematics. The integer codes present a convenient tool for primary FSS classification, synthesis of new arrangements of working machines, synthesis of the basic mathematical models such as form-shaping function (FSF), and so on.

Formulation of the FSS code uses interpretation of a link as a rigid body and the manipulation system as a chain of the rigid bodies, whereat the target element (e.g., a workpiece) is attached to the initial link S_0, and the active element (e.g., a working tool) is attached to the ending link, whereat S_n (Fig. 3.13). Since the rigid body has no more than six DOF in relative motion with respect to the reference one, its arbitrary motion can be described as a sequence of six elementary motions listed in Table 3.1.

The coordinate code k of the FSS identifies the quantity of the links, their sequence, and relative motions. The code presents an n-digit ordered list of identifiers of motions,

$$k = k_1 k_2 \ldots k_n \tag{3.81}$$

where n is, as before, the number of the FSS links, $n \geq 2$; i is the ordered number of the link ($i = 1, 2,\ldots,n$); and k_i is the identifier of motion of the ith link defined by either one natural number from 1 to 6 or an l_i-length list of these numbers (l_i is the DOF number of the ith link).

If the ith kinematic link has one DOF in its motion relative to the (i - 1)th link, $l_i = 1$, the identifier of motion is one of six symbols listed in Table 3.1:

$$k_i = \begin{cases} 1, \text{ for translation along the } X_{i-1}\text{-axis} \\ 2, \text{ for translation along the } Y_{i-1}\text{-axis} \\ 3, \text{ for translation along the } Z_{i-1}\text{-axis} \\ 4, \text{ for rotation around the } X_{i-1}\text{-axis} \\ 5, \text{ for rotation around the } Y_{i-1}\text{-axis} \\ 6, \text{ for rotation around the } Z_{i-1}\text{-axis} \end{cases} \tag{3.82}$$

In code theory terms, the coordinate code k, Eq. (3.81), presents an n-length *word*. The identifier k_i, i.e., one of the six first natural numbers 1, 2,...,6, constituting the word is a *character* of the *sextuple alphabet*. The identifier describes either a real motion of the ith link when it is the kinematic link or a constant distance when it is the geometric link. To fix this difference, the general approach applied in robotics is used: the motion symbols of the kinematic links are printed in bold face (**1**, **2**,...,**6**), while displacements associated with geometrical links are denoted by plain numbers (1, 2,...,6).

If the link has more than 1-DOF in relative motion, $2 \leq l_i \leq 6$, the identifier presents a bracketed list of l_i ordered symbols each associated with an elementary motion:

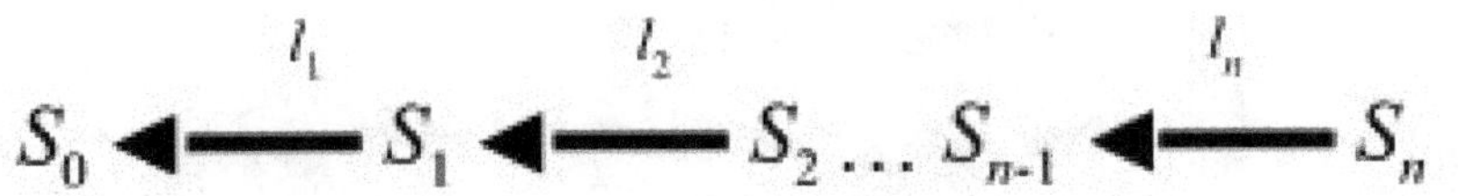

Figure 3-13 Link-motion graph of the FSS: $S_0, S_1, S_2,\ldots,S_n$ are the coordinate frames associated with the FSS links; $l_1, l_2,\ldots,l_n$ are the DOF numbers of the links.

$$k_i = \begin{cases} (k_{i1}\, k_{i2}), \text{ for } l_i = 2 \\ \ldots\ldots\ldots\ldots\ldots\ldots\ldots\ldots\ldots\ldots \\ (k_{i1}\, k_{i2}\, k_{i3}\, k_{i4}\, k_{i5}\, k_{i6}), \text{ for } l_i = 6 \end{cases} \tag{3.83}$$

where k_{ij} ($i = 1, 2,\ldots,n; j = 1, 2,\ldots,l_i$) is the identifier, Eq. (3.82), describing one of l_i elementary motions of the parallel-kinematics link.

Examples of identifiers for the parallel-kinematics links are as follows:

$$k_i = (\mathbf{123}) \tag{3.84}$$

$$k_i = (\mathbf{345}) \tag{3.85}$$

$$k_i = (\mathbf{123456}) \tag{3.86}$$

where Eq. (3.84) describes a parallel-kinematics link performing 3-D translation; Eq. (3.85) describes a parallel-kinematics link performing translation along the *Z*-axis and rotations around the *X*-axis and *Y*-axis in the given subsequence; and Eq. (3.86) describes a parallel-kinematics link performing 6-D motion in the given subsequence (for example, the Gough-Stewart platform).

The same technique is applied to describe a geometric link. For example, the 3-D translation *within a single rigid body* is denoted by bracketed three plain letters:

$$k_i = (123) \tag{3.87}$$

Such transformation takes place when there is a need to transfer the origin of the coordinate frame without necessity to change the axes' orientations. Typical examples are as follows: (a) a transfer from the *home position* (*zero point*) of the CNC-machine to the origin of the workpiece coordinate system (Fig. 3.14); (b) three axes of the guideways for the translational motions along three orthogonal directions located at three different levels; (c) coordinate systems of a base of the industrial robot and its target element are usually different, but associated with a single rigid body – the robot base.

The coordinate code *k*, Eq. (3.81), defines the specified FSS up to the spatial location and orientation of the axes of the stationary coordinate frame. The coordinate code includes the symbols of all (without exception) relative motions of the kinematic links and *necessary* geometry links. Two versions of coordinate code can be distinguished:

- The (pure) kinematic code, in which only kinematic links take place, and there are no geometric links, $n = l$. Hence all code elements are defined in bold.

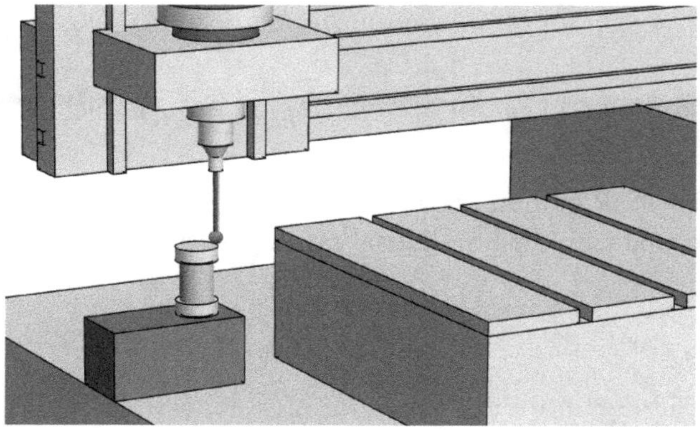

Figure 3-14 Home position touchsetter in CNC machines.

- The structure-kinematic code, in which both kinematic land geometric links take place, $n > l$, and the code elements associated with the geometric links are written as plain letters. As may be required, this version will be specified by additional subscript k_{st}.

The code allows effective and compact kinematic description of the robots and machines with serial-, parallel-, and hybrid-kinematics using combinations of six natural numbers 1, 2,...,6. In turn, this allows a relatively simple *synthesis of the mathematical equations* describing the kinematics of mechanical systems of any complexity. For example, links of the orthogonal serial-kinematics robots and machines has one DOF in motion relative to the neighbor link of the manipulation chain, their coordinate codes present the n-ordered numbers, Eq. (3.81), where the identifiers k_i are defined through Eq. (3.82). Another situation takes place in hybrid-kinematics systems. For example, the n-links-length FSS combining *a single* parallel-kinematics 3-DOF link and (n - 1) kinematic and/or geometric 1-DOF links has the following coordinate code:

$$k = k_1 \ldots k_{i-1} (k_{i1}\, k_{i2}\, k_{i3})\, k_{i+1} \ldots k_n \tag{3.88}$$

To minimize the code length, the following features of mechanical motions are used: a direction of the translation is described by a coordinate axis whose unit vector can be free displaced parallel to itself in the space, and a direction of the rotation axis is described by a coordinate axis whose unit vector can slide along the given straight line. Hence, the origin of the axis can be chosen in any point of the line.

3.2.1.3 Arrangement code

An arrangement code K presents an extension of the coordinate code k and additionally fixes an ordered position of the unique link of the FSS, which is stationary in its *absolute* motion such as machine bed, frame, housing, column, and so on. If the ith link is the fixed one, then it is sandwiched between the previous and subsequent links executing absolute motions relative to the fixed link. In the arrangement code, the stationary link is symbolized through zero located between k_i and k_{i+1}:

$$K = k_1 k_2, \ldots, k_i 0 k_{i+1}, \ldots, k_n, \tag{3.89}$$

where $k_1, k_2, \ldots, k_n$ are the identifiers, Eq. (3.82).

For the given coordinate code k, Eq. (3.81), there are $l + 1$ possible variants for the arrangement of the bed in the FSS (recall that l is the number of *kinematic* links of the FSS). The corresponding arrangements will differ by location of the digit 0 within the code.

The industrial robots (IR) present a specific case, in which the stationary link is, as a rule, the initial link of the kinematic chain. Hence, the arrangement code of the IR is

$$K = 0k_1k_2,\ldots,k_n \tag{3.90}$$

The same form has the arrangement code in automatic transfer lines for milling and boring operations, where the workpiece is stationary based and machining processes are carried out by motions of the cutting tools. Contrary to that, the fixed link (the bed, column, portal, and so on) of machine tools and CMMs is usually located in an intermediate place of the kinematic chain, and both the initial link (target elements, element to be processed, etc.) and ending link (working tool) work out an actual motion relative to the bed.

3.2.1.4 Example: Codes of serial-type machines

The numerical code-based approach becomes particularly attractive for serial-kinematics machines and robots and initially was developed just for this application. The reason is that the serial-type machines and robots are usually designed as an orthogonal system, wherein the pairs of successive axes of motions are either orthogonal or parallel to one another.

In the lathe (Fig. 3.15a), the spindle axis is parallel to direction of longitudinal feed, and two feed directions are orthogonal to each other. Hence, the three axes in the whole combine the regular cylindrical coordinate frame. Terminal elements of the kinematic chain of the lathe are the workpiece (the target element) rotating in common with the spindle relative to the bed and the cutting tool (the active element) performing two translations: parallel and perpendicular to the spindle axis in common with longitudinal carriage and cross-carriage, respectively.

Hence, the FSS of the lathe includes three movable links ($n = 3$) and combines four units: the bed associated with coordinate frame $S_1 \equiv O_1X_1Y_1Z_1$ and three mobile units of a regular turning machine – spindle (frame $S_0 \equiv O_0X_0Y_0Z_0$), longitudinal carriage (frame $S_2 \equiv O_2X_2Y_2Z_2$), and cross-carriage (frame $S_3 \equiv O_3X_3Y_3Z_3$) – provide three DOF in the functional motion of the cutting tool with respect to the workpiece.

The link-motion graph of this kinematic chain is shown in Figure 3.15b. The workpiece in common with the chuck and the main spindle presents a link associated with coordinate frame S_0, which is always assumed stationary. Despite the fact that the bed is stationary in the absolute motion, it is modeled as the link rotating around the Z_0-axis in a direction opposite to spindle rotation since only a motion relative to the previous link is considered in the FSS. According to names of the coordinate axes signed in Figure 3.15a and b, the *coordinate code k*, Eq. (3.81), is a three-digit number,

$$k = \mathbf{631} \tag{3.91}$$

where symbol $k_1 = \mathbf{6}$ corresponds to rotation around the Z-axis; $k_2 = \mathbf{3}$ describes the translation of the longitudinal carriage relative to the bed along the Z_1-axis; and finally $k_3 = \mathbf{1}$

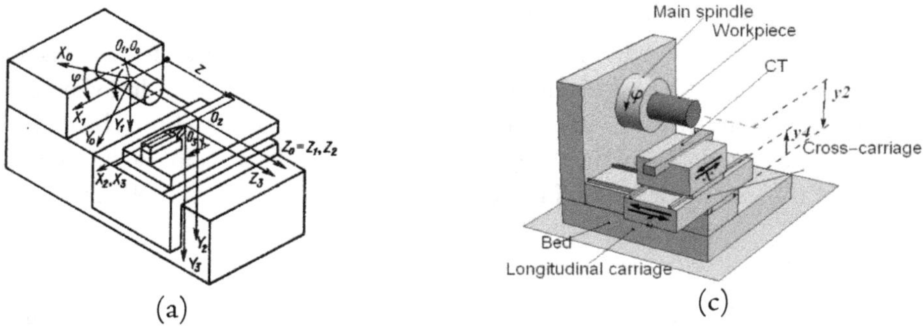

$$S_0 \xleftarrow{l_1=1} S_1 \xleftarrow{l_2=0} S_2 \xleftarrow{l_3=1} S_3$$

(b)

$$S_0 \xleftarrow{l_1=1} S_1 \xleftarrow{l_2=0} S_2 \xleftarrow{l_3=1} S_3 \xleftarrow{l_4=0} S_4 \xleftarrow{l_5=0} S_5$$

(d)

Figure 3-15 Schematics of the form-shaping system of the lathe: (a) the machine layout; (b) the link-motion graph of the kinematic chain; (c) two geometrical links entering in the structural chain; (d) the link-motion graph for the structural chain.

describes the transverse carriage translates along the X_2-axis.

Since the bed is located between the rotational spindle and longitudinal carriage, the *arrangement code* K, Eq. (3.89), of the lathe is

$$K = \mathbf{6031} \tag{3.92}$$

The *structural code* k_{st} of the lathe FSS takes into account distances between the axis of spindle rotation and guideways of both carriages (see distances y_2 and y_4 in Figure 3.15c). These distances present the Abbe offsets in accuracy calculations. The FSS becomes the six-links system (Fig. 3.15d), and its coordinate code incorporates two *geometrical* links taking into account two translation distances along the Y-axis (plain symbol 2):

$$k_{st} = \mathbf{62321} \tag{3.93}$$

The 3-DOF (**RRR**) planar robot-manipulator shown in Figure 3.16 has three kinematic links rotating around the parallel axes and three geometry links ($n = 6$). The coordinate codes k and the arrangement code K are as follows:

$$k = \mathbf{616161} \tag{3.94}$$

$$K = 0616161 \tag{3.95}$$

where $k_1 = k_3 = k_5 = \mathbf{6}$ correspond to kinematic links rotating around the Z-axes (perpendicular to the drawing plane); $k_2 = k_4 = 1$ correspond to geometrical links – the distances between to neighboring axes; $k_6 = 1$ describes the geometric link – distance between the rotation axis and end-effector point; and digit 0 in code K fixes the ordered location of the base link.

The DOF numbers corresponding to each kinematic pair are marked in the chain-type graph: $l_1 = l_3 = l_5 = 1$ and $l_2 = l_4 = l_6 = 0$. Note that the kinematics, where three neighboring

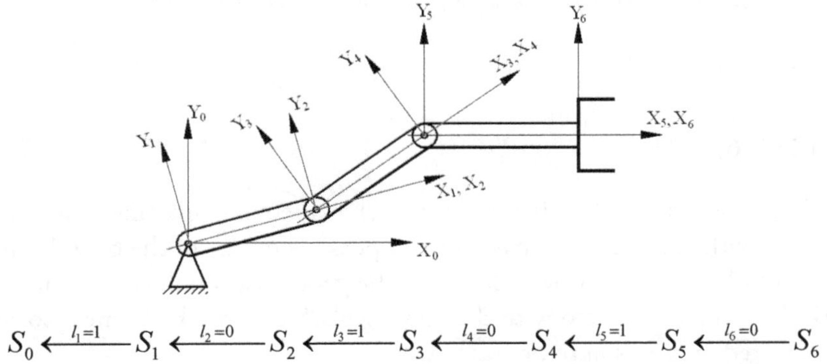

Figure 3-16 FSS of planar 3-DOF (**RRR**) manipulator.

links rotated around three parallel is typical for articulated robots.

The FSS of the reciprocating table horizontal spindle surface-grinding machine is shown in Figure 3.17. Machine table 3 with mounted workpiece 4 and cross saddle 2 moved along guideways 2 of the bed 10 (the stationary link in absolute motion) are associated with coordinate frames S_0 and S_1, respectively, while the bed with rigidly bounded column 7 is modeled by frame S_2. The wheel head 6 (frame S_3) carries out the vertical reciprocating motion along the column guideways, and the grinding spindle with grinding wheel 5 rotates around the horizontal axis (frame S_4). Hence, the machine comprises four movable links and four mobility degrees ($n = l = 4$), and its coordinate code is

$$k = \mathbf{1235} \tag{3.96}$$

3.2.1.5 *Example: Codes of parallel- and hybrid-kinematics machines*

The kinematics of the training machine shown in Figure 3.18 [2] is described using three coordinate frames associated with the base, the moving platform, and the pilot seat model,

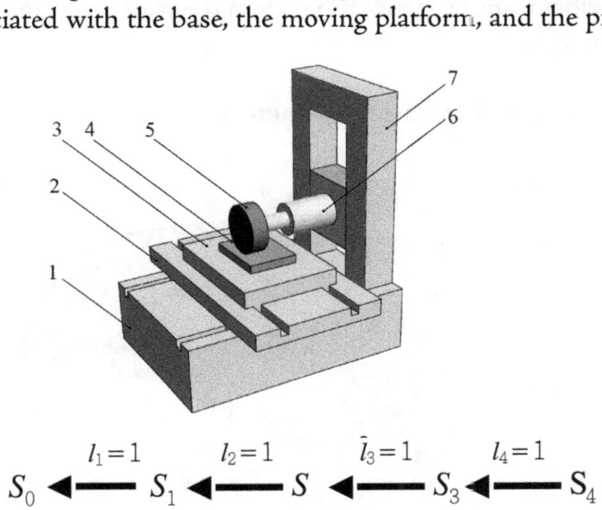

Figure 3-17 Reciprocating table horizontal spindle surface-grinding machine.

respectively. Thus, the coordinate and arrangement codes of the FSS are

$$k = (\mathbf{123456})\,(123) \tag{3.97}$$

$$K = 0\,(\mathbf{123456})\,(123) \tag{3.98}$$

where $k_1 = (\mathbf{123456})$ is the identifier relating to the kinematic link (moving platform) with 6-DOF motion with respect to the base (one of possible orders of the 6-DOF motion definitions); $k_2 = (123)$ is the identifier relating to the geometrical link connecting two coordinate frames: the frame associated with the moving platform and the frame associated with a subject to be tested or a person to be trained.

The FSS of the *tripod-based hybrid-kinematics machine* (Fig. 3.19) comprises four kinematic links ($n = 4$): upper table 1 and saddle 2 performing X- and Y-motion in the horizontal plane, the 3-DOF mobile platform, and the spindle unit. The kinematic chain is started from the home position located at the upper table. The next link is the saddle moving relative to the base. Three sliders 3, 5, and 7 control the 3-DOF motion of the moving platform 9 through connecting rods 4, 6, and 8. The sliders are moved along the vertical guideways of the base. The last link is associated with the spindle 10 rotating in common with the cutting tool. The coordinate code k and the arrangement code K are

$$k = (123)\,\mathbf{12}\,(\mathbf{345})\,\mathbf{6}3 \tag{3.99}$$

$$K = (123)\,\mathbf{12}\,0\,(\mathbf{345})\,\mathbf{6}3 \tag{3.100}$$

where $k_1 = (123)$ is the identifier of the geometrical link transferring the frame S_0 in frame S_1; identifiers $k_2 = \mathbf{1}$ and $k_3 = \mathbf{2}$ correspond to motions along the X_1- and Y_2-axis, respectively; $k_4 = (\mathbf{345})$ is the identifier associated with the 3-DOF moving platform (translation and two rotations); $k_5 = \mathbf{6}$ is the identifier associated with the spindle rotation around the Z_4-axis; and $k_6 = 3$ describes transfer (translation) from the platform plane to the fixed point of the spindle along the Z_5-axis.

3.2.2 Geometric and kinematic parameters

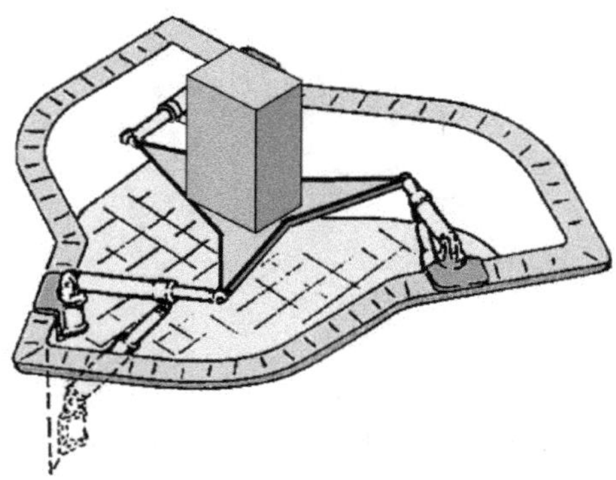

Figure 3-18 The training stand [2].

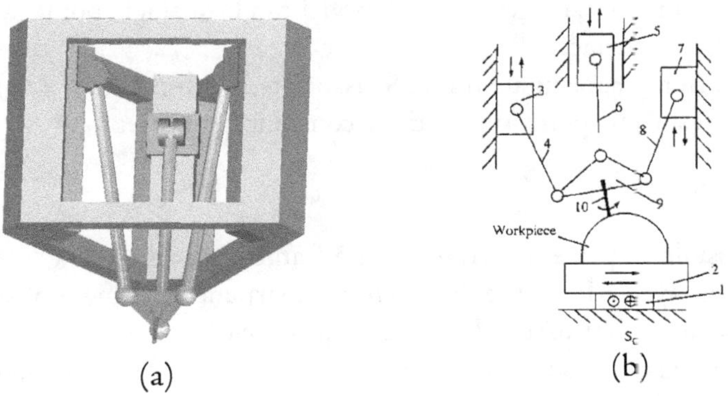

Figure 3-19 The tripod-based hybrid-kinematics machine [3]: (a) structure; (b) kinematics.

3.2.2.1 *Parameters of relative position and orientation*

To complete the formal description of the FSS, it is also necessary to consider a set of parameters defining relative position and orientation of the pair of neighboring links or their coordinate frame S_{i-1} and S_i ($i = 1, 2,\ldots,n$). The parameters of the motion (or the fixed displacement) of the ith link can be classified as follows:

- Parameter q_j of the geometrical link is

$$q_i = \begin{cases} a_i, \text{ if } q_i \text{ is the constant frame-to-frame distance} \\ \theta_i, \text{ if } q_i \text{ is the constant frame-to-frame angle} \end{cases} \quad i = 1,\ldots,n \tag{3.101}$$

- Parameter of the ith kinematic link with the DOF number $l_i = 1$ in motion relative to the S_{i-1} frame with its time history $q_i = q_j(t)$,

$$q_i = \begin{cases} a_i(t), \text{ if } q_i \text{ is the time-dependent frame-to-frame distance} \\ \theta_i(t), \text{ if } q_i \text{ is the time-dependent frame-to-frame angle} \end{cases} \quad i = 1,\ldots,n \tag{3.102}$$

Generally, a set of the parameters q_j describing the motion of the jth link with l_j-DOF is

$$q_j = \begin{cases} q_j, \text{ for } l_j = 0 \\ q_j(t), \text{ for } l_j = 1 \\ \{q_j(t), q_{j+1}(t)\}, \text{ for } l_j = 2 \\ \cdots\cdots\cdots\cdots\cdots\cdots \\ \{q_j(t), q_{j+1}(t), \ldots, q_{j+6}(t)\}, \text{ for } l_i = 6 \end{cases}, \quad (j = 1,\ldots,n) \tag{3.103}$$

3.2.2.2 *Model of active element (tool)*

The active element (AE) of the FSS is a cutting tool in the machine tool, a measuring probe in the CMM, an electrode attached to the end-effector of the welding robot, a spray gun of the painting robot, etc. In what follows, the AE will be considered as a geometrical object.

Depending on physical nature, its geometrical model can be a single point (a set of points), a line (a set of lines), or a surface.

- *Point-type AE*. In the coordinate frame S_n associated with the last (n-th) FSS link, the three coordinates of the point-type AE are constant quantities

$$x_n = a_n;\ y_n = b_n;\ \text{and}\ z_n = c_n \tag{3.104}$$

Usually, it is possible to place the origin of the frame S_n at this point, $x_n = y_n = z_n = 0$. Examples of the point-type AE are the tip of a finely sharpened cutting tool, central point of the electrode end during robot-assisted spot resistance welding, and so on.

- *Line-type AE*. Coordinates of a line-type working segment can be expressed through a function of one independent variable U,

$$x_n = x_n(U);\ y_n = y_n(U);\ \text{and}\ z_n = z_n(U) \tag{3.105}$$

with $U_{min} \le U \le U_{max}$

Consider as an example the robot-assisted spray painting. The spray paints include a spray nozzle (cap). The shape of the nozzle slit defines the line on the surface to be painted. The straight-line slit of length $2a$ can be described as follows:

$$x_n = U;\ y_n = 0;\ z_n = 0\ (-a \le U \le a) \tag{3.106}$$

As other examples, some geometrical models of the blade cutting tools are considered. A model of a single-blade cutter (Fig. 3.20) is segment AB of the straight line. For simplicity, it is assumed that the blade lies in plane YZ; hence, coordinates of section AB are

$$x_n = U \sin\theta;\ y_n = -U \cos\theta;\ \text{and}\ z_n = 0 \tag{3.107}$$

with $0 \le U \le |AB|$

where U is the current distance BO_2; and $|AO_2|$ is the length of section AB.

A model of the multiple-blade tool is shown in Figure 3.20b: cylindrical milling cutter with helical teeth. In coordinate frame S_n, Eqs. (3.105) become

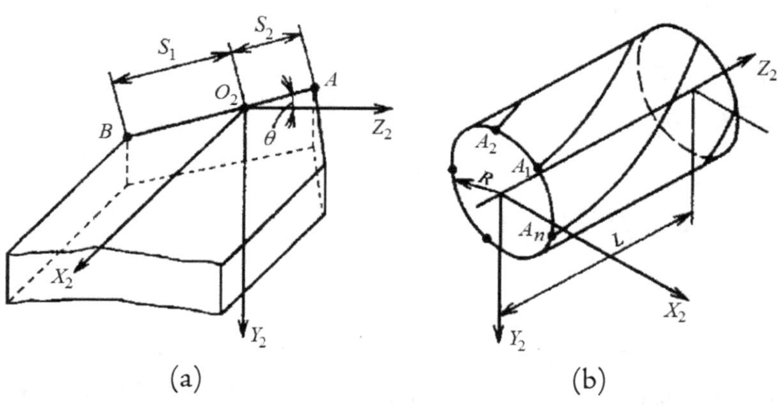

Figure 3-20 Geometrical models of blade tool: (*a*) single-blade cutter; (*b*) multi-blade cutter with helical teeth.

$$x_n = R\cos(\varphi_i + z/p);\ y_n = R\sin(\varphi_i + z/p);\ \text{and } z_n = z \tag{3.108}$$

with $0 \le z \le L$; and $i = 1,..,m$

where z is the independent variable having the meaning of linear displacement along the Z_n-axis; i is the number of the blade; m is the number of blades on the milling cutter; φ_i is the polar angle defining the initial blade position on the frontal face (with $U = 0$) of the milling cutter; R is the radius of the milling cutter; p is the screw parameter linked with pitch $H = 2\pi p$.

In the case $p \to \infty$, Eqs. (3.105) of the blades of the cylindrical milling cutter become:

$$X_n = R\cos\varphi_i;\ y_n = R\sin\varphi_i;\ \text{and } n = z \tag{3.109}$$

with $0 \le z \le L$; and $i = 1,..,m$

- *Surface-type AE.* Coordinates of a surface can be expressed as a function of two independent variables U and V,

$$x_n = x_n(U, V);\ y_n = y_n(U, V);\ \text{and } z_n = z_n(U, V), \tag{3.110}$$

with $U_{min} \le U \le U_{max}$ and $V_{min} \le V \le V_{max}$.

Consider some examples of surface-type *AE*s:

(a) Cylindrical grinding wheel of radius R and thickness b (Fig. 3.21a)

$$x_n = R\cos\varphi_t;\ y_n = R\sin\varphi_t;\ \text{and } z_n = z_t \tag{3.111}$$

with $0 \le z_t \le b$ and $0 \le \varphi_t \le 2\pi$

(b) Conical grinding wheel (Fig. 3.21b):

$$x_n = \rho\cos\varphi;\ y_n = \rho\sin\varphi;\ \text{and } z_n = \rho\,\text{cotan}\,\theta - h \tag{3.112}$$

$0 \le \varphi \le 2\pi$ and $h\tan\theta \le \rho \le (h + b)\tan\theta$

where φ is the polar angle; ρ is the polar radius of a current cross-section; b is the wheel height; θ is the half cone vertex angle; and h is the distance between the cone vertex and small end face.

(c) Spherical segments (Fig. 3.21c) are used for the measuring probe ends in the CMM and grinding wheels. Coordinates of the spherical segment are

$$x_n = R\sin\theta\cos\varphi;\ y_n = R\sin\theta\sin\varphi;\ \text{and } z_n = R\cos\theta \tag{3.113}$$

with $0 \le \varphi \le 2\pi$ and $\theta_{min} \le \theta \le \theta_{max}$

$$\theta = \begin{cases} \arcsin\dfrac{h}{R}, & \text{for } h < R \\[2ex] \pi/2 + \arcsin\dfrac{h-R}{R}, & \text{for } h \ge R \end{cases}$$

where φ and θ are the latitude and altitude angles, respectively; R is the sphere radius; h is the distance of the segment plane from the north pose.

Thus, the number of geometrical parameters M defining the FSS active element depends on its dimensionality:

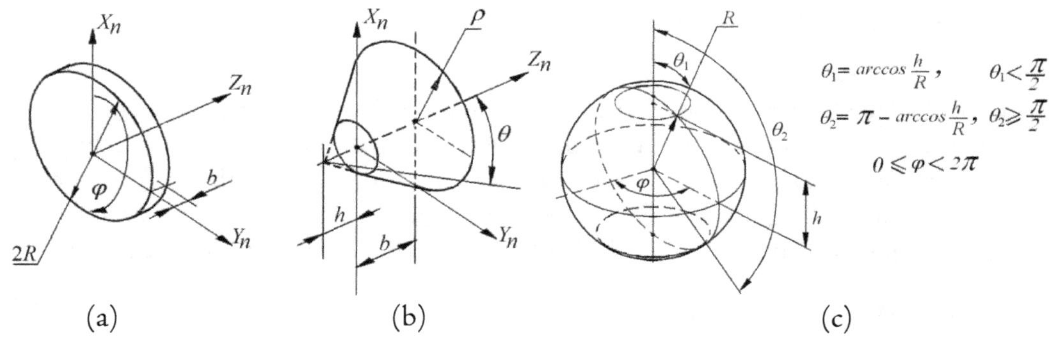

Figure 3-21 Geometrical model of surface-type tools: (a) cylindrical working surface (b) conical surface; (c) spherical segment.

$$M = \begin{cases} 0, & \text{for point-type tool} \\ 1, & \text{for line-type tool} \\ 2, & \text{for surface-type tool} \end{cases} \tag{3.114}$$

3.2.2.3 *Model of the target elements and the AE-TE contact*

By analogy with the active element (AE) of the FSS, the target (TE) is considered as a geometrical subject, whose model can be a single point (a set of separate points), a line (a set of separate lines), or a surface. Correspondingly, the dimensionality m of the TE is:

$$m = \begin{cases} 0, & \text{for point-type object} \\ 1, & \text{for line-type object} \\ 2, & \text{for surface-type object} \end{cases} \tag{3.115}$$

The resistance spot-welding of the thin parts can be considered as an example of the zero-dimension target element, $m = 0$. The butt and fillet welds as well as resistance seam-welds present examples of the line-type objects, $m = 1$. Obviously, the consideration of the welds as a point or a line is only a kinematic-related interpretation since actual welds have specific volumetric forms. Finally, the machined plane, spheres, cones, and free-form surfaces present examples of the surface-type target elements, $m = 2$.

The exposure of an active element of the FSS to the target one can have different physical natures. It may be a power mechanical contact between the AE and TE during machining and assembly operations and a non-mechanical contact through electrical sparks between AE and TE as a pair of electrical electrodes during electric discharge wire cutting or arc welding, through other active physical agents such as a gas abrasive stream during abrasive jet cutting, a paint stream during spray painting, and combination of a large current with a mechanical pressure during spot welding.

In what follows, all interrelations "AE-TE" will be considered as a local contact (hereafter "contact") characterized by two parameters: geometry of the contact and its continuity in time. Contact area presents a set of points common to the tool and workpiece at a current moment of time. A general characteristic of the contact geometry is the dimensionality

$$\mu = \begin{cases} 0, \text{ for a single-point or multipoint contact area} \\ 1, \text{ for a single-line or multiline contact area} \\ 2, \text{ for a surface-type contact area} \end{cases} \quad (3.116)$$

Obviously the dimensionality μ of the contact area is less than both dimensionality M, Eq. (3.114), of the AE (tool) and dimensionality of the TE m, Eq. (3.115),

$$\mu \leq (M, m) \quad (3.117)$$

Six possible combinations of the dimensionalities μ, Eq. (3.115), and m, Eq. (3.115), are shown in Table 3.2. In Table 3.3, 14 possible combinations of M, m, and μ are shown.

Apart from the contact dimensionality, contact conditions are characterized by contact continuity defined through the binary continuity factor

$$\tau = \begin{cases} 0, \text{ for the interrupt contact} \\ 1, \text{ for the continous contact} \end{cases} \quad (3.118)$$

Table 3-2 Possible combinations of dimensionalities μ and m.

m	Dimensionality ν of the contact area		
	μ=0	μ=1	μ=2
m = 0	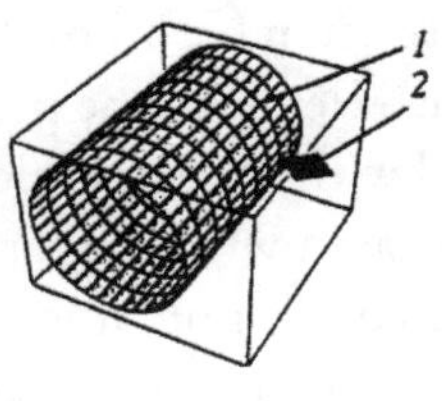		
m = 1	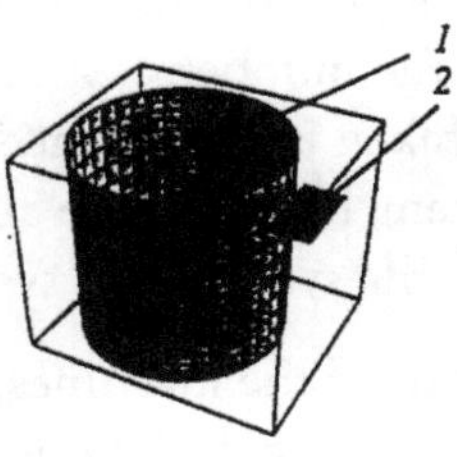	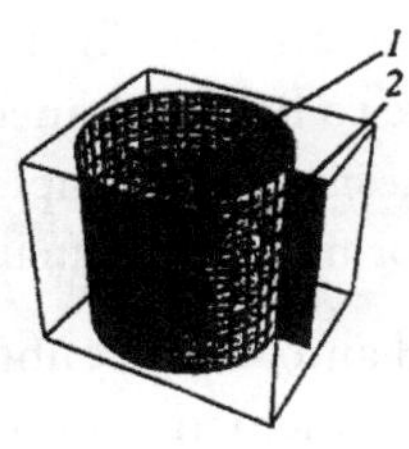	
m = 2	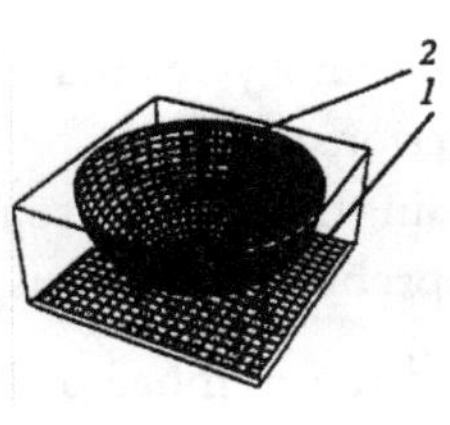	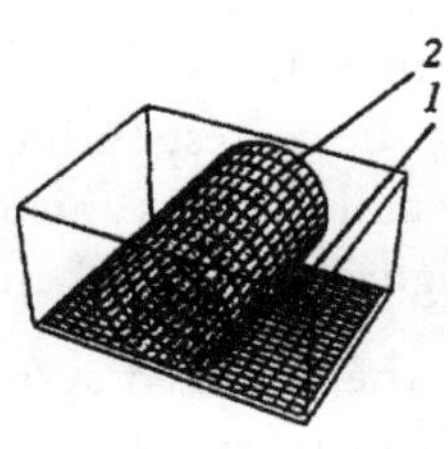	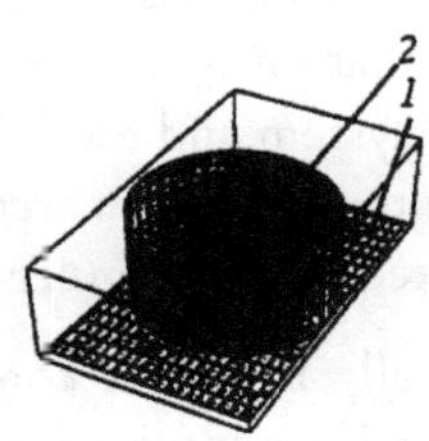

Table 3-3 Topology of "AE-TE" contact.

m	$M = 0$	$M = 1$		$M = 2$		
0	$\mu = 0$	$\mu = 0$	–	$\mu = 0$	–	–
1	$\mu = 0$	$\mu = 0$	$\mu = 1$	$\mu = 0$	$\mu = 1$	–
2	$\mu = 0$	$\mu = 0$	$\mu = 1$	$\mu = 0$	$\mu = 1$	$\mu = 2$

In machining on machine tools, an intermittent contact is typical for blade machining, such as milling, fly hobbing, and slotting, while machining by the surface is performed by continuous contact. In Table 3.4, examples of form-shaping processes of different topological classes are defined according to quartet of parameters:

$$C = m\, M\, \mu\, \tau \qquad (3.119)$$

For example, the fillet welding carried out by means of an electrode (m = 1 and M = 0) with point-type continuous contact (μ = 0 and τ = 1) relates to class 1001. Grinding the sculpture surface carried out by means of the spherical grinding wheel (m = 2 and M = 2) with point-type continuous contact (μ = 0 and τ = 1) relates to class 2201, and so on.

3.3 Form-shaping function formulation

3.3.1 Mathematical technique

3.3.1.1 Introduction

The main model of the FSS theory is the form-shaping function (FSF) of the machine. The FSF connects the position-orientation vectors of the active element (the processing tool) in two coordinate frames: in the coordinate frame associated with the active element AE of the quasi-kinematic chain and in the coordinate frame associated with the target element TE (the element to be operated) of the FSS. To describe the connection, it is essential to carry out relatively cumbersome coordinate transformations, which are effectively formulized by means of matrix calculus. The matrix techniques are developed for analysis and synthesis of complex systems, whose properties are described by a large number of kinematic and geometric characteristics, which have to be simultaneously taken into account. The techniques are not only convenient for synthesis of the complex system, but they also allow application of the formal matrix properties for non-trivial analysis of the synthesized system:

- Calculations of the nominal and actual configurations of the machines;
- Evaluation of the geometric kinematic, static, and dynamic deviations in any point of the machine working space along any given direction;
- Evaluation of the critical (maximal and minimal) accuracy-related characteristics of the system and establishing the corresponding directions;
- Estimation of the accuracy and stiffness anisotropy in the working space;
- Detection of singular configurations and their comprehensive analysis, etc.

Generally, the more complex the analyzed system and the number of its elements, the more effective is application of the matrix methods. The matrix techniques can be considered

Table 3-4 Topological classes of the form-shaping processes.

m	M	μ	τ	Class	Example
0	0	0	0	0000	Spot-welding at an isolated point by a point-type electrode
0	1	0	0	0100	
0	2	0	0	0200	Measurement of position of an isolated point by a spherical probe
1	0	0	0	1000	Spot-welding of an intermittent weld by a point-type electrode
1	0	0	1	1001	Spot- or arc-welding of a line-type weld by a point-type electrode. Engraving by a point-type tool
1	1	0	0	1100	
1	1	0	1	1101	Spot welding of a line-type weld by a disk-shape electrode.
1	1	1	0	1110	
1	1	1	1	1111	Shearing
1	2	0	0	1200	Point-to-point measurement of a line by a spherical probe
1	2	0	1	1201	Measurement of a trajectory by a spherical probe
1	2	1	0	1210	
1	2	1	1	1211	
2	0	0	0	2000	Turning by a fly-hob
2	0	0	1	2001	Turning by a single-point edge
2	1	0	0	2100	Milling by a straight tooth milling cutter.
2	1	0	1	2101	Milling by a helical tooth milling cutter
2	1	1	0	2110	Milling by a spherical tooth milling cutter
2	1	1	1	2111	Turning by a form edge tool. Broaching. Core drill
2	2	0	0	2200	Measurement of a surface by a spherical probe
2	2	0	1	2201	Grinding by a spherical wheel
2	2	1	1	2211	Grinding by a wheel periphery. Longitudinal rolling. Roller painting.
2	2	2	0	2220	Grinding by a wheel face. Die forging, die forming

as a natural bridge between main sources of machine errors – geometrical deviations, kinematic errors, and compliance-caused deformations. Effectiveness of matrix methods has drastically increased with proliferation of computers in design and manufacturing.

Synthesis of the FSF presents a solution of the "direct kinematics" problem [4], calculating unknown parameters of the relative position/orientation of the AE (technological tool) with respect to the TE (part to be processed) when the programmed motions of the drives and actuators are given. This problem has simple solution for serial-kinematics machines (SKM) and just this fact is used for synthesis of their FSF. Namely, the procedure is reduced to matrix multiplication and the solution has an explicit analytical form, and for machines with orthogonal kinematics, the procedure is simplified to a product of the matrices of elementary motions according to coordinate code k, Eq. (3.81).

However, the majority of real-life applications, especially the machine setup, require solution of the "inverse kinematics" problem, i.e., definition of the drive motions when the pre-programmed pose (position and orientation) of the AE relative to the TE is given. In some cases, the inverse kinematics problem for SKMs is not so complicated, but as a rule, the analysis presents the solution of a set of the simultaneous non-linear equations. There are many cases, in which analytical solutions are found out, and in some cases, only the *numerical* solution is possible.

The situation is in complete contrast to solution of "inverse kinematics" problem for PKMs, which usually has a simple analytical solution, and this fact is effectively used for synthesis of the FSF of parallel- and hybrid-kinematics machines.

In this book, the universal tool – Mathematica® software system [5] – is applied for accuracy-related calculations. This software system substantially simplifies analytical transformations and numerical procedures needed for solving design problems in a user-friendly environment. The Mathematica software enables numerical and symbolic operations, and graphical presentations in a unified way – through direct interpretation of standard commands with a set of parameters, without preliminary declaration of parameter types. The involved procedures closely resemble use of a hand-held calculator, but this calculator has a huge computing power and provides a knowledge base. This statement is valid for most of the operations including those so complicated and cumbersome as linear programming, analytical and numerical differentiation and integration, solving linear, polynomial, and transcendental equations, solving the eigenvalue problems, building 2-D and 3-D graphics, etc., needed for solution of numerous complicated and sophisticated problems.

3.3.1.2 Homogeneous coordinates and their transformations

The special matrix tool for analysis, synthesis, programming, and optimization of machine and robot models becomes the homogeneous transformation matrix (HTM) technique, which operates with the matrices and vectors of *fourth order* for coordinate transformations in 3-D space [3, 7-9]. In particular, it allows a simple transfer from the abovementioned codes, Eqs. (3.81), (3.88), and (3.89), to synthesis of the FSF for a broad spectrum of machines, accuracy calculations, and so on. The main feature of this homogeneous transformation technique is in the fact that any coordinate transformations are expressed through a single mathematical operation – matrix multiplication. Contrary to this, upon using matrices and vectors of the third order, two different operations are needed: rotation of the coordinate system is modeled by multiplying the vector on the matrix while translation is modeled by addition of the third-order vectors.

Two types of homogeneous vectors of the fourth order in 3-D space are applied: position and non-position vectors. *The position vector* of a point is

$$\mathbf{r} = \begin{bmatrix} x \\ y \\ z \\ 1 \end{bmatrix} \tag{3.120}$$

where x, y, and z are the regular Cartesian coordinates of the point; x, y, z, and 1 are the *homogeneous coordinate* of the same point.

By definition, the origin of the *position vector*, Eq. (3.120), is always fixed with the origin of the coordinate frame (Fig. 3.1). For illustration, vectors $\mathbf{r}_Q = OQ$ and $\mathbf{r}_p = OO_p$ (Fig. 3.2) are the position vectors as referred to coordinate frame $OXYZ$, and vector ρ is also the position vectors but as referred to coordinate frame $O_pX_pY_pZ_p$. The distance of the point from the origin is

$$d = (x^2 + y^2 + z^2)^{1/2} \tag{3.121}$$

The non-position vector, whose three first coordinates are real numbers and the fourth coordinate equals zero, is

$$a = [a_x, a_y, a_z, 0]^{\mathrm{T}} \tag{3.122}$$

For illustration, the vector ρ (Fig. 3.2) is the non-position vector as referred to coordinate frame $OXYZ$. The magnitude (length) of the non-position vector is defined as that for position vector, Eq. (3.121),

$$||\boldsymbol{a}|| = (a_x^{\ 2} + a_y^{\ 2} + a_z^{\ 2})^{1/2} \tag{3.123}$$

Some typical non-position vectors are as follows:

- Difference vector describing displacement of a point from the initial 1 to final 2 positions,

$$\mathbf{d} = \mathbf{r}_2 - \mathbf{r}_1 = [d_x, d_y, d_z, 0]^{\mathrm{T}} \tag{3.124}$$

$$d_x = x_2 - x_1;\ d_y = y_2 - y_1;\ d_z = z_2 - z_1$$

where x_1, y_1, z_1 and x_2, y_2, z_2 are the coordinates of position vectors $\mathbf{r}_1$ and $\mathbf{r}_2$, respectively.

- A derivative with respect to parameter t of parameter-dependent vector $\mathbf{r} = \mathbf{r}(t)$, e.g., the velocity vector

$$\mathbf{v} = [v_x, v_y, v_z, 0]^{\mathrm{T}} \tag{3.125}$$

where $v_x = dx/dt$; $v_y = dy/dt$; and $v_z = dz/dt$.

Other examples of the derivative-related *non-positional vectors* are the acceleration vector; tangent, normal, and binormal vectors at a curve; tangents and normal at a surface; and the unit vector $\mathbf{e}$ comprising three directional cosines, e.g., the unit vector of the direction defined by non-position vector $\boldsymbol{a}$, Eq. (3.123),

$$\mathbf{e} = [e_x, e_y, e_z, 0]^{\mathrm{T}} = \boldsymbol{a}/||\boldsymbol{a}|| \tag{3.126}$$

where $\boldsymbol{e}_x = a_x/||a||$, $e_y = a_y/||a||$, and $e_z = a_z/||a||$ are the directional cosines; and $||\boldsymbol{a}||$ is the magnitude of vector $\boldsymbol{a}$, Eq. (3.122).

In terms of homogeneous coordinates, four unit vectors of the Cartesian coordinate frame are defined: three non-position unit vectors $\mathbf{e}^1$, $\mathbf{e}^2$, $\mathbf{e}^3$ associated with the X-, Y-, and Z-axis, respectively, and the position vector $\mathbf{e}^4$ associated with the origin of the coordinates,

$$\mathbf{e}^1 = \begin{bmatrix} 1 \\ 0 \\ 0 \\ 0 \end{bmatrix}; \mathbf{e}^2 = \begin{bmatrix} 0 \\ 1 \\ 0 \\ 0 \end{bmatrix}; \mathbf{e}^3 = \begin{bmatrix} 0 \\ 0 \\ 1 \\ 0 \end{bmatrix}; \text{ and } \mathbf{e}^4 = \begin{bmatrix} 0 \\ 0 \\ 0 \\ 1 \end{bmatrix} \tag{3.127}$$

Position and non-position vectors can be represented through the unit vectors

$$\mathbf{r} = x\,\mathbf{e}^1 + y\,\mathbf{e}^2 + z\,\mathbf{e}^3 + 1\,\mathbf{e}^4$$

$$\boldsymbol{a} = a_x\,\mathbf{e}^1 + a_y\,\mathbf{e}^2 + a_z\,\mathbf{e}^3 + 0\,\mathbf{e}^4$$

The basic mathematical operations

- *Addition*: The sum of two position vectors is *undefined*. The addition is defined for two vectors where one or both of them are non-position,

$$\mathbf{r} + \mathbf{d} = \begin{bmatrix} x \\ y \\ z \\ 1 \end{bmatrix} + \begin{bmatrix} d_x \\ d_y \\ d_z \\ 0 \end{bmatrix} = \begin{bmatrix} x + d_x \\ y + d_y \\ z + d_z \\ 1 \end{bmatrix} \text{and } \boldsymbol{a}_1 + \boldsymbol{a}_2 = \begin{bmatrix} a_{x1} + a_{x2} \\ a_{y1} + a_{y2} \\ a_{z1} + a_{z2} \\ 0 \end{bmatrix}$$

- *Multiplication* by a scalar is defined only for non-position vectors,

$$p\,\boldsymbol{a} = p \begin{bmatrix} a_x \\ a_y \\ a_z \\ 0 \end{bmatrix} = \begin{bmatrix} pa_x \\ pa_y \\ pa_z \\ 0 \end{bmatrix}$$

- *The scalar (dot) product* of two non-position vectors or a pair composing a position and a non-position vector is a scalar,

$$\mathbf{a} \cdot \mathbf{b} = \begin{bmatrix} a_x \\ a_y \\ a_z \\ 0 \end{bmatrix} \cdot \begin{bmatrix} b_x \\ b_y \\ b_z \\ 0 \end{bmatrix} = a_x b_x + a_y b_y + a_z b_z$$

$$\mathbf{r}.\mathbf{a} = \begin{bmatrix} x \\ y \\ z \\ 1 \end{bmatrix} \cdot \begin{bmatrix} a_x \\ a_y \\ a_z \\ 0 \end{bmatrix} = a_x x + a_y y + a_z z \tag{3.128}$$

The regular geometric interpretation of the scalar product remains valid: it is projection of one vector on a direction of the second one. In these terms, $\mathbf{a} \cdot \mathbf{b} = ||\mathbf{a}|| \, ||\mathbf{b}|| \cos \alpha$, where α is the angle between vectors **a** and **b**; $||\mathbf{a}||$ and $||\mathbf{b}||$ are their magnitudes. In particular, if **e** is a unit vector, whose components are the direction cosines, $||\mathbf{e}|| = 1$, the product $\mathbf{e} \cdot \mathbf{b}$ presents projection of **b** onto the direction defined through vector **e**.

- *The vector product* of two non-position vectors or a pair composing a position and a non-position vector presents a new non-position vector,

$$\mathbf{r} \times \boldsymbol{a} = \begin{bmatrix} x \\ y \\ z \\ 1 \end{bmatrix} \times \begin{bmatrix} a_x \\ a_y \\ a_z \\ 0 \end{bmatrix} = \begin{bmatrix} ya_z - za_y \\ za_x - xa_z \\ xa_y - ya_x \\ 0 \end{bmatrix}$$

The regular geometric interpretation of the vector product remains valid: it is a moment of the non-position vector a relative to the initial point of the position vector **r**. In these terms, $\mathbf{r} \times a$ is a vector perpendicular to **r** and a and of magnitude $||\mathbf{r} \times a|| = ||\mathbf{r}|| \, ||a|| \sin \alpha$, where α is the angle between $||\mathbf{r}||$ and $||a||$.

- *The scalar triple product* (box product) $[abc]$ is defined for three vectors among which at least one is a non-position vector,

$$[abc] \equiv a \cdot [b \times c] \equiv [a \times b] \cdot c$$

$$[abc] = [b\,c\,a] = [c\,a\,b] = -[b\,a\,c] = -[c\,b\,a] = -[a\,c\,b]$$

$[abc] > 0$ if vectors a, b, c form a right-hand triplet

The product $[abc]$ is equal to the determinant of the 3×3 matrix whose rows (or columns) compose the Cartesian coordinates of the vectors (the fourth homogeneous coordinates are ignored!). For example, if $a = (a_x, a_y, a_z, 0)^{\mathrm{T}}$; $b = (b_x, b_y, b_z, 0)^{\mathrm{T}}$; and $c = (c_x, c_y, c_z, 0)^{\mathrm{T}}$, then

$$[\boldsymbol{abc}] = \begin{vmatrix} a_x & a_y & a_z \\ b_x & b_y & b_z \\ c_x & c_y & c_z \end{vmatrix}$$

To simplify transfer from homogeneous to Cartesian coordinates, one can use the 3×4 matrix:

$$\mathbf{T}_{4to3} = \begin{bmatrix} 1 & 0 & 0 & 0 \\ 0 & 1 & 0 & 0 \\ 0 & 0 & 1 & 0 \end{bmatrix} \tag{3.129}$$

Actually,

$$\mathbf{T}_{4to3}\mathbf{r} = \begin{bmatrix} 1 & 0 & 0 & 0 \\ 0 & 1 & 0 & 0 \\ 0 & 0 & 1 & 0 \end{bmatrix} \begin{bmatrix} x \\ y \\ z \\ 1 \end{bmatrix} = \begin{bmatrix} x \\ y \\ z \end{bmatrix} \text{and } \mathbf{T}_{4to3}\,\boldsymbol{a} = \begin{bmatrix} 1 & 0 & 0 & 0 \\ 0 & 1 & 0 & 0 \\ 0 & 0 & 1 & 0 \end{bmatrix} \begin{bmatrix} a_x \\ a_y \\ a_z \\ 0 \end{bmatrix} = \begin{bmatrix} a_x \\ a_y \\ a_z \end{bmatrix}$$

3.3.1.3 *Homogeneous transformation matrices (HTM)*

Consider two rigid bodies associated with coordinate frames S_{i-1} and S_i, respectively. The same point in space has different coordinates in these frames with the exception of the trivial case when S_{i-1} and S_i are identical. Denote by $\mathbf{r}_{i-1}$ and $\mathbf{r}_i$, the position vectors, Eq. (3.120), of a point in two coordinate frames. They are linked by the matrix relation

$$\mathbf{r}_{i-1} = {}^{i-1}\mathbf{A}_i\,\mathbf{r}_i \tag{3.130}$$

where ${}^{i-1}\mathbf{A}_i$ is the 4 × 4 *homogeneous transformation matrices (HTM)* transforming homogeneous coordinates of vector $\mathbf{r}_i$ from frame S_i into frame S_{i-1},

$$ {}^{i-1}\mathbf{A}_i = \begin{bmatrix} a_{11} & a_{12} & a_{13} & a_{14} \\ a_{21} & a_{22} & a_{23} & a_{24} \\ a_{31} & a_{32} & a_{33} & a_{34} \\ 0 & 0 & 0 & 1 \end{bmatrix}_i \tag{3.131}$$

In this matrix, the first three coordinates a_{14}, a_{24}, and a_{34} of the fourth column are Cartesian coordinates of the origin O_i of system S_i in system S_{i-1}, i.e.,

$$\overline{O_{i-1}O_i} \equiv \mathbf{p}_i = [a_{14}, a_{24}, a_{34}]_i^{\mathrm{T}} \tag{3.132}$$

Components of vector $\mathbf{p}_i$ are the vector of translation of the ith link relative to the $(i - 1)$th one, whose components $p_{xi} = a_{14i}$, $p_{yi} = a_{24i}$, and $p_{zi} = a_{34i}$ represent linear displacements of the ith link along the X_{i-1}-, Y_{i-1}-, and Z_{i-1}-axis, respectively. Furthermore, the upper left 3 × 3 block ${}^{i-1}\mathbf{R}_i$ describes rotation of system S_i relative to its origin O_i, i.e., after pure translation described vector $\mathbf{p}_i$ when the analogous axes of systems S_i and S_{i-1} become parallel and similarly directed,

$$ {}^{i-1}\mathbf{R}_i = \begin{bmatrix} a_{11} & a_{12} & a_{13} \\ a_{21} & a_{22} & a_{23} \\ a_{31} & a_{32} & a_{33} \end{bmatrix}_i \tag{3.133}$$

The columns of the rotation block-matrix present the directional cosines of the X_i-, Y_i-, and Z_i-axes with respect to the X_{i-1}-, Y_{i-1}-, and Z_{i-1}-axes. Hence, the block ${}^{i-1}\mathbf{R}_i$ is the orthogonal matrix, i.e., for all i, k = 1, 2, 3,

$$\sum_{j=1}^{3} a_{ij}a_{kj} = \sum_{j=1}^{3} a_{ji}a_{jk} = \begin{cases} 0, & \text{for } i \neq k \\ 1, & \text{for } i = k \end{cases},$$

$$\det({}^{i-1}\mathbf{R}_i) = \begin{vmatrix} a_{11} & a_{12} & a_{13} \\ a_{21} & a_{22} & a_{23} \\ a_{31} & a_{32} & a_{33} \end{vmatrix}_i = 1$$

$${}^{i}\mathbf{R}_{i-1} = ({}^{i-1}\mathbf{R}_i)^{-1} = ({}^{i-1}\mathbf{R}_i)^{\mathrm{T}} \tag{3.134}$$

i.e., the determinant of matrix ${}^{i-1}\mathbf{R}_i$ is equal to one, and the inverse rotation matrix $({}^{i-1}\mathbf{R}_i)^{-1}$ equals the transposed matrix.

In terms of Eqs. (3.132) and (3.134), matrix ${}^{i-1}\mathbf{A}_i$ can be rewritten in the following compact form:

$$
{}^{i-1}\mathbf{A}_i = \begin{bmatrix} {}^{i-1}\mathbf{R}_i & \mathbf{p}_i \\ \mathbf{0} & 1 \end{bmatrix} \tag{3.135}
$$

where $\mathbf{0} = [0, 0, 0]$.

Since the following material is dedicated to motions of the FSS links, the pair of coordinate frame S_{i-1} and S_i can be considered as those associated with a stationary and a mobile link, respectively, i.e., the motion of the ith link relative to the $(i-1)$th one is considered, irrespective of the fact that either of the links is stationary in *absolute* motion.

3.3.1.4 HTMs of elementary motions

Among the HTMs, there are six matrices of elementary motions $\mathbf{A}^j$, where j is the motion symbol according to Tables 3.1 and 3.5, $j = 1, 2,\ldots,6$. If systems S_{i-1} and S_i are tied with two FSS links, Eq. (3.130) can be represented as follows:

$$
\mathbf{r}_{i-1} = \mathbf{A}^j \, \mathbf{r}_i, \tag{3.136}
$$

where $\mathbf{A}^j$ is one of six matrices describing one of six elementary motions of the system S_i relative to system S_{i-1}: either translation along or rotation around an axis X_{i-1}, Y_{i-1}, or Z_{i-1}; matrix $\mathbf{A}^j$ depends on a single parameter q^j ($j = 1, 2,\ldots,6$), which has the meaning of an elementary displacement of a rigid body – translational distance or angle of rotation,

$$
\mathbf{A}^j = \mathbf{A}^j(q^j), \text{ with } j = 1, 2,\ldots,6 \tag{3.137}
$$

$$
{}^{i-1}\mathbf{A}_i\,(q_i) = \begin{cases} \mathbf{A}^1(x_i), \text{ for translation along the } X_{i-1}\text{-axis} \\ \mathbf{A}^2(y_i), \text{ for translation along the } Y_{i-1}\text{-axis} \\ \mathbf{A}^3(z_i), \text{ for translation along the } Z_{i-1}\text{-axis} \\ \mathbf{A}^4(\theta_i), \text{ for rotation around the } X_{i-1}\text{-axis} \\ \mathbf{A}^5(\psi_i), \text{ for rotation around the } Y_{i-1}\text{-axis} \\ \mathbf{A}^6(\varphi_i), \text{ for rotation around the } Z_{i-1}\text{-axis} \end{cases} \tag{3.138}
$$

Hence, matrices $\mathbf{A}^1$, $\mathbf{A}^2$, and $\mathbf{A}^3$ describe axial translational displacements of the system S_i along one of the axes of the system S_{i-1}, and matrices $\mathbf{A}^4$, $\mathbf{A}^5$, and $\mathbf{A}^6$ describe rotation of the system S_i around one of the axes of the system S_{i-1} (Table 3.5):

$$
\mathbf{A}^1(x_i) = \begin{bmatrix} 1 & 0 & 0 & x_i \\ 0 & 1 & 0 & 0 \\ 0 & 0 & 1 & 0 \\ 0 & 0 & 0 & 1 \end{bmatrix}; \mathbf{A}^2(y_i) = \begin{bmatrix} 1 & 0 & 0 & 0 \\ 0 & 1 & 0 & y_i \\ 0 & 0 & 1 & 0 \\ 0 & 0 & 0 & 1 \end{bmatrix}; \mathbf{A}^3(z_i) = \begin{bmatrix} 1 & 0 & 0 & 0 \\ 0 & 1 & 0 & 0 \\ 0 & 0 & 1 & z_i \\ 0 & 0 & 0 & 1 \end{bmatrix} \tag{3.139}
$$

$$\mathbf{A}^4(\theta_i)=\begin{bmatrix}1&0&0&0\\0&\cos\theta_i&-\sin\theta_i&0\\0&\sin\theta_i&\cos\theta_i&0\\0&0&0&1\end{bmatrix};\mathbf{A}^5(\psi_i)=\begin{bmatrix}\cos\psi_i&0&\sin\psi_i&0\\0&1&0&0\\-\sin\psi_i&0&\cos\psi_i&0\\0&0&0&1\end{bmatrix};\mathbf{A}^6(\varphi_i)=\begin{bmatrix}\cos\varphi_i&-\sin\varphi_i&0&0\\\sin\varphi_i&\cos\varphi_i&0&0\\0&0&1&0\\0&0&0&1\end{bmatrix} \quad (3.140)$$

Note that the left superscript $(i - 1)$ and right subscript i of the matrices are omitted, and their association with the ith link is identified by means of the subscript i of the unique parameter of the matrix. The geometrical illustrations of the transformations by formula (3.136) are shown in Figure 3.22. The designations of the axes of motions in Figure 3.22 are as follows: since the axes of translations are displaced along each other, their designations are comma-separated, e.g.,"X_{i-1}, X_i", and since the axes of rotations are identical, they are written with identity symbol, e.g.,"$X_{i-1} \equiv X_i$". Each of matrices $\mathbf{A}^1, \mathbf{A}^2,\ldots,\mathbf{A}^6$ depends on a single parameter.

All the coordinate systems are right-sided, *i.e.*, rotation from the X-axis to the Y-axis must take place counterclockwise when looking from the end of the Z-axis. The rule of signs for arguments q^j of matrix $\mathbf{A}^j(q^j)$: argument q^j is assumed a positive if the system S_i moves in positive direction relative to the axes of system S_{i-1}:

(a) Linear displacements $x = q^1$, $y = q^2$, and $z = q^3$ are assumed positive if system S_i moves in a positive direction to the corresponding axis of system S_{i-1};

(b) Angles of rotation $\theta = q^4$, $\psi = q^5$, and $\varphi = q^6$ are assumed positive if system S_i rotates counterclockwise (when looking from positive end of the axis of rotation) around the corresponding axis of system S_{i-1}.

The important feature of the HTM technique is that any coordinate transformations (both relative translations and relative rotations) can be described with aid of the unique mathematical operation — matrix multiplication, whereas upon using vectors and matrices of the third order, two operations (addition and multiplication) are needed. Specifically, the translation is modeled by vector addition, while the rotation is modeled by multiplying the matrix. Actually, two subsequent motions of a rigid body are described by matrix ${}^{i-1}\mathbf{A}_{i+1}$ presenting the product of matrices ${}^{i-1}\mathbf{A}_i$ and ${}^{i}\mathbf{A}_{i+1}$. Namely, let $\mathbf{r}_{i-1} = {}^{i-1}\mathbf{A}_i\,\mathbf{r}_i$ and $\mathbf{r}_i = {}^{i}\mathbf{A}_{i+1}\,\mathbf{r}_{i+1}$. Hence,

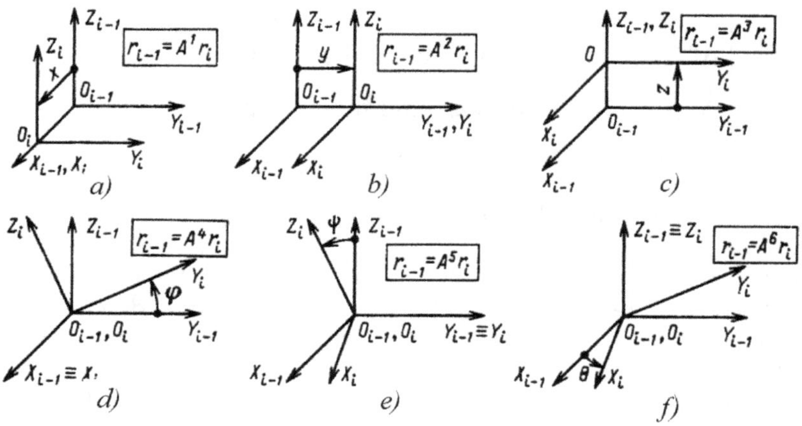

Figure 3-22 Geometrical interpretation of the elementary displacement of the rigid body: *a, b, c* – translations along X-, Y-, and Z-axis, respectively; *d, e, f* – rotations around X-, Y-, and Z-axis, respectively.

Table 3-5 Matrices of elementary motions or displacements.

Relative movement	Matrices modeling motion relative to axis X	Y	Z
Translation along axis	$\mathbf{A}^1 = \begin{pmatrix} 1 & 0 & 0 & x \\ 0 & 1 & 0 & 0 \\ 0 & 0 & 1 & 0 \\ 0 & 0 & 0 & 1 \end{pmatrix}$	$\mathbf{A}^2 = \begin{pmatrix} 1 & 0 & 0 & 0 \\ 0 & 1 & 0 & y \\ 0 & 0 & 1 & 0 \\ 0 & 0 & 0 & 1 \end{pmatrix}$	$\mathbf{A}^3 = \begin{pmatrix} 1 & 0 & 0 & 0 \\ 0 & 1 & 0 & 0 \\ 0 & 0 & 1 & z \\ 0 & 0 & 0 & 1 \end{pmatrix}$
Rotation around axis	$\mathbf{A}^4 = \begin{pmatrix} 1 & 0 & 0 & 0 \\ 0 & \cos\theta & -\sin\theta & 0 \\ 0 & \sin\theta & \cos\theta & 0 \\ 0 & 0 & 0 & 1 \end{pmatrix}$	$\mathbf{A}^5 = \begin{pmatrix} \cos\psi & 0 & \sin\psi & 0 \\ 0 & 1 & 0 & 0 \\ -\sin\psi & 0 & \cos\psi & 0 \\ 0 & 0 & 0 & 1 \end{pmatrix}$	$\mathbf{A}^6 = \begin{pmatrix} \cos\varphi & -\sin\varphi & 0 & 0 \\ \sin\varphi & \cos\varphi & 0 & 0 \\ 0 & 0 & 1 & 0 \\ 0 & 0 & 0 & 1 \end{pmatrix}$

The generalized coordinates θ, ψ, φ are the angles of rotation relative to the axes; *x, y, z* are linear displacements along the axes; a superscript in the notation of matrix **A** is the number of the generalized coordinate.

$$\mathbf{r}_{i\text{-}1} = {}^{i\text{-}1}\mathbf{A}_i\,{}^{i}\mathbf{A}_{i+1}\,\mathbf{r}_{i+1} = {}^{i\text{-}1}\mathbf{A}_{i+1}\,\mathbf{r}_{i+1}$$

$${}^{i\text{-}1}\mathbf{A}_{i+1} = {}^{i\text{-}1}\mathbf{A}_i\,{}^{i}\mathbf{A}_{i+1}$$

As a result, the HTM-based technique allows significant simplification of description of complicated motions of the bodies in 3-D space, which is reduced to specified multiplication of the HTMs in the predetermined order. Five specific properties of matrices $\mathbf{A}^j(q^j)$ listed in Table 3.5 and their mechanical nature are as follows:

Property 1. Matrix $\mathbf{A}^j$ of the sum of arguments is equal to the product of the same type of matrices of the addenda (for all $j = 1,\ldots,6$),

$$\mathbf{A}^j(q_1)\,\mathbf{A}^j(q_2) = \mathbf{A}^j(q_2)\,\mathbf{A}^j(q_1) = \mathbf{A}^j(q_1 + q_2) \tag{3.141}$$

Generally,

$$\prod_{i=1}^{n} \mathbf{A}^j(q_i) = \mathbf{A}^j(q_1 + q_2 + \ldots + q_n) \tag{3.142}$$

$$\mathrm{A}^j(nq) = [\mathrm{A}^j(q)]^n$$

Equation (3.141) can be proven through direct multiplication of the elementary matrices. For example, for $j = 1$ and $j = 6$,

$$\mathbf{A}^1(x_1)\mathbf{A}^1(x_2) = \begin{bmatrix} 1 & 0 & 0 & x_1 \\ 0 & 1 & 0 & 0 \\ 0 & 0 & 1 & 0 \\ 0 & 0 & 0 & 1 \end{bmatrix}\begin{bmatrix} 1 & 0 & 0 & x_2 \\ 0 & 1 & 0 & 0 \\ 0 & 0 & 1 & 0 \\ 0 & 0 & 0 & 1 \end{bmatrix} = \begin{bmatrix} 1 & 0 & 0 & x_1 + x_2 \\ 0 & 1 & 0 & 0 \\ 0 & 0 & 1 & 0 \\ 0 & 0 & 0 & 1 \end{bmatrix} = \mathrm{A}^1(x_1 + x_2)$$

$$\mathrm{A}^6(\varphi_1)\,\mathrm{A}^6(\varphi_2) =$$

$$\begin{bmatrix} \cos\varphi_1 & -\sin\varphi_1 & 0 & 0 \\ \sin\varphi_1 & \cos\varphi_1 & 0 & 0 \\ 0 & 0 & 1 & 0 \\ 0 & 0 & 0 & 1 \end{bmatrix}\begin{bmatrix} \cos\varphi_2 & -\sin\varphi_2 & 0 & 0 \\ \sin\varphi_2 & \cos\varphi_2 & 0 & 0 \\ 0 & 0 & 1 & 0 \\ 0 & 0 & 0 & 1 \end{bmatrix} = \begin{bmatrix} \cos(\varphi_1+\varphi_2) & -\sin(\varphi_1+\varphi_2) & 0 & 0 \\ \sin(\varphi_1+\varphi_2) & \cos(\varphi_1+\varphi_2) & 0 & 0 \\ 0 & 0 & 1 & 0 \\ 0 & 0 & 0 & 1 \end{bmatrix} = A^6(\varphi_1+\varphi_2)$$

Property 1 means that displacement on distance $q_1 + q_2$ can be carried out as the sequence of two displacements on distances q_1 and q_2 (or q_2 and q_1) along or around the same axis.

Property 2. Matrix $\mathbf{A}^j$ of zero argument ($q = 0$) is the 4 × 4 identity matrix (for all j = 1,...,6),

$$\mathbf{A}^j(0) = \mathbf{I}_{4\times4} \tag{3.143}$$

Property 2 means that displacement on zero distance (angle) leaves the body position and orientation unchanged.

Property 3. The inverse matrix is a matrix of the argument with an opposite sign,

$$[\mathbf{A}^j(q)]^{-1} = \mathbf{A}^j(-q), \text{ with } j = 1,\ldots,6 \tag{3.144}$$

Proof: According to *Properties 1 and 2*, $\mathbf{A}^j(0) = \mathbf{A}^j(q - q) = \mathbf{A}^j(q)\,\mathbf{A}^j(-q) = \mathbf{I}$. Thus, according to definition of the inverse matrix, Eq. (3.144) is true.

Property 3 means that inverse displacement presents displacement along the same axis at the same distance in the inverse direction.

Property 4. The inverse matrix of the product of any matrices of type $\mathbf{A}^j$ is equal to the product of the same matrices of the argument with an opposite sign taken in the inverse order,

$$\left[\prod_{i=1}^{N} \mathbf{A}^{j_i}(q_i)\right]^{-1} = \prod_{i=N}^{1} \mathbf{A}^{j_i}(-q_i) \tag{3.145}$$

Mechanical interpretation of the *Property 4* results from generalization of *Property 3*.

Property 5. Generally, multiplication of $\mathbf{A}^j$ (j = 1,...,6) matrices is not-commutative, $\mathbf{A}^j\mathbf{A}^k \neq \mathbf{A}^k\mathbf{A}^j$. For example,

$$\mathbf{A}^1(q)\mathbf{A}^5(q) = \begin{pmatrix} 1 & 0 & 0 & a \\ 0 & 1 & 0 & 0 \\ 0 & 0 & 1 & 0 \\ 0 & 0 & 0 & 1 \end{pmatrix}\begin{bmatrix} \cos\psi & 0 & \sin\psi & 0 \\ 0 & 1 & 0 & 0 \\ -\sin\psi & 0 & \cos\psi & 0 \\ 0 & 0 & 0 & 1 \end{bmatrix} = \begin{bmatrix} \cos\psi & 0 & \sin\psi & 0 \\ 0 & 1 & 0 & 0 \\ -\sin\psi & 0 & \cos\psi & 0 \\ 0 & 0 & 0 & 1 \end{bmatrix};$$

$$\mathbf{A}^5(q)\mathbf{A}^1(q) = \begin{pmatrix} \cos\psi & 0 & \sin\psi & 0 \\ 0 & 1 & 0 & 0 \\ -\sin\psi & 0 & \cos\psi & 0 \\ 0 & 0 & 0 & 1 \end{pmatrix}\begin{pmatrix} 1 & 0 & 0 & a \\ 0 & 1 & 0 & 0 \\ 0 & 0 & 1 & 0 \\ 0 & 0 & 0 & 1 \end{pmatrix} = \begin{pmatrix} \cos\psi & 0 & \sin\psi & a\cos\psi \\ 0 & 1 & 0 & 0 \\ -\sin\psi & 0 & \cos\psi & -a\sin\psi \\ 0 & 0 & 0 & 1 \end{pmatrix}$$

i.e., $\mathbf{A}^1(q)\,\mathbf{A}^5(q) \neq \mathbf{A}^5(q)\,\mathbf{A}^1(q)$.

However, there are three special cases, when the multiplication of $\mathbf{A}^j$ matrices is commutative. Namely, $\mathbf{A}^j\,\mathbf{A}^k = \mathbf{A}^k\,\mathbf{A}^j$ if and only if any of the following three conditions is fulfilled:

(5a) $j = 1, 2, 3$ and $k = 1, 2, 3$; i.e., the displacements of two consecutive links are translations;

(5b) $j = k$; i.e., the displacements of two consecutive links present translation along or rotation around the same axis (see *Property 1*);

(5c) $|j - k| = 3$; i.e., the displacements of two consecutive links form a cylindrical pair: either of two bodies translates and the other one rotates relative to a common axis.

Properties 1-5 have great importance not only for computations, but are also used when analyzing the configurations of machines.

To simplify transfer of HTM rotation matrices from homogeneous to regular 3 × 3 form, one can use the 3 × 4 matrix $\mathbf{T}_{4to3}$, Eq. (3.129). For example,

$$\mathbf{T}_{4to3}\,\mathbf{A}^5\,(\mathbf{T}_{4to3})^{\mathrm{T}} = \begin{bmatrix} 1 & 0 & 0 & 0 \\ 0 & 1 & 0 & 0 \\ 0 & 0 & 1 & 0 \end{bmatrix} \begin{bmatrix} \cos\psi & 0 & \sin\psi & 0 \\ 0 & 1 & 0 & 0 \\ -\sin\psi & 0 & \cos\psi & 0 \\ 0 & 0 & 0 & 1 \end{bmatrix} \begin{bmatrix} 1 & 0 & 0 \\ 0 & 1 & 0 \\ 0 & 0 & 1 \\ 0 & 0 & 0 \end{bmatrix} = \begin{bmatrix} \cos\psi & 0 & \sin\psi \\ 0 & 1 & 0 \\ -\sin\psi & 0 & \cos\psi \end{bmatrix}$$

Finally, let us consider the effect of multiplying the HTM of elementary motions by position vectors $\mathbf{r}$ and non-position vectors a and b:

$$\mathbf{A}^j\,a = a, \text{ for } j = 1, 2, 3;$$

$$\mathbf{A}^j\,a \cdot \mathbf{A}^j\,b = a \cdot b, \text{ for } j = 1, 2, \ldots, 6;$$

$$\mathbf{A}^j\,\mathbf{r} \cdot \mathbf{A}^j\,a = \mathbf{r} \cdot a, \text{ for } j = 4, 5, 6;$$

$$[\mathbf{A}^j\,a \times \mathbf{A}^j\,b] = [a \times b], \text{ for } j = 1, 2, 3;$$

$$[\mathbf{A}^j\,a \times \mathbf{A}^j\,a] = \mathbf{A}^j[a \times b], \text{ for } j = 4, 5, 6;$$

$$[\mathbf{A}^j\,\mathbf{r} \times \mathbf{A}^j\,a] = \mathbf{A}^j\,[\mathbf{r} \times a], \text{ for } j = 4, 5, 6$$

3.3.1.5 *Multi-parametric matrix: Expression through elementary HTMs*

Robots and machines with parallel and hybrid kinematics have at least one link with the DOF number more than one. A coordinate transformation associated with the *i*th multi-DOF kinematic link depends on l_i parameters, $2 \le l_i \le 6$. The general case of the spatial motion of the rigid body can be presented as six subsequent elementary motions with translation parameters x_i, y_i, z_i and three Euler angles $\theta_i, \psi_i, \varphi_i$ and described through the six-parametric ${}^{i\text{-}1}\mathbf{A}_i = {}^{i\text{-}1}\mathbf{A}_i(x_i, y_i, z_i, \theta_i, \psi_i, \varphi_i)$, where *i* is the link number. This matrix can be represented as the product of six different matrices $\mathbf{A}^j(q_j)$, Eq. (3.137), listed in the given order. The product value depends on the matrix sequence. Hereafter, the following sequence will be used:

$${}^{i\text{-}1}\mathbf{A}_i = {}^{i\text{-}1}\mathbf{A}_i(x_p, y_p, z_p, \theta_i, \psi_i, \varphi_i) = \mathbf{A}^1(x_p)\,\mathbf{A}^2(y_p)\,\mathbf{A}^3(z_p)\,\mathbf{A}^4(\theta_i)\,\mathbf{A}^5(\psi_i)\,\mathbf{A}^6(\varphi_i) \tag{3.146}$$

Substitution of matrices of elementary motions, Eq. (3.138), into the right-hand side of Eq. (3.146) yields the six-parametric matrix $\mathbf{A}_{Eu}$ referred to as the "extended Euler matrix":

$$^{i\text{-}1}A_i(x_p, y_p, z_p, \theta_i, \psi_i, \varphi_i) = \mathbf{A}_{Eu}(x_{pi}, y_{pi}, z_{pi}, \theta_i, \psi_i, \varphi_i) =$$

$$\begin{bmatrix} \cos\varphi_i \cos\psi_i & -\sin\varphi_i \cos\psi_i & \sin\psi_i & x_{pi} \\ \cos\varphi_i \sin\psi_i \sin\theta_i + \sin\varphi_i \cos\theta_i & \cos\varphi_i \cos\theta_i - \sin\varphi_i \sin\psi_i \sin\theta_i & -\cos\psi_i \sin\theta_i & y_{pi} \\ -\cos\varphi_i \sin\psi_i \cos\theta_i + \sin\varphi_i \sin\theta_i & \sin\varphi_i \sin\psi_i \cos\theta_i + \cos\varphi_i \sin\theta_i & \cos\psi_i \cos\theta_i & z_{pi} \\ 0 & 0 & 0 & 1 \end{bmatrix} \quad (3.147)$$

where x_{pi}, y_{pi}, z_{pi} are the distances of the ith body translation along the $X_{i\text{-}1}$-, $Y_{i\text{-}1}$-, and $Z_{i\text{-}1}$-axes, respectively; θ_i, ψ_i, φ_i are the Euler angles (Fig. 3.4): θ_i is the roll angle of rotations around the X-axis, ψ_i is the pitch angle of rotations around the Y-axis, and φ_i is the yaw angle of rotations around the Z-axis.

Note that Eq. (3.147) relates to the case, in which three rotations are performed around the X-, Y-, Z-axis in series. It is not a unique presentation of the general rotation: this is only one of 24 versions [1, 4] of the possible combinations and sequences of the Euler angles forming different expressions for matrix $\mathbf{A}_{Eu}$.

There are two important partial cases of the parallel-kinematics links when the ith link performs a pure translation or a pure rotation:

- Translation motion is described by the 4 × 4 translation matrix

$$\mathbf{A}_{tr} = \mathbf{A}_{tr}(x_{pi}, y_{pi}, z_{pi}) = \mathbf{A}^1(x_{pi})\mathbf{A}^2(y_{pi})\mathbf{A}^3(z_{pi}) = \begin{bmatrix} 1 & 0 & 0 & x_{pi} \\ 0 & 1 & 0 & y_{pi} \\ 0 & 0 & 1 & z_{pi} \\ 0 & 0 & 0 & 1 \end{bmatrix} \quad (3.148)$$

- Rotation is described by one of the 4 × 4 Euler matrices

$$\mathbf{R}_{Eu}(\theta_i, \psi_i, \varphi_i) = \mathbf{A}^4(\theta_i)\,\mathbf{A}^5(\psi_i)\,\mathbf{A}^6(\varphi_i) =$$

$$\begin{bmatrix} \cos\varphi_i \cos\psi_i & -\sin\varphi_i \cos\psi_i & \sin\psi_i & 0 \\ \cos\varphi_i \sin\psi_i \sin\theta_i + \sin\varphi_i \cos\theta_i & \cos\varphi_i \cos\theta_i - \sin\varphi_i \sin\psi_i \sin\theta_i & -\cos\psi_i \sin\theta_i & 0 \\ -\cos\varphi_i \sin\psi_i \cos\theta_i + \sin\varphi_i \sin\theta_i & \sin\varphi_i \sin\psi_i \cos\theta_i + \cos\varphi_i \sin\theta_i & \cos\psi_i \cos\theta_i & 0 \\ 0 & 0 & 0 & 1 \end{bmatrix} \quad (3.149)$$

According to Chasles' theorem, the general displacement of a rigid body in 3-D space can be represented through rotation and translation. In matrix terms, the extended Euler matrix $\mathbf{A}_{Eu}$, Eq. (3.147), is expressed as a product of the translation matrix $\mathbf{A}_{tr}$, Eq. (3.148), and the Euler matrix $\mathbf{A}_{rot}$, Eq. (3.149):

$$\mathbf{A}_{Eu}(x_{pi}, y_{pi}, z_{pi}, \theta_i, \psi_i, \varphi_i) = \mathbf{A}_{tr}(x_{pi}, y_{pi}, z_{pi})\,\mathbf{R}_{Eu}(\theta_i, \psi_i, \varphi_i)$$

Multiplication and inversion of the HTMs leave their structure, Eq. (3.135), unchanged. Actually, this is true for two HTM matrices:

$$^{i-1}\mathbf{A}_{i+1} = {}^{i-1}\mathbf{A}_i\, {}^{i}\mathbf{A}_{i+1} = \begin{bmatrix} ^{i-1}\mathbf{R}_i & \mathbf{p}_i \\ \mathbf{0} & 1 \end{bmatrix} \begin{bmatrix} ^{i}\mathbf{R}_{i+1} & \mathbf{p}_{i+1} \\ \mathbf{0} & 1 \end{bmatrix} =$$

$$\begin{bmatrix} ^{i-1}\mathbf{R}_i\, {}^{i}\mathbf{R}_{i+1} & \mathbf{p}_i + {}^{i-1}\mathbf{R}_i \mathbf{p}_{i+1} \\ \mathbf{0} & 1 \end{bmatrix} = \begin{bmatrix} ^{i-1}\mathbf{R}_{i+1} & \mathbf{p}_{i,i+1} \\ \mathbf{0} & 1 \end{bmatrix} \tag{3.150}$$

where $^{i-1}\mathbf{R}_{i+1}$ is the orthogonal matrix since it is the product of two orthogonal cofactor matrices $^{i-1}\mathbf{R}_i$ and $^{i-1}\mathbf{R}_{i+1}$, and hence, presents a rotation matrix; and $\mathbf{p}_{i,i+1}$ is the translation vector as the sum of two regular translation vectors.

The form-preserving feature for a product of n HTMs can be proven by induction on n,

$$\mathbf{A} = \prod_{i=1}^{n} {}^{i-1}\mathbf{A}_i = \begin{bmatrix} ^{0}\mathbf{R}_n & \mathbf{p}_{0,n} \\ \mathbf{0} & 1 \end{bmatrix} \tag{3.151}$$

Inversion of the HTM also leaves its form unchanged:

$$\mathbf{A}^{-1} = \begin{bmatrix} \mathbf{R} & \mathbf{p} \\ \mathbf{0} & 1 \end{bmatrix}^{-1} = \begin{bmatrix} \mathbf{R}^{-1} & -\mathbf{R}^{-1}\mathbf{p} \\ \mathbf{0} & 1 \end{bmatrix} \tag{3.152}$$

Actually, the multiplication shows that $\mathbf{AA}^{-1} = \mathbf{A}^{-1}\mathbf{A} = \mathbf{I}$.

3.3.2 Synthesis and applications of the Form-Shaping Function

3.3.2.1 *Manipulation matrix*

In the general case, the set of motions of the FSS links is modeled by a manipulation matrix representing an ordered product of n HTMs:

$$^{0}\mathbf{A}_n = \prod_{i=1}^{n} {}^{i-1}\mathbf{A}_i \tag{3.153}$$

where cofactor $^{i-1}\mathbf{A}_i$ is the 4 × 4 HTM, Eq. (3.131), describing transformation of homogeneous coordinates from frame S_i into frame S_{i-1}; and n is the number of the FSS links.

Obviously, the manipulation matrix $^{0}\mathbf{A}_n$ possesses all properties described in the previous section. Each of n cofactor-matrices in the product, Eq. (3.153), depends on one or more (no more than six) kinematic and/or geometric parameters,

$$^{i-1}\mathbf{A}_i = {}^{i-1}\mathbf{A}_i(\mathbf{q}_i), \text{ with } i = 1,\ldots,n \tag{3.154}$$

$$\mathbf{q}_i = \begin{cases} q_i, \text{for 0 - or 1-DOF link} \\ (q_{1i}, q_{2i}), \text{for 2 -DOF link} \\ \ldots\ldots\ldots\ldots \\ (q_{1i}, q_{2i}, \ldots, q_{6i}), \text{for 6 -DOF link} \end{cases}$$

where $\mathbf{q}_i$ is either a single parameter or a list of parameters.

Graphically, the nominal FSS can be represented as a chain of coordinate frames

$$S_0 \xleftarrow{{}^{0}\mathbf{A}_1(\mathbf{q}_1)} S_1 \xleftarrow{{}^{1}\mathbf{A}_2(\mathbf{q}_2)} \cdots\cdots \xleftarrow{{}^{i-1}\mathbf{A}_i(\mathbf{q}_i)} S_i \xleftarrow{{}^{i}\mathbf{A}_{i+1}(\mathbf{q}_{i+1})} \cdots\cdots \xleftarrow{{}^{n-1}\mathbf{A}_n(\mathbf{q}_n)} S_n$$

Figure 3-23 The FSS graph: each pair of the frames S_i and S_{i-1} is interrelated through the coordinate transformation matrix ${}^{i-1}\mathbf{A}_i(\mathbf{q}_i)$.

S_i (i = 0, 1,...,n) connected by the matrices ${}^{i-1}\mathbf{A}_i(\mathbf{q}_i)$ (Fig. 3.23), where $\mathbf{q}_i$ presents a set of parameters defining motion of the ith link relative to the (i – 1)th one.

The single-parametric cofactor HTM associated with the ith link has the form given by Eq. (3.138) and describes the transformation relating to either a 1-DOF kinematic link (l_i = 1) or a zero-DOF geometric link (l_j = 0),

$$^{i-1}\mathbf{A}_i = \mathbf{A}^{k_i}(q_i) \tag{3.155}$$

where $\mathbf{A}^{k_i}(q_i)$ is one of six matrices (Table 3.5); and k_i is one of the integers 1, 2,..,6 associated with six elementary motions.

Multiplication matrix ${}^0\mathbf{A}_n$ depends on l kinematic parameters of relative motions and (n - l) geometric parameters of the FSS kinematic chain, where l is the mobility number, Eq. (3.79),

$$^0\mathbf{A}_n = {}^0\mathbf{A}_n(q_1, q_2,\ldots,q_l, a_1, a_2,\ldots,a_{n-l}) \tag{3.156}$$

Note that matrix ${}^0\mathbf{A}_n$ for the *serial*-kinematics machines presents a product of *the single-parametric HTM* and is uniquely constructed from coordinate code k and co-factor matrices ${}^{i-1}\mathbf{A}_i$, (3.155). If $k = k_1 k_2 \ldots k_n$, then

$$^0\mathbf{A}_n = \prod_{i=1}^{n} \mathbf{A}^{k_i}(q_i) = \mathbf{A}^{k_1}(q_1)\mathbf{A}^{k_2}(q_2)\ldots\mathbf{A}^{k_n}(q_n) \tag{3.157}$$

where $k_1, k_2,\ldots, k_n$ are ordered integers composing coordinate code k.

Graphically, the FSS of the serial-type machine can be represented as a chain of coordinate frames S_i (i = 0, 1,...,n) connected by the HTMs ${}^{i-1}\mathbf{A}_i(q_i)$ (Fig. 3.24). In what follows, the list of the HTMs composing the manipulation matrix ${}^0\mathbf{A}_n$ associated with the described serial-type machine or robot will be represented in the form of Table 3.6.

The multi-parametric cofactor HTM is associated with the ith link having more than one DOF. The parallel- and hybrid-type FSS compose not only the single-parametric links, but also consists of at least one link with from two to six DOF ($2 \le l_i \le 6$):

$${}^{i-1}\mathbf{A}_i(q_{i1}, q_{i2});\ {}^{i-1}\mathbf{A}_i(q_{i1}, q_{i2}, q_{i3}),\ldots,\ {}^{i-1}\mathbf{A}_i(q_{i1}, q_{i2},\ldots,q_{i6})$$

$$S_0 \xleftarrow{\mathbf{A}^{k_1}(q_1)} S_1 \xleftarrow{\mathbf{A}^{k_2}(q_2)} \cdots\cdots \xleftarrow{\mathbf{A}^{k_i}(q_i)} S_i \xleftarrow{\mathbf{A}^{k_{i+1}}(q_{i+1})} \cdots\cdots \xleftarrow{\mathbf{A}^{k_n}(q_n)} S_n$$

Figure 3-24 The FSS graph of the serial-type machine with orthogonal kinematics.

Table 3-6 Form-shaping system of orthogonal-kinematics machine.

S_i	Link	Motion relative to the previous link	Matrix ${}^{i-1}\mathbf{A}_i$
S_0	Target element	Conventionally fixed	–
S_1	The 1st moving link	Rotation or translation with parameter q_1	${}^0\mathbf{A}_1(q_1)$
....	....	....	....
S_i	The *i*th link name	Rotation or translation with parameter q_i	${}^{i-1}\mathbf{A}_i(q_i)$
....	....	....	....
S_n	The *n*th link name	Rotation or translation with parameter q_n	${}^{n-1}\mathbf{A}_n(q_n)$

Correspondingly, their manipulation matrix ${}^0\mathbf{A}_n$, Eq. (3.153), can include both some single-parametric cofactor matrices and at least one multi-parametric cofactor matrix depending on l_i parameters. In the general case, the multi-parametric matrix can be described by the extended Euler matrix, Eq. (3.147), and some constraints.

3.3.2.2 FSF synthesis

The FSF transforms the position and orientation vectors of the AE from the ending coordinate frame S_n of the kinematic chain into frame S_0 associated with the TE of the FSS. The *position form-shaping function* (PosFSF) connects two AE-associated position vectors — $\mathbf{r}_n$ expressed in frame S_n and $\mathbf{r}_0$ expressed in frame S_0 – by means of the manipulation matrix ${}^0\mathbf{A}_n$, Eq. (3.153):

$$\mathbf{r}_0 = {}^0\mathbf{A}_n\,\mathbf{r}_n \tag{3.158}$$

with

$$\mathbf{r}_0 = \begin{bmatrix} x_0 \\ y_0 \\ z_0 \\ 1 \end{bmatrix} \text{and } \mathbf{r}_n = \begin{bmatrix} x_n \\ y_n \\ z_n \\ 1 \end{bmatrix} \tag{3.159}$$

where x_0, y_0, and z_0 are the coordinates of an AE point referring to the frame S_0; and x_n, y_n, and z_n are those referring to the frame S_n.

The *orientation form-shaping function* (OrFSF) involves the same 4 × 4 manipulation matrix ${}^0\mathbf{A}_n$, Eq. (3.153) and a pair of the 4 × 1 non-position vectors $\mathbf{c}_0$ and $\mathbf{c}_n$:

$$\mathbf{c}_0 = {}^0\mathbf{A}_n\,\mathbf{c}_n \tag{3.160}$$

with

$$\mathbf{c}_0 = \begin{bmatrix} c_{0x} \\ c_{0y} \\ c_{0z} \\ 0 \end{bmatrix} \text{and } \mathbf{c}_n = \begin{bmatrix} c_{nx} \\ c_{ny} \\ c_{nz} \\ 0 \end{bmatrix} \tag{3.161}$$

where c_{0x}, c_{0y}, and c_{0z} are the direction cosines referring to frame S_0; and c_{nx}, c_{ny}, and c_{nz} are those referring to the frame S_n.

Note that both $\mathbf{c}_0$ and $\mathbf{c}_n$ are the unit vectors, $||\mathbf{c}_0|| = ||\mathbf{c}_n|| = 1$. Hence, Eq. (3.160) can result in only two of the three independent scalar equations.

Since both PosFSF and OrFSF consist of the same manipulation matrix, the FSF describing the kinematic chain of a machine can be represented in a compact form

$$[\mathbf{r}_0\ \mathbf{c}_0] = {}^0\mathbf{A}_n\ [\mathbf{r}_n\ \mathbf{c}_n] \tag{3.162}$$

i.e.,

$$\begin{bmatrix} x_0 & c_{0x} \\ y_0 & c_{0y} \\ z_0 & c_{0z} \\ 1 & 0 \end{bmatrix} = {}^0\mathbf{A}_n \begin{bmatrix} x_n & c_{nx} \\ y_n & c_{ny} \\ z_n & c_{nz} \\ 1 & 0 \end{bmatrix} \tag{3.163}$$

Position vector $\mathbf{r}_n$, Eq. (3.159), and orientation unit-vector $\mathbf{c}_n$, Eq. (3.161), of the AE (processing tool) depend on M parameters defined by Eq. (3.114). Taking into account that matrix ${}^0\mathbf{A}_n$ depends on n parameters, one can conclude that the FSF presents the function of $n+M$ parameters: $\mathbf{r}_0 = \mathbf{r}_0(q_1, q_1,\ldots,q_{n+M})$ and $\mathbf{c}_0 = \mathbf{c}_0(q_1, q_1,\ldots,q_{n+M})$.

3.3.2.3 *FSF features*

- Axis and angle of rotation: The FSF manipulation matrix ${}^0\mathrm{A}_n$, Eq. (3.157), whether it be the matrix associated with a serial, parallel, or hybrid kinematics machine, holds the general structure given by Eq. (3.131):

$${}^0\mathbf{A}_n = \begin{bmatrix} a_{11} & a_{12} & a_{13} & a_{14} \\ a_{21} & a_{22} & a_{23} & a_{24} \\ a_{31} & a_{32} & a_{33} & a_{34} \\ 0 & 0 & 0 & 1 \end{bmatrix} \tag{3.164}$$

In this matrix, the upper left 3 × 3 describes the rotation of system S_i of the nth rigid body as referring to the coordinate frame S_0, where a_{1k}, a_{2k}, and a_{3k} ($k = 1, 2, 3$) present the directional cosines of the X_n-, Y_n-, and Z_n-axes with respect to X_0-, Y_0-, and Z_0-axes. According to Chasles' theorem, there is an axis or rotation $\mathbf{w} = [w_x, w_y, w_z, 0]$ and angle α of rotation around this axis. In terms of matrix ${}^0\mathbf{A}_n$, Eq. (3.164), they are defined as follows:

$$\alpha = \arccos\frac{a_{11}+a_{22}+a_{33}-1}{2} \tag{3.165}$$

$$\mathbf{w} = \frac{1}{2\sin\alpha}\begin{bmatrix} a_{32}-a_{23} \\ a_{13}-a_{31} \\ a_{21}-a_{12} \\ 0 \end{bmatrix} \tag{3.166}$$

- Inversion of the manipulation matrix ${}^0\mathbf{A}_n$, Eq. (3.157), yields an equation of the active element (tool) when the target element is known:

$$(\mathbf{r}_n\ \mathbf{c}_n) = {}^n\mathbf{A}_0\ (\mathbf{r}_0\ \mathbf{c}_0) \tag{3.167}$$

with

$${}^n\mathbf{A}_0 = ({}^0\mathbf{A}_n)^{-1} = ({}^0\mathbf{A}_1\ {}^1\mathbf{A}_2 \times\ldots\times {}^{n-1}\mathbf{A}_n)^{-1} = {}^n\mathbf{A}_{n-1} \times\ldots\times {}^2\mathbf{A}_1\ {}^1\mathbf{A}_0 \tag{3.168}$$

Recall that according to *Property* 3 of the one-parametric HTM, ${}^i\mathbf{A}_{i-1} = {}^{i-1}\mathbf{A}_i(-q_i)$.

- Using another coordinate frame: In some cases, the FSF, Eq. (3.158), can be simplified through transferring to another coordinate frame. For this purpose, both sides of the FSF are multiplied by a matrix ${}^j\mathrm{A}_0$, where j is the number of the coordinate frame of interest:

$${}^j\mathbf{A}_0\ [\mathbf{r}_0\ \mathbf{c}_0] = {}^j\mathbf{A}_0{}^0\mathbf{A}_n\ [\mathbf{r}_n\ \mathbf{c}_n] \tag{3.169}$$

- FSF and kinematics problems: The FSF formulates the so-called *forward (or direct) kinematics problem* when parameters of link motions are given. Using this information the FSF defines the position and orientation of the nth link associated with the active element of the FSS with respect to the initial link associated with the target element through parameters of the link motions. The FSF synthesis is usually a simple analytical problem as applied to the serial-type kinematics machine taking into account that the majority of them have orthogonal kinematics. The same problem for parallel- and hybrid-kinematics machines is considerably more complicated, and more often than can be solved only in numerical form.
- Why is the D-H convention not used? A mechanical linkage can consist of links in which the input and output axes present a pair of skew lines crossing by an arbitrary angle. Coordinate transformation between these axes can be reduced to two successive screw transformations relative to the axes [6]. Denavit and Hartenberg used this fact to apply to linkage kinematics problems in the form of so-called the *D-H matrices* and *D-H convention*. The D-H matrices present four-parametric HTMs obtained by the product of four one-parametric HTMs, with fixed choice of the coordinate axes X and Z for the input and output axes, respectively. This approach becomes the regular one for the robot modeling [8]. However, orthogonal kinematic chains of machine tools, CMM, and IR can be successfully analyzed without application of the D-H convention [7-10], and what is more important, the D-H matrices are not convenient for solution of *accuracy problems* since there are some significant difficulties in calculation of geometric and kinematic errors when two or more links move along the parallel axes [11, 12]. Taking into account that this book is dedicated to accuracy calculation problems, the technique of one-parametric HTMs is used here as a basic approach.

3.2.3.4 *FSF-based choosing machine architectures*

Synthesis of the machine tool architecture is one of the most important problems occurring and solved in early design stages to choice and compare new arrangements under conditions of poor information. For this purpose, the integer or alphabetic codes with a moderate number of identifiers associated with kinematic and structural FSS components are used, and these codes are usually considered in common with simple-type graphs. To model different architectures, regular permutations and combinations of the code elements are synthesized and evaluated for performance indexes. One of the first fundamental works was written by Vragov [13].

Intensive investigations are dedicated to analysis of the 5-axis machines: the FSF theory connected analysis of layouts with the different orders of translations and rotations units with comparison of output errors [14, 15]; application of tree-type graphs for configuration synthesis of machining centers with and without tool change mechanisms [16, 17]; application of the FSF theory for design evaluation of miniature machine tools entering into flexible manufacturing systems [18]; using the neural nets for design of gear manufacturing machines [19].

To be specific, let us consider serial type machines with orthogonal kinematics. In these machines, there is one-to-one correspondence between the type and order of link motions and position of the motion symbol in the coordinate k, Eq. (3.81). The physical rearrangement of two links of the FSS is modeled by permutation of two matrices $\mathbf{A}^j$ ($j = 1, 2, \ldots, 6$) in the FSF. After the permutation, the result of a physical operation can change or stay unchanged depending on whether the corresponding matrices in FSF are commutative or not. All the allowed rearrangements that do not change the FSF are reduced to three cases 5a, 5b, and 5c that are listed in Property 5 of homogeneous transformation matrices $\mathbf{A}^j$ (see *subsection 3.3.1.3*).

It follows from this that the following four transformations (and only such) or any of their combinations do not alter the resulting function of the form-shaping system:

Transformation 1. Rearrangement of two adjacent links of the chain executing translational motions relative to the preceding link.

Transformation 2. Rearrangement of two adjacent links one of which executes translation and the other rotation around the one axis (two links comprising a cylindrical pair).

Transformation 3. Unification of several links, executing identical motions into a single link.

Transformation 4. Partitioning a link into several links executing the same motion.

Eighteen formal operations in the machine configuration realizing the listed transformations are given in Table 3.7. For illustration, two schematics of form-shaping of flat cams are shown in Figure 3.25. In the layout depicted in Figure 3.25b, tool rotation and translation present the cylindrical pair. According to Transformation 2, they can be permitted without

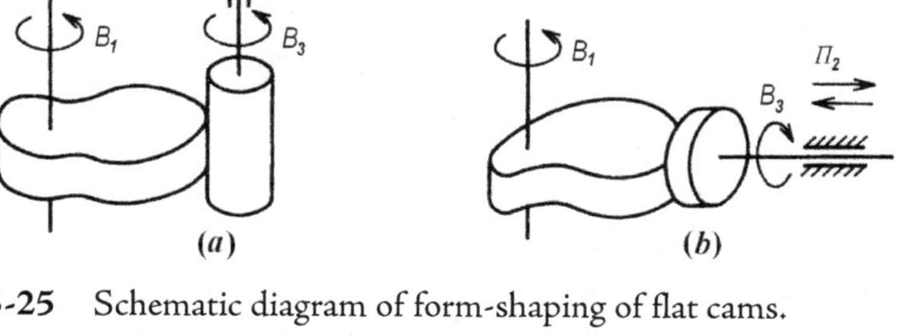

Figure 3-25 Schematic diagram of form-shaping of flat cams.

Table 3-7 Allowable formal operations in coordinate code.

Number	Symbolization	Description
Transformation 1	12 = 21, 13 = 31, 23 = 32	Translation units can be permutated
Transformation 2	14 = 41, 25 = 52, 36 = 63	Order in cylindrical pair can be permutated
Transformation 3	11 = 1, 22 = 2, 33 = 3	Identically moving units can be combined
	44 = 4, 55 = 5, 66 = 6	
Transformation 4	1 = 11, 2 = 22, 3 = 33	A unit can be broken up into two identically moving units
	4 = 44, 5 = 55, 66 = 66	

changing the result of operation. Contrary to that, in the layout depicted in Figure 3.25a, no rearrangement of links is permitted.

In the arrangement codes K, Eq. (3.89), the bed position is also figured. There is a possibility to analyze different machine layouts using combinations of the seven integers (0, 1,…,6) used for the arrangement codes. In Table 3.8, some lathe configurations are considered. A thread-cutting lathe with horizontal spindle and horizontal bed position has the coordinate code k = **631** and arrangement code K = **6031**. According to Transformations 3 and 1, the permutation in pairs 63 and 31, respectively, are allowed. Configurations 3601 and 3610 consist of the tail spindle, configurations from the last column can be used in the transfer lines, and configurations shown in the first column convert the machine in an industrial robot.

Table 3-8 Possible combinations of dimensionalities μ and m.

Form-shaping chain		Configuration for bed position			
Number	Formula	$0k_1k_2k_3$	$k_10k_2k_3$	$k_1k_20k_3$	$k_1k_2k_30$
1	631	0631 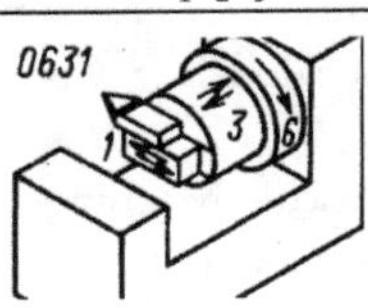	6031 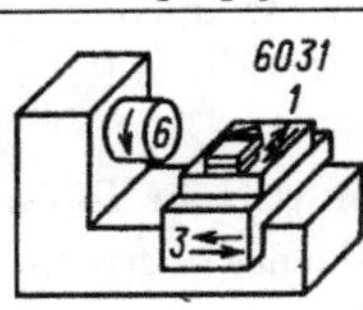	6301 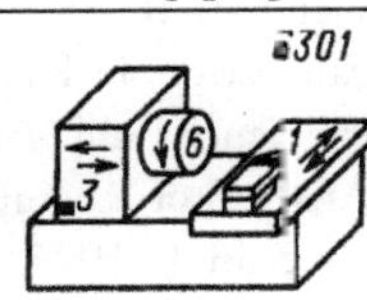	6310 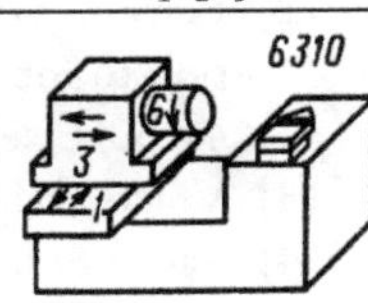
2	613	0613 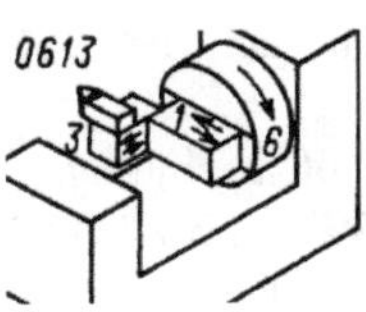	6013 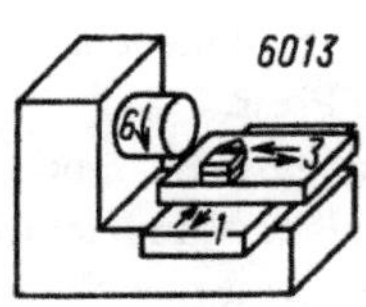	6103 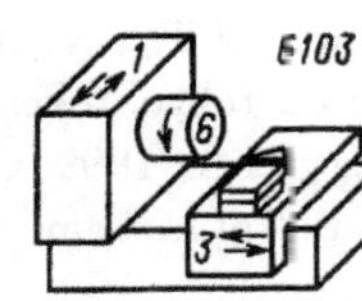	6130 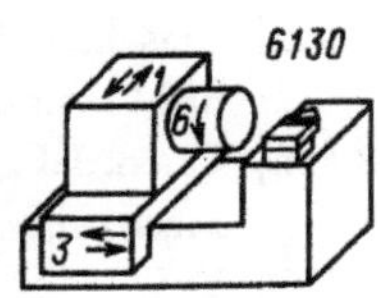
3	361	0361 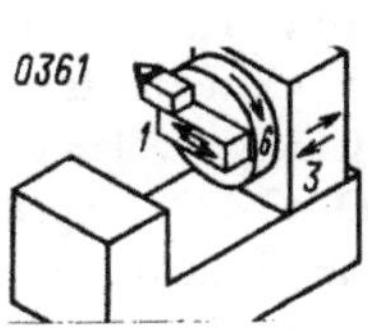	3061 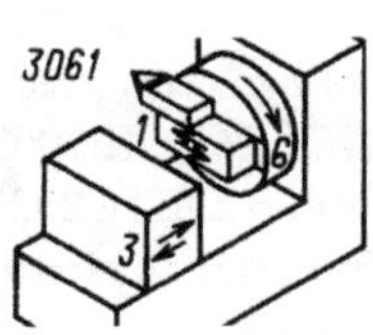	3601 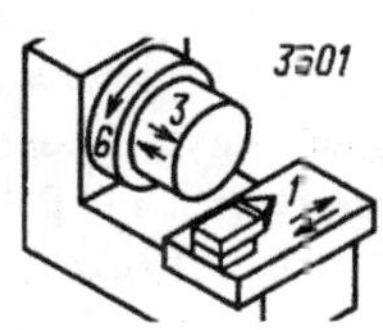	3610 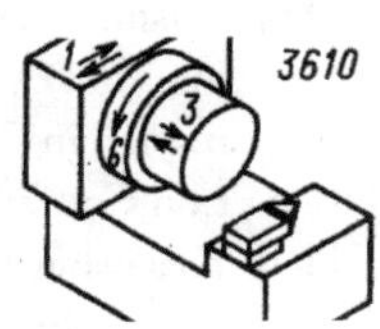

Note that rotation of the bed around the Z-axis (e.g., $K = \mathbf{60631}$) results in the arrangements with slant bed (see Figures 4.9 and 4.8 in the next chapter).

3.4 Conclusion

The form-shaping system (FSS) and its mathematical model – the form-shaping function (FSF) – are considered. Mathematical technique of the homogeneous transformation matrices is described and applied to the FSF formulation.

References

[1] Korn, G., Korn, T., 1968, *Mathematical Handbook for Scientists and Engineers*, McGraw-Hill, NY, p. 472.

[2] Stewart, D., 1965, A platform with six degrees of freedom, *Proceedings 1965/66 of the Institute of Mechanical Engineers, Part 1,* 180 (15), 371–386.

[3] Shneor, Y., Portman, V.T., 2010, Stiffness of 5-Axis Machines with Serial, Parallel, and Hybrid Kinematics: Evaluation and Comparison, *CIRP Annals - Manufacturing Technology*, **59**/1, 409–412.

[4] Tsai, L.-W., 1999, *Robot Analysis: The Mechanics of Serial and Parallel Manipulators*, John Wiley, New York, NY.

[5] Wolfram, S., 2003, *The Mathematica Book*, 5th ed., Wolfram Media Inc., 1488 pp.

[6] Denavit, J., Hartenberg, R.S., 1955, A kinematic notation for lower-pair mechanisms based on matrices". Trans ASME J. Appl. Mech., 23, pp. 215–221.

[7] Litvin, F.L.,1972, Theory of Gearing. 2nd ed., Nauka, Moscow (*in Russian*).

[8] Paul, R., 1981, Robot manipulators: mathematics, programming, and control, Cambridge, MA, MIT Press. ISBN 978-0-262-16082-7.

[9] Portman, V.T., 1981, Universal Method for Calculating the Accuracy of Mechanical Devices, *Soviet Engineering Research*, 1, No. 7, 11–15.

[10] Reshetov, D.N., Portman, V.T., 1988, Accuracy of Machine Tools, ASME Press, New York, NY.

[11] Mooring, B.W., Roth, Z. S., Driels, M.S., 1991, Fundamentals of Manipulator Calibration, John Wiley, New York, NY.

[12] Hayati, S.A., Gerald, P.R., 1986, Inverse kinematics solution for near simple robots and application to robot calibration, in Resent Trends in Robotics: Modeling, Control, and Education, pp. 41-50, Elsevier Science Publ., Amsterdam.

[13] Vragov, Y.D., Analysis of Layouts of Machine Tools, Moscow, Machinostroenie, 1978, 208 pp.

[14] Sakamoto, S., Inasaki, I., 1996, Influence of structure errors in five—axis machining centers on the machining accuracy, Transactions of the Japan Society of Mechanical Engineers, Part C, Vol. 62, Issue 594, pp. 748–753.

[15] Inasaki, I., Kishinami, T., Sakamoto, S., Takeuchi, Y., Tanaka, F., 1997, Shape Generation Theory of Machine Tools, Tokyo, Yokendo Press.

[16] Chen, F.C., Yan, H.S., 1999, Configuration synthesis of machining centers with tool change mechanisms, Int. J. of Machine Tools & Manufacture, 39(5), pp. 823-837.

[17] Yan, H.-S., Chen, F,C., 1998, Configuration Synthesis of Machining Centers Without Tool Change Arm, Mechanism and Machine Theory, 33 (1–2), pp. 197-212.

[18] Mishima, N., 2003, Design Evaluation Method for Miniature Machine Tools Utilizing Form-Shaping Theory, ASME 2003 Int. Design Engineering Technical Conferences and Computers and Information in Engineering Conference (IDETC/CIE2003), Chicago, Volume 2: 29th Design Automation Conference, Parts A and B, Paper no. DETC2003/DAC-48787, pp. 815–821.

[19] Krivosheya, A., Danilchenko, J., Storchak, M., Pasternak, S., 2016, Design of shaping machine and tooling systems for gear manufacturing, Mechanisms and Machine Science 34, pp. 425-450, Switzerland.

4 FSF of serial-kinematics systems

4.1 FSF and position analysis

Pose coordinates of a rigid body define the position and orientation of the body in space, and relationships between a pose of the output link of machine or robot and corresponding parameters of the actuators presents a key-step of position analysis.Vector **X** describing the pose has six or three elements depending on whether the considered problem is spatial or planar, respectively,

$$\mathbf{X} = [x_0, y_0, z_0, \theta, \psi, \varphi]^{\mathrm{T}}, \text{ in 3-D space} \tag{4.1}$$

$$\mathbf{X} = [x_0, y_0, \varphi]^{\mathrm{T}}, \text{ in 2-D space} \tag{4.2}$$

where x_0, y_0, z_0 are Cartesian coordinates of the position of a point of the rigid body; and θ, ψ, φ are the angles completely defining orientation of the body.

Actuated coordinates is a set of controlled parameters defining the machine structure and its drives and actuators. For definiteness, vector **Q** of the actuated coordinates is considered a vector composing just either six or three elements,

$$\mathbf{Q} = (q_1,\ldots,q_n)^{\mathrm{T}} \tag{4.3}$$

$n = 6$ or $n = 3$

Two alternative kinematics problems can be formulated when a finite (not small) motion of the body is considered:

- *Direct (or forward) kinematics* problem when the actuated parameters **Q**, Eq. (4.3), are given and the corresponding pose must be calculated,

$$\mathbf{X} = \mathbf{X}(\mathbf{Q}) \tag{4.4}$$

- *Inverse kinematics* problem,

$$\mathbf{Q} = \mathbf{Q}(\mathbf{X}) \tag{4.5}$$

Both these problems are directly connected with the FSF. Actually, the FSF formulates the forward (direct) kinematics problem, whose solution is immediately obtained in the form of Eq. (4.4) from Eq. (3.162). In turn, the inverse kinematics is obtained by solution of a system of simultaneous equations expressed through the FSF,

$$[\mathbf{r}_0\mathbf{c}_0] = [\mathbf{r}_{TE}\mathbf{c}_{TE}] \tag{4.6}$$

where $\mathbf{r}_0$ and $\mathbf{c}_0$ are defined through FSF, Eq. (3.162); and $\mathbf{r}_{TE}$ and $\mathbf{c}_{TE}$ are the vectors given by the target elements needed for processing.

The inverse kinematics solution presents a key-part of *machine setups* and will be considered in Chapter 6. Solution of the inverse kinematics allows observation of current positions of the robot links and time-history for all actuators. Note that solution of the inverse kinematics problems for the serial-kinematics systems is usually a non-trivial problem. However, active investigations in this area result in the fact that in the majority of the cases the solutions can be obtained in the closed-loop form.

4.2 Serial-kinematics robots

4.2.1 Planar articulated manipulators

4.2.1.1 FSF of the 2-DOF manipulator

The planar 2-DOF articulated (**RR**) robot (Fig. 4.1) consists of four links (n = 4), among them two kinematic links rotating around the parallel axes and two geometric links separating the parallel axes. The coordinate and arrangement codes of the manipulator are k = **6161** and K = 0**6161**, respectively. The FSS links and their motions are listed in Table 4.1.

The manipulation matrix is a product of the matrices shown in the last column of Table 4.1. Since the planar case is considered, the 3 × 3 matrices can be used:

$$^{0}\mathbf{A}_4 = \prod_{i=1}^{4} {}^{i-1}\mathbf{A}_i(q_i) = \mathbf{A}^6(\varphi_1)\mathbf{A}^1(a_1)\mathbf{A}^6(\varphi_2)\mathbf{A}^1(a_2) \tag{4.7}$$

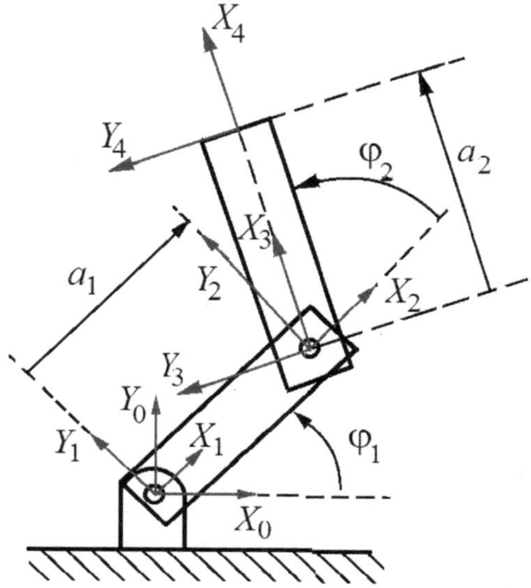

Figure 4-1 Planar 2-DOF (RR) manipulator

Table 4-1 FSS of the planar **RR** robot, $k = 6161$.

S_i	Physical body	Motion relative to the previous link	Matrix $^{i-1}\mathbf{A}_i$
S_0	Base	Fixed link	–
S_1	Link 1	Rotation around the Z_0-axis at angle φ_1	$^0\mathbf{A}_1 = \mathbf{A}^6(\varphi_1)$
S_2	Link 1	Translation along the X_1-axis at distance a_1	$^1\mathbf{A}_2 = \mathbf{A}^1(a_1)$
S_3	Link 2	Rotation around the Z_2-axis at angle φ_2	$^2\mathbf{A}_3 = \mathbf{A}^6(\varphi_2)$
S_4	Link 2	Translation along the X_3-axis at distance a_2	$^3\mathbf{A}_4 = \mathbf{A}^1(a_2)$

Parameters: φ_i $(i = 1, 2)$ is the angle of rotation of the ith kinematic link around the previous Z-axis; ai is the distance between two neighboring axes (the length of the geometrical link).

$$\mathbf{A}^6(\varphi_i) = \begin{bmatrix} \cos\varphi_i & -\sin\varphi_i & 0 \\ \sin\varphi_i & \cos\varphi_i & 0 \\ 0 & 0 & 1 \end{bmatrix} (i = 1,\ 2) \text{ and } \mathbf{A}^1(a_i) = \begin{bmatrix} 1 & 0 & a_i \\ 0 & 1 & 0 \\ 0 & 0 & 1 \end{bmatrix} (i = 1,\ 2) \tag{4.8}$$

We direct the reader's attention to the fact that the subsequence of the superscripts in Eq. (4.7) is just the coordinate code k for the considered robot. The manipulation matrix $^0\mathbf{A}_4$ and FSF of the robot are as follows:

$$^0\mathbf{A}_4 = \begin{bmatrix} \cos(\varphi_1+\varphi_2) & -\sin(\varphi_1+\varphi_2) & a_1\cos\varphi_1 + a_2\cos(\varphi_1+\varphi_2) \\ \sin(\varphi_1+\varphi_2) & \cos(\varphi_1+\varphi_2) & a_1\sin\varphi_1 + a_2\sin(\varphi_1+\varphi_2) \\ 0 & 0 & 1 \end{bmatrix}, \tag{4.9}$$

$$[\mathbf{r}_0\ \mathbf{c}_0] = {}^0\mathbf{A}_4\ [\mathbf{r}_4\ \mathbf{c}_4] \tag{4.10}$$

or

$$\mathbf{r}_0 \equiv \begin{bmatrix} x_0 \\ y_0 \\ 1 \end{bmatrix} = \begin{bmatrix} \cos(\varphi_1+\varphi_2) & -\sin(\varphi_1+\varphi_2) & a_1\cos\varphi_1 + a_2\cos(\varphi_1+\varphi_2) \\ \sin(\varphi_1+\varphi_2) & \cos(\varphi_1+\varphi_2) & a_1\sin\varphi_1 + a_2\sin(\varphi_1+\varphi_2) \\ 0 & 0 & 1 \end{bmatrix} \begin{bmatrix} x_4 \\ y_4 \\ 1 \end{bmatrix}$$

$$= \begin{bmatrix} x_4\cos(\varphi_1+\varphi_2) - y_4\sin(\varphi_1+\varphi_2) + a_1\cos\varphi_1 + a_2\cos(\varphi_1+\varphi_2) \\ y_4\cos(\varphi_1+\varphi_2) + x_4\sin(\varphi_1+\varphi_2) + a_1\sin\varphi_1 + a_2\sin(\varphi_1+\varphi_2) \\ 1 \end{bmatrix} \tag{4.11}$$

$$\mathbf{c}_0 \equiv \begin{bmatrix} c_{0x} \\ c_{0y} \\ 0 \end{bmatrix} = \begin{bmatrix} \cos(\varphi_1+\varphi_2) & -\sin(\varphi_1+\varphi_2) & a_1\cos\varphi_1 + a_2\cos(\varphi_1+\varphi_2) \\ \sin(\varphi_1+\varphi_2) & \cos(\varphi_1+\varphi_2) & a_1\sin\varphi_1 + a_2\sin(\varphi_1+\varphi_2) \\ 0 & 0 & 1 \end{bmatrix} \begin{bmatrix} c_{x4} \\ c_{y4} \\ 0 \end{bmatrix}$$

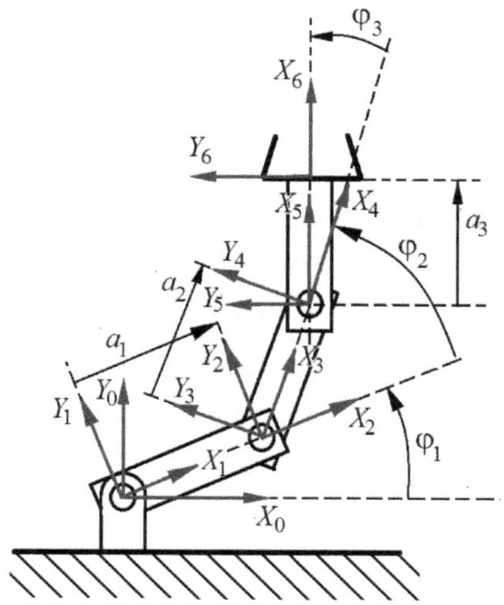

Figure 4-2 Planar 3-DOF (**RRR**) manipulator.

$$= \begin{bmatrix} c_{x4}\cos(\varphi_1+\varphi_2) - c_{y4}\sin(\varphi_1+\varphi_2) \\ c_{x4}\cos(\varphi_1+\varphi_2) + c_{y4}\sin(\varphi_1+\varphi_2) \\ 0 \end{bmatrix} \tag{4.12}$$

where $\mathbf{r}_0$ and $\mathbf{r}_4$ are the position vectors of a point of the robot target element of the FSS in coordinate frames S_0 and S_4, respectively; $\mathbf{c}_0$ and $\mathbf{c}_4$ are the unit vectors ($||\mathbf{c}_0|| = ||\mathbf{c}_4|| = 1$) of the active element (technological tool) orientation in coordinate frames S_0 and S_4, respectively.

Since the manipulator has only two DOF, it can be of limited utility, such as transfer a chosen point of the last link in a given position or motion along a given trajectory. Let us consider the motion of point O_4– the origin of the coordinate frame associated with the end-effector (Fig. 4.1). For this point, $x_4 = y_4 = 0$ and $\mathbf{r}_4 = [0, 0, 1]^T$. Hence,

$$\mathbf{r}_0 = {}^0\mathbf{A}_4 \begin{bmatrix} 0 \\ 0 \\ 1 \end{bmatrix} = \begin{bmatrix} a_1\cos\varphi_1 + a_2\cos(\varphi_1+\varphi_2) \\ a_1\sin\varphi_1 + a_2\sin(\varphi_1+\varphi_2) \\ 1 \end{bmatrix} \tag{4.13}$$

4.2.1.2 FSF of the 3-DOF manipulator

The planar 3-DOF articulated (**RRR**) robot (Fig. 4.2) consists of six links ($n = 6$), among them three kinematic links rotating around the parallel axes and three geometric links, two of which separate the parallel axes. The coordinate and arrangement codes are k = **616161**, Eq. (3.94), and K = **0616161**, Eq. (3.95), respectively. The FSS links are listed in Table 4.2.

According to Table 4.2, the 3 × 3 manipulation matrix is a product of the matrices listed in the last column,

Table 4-2 FSS of the planar **RRR** robot, k = **616**161.

S_i	Physical body	Motion relative to previous link	Matrix $^{i-1}\mathbf{A}_i$
S_0	Base	Fixed link	–
S_1	Link 1	Rotation around the Z_0-axis at angle φ_1	$^0\mathbf{A}_1 = \mathbf{A}^6(\varphi_1)$
S_2	Link 1	Translation along the X_1-axis at distance a_1	$^1\mathbf{A}_2 = \mathbf{A}^1(a_1)$
S_3	Link 2	Rotation around the Z_2-axis at angle φ_2	$^2\mathbf{A}_3 = \mathbf{A}^6(\varphi_2)$
S_4	Link 2	Translation along the X_3-axis at distance a_2	$^3\mathbf{A}_4 = \mathbf{A}^1(a_2)$
S_5	Link 3	Rotation around the Z_5-axis at angle φ_3	$^4\mathbf{A}_5 = \mathbf{A}^6(\varphi_3)$
S_6	End-effector	Translation along the X_5-axis at distance a_3	$^5\mathbf{A}_6 = \mathbf{A}^1(a_3)$

Parameters: φ_i (i = 1, 2, 3) is the angle of rotation of the ith kinematic link around the previous Z-axis; ai is the distance between axes possessing by the ith link.

$$^0\mathbf{A}_6 = \prod_{i=1}^{6} {}^{i-1}\mathbf{A}_i(q_i) = \mathbf{A}^6(\varphi_1)\,\mathbf{A}^1(a_1)\,\mathbf{A}^6(\varphi_2)\,\mathbf{A}^1(a_2)\,\mathbf{A}^6(\varphi_3)\,\mathbf{A}^1(a_3) \tag{4.14}$$

$$\mathbf{A}^6(\varphi_i) = \begin{bmatrix} \cos\varphi_i & -\sin\varphi_i & 0 \\ \sin\varphi_i & \cos\varphi_i & 0 \\ 0 & 0 & 1 \end{bmatrix} \text{and } \mathbf{A}^1(a_i) = \begin{bmatrix} 1 & 0 & a_i \\ 0 & 1 & 0 \\ 0 & 0 & 1 \end{bmatrix},\ i = 1,\ 2,\ 3 \tag{4.15}$$

Hence,

$$^0\mathbf{A}_6 = \begin{bmatrix} \cos(\varphi_1+\varphi_2+\varphi_3) & -\sin(\varphi_1+\varphi_2+\varphi_3) & a_{14} \\ \sin(\varphi_1+\varphi_2+\varphi_3) & \cos(\varphi_1+\varphi_2+\varphi_3) & a_{24} \\ 0 & 0 & 1 \end{bmatrix} \tag{4.16}$$

$$a_{14} = a_1 \cos\varphi_1 + a_2 \cos(\varphi_1 + \varphi_2) + a_3 \cos(\varphi_1 + \varphi_2 + \varphi_3)$$

$$a_{24} = a_1 \sin\varphi_1 + a_2 \sin(\varphi_1 + \varphi_2) + a_3 \sin(\varphi_1 + \varphi_2 + \varphi_3) \tag{4.17}$$

The FSF of the robot is

$$[\mathbf{r}_0\mathbf{c}_0] = {}^0\mathbf{A}_6[\mathbf{r}_6\mathbf{c}_6] \tag{4.18}$$

$$\mathbf{r}_0 = [x_0, y_0, 1]^T \text{ and } \mathbf{r}_6 = [x_6, y_6, 1]^T \tag{4.19}$$

$$\mathbf{c}_0 = [c_{0x}, c_{0y}, 0]^T \text{ and } \mathbf{c}_6 = [c_{6x}, c_{6y}, 0]^T \tag{4.20}$$

where $\mathbf{r}_0$ and $\mathbf{r}_6$ are the position vectors of a point of the robot target element in coordinate frames S_0 and S_6, respectively; and $\mathbf{c}_0$ and $\mathbf{c}_6$ are the unit vectors ($||\mathbf{c}_0|| = ||\mathbf{c}_6|| = 1$) of the active element (technological tool) orientation referring to the frames S_0 and S_6, respectively.

Solution of the forward kinematics problem is obtained by multiplication of the manipulation matrix and vectors $\mathbf{r}_6$ and $\mathbf{c}_6$ in the listed order:

$$\begin{bmatrix} x_0 & c_{0x} \\ y_0 & c_{0y} \\ 1 & 0 \end{bmatrix} = \begin{bmatrix} \cos(\varphi_1+\varphi_2+\varphi_3) & -\sin(\varphi_1+\varphi_2+\varphi_3) & a_{14} \\ \sin(\varphi_1+\varphi_2+\varphi_3) & \cos(\varphi_1+\varphi_2+\varphi_3) & a_{24} \\ 0 & 0 & 1 \end{bmatrix} \begin{bmatrix} x_6 & c_{6x} \\ y_6 & c_{6y} \\ 1 & 0 \end{bmatrix}$$

$$= \begin{bmatrix} x_6\cos(\varphi_1+\varphi_2+\varphi_3) - y_6\sin(\varphi_1+\varphi_2+\varphi_3) + a_{14} & c_{6x}\cos(\varphi_1+\varphi_2+\varphi_3) - c_{6y}\sin(\varphi_1+\varphi_2+\varphi_3) \\ x_6\sin(\varphi_1+\varphi_2+\varphi_3) + y_6\cos(\varphi_1+\varphi_2+\varphi_3) + a_{24} & c_{6x}\sin(\varphi_1+\varphi_2+\varphi_3) + c_{6y}\cos(\varphi_1+\varphi_2+\varphi_3) \\ 1 & 0 \end{bmatrix} \quad (4.21)$$

where a_{14} and a_{24} are defined through Eqs. (4.17).

Note that the 3-DOF set of **RRR** links forming the robots kinematics presents a typical structural part entering in two serial-kinematics robots – SCARA arm and Scorbot robot considered in two following subsections.

4.2.1.3 FSF of the SCARA arm robot

The 4-DOF SCARA arm (Fig. 4.3) has **RRPR** kinematics. In this robot, there are three articulated links as in the **RRR** planar manipulator and an actuated translational link (guide rod) sliding along the last axis of rotation. Since the arm has only four DOF, direction of the

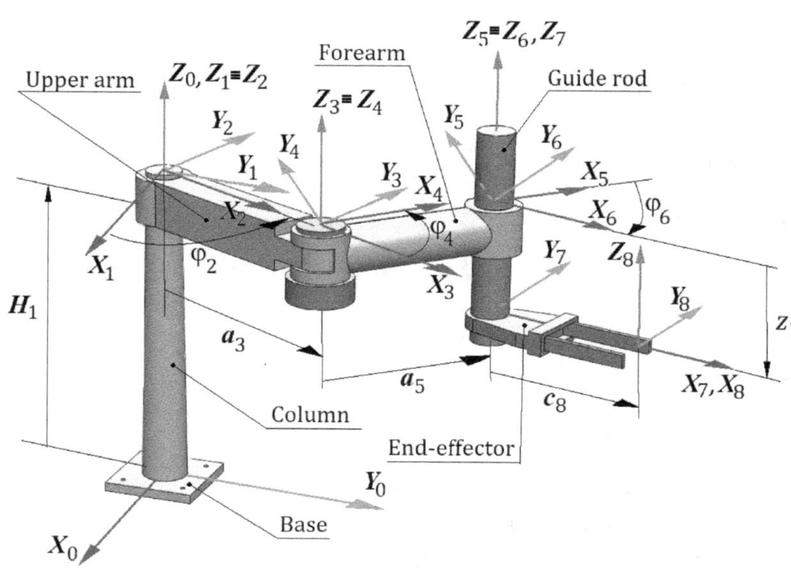

Figure 4-3 The SCARA arm: the kinematic scheme and coordinate frames.

end-effector translation is always oriented at right angle to the planes of the rotating links. Nevertheless, presence of the translational link allows useful applications of the robot such as assembling plain PC boards.

The arm kinematics includes four kinematic links (three rotations and one translation) and four geometric links. Coordinate code of the SCARA arm is k = 36262362. The FSS links are listed in Table 4.3. The manipulation matrix is as follows:

$$ {}^{0}\mathbf{A}_8 = \prod_{i=1}^{8} {}^{i-1}\mathbf{A}_i(q_i) = \mathbf{A}^3(H_1)\,\mathbf{A}^6(\varphi_2)\,\mathbf{A}^1(a_3)\,\mathbf{A}^6(\varphi_4)\,\mathbf{A}^1(a_5)\,\mathbf{A}^3(z_6)\,\mathbf{A}^6(\varphi_7)\,\mathbf{A}^1(a_8) \quad (4.22) $$

$$ \mathbf{A}^6(\varphi_i) = \begin{bmatrix} \cos\varphi_i & -\sin\varphi_i & 0 & 0 \\ \sin\varphi_i & \cos\varphi_i & 0 & 0 \\ 0 & 0 & 1 & 0 \\ 0 & 0 & 0 & 1 \end{bmatrix} (i = 2, 4, 7); $$

$$ \mathbf{A}^1(a_i) = \begin{bmatrix} 1 & 0 & 0 & a_i \\ 0 & 1 & 0 & 0 \\ 0 & 0 & 1 & 0 \\ 0 & 0 & 0 & 1 \end{bmatrix} (i = 3, 5, 8); $$

$$ \mathbf{A}^3(H_1) = \begin{bmatrix} 1 & 0 & 0 & 0 \\ 0 & 1 & 0 & 0 \\ 0 & 0 & 1 & H_1 \\ 0 & 0 & 0 & 1 \end{bmatrix}; \text{ and } \mathbf{A}^3(z) = \begin{bmatrix} 1 & 0 & 0 & 0 \\ 0 & 1 & 0 & 0 \\ 0 & 0 & 1 & z_6 \\ 0 & 0 & 0 & 1 \end{bmatrix} \quad (4.23) $$

Hence,

$$ {}^{0}\mathbf{A}_8 = \begin{bmatrix} \cos(\varphi_2+\varphi_4+\varphi_7) & -\sin(\varphi_2+\varphi_4+\varphi_7) & 0 & a_{14} \\ \sin(\varphi_2+\varphi_4+\varphi_7) & \cos(\varphi_2+\varphi_4+\varphi_7) & 0 & a_{24} \\ 0 & 0 & 1 & H_1+z_6 \\ 0 & 0 & 0 & 1 \end{bmatrix}, \quad (4.24) $$

$$ a_{14} = a_3 \cos\varphi_2 + a_5 \cos(\varphi_2+\varphi_4) + a_8 \cos(\varphi_2+\varphi_4+\varphi_7) \quad (4.25) $$

$$ a_{24} = a_3 \sin\varphi_2 + a_5 \sin(\varphi_2+\varphi_4) + a_3 \sin(\varphi_2+\varphi_4+\varphi_7) $$

Table 4-3 FSS of the planar SCARA arm, K = 036161361.

S_i	Body	Motion relative to previous link	Matrix $^{i-1}\mathbf{A}_i$
S_0	Base	Fixed link	–
S_1	Column	Translation along the Z_0-axis at distance H	$^0\mathbf{A}_1 = \mathbf{A}^3(H_1)$
S_2	Upper arm	Rotation around the Z_1-axis at angle φ_2	$^1\mathbf{A}_2 = \mathbf{A}^6(\varphi_2)$
S_3	Upper arm	Translation along the X_2-axis at distance a_3	$^2\mathbf{A}_3 = \mathbf{A}^1(a_3)$
S_4	Forearm	Rotation around the Z_3-axis at angle φ_4	$^3\mathbf{A}_4 = \mathbf{A}^6(\varphi_4)$
S_5	Forearm	Translation along the X_4-axis at distance a_5	$^4\mathbf{A}_5 = \mathbf{A}^1(a_5)$
S_6	Guide rod	Translation around the Z_5-axis at distance z_6	$^5\mathbf{A}_6 = \mathbf{A}^3(z_6)$
S_7	End-effector	Rotation around the Z_6-axis at angle φ_7	$^6\mathbf{A}_7 = \mathbf{A}^6(\varphi_7)$
S_8	End-effector	Translation along the X_7-axis at distance a_8	$^7\mathbf{A}_8 = \mathbf{A}^1(a_8)$

Parameters: φ_i (i = 2, 4, 7) is the actuated angle of rotation of the ith kinematic link around the Z_{i-1}-axis; a_3 and a_5 are the constant distance between neighboring axes of rotation possessing by the ith link; z_6 is the actuated displacement of the guide rod; and a_8 is the length within the EE.

The FSF of the SCARA arm robot is

$$[\mathbf{r}_0\mathbf{c}_0] = {}^0\mathbf{A}_8\,[\mathbf{r}_8\mathbf{c}_8] \tag{4.26}$$

with

$$\mathbf{r}_0 = [x_0, y_0, z_0, 1]^{\mathrm{T}} \text{ and } \mathbf{r}_8 = [x_8, y_8, z_8, 1]^{\mathrm{T}} \tag{4.27}$$

$$\mathbf{c}_0 = [c_{0x}, c_{0y}, 0, 0]^{\mathrm{T}} \text{ and } \mathbf{c}_8 = [c_{8x}, c_{8y}, 0, 0]^{\mathrm{T}} \tag{4.28}$$

where $\mathbf{r}_0$ and $\mathbf{r}_8$ are the position vectors of a point of the robot target element in coordinate frames S_0 and S_8, respectively; and $\mathbf{c}_0$ and $\mathbf{c}_8$ are the unit vectors ($||\mathbf{c}_0|| = ||\mathbf{c}_8|| = 1$) of the active element (technological tool) orientation referring to the frames S_0 and S_8, respectively.

Note that Eq. (4.28) describes the case shown in Figure 4.3: the three parallel rotation axes are denoted the Z-axis, and thus, the z-coordinates of the orientation vectors $\mathbf{c}_0$ and $\mathbf{c}_8$ are equal to zero. Solution of the direct kinematics problem is obtained by substitution of Eqs. (4.27) and (4.28) in Eq. (4.26):

$$\begin{bmatrix} x_0 & c_{0x} \\ y_0 & c_{0y} \\ 0 & 0 \\ 1 & 0 \end{bmatrix} = \begin{bmatrix} \cos(\varphi_2+\varphi_4+\varphi_7) & -\sin(\varphi_2+\varphi_4+\varphi_7) & 0 & a_{14} \\ \sin(\varphi_2+\varphi_4+\varphi_7) & \cos(\varphi_2+\varphi_4+\varphi_7) & 0 & a_{24} \\ 0 & 0 & 1 & H_1+z_6 \\ 0 & 0 & 0 & 1 \end{bmatrix} \begin{bmatrix} x_8 & c_{8x} \\ y_8 & c_{8y} \\ 0 & 0 \\ 1 & 0 \end{bmatrix}$$

$$
= \begin{bmatrix} x_8 \cos(\varphi_2+\varphi_4+\varphi_7) - y_8 \sin(\varphi_2+\varphi_4+\varphi_7) + a_{14} \\ x_8 \sin(\varphi_2+\varphi_4+\varphi_7) + y_8 \cos(\varphi_2+\varphi_4+\varphi_7) + a_{24} \\ H_1 + z_6 \\ 1 \end{bmatrix.
$$

$$
\begin{matrix} c_{8x} \cos(\varphi_2+\varphi_4+\varphi_7) - c_{8y} \sin(\varphi_2+\varphi_4+\varphi_7) \\ c_{8x} \sin(\varphi_2+\varphi_4+\varphi_7) + c_{8y} \cos(\varphi_2+\varphi_4+\varphi_7) \\ 0 \\ 0 \end{bmatrix} \quad (4.29)
$$

where a_{14} and a_{24} are defined through Eqs. (4.25).

4.2.1.4 FSF of the Scorbot robot

In the 5-DOF Scorbot robot (Fig. 4.4), the column rigidly mounted to the pedestal (fixed link) bears the vertical axis (the Z_0-axis) for rotation around it. The upper arm, forearm and gripper rotate around the horizontal axes relative to foregoing links. Additionally, the wrist allows rotation of the gripper relative to the forearm around two mutually perpendicular (pitch-and-roll) axes [1]. The coordinate code k and arrangement code K of the Scorbot robot are

$$k = 615151541$$

$$K = 0615151541 \quad (4.30)$$

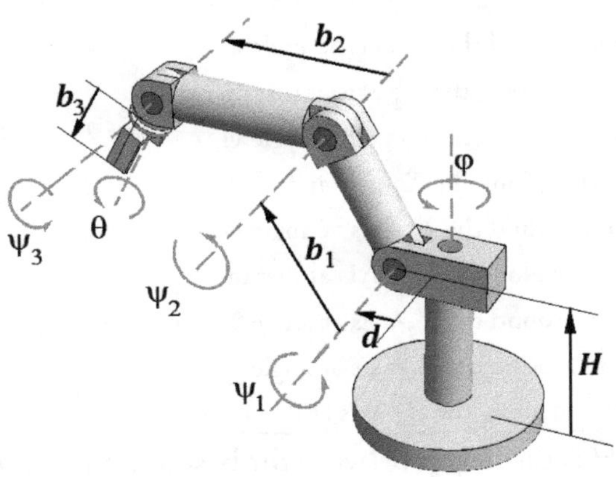

Figure 4-4 Scorbot ER-VII Robot Arm: kinematic scheme.

One more note that the arrangement code K of all considered serial-kinematics robots begins with symbol 0. This clearly demonstrates the difference between arrangements of machine tools and robots: all mobile links of the robot kinematic chain lie on one side of the bed, while in machine tools the bed divides the kinematic chain into tool- and work-associated sub-chains.

The structural code differs by addition of a geometrical link,

$$k = 3615151541 \tag{4.31}$$

where symbol $k_1 = 3$ corresponds to geometric links associated with the distance between the base plane and the point of intersection the Z_0-axis with the common perpendicular to two first axes of rotation. The FSS links of the Scorbot robot (Fig. 4.4) and associated coordinate frames are listed in Table 4.4.

The manipulation matrix ${}^0\mathbf{A}_{10}$ is the product of the ten 4 × 4 matrices listed in the last column of Table 4.4:

$$\begin{aligned} {}^0\mathbf{A}_{10} &= \left[\prod_{i=1}^{10} {}^{i\text{-}1}\mathbf{A}_i(q_i)\right] = \left[\prod_{i=1}^{10} \mathbf{A}^{j_i}(q_i)\right] \\ &= \mathbf{A}^3(H)\,\mathbf{A}^6(\varphi)\,\mathbf{A}^1(d)\,\mathbf{A}^5(\psi_1)\,\mathbf{A}^1(\mathrm{b}_1)\,\mathbf{A}^5(\psi_2)\,\mathbf{A}^1(\mathrm{b}_2)\,\mathbf{A}^5(\psi_3)\,\mathbf{A}^4(\theta)\,\mathbf{A}^1(\mathrm{b}_3) \end{aligned} \tag{4.32}$$

with

Table 4-4 Form-shaping system Scorbot robot, k = 3615151541

S_i	Link	Motion relative to previous link	Matrix ${}^{i\text{-}1}\mathbf{A}_i$
S_0	Base	Fixed link	–
S_1	Column	Translation along the Z_0-axis at height H	${}^0\mathbf{A}_1 = \mathbf{A}^3(H)$
S_2	Body	Rotation around the Z_1-axis at angle φ	${}^1\mathbf{A}_2 = \mathbf{A}^6(\varphi)$
S_3	Body	Translation along the X_2-axis at distance d	${}^2\mathbf{A}_3 = \mathbf{A}^1(d)$
S_4	Upper arm	Rotation around the Y_3-axis at angle ψ_1	${}^3\mathbf{A}_4 = \mathbf{A}^5(\psi_1)$
S_5	Upper arm	Translation along the X_4-axis at distance b_1	${}^4\mathbf{A}_5 = \mathbf{A}^1(b_1)$
S_6	Forearm	Rotation around the Y_5-axis at angle ψ_2	${}^5\mathbf{A}_6 = \mathbf{A}^5(\psi_2)$
S_7	Forearm	Translation along the X_6-axis at distance b_2	${}^6\mathbf{A}_7 = \mathbf{A}^1(b_2)$
S_8	Gripper	Rotation around the Y_7-axis at angle ψ_3	${}^7\mathbf{A}_8 = \mathbf{A}^5(\psi_3)$
S_9	Gripper	Rotation around the X_8-axis at angle θ	${}^8\mathbf{A}_9 = \mathbf{A}^4(\theta)$
S_{10}	Gripper	Translation along the X_9-axis at distance b_3	${}^9\mathbf{A}_{10} = \mathbf{A}^1(b_3)$

Geometrical parameters: H is the height between the base and the common perpendicular to the two first rotation axes; d is the shortest distance between the axes of the column and upper arm; and b_1, b_2, b_3 are axis-to-axis distance between the axes of the shoulder, elbow, and wrist.

$$
{}^{0}\mathbf{A}_1 = \mathbf{A}^3(H) = \begin{bmatrix} 1 & 0 & 0 & 0 \\ 0 & 1 & 0 & 0 \\ 0 & 0 & 1 & H \\ 0 & 0 & 0 & 1 \end{bmatrix};\ {}^{1}\mathbf{A}_2 = \mathbf{A}^6(\varphi) = \begin{bmatrix} \cos\varphi & -\sin\varphi & 0 & 0 \\ \sin\varphi & \cos\varphi & 0 & 0 \\ 0 & 0 & 1 & 0 \\ 0 & 0 & 0 & 1 \end{bmatrix};
$$

$$
{}^{2}\mathbf{A}_3 = \mathbf{A}^1(d) = \begin{bmatrix} 1 & 0 & 0 & d \\ 0 & 1 & 0 & 0 \\ 0 & 0 & 1 & 0 \\ 0 & 0 & 0 & 1 \end{bmatrix};\ {}^{3}\mathrm{A}_4 = \mathbf{A}^5(\psi_1) = \begin{pmatrix} \cos\psi_1 & 0 & \sin\psi_1 & 0 \\ 0 & 1 & 0 & 0 \\ -\sin\psi_1 & 0 & \cos\psi_1 & 0 \\ 0 & 0 & 0 & 1 \end{pmatrix};
$$

$$
{}^{4}\mathbf{A}_5 = \mathbf{A}^1(b_1) = \begin{bmatrix} 1 & 0 & 0 & b_1 \\ 0 & 1 & 0 & 0 \\ 0 & 0 & 1 & 0 \\ 0 & 0 & 0 & 1 \end{bmatrix};\ {}^{5}\mathbf{A}_6 = \mathbf{A}^5(\psi_2) = \begin{pmatrix} \cos\psi_2 & 0 & \sin\psi_2 & 0 \\ 0 & 1 & 0 & 0 \\ -\sin\psi_2 & 0 & \cos\psi_2 & 0 \\ 0 & 0 & 0 & 1 \end{pmatrix};
$$

$$
{}^{6}\mathbf{A}_7 = \mathbf{A}^1(b_2) = \begin{bmatrix} 1 & 0 & 0 & b_2 \\ 0 & 1 & 0 & 0 \\ 0 & 0 & 1 & 0 \\ 0 & 0 & 0 & 1 \end{bmatrix};\ {}^{7}\mathbf{A}_8 = \mathbf{A}^5(\psi_3) = \begin{pmatrix} \cos\psi_3 & 0 & \sin\psi_3 & 0 \\ 0 & 1 & 0 & 0 \\ -\sin\psi_3 & 0 & \cos\psi_3 & 0 \\ 0 & 0 & 0 & 1 \end{pmatrix};
$$

$$
{}^{8}\mathbf{A}_9 = \mathbf{A}^4(\theta) = \begin{pmatrix} 1 & 0 & 0 & 0 \\ 0 & \cos\theta & -\sin\theta & 0 \\ 0 & \sin\theta & \cos\theta & 0 \\ 0 & 0 & 0 & 1 \end{pmatrix};\ \text{and } {}^{9}\mathbf{A}_{10} = \mathbf{A}^1(b_3) = \begin{bmatrix} 1 & 0 & 0 & b_3 \\ 0 & 1 & 0 & 0 \\ 0 & 0 & 1 & 0 \\ 0 & 0 & 0 & 1 \end{bmatrix}.
$$

Formal multiplication yields

$$
{}^{0}\mathbf{A}_{10} = \begin{bmatrix} a_{11} & a_{12} & a_{13} & a_{14} \\ a_{21} & a_{22} & a_{23} & a_{24} \\ a_{31} & a_{32} & a_{33} & a_{34} \\ 0 & 0 & 0 & 1 \end{bmatrix} \tag{4.33}
$$

$$a_{11} = \cos\varphi \cos\psi_{123};\ a_{12} = \cos\varphi \sin\theta \sin\psi_{123} - \sin\varphi \cos\theta$$

$$a_{13} = \cos\varphi \cos\theta \sin\psi_{123} + \sin\varphi \sin\theta;\ a_{14} = \cos\varphi\,(d + b_1 \cos\psi_1 + b_2 \cos\psi_{12} + b_3 \cos\psi_{123})$$

$$a_{21} = \sin\varphi \cos\psi_{123};\ a_{22} = \sin\varphi \sin\theta \sin\psi_{123} + \cos\varphi \cos\theta$$

$$a_{23} = \sin\varphi \cos\theta \sin\psi_{123} - \cos\varphi \sin\theta;\ a_{24} = \sin\varphi\,(d + b_1 \cos\psi_1 + b_2 \cos\psi_{12} + b_3 \cos\psi_{123})$$

$$a_{31} = -\sin\psi_{123};\ a_{32} = \sin\theta \cos\psi_{123};\ a_{33} = \cos\theta \cos\psi_{123};\ \text{and,}$$

$$a_{34} = H - b_1 \sin\psi_1 - b_2 \sin\psi_{12} - b_3 \sin\psi_{123} \tag{4.34}$$

4.2.1.5 FSF of the spherical robot

Three controlled variable parameters of the spherical robot (Fig. 4.5) are as follows: the latitude angle φ giving a position of the rotation axle around the fixed column; the altitude angle θ of tilting telescopic guideways, and variable length ρ of the arm carrying the controlled point of the end-effector. If length ρ is fixed, the end-effector points with varying angles φ and θ moves over the spherical surface of radius ρ, hence follows the name of the mechanism. Varying of radius ρ extends the robot working space, which lies between two spherical surfaces corresponding to maximal and minimal radii ρ. Of course, a part of the spherical surfaces cannot be reached by the end-effector because of structural obstructions. The structural code of the 3-DOF spherical robot is k_{st} = **3642.** The robot links and their motions are listed in Table 4.5.

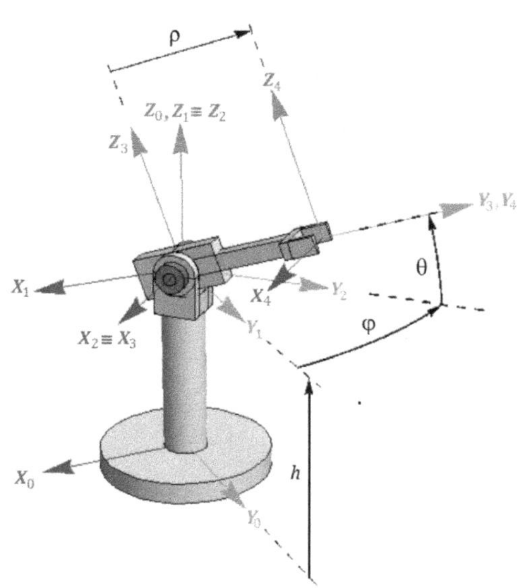

Figure 4-5 Spherical robot: links and associated coordinate frames.

Table 4-5 Form-shaping system of the spherical robot, k_{st} = 3642.

S_j	Link	Motion relative to previous link	Matrix $^{i-1}\mathbf{A}_i$
S_0	Column	Fixed link	–
S_1	Base of rotation	Transfer along the X_0-axis at height h	$^0\mathbf{A}_1 = \mathbf{A}^3(h)$
S_2	Elevated pivot	Rotation around the Z_1-axis at angle φ	$^1\mathbf{A}_2 = \mathbf{A}^6(\varphi)$
S_3	Telescoping arm	Rotation around the X_2-axis at angle θ	$^2\mathbf{A}_3 = \mathbf{A}^4(\theta)$
S_4	Sliding sleeve	Translation along the Y_3-axis at distance ρ	$^3\mathbf{A}_4 = \mathbf{A}^2(\rho)$

Parameters: h is the distance between the column pedestal and axis of rotation of the elevated pivot; φ is the angle of rotation of the elevated pivot; θ is the angle of rotation of the telescoping arm; and ρ is the current extension of the sliding sleeve.

The manipulation matrix is

$$^0\mathbf{A}_4 = \prod_{j=1}^{4} {}^{j-1}\mathbf{A}_j(q_j) = \mathbf{A}^3(h)\,\mathbf{A}^6(\varphi)\,\mathbf{A}^4(\theta)\,\mathbf{A}^2(\rho)$$

$$= \begin{bmatrix} 1 & 0 & 0 & 0 \\ 0 & 1 & 0 & 0 \\ 0 & 0 & 1 & h \\ 0 & 0 & 0 & 1 \end{bmatrix} \begin{pmatrix} \cos\varphi & -\sin\varphi & 0 & 0 \\ \sin\varphi & \cos\varphi & 0 & 0 \\ 0 & 0 & 1 & 0 \\ 0 & 0 & 0 & 1 \end{pmatrix} \begin{pmatrix} 1 & 0 & 0 & 0 \\ 0 & \cos\theta & -\sin\theta & 0 \\ 0 & \sin\theta & \cos\theta & 0 \\ 0 & 0 & 0 & 1 \end{pmatrix} \begin{bmatrix} 1 & 0 & 0 & 0 \\ 0 & 1 & 0 & \rho \\ 0 & 0 & 1 & 0 \\ 0 & 0 & 0 & 1 \end{bmatrix}$$

$$= \begin{bmatrix} \cos\varphi & -\cos\theta\sin\varphi & \sin\theta\sin\varphi & -\rho\cos\theta\sin\varphi \\ \sin\varphi & \cos\theta\cos\varphi & -\sin\theta\cos\varphi & \rho\cos\theta\cos\varphi \\ 0 & \sin\theta & \cos\theta & h+\rho\sin\theta \\ 0 & 0 & 0 & 1 \end{bmatrix} \tag{4.35}$$

The FSF of the spherical robot is

$$\mathbf{r}_0 = {}^0\mathbf{A}_4\,\mathbf{r}_4, \text{ with } \mathbf{r}_0 = \begin{bmatrix} x_0 \\ y_0 \\ z_0 \\ 1 \end{bmatrix}; \mathbf{r}_4 = \begin{bmatrix} x_4 \\ y_4 \\ z_4 \\ 1 \end{bmatrix} \tag{4.36}$$

If, for example, the active element model presents a single point ($x_4 = y_4 = z_4 = 0$), we obtain

$$\mathbf{r}_0 = {}^0\mathbf{A}_4\,\mathbf{r}_4 = \begin{bmatrix} -\rho\cos\theta\sin\varphi \\ \rho\cos\theta\cos\varphi \\ h+\rho\sin\theta \\ 1 \end{bmatrix} \tag{4.37}$$

4.3 Serial-kinematics machine tools

4.3.1 Turning machines

4.3.1.1 FSF of the lathe

The kinematic model of the FSS (Fig. 4.6a) has coordinate code k = **631**, Eq. (3.91). A set of the coordinate frames $S_i = X_iY_iZ_i$ (i = 0, 1, 2, 3) associated with the ith link are shown in the Figure 4.6b. The FSS links and their motions are listed in Table 4.6. The manipulation matrix ${}^0\mathbf{A}_3$ is synthesized according to coordinate code (see also Appendix *A2.2.1*),

$${}^0\mathbf{A}_3 = {}^0\mathbf{A}_1\,{}^1\mathbf{A}_2\,{}^2\mathbf{A}_3 = \mathbf{A}^6(\varphi)\;\mathbf{A}^3(z)\;\mathbf{A}^3(x) = \tag{4.38}$$

$$\begin{bmatrix} \cos\varphi & -\sin\varphi & 0 & 0 \\ \sin\varphi & \cos\varphi & 0 & 0 \\ 0 & 0 & 1 & 0 \\ 0 & 0 & 0 & 1 \end{bmatrix} \begin{bmatrix} 1 & 0 & 0 & 0 \\ 0 & 1 & 0 & 0 \\ 0 & 0 & 1 & z \\ 0 & 0 & 0 & 1 \end{bmatrix} \begin{bmatrix} 1 & 0 & 0 & x \\ 0 & 1 & 0 & 0 \\ 0 & 0 & 1 & 0 \\ 0 & 0 & 0 & 1 \end{bmatrix} = \begin{bmatrix} \cos\varphi & -\sin\varphi & 0 & x\cos\varphi \\ \sin\varphi & \cos\varphi & 0 & x\sin\varphi \\ 0 & 0 & 1 & z \\ 0 & 0 & 0 & 1 \end{bmatrix}$$

The FSF is as follows:

$$\mathbf{r}_0 = {}^0\mathbf{A}_3\mathbf{r}_3, \tag{4.39}$$

$$\mathbf{r}_0 \equiv \begin{bmatrix} x_0 \\ y_0 \\ z_0 \\ 1 \end{bmatrix} \text{ and } \mathbf{r}_3 \equiv \begin{bmatrix} x_3 \\ y_3 \\ z_3 \\ 1 \end{bmatrix};$$

Hence,

$$\mathbf{r}_0 = {}^0\mathbf{A}_3\mathbf{r}_3 = \begin{bmatrix} \cos\varphi & -\sin\varphi & 0 & x\cos\varphi \\ \sin\varphi & \cos\varphi & 0 & x\sin\varphi \\ 0 & 0 & 1 & z \\ 0 & 0 & 0 & 1 \end{bmatrix} \begin{bmatrix} x_3 \\ y_3 \\ z_3 \\ 1 \end{bmatrix} = \begin{bmatrix} (x+x_3)\cos\varphi - y_3\sin\varphi \\ (x+x_3)\sin\varphi + y_3\cos\varphi \\ z+z_3 \\ 1 \end{bmatrix} \tag{4.40}$$

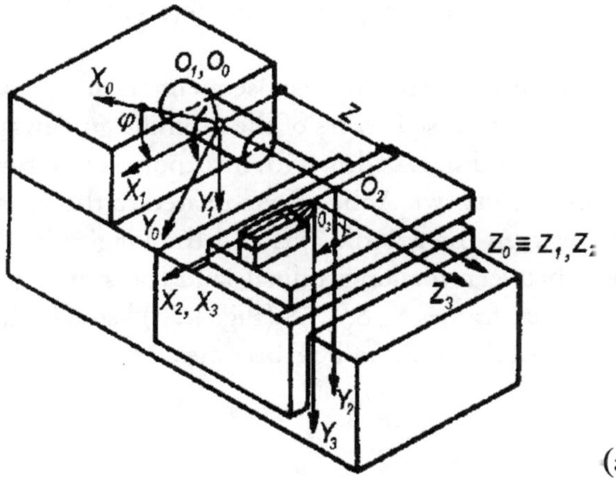

(a)

$$S_0 \xleftarrow{\;^0\mathbf{A}_1 = \mathbf{A}^6(\varphi)\;} S_1 \xleftarrow{\;^1\mathbf{A}_2 = \mathbf{A}^3(z)\;} S_2 \xleftarrow{\;^2\mathbf{A}_3 = \mathbf{A}^1(x)\;} S_3$$

(b)

Figure 4-6 Form-shaping system of the lathe: (a) the kinematic chain (b) the chain graph.

Let, for example, a machining operation is carried out by a point-type cutter, and working point O_3(Fig. 4.6a) – the apex of the *CT* – has coordinates $x_3 = y_3 = z_3 = 0$. Hence, the position FSF for the point-type turning is (see also Appendix *A2.2.2*)

$$\mathbf{r}_0 = \mathbf{r}_0(x, \varphi, z) = \begin{bmatrix} x \cos \varphi \\ x \sin \varphi \\ z \\ 1 \end{bmatrix} \tag{4.41}$$

Equations (4.38-41) define the relative motion of point O_3 with respect to frame S_0 as a function of three unknowns x, φ, and z, where x and φ present the polar radius and polar angle, respectively, and z is the longitudinal coordinate defining a cylindrical coordinate frame.

Table 4-6 Form-shaping system of the lathe, kinematic code k = **631.**

S_i	**Link**	**Motion relative to previous link**	**Matrix $^{i-1}\mathbf{A}_i$**
S_0	Main spindle	Conventionally fixed link	–
S_1	Bed	Rotation around the Z_0-axis at angle φ	$^0\mathbf{A}_1 = \mathbf{A}^6(\varphi)$
S_2	Longitudinal carriage	Translation along the Z_1-axes at distance z	$^1\mathbf{A}_2 = \mathbf{A}^3(z)$
S_3	Cross-carriage	Translation along the X_2-axes at distance x	$^2\mathbf{A}_3 = \mathbf{A}^1(x)$

Parameters: x and z are the positions of the longitudinal carriage and cross-carriage, respectively; φ is the angle of the spindle rotation.

Obviously, Eq. (4.41) describes the spatial geometrical objects in the cylindrical coordinate frame.

The structural code k_{st} is more suitable for a subsequent accuracy analysis since it automatically includes the Abbe offset caused errors of the machine guideways. The code k_{st} consists of all the symbols of the coordinate code k with incorporation of two geometric links y_2 and y_4 (Fig. 4.7). These links define two Abbe offsets related to the considered architecture of the lathe: distance y_2 between the spindle axis and the longitudinal-guideways-associated plane, and distance y_4 between the longitudinal- and transverse-guideways-associated planes. For the set of coordinate frames $S_0, S_1,\dots,S_5$(Fig. 4.7b) listed in Table 4.7, the code k_{st} is given through Eq. (3.93), and the position FSF becomes

$$k_{st} = \mathbf{62321} \tag{4.42}$$

$$\mathbf{r}_0 = {}^0\mathbf{A}_5\mathbf{r}_5 \tag{4.43}$$

$$\mathbf{r}_0 = [x_0, y_0, z_0, 1]^{\mathrm{T}};\ \mathbf{r}_5 = [x_5, y_5, z_5, 1]^{\mathrm{T}};$$

$$^0\mathbf{A}_5 = \mathbf{A}^6(\varphi)\mathbf{A}^2(y_2)\mathbf{A}^3(z)\mathbf{A}^2(-y_4)\mathbf{A}^1(x)$$

$$= \begin{bmatrix} \cos\varphi & -\sin\varphi & 0 & x\cos\varphi-(y_2-y_4)\sin\varphi \\ \sin\varphi & \cos\varphi & 0 & x\sin\varphi+(y_2-y_4)\cos\varphi \\ 0 & 0 & 1 & z \\ 0 & 0 & 0 & 1 \end{bmatrix} \tag{4.44}$$

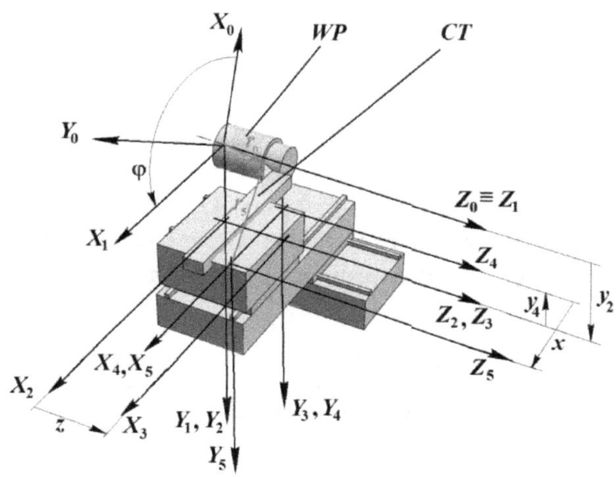

Figure 4-7 The quasi-kinematic chain of the lathe.

Table 4-7 Form-shaping system of the lathe, coordinate code $k_{st} = 62321$.

S_i	Link	Motion relative to previous link	Matrix $^{i-1}\mathbf{A}_i$
S_0	Main spindle	Conventionally fixed link	–
S_1	Bed	Rotation around the Z_0-axis at angle φ	$^0\mathbf{A}_1 = \mathbf{A}^6(\varphi)$
S_2	Longitudinal guide	Translation along the Y_1-axes at distance y_2	$^1\mathbf{A}_2 = \mathbf{A}^2(y_2)$
S_3	Longitudinal carriage	Translation along the Z_2-axes at distance z	$^2\mathbf{A}_3 = \mathbf{A}^3(z)$
S_4	Cross-guideways	Translation along the Y_3-axes at distance y_4	$^3\mathbf{A}_4 = \mathbf{A}^2(y_4)$
S_5	Cross-carriage	Translation along the X_4-axes at distance x	$^4\mathbf{A}_5 = \mathbf{A}^1(x)$

Substituting vector $\mathbf{r}_5$ and $^0\mathbf{A}_5$ into Eq. (4.42) yields the position FSF of the lathe corresponding to the code, Eq. (4.42),

$$\mathbf{r}_0 = {}^0\mathbf{A}_5\,\mathbf{r}_5 = \begin{bmatrix} (x+x_5)\cos\varphi-(y_2-y_4+y_5)\sin\varphi \\ (x+x_5)\sin\varphi+(y_2-y_4+y_5)\cos\varphi \\ z \\ 1 \end{bmatrix} \tag{4.45}$$

Since Eqs. (4.40) and (4.45) describe identical setup, there are two necessary constraints between parameters of two versions:

$$x_3 = x_5 \text{ and } y_3 = y_2 - y_4 + y_5 \tag{4.46}$$

where x_3 and y_3 are the coordinates used in Eq. (4.40), and x_5, y_2, y_4, and y_5 are the coordinates used in Eq. (4.45).

4.3.1.2 *FSF of the turning center*

Some typical layouts of high-precision turning centers (Hardinge Inc., USA) are shown in Figure 4.8 [2]. All structures have a slant bed with different tilt angles and spatial arrangement of the X- and Y-axis relative to the main spindle-associated Z-axis. Six possible layouts are shown in Figure 4.9. Compared with the regular lathe the turning center additionally includes a milling-boring axis (the Y-axis), which is orthogonal to the Z-axis, but can make different angles with the cross-carriage axis (the X-axis).

As an example, the form-shaping system of the turning center, layout (*a*) in Figure 4.9 is considered (Fig. 4.10). The turning center works in two modes: a turning mode and a milling-boring mode (Table 4.8). When the machine works in *the turning mode*, its kinematics includes only three motions: cutting motion of the main spindle with the workpiece and two feed motions – translations along the Z- and X-axis. In the milling-boring mode, the FSS combines all links participating in the turning mode and additionally includes the feed motion along the Y-axis and the main (cutting) motion – rotation around the Y-axis. Obviously the main cutting motion in the turning mode (rotation around the Z-axis) becomes either feed motion or fixed angular position.

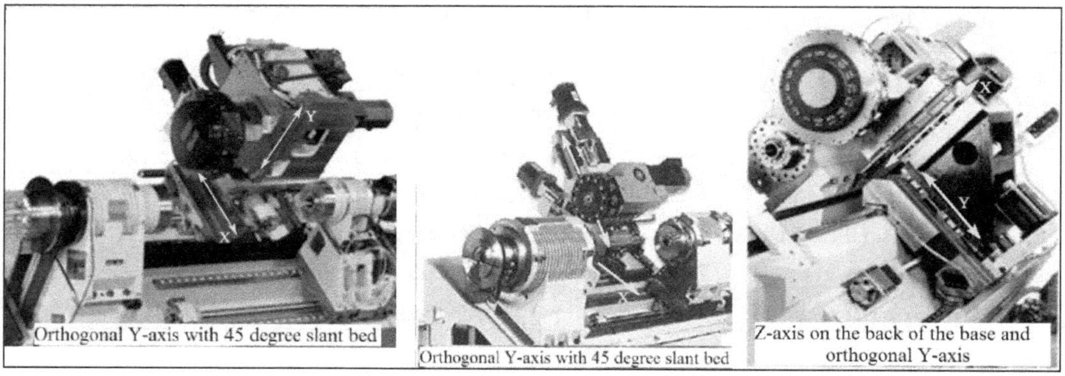

(a) (b) (c)

Figure 4-8 Typical layouts of turning centres: by Kushnir, et al. (a) the orthogonal *Y*-axis with 45 degree slant bed; (b) the 30 degree wedge on 30 degree slant base, the *X*-axis is under the wedge; (c) the *Z*-axis on the back of the base and the orthogonal *Y*-axis, the *X*-axis is on top.

The manipulation matrix for turning mode presents a product ${}^0\mathbf{A}_7$ of the first seven matrices in Table 4.8 in the given order:

$$
{}^0\mathbf{A}_7 = \prod_{i=1}^{7} {}^{i-1}\mathbf{A}_i = \mathbf{A}^6(-\varphi)\mathbf{A}^2(H)\mathbf{A}^3(z)\mathbf{A}^2(y^3)\mathbf{A}^1(x^5)\mathbf{A}^2(y^6)
$$

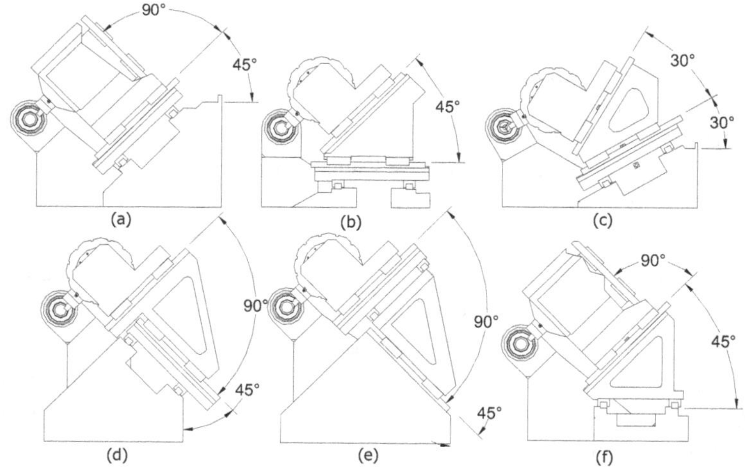

Figure 4-9 (a) Orthogonal Y-axis with 45 degree slant bed; (b) 45 degree wedge on flat base, the *X*-axis is on top of the wedge; (c) 30 degree wedge on 30 degree slant base, the *X*-axis is under the wedge; (d) *Z*-axis on the back of the base with and orthogonal *Y*-axis, X-axis is on top; (e) "*Z* on top of Y"; (f) Flat base with 45 degree carriage and orthogonal *Y*-axis

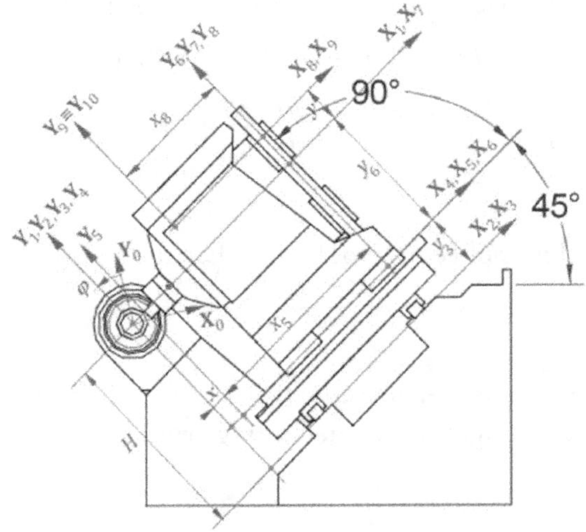

Figure 4-10 Machine links with associated coordinate systems. See layout in Figure 4.9a.

Table 4-8 Form-shaping system of the turning center (Fig. 4.10).

S_i	Link name	Motion relative to the previous link: Type, axis	Parameter	Mode: Turning	Mode: Milling	Matrix ${}^{i-1}\mathbf{A}_i = \mathbf{A}^j(q_i)$
S_0	MTS	Conventionally fixed	–	–	–	–
S_1	Bed	**Rotation, Z_0-axis**	Angle φ	Cutting	φ-feed	${}^0\mathbf{A}_1 = \mathbf{A}^6(-\varphi)$
S_2	Bed	Transfer, Y_1-axis	Distance H	Fixed	Fixed	${}^1\mathbf{A}_2 = \mathbf{A}^2(H)$
S_3	LC	**Translation, Z_2-axis**	Distance z	z-feed	z-feed	${}^2\mathbf{A}_3 = \mathbf{A}^3(z)$
S_4	LC	Transfer, Y_3-axis	Distance y_3	Fixed	Fixed	${}^3\mathbf{A}_4 = \mathbf{A}^2(y_3)$
S_5	CC	**Translation, X_4-axis**	Distance x	x-feed	x-feed	${}^4\mathbf{A}_5 = \mathbf{A}^1(x)$
S_6	CC	Transfer, X_5-axis	Distance x_5	Fixed	Fixed	${}^5\mathbf{A}_6 = \mathbf{A}^1(x_5)$
S_7	T	Transfer, Y_6-axis	Distance y_6 (1)	Fixed	Fixed	${}^6\mathbf{A}_7 = \mathbf{A}^2(y_6)$
S_8	T	**Translation, Y_7-axis**	Distance y	–	y-feed	${}^7\mathbf{A}_8 = \mathbf{A}^2(y)$
S_9	TS	Transfer, X_8-axis	Distance x_8	–	Fixed	${}^8\mathbf{A}_9 = \mathbf{A}^1(x_8)$
S_{10}	MMS	**Rotation, Y_9-axis**	Angle ψ	–	Cutting	${}^9\mathbf{A}_{10} = \mathbf{A}^5(\psi)$

Parameters: z, x, and y are the current displacements of the longitudinal carriage, cross-carriage, and turret, respectively; φ and θ are the angles of the spindle rotation; H is the center height; y_3 is the distance between longitudinal and cross guideways; x_5 is the distance between cross and Y-axis guideways; y_6 is the distance from cross guideways to the center line; x_8 is the distance between the turret guideways and the axis of the milling-boring spindle. MTS – main turning spindle; MMS – main milling spindle; LC – longitudinal carriage; CC – cross-carriage; T – turret; TS – turret station.

(1) y_6 is chosen according to condition $H + y_3 + y_6 = 0$.

$$= \begin{bmatrix} \cos\varphi & -\sin\varphi & 0 & (x_5+x)\cos\varphi-(H+y_3+y_6)\sin\varphi \\ \sin\varphi & \cos\varphi & 0 & (H+y_3+y_6)\cos\varphi+(x_5+x)\sin\varphi \\ 0 & 0 & 1 & z \\ 0 & 0 & 0 & 1 \end{bmatrix} \tag{4.47}$$

When a single point tool is applied, the tool vector has the form

$$\mathbf{r}_7 = [L, 0, 0, 1]^{\mathrm{T}} \tag{4.48}$$

where L is the distance from the Y-axis guideways to tip of the tool edge; taking into account the equation for the single-point machining from Table 4.8 ($H + y_3 + y_6 = 0$), we obtain:

$$\mathbf{r}_0 = {}^0\mathbf{A}_7\mathbf{r}_7 = \begin{bmatrix} \cos\varphi & -\sin\varphi & 0 & (x_5+x)\cos\varphi \\ \sin\varphi & \cos\varphi & 0 & (x_5+x)\sin\varphi \\ 0 & 0 & 1 & z \\ 0 & 0 & 0 & 1 \end{bmatrix} \begin{bmatrix} L \\ 0 \\ 0 \\ 1 \end{bmatrix} = \begin{bmatrix} (L+x_5+x)\cos\varphi \\ (L+x_5+x)\sin\varphi \\ z \\ 1 \end{bmatrix} \tag{4.49}$$

When the machine works in *the milling-boring mode*, its kinematics includes all kinematic motions — cutting motion of the milling spindle with a milling cutter installed in a turret station, and four feed motions — translations along the Z-, X-, and Y-axis and rotation of the main spindle (note that the main spindle motion becomes a rotational feed motion). The manipulation matrix presents a product of matrices listed in Table 4.8 in the given order:

$${}^0\mathbf{A}_{10} = \prod_{i=1}^{10} {}^{i-1}\mathbf{A}_i \tag{4.50}$$

For a cylindrical milling cutter of radius R oriented parallel to the Y-axis, the FSF becomes

$$[\mathbf{r}_0\,\mathbf{e}_0] = {}^0\mathbf{A}_{10}[\mathbf{r}_{10}\,\mathbf{e}_{10}] = \begin{bmatrix} (x+x_5+x_8+R\cos\psi)\cos\varphi - y\sin\varphi & -\sin\varphi \\ y\cos\varphi+(x+x_5+x_8+R\cos\psi)\sin\varphi & \cos\varphi \\ z-R\sin\psi & 0 \\ 1 & 0 \end{bmatrix} \tag{4.51}$$

4.3.2 Multi-axis machines

4.3.2.1 *XYZ triplet in 3-axis machines*

The Cartesian linkage, i.e., the 3-DOF set of three actuated orthogonal translational links (**PPP**), subsequently referred to as "*XYZ* translational triplet",is widely used as a main part of kinematic systems of different machines. Among them are:

- Cartesian robots and coordinate measuring machines (CMM),where the *XYZ* translational triplet presents the basic part of the machine providing its target technological function;
- Machine tools such as milling machines and contour-cutting machines, where the *XYZ* triplet is accompanied with the regular cutting motion – rotation of the main spindle;
- Engraving machines, where the cutting process is actuated without rotation of cutting tool;
- Machines for non-traditional machining methods such as electric discharge machines (EDM) for wire cutting, laser cutting and water cutting machines, etc.
- 3D printers.

Possible structures of the Cartesian linkages are shown in Figure 4.11, where in the arrangement codes, as before, symbol 0 corresponds to a fixed frame (base, bed, column, etc.), and 1, 2, 3 are symbolized translations along the *X*-, *Y*-, and *Z*-axis, respectively.

Additionally, the Cartesian linkage makes an integral part of the kinematic system of the majority of 4-, 5- and 6-axis machines, where apart from the 3-DOF linkage there are one, two, or three rotations for controlled inclination of the *AE* and *TE* relative to each other.

The FSS of the *XYZ* Cartesian robot (Fig. 4.12) includes an end-effector (*EE*) mounted rigidly to the *Z*-translation link performing a pre-designed trajectory with respect to a target element (*TE*) attached to the base.The FSS of the laser cutting machine for cutting sheets (Fig. 4.13) has a laser cutter (*LC*) fitted to the *XYZ*-translation triplet and the bed-mounted workpiece (*WP*). Similar kinematics is used on machines using other physical cutting means, for example, in ultrasonic and water jet cutting machines.

In the CMM (Fig.4.14), measuring probe (*PR*) attached to the vertically moving link performs a pre-designed trajectory with respect to the measured *WP* attached to the fixed table. Both gantry- and bridge-type CMMs have identical arrangement codes K = **0123**. Taking into account specific requirements for the CMM accuracy, their structures are significantly

(a) ***TE* ← 0123 ← *EE***
(b) ***WP* ← 0123 ← *LC***
(c) ***WP* ← 0123 ← *PR***
(d) ***WP* ← 1023 ← *PR***
(e) ***WP* ← 1230**(feed motions)← **6**(cutting motion) ← ***CT***
(f) ***WP* ← 1023** (feed motions)← **6**(cutting motion) ← ***CT***
(g) ***WP* ← 1203** (feed motions)← **6**(cutting motion) ← ***CT***

Figure 4.11. FSS graphs of the 3-axis machines: (a) Cartesian robot: *PE* is the passive element, *EE* is the end-effector; (b) laser cutting machine: *LC* is the laser cutter, *WP* is the workpiece; (c) and (d) *FSS* of the CMMs: *WP* is the workpiece, *PR* is the probe; (e), (f) and (g) are FSSs of machine tools: *CT* is the cutting tool.

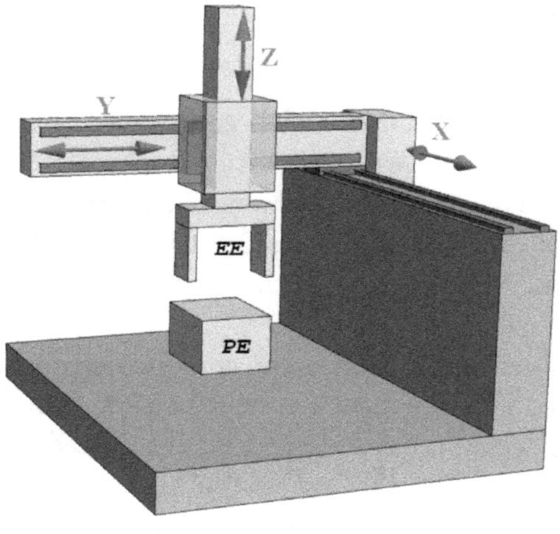

Figure 4-12 *XYZ* Cartesian robot, k = **0123**.

stiffer than those of the Cartesian robot and laser cutting machine. The CMM with a moving table (Fig. 4.15), arrangement code K = **1023**, is also widely used.

The manipulation matrix of the *XYZ* translational triplet is obtained through multiplication of three matrices of translational motions,

$$
{}^{0}\mathbf{A}_3 = \prod_{i=1}^{3} {}^{i-1}\mathbf{A}_i(q_i) = \mathbf{A}^1(x)\ \mathbf{A}^2(y)\mathbf{A}^3(z)
$$

$$
= \begin{bmatrix} 1 & 0 & 0 & x \\ 0 & 1 & 0 & 0 \\ 0 & 0 & 1 & 0 \\ 0 & 0 & 0 & 1 \end{bmatrix} \begin{bmatrix} 1 & 0 & 0 & 0 \\ 0 & 1 & 0 & y \\ 0 & 0 & 1 & 0 \\ 0 & 0 & 0 & 1 \end{bmatrix} \begin{bmatrix} 1 & 0 & 0 & 0 \\ 0 & 1 & 0 & 0 \\ 0 & 0 & 1 & z \\ 0 & 0 & 0 & 1 \end{bmatrix} = \begin{bmatrix} 1 & 0 & 0 & x \\ 0 & 1 & 0 & y \\ 0 & 0 & 1 & z \\ 0 & 0 & 0 & 1 \end{bmatrix} \tag{4.52}
$$

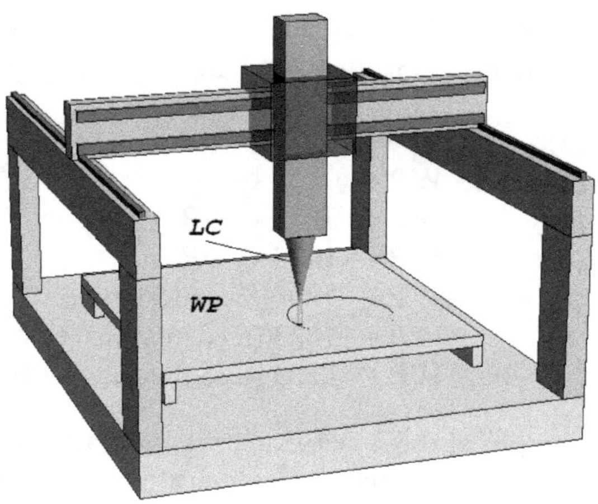

Figure 4-13 Laser cutting machine, K = 0123.

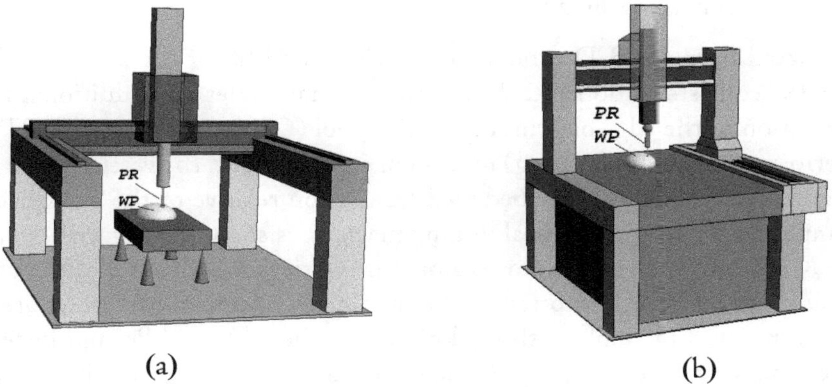

Figure 4-14 Coordinate measuring machines (CMM), K = 0123: (a) gantry-type, (b) bridge-type with moving column.

The FSF of all machines shown in Figures 4.12-15 are

$$\mathbf{r}_0 = {}^0\mathbf{A}_3\mathbf{r}_3 = \begin{bmatrix} x + x_3 \\ y + y_3 \\ z + z_3 \\ 1 \end{bmatrix}, \tag{4.53}$$

where x_3, y_3, and z_3 are the coordinate of the position vector $\mathbf{r}_3$ associated with the working point of the AE (tool or measuring probe).

For convenience, matrix ${}^0\mathbf{A}_3$, Eq. (4.52), describing the XYZ translations will also designated

$${}^0\mathbf{A}_3 \equiv \mathbf{A}_{tr} \equiv \mathbf{A}(x,y,z) \tag{4.54}$$

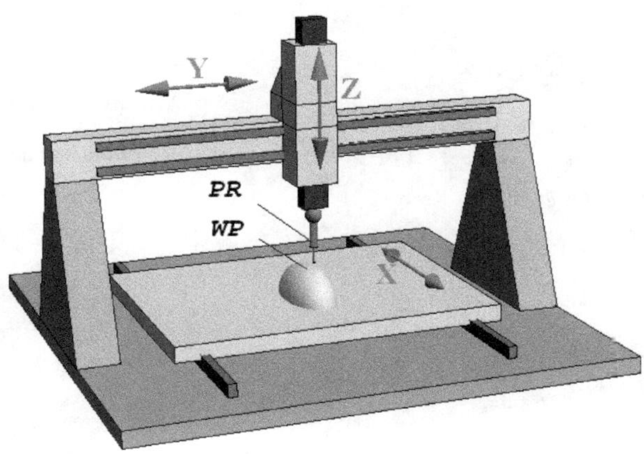

Figure 4-15 CMM with a movable table, K = **1023.**

4.3.2.2 *FSF of the 3-axis machine tools*

As differentiated from the 3-D robots and CMMs, the FSS of the *3-axis machine tools* (Figs. 4.16-19) consists of both the *XYZ* translational triplet and additional kinematic motion –rotation of the edge or grinding cutting tool (*CT*). As a result, the *CT* simultaneously performed the main (cutting) motion in common with the vertically- or horizontally-oriented spindle and a prescribed feed translation relative to the workpiece (WP). As an example, the knee-type vertical milling machine is shown in Figure 4.16 (arrangement code $K = 12306$). The knee moves along the vertical guideways of the column-type bed that has guideways for *Y*-motion of the saddle whose upper guideways are used for *X*-motion of the machine table with workpiece, i.e., the table, saddle, and knee perform three orthogonal feeds providing a pre-designed trajectory of the *CT* relative to the *WP* fitted to the machine table.

Other versions of the *XYZ* translational systems are realized in knee-less milling machines. An example is the 3-axis machining center with the crossed motion (*XY*) table (Fig. 4.17, code $K =$ **12036**). According to the code, the fixed link (the bed in common with a column) is located between two successive translational units that results in higher stiffness compared with the knee-type machine. This architecture is typical for high-accuracy machines, such as jig-boring machine tools, CNC milling and machining centers. Another version is the vertical portal-type CNC machining center (Fig. 4.18), code $K = 10236$, where the workpiece has a single motion relative to the bed compared to the knee-type and the milling center with three and two motions relative to the bed, respectively.

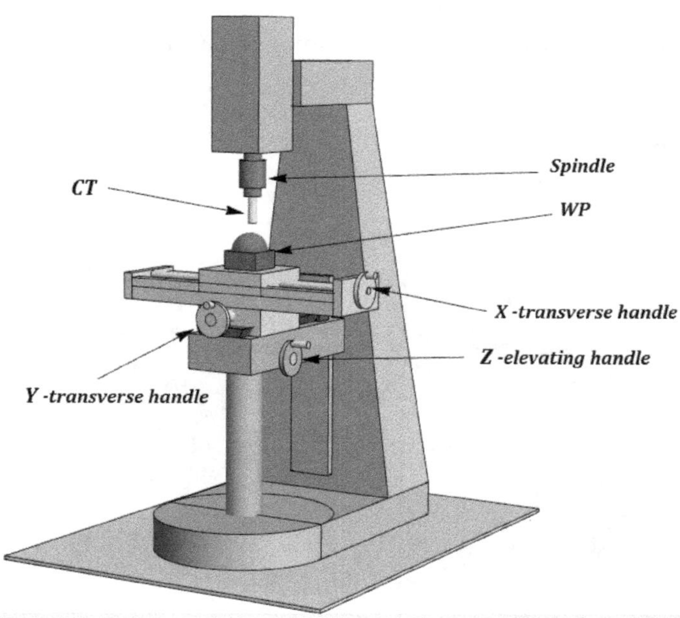

Figure 4-16 Vertical knee-type milling machine, $k =$ **12306**

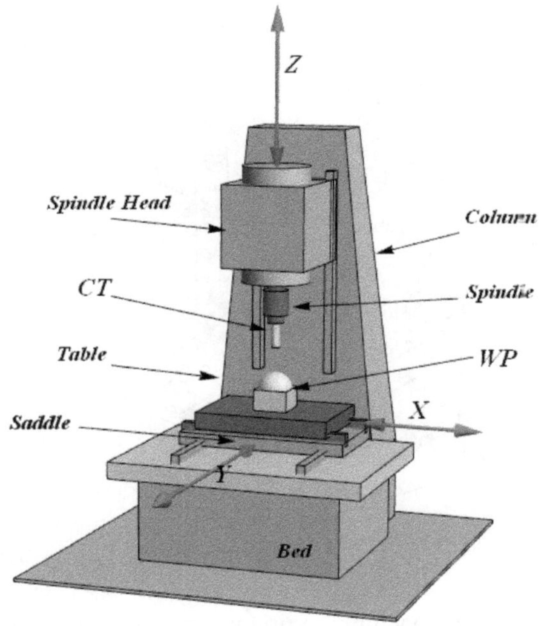

Figure 4-17 The 3-axis machining center with the XY-table, k = **12036.**

The manipulation matrix of the milling machines (Fig. 4.16-18), is obtained by multiplication of matrix $\mathbf{A}_{tr}$, Eq. (4.54), and a matrix of the main motion – cutting tool rotation. Apart from layouts with the vertical cutting tool spindle (coordinate code **1236**), layouts with horizontal axis (code **1235**) such as the horizontal boring and milling machines are widely applied.

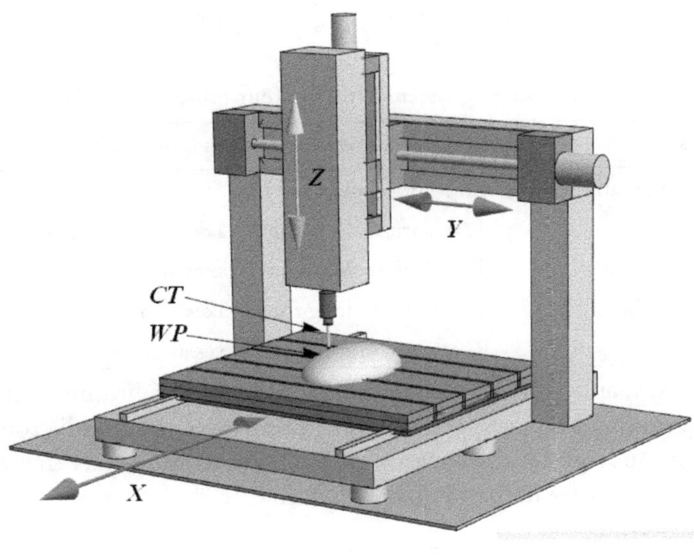

Figure 4-18 The 3-axis portal-type machine, k = **10236.**

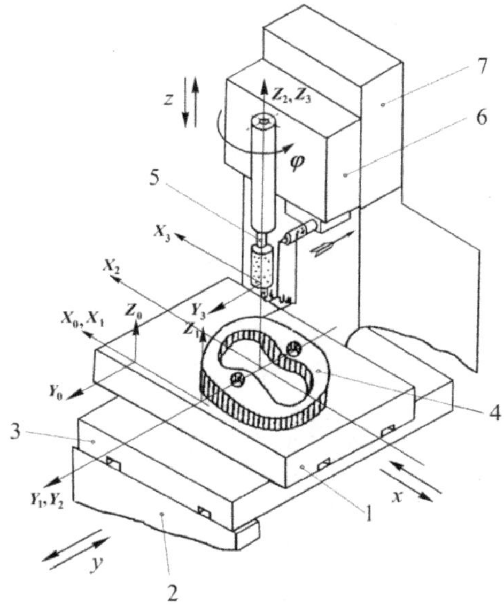

Figure 4-19 Contour-grinding machine: setup for machining of an epitrochoidal cylinder.

As an example, the FSS of the contour grinding machine (the NC semi-automatic profile grinding machine MA396F3 [3]) with a cylindrical wheel attached to the vertical spindle is considered (Fig. 4.19). The machine has a vertical layout with horizontal cross-type table and slides. The machine is designed for grinding open and enclosed internal and external faces of complex shape (for example, the epitrochoidal contour) with rectilinear generating-lines, or curvilinear faces (for example hydraulic pump stators, cams, templets, dies, etc.). The machine can also grind faces with a non-rectilinear generating-line (with the wheel dressed with a form tool) and perform jig-grinding jobs.

Table 4-9 Form-shaping system of the contour-grinding machine, K = **12306**

S_i	**Link**	**Motion relative to previous link**	**Matrix $^{i-1}\mathbf{A}_i$**
S_0	Machine table	Conventionally fixed link	–
S_1	Saddle	Translation along the X_0-axes at distance $-x$	${}^0\mathbf{A}_1 = \mathbf{A}^1(-x)$
S_2	Bed	Translation along the Y_1-axes at distance $-y$	${}^1\mathbf{A}_2 = \mathbf{A}^2(-y)$
S_3	Spindle head	Translation along the Z_2-axes at distance z	${}^2\mathbf{A}_3 = \mathbf{A}^3(z)$
S_4	Spindle	Rotation around the Z_3-axes at angle φ	${}^3\mathbf{A}_4 = \mathbf{A}^6(\varphi)$

Parameters: x, y, and z are the current positions of the longitudinal carriage, cross-carriage, and grinding head, respectively; φ is the angle of the spindle rotation.
Note: The minus sign for distances x and y follows from the fact that relative motion of the saddle and bed are in the opposite directions to the actual rotation of the table and saddle, respectively.

The initial link is associated with home position (coordinate frame S_0). The machine tool composes four movable links (n = 4): table 1 with mounted workpiece 4 is associated with coordinate frame S_1, the cross-carriage 3 and the bed(the stationary link in its absolute motion) are associated with coordinate frames S_2 and S_3, respectively. The grinding head 6 carries out the vertical reciprocating motion along the Z_3-axis, and the grinding spindle 5 with grinding wheel rotates around the vertical Z_4-axis. The coordinate and arrangement codes of the FSS of the contour-grinding machine are as follows:

k = (123)**1323326** and K = **12306**

where symbol k_3 = 3 in coordinate code k describes the geometric link associated with distance between the upper basic plate of the machine table and the guideways of the X-motion; symbol k_5 = 3 describes the geometric link associated with the shortest distance between the guideways of X- and Y-motions; and symbol k_7 = 2 describes the geometric link associated with the shortest distance between the guideways of Y- and Z-motions.

The FSS links and their motions are listed in Table 4.9, and the FSF is as follows:

$$\mathbf{r}_0 = {}^0\mathbf{A}_4\mathbf{r}_4 \tag{4.55}$$

with

$$\mathbf{r}_0 = [x_0, y_0, z_0, 1]^{\mathrm{T}};\ \mathbf{r}_4 = [x_4, y_4, z_4, 1]^{\mathrm{T}}; \tag{4.56}$$

$${}^0\mathbf{A}_4 = \mathbf{A}^1(-x)\,\mathbf{A}^2(-y)\,\mathbf{A}^3(z)\,\mathbf{A}^6(\varphi) = \mathbf{A}(-x,\ -y, z)\,\mathrm{A}^6(\varphi)$$

$$= \begin{bmatrix} 1 & 0 & 0 & -x \\ 0 & 1 & 0 & -y \\ 0 & 0 & 1 & z \\ 0 & 0 & 0 & 1 \end{bmatrix} \begin{bmatrix} \cos\varphi & -\sin\varphi & 0 & 0 \\ \sin\varphi & \cos\varphi & 0 & 0 \\ 0 & 0 & 1 & 0 \\ 0 & 0 & 0 & 1 \end{bmatrix} = \begin{bmatrix} \cos\varphi & -\sin\varphi & 0 & -x \\ \sin\varphi & \cos\varphi & 0 & -y \\ 0 & 0 & 1 & z \\ 0 & 0 & 0 & 1 \end{bmatrix} \tag{4.57}$$

Substituting matrix ${}^0\mathbf{A}_4$, Eq. (4.53) and vector $\mathbf{r}_4$ in Eq. (4.52) yields the position FSF:

$$\mathbf{r}_0 = \begin{bmatrix} \cos\varphi & -\sin\varphi & 0 & -x \\ \sin\varphi & \cos\varphi & 0 & -y \\ 0 & 0 & 1 & z \\ 0 & 0 & 0 & 1 \end{bmatrix} \begin{bmatrix} x_4 \\ y_4 \\ z_4 \\ 1 \end{bmatrix} = \begin{bmatrix} -x + x_4\cos\varphi - y_4\sin\varphi \\ -y + x_4\sin\varphi + y_4\cos\varphi \\ z + z_4 \\ 1 \end{bmatrix} \tag{4.58}$$

In this machine, the grinding wheel is of cylindrical form:

$$\mathbf{r}_5 = [R\cos\varphi_t; R\sin\varphi_t, z_t, 1]^{\mathrm{T}}, \text{ with } 0 \le z_t \le h \text{ and } 0 \le \varphi_t \le 2\pi \tag{4.59}$$

where R and h are the radius and height of the grinding wheel, respectively (Fig. 3.2.15a); and φ_t and z_t are polar angle and longitudinal coordinates on the cylindrical surface of the grinding wheel, respectively.

Hence, the FSF becomes

$$
\mathbf{r}_0 = \begin{bmatrix} \cos\varphi & -\sin\varphi & 0 & -x \\ \sin\varphi & \cos\varphi & 0 & -y \\ 0 & 0 & 1 & z \\ 0 & 0 & 0 & 1 \end{bmatrix} \begin{bmatrix} R\cos\varphi_t \\ R\sin\varphi_t \\ z_t \\ 1 \end{bmatrix} = \begin{bmatrix} -x + R\cos(\varphi+\varphi_t) \\ -y + R\sin(\varphi+\varphi_t) \\ z + z_t \\ 1 \end{bmatrix} \tag{4.60}
$$

Correspondingly, the FSF of the machines with the horizontally oriented main sipndle axis (e.g., Figure 3.17, k = **1235**) is as follows:

$$
\mathbf{r}_0 = \begin{bmatrix} \cos\varphi & 0 & \sin\varphi & -x \\ 0 & 1 & 0 & -y \\ -\sin\varphi & 0 & \cos\varphi & z \\ 0 & 0 & 0 & 1 \end{bmatrix} \begin{bmatrix} -R\cos\varphi_t \\ y_t \\ R\sin\varphi_t \\ 1 \end{bmatrix} = \begin{bmatrix} -x - R\cos(\varphi+\varphi_t) \\ -y + y_t \\ z + R\sin(\varphi+\varphi_t) \\ 1 \end{bmatrix} \tag{4.61}
$$

4.3.2.3 FSF of the 5-axis machine tools

Similarly to the three-axis machines, five-axis machines consist of the *XYZ* translational triplet and additionally combine two rotary axes, which allow currently specified orientations of the cutting tool relative to the workpiece. In Figure 4.20, orientations of the spherical milling tool axis during 3- and 5-axis machining are compared. During machining on the 3-axis machine, the spherical tool is applied and orientation of its axis relative the machine table is unchanged (Fig. 4.20a).With the constant frequency of spindle rotation,as the tool moves away from the semi-sphere base, the cutting velocity decreases since it is proportional to the radius of the tool working point circle.

A controlled tilt of the cutting tool and/or machine table allows the most effective contact conditions of the tool with the machined surface. In Figure 4.20b and c, two possible versions of the 5-axis machining are schematized: (b) with one-tilting spindle axis and rotary machine table, and (c) with double-pivot cutting tool. In both cases, the tool edges are located on a cylinder of constant diameter (e.g., end milling cutter) and cutting velocity depends only on the frequency of spindle rotation.

In modern manufacturing, the 5-axis machines become an effective tool for increasing the productivity thanks to, for example, significant reduction of cycle and door-to-door time since some machining operations can be performed on one machine in one setup, and thus, resetting operations are eliminated. Owing to this:

- there is no need to design and machine additional locating elements on the workpiece;
- the most convenient orientation of the cutting tool can be chosen that allows permitting higher removable rates than 3-axis machining;
- the accuracy of machined parts is potentially increased.

These are the reasons that the 5-axis machines are extensively used in die-mold machining, in aerospace and shipbuilding industries, and similar cases for processing parts of complicated

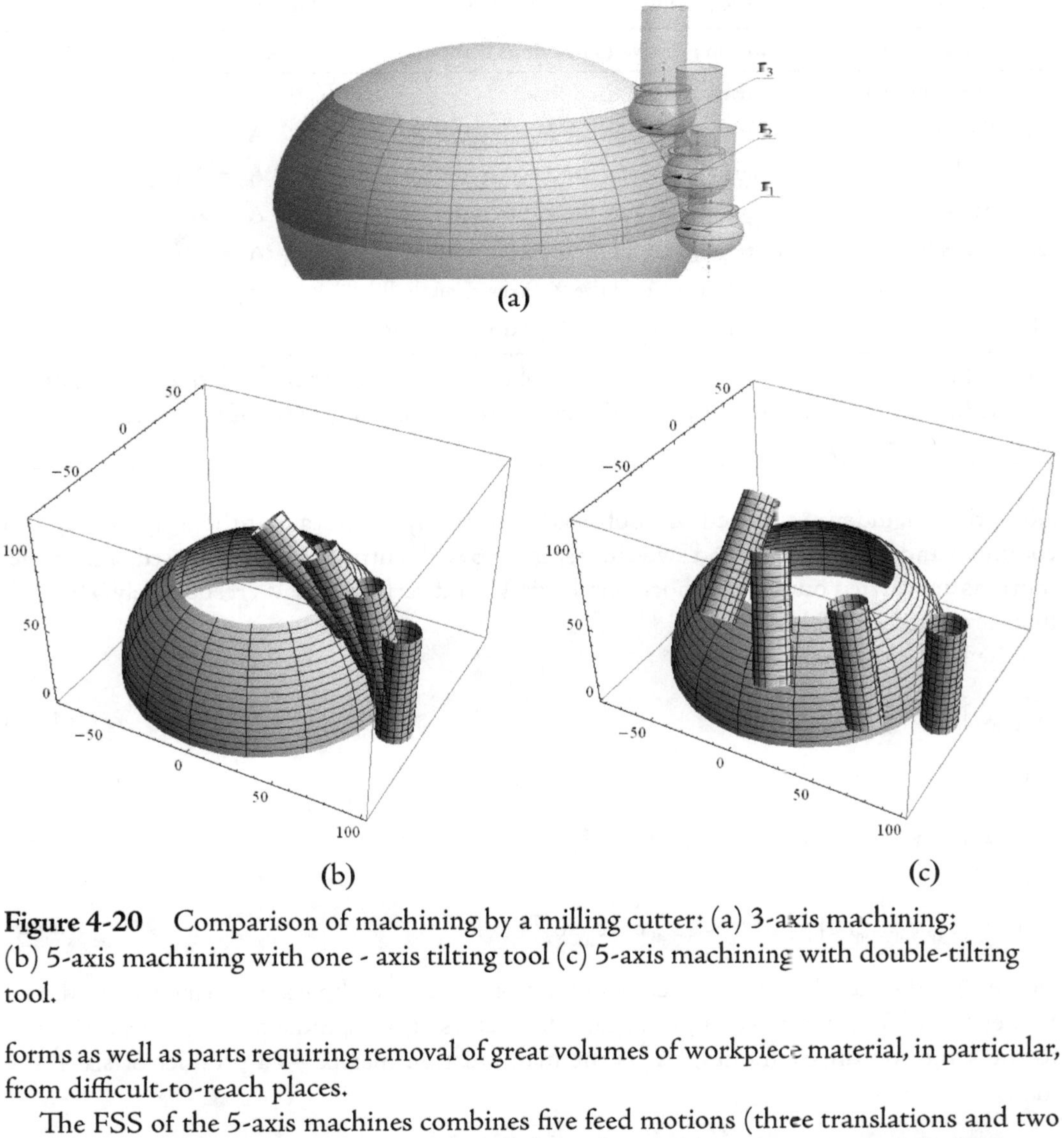

Figure 4-20 Comparison of machining by a milling cutter: (a) 3-axis machining; (b) 5-axis machining with one - axis tilting tool (c) 5-axis machining with double-tilting tool.

forms as well as parts requiring removal of great volumes of workpiece material, in particular, from difficult-to-reach places.

The FSS of the 5-axis machines combines five feed motions (three translations and two rotations) and a main motion (cutting tool rotation). Possible architectures of the FSS consist of the *XYZ* triplet describing translations along the *X*-, *Y*-, and *Z*-axis, and two feed rotations varying in place relative to *XYZ* triplet. In all the cases, we assume that main (cutting) motion is rotation around *Z*-axis (code **6**) as is obtained in CNC machining centers. The feed rotations can be performed: (a) both by the tool head (Fig. 4.21a [4]), with arrangement code K = **1203646**; (b) both by the workpiece table (Fig. 4.21b), with arrangement code K = **6412036** called "trunnion type" machine; and (c) one by the tool head and another by the workpiece table (Fig. 4.21c), with arrangement code K = **6120346**.

In Figure 4.22, a layout of the 5-axis single-column vertical machine tool [5], where tilting and swiveling tool motions are schematized in Figure 4.21a, is shown. Machine table 1 with workpiece *W* moves along the *X*-guideways machined on carriage 2, which, in turn, moves

Table 4-10 FSS of the 5-axis machine tilt-swivel tool, K = **1203646.**

S_i	**Link**	**Motion relative to previous link**	**Matrix** $^{i-1}\mathbf{A}_i$
S_0	Machine table	Conventionally fixed link	–
S_1	Saddle	Translation along the X_0-axes at distance $-x$	$^0\mathbf{A}_1 = \mathbf{A}^1(-x)$
S_2	Bed	Translation along the Y_1-axes at distance $-y$	$^1\mathbf{A}_2 = \mathbf{A}^2(-y)$
S_3	Spindle head	Translation along the Z_2-axes at distance z	$^2\mathbf{A}_3 = \mathbf{A}^3(z)$
S_4	Swiveling head	Rotation around the Z_3-axes at angle φ_{sw}	$^3\mathbf{A}_4 = \mathbf{A}^6(\varphi_{sw})$
S_5	Tilting head	Rotation around the X_4-axes at angle θ	$^4\mathbf{A}_5 = \mathbf{A}^4(\theta)$
S_6	Spindle	Rotation around the Z_6-axes at angle φ_s	$^5\mathbf{A}_6 = \mathbf{A}^6(\varphi_s)$

Note: The sign minus for distances x and y follows from the fact that relative motion of the saddle and bed is in the opposite directions to the actual motion of the table and saddle, respectively.

along the Y-guideways of bed 3. Tool head 4 vertically moves along the Z-guideways on column 5 and carries tilting and swiveling spindle 6 with cutting tool CT. The tilt and swivel motions are carried out around horizontal axis V_x and vertical axis V_z, respectively. The FSS links are listed in Table 4.10.

The FSF ofthe machine is

$$[\mathbf{r}_0\mathbf{c}_0] = {}^0\mathbf{A}_6\,[\mathbf{r}_6\mathbf{c}_6] \tag{4.62}$$

with

$$\mathbf{r}_0 = [x_0, y_0, z_0, 1]^{\mathrm{T}} \text{ and } \mathbf{r}_6 = [x_6, y_6, z_6, 1]^{\mathrm{T}} \tag{4.63}$$

$$\mathbf{c}_0 = [c_{x0}, c_{y0}, c_{z0}, 0]^{\mathrm{T}} \text{ and } \mathbf{c}_6 = [c_{x6}, c_{y6}, c_{z6}, 0]^{\mathrm{T}} \tag{4.64}$$

where $\mathbf{r}_0$ and $\mathbf{r}_6$ are the position vectors of a point of the workpiece referring to coordinate frames S_0 and S_6, respectively; $\mathbf{c}_0$ and $\mathbf{c}_6$ are the vectors of the spindle orientation as referring to the same coordinate frames; $^0\mathbf{A}_6$ is the manipulation matrix — a product of six 4 × 4 matrices,

$$^0\mathbf{A}_1 = \mathbf{A}^1(-x) = \begin{pmatrix} 1 & 0 & 0 & -x \\ 0 & 1 & 0 & 0 \\ 0 & 0 & 1 & 0 \\ 0 & 0 & 0 & 1 \end{pmatrix};\ {}^1\mathbf{A}_2 = \mathbf{A}^2(-y) = \begin{pmatrix} 1 & 0 & 0 & 0 \\ 0 & 1 & 0 & -y \\ 0 & 0 & 1 & 0 \\ 0 & 0 & 0 & 1 \end{pmatrix};$$

$$^2\mathbf{A}_3 = \mathbf{A}^3(z) = \begin{pmatrix} 1 & 0 & 0 & 0 \\ 0 & 1 & 0 & 0 \\ 0 & 0 & 1 & z \\ 0 & 0 & 0 & 1 \end{pmatrix};\ {}^3\mathbf{A}_4 = \mathbf{A}^6(\varphi_{sw}) = \begin{bmatrix} \cos\varphi_{sw} & -\sin\varphi_{sw} & 0 & 0 \\ \sin\varphi_{sw} & \cos\varphi_{sw} & 0 & 0 \\ 0 & 0 & 1 & 0 \\ 0 & 0 & 0 & 1 \end{bmatrix};$$

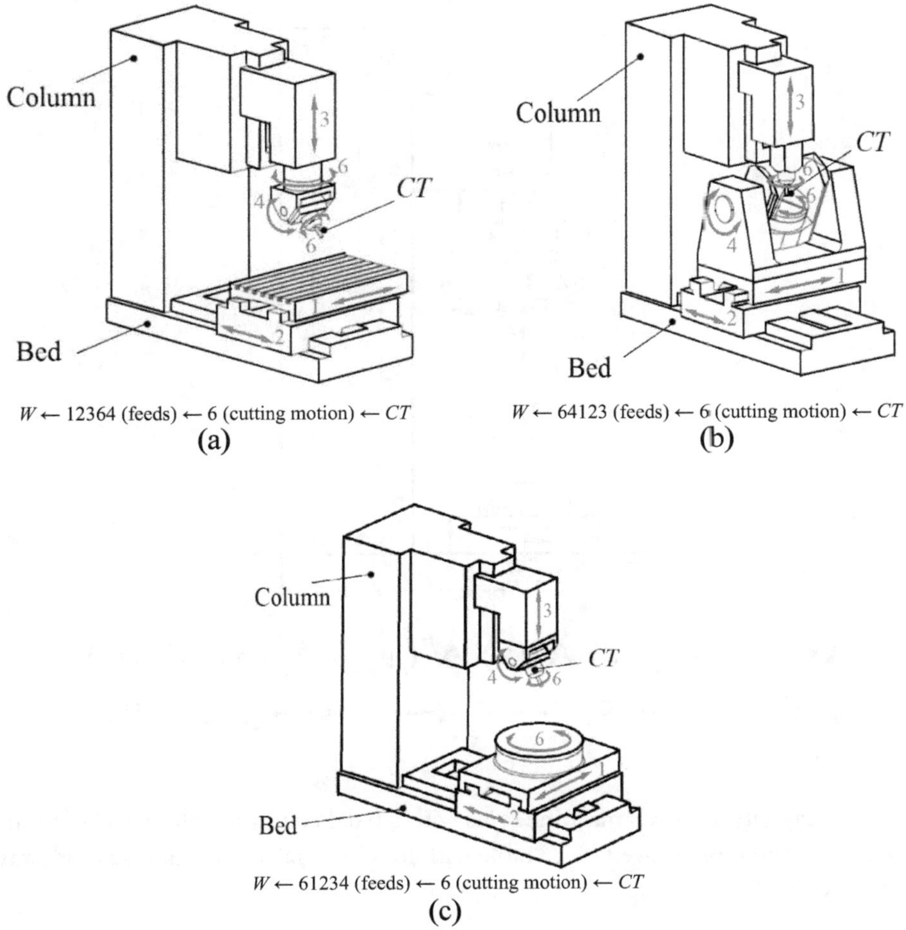

Figure 4-21 FSS versions of the 5-axis machine tools [4]: (*a*) double-tilting CT; (*b*) tilt-swivel workpiece table; (*c*) tilting CT and swivel workpiece table.

$$
{}^{4}\mathbf{A}_{5} = \mathbf{A}^{4}(\theta) = \begin{pmatrix} 1 & 0 & 0 & 0 \\ 0 & \cos\theta & -\sin\theta & 0 \\ 0 & \sin\theta & \cos\theta & 0 \\ 0 & 0 & 0 & 1 \end{pmatrix}; \text{ and } {}^{5}\mathbf{A}_{6} = \mathbf{A}^{6}(\varphi) = \begin{pmatrix} \cos\varphi & -\sin\varphi & 0 & 0 \\ \sin\varphi & \cos\varphi & 0 & 0 \\ 0 & 0 & 1 & 0 \\ 0 & 0 & 0 & 1 \end{pmatrix} \tag{4.65}
$$

Hence, the manipulation matrix is as follows:

$$
{}^{0}\mathbf{A}_{6} = \mathbf{A}^{1}(-x)\,\mathbf{A}^{2}(-y)\,\mathbf{A}^{3}(z)\,\mathbf{A}^{6}(\varphi_{sw})\mathbf{A}^{4}(\theta)\mathbf{A}^{6}(\varphi)
$$

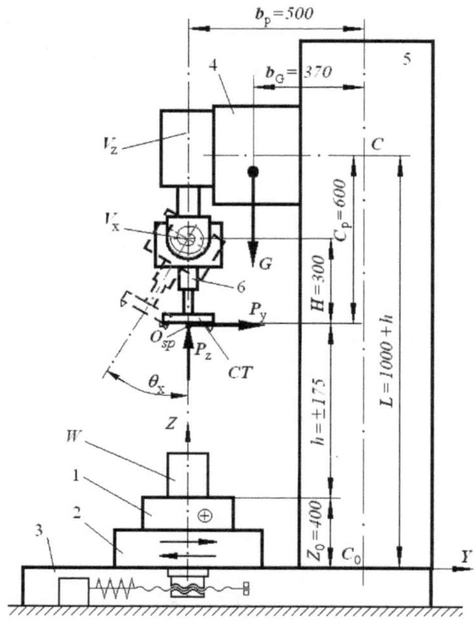

(a)

$$\mathbf{A}^1(-x)\ \mathbf{A}^2(-y)\ \mathbf{A}^3(z)\ \mathbf{A}^6(\varphi_{sw})\ \mathbf{A}^4(\theta)\ \mathbf{A}^6(\varphi_s)$$

$$S_0 \leftarrow S_1 \leftarrow S_2 \leftarrow S_3 \leftarrow S_4 \leftarrow S_5 \leftarrow S_6$$

(b)

Figure 4-22 Layouts of 5-axis machine tools with a fixed column, a tilt-swivel vertically moved tool head, and cross moved (XY) table with longitudinal and transverse guideways: (a) general scheme; (b) FSS graph.

$$= \mathbf{A}_{tr}\left(-x,\ -y, z\right)\mathbf{A}^6(\varphi_s)\mathbf{A}^4(\theta)\mathbf{A}^6(\varphi) = \begin{bmatrix} a_{11} & a_{12} & a_{13} & -x \\ a_{21} & a_{22} & a_{23} & -y \\ a_{31} & a_{32} & a_{33} & z \\ 0 & 0 & 0 & 1 \end{bmatrix} \tag{4.66}$$

where

$a_{11} = \cos\varphi\cos\varphi_{sw} - \cos\theta\sin\varphi\sin\varphi_{sw}$

$a_{12} = -\sin\varphi\cos\varphi_{sw} - \cos\varphi\cos\theta\sin\varphi_{sw}$; $a_{13} = \sin\theta\sin\varphi_{sw}$

$a_{21} = \cos\varphi\sin\varphi_{sw} + \cos\theta\sin\varphi\cos\varphi_{sw}$; $a_{22} = \cos\theta\cos\varphi\cos\varphi_{sw} - \sin\varphi\sin\varphi_{sw}$

$a_{23} = -\sin\theta\cos\varphi_{sw}$; $a_{31} = \sin\theta\sin\varphi$; $a_{32} = \sin\theta\cos\varphi$; and $a_{33} = \cos\theta$

In Figure 4.23, two layouts of the portal-type 5-axis machines with the same pair of the tool motions as that in Figure 4.22a are shown. In turn, they are different from one another by the units performing the longitudinal (along the X-axis) motion: the table (Fig. 4.23a)

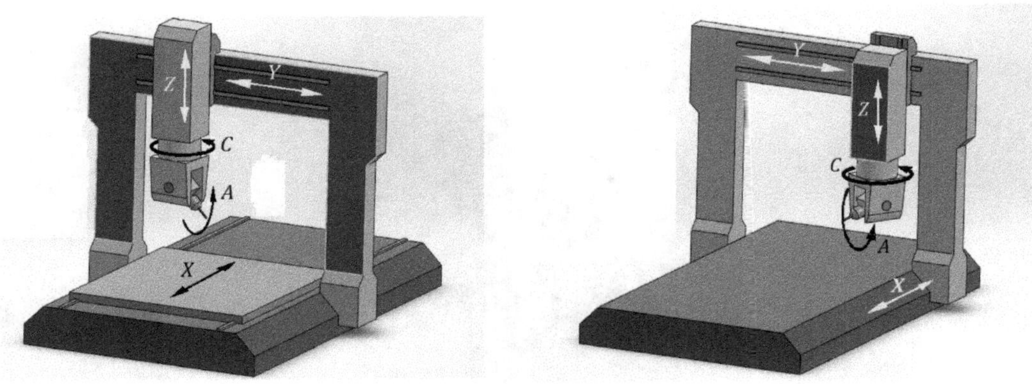

(a) (b)

Figure 4-23 Layouts of 5-axis double column machine tools with a tilt-swivel tool and transversally and vertically moved milling head (*Y*- and *Z*-axis): (a) with a fixed column and an *X*-axis-moved table; (b) with an *X*-axis-moved portal and a fixed table.

and the portal (Fig. 4.23b). The 5-axis machine shown in Figure 4.23b is dedicated for machining of the large capacity workpieces fitted on the fixed table. A double column portal works out the longitudinal (*X*-axis) movement, and has a cross rail with transversally-and vertically-moving milling head (*Y*- and *Z*-axis, respectively). The machines shown in Figures 4.23a and b have the same coordinate code k = **123646**, but different arrangement codes: K = **1023646** and **0123646**, respectively. For example, the manipulation matrix ${}^0\mathbf{A}_6$ of the machine tool shown in Figure 4.23b is similar to that expressed by Eq. (4.66) with opposite signs of coordinates x and y:

$$ {}^0\mathbf{A}_6 = \mathbf{A}^1(x)\mathbf{A}^2(y)\mathbf{A}^3(z)\mathbf{A}^6(\varphi_{sw})\mathbf{A}^4(\theta)\mathbf{A}^6(\varphi) = \begin{bmatrix} a_{11} & a_{12} & a_{13} & x \\ a_{21} & a_{22} & a_{23} & y \\ a_{31} & a_{32} & a_{33} & z \\ 0 & 0 & 0 & 1 \end{bmatrix} \tag{4.67} $$

where a_{11}; a_{12}; a_{13}; a_{21}; a_{22};a_{23};a_{31}; a_{32}; and a_{33} are defined through Eq. (4.66).

In Figure 4.24, the layout of the CNC machining center of type shown in Figure 4.21 b is shown. The machine has the swivel head and the built-in rotary table making reversible compete turn (±360°). Its arrangement code is **6021356**.

In Figure 4.25, two layouts of the 5-axis trunnion-rotary table (tilt-swivel workpiece table) machines, whose arrangement codes are **6412036** and **6512036** (they compared with two double-tilting tables layouts **541203** and **451203**), are displayed. Correspondingly, manipulation matrices ${}^0\mathbf{A}_6$ for the two layouts differ from that given by Eq.(4.67) by positions of the rotation feed motions:

$$ {}^0\mathbf{A}_6 = \mathbf{A}^6(-\varphi_{sw})\,\mathbf{A}^5(-\theta)\,\mathbf{A}^1(-x_{tr})\,\mathbf{A}^2(-y_{tr})\,\mathbf{A}^3(z_{tr})\,\mathbf{A}^6(\varphi) \tag{4.68} $$

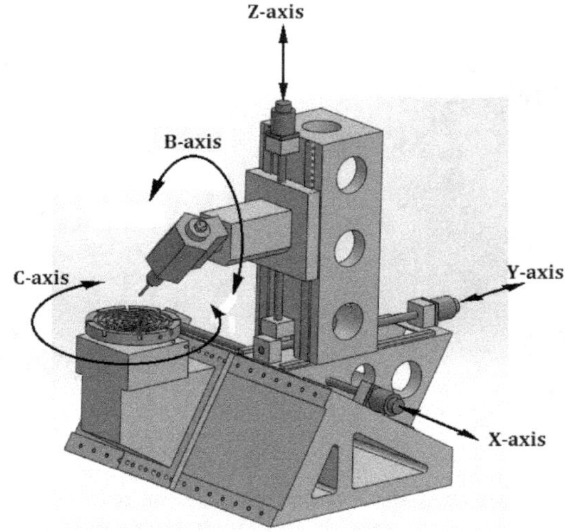

Figure 4-24 5-axis CNC machining center with the swivel head and the ±360° rotary table (similar to Willemin-Macodel SA machine).

$$ {}^{0}\mathbf{A}_6 = \mathbf{A}^6(-\varphi_{sw})\,\mathbf{A}^4(-\psi)\,\mathbf{A}^1(-x_{tr})\,\mathbf{A}^2(-y_{tr})\,\mathbf{A}^3(z_{tr})\,\mathbf{A}^6(\varphi) \tag{4.69} $$

Two other rotation pairs, with arrangement codes **5412036** and **5612036**, are also possible. Their FSFs are as follows:

$$ {}^{0}\mathbf{A}_6 = \mathbf{A}^5(-\psi)\mathbf{A}^4(-\theta)\,\mathbf{A}^1(-x_{tr})\,\mathbf{A}^2(-y_{tr})\,\mathbf{A}^3(z_{tr})\,\mathbf{A}^6(\varphi) \tag{4.70} $$

$$ {}^{0}\mathbf{A}_6 = \mathbf{A}^4(-\psi)\mathbf{A}^5(-\varphi)\,\mathbf{A}^1(-x_{tr})\,\mathbf{A}^2(-y_{tr})\,\mathbf{A}^3(z_{tr})\,\mathbf{A}^6(\varphi) \tag{4.71} $$

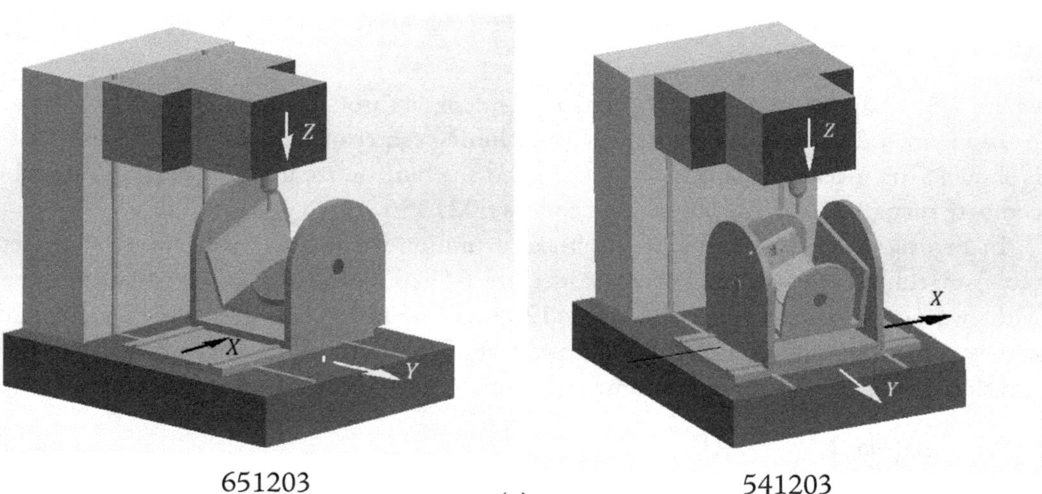

(a)

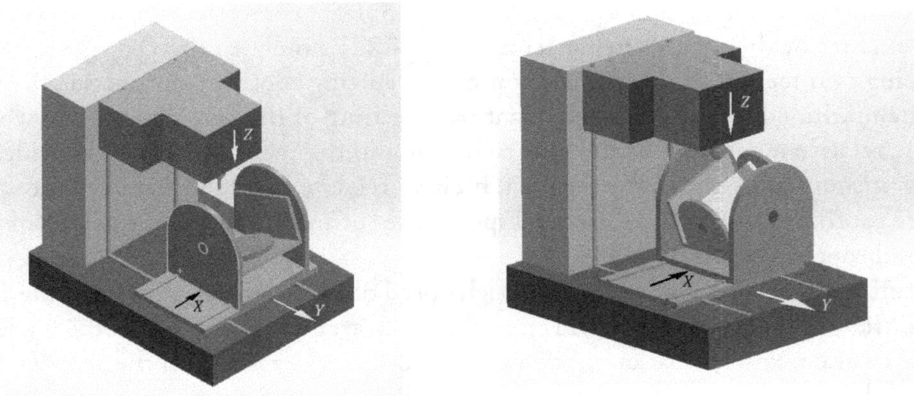

641203 451203

(b)

Figure 4-25 The layouts of the 5-axis trunnion-type machine (with the tilt-swivel work-piece table and vertical cutting tool axis).

Note that the four manipulation matrices are different from each other since the rotation matrices $\mathbf{A}^4(\theta)$, $\mathbf{A}^5(\psi)$, and $\mathbf{A}^6(\varphi)$ are non-commutative.

The layouts shown in Figures 4.21b and c are effectively used for machining impellers (Fig. 4.26): removal of material between a pair of neighboring blades can be performed in the 5-axis mode followed by periodic rotation on the angular pitch. Hence, the table rotation extends the space for the 3-axis operations and, additionally, is typically preferred for machining heavier workpieces.

4.3.3 Gearing machines for spur shaping

4.3.3.1 FSF of the rack-cutter machine

The active element of the rack-cutter machine is the rack with 3-7 teeth. The rack-generating method for cutting the involute spur gear teeth is the most accurate since the

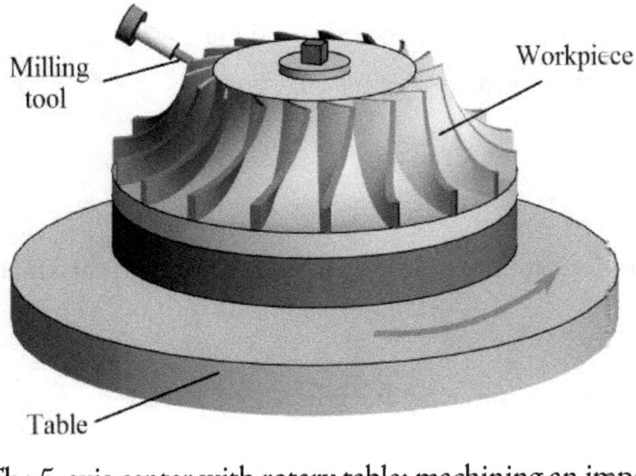

Figure 4-26 The 5-axis center with rotary table: machining an impeller.

main working part of a rack tooth profile has the simplest geometrical form – a straight line. Kinematics of the rack-cutter machine (Fig. 4.27) simulates the rack-gear meshing by using two feed motions: translation of the cutting tool and proportional rotation of the machined pinion imitating a natural meshing of the cutting tool and workpiece. The relative movement of the gear to be cut and the rack-tool can be considered as rolling without slipping of the gear pitch circle relative to the rack pitch-line. The workpiece tooth profile presents an envelope of a set of instantaneous positions of the cutting tool tooth.

The cutting process is realized by the high-speed axial reciprocal motion of the tool relative to the workpiece. In Table 4.11, the FSS links are listed, where φ is the angle of workpiece rotation, x is translation motion of the workpiece agreed upon the angle φ, E is the distance between the gear pitch circle and rack pitch-line, and z is the current positions of the tool ram during cutting reciprocated motion.

The FSF is as follows:

$$\mathbf{r}_0 = {}^0\mathbf{A}_4\,\mathbf{r}_4 = \mathbf{A}^6(-\varphi)\mathbf{A}^1(-x)\mathbf{A}^2(E)\,\mathbf{A}^3(z)\,\mathbf{r}_4$$

$$= \begin{bmatrix} \cos\varphi & \sin\varphi & 0 & 0 \\ -\sin\varphi & \cos\varphi & 0 & 0 \\ 0 & 0 & 1 & 0 \\ 0 & 0 & 0 & 1 \end{bmatrix} \begin{bmatrix} 1 & 0 & 0 & -x \\ 0 & 1 & 0 & 0 \\ 0 & 0 & 1 & 0 \\ 0 & 0 & 0 & 1 \end{bmatrix} \begin{bmatrix} 1 & 0 & 0 & 0 \\ 0 & 1 & 0 & E \\ 0 & 0 & 1 & 0 \\ 0 & 0 & 0 & 1 \end{bmatrix} \begin{bmatrix} 1 & 0 & 0 & 0 \\ 0 & 1 & 0 & 0 \\ 0 & 0 & 1 & z \\ 0 & 0 & 0 & 1 \end{bmatrix} \begin{bmatrix} x_4 & c_{x4} \\ y_4 & c_{y4} \\ z_4 & c_{z4} \\ 1 & 0 \end{bmatrix}$$

$$= \begin{bmatrix} \cos\varphi & \sin\varphi & 0 & E\sin\varphi - x\cos\varphi \\ -\sin\varphi & \cos\varphi & 0 & E\cos\varphi + x\sin\varphi \\ 0 & 0 & 1 & z \\ 0 & 0 & 0 & 1 \end{bmatrix} \begin{bmatrix} x_4 & c_{x4} \\ y_4 & c_{y4} \\ z_4 & c_{z4} \\ 1 & 0 \end{bmatrix}$$

$$= \begin{bmatrix} (x_4 - x)\cos\varphi + (y_4 + E)\sin\varphi & c_{x4}\cos\varphi + c_{y4}\sin\varphi \\ -(x_4 - x)\sin\varphi_1 + (y_4 + E)\cos\varphi & -c_{x4}\sin\varphi + c_{y4}\cos\varphi \\ z + z_4 & c_{z4} \\ 1 & 0 \end{bmatrix} \qquad (4.72)$$

The synchronized rotation and translation is expressed through the kinematic constraint:

$$x = \varphi\, R_p \qquad (4.73)$$

where R_p is the radius of the workpiece pitch circle.

Table 4-11 FSS of the rack-gearing machine, k = **6132,** K = **60132**

S_i	**Link**	**Motion relative to previous link**	**Matrix** $^{i\text{-}1}\mathbf{A}_i$
S_0	Workpiece spindle	Conventionally fixed link	–
S_1	Slider	Rotation around the Z_0-axis at angle$-\varphi$	$^0\mathbf{A}_1 = \mathbf{A}^6(-\varphi)$
S_2	Bed	Translation along the X_1 axis at distance x	$^1\mathbf{A}_2 = \mathbf{A}^1(-x)$
S_3	Bed	Translation along the Y_2-axis at distance E	$^2\mathbf{A}_3 = \mathbf{A}^2(E)$
S_4	Ram	Translation along the Z_3 axis at distance z	$^3\mathbf{A}_4 = \mathbf{A}^3(z)$

The minus sign for angle φ and distances x and E follow from the fact that relative motions of the slider and bed are opposite to the both actual meshing motions.

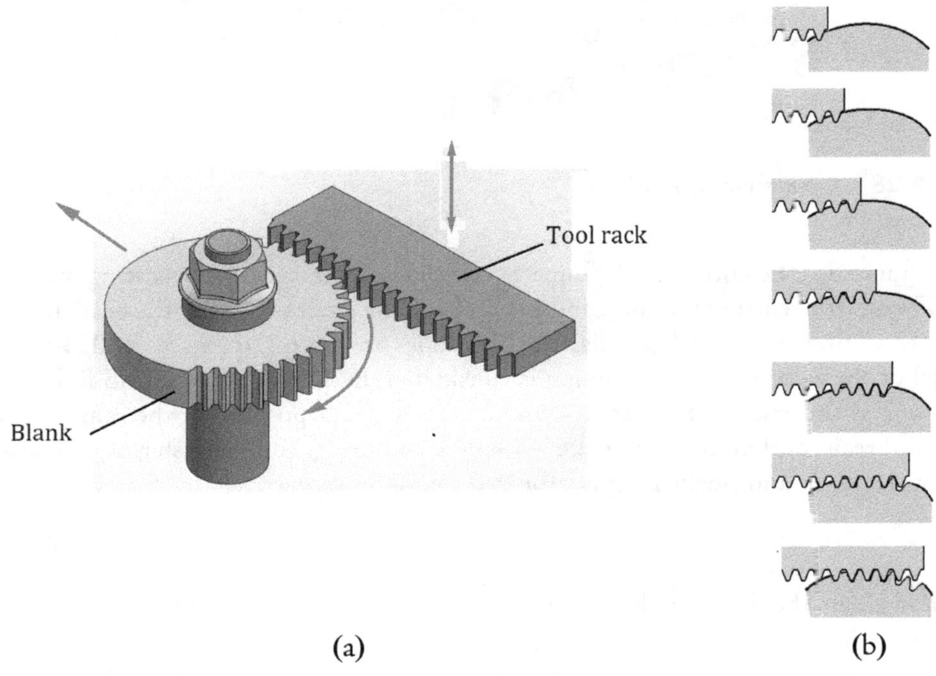

Figure 4-27 Rack-gearing process: (a) tool-workpiece meshing; (b) relative motions of the tool and workpiece.

4.3.3.2 FSF of the gear-shaping machine for spur shaping

The FSS of gearing machine tools using enveloping principle for the gear teeth machining is considered. This method forms the workpiece tooth profile as an envelope of a set of instantaneous positions of the cutting tool tooth in the process of the joint motions of the cutting tool and workpiece imitating their natural meshing.

The FSS of the gear-shaping machine (Fig. 4.28) for spur shaping composes four movable links (n = 4). The table *1* in common with the mounted workpiece associated with coordinate frame S_0 rotates around vertical axis. The carriage (coordinate frame S_1) performs an in-feed motion along the axis-to-axis line fitting the current distance E between the workpiece and the gear shaper cutter. The bed (the stationary link in absolute motion) and a column with attached vertical guideways for the gear cutter translational motion is associated with

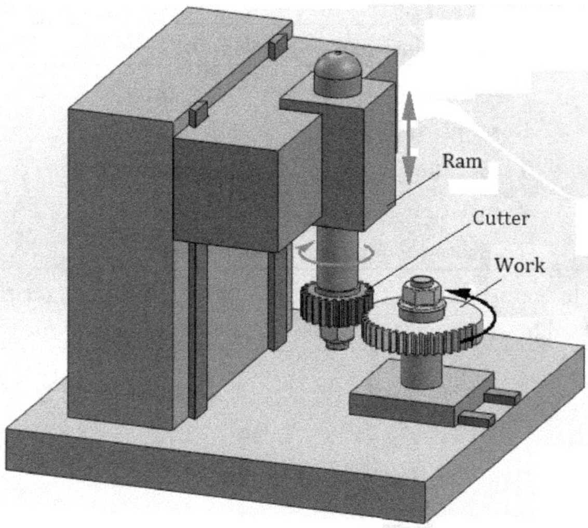

Figure 4-28 Gear-shaping machine.

coordinate frame S_2. The cutter head (frame S_3) carries out the axial reciprocating motion along the Z_2-axis (the cutting motion), and the tool spindle (frame S_4) with the gear shaper cutter rotates around the vertical Z_3-axis. The FSS codes are k = **6236** and K = **60236**.

A model of the external gear machining is considered. In Table 4.12, the FSS links are listed, where E is the axis-to-axis distance, and z is the current position of the ram during cutting (main) reciprocal motion. Here the workpiece and the cutting tool (shaper) perform the counter-clockwise and clockwise synchronized rotations, respectively:

$$\varphi_1 = -(Z_2/Z_1)\,\varphi_2 \tag{4.74}$$

where Z_1 and Z_2 are the teeth number of the tool and workpiece, respectively.

The FSF is

$$[\mathbf{r}_0\mathbf{c}_0] = {}^0\mathbf{A}_4[\mathbf{r}_4\mathbf{c}_4] \tag{4.75}$$

with

$$^0\mathbf{A}_4 = \mathbf{A}^6(-\varphi_1)\,\mathbf{A}^2(E)\,\mathbf{A}^3(z)\,\mathbf{A}^6(-\varphi_2)$$

$$= \begin{bmatrix} \cos\varphi_1 & \sin\varphi_1 & 0 & 0 \\ -\sin\varphi_1 & \cos\varphi_1 & 0 & 0 \\ 0 & 0 & 1 & 0 \\ 0 & 0 & 0 & 1 \end{bmatrix} \begin{bmatrix} 1 & 0 & 0 & 0 \\ 0 & 1 & 0 & E \\ 0 & 0 & 1 & 0 \\ 0 & 0 & 0 & 1 \end{bmatrix} \begin{bmatrix} 1 & 0 & 0 & 0 \\ 0 & 1 & 0 & 0 \\ 0 & 0 & 1 & z \\ 0 & 0 & 0 & 1 \end{bmatrix} \begin{bmatrix} \cos\varphi_2 & \sin\varphi_2 & 0 & 0 \\ -\sin\varphi_2 & \cos\varphi_2 & 0 & 0 \\ 0 & 0 & 1 & 0 \\ 0 & 0 & 0 & 1 \end{bmatrix}$$

Table 4-12 Form-shaping system of the gear-shaping machine, K = **60236.**

S_i	Link	Motion relative to previous link	Matrix ${}^{i-1}\mathbf{A}_i$
S_0	Workpiece spindle	Conventionally fixed link	–
S_1	Bed	Rotation around the Z_0-axis at angle $-\varphi_1$	${}^0\mathbf{A}_1 = \mathbf{A}^6(-\varphi_1)$
S_2	Bed	Translation along the Y_2-axis at distance E	${}^1\mathbf{A}_2 = \mathbf{A}^2(E)$
S_3	Ram	Translation along the Z_3 axis at distance z	${}^2\mathbf{A}_3 = \mathbf{A}^3(z)$
S_4	Shaper	Rotation around the Z_2 axis at angle $-\varphi_2$	${}^3\mathbf{A}_4 = \mathbf{A}^6(-\varphi_2)$

The minus sign for angle φ_1 follows from the fact that relative motions of the bed are opposite to the actual workpiece rotation.

$$= \begin{bmatrix} \cos(\varphi_1+\varphi_2) & \sin(\varphi_1+\varphi_2) & 0 & E\sin\varphi_1 \\ -\sin(\varphi_1+\varphi_2) & \cos(\varphi_1+\varphi_2) & 0 & E\cos\varphi_1 \\ 0 & 0 & 1 & z \\ 0 & 0 & 0 & 1 \end{bmatrix} \tag{4.76}$$

After substitution of matrix ${}^0\mathbf{A}_4$, Eq. (4.76), in Eq. (4.75), the FSF becomes:

$$[\mathbf{r}_0\mathbf{c}_0] = \begin{bmatrix} \cos(\varphi_1+\varphi_2) & \sin(\varphi_1+\varphi_2) & 0 & E\sin\varphi_1 \\ -\sin(\varphi_1+\varphi_2) & \cos(\varphi_1+\varphi_2) & 0 & E\cos\varphi_1 \\ 0 & 0 & 1 & z \\ 0 & 0 & 0 & 1 \end{bmatrix} \begin{bmatrix} x_4 & c_{x4} \\ y_4 & c_{y4} \\ z_4 & c_{z4} \\ 1 & 0 \end{bmatrix}$$

$$= \begin{bmatrix} x_4\cos(\varphi_1+\varphi_2)+y_4\sin(\varphi_1+\varphi_2)+E\sin\varphi_1 & c_{x4}\cos(\varphi_1+\varphi_2)+c_{y4}\sin(\varphi_1+\varphi_2) \\ -x_4\sin(\varphi_1+\varphi_2)+y_4\cos(\varphi_1+\varphi_2)+E\cos\varphi_1 & -c_{x4}\sin(\varphi_1+\varphi_2)+c_{y4}\cos(\varphi_1+\varphi_2) \\ z+z_4 & c_{z4} \\ 1 & 0 \end{bmatrix} \tag{4.77}$$

4.4 Conclusion

Some examples of synthesis of the form-shaping functions (FSF) of machines and robots with serial kinematics are considered. Independently of the type of machine, the synthesis procedure is carried out in the standard form: composing the coordinate and arrangement codes, preparing the table combining the form-shaping system links, and proper synthesis of the FSF.

References

[1] http://www.intelitek.com/robots/ Scorbot-ER 4u, Intelitek, April, 2017.

[2] Kushnir, E., Portman, V.T., Aguilar, A., Clark, W., 2017, Layout evaluation at earlier stages of machine tool design: form-shaping function-based approach, The International Journal of Advanced Manufacturing Technology, Vol. 90 (9), pp. 3333–3346.

[3] Reibakh, Yu.S., Levin, A.A., 1977, MA396F3 NC Semi-Automatic Profile Grinding Machine, Machines and Tooling, Vol. 48, Issue 7, pp. 30-35.

[4] Yoshida, A., Sugimura, N., Takahashi, A., Thasana, W., Tanimizu, Y., and Iwamura, K., 2014, A Study on Tolerance Design for Kinematic Motion Deviations of Machining Centers Based on Geometric Tolerances, The 5th TSME International Conference on Mechanical Engineering, AMM028, 17-19th December, The Empress, Chiang Mai.

[5] Shneor, Y. and Portman, V.T, 2010, Stiffness of 5-Axis Machines with Serial, Parallel, and Hybrid Kinematics: Evaluation and Comparison, CIRP Annals - Manufacturing Technology, 59/1, pp. 409-412.

5. FSF of parallel- and hybrid-kinematics machines

5.1 General description

5.1.1 Main features

In the two past decades, the parallel kinematics mechanisms (PKMs) have received wide acceptance for the design of robots and machines of different types. In these mechanisms, the output link – end-effector (EE) – usually presents a multi-degree-of-freedom platform (Fig. 5.1), which is directly and simultaneously (in-parallel) connected with another link (e.g., a fixed base) through several independent kinematic chains. Two parallel chains together with the base and platform form a kinematic closed loop. This fact explains the presently accepted definition of a generalized parallel manipulator, *a closed-loop kinematic chain mechanism whose end-effector is linked to the base by several independent kinematic chains* [1]. It should be emphasized that introducing the PKMs into real engineering projects has been made possible only by the simultaneous emergence of the modern computerized control systems.

Distinguishing features of the PKMs and their dissimilarity from the serial-kinematics machines (SKMs) can be demonstrated by means of the following examples. In Figure 5.2, a pair of different mechanisms with two degrees of freedom (DOF) is shown. Both planar linkages consist of links connected only through revolute joints and both have 2-DOF. Each mechanism has the EE driven by two active (actuated) joints, which are shown by white circles, while passive joints are symbolized by black circles (Fig. 5.2b). Angular parameters of their actuated motions are labeled φ_1 and φ_2. The DOF number can be calculated through the Grübler-Kutzbach equation [2]:

$$F = \lambda\,(n - j - 1) + (f_1 + \ldots + f_j) \tag{5.1}$$

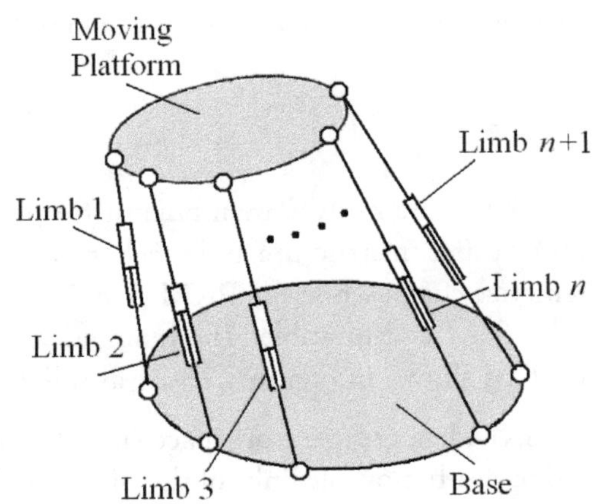

Figure 5-1 Parallel mechanism sketch.

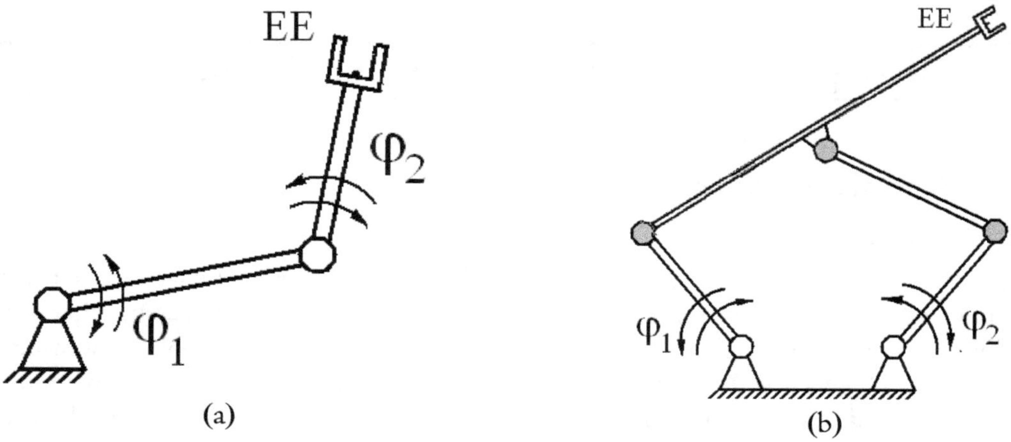

Figure 5-2 Mechanisms with two degrees of freedom: (a) serial kinematics; (b) parallel kinematics.

$$\lambda = \begin{cases} 3, \text{ for planar or spherical mechanisms} \\ 6, \text{ for spatial mechanism} \end{cases} \tag{5.2}$$

where n is the number of links in the mechanism, including the fixed link; j is the number of joints assuming that all the joints are connected only by a pair of links; and f_i (i = 1,...,j) is the DOF number permitted by ith joint.

Parameters entering into Eq. (5.1) for the SKM shown in Figure 5.2a are as follows: λ = 3 since the mechanism is planar; n = 3 since the mechanism has three (two movable and one fixed) links; j = 2 since the mechanism has two joints; and $f_1 = f_2 = 1$ since a revolute joint permits 1-DOF. Hence, the calculation of the DOF number by Eq. (5.1) yields

$$F = 3\,(3 - 2 - 1) + (1 + 1) = 2 \tag{5.3}$$

Correspondingly, the parameters relating to the PKM shown in Figure 5.2b are: λ = 3; n = 5; j = 5; $f_1 = ... = f_5 = 1$. Hence,

$$F = 3\,(5 - 5 - 1) + 1 \times 5 = 2 \tag{5.4}$$

Both mechanisms can be used as the 2-DOF manipulator for planar positioning a given point of the EE. However, they differ in structure and mechanical features. The SKM (Fig. 5.2a) presents an open kinematics chain, while the PKM (Fig. 5.2b) has a closed kinematic loop formed by all links, both fixed and movable. The main advantages of the SKM (Fig. 5.2a) as compared with the PKM shown in Figure 5.2b are as follows:

- With similar sizes, the SKM has greater workspace (i.e., the area its end-effector can reach) than the PKM since both movable links of the SKM can be rotated by complete revolution, while the angles of rotation of the PKM links are limited.

- Generally, there is a practically unhindered approach to the SKM workspace points.
- There are no configurations of the SKM with zero stiffness value, while this can take place in the PKM-based structures. Note that the SKM can have configurations with zero compliance value (infinite stiffness value).

In turn, the PKM has some significant advantages explaining great scientific and manufacturing interest in the PKMs and their widespread industrial applications. With identical accuracy of manufacturing, the PKM is more accurate and stiffer than the SKM. As seen from Figure 5.2b, a load acting on the EE of the PKM is distributed over two kinematic sub-chains (limbs), while the SKM has only one supporting chain. For the same reasons, the PKM usually has better dynamic characteristics.

The listed features are significantly more pronounced for manipulators with more than 2-DOF. In Figure 5.3, two 6-DOF mechanisms are shown: the SKM – the Stanford arm (Fig. 5.3a) and the PKM (Fig. 5.3b) – the Gough-Stewart platform (GSP). The Stanford arm [3] presents the robot with six mobile links connected by six kinematic pairs, whose motions are driven and controlled by six drives. The arm contains a fixed link (column 1) and six movable links: link 2 rotates around the vertical axis of the column; link 3 rotates around the horizontal axis intersecting the vertical axis orthogonal to it; an axis of link 4 intersects the previous one perpendicularly; links 3 and 4 compose a sliding pair and link 4 carries out a translational motion relative to link 3; link 5 rotates about the axis, which is in line with the axis of translation; the last-listed axis and two next rotation axes (they are not shown) intersect one another perpendicularly at a single point – the wrist center of the hand.

Hence, the robot has seven links ($n = 7$), one prismatic and five revolution joints ($j = 6$) with 1-DOF each ($f_1 = \ldots = f_6 = 1$), and the calculation of the DOF number by Eq. (5.1) yields:

$$F = 6(7 - 6 - 1) + 6 \times 1 = 6 \tag{5.5}$$

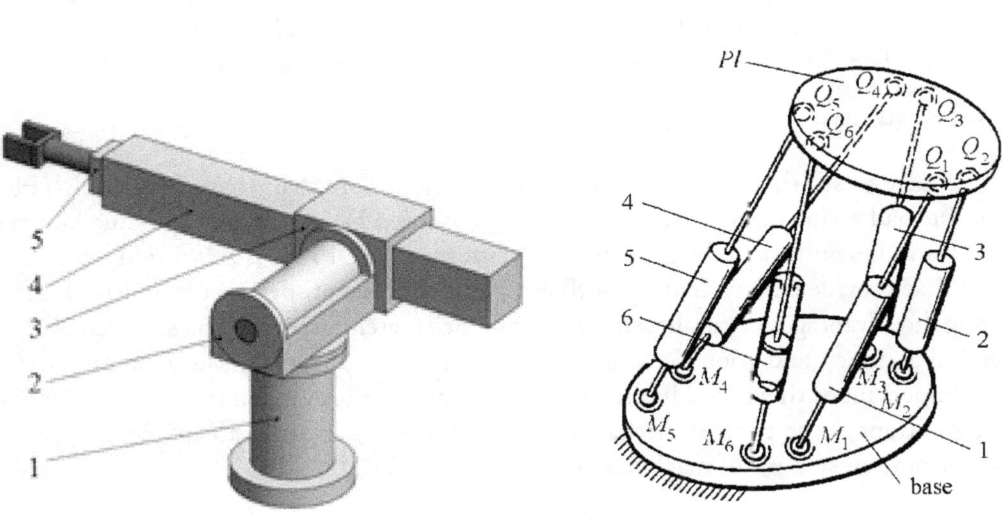

Figure 5-3 The 6-DOF mechanisms: (a) serial-type – the Stanford hand; (b) parallel-type – the Gough-Stewart platform.

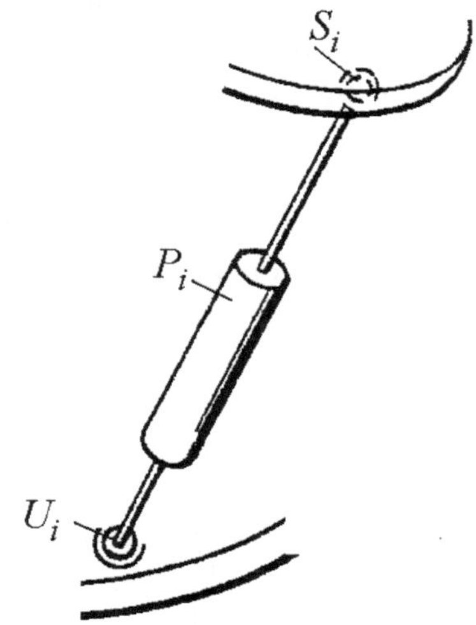

Figure 5-4 Kinematic joints associated with the GSP limb.

The structure of the Gough-Stewart platform [4] is shown in Figure 5.3b. A moving platform *Pl* is connected with a fixed base by six extensible limbs 1, 2,...,6 of controlled-variable lengths. This kinematics of the limb is designed as U**P**S, where universal, prismatic, and spherical joints are symbolized by U, **P**, and S, respectively (Fig 5.4). Symbol **P** in bold denotes the actuating pair. The built-in limb translation drive combines supporting and kinematic functions for moving platform *Pl* with respect to the base.

In the GSP mechanism, the number of links is $n = 14$: the platform, base, six upper and six lower parts of the limbs. The limb is connected with the base and the moving platform by the 2-DOF universal joint and 3-DOF spherical joints, respectively. Hence, the mechanism consists of 18 joints: six translational joints with DOF numbers $f_1 = ... = f_6 = 1$, six universal joints with $f_7 = ... = f_{12} = 2$, and six joints with $f_{13} = ... = f_{18} = 3$. Thus, Eq. (5.1) yields:

$$F = 6\,(14 - 18 - 1) + 6 \times 1 + 6 \times 2 + 6 \times 3 = 6 \tag{5.6}$$

It is easy to see that the workspace of the SKM is more than that of the PKM. However, the load acting on the EE of the SKM is absorbed by one support (unique kinematic chain), and in contrast to this, the moving platform of the PKM is supported on six limbs (legs) that provide better accuracy, stiffness, and dynamics properties of the PKM. Hence, among the advantages of parallel structures are their potential high stiffness, acceptable geometrical accuracy, and good dynamic properties. In spite of some disadvantages – and the most important of them are a limited working space and the availability of multiple singular configurations – the parallel mechanisms have gained widespread acceptance for position and orientation systems of machines and robots [1, 2, 5–7].

5.1.2 PKM examples

The PKMs are used over a wide load range. For example, the flight simulator (Fig. 5.5) described by Stewart [4] works out the 6-DOF spatial motion using six simple drives. The platform center is displaced along a given direction, and simultaneously the platform is rotated around three coordinate axes. The similar mechanism for tear testing (Fig. 5.6) was reported by Gough [8] in 1956. From the early nineties, some machine tools with parallel kinematics were developed using the GSP-based manipulation systems: the machine tool of the non-traditional configuration [9], VARIAX® of the Giddings & Lewis Machine Tools, Wisconsin, Octahedral-Hexapod machine tool for five-axis machining, Ingersol Milling Machine, machine tools of Toyota Co. [1, 10]. Milling centers with parallel kinematics for machining free-form surfaces are shown in Figure 5.7.

The main reason to apply the parallel kinematics for machine tools was in the effort to enhance the dynamic properties of the machine mechanical system by reducing the masses of the moving parts, in particular for high-speed machining conditions. Actually, the GSP-based 5-axis machine has a single massive moving part. Contrary to this, in serial-kinematics machines, no less than three of five units move in series relative to the base (Fig. 5.8).The seemingly effective solution of this problem is application of PKM-based structures.

However, there are some restrictions inherent in PKM structure and kinematics to meet the typical machine tool requirements. The most serious of them are the stiffness properties of the PKM, which are significantly configuration-dependent (vary from one configuration to another), and moreover, the PKM can have singular configurations with *zero stiffness* values. Additionally, the workspace of the PKM is limited, and in particular, the possible tilting angles of the moving platform are significantly limited compared with traditional 5-axis machines. Since machining of complicated geometric forms in the PKMs is accompanied with greater limb reverses, the machined profiles can obtain the corresponding number of typical clearance-caused errors. All these facts have been hindered with further commercialization of PKM-based machine tools.

Two micro-positioning manipulation systems (the term "micro" relates to a range and positioning accuracy) are shown in Figure 5.9. The 220 × 230 mm platform of the manipulator (Fig. 5.9a) can move in the range ±6 mm and ±5 deg with load capacity 2 kg, with

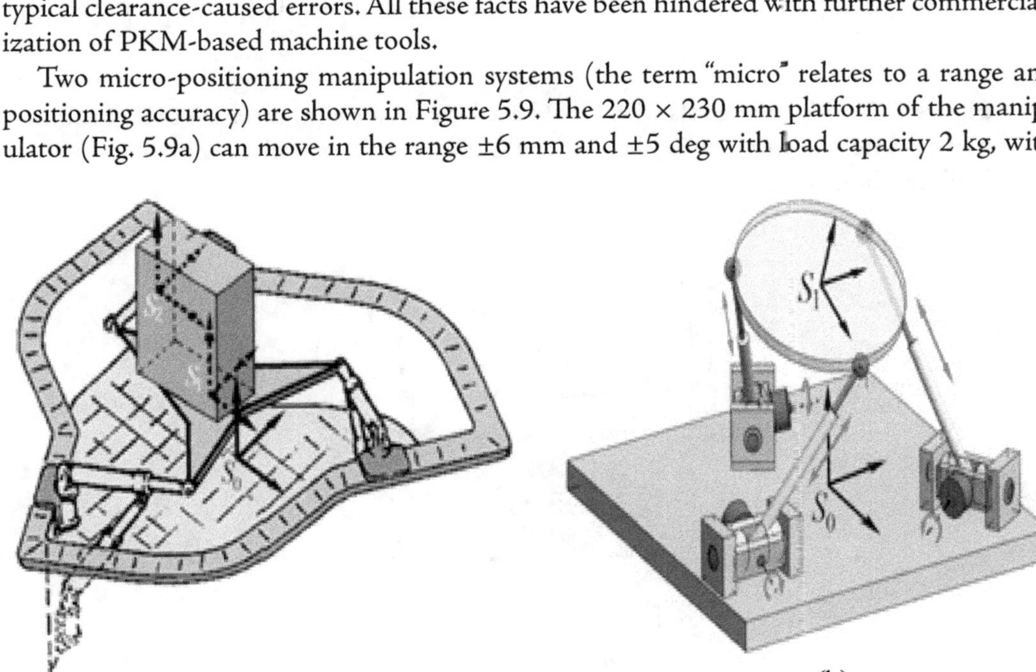

Figure 5-5 Flight simulator [4]. S_0 - base frame, S_1 - platform frame, and S_2 - seat frame.

Figure 5-6 Mechanism for tire testing [7].

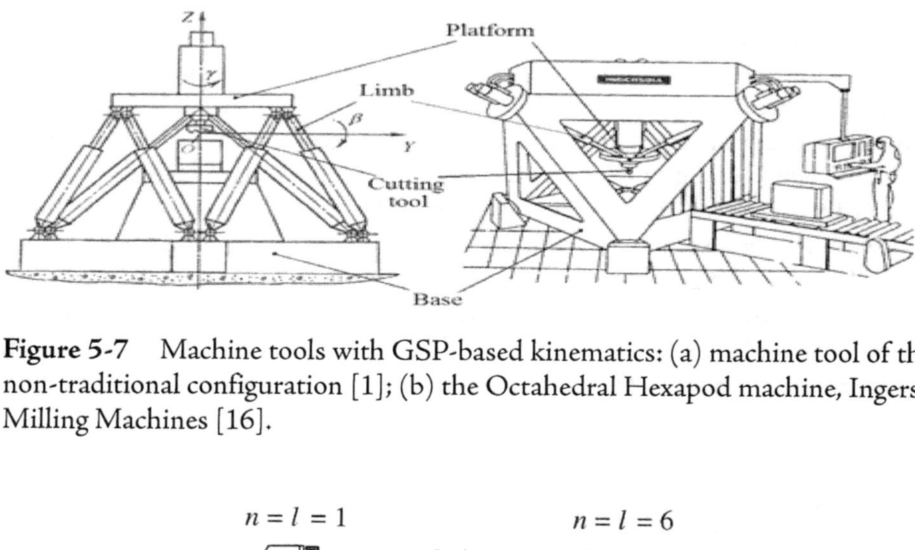

Figure 5-7 Machine tools with GSP-based kinematics: (a) machine tool of the non-traditional configuration [1]; (b) the Octahedral Hexapod machine, Ingersoll Milling Machines [16].

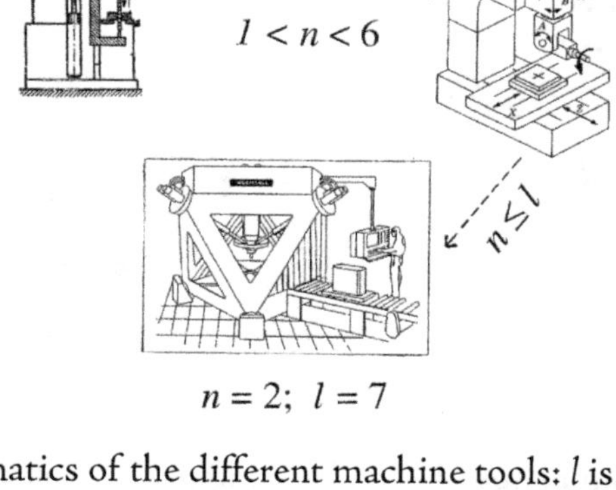

Figure 5-8 Kinematics of the different machine tools: l is the DOF number (including the main motion); n is the number of the form-shaping links.

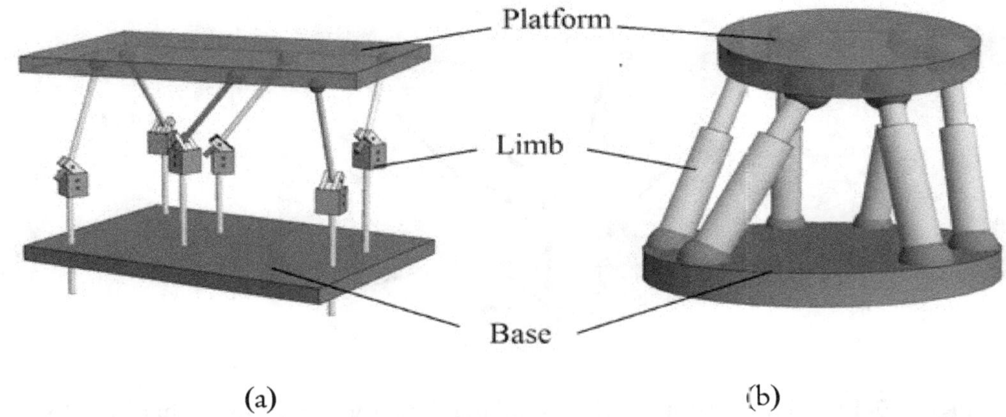

Figure 5-9 Micro positioning manipulation systems produced by Physical Instrument (PI) [11].

0.1 μm and 0.4 arc sec of minimum incremental motions. The robot with the moving platform of 200 mm diameter (Fig. 5.9b) is developed for aligning optics in astronomical telescopes [11]. It has variable-length legs providing the following working range: platform motion in the platform plane is ±50 mm,in the axial (normally to the plane) direction is ±25 mm, the tilting angles are ±15°; and rotation around the axis ±30°.Load capacity is 25÷100 kg depending on platform orientation; minimum incremental linear motion is 1÷3 μm with repeatability of ±(0.4÷0.5) μm; and minimum incremental angular motion is 5 μrad with repeatability of ±(7÷12) μrad. There are also the nano-resolution manipulators using piezoelectric (PZT) actuators with working range of 0.8 mm and 10 mrad and very high resolution.

In contrast to the micron and submicron devices, the manipulation system shown in Figure 5.10 presents the orientation system of a huge radio telescope – a square kilometer array [12]. In this system, six large span cables are used to tune the feed to implement tracking with six big power servomechanisms, with a load capacity of tens of tons.

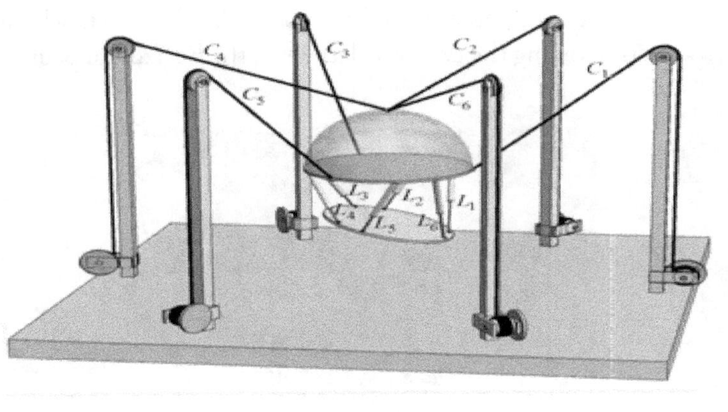

Figure 5-10 An orientation system of the huge radio telescope [12]. $C_1, C_2, \ldots, C_6$ – the tendon cables; $L_1, L_2, \ldots, L_6$ – the limbs of the GSP-type fine turning system.

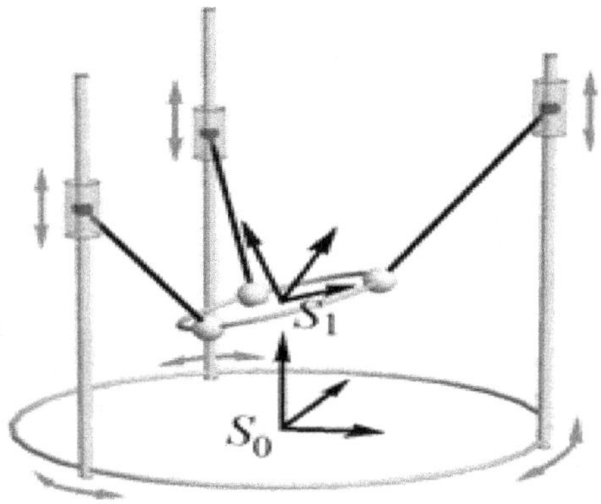

Figure 5-11 The 6-DOF spatial tripod manipulator – Eclipse mechanism [13].

The specific structure and exterior of the 6-DOF platform mechanisms with six limbs have resulted in the entomology-related name *hexapod*. Correspondingly, the name *tripod* is applied to mechanisms having just *three* limbs. In Figure 5.10, the 6-DOF spatial tripod-type manipulator, called the Eclipse mechanism, with the **P**RS limbs is shown [13]. The kinematics of this mechanism is as follows: three vertical columns can slide along the horizontal-base-located circular guideways and three sliders can translate along the columns. One end of the limb is connected to the mobile platform through the spherical joint, while the opposite end is connected through the revolute joint to the slider, which translates along the fixed guideways on the columns.

Another tripod-type mechanism is shown in Figure 5.12 [14], where an end-effector – platform 1 – has 3-DOF in its motion relative to the base. The mechanism shown in Figure 5.12 consists of the same limb kinematics as that in Figure 5.11, but the vertical translation guideways are fixed with respect to base 8. Vertical position and orientation of platform 1 result from active motions of sliders 2, 3, 4 along three vertical guideways. Connecting rods 5, 6, 7 are attached to the platform and sliders through spherical and revolute joints, respectively. The *i*th rod ($i = 1, 2, 3$) moves in its vertical plane.

At the upper left of Figure 5.12, the sketch of the 5-axis tripod-based machine tool is shown. In this machine, translations along the two translations in the horizontal plane are actuated by

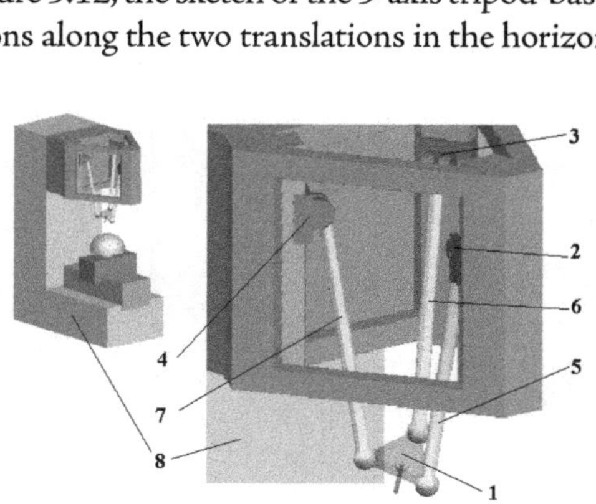

Figure 5-12 Tripod-based mechanism [14].

Figure 5-13 The basic hexaglide principle [15]

machine table and saddle, while vertical and tilting motion of the platform-mounted spindle are carried out by means of the platform manipulation. This combination of the parallel- and serial-type manipulations characterizes the hybrid-type kinematics machine.

Contrary to the GSP-based mechanism, where manipulating motions are worked out using the *limb extension*, both mechanisms shown in Figures 5.11 and 5.12 consist of limbs of *non-variable* lengths. Additional examples of such structures are shown in Figure 5.13. Controlled translations of the opposite ends result in desirable motions of the mobile platform. Six sliders of a mechanism called a hexaglide move along six parallel guideways aimed at the 6-DOF manipulation of the platform. The platform is attached to the limbs connecting the sliders and platform through universal and spherical joints, respectively [15].

Apart from sliding pairs applied in the actuated chains with extensible and non-variable limb lengths, the kinematic chains with articulated linkages are also developed. In Figure 5.14, a 3-DOF planar manipulator is driven by a triplet of two-linkage chains consisting of three revolute joints (3**R**RR kinematics). In this manipulator, rotations of cranks 2, 2', and 2" at angles φ_1, φ_2, and φ_3, respectively, are transformed through intermediate bars 3, 3', and 3"

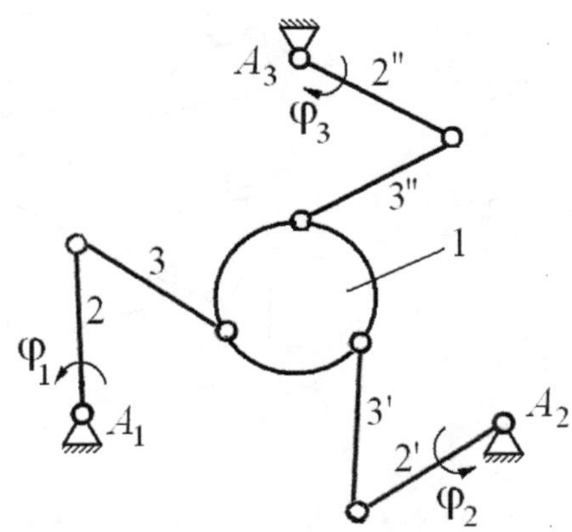

Figure 5-14 Planar 3RRR manipulator.

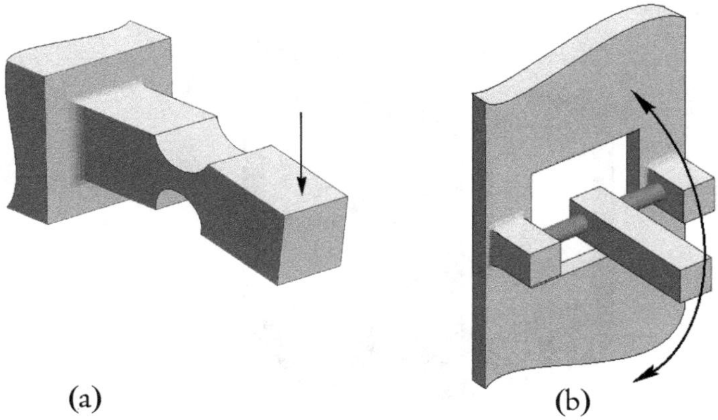

Figure 5-15 (a) Flexure revolute hinge [17]; (b) torsional elastic hinge [16].

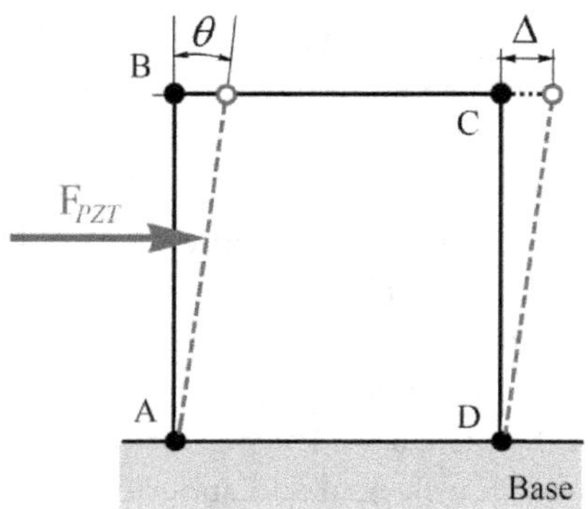

Figure 5-16 Parallelogram with flexure joints for micro- and nano-mechanisms.

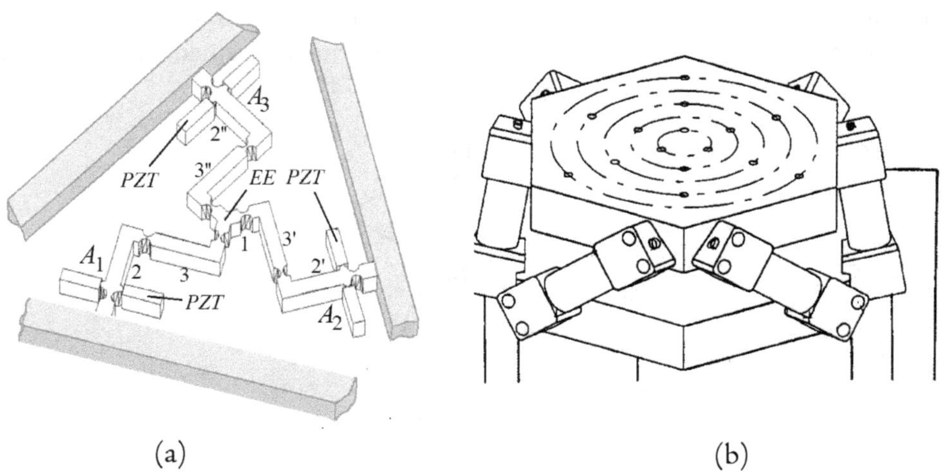

Figure 5-17 Micromanipulators with flexure hinges with flexure joints: (a) 3-DOF planar robot [18], designation are similar to those in Figure 5.14; (b) 6-DOF robot with GSP-type structure [19].

in planar motion (displacements in the plane and rotation around the axis perpendicular to the plane) of platform 1. All joints of the mechanism are revolute.

A specific class of the PKM presents the structures with elastic hinges in which the PZT actuated motions are carried out within a relatively small dimensional range with very high (submicron) accuracy. For this purpose, the flexure rotational and revolute hinges [16, 17] are widely used (Fig. 5.15). Contrary to regular kinematic hinges, the elastic hinges featur very high accuracy of movements thanks to luft absence and minimal elastic hysteresis. Note that robot links connected by elastic hinges are immediately returned to initial position after reloading the active forces. Hence, any other position requires application of a permanent external fixation. As an example, the 1-DOF parallelogram (Fig. 5.16) with elastic hinges used for micro- and nano-mechanisms is shown.

In Figure 5.17a, the PZT-actuated planar 3-DOF parallel kinematics manipulator [18] is shown. It has nine fluxer hinges and presents a full analog of the planar 3**R**RR manipulator (Fig. 5.14). To demostrate this similarity in kinematics and difference in hinges, the identical designations for hinges and links are used for both robots. The linear and angular workspace of the robot are 0÷100 μm and 0÷0.1 μrad, respectively. The robot stiffness value along the linear and around the angular displacements are 9400-9500 N/m and 0.062 N-m/deg, respectively.

In Figure 5.17b, the 6-DOF spatial robot is shown in which 12 elastic hinges are used instead of the regular six universal and six spherical hinges [19]. The robot kinematics is similar to the reguar Gough-Stewart platform. The micromoving platform obtains displacements in the range ±0.04 mm with experimentally measured accuracy of about 20 nm.

5.2 Position analysis of the parallel-kinematics systems

5.2.1 Layout and inverse kinematics

Similar to serial-kinematics systems, position analysis of the parallel-kinematics machines includes both direct kinematics $\mathbf{X} = \mathbf{X}(\mathbf{Q})$, Eq. (4.4), and inverse kinematics $\mathbf{Q} = \mathbf{Q}(\mathbf{X})$, Eq. (4.5), where vector **X**, Eq. (4.1), composes the pose coordinates and vector **Q**, Eq. (4.3), composes the actuator parameters. It should be particularly emphasized that, as distinguished from the serial-kinematics machines, the inverse kinematics problem for the parallel-kinematics mechanism usually has a relatively simple solution. Contrary to that, the direct kinematics usually presents a sophisticated problem, whose solution can be obtained in the majority of the cases through *inversion of the inverse kinematics* problem.

Since the FSS of parallel- and hybrid-kinematics machines consists of a multi-DOF link, the corresponding FSF consists of the extended Euler's matrix $\mathbf{A}_{Eu}$, Eq. (3.147). To define motions of the actuated limbs, the six parameters of the pose have to be expressed through another set of six parameters. $q_{i_1}, q_{i_2}, \ldots, q_{i_6}$ The latter sextet (or its part) presents parameters of the actuated limbs. The two sextets are interrelated through six scalar simultaneous equations

$$F_1(x_p, y_p, z_p, \theta_p, \psi_p, \varphi_p, q_{i1}, q_{i2}, \ldots, q_{i6}) = 0$$

$$F_2(x_p, y_p, z_p, \theta_p, \psi_p, \varphi_p, q_{i1}, q_{i2}, \ldots, q_{i6}) = 0$$

....

$$F_6(x_p, y_p, z_p, \theta_i, \psi_i, \varphi_i, q_{i1}, q_{i2}, \ldots, q_{i6}) = 0 \tag{5.7}$$

Physical interpretation of system (5.7) and solution procedure depend on the DOF number l_i of the link associated with matrix $^{i-1}\mathbf{A}_i(x_p, y_p, z_p, \theta_i, \psi_i, \varphi_i)$.

(a) $l_i = 6$ and the functional matrix of system (5.7) is non-zero,

$$\mathbf{J}_i = \frac{\partial(F_1, F_2, \ldots, F_6)}{\partial(q_{i_1}, q_{i_2}, \ldots, q_{i_6})} \neq 0 \tag{5.8}$$

Hence, the simultaneous equations (5.7) can be solved with respect to the six parameters, $q_{i1}, q_{i2}, \ldots, q_{i6}$ which are expressed in terms of pose parameters $x_i, y_i, z_i, \theta_i, \psi_i$, and φ_i,

$$q_{i1} = q_{i1}(x_p, y_p, z_p, \theta_i, \psi_i, \varphi_i)$$

$$q_{i2} = q_{i2}(x_p, y_p, z_p, \theta_i, \psi_i, \varphi_i)$$

....

$$q_{i6} = q_{i6}(x_p, y_p, z_p, \theta_i, \psi_i, \varphi_i) \tag{5.9}$$

Equations (5.9) formulate a solution of the inverse kinematics problem for the parallel kinematics link, i.e., fix the six necessary displacements of the actuated elements resulting in the given current parameter of the platform pose. For example, the pose of the Gough-Stewart platform (Fig. 5.3) as a mechanism with 6-DOF is defined through six parameters: three coordinates of the position of a platform point x_p, y_p, z_p and three Euler angles θ, ψ, and φ of the orientation of the moving platform, $\mathbf{X} = (x_p, y_p, z_p, \theta, \psi, \varphi)$. The sextet of the robot pose parameters is controlled by current lengths of six limbs is $\mathbf{Q} = (L_1, L_2, L_3, L_4, L_5, L_6)$, and solution of the inverse kinematics becomes:

$$L_1 = L_1(x_p, y_p, z_p, \theta, \psi, \varphi)$$

$$L_2 = L_2(x_p, y_p, z_p, \theta, \psi, \varphi)$$

....

$$L_6 = L_6(x_p, y_p, z_p, \theta, \psi, \varphi) \tag{5.10}$$

(b) $1 < l_i < 6$, and a set of $(6 - l_i)$ *layout constraints* reduces the DOF number from six to l_i. For the correctly designed structure, this system must be solved with respect to l_i of six pose parameters. If for example $l_i = 3$, these equations have the form

$$F_1(x_p, y_p, z_p, \theta_i, \psi_i, \varphi_i) = 0$$

$$F_2(x_p, y_p, z_p, \theta_i, \psi_i, \varphi_i) = 0$$

$$F_3(x_p, y_p, z_p, \theta_i, \psi_i, \varphi_i) = 0 \tag{5.11}$$

(c) $1 < l_i < 6$, and a set of l_i simultaneous equations formulates the inverse kinematics problem. For $l_i = 3$, this group has the form:

$$G_1(x_p, y_p, z_p, \theta_i, \psi_i, \varphi_i, q_{i1}, q_{i2}, q_{i3}) = 0$$

$$G_2(x_p, y_p, z_p, \theta_i, \psi_i, \varphi_i, q_{i1}, q_{i2}, q_{i3}) = 0$$

$$G_3(x_p, y_p, z_p, \theta_i, \psi_i, \varphi_i, q_{i1}, q_{i2}, q_{i3}) = 0 \tag{5.12}$$

with

$$\left|\frac{\partial(G_1, G_2, G_3)}{\partial(q_{i_1}, q_{i_2}, q_{i_3})}\right| \neq 0 \tag{5.13}$$

where the sextet $x_p, y_p, z_p, \theta_i, \psi_i, \varphi_i$ obeys Eqs. (5.12).

In all the cases, the inverse kinematics defining the controlled positions and motions of the robot drives and actuators is the crucial step of the calculations relating to the parallel- and hybrid-kinematics systems. Equations for solution the inverse kinematics are synthesized using the fact that the parallel mechanisms consists of kinematic loops (closed kinematic chains) with built-in actuators. Thus, an intermediate element of the loop can be simultaneously expressed through two different half-chains: one half-chain uses the given information about the end-effector pose parameters x, y, z, θ, ψ, φ, and another half-chain consists of the required information about a current position q_j of the built-in-limb actuator. Equating these expressions for each actuator of the ith link yields:

$$g_j(q_j) = f_j(x_p, y_p, z_p, \theta_i, \psi_i, \varphi_i), \text{ with } j = 1, 2, \ldots, l_i \tag{5.14}$$

where l_i is the DOF number of the ith link.

Since the pose parameters are given, the right-hand part of Eq. (5.14) is known, and the equation can be solved with respect to q_j. For example, point Q_i of the upper end of the ith limb of the GSP (Fig. 5.3b) can be expressed through two position vectors,

$$\mathbf{r}_{Qi}(L_i) = \mathbf{r}_{Qi}(x_p, y_p, z_p, \theta, \psi, \varphi), \text{ with } i = 1, 2, \ldots, 6 \tag{5.15}$$

where $l_i = 6$ and L_i is the length of the ith link, $L_i = |M_iQ_i|$.

5.2.2 Planar 3-RRR manipulator

5.2.2.1 *Layout and geometry*

A planar 3-DOF parallel manipulator is shown in Figure 5.18. The moving platform $Q_1Q_2Q_3$ moves in the drawing plane driven through three structure-identical limbs. The ith limb

(i = 1, 2, 3) contains two links of lengths a_i and b_i connected with one another through a revolute joint D_i. In turn, the limb links are connected with the platform at point B_i and with fixed pivots A_i. Hence, parameters in the Grübler-Kuztbach equation are: the manipulator is planar ($\lambda = 3$), it consists of $n = 8$ links and $j = 9$ joints, each link permitted 1-DOF ($f_j = 1$). According to Eqs.(5.1-2), the number F of the DOF is

$$F = \lambda\,(n - j - 1) + (f_1 + \ldots + f_j) = 3(8 - 9 - 1) + 9 = 3 \tag{5.16}$$

Triangles $A_1A_2A_3$ and $B_1B_2B_3$ are equilateral:

$$|A_1A_2| = |A_2A_3| = |A_1A_3| = l_A \text{ and } |B_1B_2| = |B_2B_3| = |B_1B_3| = l_B \tag{5.17}$$

The mechanism contains three **R**RR limbs, each of which connects the moving platform with the base using three revolute joints (symbol R).The first **R**(in bold) means that the first joint is actuated. With the same sizes of the links, the manipulator can have eight versions of its assembly (Fig. 5.19), among which only two (111 and 222) are useful in practice. A pose of the manipulator is defined by position of the end-effector (moving platform $Q_1Q_2Q_3$), i.e., coordinates x_0, y_0 of its central point O_p, and by orientation of the moving platform, i.e., angle φ between corresponding axes of the triangles $A_1A_2A_3$ and $Q_1Q_2Q_3$. The current pose of the platform is provided through the control actuated angles α_1, α_2, and α_3, while angles β_1, β_2, and β_3 associated with relative orientation of links a_i and b_i are passive. Hence, the articular coordinates **Q** and a set of the desired coordinates of the robot pose **X** are

$$\mathbf{Q} = (\alpha_1, \alpha_2, \alpha_3) \text{ and } \mathbf{X} = (x_0, y_0, \varphi) \tag{5.18}$$

Coordinates of points A_1, A_2, and A_3 in the fixed coordinate frame OXY are

$$\mathbf{r}_{A1} = l_A\begin{bmatrix} 1/2 \\ -\sqrt{3}/6 \end{bmatrix};\ \mathbf{r}_{A2} = l_A\begin{bmatrix} 0 \\ \sqrt{3}/3 \end{bmatrix};\ \text{and } \mathbf{r}_{A3} = l_A\begin{bmatrix} -1/2 \\ -\sqrt{3}/6 \end{bmatrix} \tag{5.19}$$

Coordinates of points Q_1, Q_2, and Q_3 in coordinate frame $O_pX_pY_p$ associated with the moving platform are

$$\rho_1 = l_B\begin{bmatrix} 1/2 \\ -\sqrt{3}/6 \end{bmatrix};\ \rho_2 = l_B\begin{bmatrix} 0 \\ \sqrt{3}/3 \end{bmatrix};\ \text{and } \rho_3 = l_B\begin{bmatrix} -1/2 \\ -\sqrt{3}/6 \end{bmatrix} \tag{5.20}$$

After translation of the moving platform along vector $\mathbf{r}_0 = OO_p$ and rotation at angle φ, coordinates of points A_1, A_2, and A_3 are unchanged. The coordinates of points Q_1, Q_2, and Q_3 in the fixed coordinate frame OXY depend on displacement and rotation of the moving platform as follows:

$$\mathbf{r}_i = \mathbf{r}_0 + \mathbf{A}(\varphi)\,\rho_i\ (i = 1, 2, 3) \tag{5.21}$$

where ρ_i is given by Eqs. (5.20); $\mathbf{r}_0$ is the translation vector; and $\mathbf{A}(\varphi)$ is the rotation matrix,

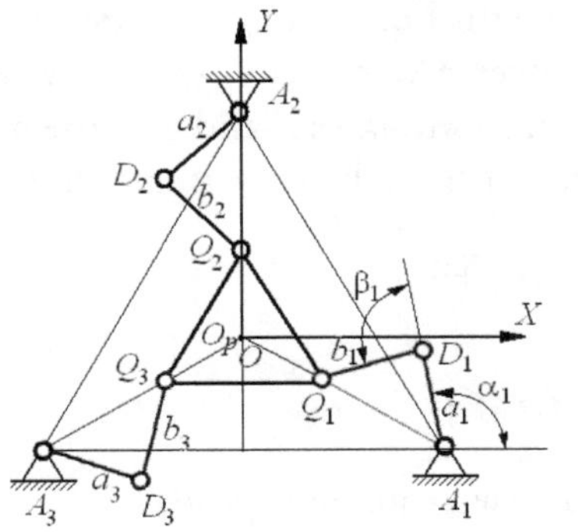

Figure 5-18 The 3-RRR planar parallel manipulator.

$$\mathbf{r}_0 = \begin{bmatrix} x_0 \\ y_0 \end{bmatrix} \text{and } \mathbf{A}(\varphi) = \begin{bmatrix} \cos\varphi & -\sin\varphi \\ \sin\varphi & \cos\varphi \end{bmatrix}$$

For example, vector $\mathbf{r}_1$ of point B_1 and its coordinates x_1 and y_1 in the fixed frame are

$$\mathbf{r}_1 = \mathbf{r}_0 + \mathbf{A}(\varphi)\boldsymbol{\rho}_1 = \begin{bmatrix} x_0 \\ y_0 \end{bmatrix} + l_B \begin{bmatrix} \cos\varphi & -\sin\varphi \\ \sin\varphi & \cos\varphi \end{bmatrix} \begin{bmatrix} 1/2 \\ -\sqrt{3}/6 \end{bmatrix}$$

Hence,

$$x_1 = x_0 + l_B[(1/2)\cos\varphi + (\sqrt{3}/6)\sin\varphi] \tag{5.22}$$

$$y_1 = y_0 + l_B[(1/2)\sin\varphi - (\sqrt{3}/6)\cos\varphi] \tag{5.23}$$

5.2.2.2 *Inverse kinematics*

For the considered manipulator, the articular coordinates **Q** and the robot pose coordinates **X** are expressed through Eqs. (5.20). Thus, the inverse kinematics is formulated as follows:

$$\alpha_i = \alpha_i(x_0, y_0, \varphi), \text{ with } i = 1, 2, 3 \tag{5.24}$$

where x_0, y_0, and φ are the set of parameters defining the pose of the moving platform.

The coordinates of points Q_1, Q_2, and Q_3 are connected with the coordinates of points A_1, A_2, and A_3 as follows:

$$x_i = x_{Ai} + a_i \cos\alpha_i + b_i \cos(\alpha_i + \beta_i), i = 1, 2, 3$$

$$y_i = y_{Ai} + a_i \sin\alpha_i + b_i \sin(\alpha_i + \beta_i), i = 1, 2, 3 \tag{5.25}$$

Equations (5.25) together with Eqs. (5.21) describe the mechanism geometry and present a base for solution of the inverse kinematics, when x_0, y_0 and φ are given and angles α_i and β_i are to be found. In this case, coordinates x_i and y_i of the moving platform are immediately defined through Eqs. (5.25), which can be rewritten as follows:

$$x_i - x_{Ai} = a_i \cos\alpha_i + b_i \cos(\alpha_i + \beta_i) \tag{5.26}$$

$$y_i - y_{Ai} = a_i \sin\alpha_i + b_i \sin(\alpha_i + \beta_i) \tag{5.27}$$

Summing the square of the two equations yields

$$(x_i - x_{Ai})^2 + (y_i - y_{Ai})^2 = a_i^2 + b_i^2 + 2a_ib_i \cos\beta_i \tag{5.28}$$

Hence,

$$\beta_i = \cos^{-1}\frac{(x_i - x_{Ai})^2 + (y_i - y_{Ai})^2 - a_i^2 - b_i^2}{2a_ib_i}, i=1,2,3 \tag{5.29}$$

Further, multiplying Eq. (5.26) by -sin α_i and Eq. (5.27) by cos α_i with the following summing with one another yields

$$-(x_i - x_{Ai})\sin\alpha_i + (y_i - y_{Ai})\cos\alpha_i = b_i \sin\beta_i \tag{5.30}$$

The left-hand side of Eq. (5.30) presents the general sinusoidal function [20]

$$C_i \cos(\alpha_i + \lambda_i) = b_i \sin\beta_i \tag{5.31}$$

with

$$C_i = \sqrt{(x_i - x_{Ai})^2 + (y_i - y_{Ai})^2} \text{ and } \lambda_i = \tan^{-1}\frac{x_i - x_{Ai}}{y_i - y_{Ai}} \tag{5.32}$$

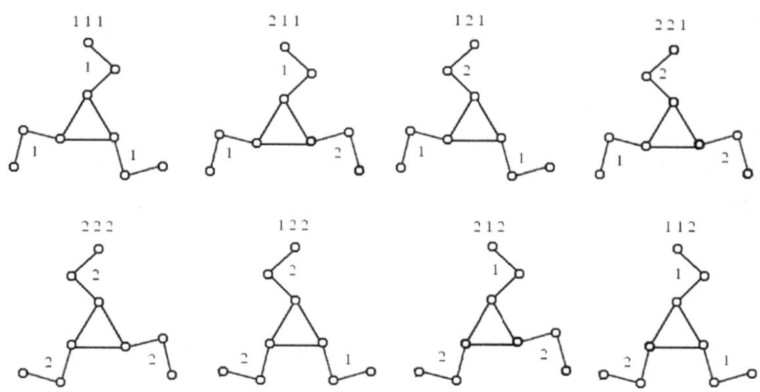

Figure 5-19 Versions of the manipulator assembly.

Solution of Eq. (5.30) with respect of angle α_i is

$$\alpha_i = \cos^{-1}\frac{b_i \sin\beta_i}{C_i} - \lambda_i \tag{5.33}$$

Since each of angles α_i, Eq. (5.33), and β_i, Eq. (5.29), has two different values, the inverse kinematics problem has eight solutions, if any. The eight versions of assembly corresponding to the eight solutions are shown in Figure 5.19. Note that usually angles α_i present controlled (actuated) parameters of robot motions while angles β_i are passive parameters. The set of Eqs. (5.29) and (5.33) gives complete solution of the inverse kinematics problem.

Numerical Example 5.1.
Given: Dimensions of the mechanism (Fig. 5.18) are: l_A = 1.0 m; l_B = 0.5 m; and $a_i = b_i$ = 0.25 m (i = 1, 2, 3). The center of the moving platform is located at point x_0 = 0.1, y_0 = 0.15 m; the platform is oriented according to angle $\varphi = \pi/12$. **Required**: Solution of the inverse kinematics problem.

Solution:

(1) Position vectors of points A_1, A_2, and A_3 (in m) in the fixed coordinates frame OXY are

$$\mathbf{r}_{A1} = l_A\begin{bmatrix} 1/2 \\ -\sqrt{3}/6 \end{bmatrix} = \begin{bmatrix} 0.5 \\ -0.288675 \end{bmatrix}; \mathbf{r}_{A2} = l_A\begin{bmatrix} 0 \\ \sqrt{3}/3 \end{bmatrix} = \begin{bmatrix} 0 \\ 0.57735 \end{bmatrix};$$

$$\mathbf{r}_{A3} = l_A\begin{bmatrix} -1/2 \\ -\sqrt{3}/6 \end{bmatrix} = \begin{bmatrix} -0.5 \\ -0.288675 \end{bmatrix};$$

(2) Position vectors of points Q_1, Q_2, and Q_3 (in m) in the coordinates frame $O_pX_pY_p$:

$$\rho_1 = l_B\begin{bmatrix} 1/2 \\ -\sqrt{3}/6 \end{bmatrix} = \begin{bmatrix} 0.25 \\ -0.144338 \end{bmatrix}; \rho_2 = l_B\begin{bmatrix} 0 \\ \sqrt{3}/3 \end{bmatrix} = \begin{bmatrix} 0 \\ 0.288675 \end{bmatrix};$$

$$\rho_3 = l_B\begin{bmatrix} -1/2 \\ -\sqrt{3}/6 \end{bmatrix} = \begin{bmatrix} -0.25 \\ -0.144338 \end{bmatrix}$$

(3) Coordinates of points Q_1, Q_2, and Q_3 in the fixed coordinate frame OXY are

$$\mathbf{r}_i = \mathbf{r}_0 + \mathbf{A}(\varphi)\rho_i = \begin{bmatrix} 0.10 \\ 0.15 \end{bmatrix} + \begin{bmatrix} \cos(\pi/12) & -\sin(\pi/12) \\ \sin(\pi/12) & \cos(\pi/12) \end{bmatrix}\rho_i, \text{ with } i = 1,\ 2,\ 3$$

Substituting position vectors ρ_1, ρ_2, and ρ_3 yields

$x_1 = 0.1 + 0.25 \cos(\pi/12) + 0.144338 \sin(\pi/12) = 0.378839$

$y_1 = 0.15 + 0.25 \sin(\pi/12) - 0.144338 \cos(\pi/12) = 0.0752854$

$x_2 = 0.1 - 0.288674 \sin(\pi/12) = 0.0252854$

$y_2 = 0.15 + 0.288674 \cos(\pi/12) = 0.428839$

$x_3 = 0.1 - 0.25 \cos(\pi/12) + 0.144338 \sin(\pi/12)] = -0.104124$

$y_3 = 0.15 - 0.25 \sin(\pi/12) - 0.144338 \cos(\pi/12)] = -0.0541241$

(4) Angles β_i ($i = 1, 2, 3$) computed through Eq. (5.29)are

$\beta_1 = \pm 1.39268 = \pm 79.795^\circ$; $\beta_2 = \pm 3.52949 = \pm 144.93^\circ$; $\beta_3 = \pm 0.80397 = \pm 46.064^\circ$

(5) Angle α_i ($i = 1, 2, 3$) are calculated through Eq. (5.33), where $b_1 = b_2 = b_3 = 0.25$ m. Two triplets of the angle values are

$\alpha_{11} = \pm 3.58849 = \pm 148.31^\circ$ $\qquad \alpha_{12} = \pm 1.19582 = \pm 68.5152^\circ$

$\alpha_{21} = \pm 3.00418 = \pm 173.127^\circ$ $\qquad \alpha_{22} = \pm 0.474694 = \pm 27.198^\circ$

$\alpha_{31} = \pm 0.93686 = \pm 53.6781^\circ$ $\qquad \alpha_{32} = \pm 0.132892 = \pm 7.61417^\circ$

The solution of the inverse kinematics problem is completed. Note that there are 64 combinations of angles α_i, while the inverse kinematics has only eight solutions corresponding to the eight assembly versions shown in Figure 5.19. In Figure 5.20, the results of calculations for one of the eight versions are detailed: $\alpha_1 = 68.5152^\circ$; $\alpha_2 = 27.198^\circ$; $\alpha_3 = 7.61417^\circ$; $\beta_1 = 79.7945^\circ$; $\beta_2 = 144.929^\circ$; and $\beta_3 = 46.064^\circ$.

5.2.3 Gough-Stewart platform: inverse kinematics problem

5.2.3.1 *Algorithm*

The pose of the Gough-Stewart platform depends on and is controlled by current lengths of the six limbs. According to Eq. (5.10), the solution of the inverse kinematics, $\mathbf{Q} = \mathbf{Q}(\mathbf{X})$, is

$$L_1 = L_1(x_p, y_p, z_p, \theta, \psi, \varphi),$$

$$L_2 = L_2(x_p, y_p, z_p, \theta, \psi, \varphi),$$

....

$$L_6 = L_6(x_p, y_p, z_p, \theta, \psi, \varphi) \tag{5.34}$$

where x_p, y_p, z_p are the coordinates of the position of the platform center point O_p (Fig. 5.21); θ, ψ, and φ are the Euler angles of the moving platform orientation; and l_i is the current

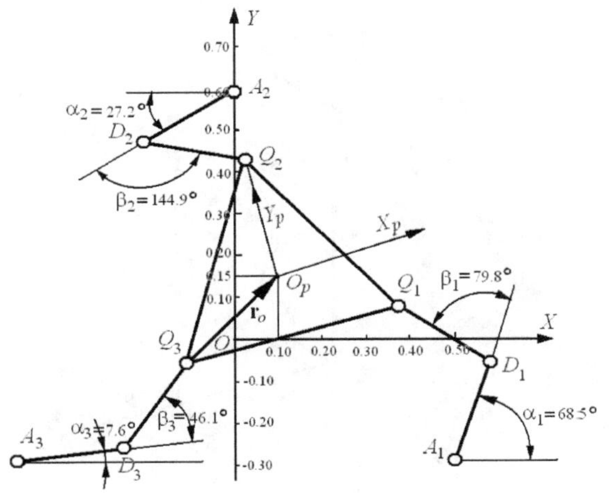

Figure 5-20 Configuration of the manipulator according to Numerical Example 5.1.

lengths of the *i*th limb, $l_i = |M_iQ_i|$ depending on a current position of the built-in-limb translational drive.

Actually, if the pose of the mobile platform is known, the limb length represents the distance between the upper and lower ends of the limb and this distance can be calculated from regular geometrical consideration. For this purpose, vector $\mathbf{r}_i$ of the *i*th limb ($i = 1,\ldots,6$) is defined as the difference of the vectors of the upper and lower ends of the limbs

$$\mathbf{r}_i = [x_i, y_i, z_i, 0]^T = \mathbf{r}_{Qi} - \mathbf{r}_{Mi} \; (i = 1,\ldots,6) \tag{5.35}$$

where x_i, y_i, z_i are the Cartesian coordinates defining the *i*th limb; $\mathbf{r}_{Mi}$ is the configuration independent position vector of point M_i associated with the fixed base; and $\mathbf{r}_{Qi}$ is the position vector corresponding to point Q_i (the upper point of the limb M_iQ_i).

The vectors entering into Eq. (5.35) are expressed through structural and pose parameters of the mechanism. The vector of the lower ends of the *i*th limb depends on radius r_1 of the base circle (Fig. 5.22a) and the polar angle α_i of point M_i,

$$\mathbf{r}_{Mi} = [r_1 \cos \alpha_i, r_1 \sin \alpha_i, 0, 1]^T, i = 1,\ldots,6 \tag{5.36}$$

Vector $\rho_i = \overline{O_pQ_i}$ of the upper ends Q_i, as referred to the moving platform associated frame $O_pX_pY_pZ_p$ (Fig. 5.22b) is configuration-independent and structure-dependent. The position vector is defined through radius r_2 and polar angle β_i,

$$\rho_i = [r_2 \cos \beta_i, r_2 \sin \beta_i, 0, 1]^T, i = 1,\ldots,6 \tag{5.37}$$

Transfer of the coordinates of the upper end Q_i from the frame $O_pX_pY_pZ_p$ to the base associated frame $OXYZ$ is carried out using the extended Euler matrix $\mathbf{A}_{Eu}(x_p, y_p, z_p, \theta, \psi, \varphi)$, Eq. (3.147):

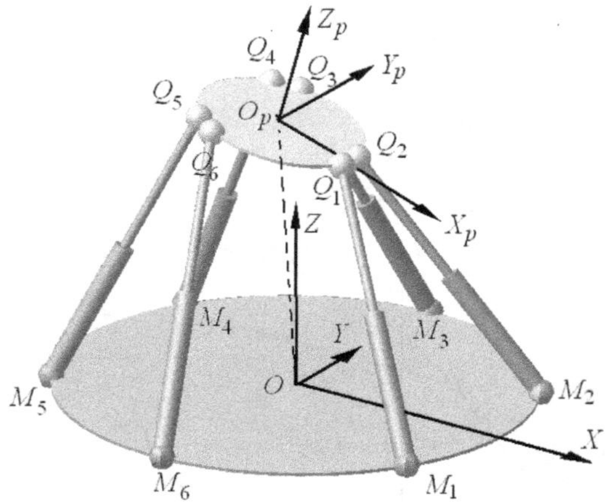

Figure 5-21 GSP configuration.

$$\mathbf{r}_{Qi} = \mathbf{A}_{Eu}(x_p, y_p, z_p, \theta, \psi, \varphi)\, \boldsymbol{\rho}_i =$$

$$\begin{bmatrix} \cos\varphi\,\cos\psi & -\sin\varphi\,\cos\psi & \sin\psi & x_p \\ \cos\varphi\,\sin\psi\,\sin\theta + \sin\varphi\,\cos\theta & \cos\varphi\cos\theta - \sin\varphi\sin\psi\,\sin\theta & -\cos\psi\,\sin\theta & y_p \\ -\cos\varphi\,\sin\psi\,\cos\theta + \sin\varphi\sin\theta & \sin\varphi\sin\psi\,\cos\theta + \cos\varphi\sin\theta & \cos\psi\,\cos\theta & z_p \\ 0 & 0 & 0 & 1 \end{bmatrix} \begin{bmatrix} r_2\cos\beta_i \\ r_2\sin\beta_i \\ 0 \\ 1 \end{bmatrix}$$

$$= \begin{bmatrix} x_p + r_2\cos\psi\cos(\varphi+\beta_i) \\ y_p + r_2[\cos\theta\sin(\varphi+\beta_i) + \sin\theta\sin\psi\cos(\varphi+\beta_i)] \\ z_p + r_2[\sin\theta\sin(\varphi+\beta_i) - \cos\theta\sin\psi\cos(\varphi+\beta_i)] \\ 1 \end{bmatrix}, \; i = 1,\ldots,6 \qquad (5.38)$$

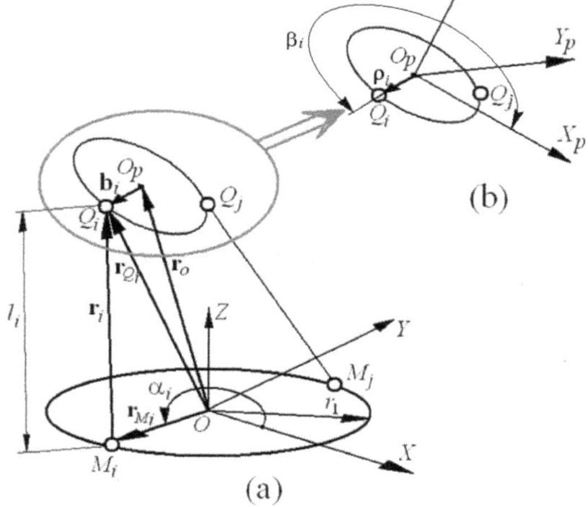

Figure 5-22 Limb location and orientation. Vector O_pQ_i is shown as referred to: (a) frame $OXYZ$ (vector $\mathbf{b}_i$); and (b) frame $O_pX_pY_pZ_p$ (vector $\boldsymbol{\rho}_i$).

where x_p, y_p, and z_p are the parameters of the ith body translation along the X-, Y-, and Z-axes, respectively; θ, ψ, and φ are the Euler angles of the ordered rotations around the X-, Y-, Z-axes, respectively.

Substitution of Eqs. (5.38) and (5.36) in Eq. (5.35) yields the non-position vector of the limb:

$$\mathbf{r}_i \equiv \begin{bmatrix} x_i \\ y_i \\ z_i \\ 0 \end{bmatrix} = \begin{bmatrix} x_p + r_2 \cos\psi \cos(\varphi+\beta_i) \\ y_p + r_2[\cos\theta \sin(\varphi+\beta_i) + \sin\theta \sin\psi \cos(\varphi+\beta_i)] \\ z_p + r_2[\sin\theta \sin(\varphi+\beta_i) - \cos\theta \sin\psi \cos(\varphi+\beta_i)] \\ 1 \end{bmatrix} - \begin{bmatrix} r_1 \cos\alpha_i \\ r_1 \sin\alpha_i \\ 0 \\ 1 \end{bmatrix}$$

$$= \begin{bmatrix} x_p + r_2 \cos\psi \cos(\varphi+\beta_i) - r_1 \cos\alpha_i \\ y_p + r_2[\cos\theta \sin(\varphi+\beta_i) + \sin\theta \sin\psi \cos(\varphi+\beta_i)] - r_1 \sin\alpha_i \\ z_p + r_2[\sin\theta \sin(\varphi+\beta_i) - \cos\theta \sin\psi \cos(\varphi+\beta_i)] \\ 0 \end{bmatrix} (i=1,\ldots,6) \quad (5.39)$$

The current length L_i of the ith limb calculated as a magnitude $||\mathbf{r}_i||$ of vector $\mathbf{r}_i$, Eq. (5.39), presents solution of the inverse-kinematics problem:

$$L_i = \|\mathbf{r}_i\| = \sqrt{x_i^2 + y_i^2 + z_i^2} =$$

$$[(x_p + r_2 \cos\psi \cos(\varphi+\beta_i) - r_1 \cos\alpha_i)^2 + (y_p + r_2 \cos\theta \sin(\varphi+\beta_i) + r_2 \sin\theta \sin\psi \cos(\varphi+\beta_i) - r_1 \sin\alpha_i)^2 + (z_p + r_2 \sin\theta \sin(\varphi+\beta_i) - r_2 \cos\theta \sin\psi \cos(\varphi+\beta_i))^2]^{1/2} \, (i=1,\ldots,6) \quad (5.40)$$

Mathematica codes for Eqs. (5.38-40) are shown in Appendix *A2.2.5*.

In what follows, representation of vector $\overline{O_pQ_i}$ in the base coordinate frame $OXYZ$ will be needed. As is clear from Figure 5.22, vector $\overline{O_pQ_i}$ can be represented in the frame $OXYZ$ by the difference of two position vectors $\mathbf{r}_{Qi}$ and $\mathbf{r}_p$, i.e., presents non-position vector:

$$\mathbf{b}_i = \mathbf{r}_{Qi} - \mathbf{r}_p, \; i = 1,\ldots,6 \quad (5.41)$$

where $\mathbf{r}_p$ is the position vector of the platform geometrical center O_p,

$$\mathbf{r}_p = (x_p, y_p, z_p, 1)^{\mathrm{T}} \quad (5.42)$$

Vectors $\boldsymbol{\rho}_i$ and $\mathbf{b}_i$ are separately shown in Figures 5.22a and 5.22b, respectively. Obviously, vectors $\boldsymbol{\rho}_i$ and $\mathbf{b}_i$ coincide graphically, since both present the same vector $\overline{O_pQ_i}$ expressed in two different coordinate frames. Substitution of Eqs. (5.38) and (5.42) in Eq. (5.41) yields

$$\mathbf{b}_i = \mathbf{r}_{Qi} - \mathbf{r}_p = \begin{bmatrix} x_p + r_2 \cos\psi\cos(\varphi+\beta_i) \\ y_p + r_2[\cos\theta\sin(\varphi+\beta_i) + \sin\theta\sin\psi\cos(\varphi+\beta_i)] \\ z_p + r_2[\sin\theta\sin(\varphi+\beta_i) - \cos\theta\sin\psi\cos(\varphi+\beta_i)] \\ 1 \end{bmatrix} - \begin{bmatrix} x_p \\ y_p \\ z_p \\ 1 \end{bmatrix}$$

$$= r_2 \begin{bmatrix} \cos\psi\cos(\varphi+\beta_i) \\ [\cos\theta\sin(\varphi+\beta_i) + \sin\theta\sin\psi\cos(\varphi+\beta_i)] \\ [\sin\theta\sin(\varphi+\beta_i) - \cos\theta\sin\psi\cos(\varphi+\beta_i)] \\ 0 \end{bmatrix} \quad (5.43)$$

5.2.3.2 *Partial cases: The 6–3 and 3–3 GSP*

In the 6–3 GSP architecture (Fig. 5.23), the lower ends $M_1, M_2, \ldots, M_6$ of the limbs lie in the base plane at *six* points located in the circle of radius r_1 in 60° intervals, while the six upper ends $Q_1, Q_2, \ldots, Q_6$ are attached in pairs to the mobile platform plane at *three* points located in the circle of radius r_2 in 120° intervals. The first example demonstrates a solution of the inverse kinematics problem for a fixed pose of the moving platform. In the second example, the limb lengths are calculated when the platform carried out a vertical motion.

Numerical Example 5.2.

Given: Dimensions of the 6–3 GSP (Fig. 5.23): $r_1 = 0.7$ m; $r_2 = 0.25$ m; polar angles of points M_i: $\alpha_i = -\pi/6 + (i-1)\pi/3$, with $i = 1,\ldots,6$; and polar angles of points Q_i: $\beta_i = \beta_{i+1} = (i-1)\pi/3$, with $i = 1, 3, 5$. Platform pose: $x_p = -0.05$ m, $y_p = 0.1$ m; $z_p = 0.8$ m; $\theta = \varphi = 0$, $\psi = \pi/12$. **Required:** Calculation of the limb lengths L_i in the given configuration.

Solution: According to Eq. (5.39) we have

$$\mathbf{r}_i = \begin{bmatrix} x_p + r_2 \cos\psi\cos(\varphi+\beta_i) - r_1\cos\alpha_i \\ y_p + r_2[\cos\theta\sin(\varphi+\beta_i) + \sin\theta\sin\psi\cos(\varphi+\beta_i)] - r_1\sin\alpha_i \\ z_p + r_2[\sin\theta\sin(\varphi+\beta_i) - \cos\theta\sin\psi\cos(\varphi+\beta_i)] \\ 0 \end{bmatrix}$$

$$= \begin{bmatrix} -0.05 + 0.25\cos(\pi/12)\cos\beta_i \text{-} 0.7\cos\alpha_i \\ 0.1 + 0.25\sin\beta_i \text{-} 0.7\sin\alpha_i \\ 0.8 \text{-} 0.25\sin(\pi/12)\cos\beta_i \\ 0 \end{bmatrix}$$

i.e.,

$x_i = -0.05 + 0.24148 \cos \beta_i - 0.7 \cos \alpha_i;$

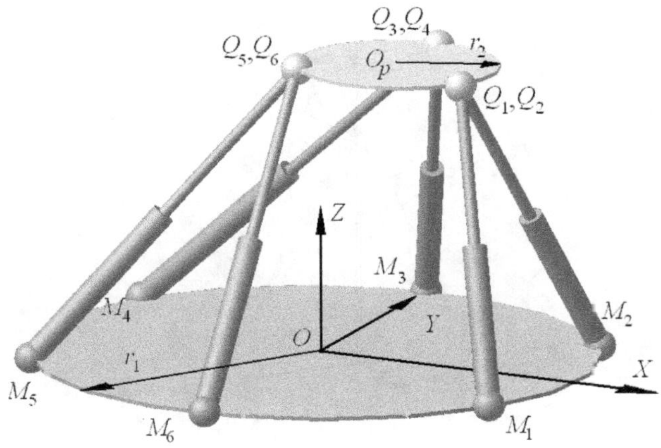

Figure 5-23 The 6-3 GSP architecture.

$y_i = 0.1 + 0.25 \sin \beta_i - 0.7 \sin \alpha_i;$

$z_i = 0.8 + 0.064705 \cos \beta_i; \ (i = 1,\ldots,6)$

According to Eq. (5.40), the current length $L_i (i = 1, 2,\ldots,6)$ is as follows:

$$L_i = \sqrt{x_i^2 + y_i^2 + z_i^2} =$$

$$\sqrt{(-0.05 + 0.24148 \cos\beta_i - 0.7 \cos\alpha_i)^2 \ + (0.1 + 0.25 \sin\beta_i - 0.7 \sin\alpha_i)^2 + (0.}$$

Results of calculations are listed in Table 5.1 and shown in Figure 5.24.

Numerical Example 5.3.
Given: Dimensions of the 6–3 GSP: $r_1 = 0.7$ m, $r_2 = 0.25$ m;polar angles of points M_i: $\alpha_i = -\pi/6 + (i - 1)\pi/3$, with $i = 1, 2,\ldots,6$; andpolar angles of points Q_i: $\beta_i = \beta_{i+1} = (i - 1)\pi/3$, with $i = 1, 3, 5$. Platform pose: $x_p = -0.05$ m, $y_p = 0.1$ m; $\theta = \varphi = 0$; $\psi = \pi/12$. **Required:** Calculation of the limb lengths during movement of the moving platform along the Z-axis with $z_p = 0.2 \div 0.8$ m.

Solution: Since all data excluding z_p value are the same as that in the previous Example, we can use, as before, Eqs. (5.39) and (5.40):

$$\mathbf{r}_i = \begin{bmatrix} x_p + r_2 \cos\psi \cos(\varphi + \beta_i) - r_1 \cos\alpha_i \\ y_p + r_2[\cos\theta \sin(\varphi + \beta_i) + \sin\theta \sin\psi \cos(\varphi + \beta_i)] - r_1 \sin\alpha_i \\ z_p + r_2[\sin\theta \sin(\varphi + \beta_i) - \cos\theta \sin\psi \cos(\varphi + \beta_i)] \\ 0 \end{bmatrix}$$

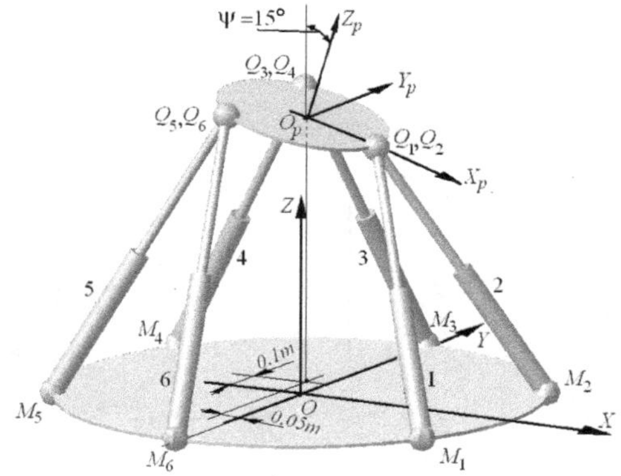

Figure 5-24 The configuration according to Numerical Example 5.2.

$$= \begin{bmatrix} -0.05+0.24148\cos\beta_i - 0.7\cos\alpha_i \\ 0.1+0.25\sin\beta_i - 0.7\sin\alpha_i \\ z_p - 0.064705\cos\beta_i \\ 0 \end{bmatrix};$$

$$L_i = \sqrt{(-0.05+0.24148\cos\beta_i - 0.7\cos\alpha_i)^2 + (0.1+0.25\sin\beta_i - 0.7\sin\alpha_i)^2 + (z_p - 0.064705\cos\beta_i)^2}$$

Results of calculations are shown in Figure 5.25. Note that the limb lengths for $z_p = 0.8$ m are obtained in the previous Example and listed in Table 5.1.

In 3–3 GSP architecture (Fig. 5.26), both lower and upper limb ends are attached in pairs to the base and the mobile platform, respectively: $\alpha_1 = \alpha_6 = -\pi/3$, $\alpha_2 = \alpha_3 = \pi/3$, $\alpha_4 = \alpha_5 = \pi$, and polar angles of point Q_i $\beta_i = \beta_{i+1} = (i - 1)\pi/3$, with $i = 1, 3, 5$. As is clear from Figures 5.6-5.9, this architecture is in most common use for different applications. The reason can be explained through analysis of the stiffness properties of the GSP when the moving platform is located in the central position and horizontally orientated. In this configuration, the 6–6 GSP, in which the limb ends arranged in hexagon vertices of the base and movingplatform, has zero stiffness value in some directions since six lines presenting continuation of the limbs meet at one point, i.e., this is a singular configuration of the mechanism. Contrary to that, the 3–3 GSP has the maximal stiffness value compared with the other symmetrical architectures (including the 6–3 GSP).

5.2.4 Direct kinematics

(DKP) determines the pose of the end-effector (EE) when the actuated coordinates of the robot, i.e., the lengths of the limbs due to current displacements of the robot's actuators are given. Solution of the DKP is applied for formulation of feedback signals in the systems controlling the EE pose and velocity.

Formally, the solution can be obtained from the set of simultaneous equations formulating the inverse kinematic problem (IKP). In these terms, synthesis of the equations of the

Table 5-1 Polar angles and lengths of the limbs.

Limb i	α_i	β_i	Length L_i
1	$-\pi/6$	0	0.956643
2	$\pi/6$	0	0.880435
3	$\pi/2$	$2\pi/3$	0.932218
4	$5\pi/6$	$2\pi/3$	0.939985
5	$7\pi/6$	$4\pi/3$	0.967972
6	$3\pi/2$	$4\pi/3$	1.03074

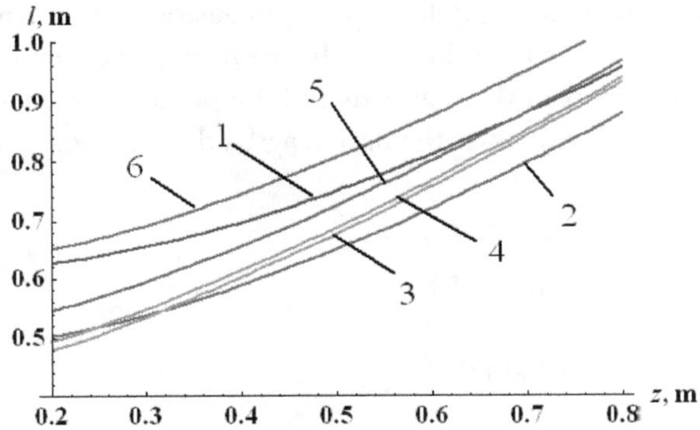

Figure 5-25 The limb lengths during motion of the moving platform along the Z-axis (*Numerical Example* 5.3).

IKP can be considered as an initial stage for solution of the DKP. As a rule, the DKP for parallel manipulators is significantly more complicated than their IKP and cannot be solved in an explicit analytical form since the initial set of equations is non-linear, of high order, and transcendental. As a result, the solution of the DKP is usually not unique, i.e., the same set

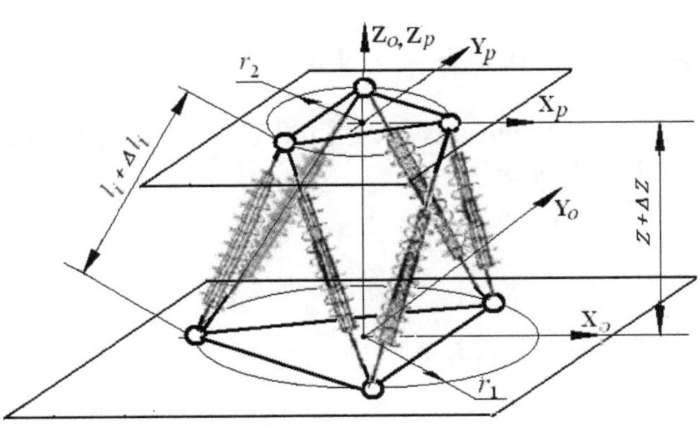

Figure 5-26 The 3–3 Gough-Stewart platform in central position and horizontal orientation of the moving platform.

of articular coordinates (positions of actuators) corresponds to some different poses of the robot. These poses are called "assembly modes".

As a relatively simple example, the DKP solution for the 3-DOF planar parallel robot shown in Figure 5.27 [1] is considered. A current pose of the moving platform in the *XY*-plane is defined by controlled lengths l_1, l_2, and l_3 of three extended limbs. The limbs are connected with the platform and the base through revolute joints, *i.e.*, the limbs are of the RPR type. The mechanism is planar ($\lambda = 3$), contains $n = 8$ links and $j = 9$ joints (6 revolute + 3 prismatic) possessing 1-DOF each, $f_1 = \ldots = f_9 = 1$. The DOF number F, Eq. (5.1), is

$$F = \lambda\,(n - j - 1) + (f_1 + \ldots + f_j) = 3\,(8 - 9 - 1) + 9 \times 1 = 3 \tag{5.44}$$

The aim of the DKP is to define a platform pose (its position and orientation) for given values of the actuated limbs l_1, l_2, and l_3. In this case, the platform position is defined through coordinates x and y of the left point of the platform and the platform orientation associated with angle Φ. Solution of the DKP expresses the limb lengths through pose parameters x, y, and Φ

$$l_1^2 = x^2 + y^2 \tag{5.45}$$

$$l_2^2 = (c_2 - x - a\cos\Phi)^2 + (y + a\sin\Phi)^2 \tag{5.46}$$

$$l_3^2 = (c_3 - x - b\cos\Phi)^2 + (y + b\sin\Phi)^2 \tag{5.47}$$

The following solution is described according to Merlet [1]. Subtracting Eq. (5.45) from Eqs. (5.46) and (5.47) results in two linear equations with respect to x and y:

$$2(a\cos\Phi - c_2)x + (2a\sin\Phi)y = 2ac_2\cos\Phi + l_2^2 - (a^2 + c_2^2 + l_1^2) \tag{5.48}$$

$$2(b\cos\Phi - c_3)x + (2b\sin\Phi)y = 2bc_3\cos\Phi + l_3^2 - (b^2 + c_3^2 + l_1^2) \tag{5.49}$$

Solutions $x = x(\Phi)$ and $y = y(\Phi)$ are substituted in Eq. (5.45). This results in a cubic equation in $T = \cos\Phi$:

$$A_3T^3 + A_2T^2 + A_1T + A_0 = 0, \text{ with } -1 \le T \le 1 \tag{5.50}$$

This equation has three real solutions T_1, T_2, and T_2. Let the solutions satisfying requirement $|T| \le 1$ be T_1 and T_2. Hence,

$$\Phi_{1,2} = \pm \arccos T_1 \text{ and } \Phi_{3,4} = \pm \arccos T_2 \tag{5.51}$$

$$x_i = x(\Phi_i) \text{ and } y_i = y(\Phi_i),\ i = 1, 2, 3, 4 \tag{5.52}$$

Numerical Example 5.4.

Let parameters of the mechanism in Figure 5.27 be: $a = 0.2$ m, $b = 0.12$ m, $c_2 = 0.28$ m, $c_3 = 0.14$ m. The current lengths of the limbs are $l_1 = 0.10$ m, $l_2 = 0.15$ m, and $l_3 = 0.12$ m.

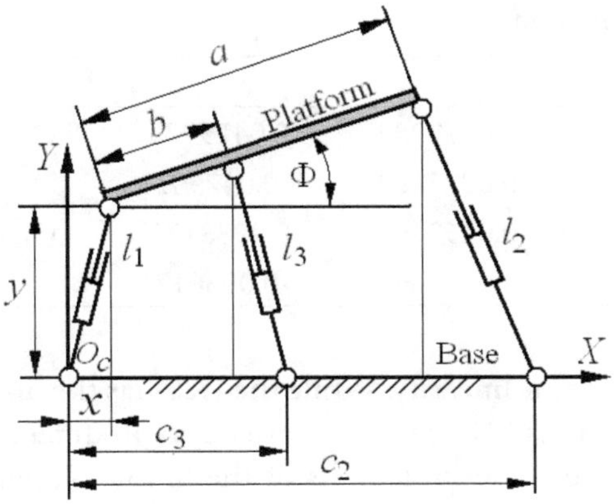

Figure 5-27 A planar 3-RPR parallel manipulator.

Required: Definition of platform pose (x_i, y_i, and Φ).
Solution: Substitution of the initial data in Eqs. (5.46) and (5.47) followed by solution of the set of two simultaneous equations with respect to x and y yields:

$$x = 0.606071 - 0.6 \cos \Phi \tag{5.53}$$

$$y = (0.58375 - 1.16607 \cos \Phi + 0.6 \cos^2 \Phi)/\sin \Phi \tag{5.54}$$

Substitution of these expressions into Eq. (5.45) yields

$$0.1^2 = (0.606071 - 0.6 \cos \Phi)^2 + (0.58375 - 1.16607 \cos \Phi + 0.6 \cos^2 \Phi)^2/\sin^2 \Phi$$

Formal transformations yield a cubic equation in the form of Eq. (5.50):

$$0.698087 - 2.08867\, T + 2.0629\, T^2 - 0.672\, T^3 = 0$$

The equation has three solutions $T_1 = 0.863905$, $T_2 = 0.98456$, and $T_3 = 1.22133$. Obviously only T_1 and T_2 can be considered as possible values of cosΦ. As a result, there are four solutions:

$$\Phi_{1,2} = \pm \arccos 0.863905 = \pm 30.2421°; \text{ and } \Phi_{3,4} = \pm \arccos 0.98456 = \pm 10.0815°.$$

Substituting the angle values into Eqs. (5.53) and (5.54) yields coordinates x and y listed in Table 5.2. In this case, there are four poses corresponding to four solutions. In Figure 5.28, two of four assembly modes corresponding to Solutions 1 and 3 are shown. The 2nd and 4th poses present the specular reflection of poses 1 and 3, respectively, relative to the X-axis in Figure 5.28.

The example demonstrates the complexity of the DKP solution for parallel kinematics machines compared with the IKP. Usually, the explicit solution in symbolic form is impossible, and a numerical solution is not easy. For example, in the case of the GSP mechanism, the direct kinematics solution expresses six position and orientation parameters through the current lengths $l_1,\ldots,l_6$ of the limbs as a set of six simultaneous transcendental equations

Table 5-2 Assembly modes.

Pose	Φ, degs	*x*, m	*y*, m	Fig. 5.26
1	30.2421	0.0877282	0.0479975	On the left
2	-30.2421	0.0877282	-0.0479975	
3	10.0815	0.0153355	0.0988171	On the right
4	-10.0815	0.0153355	-0.0988171	

characterized by high-order non-linearity. While the IKP has the unique solution, the direct one can have up to 40 solutions [21], i.e., there are up to 40 different configurations of the mechanism corresponding to six given lengths of the limbs. Note that 20 solutions differ from 20 others only by the sign of the *z*-coordinates, i.e., present specular reflections relative to the base plane.

5.3 FSF formulation

5.3.1 FSF of the hexapod-based milling machine

Synthesis of the FSF of parallel- and hybrid-kinematics FSS is a procedure whose complexity is approximately similar to that for serial-kinematics systems. Actually, the manipulating matrix results from ordered multiplication of the links-associated matrices whether it is one- or multi-parametric matrix. However, a single parameter of the one-parametric matrix is different in nature from parameters entering into a multi-parametric matrix. Namely, a single parameter q_i in one-parametric matrix ${}^{i-1}\mathbf{A}_i(q_i)$, Eq. (3.138), associated with a 1-DOF link directly describes the motion $q_i = q_i(t)$, if the *i*th link is the actuated one, or dimension q_i, if the *i*th link is the geometric one. Contrastingly, six parameters $x_i, y_i, z_i, \theta_i, \psi_i, \varphi_i$ of multi-parametric matrix $i{-}^{1}\,\mathbf{A}_{Eu}$, Eq. (3.147), describe a pose of the platform of the parallel kinematics link, while kinematic parameters – the controlled motions of the actuators (in the case of the GSP, the current lengths of the limbs) – are defined through solution of the inverse kinematics problem described in Section 5.2.1. In other terms, solution of the inverse kinematics problem can be considered as the set of six *kinematic constraints*.

Two typical structures of the hexapod-based machine tools are shown in Figure 5.29. The workspace in the architecture shown in Figure 5.29a is located between the limbs

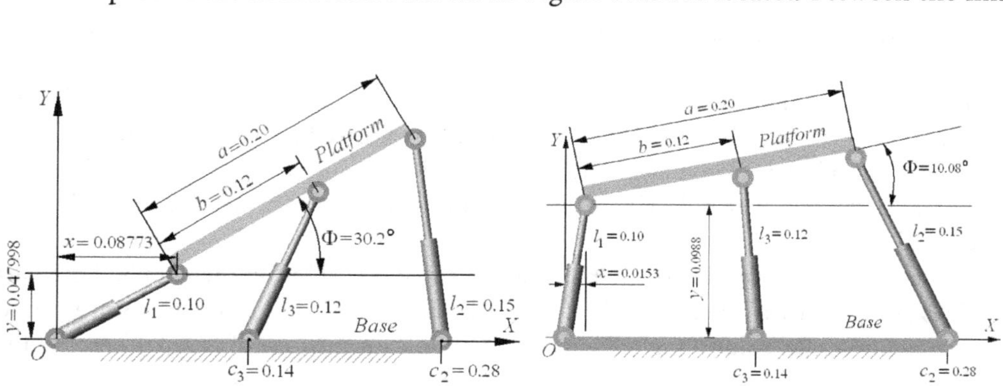

Figure 5-28 Solution of the direct kinematics problem.

(internal workspace) similarly to that in Figure 5.7a and c. Contrary to this, the limbs can be located outside of the workspace (Fig. 5.29b) as is shown in Figure 5.7b (external workspace). In both cases, the PKM consists of the moving platform having 6-DOF in motion relative to the base. The machine has six translational feed motions(six translations resulting in elongations of the limbs) and one cutting motion – rotation of the spindle around its axis. The workpiece is arranged on the stationary table associated with the base, and the moving platform bears the main spindle with milling cutter.

The FSS of the machine combines the fixed base and two kinematic links (n = 2): the 6-DOF moving platform and one-DOF rotating spindle. Hence, the mobility number is l = 7.The FSS described in Table 5.3 also includes two geometric links. The coordinate code of the FSS of the hexapod-based milling machine is k = (123) (**123456**) **6**3, and the FSF is

$$[\mathbf{r}_0\mathbf{c}_0] = {}^0\mathbf{A}_4\,[\mathbf{r}_4\mathbf{c}_4], \tag{5.55}$$

with

$$\mathbf{r}_0 = [x_0, y_0, z_0, 1]^{\mathrm{T}} \text{ and } \mathbf{r}_4 = [x_4, y_4, z_4, 1]^{\mathrm{T}} \tag{5.56}$$

$$\mathbf{c}_0 = [c_{0x}, c_{0y}, c_{0z}, 0]^{\mathrm{T}} \text{ and } \mathbf{c}_4 = [c_{4x}, c_{4y}, c_{4z}, 0]^{\mathrm{T}} \tag{5.57}$$

$${}^0\mathbf{A}_4 = \mathbf{A}_{tr}(-x_1, -y_1, -z_1)\,\mathbf{A}_{Eu}(x, y, z, \theta, \psi, \varphi)\,\mathbf{A}^6(\varphi_{sp})\,\mathbf{A}^3(h) \tag{5.58}$$

where $\mathbf{r}_0$ and $\mathbf{c}_0$ are the homogeneous position and orientation vectors, respvectively, as referred to coordinate frame S_0 associated with the target object (e.g., the workpiece); $\mathbf{r}_6$ and $\mathbf{c}_6$ are the same in coordinateframe S_6; $x, y, z, \theta, \psi, \varphi, h$, and φ_{sp} are the parameters specified in Table 5.3; $\mathbf{A}_{Eu}(x, y, z, \theta, \psi, \varphi)$ is the extended Euler's matrix, Eq. (3.147); and $\mathbf{A}^3(h)$ and $\mathbf{A}^6(\varphi_{sp})$ matrices of elementary motions (Table 3.5),

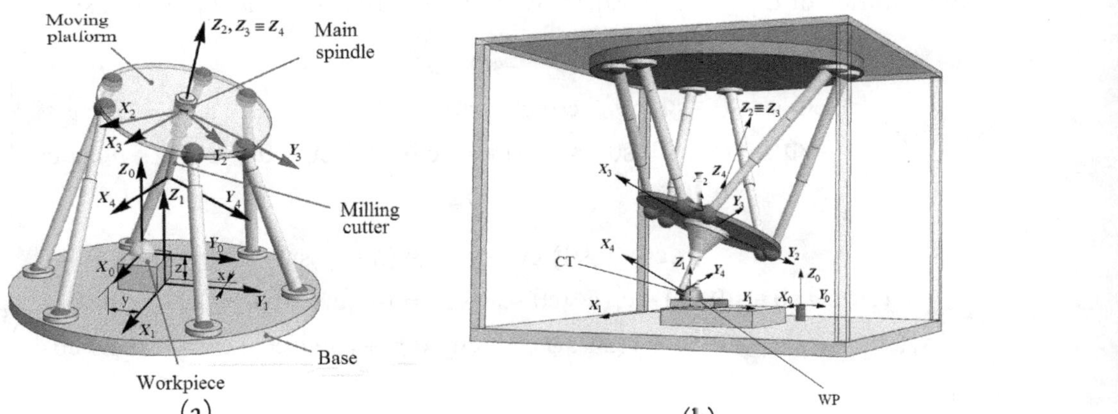

Figure 5-29 Hexapod-based machine: (a) internal workspace; (b) external workspace.

$$^{0}\mathbf{A}_1 = \mathbf{A}_{tr}\left(-x_h,-y_h,-z_h\right) = \begin{bmatrix} 1 & 0 & 0 & -x_h \\ 0 & 1 & 0 & -y_h \\ 0 & 0 & 1 & -z_h \\ 0 & 0 & 0 & 1 \end{bmatrix};$$

$$^{1}\mathbf{A}_2 = \mathbf{A}_{Eu}(x,y,z,\theta,\psi,\varphi) = \begin{bmatrix} \cos\varphi\cos\psi & -\sin\varphi\cos\psi & \sin\psi & x \\ \cos\varphi\sin\psi\sin\theta+\sin\varphi\cos\theta & \cos\varphi\cos\theta-\sin\varphi\sin\psi\sin\theta & -\cos\psi\sin\theta & y \\ -\cos\varphi\sin\psi\cos\theta+\sin\varphi\sin\theta & \sin\varphi\sin\psi\, c\theta+\cos\varphi\sin\theta & \cos\psi\cos\theta & z \\ 0 & 0 & 0 & 1 \end{bmatrix};$$

$$^{2}\mathbf{A}_3 = \mathbf{A}^6(\varphi_{sp}) = \begin{bmatrix} \cos\varphi_{sp} & -\sin\varphi_{sp} & 0 & 0 \\ \sin\varphi_{sp} & \cos\varphi_{sp} & 0 & 0 \\ 0 & 0 & 1 & 0 \\ 0 & 0 & 0 & 1 \end{bmatrix}; \text{ and } {}^{3}\mathbf{A}_4 = \mathbf{A}^3(h) = \begin{bmatrix} 1 & 0 & 0 & 0 \\ 0 & 1 & 0 & 0 \\ 0 & 0 & 1 & h \\ 0 & 0 & 0 & 1 \end{bmatrix} \tag{5.59}$$

Hence, the manipulating matrix can be represented as follows:

$$^{0}\mathbf{A}_4 = {}^{0}\mathbf{A}_1\left(-x_h,-y_h,-z_h\right)\ {}^{1}\mathbf{A}_2(x,y,z,\theta,\psi,\varphi)\,\mathbf{A}^6(\varphi_{sp})\,\mathbf{A}^3(h)$$
$$= \begin{bmatrix} c\Phi\, c\psi & -s\Phi\, c\psi & s\psi & x-x_h+h\, s\psi \\ c\Phi\, s\psi\, s\theta+s\Phi\, c\theta & c\Phi\, c\theta-s\Phi\, s\psi\, s\theta & -c\psi\, s\theta & y-y_h-h\, c\psi\, s\theta \\ -c\Phi\, s\psi\, c\theta+s\Phi\, s\theta & s\Phi\, s\psi\, c\theta+c\Phi\, s\theta & c\psi\, c\theta & z-z_h+h\, c\psi\, c\theta \\ 0 & 0 & 0 & 1 \end{bmatrix} \tag{5.60}$$

where $c\Phi \equiv \cos(\varphi + \varphi_{sp})$, $s\Phi \equiv \sin(\varphi_2 + \varphi_{sp})$, and so on.

Substitution of Eq. (5.60) in Eq. (5.55) yields position and orientation FSFs:

$$\mathbf{r}_0 = {}^{0}\mathbf{A}_4\mathbf{r}_4 = \begin{bmatrix} c\Phi\, c\psi & -s\Phi\, c\psi & s\psi & x-x_h+h\, s\psi \\ c\Phi\, s\psi\, s\theta+s\Phi\, c\theta & c\Phi\, c\theta-s\Phi\, s\psi\, s\theta & -c\psi\, s\theta & y-y_h-h\, c\psi\, s\theta \\ -c\Phi\, s\psi\, c\theta+s\Phi\, s\theta & s\Phi\, s\psi\, c\theta+c\Phi\, s\theta & c\psi\, c\theta & z-z_h+h\, c\psi\, c\theta \\ 0 & 0 & 0 & 1 \end{bmatrix} \begin{bmatrix} x_4 \\ y_4 \\ z_4 \\ 1 \end{bmatrix}$$
$$= \begin{bmatrix} (x_4 c\Phi - y_4 s\Phi)\, c\psi + x - x_h + (h+z_4) s\psi \\ x_4(c\Phi s\psi\, s\theta + s\Phi c\theta) + y_4(c\Phi c\theta - s\Phi s\psi\, s\theta) - z_4 c\psi\, s\theta + y - y_h - h\, c\psi\, s\theta \\ x_4(-c\Phi s\psi\, c\theta + s\Phi s\theta) + y(s\Phi s\psi\, c\theta + c\Phi s\theta)0 + z_4 c\psi\, c\theta + z - z_h + h\, c\psi\, c\theta \\ 1 \end{bmatrix} \tag{5.61}$$

Table 5-3 Form-shaping system of the hexapod-based machine.

S_i	Link	Motion relative to previous link	Matrix ${}^{i-1}\mathbf{A}_i$
S_0	Table with workpiece	Conventionally fixed link	-
S_1	Base	Transfer from the workpiece coordinate frame to the home position	$\mathbf{A}_{tr}(-x_1, -y_1, -z_1)$
S_2	Moving platform	3-DOF translation and 3-DOF rotation	${}^1\mathbf{A}_2 = \mathbf{A}_{Eu}(x, y, z, \theta, \psi, \varphi)$
S_3	Main spindle	Z_2-axis rotation at angle φ_{sp}	${}^2\mathbf{A}_3 = \mathbf{A}^6(\varphi_{sp})$
S_4	Milling cutter	Z_3-axis translation at distance d	${}^3\mathbf{A}_4 = \mathbf{A}^3(d)$

Parameters: x_1, y_1, z_1 are the coordinates of the origin of the workpiece coordinate system as referred to of the home position of the base-associated coordinate frame; x, y, z are the platform centroid position; θ, ψ, φ are the Euler angles of the platform rotation; φ_{sp} is the angle of the spindle rotation; and h is the offset of the cutting tool relative to the platform.

$$\mathbf{c}_0 = {}^0\mathbf{A}_4\,\mathbf{c}_4 = \begin{bmatrix} (c_{4x}\mathrm{c}\Phi - c_{4y}\mathrm{s}\Phi)\ \mathrm{c}\psi + c_{4z}\mathrm{s}\psi \\ c_{4x}(\mathrm{c}\Phi\ \mathrm{s}\psi\ \mathrm{s}\theta + \mathrm{s}\Phi\ \mathrm{c}\theta) + c_{4y}(\mathrm{c}\Phi\ \mathrm{c}\theta - \mathrm{s}\Phi\ \mathrm{s}\psi\ \mathrm{s}\theta) - c_{4z}\mathrm{c}\psi\ \mathrm{s}\theta \\ c_{4x}(-\mathrm{c}\Phi\ \mathrm{s}\psi\ \mathrm{c}\theta + \mathrm{s}\Phi\ \mathrm{s}\theta) + c_{4y}(\mathrm{s}\Phi\ \mathrm{s}\psi\ \mathrm{c}\theta + \mathrm{c}\Phi\ \mathrm{s}\theta)0 + c_{4z}\mathrm{c}\psi\ \mathrm{c}\theta \\ 0 \end{bmatrix} \tag{5.62}$$

The limb lengths $L_1, L_2, \ldots, L_6$ are defined by Eq. (5.40) using six parameters of the platform position $x, y, z, \theta, \psi, \varphi$.

5.3.2 FSF of tripod-based hybrid-kinematics machine

The FSS of the hybrid-kinematics machine presents a chain of links includes both 1-DOF and multi-DOF links. The FSF consists of links with the l_i-DOF number, where $2 \le l_i \le 6$, has to be complemented by $v = (6 - l_i)$ *layout constraints*.

The tripod-type machine consists of the 3-DOF moving platform, i.e., $l_i = 3$ and $v = 6 - 3 = 3$. The tripod-based machine tools are of considerable industrial interest as applied for 5-axis machine tools [10, 22-24]. In comparison with Stewart-platform-based structures, the tripod-based structure has superior stiffness, greater working volume, and reduced number of singular configurations. These advantages result from direct application of two sliders for reciprocating translations.

In the machine shown in Figures 5.30 and 5.31, translations along the X- and the Y-axes are actuated by individual motors, while vertical position z_p of platform 1 and two angles θ_x and θ_y defining a horizontal slope of the platform result from active motions of three sliders 2, 3, 4 along vertical guideways 5, 6, and 7. Connecting rods 8, 9, 10 are attached to the platform and sliders through spherical joints B_i and revolute joints Q_i, respectively. Platform-mounted spindle 11 carries out the rotational cutting motion (main motion) of the cutting tool 12 for milling-drilling operations. The columns with the guideways are mounted on fixed bed 13. The XY-motions of workpiece 14 are carried out by table 15 and carriage 16.

The form-shaping system of the machine includes a geometrical link fixing the home-point position and five kinematic links: longitudinal table 15 (with the workpiece), cross-carriage 16, bed 13, milling head 1, and spindle 11 with a cutting tool (Table 5.4). According to Eq. (3.99), the coordinate code of the FSS is k = (123) **12(345)63**

The FSF of the machine tool is

$$[\mathbf{r}_0\mathbf{c}_0] = {}^0\mathbf{A}_6\,[\mathbf{r}_6\mathbf{c}_6] \tag{5.63}$$

with

$$\mathbf{r}_0 = [x_0, y_0, z_0, 1]^{\mathrm{T}} \text{ and } \mathbf{r}_6 = [x_6, y_6, z_6, 1]^{\mathrm{T}}$$

$$\mathbf{c}_0 = [c_{0x}, c_{0y}, c_{0z}, 0]^{\mathrm{T}} \text{ and } \mathbf{c}_6 = [c_{6x}, c_{6y}, c_{6z}, 0]^{\mathrm{T}}$$

where $\mathbf{r}_0$ and $\mathbf{r}_6$ are the position vectors of the active point of the milling cutter in two coordinate frames: the frame S_0 associated with home position on the machine table 15 and tool-associated frame S_6, respectively; $\mathbf{c}_0$ and $\mathbf{c}_6$ are the vectors of direction cosines describing the orientation vector of the cutting tool in the coordinate frames S_0 and S_6, respectively; and ${}^0\mathbf{A}_6$ is the 4 × 4 manipulation matrix presenting product of six matrices,

$${}^0\mathbf{A}_6 = {}^0\mathbf{A}_1(-x_h, -y_h, -z_h)\,\mathbf{A}^2(-y_c)\mathbf{A}^1(-x_t)\,{}^2\mathbf{A}_3(z_p, \theta_4, \psi_4)\mathbf{A}^6(\varphi_{sp})\,\mathbf{A}^3(d) \tag{5.64}$$

with

$${}^0\mathbf{A}_1(-x_h,-y_h,-z_h) = \begin{bmatrix} 1 & 0 & 0 & -x_h \\ 0 & 1 & 0 & -y_h \\ 0 & 0 & 1 & -z_h \\ 0 & 0 & 0 & 1 \end{bmatrix};\ {}^1\mathbf{A}_2 = \mathbf{A}^2(-y_c) = \begin{bmatrix} 1 & 0 & 0 & 0 \\ 0 & 1 & 0 & -y_c \\ 0 & 0 & 1 & 0 \\ 0 & 0 & 0 & 1 \end{bmatrix};$$

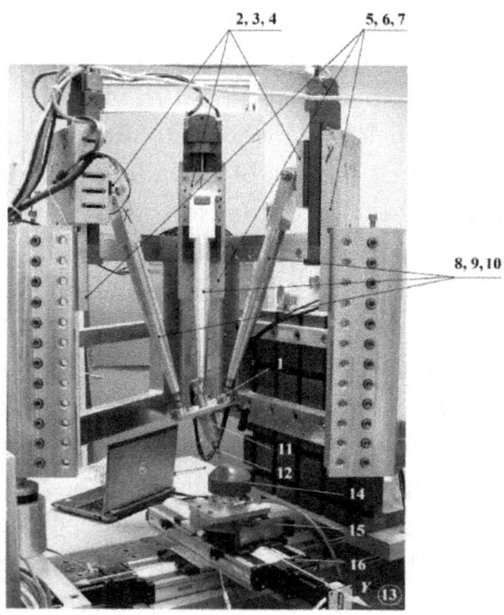

Figure 5-30 Tripod-based machine tool: General view.

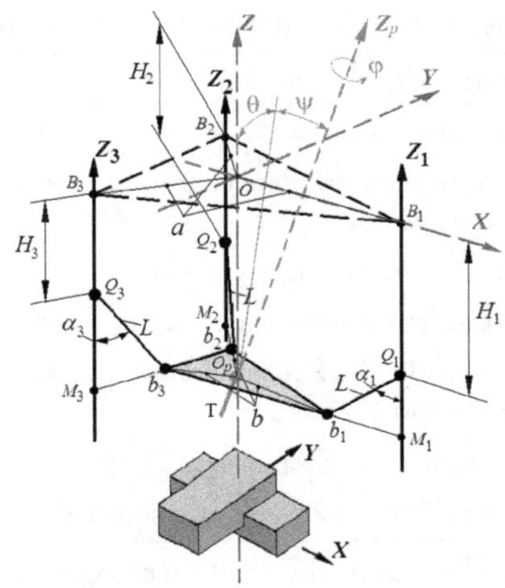

Figure 5-31 Tripod-based machine tool: Schematics of the mechanism.

$$ {}^{2}\mathbf{A}_3 = \mathbf{A}^1(-x_t) = \begin{bmatrix} 1 & 0 & 0 & -x_t \\ 0 & 1 & 0 & 0 \\ 0 & 0 & 1 & 0 \\ 0 & 0 & 0 & 1 \end{bmatrix}; \; {}^{3}\mathbf{A}_4 = {}^{3}\mathbf{A}_4(z_p, \theta_4, \psi_4); $$

$$ {}^{4}\mathbf{A}_5 = \mathbf{A}^6(\varphi_{sp}) = \begin{bmatrix} \cos\varphi_{sp} & -\sin\varphi_{sp} & 0 & 0 \\ \sin\varphi_{sp} & \cos\varphi_{sp} & 0 & 0 \\ 0 & 0 & 1 & 0 \\ 0 & 0 & 0 & 1 \end{bmatrix}; \; {}^{5}\mathbf{A}_6 = \mathbf{A}^3(d) = \begin{bmatrix} 1 & 0 & 0 & 0 \\ 0 & 1 & 0 & 0 \\ 0 & 0 & 1 & d \\ 0 & 0 & 0 & 1 \end{bmatrix}, \tag{5.65} $$

y_c and x_t are the parameters of the feed motions executing by the table and carriage; z_4 is the height position of the mobile platform; θ_4 andψ_4 are tilting angles of the platform orientation; φ_{sp} is the angle of the spindle rotation (the main motion); and ${}^{3}\mathbf{A}_3$ is the 4 × 4 matrix describing 3-DOF motion of the mobile platform expressed through extended Euler matrix $\mathbf{A}_{Eu}$ and six constraints (see Chapter 7) connecting nine current parameters of the tripod mechanism,

$$ {}^{3}\mathbf{A}_4 = {}^{3}\mathbf{A}_4(z_p, \theta_4, \psi_4) = \mathbf{A}_{Eu}(x_p, y_p, z_p, \theta_4, \psi_4, \varphi_4) $$

$$ = \begin{bmatrix} c\varphi_4\ c\psi_4 & -s\varphi_4\ c\psi_4 & s\psi_4 & x_p \\ c\varphi_4\ s\psi_4\ s\theta_4 + s\varphi_4\ c\theta_4 & c\varphi_4\ c\theta_4 - s\varphi_4\ s\psi_4\ s\theta_4 & -c\psi_4\ s\theta_4 & y_p \\ -c\varphi_4\ s\psi_4\ c\theta_4 + s\varphi_4\ s\theta_4 & s\varphi_4\ s\psi_4\ c\theta_4 + c\varphi_4\ s\theta_4 & c\psi_4\ c\theta_4 & z_p \\ 0 & 0 & 0 & 1 \end{bmatrix} \tag{5.66} $$

Table 5-4 Form-shaping system of the hybrid-kinematics machine.

S_i	Link	Motion relative to previous link	Matrix ${}^{i\text{-}1}\mathbf{A}_i$
S_0	*Y*-carriage	Conventionally fixed link	–
S_1	*Y*-carriage	Transfer from the workpiece coordinate system to the home position	${}^0\mathbf{A}_1 = {}^0\mathbf{A}_1(-x_b,-y_b,-z_b)$
S_2	*X*-carriage	Relative translation along the Y_1-axis	${}^1\mathbf{A}_2 = \mathbf{A}^2(-y_c)$
S_3	Bed	Relative translation along the X_2-axis	${}^2\mathbf{A}_3 = \mathbf{A}^1(-x_t)$
S_4	Platform	Translation along the Z_3-axis and rotations around X_3- and Y_3-axis	${}^3\mathbf{A}_4 = {}^3\mathbf{A}_4(z_p, \theta_4, \psi_4)$
S_5	Main spindle	Rotation around the Z_4-axes at angle φ_{sp}	${}^4\mathbf{A}_5 = \mathbf{A}^6(\varphi_{sp})$
S_6	Cutting tool	Translation along the Z_5-axes at distance d	${}^5\mathbf{A}_6 = \mathbf{A}^3(d)$

Parameters: x_b, y_b, z_b are the coordinates of the workpiece coordinate frame relative to the home position (zero point); x_t and y_c are the controlled motions of the *X*- and *Y*-carriage, respectively; z_p, ψ_4, and φ_4 are the controlled vertical translation and the angles of the controlled rotation of the platform; φ_{sp} the angle of the spindle rotation; and d is the offset of the cutting tool relative to the platform plane.

References

[1] Merlet, J.-P., 2006, *Parallel Robots*, 2nd ed., Springer, Heidelberg.

[2] Tsai, L.-W., 1999, *Robot Analysis*, John Wiley & Sons, Inc., New York, NY, USA.

[3] Schneiman, V., 1969, Design of Computer Controlled Manipulators, *Report AIM 92*, Stanford Artificial Intelligence Laboratory, Stanford, CA.

[4] Stewart, D., 1965, A platform with six degrees of freedom, *Proceedings 1965/66 of the Institute of Mechanical Engineers*, Part1, 180 (15), pp. 371-386.

[5] Boër, C.R., Molinari-Tosatti, L., Smith, K.S, eds., 1999, *Parallel Kinematic Machines – Theoretical Aspects and Industrial requirements*, Springer Verlag, Berlin.

[6] Ceccarelli, M., 2004, *Fundamentals of Mechanics of Robotic Manipulation*, Kluwer Academic Publishers, Dordrecht, Netherlands.

[7] Hunt, K.H., 1990, *Kinematic Geometry of Mechanisms*, Oxford Engineering Science Series, Paperback Edition, Oxford.

[8] Gough, V.E., 1956, Contribution to discussion of papers on research in automotive stability, control and tire performance, *Proceedings of Auto Division, Institute of Mechanical Engineers*, 1956-57, p. 392.

[9] Astanin, V.O., Sergienko, V.M., 1993, Investigation of the Machine Tool of the Non-Traditional Configuration, *Machine Tools and Tooling*, 3, pp. 5-8 (*in Russian*).

[10] Weck, M., Staimer, D., 2002, Parallel Kinematics Tools – Current State and Future Potential, *CIRP Annals - Manufacturing Technology*, 51(2), pp. 671-683.

[11] PI Website, 2017, http://www.physikinstrument.com/micropositioningsystems.

[12] Su, Y.X., Duan, B.Y., Peng, B, Nan, R., 2003, Mechatronics design of response enhancement of Stewart fine tuning platform for the square kilometer array, *Mechatronics*, 13, pp. 111-125.

[13] Park, F.C., Kim, J.W., 1999, Singularity Analysis of Closed Kinematic Chains, *ASME Journal of Mechanical Design*, 121, pp. 32-38.

[14] Shneor, Y., Portman, V.T., 2010, Stiffness of 5-Axis Machines with Serial, Parallel, and Hybrid Kinematics: Evaluation and Comparison, *CIRP Annals - Manufacturing Technology*, 59/1, pp. 409–412.

[15] Graham, 2017, Parallel kinematic machines, www.indoor.flyer.co.uk/kinematics.html.

[16] Oiwa, T., Matsunaga T., 1997, A Study on Torsional Elastic Hinge, *Journal of Japan Society of Precision Engineers*, Vol. 63, 10, pp.1454-1458 (in Japanese).

[17] Oiwa, T., Sugimoto, T., 2000, Shape Optimization for Flexure Hinges, *Journal of Japan Society of Precision Engineers*, 60, pp.955-959 (in Japanese).

[18] Yi, B.-J., Chung, G.B., Na, H.-Y, Kim, W.K., Suh, I.H., 2003, *Design and experiment of a 3-DOF parallel micromechanism utilizing flexure hinges, IEEE Transactions on Robotics and Automation*, Vol. 19 (4), pp. 604-612.

[19] Portman, V.T., Sandler, B.-Z., and Zahavi, E., 2000, "Rigid 6 x 6 Parallel Platform for Precision 3-D Micromanipulation: Theory and Design Application," *IEEE Transactions on Robotics and Automation*, Vol. 16, No. 6, pp. 629-643.

[20] Korn, G., Korn, T., 1968, Mathematical Handbook for Scientists and Engineers, McGraw-Hill, NY, p. 472.

[21] Raghavan, M., The Stewart Platform of General Geometry Has 40 Configurations, ASME Journal of Mechanical Design, 115, pp. 277-282.

[22] Xi, F., Zhang, D., Mechefske, C.M., Lang, Y.T., 2004, Global kinetostatic modeling of tripod-based parallel kinematic machine, Mechanism and Machine Theory, 39, 357–377.

[23] Bi, Z.M., Lang, S.Y.T., Verner, M., 2008, Dynamic modeling and validation of a tripod-based machine tool, The International Journal of Advanced Manufacturing Technology, Volume 37 (3), pp. 410–421.

[24] Wang, Y., Huang, T., Gosselin, C.M., 2003, Interpolation Algorithm and Its Error Prediction of a Tripod-Based Parallel Kinematic Machine, ASME 2003 International Design Engineering Technical Conferences and Computers and Information in Engineering Conf., Vol. 2, 29th Design Automation Conference, Parts A and B, Chicago, Illinois, USA.

6. Machine setup

6.1 Problem Formulation

6.1.1 Kinematic setup equations

The goal of the kinematic setup is to express the $(n+M)$parameters $q_1, q_2, \ldots, q_{n+M}$ of the FSF through parameters of the target element (machined workpiece, measured part, assembled parts, etc.). In robotics, this procedure is known as the *inverse kinematics problem*, whose solution has been studied in detail during past decades. Industrial robots, modern machine tools and CMMs are produced as flexible machines for multifunctional applications to provide a wide spectrum of technological operations. As is customary in high-technology equipment, two interrelated tools are applied to provide flexibility and multi-functionality:

(a) *Preprogrammed tools* realized through the *hardware*: active changeable elements such as the end-effector attached tools as applied to robots, the cutting tools as applied to machine tools, the measuring probes as applied to CMMs; excess form-shaping motions and auxiliary locating displacements; and mechanical and electro-mechanical control devices.

(b) *Programmed and reprogrammed* tools realized through the *software*: computer-based means for choosing, fixing, connecting, and constraining the FSF parameters.

Choosing the instrumentation presents the machine setup. In what follows, only geometry- and kinematics-related setup problems will be considered. In this interpretation, the setup requires solution of the inverse kinematics: a set of desired positions and orientations of the target (processed) element (*TE*) is given, while motions of the actuated joints have to be defined. Since the inverse kinematics problem is as a rule multi-variant, a suitable solution has to be chosen from some possible solutions.

Remember that the inverse kinematics of serial-kinematics machines features high complexity since synthesized equations are of both transcendental and high order, whereas with parallel kinematics-machines it is usually very straightforward. In spite of mathematical complexity, the inverse kinematics problem of serial-kinematics industrial robots, including 6-DOF robots, was solved in closed-form for the majority of the cases [1]. In what follows, the setup problem is considered using the FSF-based approach using a pair of vector equations:

$$\mathbf{r}_0 = \mathbf{r}_w \tag{6.1}$$

$$\mathbf{c}_0 = \mathbf{c}_w, \text{ with } ||\mathbf{c}_0|| = ||\mathbf{c}_w|| = 1 \tag{6.2}$$

where $\mathbf{r}_0$ is the position FSF, Eq. (3.160); $\mathbf{c}_0$ is the orientation FSF, Eq. (3.162); $\mathbf{r}_w$ and $\mathbf{c}_w$ are the known position and non-position vectors, respectively, associated with the *TE* of the FSS.

In Figure 6.1, a contact between the spherical measuring probe (*AE*) and measured surface of the workpiece (*TE*) illustrates an orientation constraint. In this case, the *AE* and *TE* contact

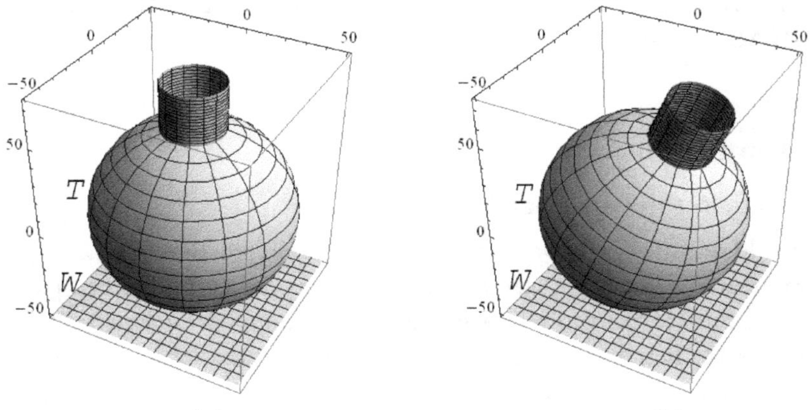

(a) (b)

Figure 6-1 Orientation constraints: T – the tool (an active element of the FSS); W – the workpiece (a target element).

in the same point of *TE*, but with different orientation of the AE, and hence, despite the fact that the position FSFs figured in Eq. (6.1) are identical in both cases, the orientation FSFs, Eq. (6.2), can differ.

6.1.2 Setup conditions

The position vector $\mathbf{r}_w$ in the right-hand sides of Eqs. (6.1) and (6.2) depends on m parameters, Eq. (3.115). Depending on the *TE* dimensionality, $m = 0, 1$, or 2. For definiteness, we consider the case in which the *TE* is a surface ($m = 2$) given by equations:

$$\mathbf{r}_w = \mathbf{r}_w(u, v) \tag{6.3}$$

$$\mathbf{c}_w = \mathbf{c}_w(u, v) \tag{6.4}$$

We consider a region where the surface (6.5) is regular, the partial derivatives $\partial\mathbf{r}_w/\partial u$ and $\partial\mathbf{r}_w/\partial v$ exist and continuous, and the normal $\mathbf{N}$ to the surface exists:

$$\mathbf{N}(u,v) = \frac{\partial \mathbf{r}_w}{\partial u} \times \frac{\partial \mathbf{r}_w}{\partial v} \neq 0 \tag{6.5}$$

$$\mathbf{n}(u, v) = \mathbf{N}/\|\mathbf{N}\| \tag{6.6}$$

where $\mathbf{n} = \mathbf{n}(u, v)$ is the unit vector of the normal $\mathbf{N}$.

In turn, the position vector $\mathbf{r}_0$ in the left-hand sides of Eqs. (6.1) and (6.2) can be represented through $n + M$ parameters, where n is the number of the FSS links and M is the number of parameters defining the *AE* (the tool). Hence, Eqs. (6.1) and (6.2) become

$$\mathbf{r}_0(q_1, q_2, \ldots, q_{n+M}) = \mathbf{r}_w(u_w, v_w) \tag{6.7}$$

$$\mathbf{c}_0(q_1, q_2, \ldots, q_{n+M}) = \mathbf{c}_w(u_w, v_w) \tag{6.8}$$

Equations(6.7) and (6.8) can be reformulated as a system of $(n + M)$ simultaneous equations

$$f_1(q_1, q_2,\ldots,q_{n+M}, u_w, v_w) = 0$$

$$\cdots\cdots\cdots\cdots\cdots\cdots$$

$$f_{n+M}(q_1, q_2,\ldots, q_{n+M}, u_w, v_w) = 0 \tag{6.9}$$

The system is solved with respect to parameters

$$q_1 = q_1(u_w, v_w)$$

$$q_{n+M} = q_{n+M}(u_w, v_w) \tag{6.10}$$

The system of simultaneous equations (6.9) combines Eqs. (6.3-4) with some added constraints. According to the theorem for existing implicit functions [2], solution of system (6.10) with respect to $(n+M)$ unknowns is possible if its Jacobi an matrix **J** at point (u, v) has the full rank,

$$\mathbf{J}(u,v) = \frac{\partial(f_1,\ldots,f_{n+M})}{\partial(q_1,\ldots,q_{n+M})} \tag{6.11}$$

$$\text{Rank}[\mathbf{J}] = n + M \tag{6.12}$$

Among solutions (6.9), two equations expressing the pair u, v of the Gaussian coordinates of the machined surface through $(l + M)$ parameters must take place:

$$u = u(q_1, q_2,\ldots,q_{n+M})$$
$$v = v(q_1, q_2,\ldots,q_{n+M}) \tag{6.13}$$

The rest of the $L = n + M - 2$ equations connect $n + M$ parameters of the FSF as follows.

$$F_1(q_1, q_2,\ldots,q_{n+M}) = 0$$

$$\cdots\cdots\cdots\cdots\cdots\cdots$$

$$F_L(q_1, q_2,\ldots, q_{n+M}) = 0 \tag{6.14}$$

with

$$L = n + M - 2 \tag{6.15}$$

Equations (6.14) formulate L conditions subsequently referred to as "constraints". The set of L equations (6.14) can be solved with respect to L setup parameters through a new pair of Gaussian coordinates if the following requirement for the rank is true:

$$\text{Rank}\left[\frac{\partial(F_1,\ldots, F_L)}{\partial(q_1,\ldots, q_{L+M})}\right] = L = n + M - 2 \tag{6.16}$$

Generalization of condition (6.15) is formulated as follows. Dimensionality m of the target element of the FSS (object to be processed), Eq. (6.1), dimensionality M of the *AE* (processing tool), Eq. (3.114), links number n (note that $n \geq l$, where l is the machine mobility number), and rank L, Eq. (6.16), of the constraints system must conform to a *Setup condition:*

$$L = n + M - m \tag{6.17}$$

In the multifunctional machines, the number of links n and the *TE* dimensionality m are usually fixed, while the type of the *AE* (processing tool) and the number of constraints (i.e., values M and L) are the *machine setup* subject. Classification of the physical sources of the different constraints and their analysis and synthesis will be detailed below. Setup of the lathe is considered as a methodological example.

In the lathe (Fig. 3.15), there are three FSS motions ($n = 3$): spindle rotation at angle φ; and longitudinal and transverse feeds at distances z and x, respectively (note that in this case, $n = l$). For example, the turning of a body of revolution ($m = 2$) by a single-point cutting tool ($M = 0$) must be accompanied by the constraints' number

$$L = n + M - m = 0 + 3 - 2 = 1 \tag{6.18}$$

During machining by single-point cutter on the lathe, only the position FSF, Eq. (6.7), is used (the point orientation is undefined):

$$\mathbf{r}_0 = \begin{bmatrix} x\cos\varphi \\ x\sin\varphi \\ z \end{bmatrix} = \begin{bmatrix} f(v)\cos u \\ f(v)\sin u \\ v \end{bmatrix}, \text{ with } f(v) \neq 0 \tag{6.19}$$

where u and v are Gaussian coordinates of the machined surface of revolution; and $f(v)$ is the given generatrix of the surface.

Vector equation (6.19) can be represented through three scalar equations

$$f_1(x, z, \varphi, u, v) = x\cos\varphi - f(v)\cos u = 0$$

$$f_2(x, z, \varphi, u, v) = x\sin\varphi - f(v)\sin u = 0$$

$$f_3(x, z, \varphi, u, v) = z - v = 0 \tag{6.20}$$

The Jacobian **J**, Eq. (6.11), and its rank Rank [**J**], Eq. (6.12), are as follows:

$$\mathbf{J} = \frac{\partial(f_1, f_2, f_3)}{\partial(x, \varphi, z, u, v)} = \begin{bmatrix} \cos\varphi & -f(v)\sin\varphi & 0 & 0 & 0 \\ \sin\varphi & f(v)\cos\varphi & 0 & 0 & 0 \\ 0 & 0 & 1 & 1 & 1 \end{bmatrix}$$

$$\text{Det}\begin{bmatrix} \cos\varphi & -f(v)\sin\varphi & 0 \\ \sin\varphi & f(v)\cos\varphi & 0 \\ 0 & 0 & 1 \end{bmatrix} = f(v) \neq 0$$

Hence, Rank [**J**] = 3. Solution of three simultaneous equations (6.20) expresses the three FSF parameters x, z and φ through the Gaussian coordinates u and v of the machined surface,

$$u = \varphi \tag{6.21}$$

$$v = z \tag{6.22}$$

$$x = f(z) \tag{6.23}$$

Equations (6.21-23) establish the new Gaussian coordinates φ and z, and Eq. (6.23) presents a single constraint $x = f(z)$ connecting these coordinates. Formally, system (6.14) consists of a single equation

$$F(x, z, \varphi) = z - v = 0$$

In the considered case, the setup of the lathe uses two different means: *hardware* for choosing the single-point cutting tool and *software* for establishing the programmable constraint, Eqs. (6.23). The detailed setup spectrum of the lathe will be considered below (see Subsection 6.3.5).

For example, setup when the constraint becomes $x = R$ results in machining the cylindrical surface, whose vector equation is

$$\mathbf{r}(u,v) = \mathbf{r}(\varphi, z) = \begin{bmatrix} R\cos\varphi \\ R\sin\varphi \\ z \end{bmatrix}$$

6.2 Constraint synthesis

6.2.1 Constraint classification

It is obvious from the setup rule, Eq. (6.17), that the constraint number $L = 0$ if $m = l + M$, i.e., the sum of the link motions and tool dimensionality equals the dimensionality of the processed (target) element. The simple example of such situation is broaching of different holes. In this process, there is a single FSS motion (translation along the hole axis), i.e., $l = 1$, and edge-type cutting tool ($M = 1$) – a broach whose last edges have the final form of the machined surface ($m = 2$). Hence, according to Eq. (6.17), $L = n + M - m = 1 + 1 - 2 = 0$.

If $l + M > m$, constraints of diverse geometrical nature take place. All L constraints, available in the machine during its functioning, can be classified as follows: L_f functional constraints; L_e enveloping constraints; and L_i implicit constraints,

$$L = L_f + L_e + L_i \tag{6.24}$$

Note that any type of constraints can be absent, i.e., a certain component entering in number L can be equal to zero. The constraints according to entered classification are formulated in Table 6.1.

6.2.2 Functional constraints

6.2.2.1 Definition

There are two groups of functional constraints: (a) *position* constraints fix the values of the FSS parameter, and (b) *kinematic* constraintsestablish relations between the parameters of the FSS links and their velocities.The position constraint defines the preprogrammed linear or angular dimension of the processed subject,

$$q_i = C_i \ (C_i \text{ is a constant, } i = 1,\ldots,n) \tag{6.25}$$

Table 6-1 Constraint formulation.

Class of constraint	Number	Formulation
Functional constraints:	L_f	
Positional		$q_j - C_j = 0\ (j = 1, \ldots, n)$
Kinematical		$f(q_1, q_2, \ldots, q_n) = 0$
Enveloping constraints:	L_e	
One-parametric		$\left\lvert \frac{\partial \mathbf{r}_0}{\partial q_j} \frac{\partial \mathbf{r}_0}{\partial q_{j+1}} \frac{\partial \mathbf{r}_0}{\partial q_{j+2}} \right\rvert = 0,$
Two-parametric		$\left\lvert \frac{\partial \mathbf{r}_0}{\partial q_j} \frac{\partial \mathbf{r}_0}{\partial q_{j+1}} \frac{\partial \mathbf{r}_0}{\partial q_{j+2}} \right\rvert = 0 \text{ and } \left\lvert \frac{\partial \mathbf{r}_0}{\partial q_j} \frac{\partial \mathbf{r}_0}{\partial q_{j+1}} \frac{\partial \mathbf{r}_0}{\partial q_{j+3}} \right\rvert = 0$
Implicit constraints:	L_i	
Collinearity		$q_i + q_{i+1} = C_i$
Coplanarity		$\left\lvert \frac{\partial \mathbf{r}_0}{\partial q_j} \frac{\partial \mathbf{r}_0}{\partial q_{j+1}} \frac{\partial \mathbf{r}_0}{\partial q_{j+2}} \right\rvert \equiv 0$

n – the number of the FSS links.

Such a constraint halts one of the motions of the FSS links and can be applied to any moving link of the machine with the exception of the link executing a principal technological motion, i.e., the cutting motion during machining. The typical position constraints were considered for the lathe, $x = R$ defines the radius of the machined cylinder, constraint $z = c$ defines the distance of the machined face from the initial cross-section, etc.

The kinematic constraint directly connects l motion parameters (remember, l is the mobility number, $l \leq n$) among $q_1,\ldots,q_n$,

$$f(q_i,\ldots,q_n) = 0 \tag{6.26}$$

Functional constraints can always be represented both as equations containing the velocities of link motions and as constraints among the coordinates of these motions. The latter form results from integrating the initial velocity dependences. Equation (6.26) describes the *kinematic constraint* encompassing two or more motion parameters. The widespread case is a linear dependence between parameters specifying the driving ratios between the velocities of motions or between corresponding displacements,

$$\sum_{i=1}^{l} a_i q_i = 0 \tag{6.27}$$

6.2.2.2 Setup of the thread-cutting kinematic chain

In the lathe, translation of the longitudinal carriage (with cutting tool) and rotation of the main spindle with the workpiece are connected, and the kinematic constraint, Eq. (6.27), becomes

$$s_w \pm L_w \varphi_{sp}/(2\pi) = 0 \tag{6.28}$$

where s_w is the translation displacement of the longitudinal carriage along the workpiece axis, in mm; φ_{sp} is angle of main spindle rotation, in rad; L_w is the lead of the screw machined on the workpiece, in mm; and the sign depends on the threading direction (minus for the right screw and plus for the left screw).

Constraint (6.28) depends on the single setup parameter L_w. In the lathe with a traditional (manual) control system, both motions in the screw-cutting kinematic chain are driven by a single electrical motor through a complicated mechanical transmission. Constraint (6.28) is realized by means of a set of gear trains connecting primary motion of the main spindle with feed motion of the leadscrew. Longitudinal displacement s_2 depends on the angle of the leadscrew rotation φ_{ls} and lead as $s_w = L_{ls}\varphi_{ls}/(2\pi)$. Thus, Eq, (6.28) becomes

$$L_{ls}\varphi_{ls} \pm L_w\varphi_{sp} = 0 \tag{6.29}$$

Hence, the "spindle-lead" transmission ratio is $i = \varphi_{sp}/\varphi_{ls} = \pm L_{ls}/L_w$ and, as a result, the functional constraint, Eq, (6.29), can be represented in the regular form of the setup procedure of the lathe

$$i_{var} = \pm L_{ls}/(i_{const} L_w) \tag{6.30}$$

where i_{const} and i_{var} are the constant and varying transfer ratios, respectively, of the screw kinematic chain of the machine.

Usually, the change gear train intended for realization of the ratio i_{var} comprises four changeable gears whose teeth numbers are *a*, *b*, *c*, *d*, and

$$i_{var} = \frac{a}{b}\cdot\frac{c}{d} \tag{6.31}$$

In the general case, the required transfer ratio, Eq.(6.31), can be established through four integers *a*,*b*, *c*, and *d* only with limited degree of exactness, usually of order 10^{-4}–10^{-5}.

In the CNC-turning machines, instead of the hardware in the form of the change gear trains the software modules connecting motions of two different electric motors are applied, and motors' velocities are programmed according to screw relation, Eq. (6.28) or (6.29). Usually, the standard software consists of a canned cycle for thread turning with prompted programming. The cycle parameters are: (a) the threading single-point tool, whose edges form the thread profile in the axial (or another) section of the workpiece; (b) the internal or mean diameter of the machined thread; and (c) the pitch of the machined thread.

6.2.2.3 Processing the sculpture (free form) surfaces

If the surface is given through a set of their points, approximation of the machined surface which can be considered as a specific form of kinematic constraint, Eq. (6.26), is applied. These surfaces are typical for the automobile, aerospace, die/mold industries, and turbo-machinery. In many cases, the free-form surfaces are numerically given in N points through their coordinates, which are used for recreation of the surface under additional requirements for its form

$$f(x_i, y_i, z_i) = 0, i = 1,\dots, N \tag{6.32}$$

Currently, curves and surfaces defined through the set of points (6.32) are represented by spline approximation of different orders using a mathematical model called non-uniform rational basis spline (NURBS)[3]. The approximation functions with NURBS curves and surfaces with free nodes are supported by Wolfram Mathematica® [4]. The command

BSplineCurve[{pt$_1$, pt$_2$, ...}] (6.33)

where $pt_i(i = 1, 2,...,N)$ is ith control point of a graphics primitive that represents a non-uniform rational B-spline curve.

By default, **BSplineCurve** *uses cubic splines. In Figure 6.2a the NURBS curve is represented for the set of ten points* pt_i $(i = 1, 2,...,10)$. The option setting **SplineDegree** $\rightarrow d$ specifies that the underlying polynomial basis should have maximal degree d. Figure 6.2b shows ghe *NURBS curve forthe same set of points for spline degreesd* $= 1, 2,...,6$. A graphics primitive that represents a non-uniform rational B-spline *surface* defined by an array of x, y, z control points is **BSplineSurface** [array].

As an example, the surface splines for default (the third degree of the approximation) and for different degrees of approximation are shown in Figure 6.2c and d, respectively, for the 25 points listed in the Figure legend. Note that if the spline degrees $d > 1$, the approximated subject is deviated from the given points, and the deviations must be no more than half of the pre-designed tolerance. These deviations are entered as part of the total errors of size, form, and position when the accuracy of the machined surface is evaluated.

The functional constraints are realized through the set of machine active axes. As applied to machine tools, depending on requirements that must be imposed upon position and orientation of the cutting tool during the surface processing on the CNC machines, there are three types of machine control processes.

2.5-axis machining allows cutting processes when the current point position can be represented through single argument functions, and the tool orientation is structurally fixed. The functional constraint, Eq. (6.33), can have implicit, explicit, or parametric form:

$$f(x, y) = 0 \tag{6.34}$$

$$y = y(x) \tag{6.35}$$

$$x = x(u); y = y(u) \tag{6.36}$$

In these cases, the CNC (or NC) machines have three controlled coordinates but only two axes are interrelated to perform the given function, Eqs. (6.34-36). Geometrically, Eqs.(6.34-36) define a planar curve, and the third coordinate is controlled by a simple cycle at each point. The 2.5-axis control allows machining of all cylinders whose generatrix is parallel to the spindle axis and a directrix lies in the plane perpendicular to the spindle axis. For example, the horizontal and vertical planes machined on the milling machines, right circular cylinders with vertical and horizontal axes machined on milling and boring machines fall into this category. It is important that the majority of the regular machine elements have surfaces of this kind, and these elements are machined by means of regular milling and drilling operations. As a result, the 2.5-axis control allows machining up to 80% of mechanical parts.

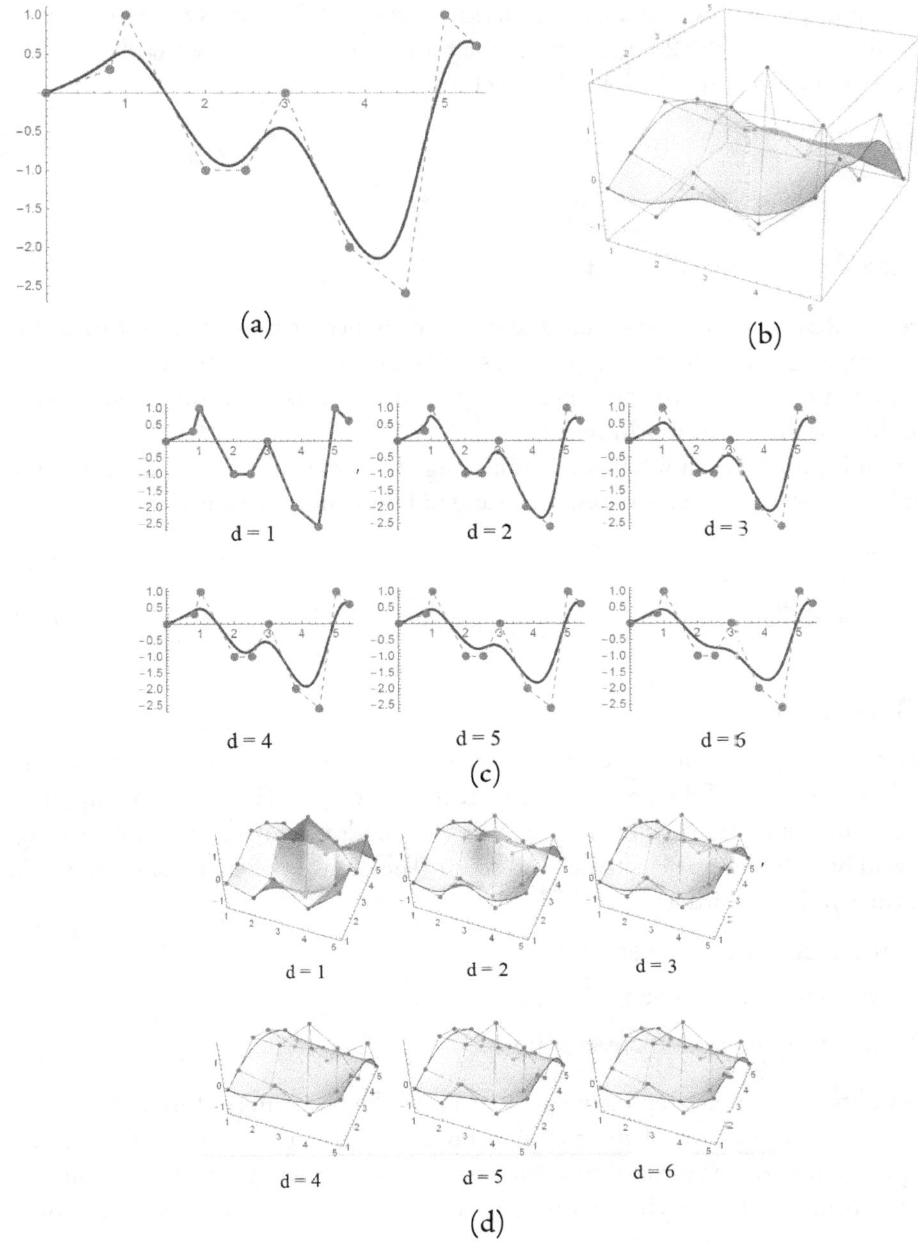

Figure 6-2 The NURBS approximation built by Wolfram Mathematica®: (a) the NURBS curve for set of points: {0, 0}, {0.8, 0.3}, {1, 1}, {2, -1}, {2.5, -1}, {3, 0}, {3.8, -2}, {4.5, -2.6}, {5, 1}, {5.4, 0.6}; (b) NURBS curve for the same points for the spline degree d = 1, 2,…,6; (c) NURBS surface for set of points:{1, 1, -0.15}, {1.2, 1.5, 0.4}, {1, 3, 0.8}, {1, 4.2, 0.4}, {1.3, 5.1, -0.1}, {{2, 1, -0.5}, {2.2, 1.9, -0.3}, {1.8, 3.5, 0.3}, {2.2, 4, 0}, {1.7, 5.4, -0.7}, {2.8, 1, 0.6}, {3.5, 2, -0.8`}, {3.2, 3, 1.9}, {2.9, 4, 0.1}, {3, 5, -0.5}, {4, 1, -0.3}, {4.5, 2.2, -0.1}, {4, 3.7, 0.6}, {3.5, 4, -0.3}, {4, 5.2, -0.7}, {5, 1, 0.7}, {5, 2, 0.9}, {5, 3, 0}, {5, 4, 0.9}, {5.1, 5.5, -1.2}; (d) NURBS surface for the same points for the spline degree d = 1, 2,…,6.

3-axis machining provides simultaneous processing when all three axes have to be moving at the same time and, as in the 2.5 machining, the tool orientation is structurally fixed. The functional constraint has either of the following forms:

$$f(x, y, z) = 0 \tag{6.37}$$

or

$$x = x(u, v);\ y = y(u, v);\ \text{and } z = z(u, v) \tag{6.38}$$

Compared to 2.5-axis machining, this allows performance of more complex operations such as the machining of free-form (sculptured) surfaces, which are commonly used in molds and airfoils industry. However, the post-processing of the 2.5-axis machining is significantly simpler than that of the 3-axis machining.

5-axis machining provides simultaneous processing of five machine axes (both point position and tool axis orientation) controlled according to following equations:

$$f(x, y, z) = 0 \tag{6.39}$$

$$\mathbf{c} = \mathbf{c}(x, y, z), \text{ with } ||\mathbf{c}|| = 1 \tag{6.40}$$

6.2.3 Enveloping constraints

In many cases, the target element presents an envelope of a family of instant positions of the working line or surface of the processing tool (active element). This method is applied in milling and grinding operations, different generating methods of gearing, the spot welding of a line-type weld by a disk-shape electrode, and so on. The number of enveloping constraints depends on the enveloping order:

$$L_e = \begin{cases} 0, & \text{if enveloping is absent} \\ 1, & \text{for one-parametric enveloping} \\ 2, & \text{for two-parametric enveloping} \end{cases} \tag{6.41}$$

The one- and two-order enveloping in Eq. (6.41) is applied in terms of Litvin [5]. According to the constraint, the vector of the common normal at a point of the tool-part contact must be perpendicular to the vector of tool velocity relative to the part at the same point. The constraints are formulated through scalar (dot-point) products of the normal at one contact surface by the tangent at the second one:

- for one-parametric enveloping ($L_e = 1$),

$$\mathbf{n}_w(u,v).\frac{\partial \mathrm{r}_t(U,V)}{\partial U} = 0 \text{ or } \mathbf{n}_w(u,v).\frac{\partial \mathrm{r}_t(U,V)}{\partial V} = 0, \tag{6.42}$$

- for two-parametric enveloping ($L_e = 2$)

$$\mathbf{n}_w(u,v).\frac{\partial \mathrm{r}_t(U,V)}{\partial U} = 0 \text{ and } \mathbf{n}_w(u,v).\frac{\partial \mathrm{r}_t(U,V)}{\partial V} = 0, \tag{6.43}$$

where $\mathbf{r}_w = \mathbf{r}_w(u, v)$ is the equation of the processed part expressed in terms of its curvilinear coordinates u and v; $\mathbf{r}_t = \mathbf{r}_t(U, V)$ is the tool equation expressed in terms of curvilinear coordinates U and V; and $\mathbf{n}_w(u, v)$ is the unit-vector of the normal $\mathbf{N}_w$ to the machined surface,

$$\mathbf{N}_w = \frac{\partial \mathbf{r}_w}{\partial u} \times \frac{\partial \mathbf{r}_w}{\partial v}$$

$$\mathbf{n}_w = \mathbf{N}_w / ||\mathbf{N}_w|| \tag{6.44}$$

In the contact point, the unit-vector of the normal $\mathbf{n}_w$ to the workpiece surface and the unit-vector of the normal $\mathbf{n}_t$ at the tool surface meet the condition

$$\mathbf{n}_w(u, v) = \pm \mathbf{n}_t(U, V) \tag{6.45}$$

This allows reformulation of constraints (6.42) and (6.43) through triple scalar products of three partial derivatives of vector $\mathbf{r}_0$ with respect to parameters q_j, q_{j+1}, q_{j+2}, respectively:

$$\left| \frac{\partial \mathbf{r}_0}{\partial q_j} \frac{\partial \mathbf{r}_0}{\partial q_{j+1}} \frac{\partial \mathbf{r}_0}{\partial q_{j+2}} \right| = 0, \text{ for } L_e = 1 \tag{6.46}$$

$$\left| \frac{\partial \mathbf{r}_0}{\partial q_j} \frac{\partial \mathbf{r}_0}{\partial q_{j+1}} \frac{\partial \mathbf{r}_0}{\partial q_{j+2}} \right| = 0 \text{ and } \left| \frac{\partial \mathbf{r}_0}{\partial q_j} \frac{\partial \mathbf{r}_0}{\partial q_{j+1}} \frac{\partial \mathbf{r}_0}{\partial q_{j+3}} \right| = 0, \text{ for } L_e = 2 \tag{6.47}$$

The triple products are calculated through determinant of the 3 × 3 matrix composed without the fourth elements of the vectors of the partial derivatives.

6.2.4 Implicit constraints

Upon duplication of identical movements by different units of the machine or – generally – upon the presence of kinematic redundancy of link motions, implicit constraints are formed between $(l + M)$ parameters of the FSF, Eq. (6.25). There are two types of the implicit constraints.

Collinearity of two successive motions, either a pair of translations along the same axis or a pair of rotations around the same axis, results in an implicit constraint of type "collinear pair". Actually, according to *Property 1* of one-parametric matrices $\mathbf{A}^k(q_j)$ Eq. (3.141), two or more motions along or around a single axis are equivalent to a motion relative to the same axis with the matrix parameter of this motion equaling the sum of parameters of initial motions, i.e.,

If ${}^{j-1}\mathbf{A}_j(q_j) = \mathbf{A}^k(q_j)$ and ${}^{j}\mathbf{A}_{j+1}(q_{j+1}) = \mathbf{A}^k(q_{j+1})$,

then

$$\mathbf{A}^k(q_j)\, \mathbf{A}^k(q_{j+1}) \equiv \mathbf{A}^k(q_\Sigma)$$

and

$$q_\Sigma = q_j + q_{j+1} \tag{6.48}$$

Note this is valid not only in the case in which two motions are actually successive, but also when the elementary co-matrices composing manipulation matrix ${}^0\mathbf{A}_n$ become successive ones after their allowable permutations listed in *Properties* 5 (see subsection 3.3.1.4).

Example: Rotation of the cylindrical wheel around its axis. Let us the axis of the cylinder be the Z-axis, *i.e.*, $\mathbf{r}_w = [R\cos\varphi_w, R\sin\varphi_w, z_w, 1]^T$, where R is the radius of the cylinder, φ_w and z_w are the polar angle and the axial coordinate of the cylinder point, respectively. Rotation of the cylinder is described by multiplication of vector $\mathbf{r}_w$ by matrix $\mathbf{A}^6(\varphi)$,

$$\mathbf{r}_0 = \mathbf{A}^6(\varphi)\mathbf{r}_w = \begin{bmatrix} \cos\varphi & -\sin\varphi & 0 & 0 \\ \sin\varphi & \cos\varphi & 0 & 0 \\ 0 & 0 & 1 & 0 \\ 0 & 0 & 0 & 1 \end{bmatrix} \begin{bmatrix} R\cos\varphi_w \\ R\sin\varphi_w \\ z_w \\ 1 \end{bmatrix} = \begin{bmatrix} R\cos(\varphi+\varphi_w) \\ R\sin(\varphi+\varphi_w) \\ z_w \\ 1 \end{bmatrix} = \begin{bmatrix} R\cos\Phi \\ R\sin\Phi \\ z_w \\ 1 \end{bmatrix}$$

where $\Phi = \varphi + \varphi_w$.

Co-planarity of three successive motions results in an implicit constraint since velocities of three successive link motions are parallel to a plane. In the Cartesian coordinate frame, three types of coplanar triplets of elementary motions are possible:

$$k = \mathbf{126} \tag{6.49}$$

$$k = \mathbf{135} \tag{6.50}$$

$$k = \mathbf{234} \tag{6.51}$$

where1, 2,...,6 are the motion identifiers (Table 3.1).

In the FSS, each triplet can be realized in six versions since any permutation of three symbols in Eqs. (6.49–50) does not disturb the co-planarity of the triplet. Hence, there are 18 possible versions of the coplanar triplets in the coordinate code of the FSS. Each constraint of type "coplanar triplet" reduces the number of independent parameters through substitution of three parameters q_j, q_{j+1}, and q_{j+2} for two parameters:

$$q_{\Sigma 1} = q_{\Sigma 1}(q_j, q_{j+1}, q_{j+2}) \text{ and } q_{\Sigma 2} = q_{\Sigma 2}(q_j, q_{j+1}, q_{j+2}) \tag{6.52}$$

The formal indication of the implicit constraints is the presence of a triplet of coplanar vectors among $(l + M)$ partial derivatives $\partial\mathbf{r}_0/\partial q_1,\ldots,\partial\mathbf{r}_0/\partial q_{n+M}$ of vector $\mathbf{r}_0$. A corresponding condition is that the triple scalar product of derivatives $\partial\mathbf{r}_0/\partial q_j$, $\partial\mathbf{r}_0/\partial q_{j+1}$, and $\partial\mathbf{r}_0/\partial q_{j+2}$ is identically equal to zero,

$$\left| \frac{\partial \mathrm{r}_0}{\partial q_j} \frac{\partial \mathrm{r}_0}{\partial q_{j+1}} \frac{\partial \mathrm{r}_0}{\partial q_{j+2}} \right| \equiv 0 \tag{6.53}$$

Note that here the triple scalar product of the derivatives *identically equals zero* (sign ≡) but does not *equal zero* (sign =) as this takes place for the enveloping constraint. As an example, the implicit constraint of type k = **126**, Eq. (6.49), is considered using a real-life representative –the surface-grinding machine (Fig. 6.3). Codes 1 and 2 correspond to the

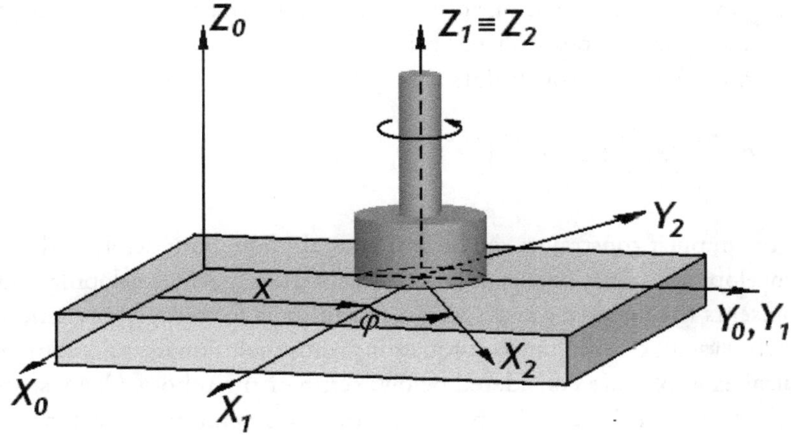

Figure 6-3 Schematic diagram of surface-grinding by wheel end face

X- and Y-feeds of the table, and code 6 corresponds to the grinding wheel rotation around the Z-axis.The FSF can be written as follows:

$$\mathbf{A}^1(x)\mathbf{A}^2(y)\mathbf{A}^6(\varphi)\mathbf{r}_w = \begin{bmatrix} 1 & 0 & 0 & x \\ 0 & 1 & 0 & 0 \\ 0 & 0 & 1 & 0 \\ 0 & 0 & 0 & 1 \end{bmatrix} \begin{bmatrix} 1 & 0 & 0 & 0 \\ 0 & 1 & 0 & y \\ 0 & 0 & 1 & 0 \\ 0 & 0 & 0 & 1 \end{bmatrix} \begin{bmatrix} \cos\varphi & -\sin\varphi & 0 & 0 \\ \sin\varphi & \cos\varphi & 0 & 0 \\ 0 & 0 & 1 & 0 \\ 0 & 0 & 0 & 1 \end{bmatrix} \begin{bmatrix} R\cos\varphi_w \\ R\sin\varphi_w \\ 0 \\ 1 \end{bmatrix} = \begin{bmatrix} u \\ v \\ 0 \\ 1 \end{bmatrix}$$

with $u = x + R\cos\Phi$; $v = y + R\sin\Phi$; and $\Phi = \varphi + \varphi_w$.

6.2.5 Setup algorithm

Setup procedure includes four steps, certain of them can be unnecessary in particular cases.

Step 1: The FSF is formulated and the constraint number L, Eq. (6.17), is calculated.

Step 2: If $L > 0$, the presence of implicit, functional, and enveloping constraints are successfully verified and constraints, if any, are formulated. Substitution of each constraint in the FSF reduces the number of the fitted parameters by one. If the number of unknown parameters becomes m, the setup process is completed; otherwise the setup process is continued.

Step 3: Equating the FSF to the target element expression,Eqs. (6.1) and (6.2), allows formulation of the enveloping constraints, if any, and additional functional constraints. Let us suppose that the target element is a surface:

$$\mathbf{r}_w(u, v) = \mathbf{r}_0(q_1, q_2, \ldots, q_{n+M}) \quad (6.54)$$

$$\mathbf{c}_w(u, v) = \mathbf{c}_0(q_1, q_2, \ldots, q_{n+M}) \quad (6.55)$$

$$\mathbf{n}_w = \frac{\dfrac{\partial \mathbf{r}_w}{\partial u} \times \dfrac{\partial \mathbf{r}_w}{\partial v}}{\left| \dfrac{\partial \mathbf{r}_w}{\partial u} \times \dfrac{\partial \mathbf{r}_w}{\partial v} \right|} \quad (6.56)$$

where $\mathbf{r}_w$ and $\mathbf{c}_w$ are the given equations of the machined surface; $\mathbf{n}_w$ is the unit vector of the normal; and $\mathbf{r}_0$ and $\mathbf{c}_0$ are the position and orientation FSF.

In this case, the enveloping constraints are

$$\mathbf{n}_w \cdot \frac{\partial \mathbf{r}_0}{\partial q_i} = 0 \text{ or} \left| \frac{\partial \mathbf{r}_w}{\partial u} \frac{\partial \mathbf{r}_w}{\partial v} \frac{\partial \mathbf{r}_0}{\partial q_i} \right| = 0 \tag{6.57}$$

Note that the implicit constraints can be shaped in the FSF both before and after formulation of each enveloping or functional constraint. A set of the implicit, enveloping, and functional constraints presents the necessary and sufficient conditions for setup formulation. The set is considered as the system of simultaneous equations whose solution fixes the setup parameters.

Some typical examples are considered below: setup of the robots, CMMs, and machine tools. As applied to serial-kinematics machines, the consideration is started with setup of the machine tools for universal operations such as milling, turning and grinding, and then machines for threading and gearing are considered. Setups of the serial-type industrial robots are illustrated by welding and painting operations. The consideration is finished with setup of the parallel- and hybrid-kinematics machines.

6.3 Setup of Multi-Axis Machines

6.3.1 Cartesian robot

The FSF of the Cartesian robot is given through Eqs. (4.52-53),

$$\mathbf{r}_0 = \mathbf{A}^1(x)\, \mathbf{A}^2(y)\, \mathbf{A}^3(z)\mathbf{r}_3 = \mathbf{A}_{tr}\mathbf{r}_t \tag{6.58}$$

where $\mathbf{A}_{tr}$ is the translation matrix, Eq. (4.54); and $\mathbf{r}_t$ is the position vector of the tool

$$\mathbf{A}_{tr} = \begin{bmatrix} 1 & 0 & 0 & x \\ 0 & 1 & 0 & y \\ 0 & 0 & 1 & z \\ 0 & 0 & 0 & 1 \end{bmatrix};$$

Let us consider application of a single-point tool mounted on the robot end-effector,

$$\mathbf{r}_t = [x_t, y_t, z_t, 1]^{\mathrm{T}}$$

Hence,

$$\mathbf{r}_0 = \mathbf{A}_{tr}\mathbf{r}_t = \begin{bmatrix} 1 & 0 & 0 & x \\ 0 & 1 & 0 & y \\ 0 & 0 & 1 & z \\ 0 & 0 & 0 & 1 \end{bmatrix} \begin{bmatrix} x_t \\ y_t \\ z_t \\ 1 \end{bmatrix} = \begin{bmatrix} x + x_t \\ y + y_t \\ z + z_t \\ 1 \end{bmatrix} = \mathbf{r}_{tr} + \mathbf{r}_t \tag{6.59}$$

$$\mathbf{r}_{tr} = [x_{tr}, y_{tr}, z_{tr}, 0]^{\mathrm{T}}$$

There are no implicit and enveloping constraints. A functional constraints can be obtained from the regular setup condition:

$$\mathbf{r}_0=\mathbf{r}_w \tag{6.60}$$

where $\mathbf{r}_w$ is the workpiece position vector.

Substitution of Eq. (6.59) in Eq. (6.60) yields

$$\mathbf{r}_{tr}= \mathbf{r}_w- \mathbf{r}_t \tag{6.61}$$

Numerical Example 6.1. A laser graver is fitted to the end-effector of the Cartesian robot. **Given:** The initial position of the laser pen-point is [0, 0, 40] mm. The Z-axis-oriented laser beam graves an ellipse in an inclined plane. The plane passes through the Y-axis and makes the angle $\pi/6$ with the XY-plane (Fig. 6.4a). The Y-oriented major and X-oriented minor semi-axes of the ellipse are 150 and 50 mm, respectively; the ellipse center is at the origin of the coordinate frame. **Required:** Synthesis of the pen-point trajectory.

Solution: Equation of the ellipse on the horizontal XY-plane is $\mathbf{r}$ = [50 cos φ, 150 sin φ, 0, 1]^T. The ellipse is brought into the given plane after rotation at the angle $\psi = -\pi/6$ around the Y-axis. Hence, the engraved ellipse equation is as follows:

$$\mathbf{r}_w = \mathbf{A}^5\mathbf{r} = \begin{bmatrix} \cos\psi & 0 & \sin\psi & 0 \\ 0 & 1 & 0 & 0 \\ -\sin\psi & 0 & \cos\psi & z \\ 0 & 0 & 0 & 1 \end{bmatrix}\begin{bmatrix} 50\cos\varphi \\ 150\sin\varphi \\ 0 \\ 1 \end{bmatrix} = \begin{bmatrix} 50\cos\varphi\cos\psi \\ 150\sin\varphi \\ -50\cos\varphi\sin\psi \\ 1 \end{bmatrix} = \begin{bmatrix} 25\sqrt{3}\cos\varphi \\ 150\sin\varphi \\ 25\cos\varphi \\ 1 \end{bmatrix}$$

According to Eq. (6.61), the trajectory of the laser beam is

$$\mathbf{r}_{tr} = \mathbf{r}_w - \mathbf{r}_t = \begin{bmatrix} 25\sqrt{3}\cos\varphi \\ 150\sin\varphi \\ 25\cos\varphi \\ 1 \end{bmatrix} - \begin{bmatrix} 0 \\ 0 \\ 10 \\ 1 \end{bmatrix} = \begin{bmatrix} 25\sqrt{3}\cos\varphi \\ 150\sin\varphi \\ 25\cos\varphi\text{-}10 \\ 0 \end{bmatrix},$$

i.e.,x = 25cos φ; y = 150 sin φ; and z = 25 cos φ -10.

In Figure 6.4, the initial position and elliptical trajectory of the laser pen-point are illustrated. For example, when the tool graves a point with φ = 0 (an end of the minor axis), the current point of the tool trajectory is x= 25 = 43.3013; y = 0; and z= 25 – 10= 15 mm.

6.3.2 CMM: measurement by probe with spherical tip

Probes with a spherical tip present a universal tool for measurement using mechanical point contact with the measured surfaces to verify actual positions and dimensions of the workpieces on machine centers before, during and after machining operations. Before processing, the initial position of the active links (tool) is verified by means of contact of the measuring tool set in the main spindle with the reference element fixed relative to the basic coordinate frame. In Figure 3.14, the home position touchsetter in the machining centers is shown.

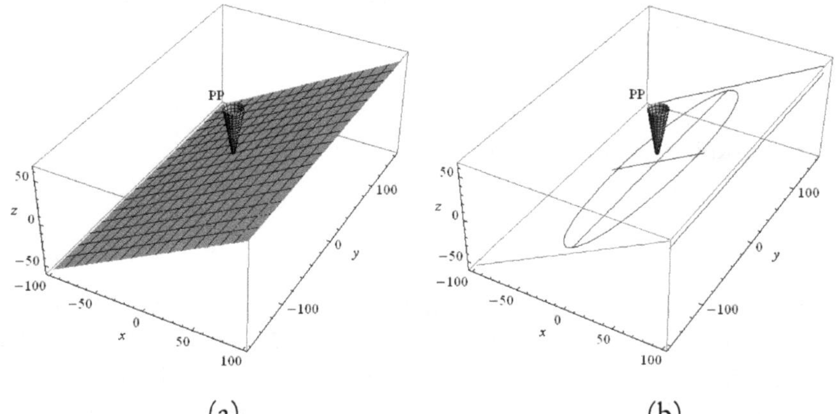

Figure 6-4 *Numerical Example 6.1*: (a) The given plane; (b) the ellipse in the plane. PP is the pen-point in the initial position.

During and after machining, the probe contacts with the processed element (Fig. 6.5). The probes with spherical tip are also in regular use for setup of CMM (Figs. 4.14-15).

The FSF of the CMM is formally identical to that of the Cartesian robot, Eq. (6.58),

$$\mathbf{r}_0 = \mathbf{A}^1(x)\,\mathbf{A}^2(y)\,\mathbf{A}^3(z)\,\mathbf{r}_3 \tag{6.62}$$

$$\mathbf{r}_3 \equiv \mathbf{r}_t = \begin{bmatrix} R\sin\theta\cos\varphi \\ R\sin\theta\sin\varphi \\ R\cos\theta \\ 1 \end{bmatrix} \tag{6.63}$$

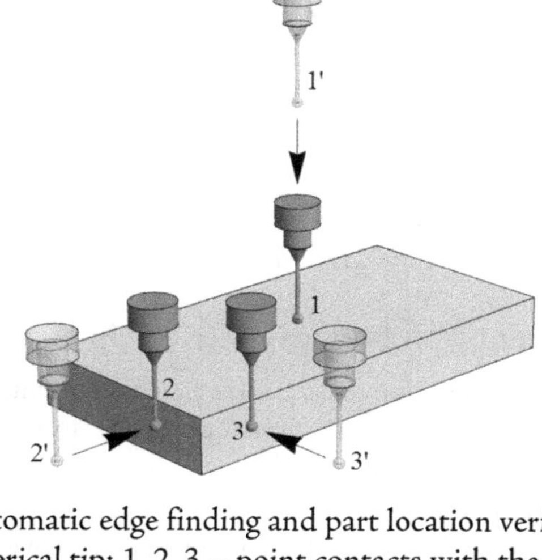

Figure 6-5 Automatic edge finding and part location verification by means of the probe with a spherical tip: 1, 2, 3 – point contacts with the part surfaces; 1', 2', 3' – probe initial positions.

where $\mathbf{r}_3$ is the vector equations of the probe as referred to their own coordinate frames; R is the probe tip radius; θ is the inclination angle counted from the Z-axis; and φ is the azimuth angle counted from the XZ-plane (Fig. 3.11).

Contact of a sphere with any surface is accompanied with two-parametric *enveloping constraints*. Equations (6.43) of the enveloping constraints are

$$\mathbf{n}_w \cdot \frac{\partial \mathbf{r}_w}{\partial \varphi} = 0 \text{ and } \mathbf{n}_w \cdot \frac{\partial \mathbf{r}_w}{\partial \theta} = 0 \tag{6.64}$$

where $\mathbf{n}_w = \mathbf{n}_w(u, v)$ is the unit vector of the normal at point (u, v) of the machined part; and $\mathbf{r}_t = \mathbf{r}_t(\varphi, \theta)$ is the tool equation, Eq. (6.63),

$$\mathbf{n}_w = \begin{bmatrix} n_x \\ n_y \\ n_z \end{bmatrix} = R \begin{bmatrix} -\sin\theta \sin\varphi \\ \sin\theta \cos\varphi \\ 0 \end{bmatrix}; \frac{\partial \mathbf{r}_w}{\partial è} = R \begin{bmatrix} \cos\theta \cos\varphi \\ \cos\theta \sin\varphi \\ -\sin\theta \end{bmatrix} \tag{6.65}$$

Hence, Eqs.(6.64) become:

$$R(-n_x \sin\theta \sin\varphi + n_y \sin\theta \cos\varphi) = 0 \tag{6.66}$$

$$R(n_x \cos\theta \cos\varphi + n_y \cos\theta \sin\varphi - n_z \sin\theta) = 0 \tag{6.67}$$

Solution of these simultaneous equations yields:

- For $\theta = 0$ or $\theta = \pi$, we have $\varphi_{1,2,3,4} = \varphi$;
- For $\theta \neq 0$ and $\theta \neq \pi$,

$$\begin{aligned} &\cos\varphi_{1,2} = \frac{n_x}{\sqrt{n_x^2 + n_y^2}}; \ \sin\varphi_{1,2} = \frac{n_y}{\sqrt{n_x^2 + n_y^2}}; \ \cos\theta_{1,2} = \pm n_z; \ \sin\theta_{1\,2} = \pm\sqrt{n_x^2 + n_y^2}; \\ &\cos\varphi_{3,4} = -\frac{n_x}{\sqrt{n_x^2 + n_y^2}}; \ \sin\varphi_{3,4} = -\frac{n_y}{\sqrt{n_x^2 + n_y^2}}; \ \cos\theta_{3,4} = \pm n_z; \ \sin\theta_{3,4} = \mp\sqrt{n_x^2 + n_y^2} \end{aligned} \tag{6.68}$$

Four solutions (6.68) result in two versions of the measured part equation,

$$\mathbf{r}_{w1,2} = \begin{bmatrix} x \pm R \sin\theta \cos\varphi \\ y \pm R \sin\theta \sin\varphi \\ z \pm R\cos\theta \\ 1 \end{bmatrix} \tag{6.69}$$

The probe trajectory equation is obtained from Eq. (6.63),

$$\mathbf{r}_{tr1,2} = \begin{bmatrix} x_w(u,v) \pm R\, n_x \\ y_w(u,v) \pm R\, n_y \\ z_w(u,v) \pm R n_z \\ 1 \end{bmatrix} = \mathbf{r}_w \pm R\, \mathbf{n}_w \tag{6.70}$$

The solutions θ = 0 and θ = π corresponds to the case in which the working point of the tool presents either of the sphere poles while the workpiece is a plane (plane-type area) whose normal is parallel to the spindle axis, $\mathbf{r}_w = [x, y, C, 1]^T$, where C is the height position of the measured plane, and $\mathbf{r}_w = [0, 0, 1, 0]^T$. Hence,

$$r_{tr3} = [x, y, C \pm R, 1]^T \tag{6.71}$$

A geometric interpretation of Eq. (6.71) is as follows: during measurements by the probe with spherical tip, the tip center trajectory is equidistant to the measured surface, and the distance is equal to radius R of the tip. Examples of measurements of the planes and convex surfaces by the probe with spherical tip are shown in Figure 6.6.

As for measurements of concave surfaces, an additional limitation takes place: radius R of the probe must be less than the minimal curvature radius $R_{c,min}$ of the measured surface,

$$R < R_{c,min} \tag{6.72}$$

Numerical example 6.2.

Given: A directrix profile of the concave parabolic cylinder is $y = 0.8x^2$.
Required: Verify whether this cylinder can be measured by probe of radius $R = 0.5$ mm.
Solution. Since the curvature of the generatrix equals zero, it is sufficient to define the curvature radius of the directrix profile

$$R_c = \frac{(1 + y'^2)^{3/2}}{y''(x)} = \frac{[1 + (1.6x)^2]^{3/2}}{1.6} = 0.625\ (1 + 2.56x^2)^{3/2}$$

$$R_{c,min} = 0.625 > R = 0.5 \text{ mm}$$

Condition (6.72) is satisfied, and thus, the measurement is possible. The scheme of measurements in accordance with conditions given in the *Example* is shown in Figure 6.7. ■

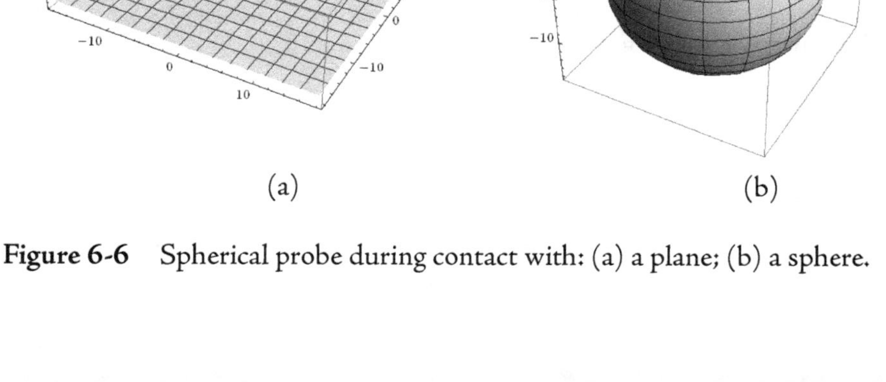

Figure 6-6 Spherical probe during contact with: (a) a plane; (b) a sphere.

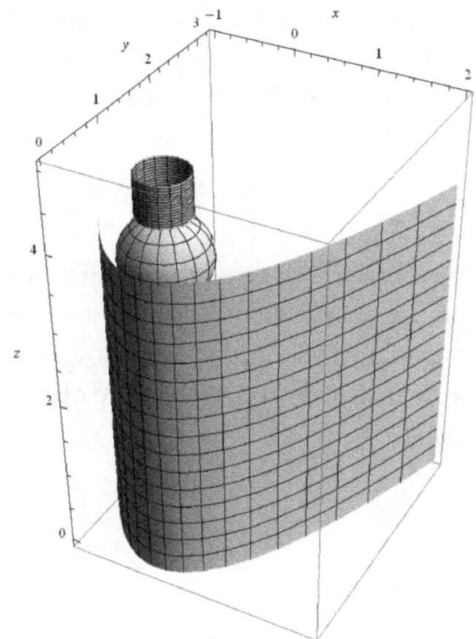

Figure 6-7 Measurement of the internal paraboloid cylinder by spherical probe.

6.3.3 CNC machine tools: 2.5-axis machining

A typical case of the 2.5-axis machining presents machining of different types of cylinders, including circular cylinders and planes, by a cylindrical cutting tool rotating around its axis (Fig. 6.8). Among the tools are grinding wheels and different types of milling cutters: helical milling cutters, end mills with cylindrical working part (not ball-end mills), key-seat and T-slot milling cutters, and so on.

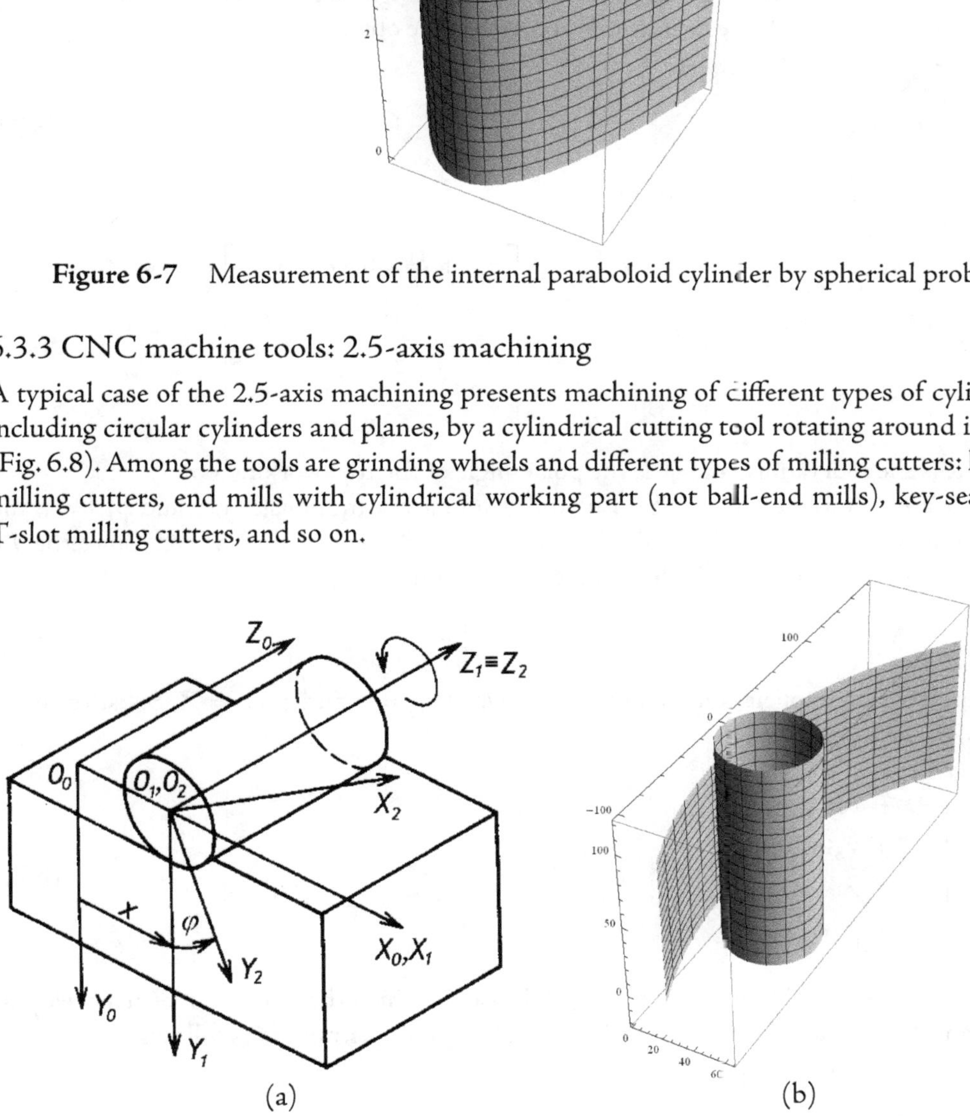

Figure 6-8 Cylindrical tool and workpiece interaction during planar profile machining: (a) a plane; (b) a concave surface.

For definiteness, the profile grinding with FSS described in Section 4.3.2.2 is considered. The FSS links are listed in Table 4.9. Since the tool axis orientation is structurally fixed, there is no need for the orientation FSF. The 5-stage setup procedure is as follows.

According to Eq. (4.55-58), the position FSF is

$$\mathbf{r}_0 = \mathbf{A}^1(-x)\mathbf{A}^2(-y)\mathbf{A}^3(z)\mathbf{A}^6(\varphi_s)\mathbf{r}_4 \tag{6.73}$$

where x, y, and z are the longitudinal feeds (translation distances along the X-, Y-, and Z-axis, respectively), the minus sign for x and y corresponds to inversion of the absolute motion since the bed is really the stationary link; and φ is the angle of the cutting tool rotation.

For the case in which the cylindrical tool is used, we obtain

$$\mathbf{r}_4 = \begin{bmatrix} R\cos\varphi_t \\ R\sin\varphi_t \\ z_t \\ 1 \end{bmatrix} \tag{6.74}$$

$$\mathbf{r}_0 = \mathbf{A}^1(-x)\mathbf{A}^2(-y)\mathbf{A}^3(z)(\varphi_t)\mathbf{r}_4 = \begin{bmatrix} -x + R\cos(\varphi_s + \varphi_t) \\ -y + R\sin(\varphi_s + \varphi_t) \\ z + z_t \\ 1 \end{bmatrix} \tag{6.75}$$

where R is the tool radius; and φ_t is the polar angle of the tool working point.

There are six variables in the FSF, Eq. (6.75): four active motions(x, y, z, and φ), $l = 4$, and two parameters (z_t and φ_t) describing the cutting tool dimensionality, $M = 2$. Since the TE (workpiece) is a surface, $m = 2$, the number of constraints, Eq. (6.17), is

$$L = n + M - m = 4 + 2 - 2 = 4 \tag{6.76}$$

There are two implicit constraints ($L_i = 2$): $\varphi_s + \varphi_t = \varphi$ and $z + z_t = Z$. Substitution in Eq. (6.75) yields

$$\mathbf{r}_0 = \begin{bmatrix} -x + R\cos\varphi \\ -y + R\sin\varphi \\ Z \\ 1 \end{bmatrix} \tag{6.77}$$

The 2.5-axis machining presents a typical case, in which the one-parametric enveloping constraint takes place ($L_e = 1$). Here, Eq. (6.42) for the constraint is as follows:

$$\mathbf{n}_w \cdot \frac{\partial \mathbf{r}_w}{\partial \ddot{o}} = \mathbf{n}_w \cdot R \begin{bmatrix} \sin\varphi \\ -\cos\varphi \\ 0 \\ 0 \end{bmatrix} = 0 \tag{6.78}$$

where $\mathbf{n}_w = \mathbf{n}_w(u, v)$ is unit vector of the normal at point (u, v) of the machined part.

The enveloping constraint becomes

$$R(n_x \sin\varphi - n_y \cos\varphi) = 0 \tag{6.79}$$

Two solutions expressed through the sine and cosine are as follows:

$$\sin\varphi_{1,2} = \pm n_y \text{ and } \cos\varphi_{1,2} = \pm n_x \tag{6.80}$$

where and $n_z = 0$.

Substitution of Eq. (6.80) in Eq. (6.78) results in the setup equation expressing the translational movement through the curvilinear coordinates u, v of the machined surface:

$$\mathbf{r}_{w1,2} = \begin{bmatrix} -x + R\cos\varphi \\ -y + R\sin\varphi \\ Z \\ 1 \end{bmatrix} = \begin{bmatrix} -x \pm Rn_x \\ -y \pm Rn_y \\ Z \\ 0 \end{bmatrix} \tag{6.81}$$

or, in the 3 × 1 vector terms,

$$\mathbf{r}_{w1,2} == \mathbf{r}_{tr} \pm R\mathbf{n}_w(u, v) \tag{6.82}$$

$$\mathbf{r}_w = \begin{bmatrix} x_w \\ y_w \\ z_w \end{bmatrix}, \mathbf{r}_{tr} = \begin{bmatrix} x_{tr} \\ y_{tr} \\ z_{tr} \end{bmatrix}, \text{ and } \mathbf{n}_w = \begin{bmatrix} n_x \\ n_y \\ n_z \end{bmatrix}$$

The setup equation expresses the translational movement equation through the curvilinear coordinates u, v of the machined surface

$$\mathbf{r}_{tr1,2} = \begin{bmatrix} x_w(u,v) \pm Rn_x \\ y_w(u,v) \pm Rn_y \\ z_w(u,v) - Z \\ 0 \end{bmatrix} \tag{6.83}$$

A geometric interpretation of Eq. (6.83) is as follows: during machining by the cylindrical tool (e.g., grinding) or by a rotating linear tool (e.g., milling), the trajectory of the tool axis is equidistant to the machined surface, and the distance equals radius R of the tool (Fig. 6.8a). The plus sign in Eq. (6.82) defines the tool trajectory performing external machining of the cylinder $\mathbf{r}_w$, and, correspondingly, the minus sign defines the tool trajectory for internal machining.

The tool moving along either of two trajectories $\mathbf{r}_{tr1}$ or $\mathbf{r}_{tr2}$ can machine not only the workpiece $\mathbf{r}_w$, but also another surface $\mathbf{r}_{w1}$ or $\mathbf{r}_{w2}$, respectively,

$$\mathbf{r}_{w1,2} = \begin{bmatrix} x_w(u,v) \\ y_w(u,v) \\ z_w(u,v) \end{bmatrix} \pm 2R\,\mathbf{n}_w(u,v) \tag{6.84}$$

In Figure 6.9a, instantaneous positions of the tool on two possible trajectories *Tr1* and *Tr2* providing machining of a cylindrical surface *W* are shown. In Figure 6.9b, the surface *W* and its equidistant *W2* machined by a single tool moved along the trajectory *Tr2* are shown.

There is a limitation for machining concave surfaces, which is identical to that of measurements: radius *R* of the tool must be less than the minimal curvature radius $R_{c,min}$ of the machined surface (Fig. 6.8b),

$$R<R_{c,min} \tag{6.85}$$

Typical applications describing 2.5 machining of surfaces on machine, whose FSF is defined through Eq. (6.84), are considered below in Numerical Examples 6.3-6.5.

Numerical example 6.3. Machining a plane by the cylindrical cutting tool. **Given:** The plane $y_w = C_w = 40$ mm is machined by cylindrical grinding wheel of radius $R = 20$ mm. **Required:** Equation of the trajectory of the tool axis during 2.5-axis machining.

Solution: The FSF, Eq. (6.84), and normal of the parallel to *Y*-axis machined plane are

$$\mathbf{r}_w(x_w, z_w) = \begin{bmatrix} x_w \\ C_w \\ z_w \end{bmatrix} \text{and } \mathbf{n}_w(z_w, \varphi_w) = \begin{bmatrix} 0 \\ 1 \\ 0 \end{bmatrix} \tag{6.86}$$

Substitution of Eq. (6.86) in Eq. (6.82) yields:

$$\mathbf{r}_{tr} = \begin{bmatrix} x_{tr} \\ y_{tr} \\ z_{tr} \end{bmatrix} = \begin{bmatrix} x_w \\ C_w \pm R \\ z_w \end{bmatrix} \tag{6.87}$$

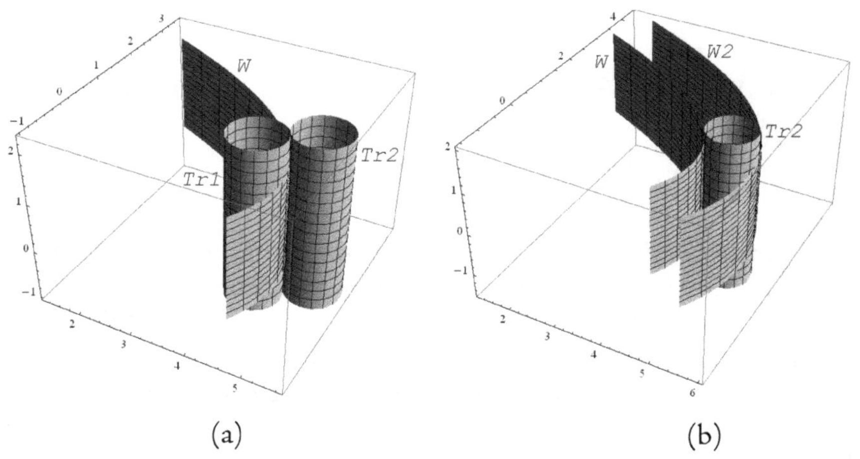

Figure 6-9 Machining a cylindrical surface *W* by cutting tool: (a) a current position of the tool on trajectories *Tr1* and *Tr2*; (b) machining a pair of surfaces *W* and *W2* by a single tool.

Hence, the solution presents two possible trajectories to machine a plane, Eq. (6.74). In Figure 6.10a, a current position of the tool on two possible trajectories of its axis for machining a plane by the cylindrical cutting tool (e.g., grinding wheel) is shown. In this case, each of two possible trajectories of the tool axis points, Eq. (6.86), presents the straight line, and these lines are parallel and located at distance $R = 20$ mm from the machined surface:

$$\mathbf{r}_{tr1} = \begin{bmatrix} x_w \\ 60 \\ z_w \end{bmatrix} \text{ and } \mathbf{r}_{tr2} = \begin{bmatrix} x_w \\ 20 \\ z_w \end{bmatrix} \text{ mm}$$

According to Eq. (6.84), the cutting tool can produce either a plane parallel to the tool trajectory or a slot (a pair of parallel planes) of width $w = 2R = 40$ mm (Fig. 6.10b), for example,

$$\mathbf{r}_{w1} = \begin{bmatrix} x_w \\ 40 \\ z_w \end{bmatrix}; \mathbf{r}_{w2} = \begin{bmatrix} x_w \\ 80 \\ z_w \end{bmatrix} \text{ mm; and } w = \left\| \mathbf{r}_{w2} - \mathbf{r}_{w1} \right\| = 40 \text{ mm}$$

Numerical example 6.4. Machining the elliptical cylinder.

Given: The FSF and normal at the elliptical cylinder with the major semi-axis a and minor semi-axis b are

$$\mathbf{r}_w(z_w, \varphi_w) = \begin{bmatrix} a\cos\varphi_w \\ b\sin\varphi_w \\ z_w \end{bmatrix} \tag{6.88}$$

$$\mathbf{n}_w(z_w, \varphi_w) = \frac{\dfrac{\partial \mathbf{r}_w}{\partial \varphi_w} \times \dfrac{\partial \mathbf{r}_w}{\partial z_w}}{\left\| \dfrac{\partial \mathbf{r}_w}{\partial \varphi_w} \times \dfrac{\partial \mathbf{r}_w}{\partial z_w} \right\|} = \frac{1}{u} \begin{bmatrix} b\cos\varphi_w \\ a\sin\varphi_w \\ 0 \end{bmatrix}, \text{ with } u = \sqrt{a^2\cos^2\varphi_w + b^2\sin^2\varphi_w} \tag{6.89}$$

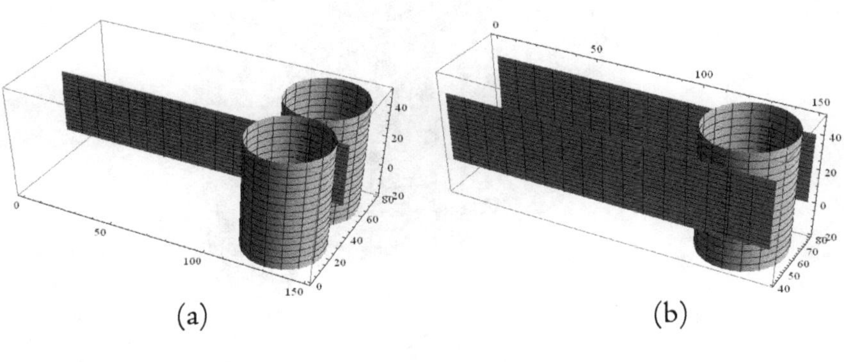

(a) (b)

Figure 6-10 *Numerical example 6.3.* Machining a plane by the cylindrical cutting tool: (a) a current position of the tool on two possible trajectories; (b) machining a straight slot by a single tool.

Required: Equation of the trajectory of the tool axis during external 2.5-axis machining of the elliptical cylinder by cylindrical tool of radius R, which is given by vector equation

$$\mathbf{r}_t = \begin{bmatrix} R\cos\varphi_t \\ R\sin\varphi_t \\ z_t \end{bmatrix}; \tag{6.90}$$

Solution: According to Eq. (6.82), the tool trajectory during external machining is

$$\mathbf{r}_{tr} = \begin{bmatrix} x_{tr} \\ y_{tr} \\ z_{tr} \end{bmatrix} = \begin{bmatrix} x_w(u,v) \\ y_w(u,v) \\ z_w(u,v) - z_t \end{bmatrix} + R\,\mathbf{n}_w = \begin{bmatrix} a\cos\varphi_w \\ b\sin\varphi_w \\ z_w - z_t \end{bmatrix} + \frac{R}{u}\begin{bmatrix} b\cos\varphi_w \\ a\sin\varphi_w \\ 0 \end{bmatrix} = \begin{bmatrix} (a+bR/u)\cos\varphi_w \\ (b+aR/u)\sin\varphi_w \\ z_w - z_t \end{bmatrix} \tag{6.91}$$

Let the semi-axis $a = 80$ mm and $b = 50$ mm, and tool radius $R = 20$ mm. In this case, the tool axis trajectory, Eq. (6.91), becomes:

$$\mathbf{r}_{tr} = \begin{bmatrix} (a+bR/u)\cos\varphi_w \\ (b+aR/u)\sin\varphi_w \\ z_w - z_t \end{bmatrix} = \begin{bmatrix} \left(80 + \dfrac{10}{\sqrt{64\cos^2\varphi_w + 25\sin^2\varphi_w}}\right)\cos\varphi_w \\ \left(50 + \dfrac{16}{\sqrt{64\cos^2\varphi_w + 25\sin^2\varphi_w}}\right)\sin\varphi_w \\ z_w - z_t \end{bmatrix} \tag{6.92}$$

In Figure 6.11, the machined workpiece of the given dimensions, the tool axis trajectory presenting equidistance to the workpiece, and four instantaneous positions of the grinding tool during external machining are shown.

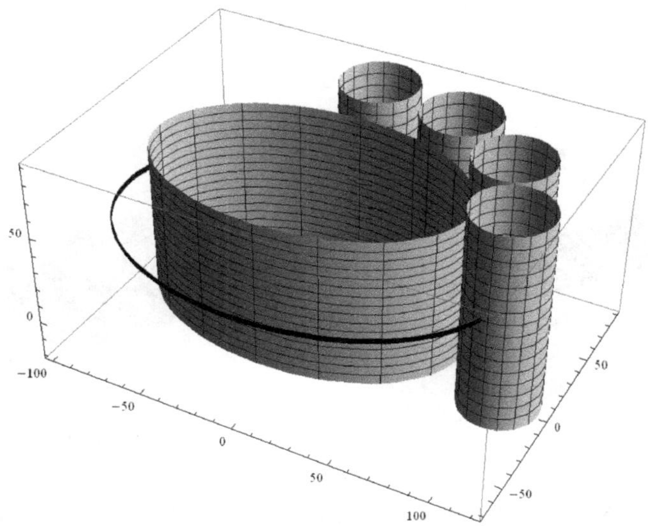

Figure 6-11 *Numerical example 6.4.* External machining of the elliptical cylinders by circular cutting tool.

Numerical example 6.5. Machining the circular cylinder.

Given: The FSF of the circular cylinder and its normal are

$$\mathbf{r}_w(z_w,\varphi_w) = \begin{bmatrix} a+R_w\cos\ \varphi_w \\ b+R_w\sin\ \varphi_w \\ z_w \end{bmatrix}; \mathbf{n}_w(z_w,\varphi_w) = \begin{bmatrix} \cos\ \varphi_w \\ \sin\ \varphi_w \\ 0 \end{bmatrix} \tag{6.93}$$

where R_w is the radius of the cylinder; a and b are coordinates of the cylinder center; and φ_w is the polar angle of cylinder point.

Required: (a) Equation of the trajectory of the tool and (b) equation of the second surface, which can be machined during movement of the tool along one of two possible trajectories.

Solution: According to Eq. (6.82), substitution of Eq. (6.93) in Eq. (6.82) yields two possible trajectories of the tool

$$\mathbf{r}_{tr1,2} = \begin{bmatrix} a+R_w\cos\varphi_w \pm R\cos\varphi_w \\ b+R_w\sin\varphi_w \pm R\sin\varphi_w \\ z_w \end{bmatrix} = \begin{bmatrix} a+(R_w \pm R)\cos\varphi_w \\ b+(R_w \pm R)\sin\varphi_w \\ z_w \end{bmatrix} \tag{6.94}$$

The second cylindrical surface additional to that given through Eq. (6.94) is calculated according to Eq. (6.84):

$$\mathbf{r}_{w1,2} = \begin{bmatrix} x_w(u,v) \\ y_w(u,v) \\ z_w(u,v) \end{bmatrix} \mp 2R\,\mathbf{n}_w(u,v) = \begin{bmatrix} a+R_w\cos\ \varphi_w \mp 2R\cos\ \varphi_w \\ b+R_w\sin\ \varphi_w \mp 2R\sin\ \varphi_w \\ z_w \end{bmatrix} = \begin{bmatrix} a-(R_w \mp 2R)\cos\ \varphi_w \\ b+(R_w \mp 2R)\sin\ \varphi_w \\ z_w \end{bmatrix} \tag{6.95}$$

Let $R_w = 50$; $a = 50$; $b = 80$; and $R = 15$ mm. Hence,

$$\mathbf{r}_w = \begin{bmatrix} a+R_w\cos\ \varphi_w \\ b+R_w\sin\ \varphi_w \\ z_w \end{bmatrix} = \begin{bmatrix} 50+50\cos\ \varphi_w \\ 80+50\sin\ \varphi_w \\ z_w \end{bmatrix};$$

$$\mathbf{r}_{tr1,2} = \begin{bmatrix} a+(R_w \pm R)\cos\ \varphi_w \\ b+(R_w \pm R)\sin\ \varphi_w \\ z_w \end{bmatrix} = \begin{bmatrix} 50+(50 \pm 15)\cos\ \varphi_w \\ 80+(50 \pm 15)\sin\ \varphi_w \\ z_w \end{bmatrix} \tag{6.96}$$

Let us consider the trajectory

$$\mathbf{r}_{tr1} = \begin{bmatrix} 50+(50+15)\cos\ \varphi_w \\ 80+(50+15)\sin\ \varphi_w \\ z_w \end{bmatrix} = \begin{bmatrix} 50+65\cos\ \varphi_w \\ 80+65\sin\ \varphi_w \\ z_w \end{bmatrix}$$

This trajectory corresponds to external machining of the cylinder by Eq. (6.96). The internal machining of the second cylindrical surface is described through equation

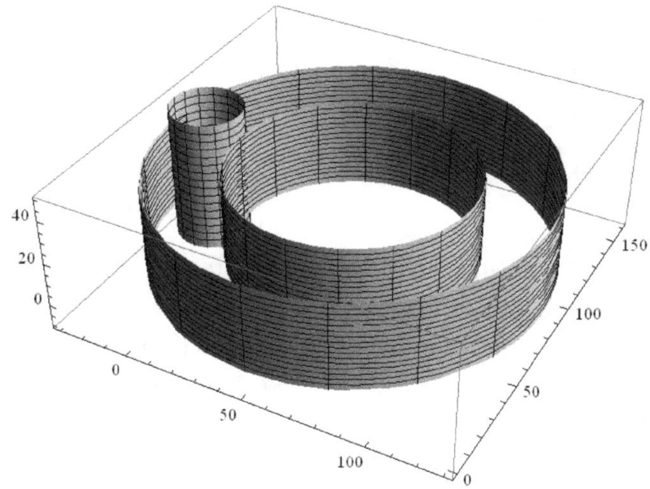

Figure 6-12 *Numerical example* 6.5. External and internal machining of the cylinders by circular cutting tool.

$$\mathbf{r}_{w2} = \begin{bmatrix} (R_w + 2R)\cos\varphi_w \\ (R_w + 2R)\sin\varphi_w \\ z_w \end{bmatrix} = \begin{bmatrix} (50+30)\cos\varphi_w \\ (50+30)\sin\varphi_w \\ z_w \end{bmatrix} = \begin{bmatrix} 50+80\cos\varphi_w \\ 80+80\sin\varphi_w \\ z_w \end{bmatrix} \quad (6.97)$$

Both surfaces (6.96) and (6.97) as well as an instantaneous position of the tool during simultaneous machining of both surfaces are shown in Figure 6.12.

6.3.4 CNC machine tools: 3-axis machining

Widening the scope of working of the machine tool from 2.5-axis to 3-axis machining is illustrated in Figure 6.13, by the example of machining plane W by a spherical cutting tool attached to the rotating spindle Sp. In both 2.5- and 3-axis machining, the spindle axis has permanent vertical or horizontal orientation depending on the machine tool type.

Examples of the cutting tool with spherical working zone are shown in Figure 6.14: (a) the 4-flute milling cutter of diameter up to 1" with ball nose and (b) the 6" × 1.5" diamond grinding wheel. In both cases, the tool working zone presents a hemisphere of upper radius $2r$, while the south pole (angle $\theta = \pi$) and its nearest area has to be eliminated from the working zone since in this area linear velocity is equal to or close to zero. The hemisphere can be full or frustum form (spherical segment, Fig. 6.14b).

Angles α_1 and α_2 (Fig. 6.15) between the spindle axis and the normals $\mathbf{n}_1$ and $\mathbf{n}_2$ at machined planes P1 and P2, respectively, are made feasible for machining is bounded as follows:

$$\theta_{min} \le \alpha_1 \le \theta_{max}; \text{ and } -\theta_{max} \le \alpha_2 \le -\theta_{min}; \text{ with } \theta \ne \pi \quad (6.98)$$

where θ_{min} and θ_{max} are the zenith angles bounding the spherical segment and defined by Eq. (3.76); angles α_1 and α_2 are defined under conditions that the spindle axis is oriented from the sphere center to the spindle, and the normal is directed from the contact point into the workpiece body.

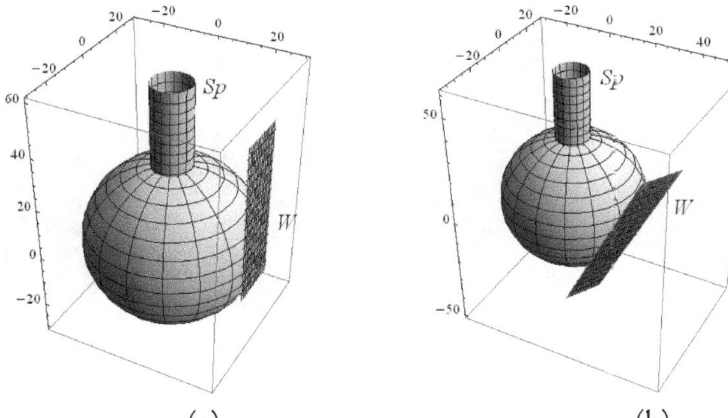

(a) (b)

Figure 6-13 Machining by a spherical cutting tool: (a) 2.5-axis machining a plane W parallel to the spindle axis; (b) 3-axis machining an oblique plane W.

In Figure 6.16, the scheme of the 3-axis machining of a sculpture surface by a rotating spherical cutting tool is shown. Since the cutting tool axis orientation is structurally fixed, the setup procedure is similar to that of 2.5 machining, but with another type of the cutting tool. The FSS links are listed in Table 6.2, where three translational links are combined in a three-parametric translational matrix $\mathbf{A}_{tr}$, Eq. (4.54). The position FSF is similar to that defined through Eq. (6.73):

$$\mathbf{r}_0 = \mathbf{A}^2(-y)\,\mathbf{A}^1(-x)\,\mathbf{A}^3(z)\,\mathbf{A}^6(\varphi_2)\,\mathbf{r}_2 = \mathbf{A}_{tr}(-x,-y,z)\,\mathbf{A}^6(\varphi_2)\,\mathbf{r}_2 \tag{6.99}$$

$$\mathbf{r}_2 = \begin{bmatrix} R\sin\theta\cos\varphi_t \\ R\sin\theta\sin\varphi_t \\ h_{sp} - R\cos\theta \\ 1 \end{bmatrix}, \quad 0 \le \varphi_t < 2\pi \text{ and } 0 < \theta_{up} \le \theta < \theta_{low} \le \pi \tag{6.100}$$

(a) (b)

Figure 6-14 Spherical cutting tools: (a) milling cutter with spherical end; (b) diamond grinding wheel 6" × 1.5" [6].

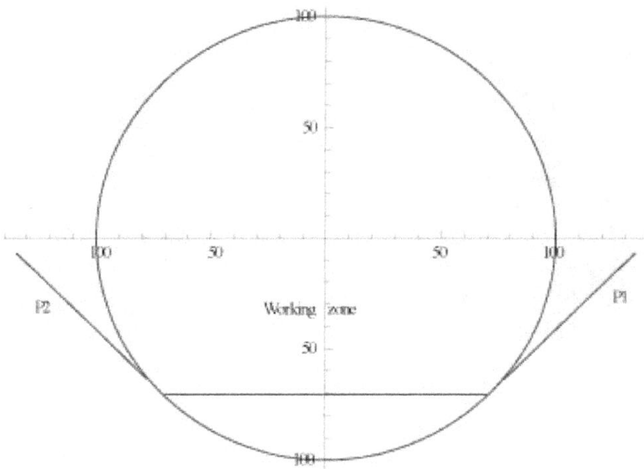

Figure 6-15 Limits of the tilting angles of the machined plane P1 and P2.

where $\mathbf{A}^1$, $\mathbf{A}^2$, and $\mathbf{A}^3$, are the 4 × 4 homogeneous matrices of links translations; x, y, and z are the linear feeds along the X-, Y-, and Z-axis, respectively, the minus sign for x and y corresponds to inversion of the absolute motion since the bed is really the stationary link; $\mathbf{A}^6$ is the 4 × 4 homogeneous matrix of the spindle rotation around the Z-axis; φ_2 is the angle of the spindle rotation; $\mathbf{r}_4$ is the position vector of the working point of the cutting tool of radius R; φ_t and θ are the azimuth and zenith angles of the tool point, respectively; and h_{sp} is the distance between tool holder and sphere center.

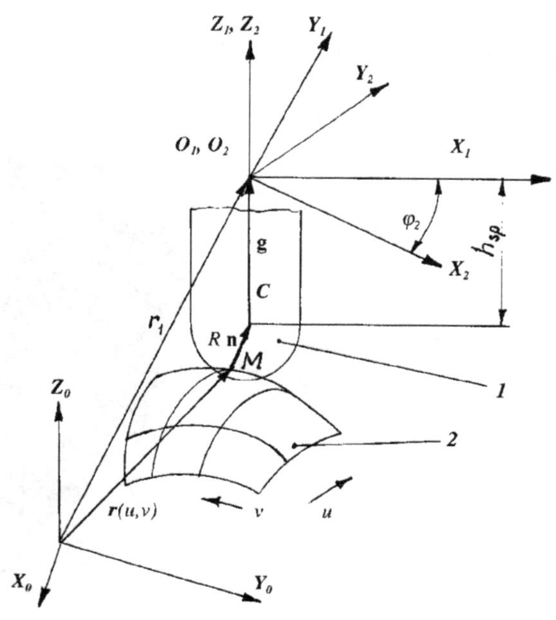

Figure 6-16 Schematic diagram of machining of free-form surface by spherical milling cutter.

Table 6-2 FSS of the milling machine, k = **1236.**

S_i	Link	Motion relative to the previous link	Matrix ${}^{i-1}\mathbf{A}_i$
S_0	Workpiece	Conventionally fixed link	–
S_1	Milling head	Translation from frame S_0 to frame S_1 at distances $-x$, $-y$, and z	${}^0\mathbf{A}_1 = \mathbf{A}_{tr}(-x, -y, z)$
S_2	Spindle	Rotation around the Z_2-axis at angle φ_2	${}^1\mathbf{A}_2 = \mathbf{A}^6(\varphi_2)$

Substituting Eq. (6.100) in Eq. (6.99) yields

$$\mathbf{r}_0 = \begin{bmatrix} 1 & 0 & 0 & -x \\ 0 & 1 & 0 & -y \\ 0 & 0 & 1 & z \\ 0 & 0 & 0 & 1 \end{bmatrix} \begin{bmatrix} \cos\varphi_2 & -\sin\varphi_2 & 0 & 0 \\ \sin\varphi_2 & \cos\varphi_2 & 0 & 0 \\ 0 & 0 & 1 & 0 \\ 0 & 0 & 0 & 1 \end{bmatrix} \begin{bmatrix} R\sin\theta\cos\varphi_t \\ R\sin\theta\sin\varphi_t \\ h_{sp} - R\cos\theta \\ 1 \end{bmatrix} = \begin{bmatrix} -x + R\sin\theta\cos(\varphi_2 + \varphi_t) \\ -y + R\sin\theta\sin(\varphi_2 + \varphi_t) \\ z + h_{sp} - R\cos\theta \\ 1 \end{bmatrix} \tag{6.101}$$

Constraint number is $L = n + M - m = 4 + 2 - 2 = 4$.
Substitution of a single *implicit constraint* ($L_i = 1$), $\varphi_2 + \varphi_t = \varphi$, yields

$$\mathbf{r}_0 = \begin{bmatrix} -x + R\sin\theta\cos\varphi \\ -y + R\sin\theta\sin\varphi \\ z + h_{sp} - R\cos\theta \\ 1 \end{bmatrix} \tag{6.102}$$

The machined surface is

$$\mathbf{r}_w = \begin{bmatrix} x(u,v) \\ y(u,v) \\ z(u,v) \\ 1 \end{bmatrix} = \begin{bmatrix} -x + R\sin\theta\cos\varphi \\ -y + R\sin\theta\sin\varphi \\ z + h_{sp} - R\cos\theta \\ 1 \end{bmatrix} \text{ and } \mathbf{n}_w = \begin{bmatrix} n_x \\ n_y \\ n_z \\ 0 \end{bmatrix} \tag{6.103}$$

where $\mathbf{r}_w = \mathbf{r}_w(u, v)$ is the position vector of the machined part; and $\mathbf{n}_w = \mathbf{n}_w(u, v)$ is the unit vector of the normal at point (u, v) of the machined part.

Contact of a sphere with any surface is accompanied with two-parametric *enveloping constraints*($L_e = 2$). Equations (6.43) of the enveloping constraints are

$$\mathbf{n}_w \cdot \frac{\partial \mathrm{r}_w}{\partial\varphi} = \begin{bmatrix} n_x \\ n_y \\ n_z \\ 0 \end{bmatrix} \cdot R \begin{bmatrix} -\sin\theta\sin\varphi \\ \sin\theta\cos\varphi \\ 0 \\ 0 \end{bmatrix} = 0 \text{ and } \mathbf{n}_w \cdot \frac{\partial \mathrm{r}_w}{\partial\theta} = \begin{bmatrix} n_x \\ n_y \\ n_z \\ 0 \end{bmatrix} \cdot R \begin{bmatrix} \cos\theta\cos\varphi \\ \cos\theta\sin\varphi \\ -\sin\theta \\ 0 \end{bmatrix} = 0 \tag{6.104}$$

Hence, Eqs.(6.104) become

$$-n_x \sin\theta\sin\varphi + n_y \sin\theta\cos\varphi = 0 \tag{6.105}$$

$$n_x \cos\theta\cos\varphi + n_y \cos\theta\sin\varphi - n_z \sin\theta = 0 \tag{6.106}$$

Solution of these simultaneous equations is considered for $\theta \neq 0$ and $\theta \neq \pi$ since solutions $\theta = 0$ and $\theta = \pi$ are not satisfied conditions (6.98):

$$\cos\varphi_{1,2} = \frac{n_x}{\sqrt{n_x^2 + n_y^2}}; \sin\varphi_{1,2} = \frac{n_y}{\sqrt{n_x^2 + n_y^2}}; \cos\theta_{1,2} = \pm n_z; \sin\theta_{1,2} = \pm\sqrt{n_x^2 + n_y^2} \tag{6.107}$$

Substitution of Eq. (6.107) in Eq. (6.103) yields

$$\mathbf{r}_w = \begin{bmatrix} -x + R\sin\theta\cos\varphi \\ -y + R\sin\theta\sin\varphi \\ z + h_{sp} - R\cos\theta \\ 1 \end{bmatrix} = \begin{bmatrix} -x \pm R\,n_x \\ -y \pm R\,n_y \\ z + h_{sp} \pm Rn_z \\ 0 \end{bmatrix} \tag{6.108}$$

The tool trajectory equation describes two possible trajectories of the sphere center

$$\mathbf{r}_{tr1,2} = \begin{bmatrix} x \\ y \\ z \\ 0 \end{bmatrix} = \mathbf{r}_w(u,v) \pm R\,\mathbf{n}_w(u,v) \tag{6.109}$$

A geometric interpretation of Eq. (6.109) is as follows: during cutting by the spherical tool, its center trajectory is equidistant to the machined surface, and the distance is equal to radius R of the tool. Examples of measurements of the planes and convex surfaces by the probe with spherical tip are shown in Figure 6.17. Note that:

1. *Limitation for machining of concave surfaces*, Eq. (6.85), is valid also for 3-axis machining: radius R of the tool must be less than the minimal curvature radius $R_{c,min}$ of the machined surface, $R<R_{c,min}$.
2. *Cutting speed.* As distinguished from the measurement mode, where the probe does not rotate, the solution must be continued to define a linear velocity of the tool rotation. During the profile machining by the spherical tool, the linear cutting speed depends on the current distance of the "tool-workpiece" contact point from the tool axis – current radius of rotation. According to definition,

$$v\,[\text{m/min}] = 2\pi R_p N/1000 \tag{6.110}$$

where N, rpm, is the rotation frequency of the cutting tool; R_p is the diameter of the current working point of the tool, $0<R_p \leq R$, mm, defined through the zenith angle θ,

$$R_p = R\sin\theta, \text{ with } \theta_{min} \leq \theta \leq \theta_{max}, \tag{6.111}$$

Hence,

$$v\,[\text{m/min}] = |\pi RN/500\,\sin\theta| \tag{6.112}$$

Numerical example 6.6. **Given:** An internal spherical zone of radius R_w = 40 mm is machined by ball-nose milling cutter of nose-diameter 2R = 10 mm. The machined spherical segment presents a hemisphere cut off by a plane spaced h_{sp} = 30 mm from the hemisphere base (Fig. 6.17). **Required:** Calculation of angular frequency of spindle rotation N, rpm, providing the permanent cutting speed v = 15 m/min.
Solution: The equation of and normal at the machined surface are as follows:

$$\mathbf{r}_w(\Phi,\Theta) = \begin{bmatrix} R_w \sin\Theta\cos\Phi \\ R_w \sin\Theta\sin\Phi \\ R_w\cos\Theta \\ 1 \end{bmatrix}; \mathbf{n}_w(\Phi,\Theta) = \pm\begin{bmatrix} \sin\Theta\cos\Phi \\ \sin\Theta\sin\Phi \\ \cos\Theta \\ 0 \end{bmatrix} \quad (6.113)$$

Note that the workpiece presents a limited spherical segment: $\pi/2 \le \Theta \le \pi$ - arctan (30/40) = 2.41886 rad. Hence, the tool trajectory lies at the sphere equidistant with the machined surface:

$$\mathbf{r}_{tr1,2} = \mathbf{r}_w(u,v) \pm R\,\mathbf{n}_w(u,v) = \begin{bmatrix} (R_w \pm R)\sin\Theta\cos\Phi \\ (R_w \pm R)\sin\Theta\sin\Phi \\ (R_w \pm R)\cos\Theta \\ 1 \end{bmatrix} \quad (6.114)$$

The enveloping requirement $\mathbf{n}_w = \pm\,\mathbf{n}_t$ results in equality of the zenith and azimuth angles of the workpiece and tool in the contact point, $\Theta = \theta$ and $\Phi = \varphi$. According Eq. (6.112),

$$v[\mathrm{m/min}] = |\pi RN\sin\theta/500| = 15 \text{ and } N = \left|v\frac{500}{\pi R\sin\theta}\right| = \left|15\frac{500}{5\pi\sin\theta}\right|$$

Hence, N = 477.5 rpm, for $\theta = \pi/2$, and N = 721.9 rpm, for θ = 2.41886.

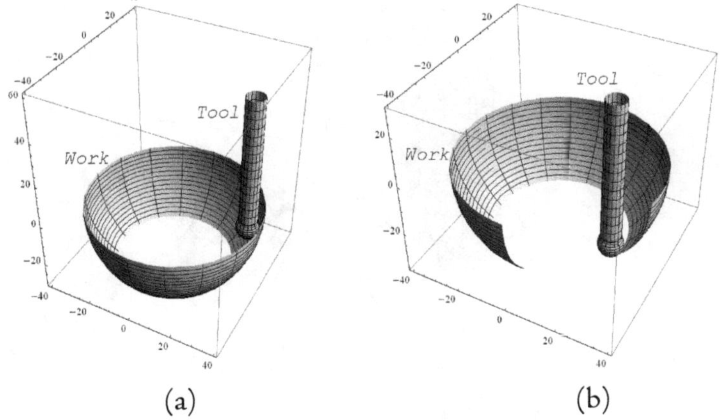

Figure 6-17 The machined sphere and two limiting instantaneous positions of the cutting tool.

6.3.5 Setup of the 5-axis machine with swivel-tilt head

6.3.5.1 General

Four possible structures of the 5-axis machines are shown in Figure 4.21. The general setup equations for 5-axis machines aim at solution with respect to five parameters of motion: three translation movements along the Cartesian coordinate axes and two angles of tool axis orientation. We begin with the solution related to the machine with tilting and swiveling motions performed by the spindle head (Fig. 6.18). According to Table 4.10, the tool axis is oriented along the Z-axis, and the FSF of the machine, Eq. (4.62), is

$$[\mathbf{r}_0\ \mathbf{c}_0] = \mathbf{A}^1(-x)\ \mathbf{A}^2(-y)\ \mathbf{A}^3(z)\ \mathbf{A}^6(\varphi_{sw})\ \mathbf{A}^4(\theta)\ \mathbf{A}^6(\varphi s)\ [\mathbf{r}_t\ \mathbf{c}_t] =$$

$$\begin{bmatrix} 1 & 0 & 0 & -x \\ 0 & 1 & 0 & -y \\ 0 & 0 & 1 & z \\ 0 & 0 & 0 & 1 \end{bmatrix} \begin{bmatrix} \cos\varphi_{sw} & -\sin\varphi_{sw} & 0 & 0 \\ \sin\varphi_{sw} & \cos\varphi_{sw} & 0 & 0 \\ 0 & 0 & 1 & 0 \\ 0 & 0 & 0 & 1 \end{bmatrix} \begin{bmatrix} 1 & 0 & 0 & 0 \\ 0 & \cos\theta & -\sin\theta & 0 \\ 0 & \sin\theta & \cos\theta & 0 \\ 0 & 0 & 0 & 1 \end{bmatrix} \begin{bmatrix} \cos\varphi_s & -\sin\varphi_s & 0 & 0 \\ \sin\varphi_s & \cos\varphi_s & 0 & 0 \\ 0 & 0 & 1 & 0 \\ 0 & 0 & 0 & 1 \end{bmatrix} \begin{bmatrix} x_t & 0 \\ y_t & 0 \\ z_t & 1 \\ 1 & 0 \end{bmatrix} \quad (6.114)$$

where x, y, z are translation movements of the machine table, saddle, and spindle head, respectively; φ_{sw} and θ are the swiveling and tilting angles, respectively, of rotation of the spindle head; and φ_s is the angle of the spindle rotation.

A cutting tool for milling-boring operations has a multi-edge structure with rectilinear or helical teeth (Fig. 6.19). Without loss of generality from the kinematics viewpoint, the setup problem can be considered using a "virtual tool" (Fig. 6.20) presenting the track of milling cutter edges of the rotating tool,

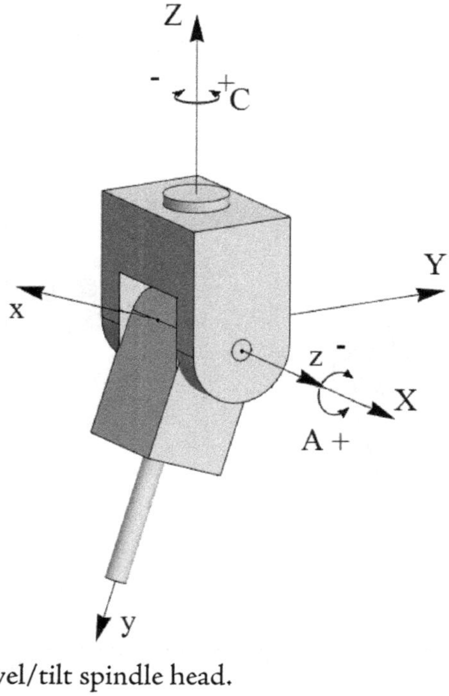

Figure 6-18 Swivel/tilt spindle head.

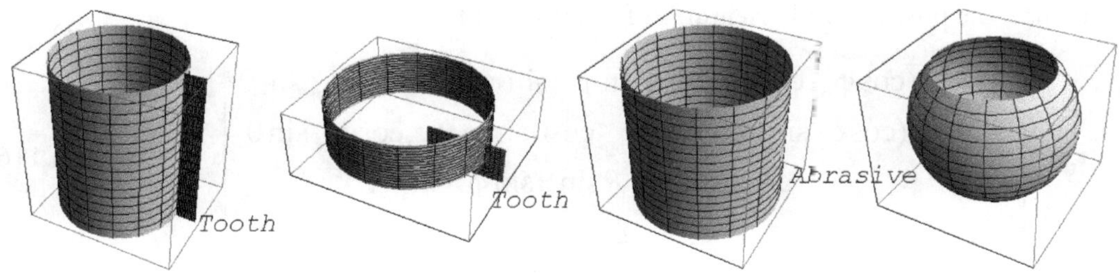

(a) (b) (c) (d)

Figure 6-19 Kinematic models of the cutting tools: (a) the one-tooth end milling cutter; (b) the one-tooth face milling cutter; (c) cylindrical grinding wheel; (d) spherical grinding wheel.

$$[\mathbf{r}_{vt}\mathbf{c}_{vt}] = \mathbf{A}^6(\varphi_s)[\mathbf{r}_t\mathbf{c}_t] = \begin{bmatrix} \cos\varphi_s & -\sin\varphi_s & 0 & 0 \\ \sin\varphi_s & \cos\varphi_s & 0 & 0 \\ 0 & 0 & 1 & 0 \\ 0 & 0 & 0 & 1 \end{bmatrix} \begin{bmatrix} x_t & 0 \\ y_t & 0 \\ z_t & 1 \\ 1 & 0 \end{bmatrix} = \begin{bmatrix} x_t\cos\varphi_s - y_t\sin\varphi_s & 0 \\ x_t\sin\varphi_s + y_t\cos\varphi_s & 0 \\ z_t & 1 \\ 1 & 0 \end{bmatrix} \quad (6.115)$$

6.3.5.2 Setup for end milling

FSF formulation

The working points of the end milling cutter of radius R (Fig. 6.19a) and tool axis are expressed through the 4 × 2 matrix

$$[\mathbf{r}_t\mathbf{c}_t] = \begin{bmatrix} R & 0 \\ 0 & 0 \\ z_t & 1 \\ 1 & 0 \end{bmatrix}; z_{fr} \le z_t \le z_{fr} + L_t$$

where R is the milling cutter radius; z_t is the current distance of the edge points from the axis of tool head tilting; z_{fr} is the coordinate of the frontal tool section; and L_t is the length of the working part of the tool.

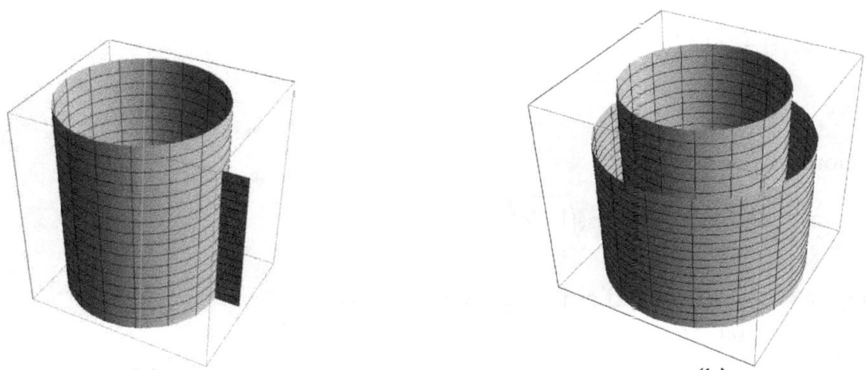

(a) (b)

Figure 6-20 Milling cutter kinematic models: (a) one-tooth cutter tool; (b) virtual tool.

Hence, the position and orientation FSF, Eq. (6.114), are

$$\mathbf{r}_0 = \begin{bmatrix} -x + R(\cos\varphi_{sw}\cos\varphi_s - \cos\theta\sin\varphi_{sw}\sin\varphi_s) + z_t\sin\varphi_{sw}\sin\theta \\ -y + R(\cos\varphi_s\sin\varphi_{sw} + \cos\varphi_{sw}\cos\theta\sin\varphi_s) - z_t\cos\varphi_{sw}\sin\theta \\ z + z_t\cos\theta + R\sin\theta\sin\varphi_s \\ 1 \end{bmatrix} \tag{6.116}$$

$$\mathbf{c}_0 = \begin{bmatrix} \sin\varphi_{sw}\sin\theta \\ -\cos\varphi_{sw}\sin\theta \\ \cos\theta \\ 0 \end{bmatrix} \tag{6.117}$$

The form-shaping system consists of six parameters of motion $x,y,z,\theta,\varphi_{sw},\varphi_s$ ($l = 6$) and one parameter of the linear-type tool z_t ($M = 1$). Constraint number L during processing a surface ($m = 2$) is

$$L = n + M - m = 6 + 1 - 2 = 5 \tag{6.118}$$

Main setup equations

Setup requirements are satisfied if the right-hand sides of Eqs. (6.116) and (6.117) are equal to the equations of the machined surface. Substitution in the left-hand sides gives

$$\mathbf{r}_w(u, v) = \mathbf{r}_0(x,y,z,\theta,\varphi_{sw},\varphi_s,z_t) \tag{6.119}$$

$$\mathbf{c}_w(u, v) = \mathbf{c}_0(\theta,\varphi_{sw}) \tag{6.120}$$

where $\mathbf{r}_w(u, v) \equiv [x_w, y_w, z_w, 1]^{\mathrm{T}}$ is the *given* position vector of the machined surface; and $\mathbf{c}_w(u, v) \equiv [c_x, c_y, c_z, 0]$ is the *given* orientation vector of the tool axis.

Substitution of the right-hand side of Eq. (6.117) in Eq. (6.120) gives three simultaneous equations with respect to angles φ_{sw} and θ (note that $\|\mathbf{c}_w\| = \|\mathbf{c}_0\| = 1$)

$$\sin\varphi_{sw}\sin\theta = c_x;$$

$$-\cos\varphi_{sw}\sin\theta = c_y;$$

$$\cos\theta = c_z \tag{6.121}$$

The solutions are

$$\cos\theta_{1,2} = c_z \text{ and } \sin\theta_{1,2} = \pm\sqrt{1-c_z^2} \tag{6.122}$$

$$\cos\varphi_{sw1,2} = \mp\frac{c_y}{\sqrt{1-c_z^2}} \text{ and } \sin\varphi_{sw1,2} = \pm\frac{c_x}{\sqrt{1-c_z^2}} \tag{6.123}$$

Note that side (profile) milling takes place only with $c_z \neq 1$. If $c_z = 1$, $c_x = c_y = 0$, the tool axis is perpendicular to the machined surface. This setup presents the face milling (not the side milling) and will be considered below (Subsection 6.3.6.2).

The *enveloping* constraints have the form:

$$\mathbf{n}_w \cdot \frac{\partial \mathbf{r}_w}{\partial \varphi_s} = 0 \tag{6.124}$$

$$\mathbf{n}_w \cdot \frac{\partial \mathbf{r}_w}{\partial z_t} = 0 \tag{6.125}$$

where $\mathbf{n}_w$ is the known unit vector of the normal at the machined surface, $\mathbf{n}_w = [n_x, n_y, n_z, 0]^T$.

Substitution of the partial derivative with respect to φ_s in Eq. (6.124) yields

$$\begin{bmatrix} n_x \\ n_y \\ n_z \\ 0 \end{bmatrix} \mathbf{A}^6(\varphi_{sw})\mathbf{A}^4(q) \begin{bmatrix} -R\sin\varphi_s \\ R\cos\varphi_s \\ 0 \\ 0 \end{bmatrix} = R[n_x(-\cos\varphi_{sw}\sin\varphi_s - \sin\varphi_{sw}\cos\theta\cos\varphi_s) +$$

$$n_y(-\sin\varphi_{sw}\sin\varphi_s + \cos\varphi_{sw}\cos\theta\cos\varphi_s) + n_z\sin\theta\cos\varphi_s] = 0 \tag{6.126}$$

After formal transformations, we obtain

$$\tan\varphi_s = \frac{-n_x\cos\theta\sin\varphi_{sw} + n_y\cos\theta\cos\varphi_{sw} + n_z\sin\theta}{n_x\cos\varphi_{sw} + n_y\sin\varphi_{sw}} \tag{6.127}$$

Equation (6.127) has two solutions

$$\varphi_{s1,2} = \begin{cases} \arctan\Lambda \\ \pi + \arctan\Lambda \end{cases}, \text{ with } \Lambda = \frac{-n_x\cos\theta\sin\varphi_{sw} + n_y\cos\theta\cos\varphi_{sw} + n_z\sin\theta}{n_x\cos\varphi_{sw} + n_y\sin\varphi_{sw}} \tag{6.128}$$

Substitution of the partial derivative with respect to z_t in Eq. (6.125) yields

$$\mathbf{n}_w \cdot \frac{\partial \mathbf{r}_w}{\partial z_t} = \mathbf{n}_w \cdot \mathbf{c}_0 \equiv 0, \tag{6.129}$$

i.e., Eq. (6.125) becomes the *implicit* (not *enveloping!*) constraint, which is formulated through Eq. (6.48) or (6.53),

$$z_t = z_t(u, v) \tag{6.130}$$

The tool trajectory design consists of two steps:

(a) *Machining step* when the tool performs regular feed motions simultaneously with the main motion – tool cutting rotation;
(b) *Idle step* when feed drives perform high-speed motions while the main motion is optional.

In both stages, it is convenient to synthesize the tool trajectory for a "virtual tool" presenting the edge in common with its cutting motion – rotation around the tool axis (Fig. 6.20),

$$\mathbf{r}_{vt} = \mathbf{A}^6(\varphi_s)\mathbf{r}_t = \begin{bmatrix} \cos\varphi_s & -\sin\varphi_s & 0 & 0 \\ \sin\varphi_s & \cos\varphi_s & 0 & 0 \\ 0 & 0 & 1 & 0 \\ 0 & 0 & 0 & 1 \end{bmatrix} \begin{bmatrix} R \\ 0 \\ z_t \\ 1 \end{bmatrix} = \begin{bmatrix} R\cos\varphi_s \\ R\sin\varphi_s \\ z_t \\ 1 \end{bmatrix} \tag{6.131}$$

Translation movement

The vector of the translation motions of the cutting tool is defined according to Eq. (6.66):

$$\mathbf{r}_{tr} \equiv \begin{bmatrix} -x \\ -y \\ z_c + z_h \\ 0 \end{bmatrix} \tag{6.132}$$

Four possible trajectories of the tool movement during the machining step are obtained through Eq. (6.71) after substitution of two functional constraints, Eq. (6.117), one enveloping constraint, Eq. (6.124), and one implicit constraint, Eq. (6.125):

$$\mathbf{r}_{tr\,1,2,3,4} = \mathbf{r}_w - \mathbf{A}^6(\varphi_{swi})\,\mathbf{A}^4(\varphi_i)\,\mathbf{A}^6(\varphi_{sk})\,\mathbf{r}_t =$$

$$\begin{bmatrix} x_w - R(\cos\varphi_{swi}\cos\varphi_{sk} - \cos\theta_j\sin\varphi_{swi}\sin\varphi_{sk}) - z_t\sin\varphi_{swi}\sin\theta_j \\ y_w - R(\cos\varphi_{sk}\sin\varphi_{swi} + \cos\varphi_{swi}\cos\theta_j\sin\varphi_{sk}) + z_t\cos\varphi_{swi}\sin\theta_j \\ z_w - z_t\cos\theta_j - R\sin\theta_j\sin\varphi_{ks} \\ 0 \end{bmatrix} \tag{6.133}$$

where angles θ_j (j = 1, 2) and φ_{swi} (i = 1, 2) are defined through Eqs. (6.122) and (6.123), respectively; angle φ_{sk} (k = 1 or 2 depending on machining mode) is defined through the enveloping constraint, Eq. (6.127); $x_w = x_w(u,v)$, $y_w = y_w(u,v)$, and $z_w = z_w(u,v)$ are the coordinates of the given position vector $\mathbf{r}_w$ of the machined surface; and z_t is expressed through Eq. (6.130).

Equating the right-hand sides of Eqs.(6.132) and (6.133) results in

$$\begin{bmatrix} x \\ y \\ z_h \\ 0 \end{bmatrix} = \begin{bmatrix} -x_w + R(\cos\varphi_{swi}\cos\varphi_{sk} - \cos\theta_j\sin\varphi_{swi}\sin\varphi_{sk}) + z_t\sin\varphi_{swi}\sin\theta_j \\ -y_w + R(\cos\varphi_{sk}\sin\varphi_{swi} + \cos\varphi_{swi}\cos\theta_j\sin\varphi_{sk}) - z_t\cos\varphi_{swi}\sin\theta_j \\ z_w - z_c - z_t\cos\theta_j - R\sin\theta_j\sin\varphi_{ks} \\ 0 \end{bmatrix} \tag{6.134}$$

Equation (6.134) presents the third kinematic constraint expressing the three translation parameters x, y, and z through two Gaussian coordinates u,v of the machined surface,

$$x = x(u,v),\; y = y(u,v),\; z = z(u,v)$$

Note that the results of calculations must satisfy the following conditions: (a) translation values x, y, and z, Eq. (6.134), must lie within the specified travel values of the machine tool, and the number of the applied constraints must satisfy Eq. (6.118). In this case, three functional constraints are given through Eqs. (6.122), (6.123), (6.134), the enveloping and implicit constraints are defined through Eqs. (6.124) and (6.125), respectively. Hence, the constraint number $L = L_f + L_e + L_i = 3 + 1 + 1 = 5$ satisfies setup condition, Eq. (6.118).

Tool trajectory: Machining step

The position vector of orbital motion of the tool, with simultaneous cutting motion (rotation of the tool around its axis) is obtained through substitution of the setup parameters in the position FSF, Eq. (6.114),

$$\mathbf{r}_{mash} = \mathbf{r}_0(u, v, \varphi_s, z_t) =$$

$$\begin{bmatrix} -x(u,v) + R(\cos\varphi_{swi}\cos\varphi_s - \cos\theta_j\sin\varphi_{swi}\sin\varphi_s) + z_t\sin\varphi_{swi}\sin\theta_j \\ -y(u,v) + R(\cos\varphi_s\sin\varphi_{swi} + \cos\varphi_{swi}\cos\theta_j\sin\varphi_s) - z_t\cos\varphi_{swi}\sin\theta_j \\ z(u,v) + z_t\cos\theta_j + R\sin\theta_j\sin\varphi_s \\ 1 \end{bmatrix} \tag{6.135}$$

where φ_s is the angle of cutting motion with frequency ω_s, $\varphi_s = \omega_s t$; and u, v are the curvilinear coordinates of the machined surface.

Tool trajectory: Idle running step

The idle running trajectory connects the *initial* and *pre-machining* positions of the virtual tool, Eq. (6.131). The *initial position* of the tool is obtained with $\theta = \varphi_{sw} = 0$ and $x = y = z = 0$. Hence, the position FSF, Eq. (6.116), becomes

$$\mathbf{r}_{0,in} = \begin{bmatrix} R\cos\varphi_s \\ R\sin\varphi_s \\ z_c + z_t \\ 1 \end{bmatrix} \tag{6.136}$$

The *pre-machining tool position* (Fig. 6.21) fixes the end of the idle running

$$\mathbf{r}_{0,pre} = \mathbf{r}_{mash}(u_b, v_b, \varphi_s, z_t) + \mathbf{r}_{pre} \tag{6.137}$$

where u_b and v_b are the values u and v in Eq. (6.137) for the tool position, from which the machining step begins; and $\mathbf{r}_{pre} = \mathbf{r}_{pre}(x_{pre}, y_{pre}, z_{pre}, 0)^T$ is the vector of the predesigned distances from the tool position immediately before beginning the machining step.

The idle tool movement from the initial position $\mathbf{r}_{0,in}$, Eq. (6.136), to the pre-machining position $\mathbf{r}_{0,pre}$, Eq. (6.137), can be established in various ways. Generally, the trajectory results from an optimization procedure, for example, subject to minimal idle time. Here, the following conditions are considered:

- Pre-machining position and orientation of the tool relative to the machined surface must be best suited for starting the machining step;
- The time of the idle movement is limited by the tilting motion velocity, which is established on the uniform maximally permissible value; and,

- If there are no obstacles, the frontal point O_t (Fig. 6.26) of the tool axis moves along a *straight-line;*
- The translation velocities are chosen in agreement with the rotation velocities.

The initial position of the point O_t is obtained from Eq. (6.136) with $R = 0$ and $z_t = z_{Ot}$,

$$\mathbf{r}_{Ot,in} = (\mathbf{r}_{0,in})_{R=0} = [0, 0, z_c + z_{Ot}, 1]^T \tag{6.138}$$

The pre-machining position of the point O_t is obtained from Eqs. (6.137) with $u = u_b$; $v = v_b$; $R = 0$; and $z_t = z_{Ot}$,

$$\mathbf{r}_{Ot,pre} = (\mathbf{r}_{mach})_{R=0} = \begin{bmatrix} -x(u_b, v_b) + x_{pre} + z_{Ot}\sin\varphi_{swi}\sin\theta_j \\ -y(u_b, v_b) + y_{pre} - z_{Ot}\cos\varphi_{swi}\sin\theta_j \\ z(u_b, v_b) + z_{pre} + z_{Ot}\cos\theta_j \\ 1 \end{bmatrix} \tag{6.139}$$

An equation of the straight line connecting the two points is

$$\mathbf{r}_{line} = \mathbf{r}_{Ot,in} + \lambda(\mathbf{r}_{Ot,pre} - \mathbf{r}_{Ot,in}) \tag{6.140}$$

where λ is the dimensionless parameter whose values $\lambda = 0$ and $\lambda = 1$ result in vector $\mathbf{r}_{Ot,pre}$ and $\mathbf{r}_{Ot,in}$, respectively.

Note that the straight line trajectory of the tool points is carried out with significantly variable velocities of the translation drives.

Numerical Example 6.7. Profiling of the horizontally-oriented bore by end mill.

Given:

- *Machine tool* is designed for machining of long-length parts, the spindle head can rotate through ±90°. around X- and Z-axis. The current height of the tilting center of the tool head is offset by $H + z$ from the base plane, where $H = 600$ mm is the initial position of the X-tilting axis and $-175 \le z \le 175$ mm is the structural travel of the spindle head along the Z-axis.
- *Workpiece:* The machined bore is of radius $r = 40$ mm and length $l = 20$ mm, the bore axis is at a height of $z_0 = 460$ mm, the front cross-section is located at $y_w = -160$ mm.
- *Cutting tool:* The end mill is of radius $R = 10$ mm and length of the working part $L = 30$ mm, the tool forward face is offset by $z_{t1} = 150$ mm from the tilting center.
- *Relative position of the tool and workpiece:*

(a) during machining, the mean cross-section of the tool ($z_t = -135$ mm) must be coincident with the mean cross-section of the machined bore ($y_w = -170$ mm);

(b) in the pre-machining position, the tool face must be offset by $y_{off} = 10$ mm from the frontal section of the machined bore (Fig. 6.21).

Required: Setup parameters: the angles φ_{sw} and θ, and the translation parameters x, y, and z.

Solution:

(1) Initial data.

(a) The machined bore position vector, orientation vector and unit vector of the normal vs.

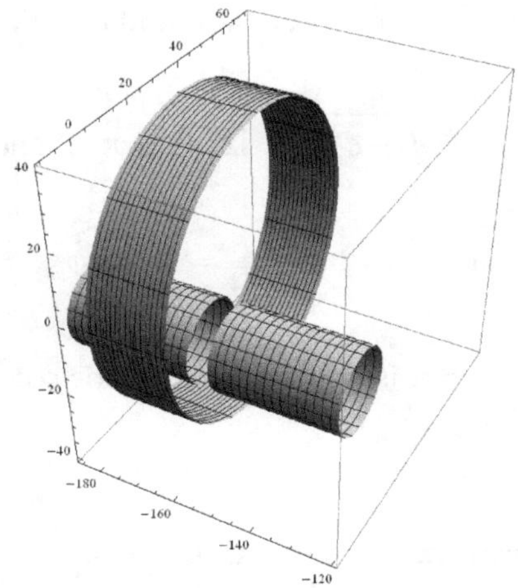

Figure 6-21 Starting and pre-machining positions of the virtual tool.

Gaussian coordinates $u \equiv y_w$ and $v \equiv \alpha$,

$$\mathbf{r}_w = \mathbf{r}_w(y_w, \alpha) = \begin{bmatrix} r\cos\alpha \\ y_w \\ z_0 + r\sin\alpha \\ 1 \end{bmatrix} = \begin{bmatrix} 40\cos\alpha \\ y_w \\ 460 + 40\sin\alpha \\ 1 \end{bmatrix}; \; -160 \le y_w \le -180 \text{ mm}; \; 0 \le \alpha < 2\pi; \qquad (6.141)$$

$$\mathbf{c}_w = \mathbf{c}_w(\alpha) = \begin{bmatrix} c_x \\ c_y \\ c_z \\ 0 \end{bmatrix} = \begin{bmatrix} 0 \\ 1 \\ 0 \\ 0 \end{bmatrix} \qquad (6.142)$$

$$\mathbf{n}_w = \mathbf{n}_w(\alpha) = \begin{bmatrix} n_z \\ n_y \\ n_z \\ 0 \end{bmatrix} = \begin{bmatrix} \cos\alpha \\ 0 \\ \sin\alpha \\ 0 \end{bmatrix} \qquad (6.143)$$

(b) The position vector of a single-edge tool:

$$\mathbf{r}_t = \begin{bmatrix} R \\ 0 \\ z_t \\ 1 \end{bmatrix} = \begin{bmatrix} 10 \\ 0 \\ z_t \\ 1 \end{bmatrix}, \text{ with } -150 \text{ mm} \le z_t \le -120 \text{ mm} \qquad (6.144)$$

The virtual tool, i.e., a cylinder presenting a trace of the rotated line (tool edge) is as follows (Fig. 6.21):

$$\mathbf{r}_{vt} = \mathbf{A}^6(\varphi_s)\begin{bmatrix} R \\ 0 \\ z_t \\ 1 \end{bmatrix} = \begin{bmatrix} R\cos\varphi_s \\ R\sin\varphi_s \\ z_t \\ 1 \end{bmatrix} = \begin{bmatrix} 10\cos\varphi_s \\ 10\sin\varphi_s \\ z_t \\ 1 \end{bmatrix}, \text{ with } -150 \le z_t \le -120 \text{ mm and } \varphi_s = \varphi_s(t) \qquad (6.145)$$

(2) *Tilting and swiveling angles.* Since $c_x = 0$, $c_y = 1$, and $c_z = 0$, Eqs. (6.122) and (6.123) yield

$$\cos\theta_{1,2} = c_z = 0 \text{ and } \sin\theta_{1,2} = \pm\sqrt{1-c_z^2} = 1; \text{ i.e.,} \theta_1 = \pi/2 \text{ and } \theta_2 = -\pi/2 \tag{6.146}$$

$$\cos\varphi_{sw1,2} = \mp\frac{c_y}{\sqrt{1-c_z^2}} = 1 \text{ and } \sin\varphi_{sw1,2} = \pm\frac{c_x}{\sqrt{1-c_z^2}} = 0; \text{ i.e.,} \varphi_{sw1} = \pi \text{ and } \varphi_{sw2} = 0 \tag{6.147}$$

Since the rotation at angle $\varphi_{sw1} = \pi$ is impossible, only the second solution, Eq. (6.147), is acceptable

$$\theta = \theta_2 = -\pi/2 \text{ and } \varphi_{sw} = \varphi_{sw2} = 0 \tag{6.148}$$

(3) *Enveloping constraints.* Substituting $n_x = \cos\alpha$, $n_y = 0$, $n_z = \sin\alpha$, $\cos\theta = -\pi/2$, and $\varphi_{sw} = 0$, in Eq. (6.128) yields:

$$\tan\varphi_{s1,2} = \frac{-n_x\cos\theta\sin\varphi_{sw} + n_y\cos\theta\cos\varphi_{sw} + n_z\sin\theta}{n_x\cos\varphi_{sw} + n_y\sin\varphi_{sw}} = -n_z/n_x = -\tan\alpha \tag{6.149}$$

$$\varphi_{s1} = -\alpha \text{ and } \varphi_{s2} = \pi - \alpha$$

Deciding between the two values (φ_{s1} or φ_{s2}) depends on whether the matter is internal or external machining.

(4) *Translational movements.* In this case, the translational movement vector, Eq. (6.132), is

$$\mathbf{r}_{tr} = \begin{bmatrix} -x \\ -y \\ H+z \\ 0 \end{bmatrix} = \begin{bmatrix} -x \\ -y \\ 600+z \\ 0 \end{bmatrix} \tag{6.150}$$

Substitution of the enveloping constraint, Eq. (e1.9), in Eq. (6.133) yields:

$$\mathbf{r}_{tr1,2} = \mathbf{r}_w - \mathbf{A}^6(0)\mathbf{A}^4(-\pi/2)\mathbf{A}^6\left(\begin{matrix}-\alpha \\ \pi-\alpha\end{matrix}\right)\mathbf{r}_t =$$

$$\begin{bmatrix} r\cos\alpha \\ y_w \\ z_0 + r\sin\alpha \\ 1 \end{bmatrix} - \begin{bmatrix} 1 & 0 & 0 & 0 \\ 0 & 0 & 1 & 0 \\ 0 & -1 & 0 & 0 \\ 0 & 0 & 0 & 1 \end{bmatrix} \begin{bmatrix} \pm\cos\alpha & \pm\sin\alpha & 0 & 0 \\ \mp\sin\alpha & \pm\cos\alpha & 0 & 0 \\ 0 & 0 & 1 & 0 \\ 0 & 0 & 0 & 1 \end{bmatrix} \begin{bmatrix} R \\ 0 \\ z_t \\ 1 \end{bmatrix} = \begin{bmatrix} (r\mp R)\cos\alpha \\ y_w - z_t \\ z_0 + (r\mp R)\sin\alpha \\ 0 \end{bmatrix} \tag{6.151}$$

Hence, the milling cutter axis trajectory presents a circle of radius $R_{tr1} = r - R$ or $R_{tr2} = r + R$. Since the *internal* profiling is considered, only $\varphi_{s1} = -\alpha$ and $R_{tr} = r - R = 40 - 10 = 30$ mm are acceptable, and numerical form of Eq. (6.151) becomes

$$\mathbf{r}_{tr} = \begin{bmatrix} 30\cos\alpha \\ y_w - z_t \\ 460 + 30\sin\alpha \\ 0 \end{bmatrix} \text{with } -180 \le y_w \le -160;\ -150 \le z_t \le -120;\ 0 \le \alpha < 2\pi \tag{6.152}$$

Equating (6.150) and (6.152) yields

$$\begin{bmatrix} -x \\ -y \\ 600+z \\ 0 \end{bmatrix} = \begin{bmatrix} 30\cos\alpha \\ y_w - z_t \\ 460+30\sin\alpha \\ 0 \end{bmatrix}, \text{ with } -180 \le y_w \le -160;\ 0 \le \alpha < 2\pi \tag{6.153}$$

This equation allows formulation of the three last setup parameters – translation values x, y, and z. The setup parameters are listed in Table 6.3.

(5) *Travel verification.* The vertical displacement z, Eq. (6.153), is available since the maximal and minimal Z-displacements $z = -140 \pm 30$ satisfy the travel data

$$-175 < -170 \le z(y_w, \alpha) \le -110 < 175 \text{ mm}$$

(6) *Tool trajectory: Machining step.* Substitution of all setup parameters listed in Table 6.3 in Eq. (6.137) results in the position vector of the tool motion: the tool axis goes around the circular orbit of radius 30 mm

$$\mathbf{r}_{mash} = \mathbf{r}_0(y_w, \alpha, \varphi_s) = \begin{bmatrix} 30\cos\alpha + 10\cos\varphi_s \\ y_w \\ 460+30\sin\alpha - 10\sin\varphi_s \\ 1 \end{bmatrix}, \ -180 \le y_w \le -160 \text{ mm} \tag{6.154}$$

where φ_s is the angle of cutting motion with frequency ω_s, $\varphi_s = \omega_s t$; α is the angle in orbital motion ($0 \le \alpha < 2\pi$); and y_w is the coordinate along the workpiece.

Taking into account the implicit constraint, Eq. (6.130), one can express Eq. (6.154) through the tool parameter z_t. Actually, there is given the requirement that the mean cross-sections of the workpiece ($y_{w,m} = -170$ mm) and the tool ($z_{t,m} = -135$ mm) must lie in the common plane during machining. Hence, the implicit constraint, Eq. (6.130), is as follows:

$$y_w = z_t - 35 \tag{6.155}$$

Thus,

Table 6-3 Setup parameters

Parameter	**Definition vs. α and y_w**	
Tilt angle	$\theta = -\pi/2$	
Swivel angle	$\varphi_{sw} = 0$	
Contact condition	$\varphi_s = -\alpha$,	with $0 \le \alpha < 2\pi$
X-axis translation	$x = -30 \cos \alpha$,	with $0 \le \alpha < 2\pi$
Y-axis translation	$y = z_t - y_w$, with $-180 \le y_w \le -160$; $-150 \le z_t \le -120$	
Z-axis translation	$z = -140 + 30 \sin \alpha$, with $0 \le \alpha < 2\pi$	

$$\mathbf{r}_{mash}(\alpha,\varphi_s,z_t) = \mathbf{r}_0(\alpha,\varphi_s,z_t) = \begin{bmatrix} 30\cos\alpha + 10\cos\varphi_s \\ z_t - 35 \\ 460 + 30\sin\alpha - 10\sin\varphi_s \\ 1 \end{bmatrix}, \quad -150 \le z_t \le -120 \text{ mm} \tag{6.156}$$

(7) *Tool trajectory: Idle step.*

(a) Initial position. With $x = y = 0$; $\theta = \varphi_{sw} = 0$ and $z = 0$, Eq. (6.138) yields:

$$\mathbf{r}_{0,in} = \begin{bmatrix} -x + R(\cos\varphi_{sw}\cos\varphi_s - \cos\theta\sin\varphi_{sw}\sin\varphi_s) + z_t\sin\varphi_{sw}\sin\theta \\ -y + R(\cos\varphi_s\sin\varphi_{sw} + \cos\varphi_{sw}\cos\theta\sin\varphi_s) - z_t\cos\varphi_{sw}\sin\theta \\ 600 + z + z_t\cos\theta + R\sin\theta\sin\varphi_s \\ 1 \end{bmatrix} = \begin{bmatrix} R\cos\varphi_s \\ R\sin\varphi_s \\ 600 + z_t \\ 1 \end{bmatrix}, \tag{6.157}$$

with $-150 \le z_t \le -120$ mm and $0 \le \varphi_s < 2\pi$.

In Figure 6.22a, the relative position of the machined bore, Eq. (6.141) and the virtual tool, Eq. (6.157), are shown. In Figure 6.22b, some instantaneous positions of the tool during its translation in the machining step, obtained through Eq. (6.155) are displayed. For example, the starting machined position of the tool with $\alpha = 0$ is

$$\mathbf{r}_0(z_t,\varphi_s) = \begin{bmatrix} 30 + 10\cos\varphi_s \\ z_t - 35 \\ 460 - 10\sin\varphi_s \\ 1 \end{bmatrix}, \text{ with } -180 \le z_t \le -160 \text{ mm}$$

(b) Pre-machining position. Assume that the initial machining position of the tool in the machining step is defined through $\alpha = 0$. According to the initial data, the distance

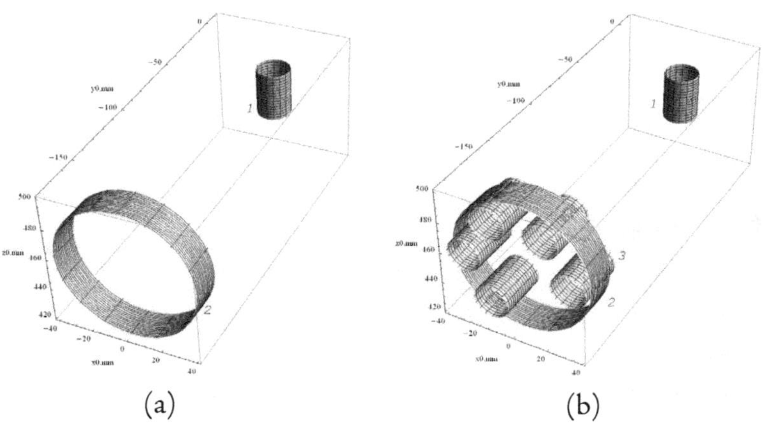

(a) (b)

Figure 6-22 Bore milling kinematics: (a) initial position of the virtual tool *1* and workpiece *2*; (b) some instantaneous positions of the virtual tool *3* during machining.

between the face sections of the tool and bore is $y_{off} = 10$ mm. Hence, the distance from the mean cross-section of the bore is $y_{pre} = l/2 + L/2 + y_{off} = 10/2 + 30/2 + 10 = 35$ mm, where $l = 20$ is the machined bore width; $L = 30$ is the lengths of the working edge (Fig. 6.22). Substitution of Eq. (6.158) in Eq. (6.137), with $\alpha = 0$, $z_t = -135$, $x_{pre} = 0$, $y_{pre} = 35$, and $z_{pre} = 0$, results in

$$\mathbf{r}_{0,pre} = \mathbf{r}_{mash} + \mathbf{r}_{pre} = \begin{bmatrix} 30 + 10\cos\varphi_s + x_{pre} \\ z_t - 35 + y_{pre} \\ 460 - 10\sin\varphi_s + z_{pre} \\ 1 \end{bmatrix} \tag{6.158}$$

Substitution of $\alpha = 0$, $z_t = -135$, $x_{pre} = 0$; $y_{pre} = 35$; and $z_{pre} = 0$ in Eq. (6.158) yields

$$\mathbf{r}_{0,pre} = \begin{bmatrix} 30 + 10\cos\varphi_s \\ z_t \\ 460 - 10\sin\varphi_s \\ 1 \end{bmatrix}, \text{ with } -150 \le y_w \le -120 \text{ mm} \tag{6.159}$$

To transfer the tool from initial position, Eq. (6.157), to its pre-machining position, Eq. (6.159), it is necessary to fix an optimal trajectory, which can be performed only by the joint action of all machine drives. For example, if the tool head begins the feed motion with the pure rotation at angle $\theta = 90°$, and only after this performs translation (Fig. 6.23), the tool moves to a relatively great distance from the workpiece. Another approach is shown in Figure 6.24a. The straight line connecting the initial and pre-machining positions of the virtual tool is proposed as a geometrically optimal trajectory - a straight line connecting two positions of a tool point, Eq. (6.140). The line connects two positions of the forward point O_t of

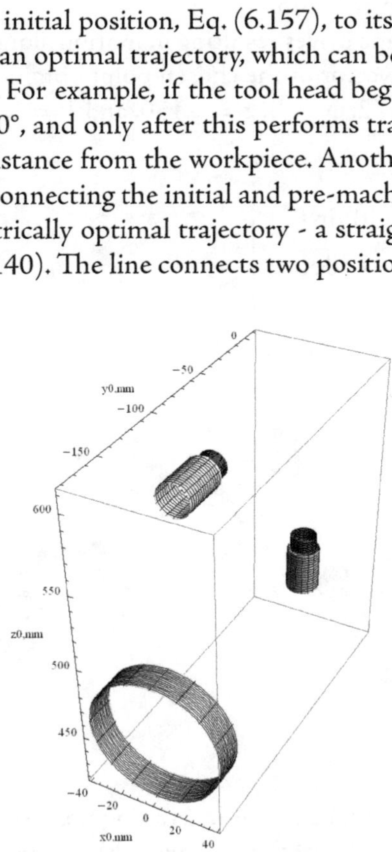

Figure 6-23 Position of the virtual tool before and after rotation, without translations.

the tool axis (the point O_t is shown in Figure 6.23):

$$\mathbf{r}_{Ot,in} = \begin{bmatrix} 0 \\ 0 \\ 450 \\ 1 \end{bmatrix} \text{and } \mathbf{r}_{Ot,pre} = \begin{bmatrix} 30 \\ -150 \\ 460 \\ 1 \end{bmatrix} \tag{6.160}$$

where the position vector $\mathbf{r}_{Ot,in}$ is obtained from Eq. (6.157) with z_t = -150 and R = 0 (instead of R = 10); and the position vector $\mathbf{r}_{Ot,pre}$ is obtained from Eq. (6.159) with z_t = -150, α = 0, and R = 0.

The line formulated using Eqs. (6.160) is

$$\mathbf{r}_{line} = \mathbf{r}_{Ot,in} + \lambda(\mathbf{r}_{Ot,fin} - \mathbf{r}_{Ot,in}) = \begin{bmatrix} 30\lambda \\ -150\lambda \\ 450+10\lambda \\ 1 \end{bmatrix} \tag{6.161}$$

where λ is time-dependent parameters whose values $\lambda = 1$ and $\lambda = 0$ result in vector $\mathbf{r}_{Ot,pre}$ and $\mathbf{r}_{Ot,in}$, respectively. Some instantaneous positions of the tool during the idle and machining motions are shown in Figure 6.24b.

The joint acts of the machine drives resulting in motion along $\mathbf{r}_{line}$ result from equation $\mathbf{r}_{line}$ = $\mathbf{r}_0$, where $\mathbf{r}_0$ is the position vector of the chosen point obtained from the FSF, Eq. (6.116), under conditions $0 \le \theta \le -\pi/2$, $\varphi_{sw} = 0$, $z_t = -150$, and $R = 0$,

$$\begin{bmatrix} 30\lambda \\ -150\lambda \\ 450+10\lambda \\ 1 \end{bmatrix} = \begin{bmatrix} -x \\ -y+150\sin\theta \\ 600+z-150\cos\theta \\ 1 \end{bmatrix} \tag{6.162}$$

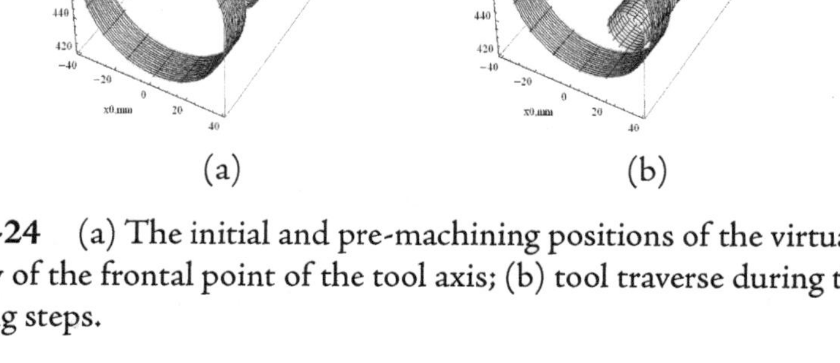

(a) (b)

Figure 6-24 (a) The initial and pre-machining positions of the virtual tool and the trajectory of the frontal point of the tool axis; (b) tool traverse during the idle and machining steps.

Hence, equations of the idle movement position vector $\mathbf{r}_{idle}$ and idle velocity are

$$\mathbf{r}_{idle} \equiv \begin{bmatrix} x \\ y \\ z \\ 0 \end{bmatrix} = \begin{bmatrix} -30\lambda \\ 150(\lambda + \sin\theta) \\ -150(1-\cos\theta) + 10\lambda \\ 0 \end{bmatrix}; \tag{6.163}$$

$$\mathbf{v}_{idle} = \begin{bmatrix} -30\dot{\lambda} \\ 150(\dot{\lambda} + \dot{\theta}\cos\theta) \\ -150\dot{\theta}\sin\theta + 10\dot{\lambda} \\ 0 \end{bmatrix} \tag{6.164}$$

Usually the idle minimal time is limited by the angular motion. The idle time needed to turn the tool through angle $\pi/2$ with uniform angular velocity $\dot{\theta}$ (without regard to acceleration and deceleration time) is $T_{idle} = \pi/(2\dot{\theta})$. To simultaneously finish all idle displacements, the X-translation velocity must be $v_x = 30\,\dot{\theta}/T_{idle} = 30\,\dot{\lambda}$, and hence, $\dot{\lambda} = 1/T_{idle} = 2\,\dot{\theta}/\pi$. In Figure 6.25, the velocity histories, Eq. (6.164), are shown with $\dot{\theta} = -3$ rpm $= -6\pi$ min^{-1}; $T_{idle} = \pi/(2\dot{\theta}) = -\pi/(2\times 6\pi) = 1/12$ min; and the translational velocities, Eq. (6.164), are

$$v_x = -60\,\dot{\theta}/\pi = 360 \text{ mm/min};$$

$$v_y = 150[2\dot{\theta}/\pi + \dot{\theta}\cos(\dot{\theta}\,t)] = -1800 - 900\pi\cos(6\pi t);$$

$$v_z = -150\,\dot{\theta}\sin(\dot{\theta}\,t) + 20\,\dot{\theta}/\pi = 900\pi\sin(6\pi t) - 120, \tag{6.165}$$

with $t = 0 \div 1/12$ min.

Note that the maximal translation velocity takes place along the Y-axis movement

$$(v_y)_{t=0} = -(1800 + 900\pi)/1000 = -4.63 \text{ m/min}.$$

This value falls in the regular range of the presently accepted idle velocities in the modern

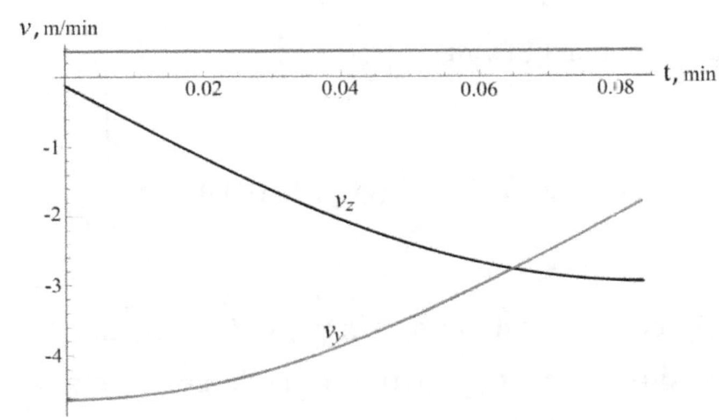

Figure 6-25 Coordinate velocities *vs.* idle running time (*Numerical Example 6.7*).

machining centers. However, conceivably the obtained mode is inconvenient since the Y-axis drive has to start the motion from the maximal velocity. If this involves difficulties, the chosen trajectory can be modified and/or the idle running time can be slightly increased.

6.3.5.3 Plane milling by face mill

The FSF of the machine has the above mentioned form given by Eq. (6.114)

$$[\mathbf{r}_0\mathbf{c}_0] = \mathbf{A}^1(-x)\,\mathbf{A}^2(-y)\,\mathbf{A}^3(z)\,\mathbf{A}^6(\varphi_{sw})\,\mathbf{A}^4(\theta)\mathbf{A}^6(\varphi_s)[\mathbf{r}_t\mathbf{c}_t] \tag{6.166}$$

where x, y, z are translation movements of the machine table, saddle, and spindle head, respectively; φ_{sw} and θ are the swiveling and tilting angles, respectively of rotation of the spindle head, and φ_s is the angle of the spindle rotation.

Without loss of generality from the kinematics viewpoint, the applied tool can be considered as a one-edge milling cutter (Fig. 6.26a),

$$[\mathbf{r}_t\,\mathbf{c}_t] = \begin{bmatrix} x_t & 0 \\ 0 & 0 \\ H_t & 1 \\ 1 & 0 \end{bmatrix}; R-l_t \le x_t \le R \tag{6.167}$$

where x_t is the current point at the edge (the edge is assumed to be oriented along the tool radius); R is the radius of the external point of the edge; l_t is the edge length; and H_t is the distance of the working tool face from the axis of tool head tilting.

A virtual tool presenting a track of the active tool points during tool rotation has the form of a ring of radius R and width l_t (Fig. 6.26b)

$$[\mathbf{r}_{vt}\,\mathbf{c}_{vt}] = \mathbf{A}^6(\varphi_s)\begin{bmatrix} x_t & 0 \\ 0 & 0 \\ H_t & 1 \\ 1 & 0 \end{bmatrix} = \begin{bmatrix} x_t\cos\varphi & 0 \\ x_t\sin\varphi & 0 \\ H_t & 1 \\ 1 & 0 \end{bmatrix}; R-l_t \le x_t \le R$$

The positionand orientation FSFs are

$$\mathbf{r}_0 \equiv \begin{bmatrix} x_0 \\ y_0 \\ z_0 \\ 1 \end{bmatrix} = \mathbf{A}^1(-x)\mathbf{A}^2(-y)\mathbf{A}^3(z)\mathbf{A}^6(\varphi_{sw})\mathbf{A}^4(\theta)\mathbf{A}^6(\varphi_s)\begin{bmatrix} x_t \\ 0 \\ H_t \\ 1 \end{bmatrix} = \tag{6.168}$$

$$\begin{bmatrix} -x + x_t(\cos\varphi_{sw}\cos\varphi_s - \cos\theta\sin\varphi_{sw}\sin\varphi_s) + H_t\sin\varphi_{sw}\sin\theta \\ -y + x_t(\cos\varphi_s\sin\varphi_{sw} + \cos\varphi_{sw}\cos\theta\sin\varphi_s) - H_t\cos\varphi_{sw}\sin\theta \\ z + x_t\sin\theta\sin\varphi_s + H_t\cos\theta \\ 1 \end{bmatrix}$$

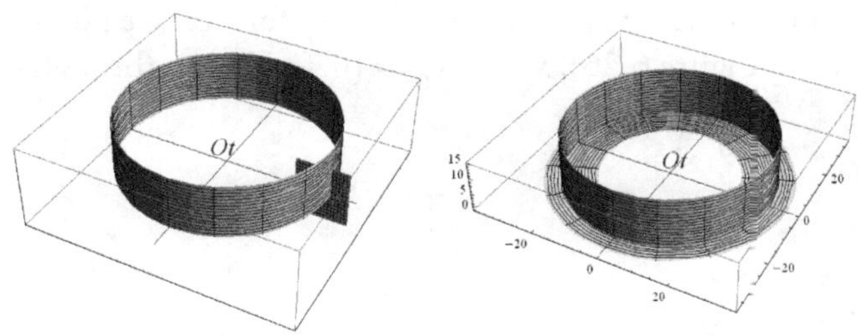

Figure 6-26 Models of the face milling cutter: (a) the one-edge tool; (b) the virtual tool.

$$\mathbf{c}_0 \equiv \begin{bmatrix} c_{0x} \\ c_{0y} \\ c_{0z} \\ 0 \end{bmatrix} = \mathbf{A}^1(-x)\mathbf{A}^2(-y)\mathbf{A}^3(z)\mathbf{A}^6(\varphi_{sw})\mathbf{A}^4(\theta)\mathbf{A}^6(\varphi_s)\begin{bmatrix} 0 \\ 0 \\ 1 \\ 0 \end{bmatrix} = \begin{bmatrix} \sin\varphi_{sw}\sin\theta \\ -\cos\varphi_{sw}\sin\theta \\ \cos\theta \\ 0 \end{bmatrix} \quad (6.169)$$

The vertical movement z of the tool head is measured from the height z_c presenting the average distance between the tilting axes and the base. Hence, $z = z_c + z_h$, where z_h is the tool head vertical traverse. The form-shaping function consists of six parameters of motion x, y, z, θ, φ_{sw}, φ ($l = 6$) and one parameter of the linear-type tool x_t ($M = 1$). Hence, the constraints number is $L = 6+1-2= 5$, as before.

Two functional constraints define the angles θ and φ_{sw}. Since the right-hand side of Eqs. (6.169) is identical with that of Eq. (6.120), the angles θ and φ_{sw} are defined through Eqs. (6.122) and (6.123).

Since the normal $\mathbf{n}_w$ at the machined plane is parallel to the tool axis during plane machining, $\mathbf{n}_w = \mathbf{c}_0$, both enveloping constraints become implicit ones. Namely, enveloping conditions result in identical zero at any x_t, φ_s, θ, and φ_{sw},

$$\mathbf{n}_w \cdot \frac{\partial \mathbf{r}_w}{\partial \varphi_s} = \mathbf{c}_0 \cdot \frac{\partial \mathbf{r}_w}{\partial \varphi_s} = \begin{bmatrix} \sin\varphi_{sw}\sin\theta \\ -\cos\varphi_{sw}\sin\theta \\ \cos\theta \\ 0 \end{bmatrix} \cdot \begin{bmatrix} x_t(-\cos\varphi_{sw}\sin\varphi_s - \cos\theta\sin\varphi_{sw}\cos\varphi_s) \\ x_t(-\sin\varphi_s\sin\varphi_{sw} + \cos\varphi_{sw}\cos\theta\cos\varphi_s) \\ x_t\sin\theta\cos\varphi_s \\ 0 \end{bmatrix} \equiv 0$$

$$\mathbf{n}_w \cdot \frac{\partial \mathbf{r}_w}{\partial x_t} = \mathbf{c}_0 \cdot \frac{\partial \mathbf{r}_w}{\partial x_t} = \begin{bmatrix} \sin\varphi_{sw}\sin\theta \\ -\cos\varphi_{sw}\sin\theta \\ \cos\theta \\ 0 \end{bmatrix} \cdot \begin{bmatrix} \cos\varphi_{sw}\cos\varphi_s - \cos\theta\sin\varphi_{sw}\sin\varphi_s \\ \cos\varphi_s\sin\varphi_{sw} + \cos\varphi_{sw}\cos\theta\sin\varphi_s \\ \sin\theta\sin\varphi_s \\ 0 \end{bmatrix} \equiv 0 \quad (6.170)$$

Translation vector $\mathbf{r}_{tr}$, Eq. (6.71), is

$$\mathbf{r}_{tr} \equiv \begin{bmatrix} -x \\ -y \\ z_c + z_h \\ 0 \end{bmatrix} \quad (6.171)$$

Current translation distances x, y, and z_h are defined for the face point of cutting tool axis O_t (point O_t is shown in Figure 6.26), $x_t = 0$, $\mathbf{r}_{Ot} = [0, 0, H_t, 1]^T$, through equation

$$\mathbf{r}_{tr} = \mathbf{r}_w - \mathbf{A}^6(\varphi_{sw1,2})\,\mathbf{A}^4(\theta_{1,2})\,\mathbf{r}_{Ot},$$

i.e.,

$$\begin{bmatrix} -x \\ -y \\ z_c + z_h \\ 0 \end{bmatrix} = \begin{bmatrix} x_w \\ y_w \\ z_w \\ 1 \end{bmatrix} - \mathbf{A}^6(\varphi_{sw1,2})\mathbf{A}^4(\theta_{1,2}) \begin{bmatrix} 0 \\ 0 \\ H_t \\ 1 \end{bmatrix} \tag{6.172}$$

Hence,

$$x_{1,2} = -x_w + H_t \sin\varphi_{sw1,2} \sin\theta_{1,2} \tag{6.173}$$

$$y_{1,2} = -y_w - H_t \cos\varphi_{sw1,2} \sin\theta_{1,2} \tag{6.174}$$

$$z_{h1,2} = z_w - z_c - H_t \cos\theta_{1,2} \tag{6.175}$$

where $\theta_{1,2}$ and $\varphi_{sw1,2}$, are expressed through Eqs. (6.122) and (6.123), respectively; and $\mathbf{r}_{Ot}$ is the position vector of the central point of the cutting tool, with $\varphi = 0$ and $x_t = 0$.

Thus, two of five constraints are implicit ones, $L_i = 2$, Eq. (6.170), there are no enveloping constraints, $L_e = 0$, and three constraints are functional, $L_f = 3$. Two functional constraints, Eqs.(6.122) and (6.123), fix tool tilting and swiveling angles φ_{sw} and θ, respectively, and the third equation fixes three translation movements through curvilinear coordinates u and v of the machined surface.

Numerical Example 6.8. Setup for plane machining by the face milling cutter.

Given:

- *Machine tool* is as in *Numerical Example 6.7*. The current height of the tilting center of the tool head is offset by $z = z_c + z_h$ from the base plane, where $z_c = 600$ mm is the mean height of the spindle head position; and $z_h = \pm\, 175$ mm is the spindle head travel along the Z-axis. Tilting and swiveling angles θ and φ_{sw} are $\pm 90°$.
- *Cutting tool*: The face milling cutter is of external diameter $D = 70$ mm and length $l_t = 15$ mm, the tool forward face is offset by $L_t = 150$ mm from the tilting center. Two applied models of the cutting tool are the single edge tool and virtual tool (Fig. 6.26).
- *Workpiece*: The machined plane piece of $B_w \times L_w = 50 \times 190$ mm (Fig. 6.27a) is turned through 60° around the Y-axis; its central point is located on the height $z_m = 300$ mm.

Required: Setup parameters: the angles φ_{sw} and θ, and the translation values x, y, and z_h.
Solution: (1) Initial data.

(a) *The machined plane* position vector $\mathbf{r}_w$, orientation vector of the cutting tool axis $\mathbf{c}_w$ and unit vector of the normal $\mathbf{n}_w$ are as follows ($\mathbf{c}_w$ is parallel with $\mathbf{n}_w$):

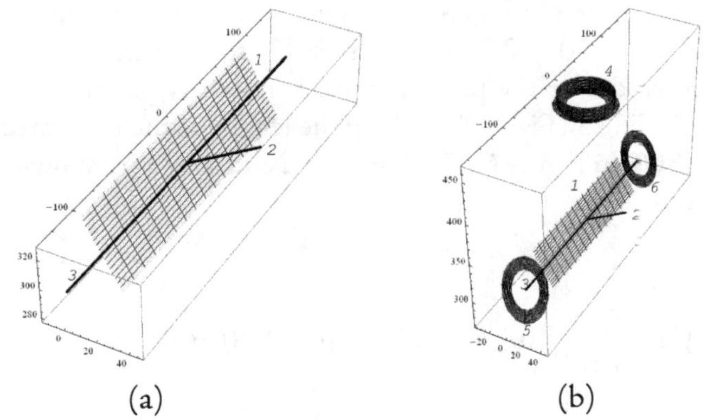

Figure 6-27 (a) Machined plane *1*, normal *2* at a plane point, and feed trajectory *3*; (b) successive positions of the virtual tool: *4* - initial position; *5* - start of cut; *6* - end of cut.

$$\mathbf{r}_w(x_p,y_p)=\begin{bmatrix} x_p\cos\beta \\ y_p \\ 300-x_p\sin\beta \\ 1 \end{bmatrix}=\begin{bmatrix} x_p/2 \\ y_p \\ 300-x_p\sqrt{3}/2 \\ 1 \end{bmatrix},\ \text{with}\ -25\le x_p\le 25; -95\le y_p\le 95\ \text{mm}$$

$$\mathbf{c}_w(x_p,y_p)=\mathbf{n}_w(x_p,y_p)=\begin{bmatrix} \sin\beta \\ 0 \\ \cos\beta \\ 0 \end{bmatrix}=\begin{bmatrix} \sqrt{3}/2 \\ 0 \\ 1/2 \\ 0 \end{bmatrix}$$

(b) *Cutting tool model.* According to Eq.(6.167), a single edge of the tool is located between $D/2 = 70/2 = 35$ and $D/2 - l_t = 35 - 15 = 20$ mm. Hence the position vector of the single edge tool (Fig. 6.26a) is

$$\mathbf{r}_t(x_t)=\begin{bmatrix} x_t \\ 0 \\ L_t \\ 1 \end{bmatrix}=\begin{bmatrix} x_t \\ 0 \\ -150 \\ 1 \end{bmatrix},\ \text{with}\ 20\le x_t\le 35\ \text{mm}$$

(c) *The virtual tool*, Eq. (6.115), presents a planar ring (Fig. 6.26b) with external and internal radii of 35 and 20 mm, respectively,

$$\mathbf{r}_{vt}(x_t,\varphi_s)=\mathbf{A}^6(\varphi_s)\begin{bmatrix} x_t \\ 0 \\ L_t \\ 1 \end{bmatrix}=\begin{bmatrix} x_t\cos\varphi_s \\ x_t\sin\varphi_s \\ -150 \\ 1 \end{bmatrix},\ \text{with}\ 20\le x_t\le 35\ \text{mm};\ 0\le\varphi_s\le 2\pi$$

(d) *The feed path*. Since the virtual tool diameter D = 70 mm is more than 80% of the workpiece width B_w = 50 = 0.71 D, it is possible to use single pass machining when the cutter is centered over the piece. Hence, a trajectory of the tool frontal point O_t presents a straight line 3 (Fig. 6.27a) along the length of the machined rectangle, with x_m = 0 and z_m = 300. In Figure 6.27b, start 5 and end 6 cuts are shown. Hence, the feed path position vector is

$$\mathbf{r}_{feed} = \mathbf{r}_w(0, y_w) = \begin{bmatrix} 0 \\ y_w \\ 300 \\ 1 \end{bmatrix}, \text{ with } -130 \le y_w \le 130 \text{ mm}$$

where the y_w-limits are fixed as $\pm(D/2 + L_w/2) = \pm(35 + 95) = \pm130$ mm.

(2) *Tilting and swiveling angles*. Since $c_x = \sqrt{3}/2$, c_y = 0, and c_z =1/2, Eqs. (6.122) and (6.123) become

$$\pm\begin{bmatrix} \sqrt{3}/2 \\ 0 \\ 1/2 \\ 0 \end{bmatrix} = \begin{bmatrix} \sin\varphi_{sw}\sin\theta \\ -\cos\varphi_{sw}\sin\theta \\ \cos\theta \\ 0 \end{bmatrix}$$

Four possible solutions are

$\theta_{1,2} = \pm\pi/3$ and $\varphi_{sw1,2} = \pm\pi/2$

$\theta_{3,4} = \pm 2\pi/3$ and $\varphi_{sw3,4} = \mp\pi/2$

Solutions 3 and 4 are excluded since angles $\theta_{3,4}$ are outside of the machine structural domain. It can be shown that only solution 1 is suited for the setup in the considered Example,

$\theta = \pi/3$ and $\varphi_{sw} = \pi/2$

(3) *Travels*. Translation displacements x, y, z are

$$x = -x_w + H_t\sin\varphi_{sw}\sin\theta = -x_p/2 - 75\sqrt{3}$$

$$y = -y_w - H_t\cos\varphi_{sw}\sin\theta = -y_p$$

$$z_h = z_w - z_c - H_t\cos\theta = -x_p\sqrt{3}/2 - 225$$

Substitution of these values in (6.168) yields

$$\mathbf{r}_0 = \begin{bmatrix} -x + x_t(\cos\varphi_{sw}\cos\varphi_s - \cos\theta\sin\varphi_{sw}\sin\varphi_s) + H_t\sin\varphi_{sw}\sin\theta \\ -y + x_t(\cos\varphi_s\sin\varphi_{sw} + \cos\varphi_{sw}\cos\theta\sin\varphi_s) - H_t\cos\varphi_{sw}\sin\theta \\ z_h + z_c + x_t\sin\theta\sin\varphi_s + H_t\cos\theta \\ 1 \end{bmatrix} = \begin{bmatrix} (x_p - x_t\sin\varphi_s)/2 \\ y_p + x_t\cos\varphi_s \\ 300 - (x_p - x_t\sin\varphi_s)\sqrt{3}/2 \\ 1 \end{bmatrix} \quad (6.176)$$

Three key-positions of the virtual tool are shown in Figure 6.27b: initial position 4, start position 5 and end position 6 of the cut. The initial position of the frontal point of tool axis O_t is obtained from Eq. (6.168) with $x = y = z_h = 0, \theta = \varphi_{sw} = 0, H_t = -150, z_c = 600, x_t = 0$, and

$$\mathbf{r}_{0,in} = \begin{bmatrix} -x + x_t(\cos\varphi_{sw}\cos\varphi_s - \cos\theta\sin\varphi_{sw}\sin\varphi_s) + H_t\sin\varphi_{sw}\sin\theta \\ -y + x_t(\cos\varphi_s\sin\varphi_{sw} + \cos\varphi_{sw}\cos\theta\sin\varphi_s) - H_t\cos\varphi_{sw}\sin\theta \\ z_h + z_c + x_t\sin\theta\sin\varphi_s + H_t\cos\theta \\ 1 \end{bmatrix} = \begin{bmatrix} 0 \\ 0 \\ 450 \\ 1 \end{bmatrix}$$

The frontal point O_t is defined from Eq. (6.176):
- in the start of cut position with $x_p = 0$ and $y_p = -130$, and $x_t = 0$, i.e.,

$$\mathbf{r}_{0,start} = \begin{bmatrix} (x_p - x_t\sin\varphi_s)/2 \\ y_p + x_t\cos\varphi_s \\ 300 - (x_p - x_t\sin\varphi_s)\sqrt{3}/2 \\ 1 \end{bmatrix} = \begin{bmatrix} 0 \\ -130 \\ 300 \\ 1 \end{bmatrix}$$

- in the end of cut position with $x_p = 0$ and $y_p = 130$, and $x_t = 0$, i.e.,

$$\mathbf{r}_{0,end} = \begin{bmatrix} (x_p - x_t\sin\varphi_s)/2 \\ y_p + x_t\cos\varphi_s \\ 300 - (x_p - x_t\sin\varphi_s)\sqrt{3}/2 \\ 1 \end{bmatrix} = \begin{bmatrix} 0 \\ 130 \\ 300 \\ 1 \end{bmatrix}$$

Idle running of the tool from the initial position to the start-of-cut position can be realized along the straight line built using dimensionless parameter λ, Eq. (6.140),

$$\mathbf{r}_{idle} = \mathbf{r}_{0,in} + \lambda(\mathbf{r}_{0,start} - \mathbf{r}_{0,in}) = \begin{bmatrix} 0 \\ -130\lambda \\ 450 - 150\lambda \\ 1 \end{bmatrix} \tag{6.177}$$

$$(\mathbf{r}_0)_{idle} = \begin{bmatrix} -x + x_t(\cos\varphi_{sw}\cos\varphi_s - \cos\theta\sin\varphi_{sw}\sin\varphi_s) + H_t\sin\varphi_{sw}\sin\theta \\ -y + x_t(\cos\varphi_s\sin\varphi_{sw} + \cos\varphi_{sw}\cos\theta\sin\varphi_s) - H_t\cos\varphi_{sw}\sin\theta \\ z_h + z_c + x_t\sin\theta\sin\varphi_s + H_t\cos\theta \\ 1 \end{bmatrix}_{x_t=0} = \begin{bmatrix} -x + H_t\sin\varphi_{sw}\sin\theta \\ -y - H_t\cos\varphi_{sw}\sin\theta \\ z_h + z_c + H_t\cos\theta \\ 1 \end{bmatrix} \tag{6.178}$$

Equating the right-hand sides of Eqs.(6.177) and (6.178) yields

$$x = -150\sin\varphi_{sw}\sin\theta$$

$$y = 130\lambda + 150\cos\varphi_{sw}\sin\theta$$

$$z_h = -150(1 + \lambda - \cos\theta) \tag{6.179}$$

Similarly to the previous *Example*, let us assume that the idle running time T_{idle} is limited by angular speed φ_{sw} of the spindle head swiveling, and the spindle head moves at permanent rates during both swiveling and tilting. Hence,

$$\varphi_{sw} = 6\pi \text{ and } \theta = 4\pi \text{ min}^{-1} \tag{6.180}$$

$$T_{idle} = \frac{\pi}{2\dot{\varphi}_{sw}} = \frac{1}{12}\text{min}$$

Differentiation of Eqs. (6.179) following substitution of the angular speeds, Eq. (6.180), yields the speeds of the idle running along the coordinate axes

$$\dot{x} = -150\,[\dot{\varphi}_{sw}\cos(\dot{\varphi}_{sw}t)\sin(\dot{\theta}t) + \dot{\theta}\sin(\dot{\varphi}_{sw}t)\cos(\dot{\theta}t)] =$$

$$-300\pi[3\cos(6\pi t)\sin(4\pi t) + 2\sin(6\pi t)\cos(4\pi t)]$$

$$\dot{y} = 150\,[-\dot{\varphi}_{sw}\sin(\dot{\varphi}_{sw}t)\sin(\dot{\theta}t) + \dot{\theta}\cos(\dot{\varphi}_{sw}t)\cos(\dot{\theta}t)] =$$

$$-300\pi[3\cos(6\pi t)\sin(4\pi t) + 2\cos(6\pi t)\cos(4\pi t)]$$

$$\dot{z}_h = -150\,\dot{\theta}\sin(\dot{\theta}t) = -600\pi\sin(4\pi t)$$

Velocities $\dot{x}$, $\dot{y}$, $\dot{z}_n$ vs. time during 1/12 min of idle running are shown in Figure 6.28.

Both considered examples show the wide spectrum of motions realized on the 5-axis machines. Actually, the 3-axis machining of the machined surfaces orientated according to the considered example requires relocation of the workpiece so that the bore and spindle axes become parallel. This procedure must be preliminary considered in the design stage and has to include designing new locating elements on the machined part, additional unloading and loading of the part onto a machine table, clamping the workpiece in another fixture, etc. Contrary to those, machining the bore on the 5-axis machine requires only tilting the spindle head at the given angles relative to the workpiece. Obviously, this significantly decreases the machining time and increases the machining accuracy.

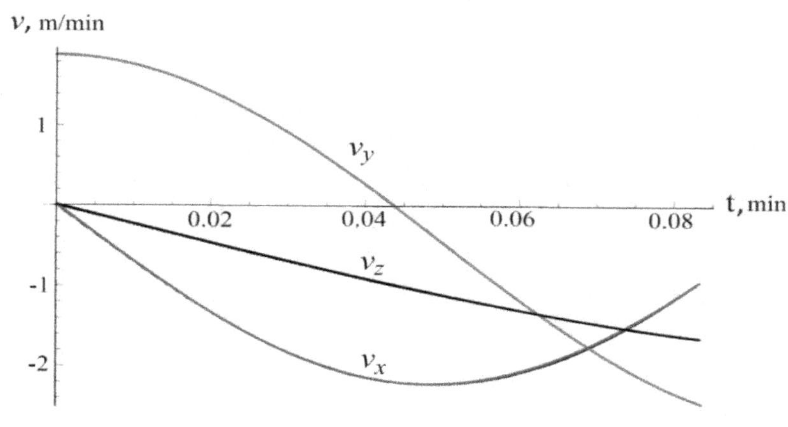

Figure 6-28 Coordinate velocities *vs.* idle running time (*Numerical Example 6.8*).

6.3.6 5-axis machines with trunnion rotary table

One of the popular configurations of the CNC 5-axis machine tools is the machine with an integrated trunnion rotary table [7]. The table with attached workpiece (Fig. 6.29) can perform a full turn around its axis and rotates around the perpendicular axis at a reasonably large angular range (±140 degs) together with the trunnion. This architecture allows machining very complicated parts, where both table rotations are required, without additional relocation and re-clamping. For example, machining of two bores in the workpiece (Fig. 6. 29) requires only programmed rotations of the table around the *Z*- and *X*-axes without any changes of workpiece positioning and clamping on the table. The same is true for machining of the planes. Hence, the table rotation extends the space for the 3-axis operations and, additionally, is typically preferred when machining heavier parts. According to Eq. (4.68), the manipulation matrix is

$$ {}^{0}\mathbf{A}_6 = \mathbf{A}^6(-\varphi_{rot})\,\mathbf{A}^4(-\theta)\,\mathbf{A}^1(-x)\,\mathbf{A}^2(-y)\,\mathbf{A}^3(z)\,\mathbf{A}^6(\varphi_{sp}) \tag{6.181} $$

where x and y are translation movements of the machine table and saddle, respectively; z is the spindle head translation, $z = z_c + z_h$; z_c is the mean height of the head position along the Z-axis; and z_h is the head displacement; φ_{rot} and θ are the rotation and tilting angles, respectively, of the rotary table motion; and φ_{sp} is the angle of the spindle rotation.

The FSF of the machine has the regular form

$$ [\mathbf{r}_0\mathbf{c}_0] = {}^{0}\mathbf{A}_6\,[\mathbf{r}_d\mathbf{c}_d] \tag{6.182} $$

In this case, the drilling operation is considered, and the applied tool is modelled as a cylinder of radius R_d. In Eq. (6.182), $\mathbf{r}_d$ is the position vector of the drill working points, and $\mathbf{c}_d$ is the vector of the drill (and spindle) axis orientation.

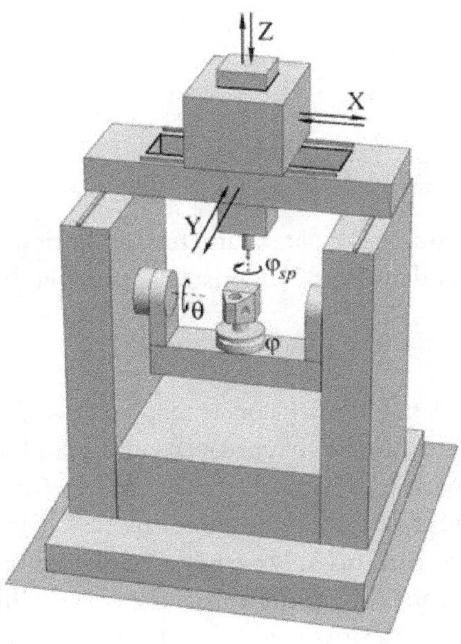

Figure 6-29 An integrated trunnion rotary table of the 5-axis machine [7].

$$[\mathbf{r}_d\ \mathbf{c}_d] = \begin{bmatrix} R_d \cos\varphi_d & 0 \\ R_d \sin\varphi_d & 0 \\ z_d & 1 \\ 1 & 0 \end{bmatrix}; H_d - l_d \le z_d \le H_d \text{ and } 0 \le \varphi_d < 2\pi \tag{6.183}$$

where z_d is the current point at the drill working part; l_d is length of the drill; and H_d is the distance of the drill face from the tool head.

A virtual tool (a track of the active tool points during tool rotation) has the form of a cylinder of radius R_d and length l_d,

$$[\mathbf{r}_{vt}\ \mathbf{c}_{vt}] = \mathbf{A}^6(\varphi_{sp}) \begin{bmatrix} R_d \cos\varphi_d & 0 \\ R_d \sin\varphi_d & 0 \\ z_d & 1 \\ 1 & 0 \end{bmatrix} = \begin{bmatrix} R_d \cos(\varphi_d + \varphi_{sp}) & 0 \\ R_d \sin(\varphi_d + \varphi_{sp}) & 0 \\ z_d & 1 \\ 1 & 0 \end{bmatrix} = \begin{bmatrix} R_d \cos\varphi & 0 \\ R_d \sin\varphi & 0 \\ z_d & 1 \\ 1 & 0 \end{bmatrix}, \tag{6.184}$$

$$\varphi = \varphi_d + \varphi_{sp} \tag{6.185}$$

Hence, the position and orientation FSFs are

$$\mathbf{r}_0 = \begin{bmatrix} (-x + R_d \cos\varphi)\cos\varphi_{rot} + [(-y + R_d \sin\varphi)\cos\theta + (z_t + z_c + z_h)\sin\theta]\sin\varphi_{rot} \\ (x - R_d \cos\varphi)\sin\varphi_{rot} + [(-y + R_d \sin\varphi)\cos\theta + (z_t + z_c + z_h)\sin\theta]\cos\varphi_{rot} \\ (z_t + z_c + z_h)\cos\theta + (y - R_d \sin\varphi)\sin\theta \\ 1 \end{bmatrix} \tag{6.186}$$

$$\mathbf{c}_0 = \begin{bmatrix} \sin\varphi_{rot} \sin\theta \\ \cos\varphi_{rot} \sin\theta \\ \cos\theta \\ 0 \end{bmatrix} \tag{6.187}$$

The form-shaping system consists of six parameters of motion $x, y, z, \theta, \varphi_{rot}, \varphi_{sp}$ ($l = 6$) and two parameters of the tool φ_d and z_d ($M = 2$). Constraint number during processing a surface is

$$L = n + M - m = 6 + 2 - 2 = 6 \tag{6.188}$$

One of six setup equations is the implicit constraint, Eq. (6.185). Two further implicit constraints follow because parameters of the virtual tool φ_d and z_d and parameters of the machined surfaces u and v must agree with one another, $u = \varphi_d$ and $v = z_d$. The remaining three constraints are functional ones and can be obtained from Eqs. (6.116) and (6.117) by equating the right-hand sides of Eqs.(6.186) and (6.187) to equations of the machined surface.

As an example, synthesis of the functional constraints for drilling of radial through-holes in a hemispherical dome (Fig. 6.30a) is considered. The given position vector $\mathbf{r}_{hs}$ of two dome hemispheres composing the unit vector $\mathbf{n}_{hs}$ of the common normal at them are

$$[\mathbf{r}_{hs}\ \mathbf{n}_{hs}]=\begin{bmatrix} x_{hs}+R_i\cos\varphi_{hs}\sin\theta_{hs} & \cos\varphi_{hs}\sin\theta_{hs} \\ y_{hs}+R_i\sin\varphi_{hs}\sin\theta_{hs} & \sin\varphi_{hs}\sin\theta_{hs} \\ z_{hs}+R_i\cos\theta_{hs} & \cos\theta_{hs} \\ 1 & 0 \end{bmatrix} \tag{6.189}$$

$$-\pi/2 \le \theta_{hs} \le \pi/2;\ 0 \le \varphi_{hs} \le 2\varphi;\ R_i = R_{in} \text{ or } R_{ex},$$

where x_{hs}, y_{hs}, z_{hs}, are the coordinate of the sphere center; R_i is the sphere radius ($R_i = R_{in}$ for internal hemisphere and $R_i = R_{ex}$ for external hemisphere); and θ_{hs} and φ_{hs} are the zenith and azimuth angles, respectively.

Since the drilled hole is directed along the sphere radius, the vector of tool axis orientation $\mathbf{c}_0$, Eq. (6.187), is identical with the normal vector at the sphere surfaces $\mathbf{n}_{hs}$, Eq. (6.189),

$$\begin{vmatrix} \sin\varphi_{rot}\sin\theta \\ \cos\varphi_{rot}\sin\theta \\ \cos\theta \\ 0 \end{vmatrix} = \begin{vmatrix} \cos\varphi_{hs}\sin\theta_{hs} \\ \sin\varphi_{hs}\sin\theta_{hs} \\ \cos\theta_{hs} \\ 0 \end{vmatrix} \tag{6.190}$$

Since $||\mathbf{c}_0|| = ||\mathbf{n}_{hs}|| = 1$, this vector equation presents three simultaneous equations with respect to angles φ_{rot} and θ. The equation has two solutions

$$\theta_{1,2} = \pm\theta_{hs};\ \varphi_{rot1,2} = \pm\pi/2 - \varphi_{hs} \tag{6.191}$$

Substitution of the angles in Eq. (6.186) results in expression of the FSF through the given angles of radial direction coinciding with tool axis orientation:

$$(\mathbf{r}_0)_{1,2}=\begin{bmatrix} \pm(-x+R_d\cos\varphi)\sin\varphi_{hs}\pm[(-y+R_d\sin\varphi)\cos\theta_{hs}\pm(z_t+z_c+z_h)\sin\theta_{hs}]\cos\varphi_{hs} \\ \pm(x-R_d\cos\varphi)\cos\varphi_{hs}\pm[(-y+R_d\sin\varphi)\cos\theta_{hs}\pm(z_t+z_c+z_h)\sin\theta_{hs}]\sin\varphi_{hs} \\ (z_t+z_c+z_h)\cos\theta_{hs}\pm(y-R_d\sin\varphi)\sin\theta_{hs} \\ 1 \end{bmatrix} \tag{6.192}$$

A trajectory along which the *frontal point of the tool axis* moves can be defined in two ways. Firstly, it is obtained from Eq. (6.192) with tool radius $R_d = 0$, i.e.,

$$(\mathbf{r}_{tr})_{1,2}=\begin{bmatrix} \mp x\sin\varphi_{hs}\pm[-y\cos\theta_{hs}\pm(z_t+z_c+z_h)\sin\theta_{hs}]\cos\varphi_{hs} \\ \pm x\cos\varphi_{hs}\pm[-y\cos\theta_{hs}\pm(z_t+z_c+z_h)\sin\theta_{hs}]\sin o_{hs} \\ y\sin\theta_{hs}\pm(z_t+z_c+z_h)\cos\theta_{hs} \\ 1 \end{bmatrix} \tag{6.193}$$

Secondly, this point must lie at the line directed along the sphere radius, i.e.,

$$\mathbf{r}_{ax}(t)=\begin{bmatrix} x_{hs}+n_x t \\ y_{hs}+n_y t \\ z_{hs}+n_z t \\ 1 \end{bmatrix}=\begin{bmatrix} x_{hs}+t\cos\varphi_{hs}\sin\theta_{hs} \\ y_{hs}+t\sin\varphi_{hs}\sin\theta_{hs} \\ z_{hs}+t\cos\theta_{hs} \\ 1 \end{bmatrix}, R_{in}-l_2 \le t \le R_{ex}+l_1, \tag{6.194}$$

where n_x, n_y, n_z are the direction cosines of the normal; t is the current parameter along the axis of the machined bore; and l_1 and l_2 are the distances of the drill tip-point relative to the external surface before hole drilling and relative to the internal surface after drilling, respectively.

Then, equating the right-hand sides of matrices (6.193) and (6.194), $\mathbf{r}_{ax}(t) = \mathbf{r}_{tr}$, gives the last setup equation which can be solved with respect to x, y, and z_h:

$$\begin{bmatrix} \mp x\sin\varphi_{hs} \pm[-y\cos\theta_{hs} \pm(z_t+z_c+z_h)\sin\theta_{hs}]\cos\varphi_{hs} \\ \pm x\cos\varphi_{hs} \pm[-y\cos\theta_{hs} \pm(z_t+z_c+z_h)\sin\theta_{hs}]\sin\varphi_{hs} \\ y\sin\theta_{hs} \pm(z_t+z_c+z_h)\cos\theta_{hs} \\ 1 \end{bmatrix} = \begin{bmatrix} x_{hs}+t\cos\varphi_{hs}\sin\theta_{hs} \\ y_{hs}+t\sin\varphi_{hs}\sin\theta_{hs} \\ z_{hs}+t\cos\theta_{hs} \\ 1 \end{bmatrix} \quad (6.195)$$

For definiteness, only the first solution considered is considered

$$x = -x_{hs}\sin\varphi_{hs} + y_{hs}\cos\varphi_{hs} \quad (6.196)$$

$$y = -x_{hs}\cos\theta_{hs}\cos\varphi_{hs} - y_{hs}\cos\theta_{hs}\sin\varphi_{hs} + z_{hs}\sin\theta_{hs} \quad (6.197)$$

$$z_h = t - z_c - z_d + x_{hs}\cos\theta_{hs}\sin\varphi_{hs} + y_{hs}\sin\theta_{hs}\sin\varphi_{hs} + z_{hs}\cos\theta_{hs} \quad (6.198)$$

Substitution of Eqs. (6.196-198) in the FSF, Eq. (6.192), results in the position vector of the machined surface

$$\mathbf{r}_0(t,\varphi) = \begin{bmatrix} x_{hs}+R_d\cos\varphi\sin\varphi_{hs}+(R_d\sin\varphi\cos\theta_{hs}+t\sin\theta_{hs})\cos\varphi_{hs} \\ y_{hs}-R_d\cos\varphi\cos\varphi_{hs}+(R_d\sin\varphi\cos\theta_{hs}+t\sin\theta_{hs})\sin\varphi_{hs} \\ z_{hs}+t\cos\theta_{hs}-R_d\sin\varphi\sin\theta_{hs} \\ 1 \end{bmatrix} \quad (6.199)$$

Additional verification is needed to establish whether the solution is acceptable.

Numerical Example 6.9. Drilling radial through-holes in a hemisphere (Fig. 6.30).

Given: The hole of diameter of 8 mm orientated along the radius given by $\theta_{hs} = \pi/12$ and $\varphi_{hs} = \pi/4$ is drilled through the hemispherical dome. The center is at point $x_{hs} = 30$, $y_{hs} = 50$, and $z_{hs} = 40$ mm, internal and external dome radii are $R_{in} = 70$ and $R_{ex} = 80$ mm, respectively. The distances of the tip-point of the drill before and after hole drilling are $l_1 = 5$ and $l_2 = 10$ mm.
Required: Pictures of the machined surface and cutting tool in the working position.
Solution:

(1) The position vector of two dome hemispheres, Eq. (6.189), is

$$\mathbf{r}_{hs} = \begin{bmatrix} 30+\begin{pmatrix}80\\70\end{pmatrix}\cos\varphi_{hs}\sin\theta_{hs} \\ 50+\begin{pmatrix}80\\70\end{pmatrix}\sin\varphi_{hs}\sin\theta_{hs} \\ 40+\begin{pmatrix}80\\70\end{pmatrix}\cos\theta_{hs} \\ 1 \end{bmatrix}; -\pi/2 \le \theta_{hs} \le \pi/2;\ 0 \le \varphi_{hs} \le 2\varphi$$

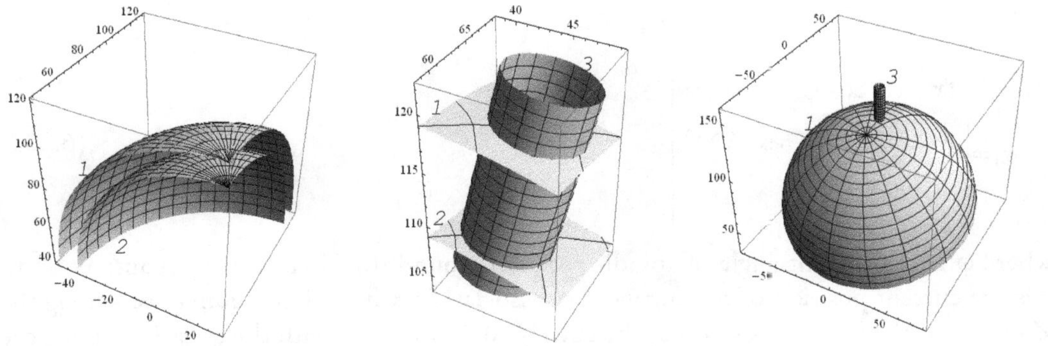

(a) (b) (c)

Figure 6-30 (a) Spherical dome octant; (b) and (c) drill position as referred to the work-piece- and bed-associated frame, respectively. *1* and *2* – the external and internal hemi-spheres of the dome; *3* – the drill in the working position.

(2) Position vector of the machined surface, Eq. (6.199), becomes

$$\mathbf{r}_0(t,\varphi) = \begin{bmatrix} x_{hs} + R_d \cos\varphi \sin\varphi_{hs} + (R_d \sin\varphi\cos\theta_{hs} + t\sin\theta_{hs})\cos\varphi_{hs} \\ y_{hs} - R_d \cos\varphi\cos\varphi_{hs} + (R_d \sin\varphi\cos\theta_{hs} + t\sin\theta_{hs})\sin\varphi_{hs} \\ z_{hs} + t\cos\theta_{hs} - R_d \sin\varphi\sin\theta_{hs} \\ 1 \end{bmatrix}$$

$$\begin{bmatrix} 30+4\cos\varphi\sin(\pi/4)+[4\sin\varphi\cos(\pi/12)+t\sin(\pi/12)]\cos(\pi/4) \\ 50-4\cos\varphi\cos(\pi/4)+[4\sin\varphi\cos(\pi/12)+t\sin(\pi/12)]\sin(\pi/4) \\ 40+t\cos(\pi/12)-4\sin\varphi\sin(\pi/12) \\ 1 \end{bmatrix}, \quad 60 \le t \le 85 \text{ mm};\ 0 \le \varphi_{hs} \le 2\varphi$$

The working position of the drill relative to the machined part is shown in Figure 6.30b. Note that in the considered machine, the kinematic chain combines two branches whose sub-FSFs are as follows:

- The branch "Tool-Bed": $[\mathbf{r}_0\mathbf{c}_0]_{TB} = \mathbf{A}^3(z)\,\mathbf{A}^6(\varphi_{sp})\,[\mathbf{r}_d\mathbf{c}_d]$
- The branch "Workpiece-Bed": $[\mathbf{r}_0\mathbf{c}_0]_{WB} = \mathbf{A}^6(\varphi_{rot})\,\mathbf{A}^4(\theta)\,\mathbf{A}^1(x)\,\mathbf{A}^2(y)\,[\mathbf{r}_{hs}\mathbf{c}_{hs}]$

Actual positions of the tool and the workpiece as referred to the bed are shown in Figure 6.30c.

6.4 Lathe setup

6.4.1 The FSF and constraints

A layout of the lathe is shown in Figure 4.6, the FSS links are listed in Table 4.6, and the position FSF, Eqs (4.38-40), is

$$\mathbf{r}_0 = \mathbf{A}^6(\varphi)\mathbf{A}^3(z)\,\mathbf{A}^1(x)\,\mathbf{r}_3 \tag{6.200}$$

i.e,

$$\mathbf{r}_0 = \begin{bmatrix} (x+x_3)\cos\varphi - y_3 \sin\varphi \\ (x+x_3)\sin\varphi + y_3\cos\varphi \\ z+z_3 \\ 1 \end{bmatrix} \tag{6.201}$$

where φ is the current angle of spindle rotation around the Z_0-axis (the cutting motion); z is the current position of the cutting tool associated with the longitudinal feed along the Z_1-axis; x is the current position of the cutting tool associated with the cross feed along the X_2-axis; and $\mathbf{r}_3$ is the vector of the cutting tool, $\mathbf{r}_3 = (x_3, y_3, z_3, 1)^T$.

In Figure 6.31, interrelationships of a cutting tool with a machined surface are shown. There are two kinematic types of the turning cutters: a single point cutter applied for machining by cylinders, faces and bodies of revolution with shaped axial section (generatrix), a linear cutter for machining screws, and bodies of revolution with different forms of generatrix.

The target element of the lathe mechanical system is the workpiece, whose nominal equations are obtained from the FSF by addition of some constraints

$$F_k = (\varphi, x, z, x_3, y_3, z_3) = 0, \text{ with } k = 1,\ldots,L \tag{6.202}$$

$$L = n + M - m \tag{6.203}$$

where $l = 3$ since there are three kinematic motions (one rotation and two translations); $m = 2$ since the target element is a machined surface; and $M = 0$, for the single-point cutting tool, or $M = 1$, for the line-type cutting tool. Hence, the number of constraints is

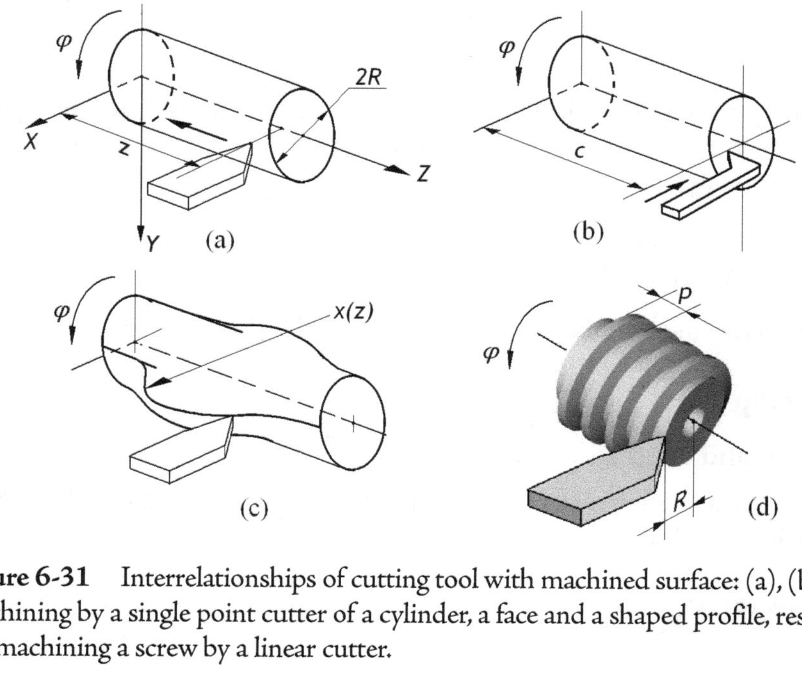

Figure 6-31 Interrelationships of cutting tool with machined surface: (a), (b) and (c) machining by a single point cutter of a cylinder, a face and a shaped profile, respectively; (d) machining a screw by a linear cutter.

$$L=\begin{cases}1, \text{ for single-point tool}\\ 2, \text{for line-type tool}\end{cases} \tag{6.204}$$

6.4.2 Machining by the single-point tool

The tool is given by $x_3 = y_3 = z_3 = 0$. The FSF, Eq. (6.200), becomes

$$\mathbf{r}_3=\begin{bmatrix}0\\0\\0\\1\end{bmatrix};\mathbf{r}_0=\begin{bmatrix}x\cos\varphi\\x\sin\varphi\\z\\1\end{bmatrix} \tag{6.205}$$

According to Eq. (6.204), there is a single functional constraint presenting the surface equation in the cylindrical coordinate frame

$$x = R(z, \varphi) \tag{6.206}$$

where R is the current radius of the machined surface.

Hence, Eq. (6.201) becomes

$$\mathbf{r}_0=\mathbf{r}_0(z,\varphi) = \begin{bmatrix}R(z,\varphi)\cos\varphi\\R(z,\varphi)\sin\varphi\\z\\1\end{bmatrix} \tag{6.207}$$

Six variants of the functional constraint listed in Table 6.4 define surfaces machined by a single point tool (Figs. 6.32-33). Setups obtained upon stopping the motions of one of the

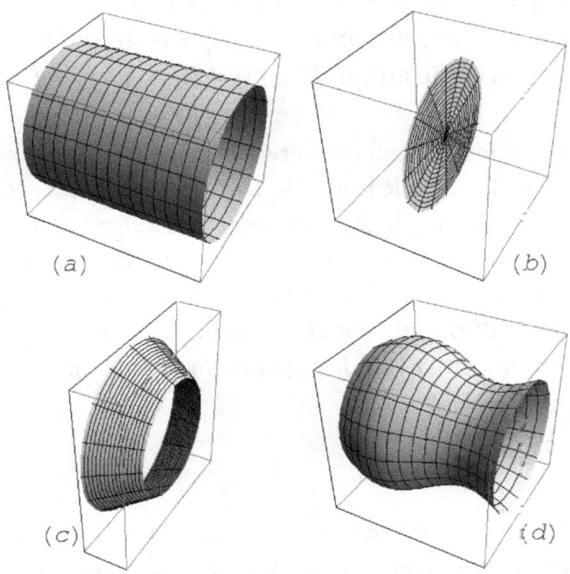

Figure 6-32 Turning bodies of revolution by a single point tool: (a) cylinder; (b) face; (c) cone; (d) shaped profile.

Table 6-4 Surfaces machined on the lathe by single-point cutter.

Machined surface	Setup condition	Surface equation	Figure
Straight cylinder of radius R	$x = R$	$\mathbf{r}_0 = \begin{pmatrix} R\cos\varphi \\ R\sin\varphi \\ z \end{pmatrix}$	6.32a
End face on distance c	$z = c$	$\mathbf{r}_0 = \begin{pmatrix} x\cos\varphi \\ x\sin\varphi \\ c \end{pmatrix}$	6.32a
Cone	$x = -az + b$	$\mathbf{r}_0 = \begin{bmatrix} (-az+b)\cos\varphi \\ (-az+b)\sin\varphi \\ z \end{bmatrix}$	6.32c
Surface of revolution	$x = R(z)$	$\mathbf{r}_0 = \begin{bmatrix} R(z)\cos\varphi \\ R(z)\sin\varphi \\ z \end{bmatrix}$	(1)6.32d
Cam with rectilinear generatrix	$x = R(\varphi)$	$\mathbf{r}_0 = \begin{bmatrix} R(\varphi)\cos\varphi \\ R(\varphi)\sin\varphi \\ z \end{bmatrix}$	(2)6.32a
Spatial cam	$x = R(z, \varphi)$	$\mathbf{r}_0 = \begin{bmatrix} R(z,\varphi)\cos\varphi \\ R(z,\varphi)\sin\varphi \\ z \end{bmatrix}$	(3)6.32b

(1)Current radii in Figure 6.32d: $R(z) = R_0 + a\sin(\pi z/L)$
(2)Current radii in Figures 6.33a and b: $R(\varphi) = R_0 + a\sin(\varphi/2)$
(3)Current radii in Figures 6.33c and d: $R(z, \varphi) = R_0 + a\sin(\varphi/2) + b z^2$

carriages at a specified position are defined in two first rows of the Table 6.4. The transverse carriage is arranged at the distance R from the workpiece axis ($x = R$) to machine the cylinder of diameter $2R$, and the longitudinal carriage is arranged at a fixed X-coordinate for end facing ($z = c$). Furthermore, constraint $x = R(z)$ is used for *contour turning* and allows machining the surface of revolution with a given form of generatrix. A constraint connecting the longitudinal feed with the current angle of the spindle rotation can have the form $x = R(\varphi)$ or $x = R(z, \varphi)$ providing machining of planar or spatial cam, respectively.

6.4.3 Machining by the profile cutter

6.4.3.1 Formulation

For definiteness, the case in which the profile cutting blade is planar and lies in the XOZ plane (Fig. 6.34) is considered. The position vector of the cutting tool and the FSF, Eq. (6.201), can be represented as follows:

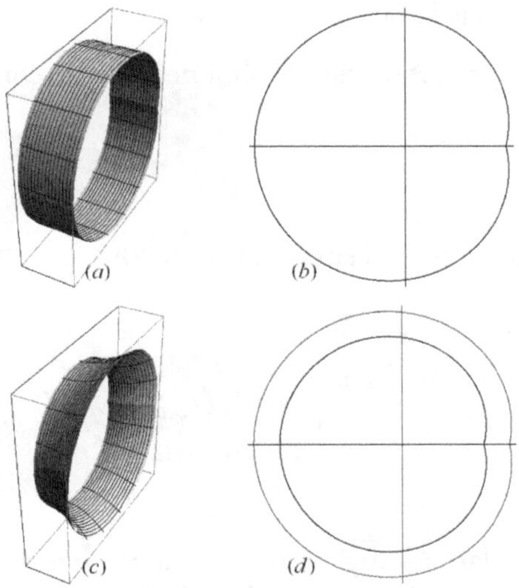

Figure 6-33 Turning cams by a single point tool: (a) and (b) cylindrical cam and its cross-section; (c) and (d) spatial came and its frontal and back faces.

$$\mathbf{r}_3 = \begin{bmatrix} x_3(U) \\ 0 \\ z_3(U) \\ 1 \end{bmatrix} \tag{6.208}$$

$$\mathbf{r}_0 = \mathbf{r}_0(x, z, \varphi, U) = \begin{bmatrix} [x + x_3(U)] \cos \varphi \\ [x + x_3(U)] \sin \varphi \\ z + z_3(U) \\ 1 \end{bmatrix} \tag{6.209}$$

where U is the current parameter defining position of an active point of the blade.

There are two groups of surfaces machined by linear-type cutter:

- Bodies of revolution when cutting tools with profile blades carry out the form turning: the tool blade shape is replicated on the workpiece generatrix;
- Helicoids when profile blade forms the helicoid profile.

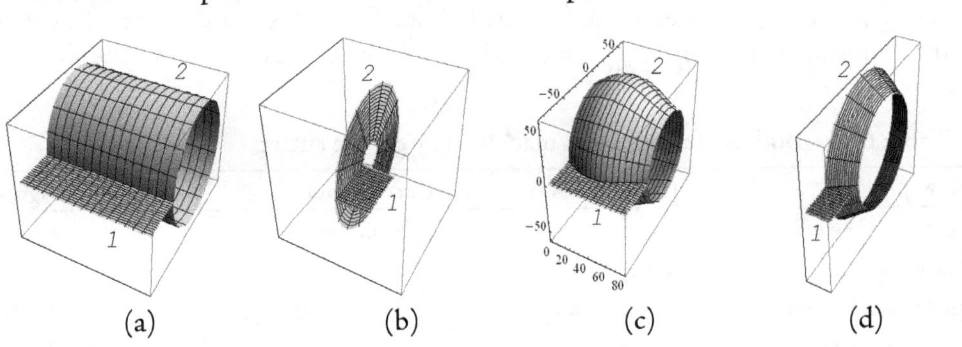

(a) (b) (c) (d)

Figure 6-34 Machining of the workpiece 2 by linear cutter *1*: (a) cylinder; (b) face, (c) shaped surface; (d) cone.

6.4.3.2 *Machining bodies of revolution*

Setup for machining a body of revolution is obtained by two constraints

$$x = x_0 \text{ and } z = z_0 \tag{6.210}$$

where x_0 and z_0 are constants.

The general form of the machined surface, Eq. (6.209), is as follows:

$$\mathbf{r} = \mathbf{r}(U,\varphi) = \begin{bmatrix} [x_0 + x_3(U)] \cos \varphi \\ [x_0 + x_3(U)] \sin \varphi \\ z_0 + z_3(U) \\ 1 \end{bmatrix}, \text{ with } U_{min} \le U \le U_{max} \tag{6.211}$$

Typical examples of the lathe setups with associated constraints and cutting edge equations are given in Table 6.5, and the machined surfaces are shown in Figure 6.34.

The setup for machining of the shaped surface of revolution (the 3rd line in the Table) is illustrated by the following example.

Numerical Example 6.10: Machining the hemispherical segment step of the multidimensional shaft (Figure 6.35a). **Given:** The hemisphere of radius R = 50 mm and centered at z_0 = 120 mm is truncated at distance h = 25 mm from the hemisphere base. **Required:** Setup parameters: $x = x_0$; $z = z_0$, and the blade profile.

Solution: (1) An equation of the hemispherical segment is as follows:

$$\mathbf{r} = \mathbf{r}(\theta,\varphi) = \begin{bmatrix} R\cos \Phi \cos\theta \\ R\sin \Phi\cos \theta \\ z_0 + R \sin \theta \\ 1 \end{bmatrix}, \text{ with } \theta_{min} \le \theta \le \theta_{max};\ 0 \le \Phi \le 2\pi \tag{6.212}$$

where R is the sphere radius; Φ is the polar angle of the sphere; θ is the blade profile angle – the azimuthal angle of the sphere; and z_0 is the sphere center.

(2) The azimuth boundaries are defined as follows: θ_{min} = 0 since this boundary defines the hemisphere base, and θ_{max} associated with the minimal diameter of the step is

Table 6-5 Bodies of revolutions machined by profile cutter.

Surface	Constraints	Cutting tool	Figure
Straight round cylinder	$x = x_0$; $z = z_0$	$\mathbf{r}_3 = [0, 0, U, 1]^T$	6.34a
End face plane	$x = x_0$; $z = z_0$	$\mathbf{r}_3 = [U, 0, 0, 1]^T$	6.34b
Shaped surface of revolution	$x = x_0$; $z = z_0$	$\mathbf{r}_3 = [x_3(U), 0, z_3(U), 1]^T$	6.34c
Straight round cone	$x = x_0$; $z = z_0$	$\mathbf{r}_3 = [U \tan\theta, 0, U, 1]^T$	6.34d

Note: x_0 and z_0 are the fixed radius and z-distance, respectively; θ is the cone half-angle.

$$\theta_{max} = \arcsin(h/R) = \arcsin(25/50) = \arcsin 0.5 = \begin{Bmatrix} \pi/6 \\ 5\pi/6 \end{Bmatrix} \tag{6.213}$$

Only solution $\theta_{max} = \pi/6$ satisfies conditions given in the *Example*.

(3) Equating the FSF, Eq. (6.211), and sphere position vector $\mathbf{r}(\vartheta, \varphi)$, Eq. (6.212), results in three simultaneous equations

$$[x + x_3(U)] \cos \varphi = R \cos \Phi \cos \theta$$

$$[x + x_3(U)] \sin \varphi = R \sin \Phi \cos \theta$$

$$z + z_3(U) = z_0 + R \cos \theta$$

Solution of the equations gives the setup parameters:

- The angle of the spindle rotation: $0 \le \varphi = \Phi \le 2\pi$;
- The final position of the tool: $x = x_0 = 0$ and $z = z_0 = 120$;
- Tool profile: $x_3(U) = R \cos \theta$ and $z_3(U) = R \sin \theta$, with $0 \le \theta \le \pi/6$,

i.e.,

$$\mathbf{r}_3 = [x_3(U), 0, z_3(U), 1]^{\mathrm{T}} = [50 \cos \theta, 0, 50 \sin \theta, 1]^{\mathrm{T}}, \text{ with } 0 \le \theta \le \pi/6$$

The setup according to the example is shown in Figure 6.35b.

6.4.3.3 Helicoid machining

Setup for turning a cylindrical single-start thread with axial lead L is obtained from Eq. (6.211) by application of two kinematic constraints

$$z = \pm P\varphi \tag{6.214}$$

$$x = d_m/2 \tag{6.215}$$

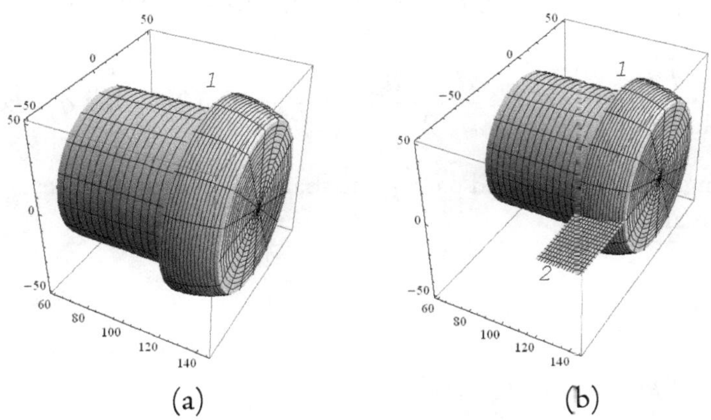

Figure 6-35 (a) A shaft with truncated hemispherical step *1*; (b) the machined step *1* with shaped tool 2 (*Numerical Example 6.10*).

where P is the thread parameter expressing through the lead, $P = L/(2\pi)$; signs + and – correspond to right- and left-hand thread, respectively; φ is the angle of the screw rotation around its axis; d_m is the mean diameter of the machined thread,

$$d_m = (d_{out} - d_{in})/2 \tag{6.216}$$

d_{out} and d_{in} are the major and minor diameters of the thread, respectively.

As an example, let us consider machining an external helical surface – Archimedes' screw (the ZA-type worm according to DIN 3975). The profile of the cutting tool blade is

$$\mathbf{r}_3 = \mathbf{r}_3(U) = \begin{bmatrix} V \\ 0 \\ U \pm L/2 \\ 1 \end{bmatrix}, \text{ with } U_1 = -\frac{d_m - d_{in}}{2q} \leq \pm U \leq U_2 = \frac{d_{out} - d_m}{2q} \tag{6.217}$$

$$q = \pm\cot(\psi/2)$$

$$V = d_m/2 + Uq \tag{6.218}$$

where U is the parameter determining the location of a current point of the cutting blade (Fig. 6.36) replicating the thread profile in the axial section of the machined thread, parameter U varies in the range between U_1 and U_2; and ψ is the profile angle.

Hence, the FSF, Eq. (6.201), for considered setup becomes

$$\mathbf{r}_0 = \mathbf{r}_0(U,\varphi) = \mathbf{A}^6(\varphi)\mathbf{A}^3(P\varphi)\mathbf{A}^1(d_m/2)\mathbf{r}_3(U) = \begin{bmatrix} R(U)\cos\varphi \\ R(U)\sin\varphi \\ U \pm L \pm P\varphi \\ 1 \end{bmatrix}, \tag{6.219}$$

with

$$R(U) = \begin{cases} R_{right} = d_m/2 + U\cot(\psi/2), \text{ for the right side profile} \\ R_{left} = d_m/2 - U\cot(\psi/2), \text{ for the left side profile} \end{cases}$$

$$U_1 = -\frac{d_m - d_{in}}{2q} \leq \pm U \leq U_2 = \frac{d_{out} - d_m}{2q};\ 0 \leq \varphi \leq l/(2\pi); \text{ and } q = \pm\cos(\psi/2).$$

where l is the length of the threaded part of the shaft to be machined.

The unit normal vector is

$$\mathbf{n} = \pm\frac{\mathbf{N}}{\|\mathbf{N}\|} \tag{6.220}$$

$$\mathbf{N} = \frac{\partial \mathbf{r}_0}{\partial U} \times \frac{\partial \mathbf{r}_0}{\partial \varphi} = \begin{bmatrix} R(U)\cos\varphi - Pq\sin\varphi \\ R(U)\sin\varphi + Pq\cos\varphi \\ -qR(U) \\ 0 \end{bmatrix} \tag{6.221}$$

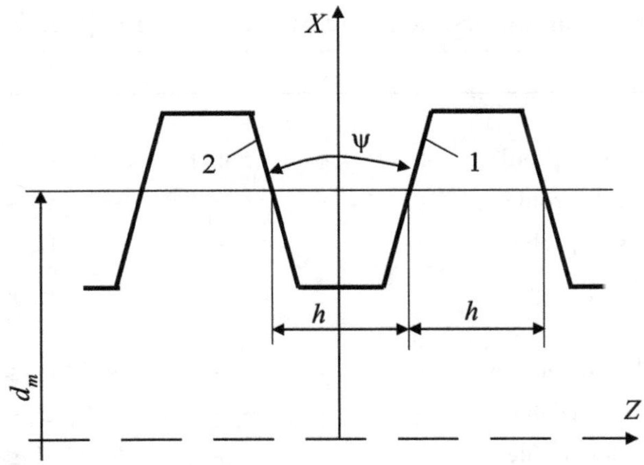

Figure 6-36 Thread profile in the screw axial section.

$$\|\mathbf{N}\| = \sqrt{P^2 q^2 + (1+q^2)R(U)^2} \tag{6.222}$$

The normal at the helicoid directed from the helicoid surface (outer-pointing normal) is considered. Its sign is plus if the normal projection at the thread axis is positive. Substitution of Eqs. (6.214) and (6.218) in Eq. (6.219) results in

$$\mathbf{r}_0 = \mathbf{r}_0(U,\varphi) = \begin{bmatrix} (d_m/2 \pm U\cot\frac{\psi}{2})\cos\varphi \\ (d_m/2 \pm U\cot\frac{\psi}{2})\sin\varphi \\ U \pm L \pm P\varphi \\ 1 \end{bmatrix} \tag{6.223}$$

$$U_1 = -\frac{d_m - d_{in}}{2q} \le \pm U \le U_2 = \frac{d_{out} - d_m}{2q};\ 0 \le \varphi \le l/(2\pi)$$

By virtue of the fact that Eq. (6.219) includes some plus-minus signs (including sign of angle ψ), possible combinations of the signs in Eq. (6.223) are listed in Table 6.6.

Numerical Example 6.11. Turning the single-start thread with profile shown in Figure 6.36. **Given:** Lead $L = 4$ mm; profile angle $\psi = 60°$; mean diameter $d_m = 36.5$ mm; external diameter $d_{ex} = 38$ mm; and internal diameter $d_{in} = 35$ mm. **Required:** Equations of profiles of two sides of right- and left-hand helicoids.
Solution: Absolute values of the thread parameters are

$P = L/(2\pi) = 2/\pi$ mm

$h = L/2 = 2$ mm

$\cot(\psi/2) = \cot(\pi/6) = \sqrt{3}$.

Table 6-6 Thread parameters: signs and boundaries in Eq. (6.223).

Part and Helicoid	p	q	h	n	U boundaries
Screw					
Right-hand thread. Right side profile	+	+	+	–	$U_1 \le U \le U_2$
Right-hand thread. Left side profile	+	–	–	+	$U_2 \le U \le U_1$
Left-hand thread. Right side profile	–	+	+	–	$U_1 \le U \le U_2$
Left-hand thread. Left side profile	–	–	–	+	$U_2 \le U \le U_1$
Nut					
Right-hand thread. Right side profile	+	+	+	–	$U_1 \le U \le U_2$
Right-hand thread. Left side profile	+	–	–	+	$U_2 \le U \le U_1$
Left-hand thread. Right side profile	–	+	+	–	$U_1 \le U \le U_2$
Left-hand thread. Left side profile	–	–	–	+	$U_2 \le U \le U_1$

$$U_1 = -\frac{d_m - d_{in}}{2q} = -\frac{36.5-35}{2\cot(\psi/2)} = -1.299$$

$$U_2 = \frac{d_{out} - d_m}{2q} = \frac{38-36.5}{2\cot(\psi/2)} = 0.75/\sqrt{3}$$

Substitution of the parameters in Eq. (6.223) results in four structural versions of the FSS:

- *Right side. Right-hand thread*

$$\mathbf{r}_0 = \mathbf{r}_0(U,\varphi) = \begin{bmatrix} R_{right}\cos\varphi \\ R_{right}\sin\varphi \\ U+h+P\varphi \\ 1 \end{bmatrix} = \begin{bmatrix} (18.25+U\sqrt{3})\cos\varphi \\ (18.25+U\sqrt{3})\sin\varphi \\ U+4 \\ 1 \end{bmatrix} \tag{6.224}$$

$R_{right} = d_m/2 + U\cos(\psi/2)$; $-1.299 \le U \le 1.299$.

- *Left side. Right-hand thread*

$$\mathbf{r}_0 = \mathbf{r}_0(U,\varphi) = \begin{bmatrix} R_{left}\cos\varphi \\ R_{left}\sin\varphi \\ U-h+P\varphi \\ 1 \end{bmatrix} = \begin{bmatrix} (18.25-U\sqrt{3})\cos\varphi \\ (18.25-U\sqrt{3})\sin\varphi \\ U \\ 1 \end{bmatrix} \tag{6.225}$$

with $R_{left} = d_m/2 - U\cos(\psi/2)$; $-1.299 \le U \le 1.299$.

- *Right side, left-hand thread*

$$\mathbf{r}_0 = \mathbf{r}_0(U,\varphi) = \begin{bmatrix} R_{left}\cos\varphi \\ R_{left}\sin\varphi \\ U+h-P\varphi \\ 1 \end{bmatrix} = \begin{bmatrix} (18.25+U\sqrt{3})\cos\varphi \\ (18.25+U\sqrt{3})\sin\varphi \\ U \\ 1 \end{bmatrix} \tag{6.226}$$

- *Left side, left-hand thread*

$$\mathbf{r}_0 = \mathbf{r}_0(U,\varphi) = \begin{bmatrix} R_{left}\cos\varphi \\ R_{left}\sin\varphi \\ U-h-P\varphi \\ 1 \end{bmatrix} = \begin{bmatrix} (18.25-U\sqrt{3})\cos\varphi \\ (18.25-U\sqrt{3})\sin\varphi \\ U-4 \\ 1 \end{bmatrix} \tag{6.227}$$

As an example, the unit vector of the normal at the right side of right-hand helicoid, Eq. (6.224), is

$$\mathbf{n} = \pm\frac{1}{\|\mathbf{N}\|}\begin{bmatrix} R_{right}\cos\varphi\text{-}pq\sin\varphi \\ R_{right}\sin\varphi + pq\sin\varphi \\ -R_r q \\ 0 \end{bmatrix} = \pm\frac{1}{\|\mathbf{N}\|}\begin{bmatrix} -31.61-3U \\ (18.25+1.732U)\cos\varphi\text{-}1.1027\sin\varphi \\ (18.25+1.732U)\sin\varphi+1.1027\cos\varphi \\ 0 \end{bmatrix} \tag{6.228}$$

$$\|\mathbf{N}\| = \sqrt{R_{right}^2(1+q^2)+q^2p^2} = \sqrt{1333.5+252.88U+12U^2}\text{; and } q = \cos(\psi/2).$$

The sign of the normal depends on what part – screw or nut – describes the helicoid. A pair composing the right and left sides of the right-hand threads (at the one pitch length) is shown in Figure 6.37. In Figure 6.38, two pairs of the right- and left-hand helicoids machined by a straight line blade are shown. The left and right sides of the threads are presented in Figures 6.39a and b, respectively.

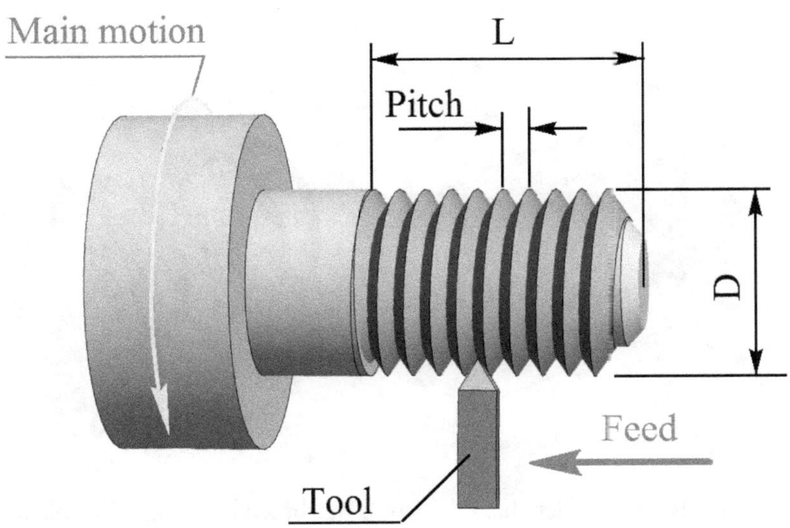

Figure 6-37 Setup for single-start thread machining (L is the thread length).

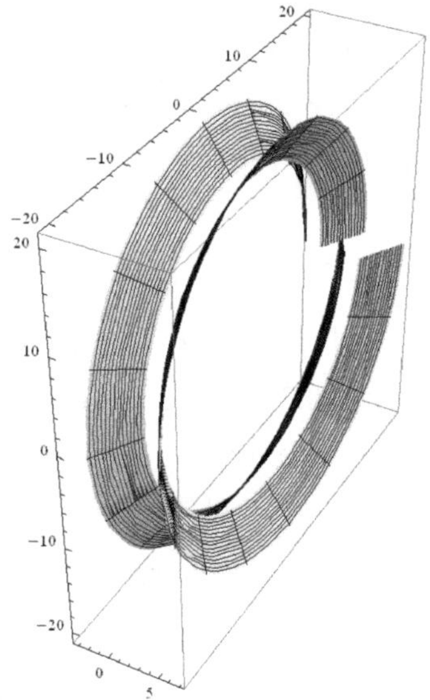

Figure 6-38 Two sides of the thread at the one pitch length.

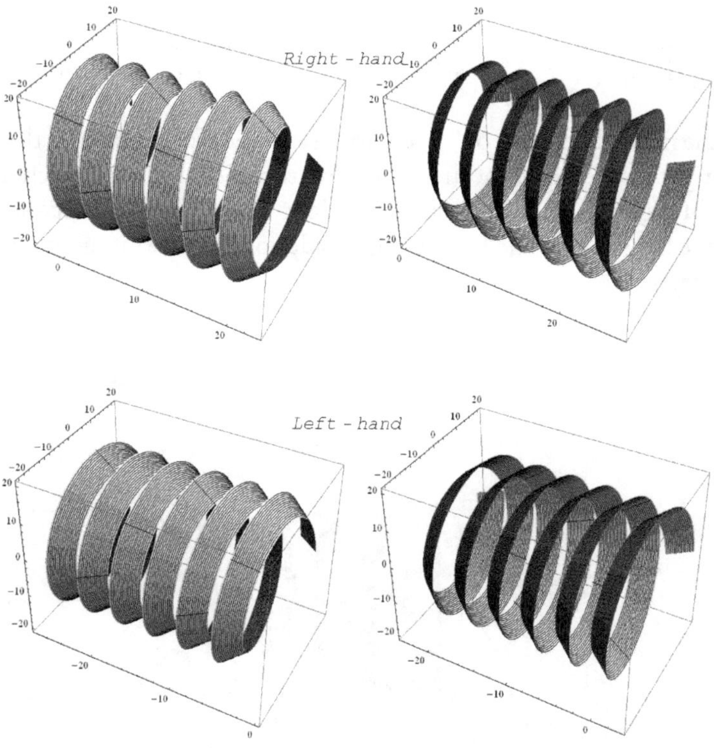

Figure 6-39 Right- and left-hand helicoids machined by straight line blade: (a) the left side of the thread; (b) the right side of the thread.

6.5 Setup of the gearing and threading machines

6.5.1 Setup of rack-cutter machine

The FSS links of the rack-cutter machine are listed in Table 4.11, and the FSF, Eq. (4.72) is

$$\mathbf{r}_0 = {}^0\mathbf{A}_4\,\mathbf{r}_4 = \mathbf{A}^6(-\varphi)\mathbf{A}^1(-x)\,\mathbf{A}^2(E)\,\mathbf{A}^3(z)\,\mathbf{r}_4 =$$

$$\begin{bmatrix} \cos\varphi & \sin\varphi & 0 & 0 \\ -\sin\varphi & \cos\varphi & 0 & 0 \\ 0 & 0 & 1 & 0 \\ 0 & 0 & 0 & 1 \end{bmatrix} \begin{bmatrix} 1 & 0 & 0 & -x \\ 0 & 1 & 0 & 0 \\ 0 & 0 & 1 & 0 \\ 0 & 0 & 0 & 1 \end{bmatrix} \begin{bmatrix} 1 & 0 & 0 & 0 \\ 0 & 1 & 0 & E \\ 0 & 0 & 1 & 0 \\ 0 & 0 & 0 & 1 \end{bmatrix} \begin{bmatrix} 1 & 0 & 0 & 0 \\ 0 & 1 & 0 & 0 \\ 0 & 0 & 1 & z \\ 0 & 0 & 0 & 1 \end{bmatrix} \begin{bmatrix} x_4 \\ y_4 \\ z_4 \\ 1 \end{bmatrix}$$

$$= \begin{bmatrix} \cos\varphi & \sin\varphi & 0 & E\sin\varphi - x\cos\varphi \\ -\sin\varphi & \cos\varphi & 0 & E\cos\varphi + x\sin\varphi \\ 0 & 0 & 1 & z \\ 0 & 0 & 0 & 1 \end{bmatrix} \begin{bmatrix} x_4 \\ y_4 \\ z_4 \\ 1 \end{bmatrix} = \begin{bmatrix} (x_4 - x)\cos\varphi + (y_4 + E)\sin\varphi \\ -(x_4 - x)\sin\varphi + (y_4 + E)\cos\varphi \\ z + z_4 \\ 1 \end{bmatrix} \quad (6.229)$$

We consider a setup for machining the left side of the tooth of the external involute spur gear with the pressure angle $\alpha = 20°$ by rack-cutter (Fig. 6.40). The active side of the cutter presents a straight-line segment tilted 20° with respect to the Y-axis. A parallel to the X-axis straight line, along which the tooth thickness and width space are equal to one another, presents the rack-cutter centrode. Note that the external part of the standard tooth addendum, which generates the gear fillet, has an arc profile and is not detailed here. During machining the centrode is rolled without slipping relative to the pitch circle of the machined gear.

The position vector of the rack-cutter in its own coordinate frame S_4 (Table 4.11) is

$$\mathbf{r}_4 = \begin{bmatrix} x_4 \\ y_4 \\ z_4 \\ 1 \end{bmatrix} = \begin{bmatrix} u\sin\alpha \\ u\cos\alpha \\ z_4 \\ 1 \end{bmatrix}; \; -h_a m/\cos\alpha \le u \le h_a m/\cos\alpha; \; 0 \le z_4 \le L, \quad (6.230)$$

where z_4 is the current coordinate along the profile length; L is the rack tooth length; h_a is the tooth addendum coefficient; m is the gear module.

In international standards, module m [mm] is expressed through the circular pitch p and diameter of pitch circle d as $m = p/\pi = d/N$ (N is the number of gear teeth) and $d = mN$. In the USA, the system in inches is customary used, in which diametrical pitch P [1/in] = N/d and pressure angle $\alpha = 14.5°$ are standardized. In this system, the diametrical pitch P and circular pitch p are connected with pitch diameter d as follows:

$$Pp = \pi \text{ and } d = N/P \quad (6.231)$$

The FSF is obtained through substitution of Eq. (6.230) in Eq. (6.229)

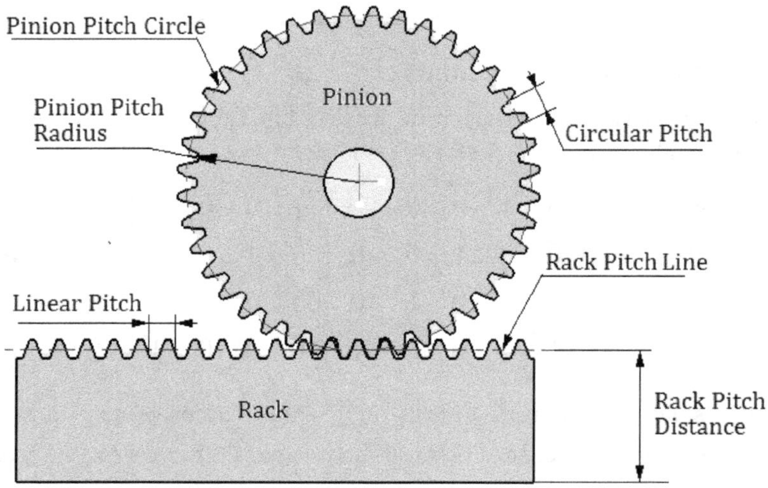

Figure 6-40 Geometry of the rack-gear meshing.

$$\mathbf{r}_0 = \begin{bmatrix} (x_4 - x)\cos\varphi + (y_4 + E)\sin\varphi \\ -(x_4 - x)\sin\varphi + (y_4 + E)\cos\varphi \\ z + z_4 \\ 1 \end{bmatrix} = \begin{bmatrix} -x\cos\varphi + u\sin(\alpha + \varphi) + E\sin\varphi \\ x\sin\varphi + u\cos(\alpha + \varphi) + E\cos\varphi \\ z + z_4 \\ 1 \end{bmatrix} \quad (6.232)$$

Hence, the FSF, Eq. (6.231), depends on six parameters: φ, x, u, z, z_4, and E. Since the number of actuated movements is $l = 3$ (tool translation, workpiece rotation, and translational cutting motion) and the dimensionalities of the tool and workpiece teeth surfaces are $M = m = 2$, the constraint number is $L = n + M - m = 4 + 2 - 2 = 4$. The constraints are specified as follows:

- Single implicit constraint: $z + z_4 = Z$;
- Two functional constraints: kinematic constraint $x = \varphi R_p$ and geometric constraint $E = R_p$, where $R_p = d/2$ is the radius of the workpiece pitch circle.

Substitution of these constraints in the FSF, Eq. (6.232), results in a set of instantaneous positions of the rack-cutter tooth

$$\mathbf{r}_0 = \mathbf{r}_0(u; \varphi; Z) = \begin{bmatrix} -\varphi R_p\cos\varphi + u\sin(\alpha + \varphi) + R_p\sin\varphi \\ \varphi R_p\sin\varphi + u\cos(\alpha + \varphi) + R_p\cos\varphi \\ Z \\ 1 \end{bmatrix} \quad (6.233)$$

- An enveloping constraint

$$[\frac{\partial \mathbf{r}_0}{\partial Z} \times \frac{\partial \mathbf{r}_0}{\partial h}] \cdot \frac{\partial \mathbf{r}_0}{\partial \varphi} = 0 \quad (6.234)$$

with

$$\frac{\partial \mathbf{r}_0}{\partial \varphi} = \begin{bmatrix} u\cos(\alpha+\varphi)+\varphi R_p \sin\varphi \\ -u\sin(\alpha+\varphi)+\varphi R_p\cos\varphi \\ 0 \\ 0 \end{bmatrix};\ \frac{\partial \mathbf{r}_0}{\partial u} = \begin{bmatrix} \sin(\alpha+\varphi) \\ \cos(\alpha+\varphi) \\ 0 \\ 0 \end{bmatrix};\ \text{and}\ \frac{\partial \mathbf{r}_0}{\partial Z} = \begin{bmatrix} 0 \\ 0 \\ 1 \\ 0 \end{bmatrix} \tag{6.235}$$

Substitution of Eqs. (6.235) in Eq. (6.234) yields

$$\begin{vmatrix} u\cos(\alpha+\varphi)+\varphi R_p \sin\varphi & \sin(\alpha+\varphi) & 0 \\ -u\sin(\alpha+\varphi)+\varphi R_p\cos\varphi & \cos(\alpha+\varphi) & 0 \\ 0 & 0 & 1 \end{vmatrix} = -u+\varphi R_p \sin\alpha = 0$$

i.e.,

$$u = \varphi R_p \sin\alpha \tag{6.236}$$

Substitution of Eq. (6.236) in Eq. (6.233) results in an equation of the machined gear tooth profile presenting the envelope of the set of instantaneous position of the rack-cutter tooth:

$$\mathbf{r}_0 = \mathbf{r}_0(\varphi; Z) = \begin{bmatrix} -\varphi R_p \cos\varphi + \varphi R_p \sin\alpha \sin(\alpha+\varphi) + R_p \sin\varphi \\ \varphi R_p \sin\varphi + \varphi R_p \sin\alpha\cos(\alpha+\varphi) + R_p\cos\varphi \\ Z \\ 1 \end{bmatrix} \tag{6.237}$$

After some trigonometric transformation, Eq. (6.237), we obtain

$$\mathbf{r}_0 = \mathbf{r}_0(\varphi; Z) = \begin{bmatrix} x_0(\varphi) \\ y_0(\varphi) \\ Z \\ 1 \end{bmatrix} = \begin{bmatrix} R_p \sin\varphi - \varphi R_p \cos\alpha\cos(\alpha+\varphi) \\ R_p\cos\varphi + \varphi R_p\cos\alpha\sin(\alpha+\varphi) \\ Z \\ 1 \end{bmatrix} \tag{6.238}$$

$\varphi_b \le \varphi \le \varphi_{add}$ and $0 \le Z \le Z_{max}$

where Z_{max} is the gear tooth length; φ_b and φ_{add} are the values of angle φ corresponding to the boundaries of the involute profile – base radius R_b and addendum radius R_{add}, respectively,

$$R_b = R_p \cos\alpha \tag{6.239}$$

$$R_{add} = R_p + m \tag{6.240}$$

Current radius R ($R > 0$) associated with angle φ is defined through $x_0(\varphi)$ and $y_0(\varphi)$ components of vector $\mathbf{r}_0$, Eq. (6.238), as follows:

$$R=\sqrt{x_0^2+y_0^2}=\sqrt{[R_p\sin\varphi-\varphi R_p\cos\alpha\cos(\alpha+\varphi)]^2+[R_p\cos\varphi+\varphi R_p\cos\alpha\sin(\alpha+\varphi)]^2}=$$
$$R_p\sqrt{1+\varphi^2\frac{(1+\cos 2\alpha)}{2}+2\varphi\sin 2\alpha} \tag{6.241}$$

Note that with $\varphi = 0$, Eq. (6.241) yields $R = R_p$. Correspondingly, the boundary angles φ_b and φ_{add} of the involute segment of the gear tooth profile represent solutions of equations obtained by equating the right-hand parts of Eqs. (6.239) and (6.240) to the right-hand part of Eq. (6.241) with substitution of $\varphi = \varphi_b$ and $\varphi = \varphi_{add}$, respectively:

$$\sqrt{1+\varphi_b^2\frac{(1+\cos 2\alpha)}{2}+2\varphi_b\sin 2\alpha}=\cos\alpha \tag{6.242}$$

$$\sqrt{1+\varphi_{add}^2\frac{(1+\cos 2\alpha)}{2}+2\varphi_{add}\sin 2\alpha}=1+m/R_p \tag{6.243}$$

As is known [5], Eq.(6.238) represents an involute profile relative to the base circle of radius $R_b = R_p\cos\alpha$.

Numerical Example 6.12. An involute profile of the spur gear (m = 4 mm, N = 24, α = 20°) of length L = 50 mm is machined on rack-cutter machine. The tooth addendum and dedendum coefficients are h_a = 1.0 and h_d = 1.25, respectively. **Required:** Schematics of the involute segment of the gear profile in common with the rack-cutter tooth with $\varphi = 0$.

Solution:

(1) *Working side of the rack-cutter:* The position vector $\mathbf{r}_4$, Eq. (6.230), is

$$\mathbf{r}_4=\begin{bmatrix}x_4\\y_4\\z_4\\1\end{bmatrix}=\begin{bmatrix}u\sin 20^\circ\\u\cos 20^\circ\\z_4\\1\end{bmatrix};u_{min}\le u\le u_{max},\ 0\le z_4\le z_{rack},$$

$u_{min} = -h_a m/\cos\alpha = -1.25 \times 4/\cos 20° = -5.32089$ mm

$u_{max} = h_a m/\cos\alpha = 1.0 \times 4/\cos 20° = 4.2567$ mm

z_{rack} is the rack-cutter tooth length.

(2) *Constraints:*

- Kinematic constraint: $x = \varphi R_p = \varphi\, m\, N/2 = 4\varphi \times 24/2 = 48\varphi$
- Geometric constraint: $E = R_p = 48$ mm
- Enveloping constraint, Eq. (6.236): $u = \varphi R_p \sin\alpha = 48\varphi\sin 20° = 16.41697\varphi$.

(3) *Equation of the machined profile:* Substitution of vector $\mathbf{r}_4$ and constraints in Eq. (6.238) yields

$$\mathbf{r}_0 = \begin{bmatrix} R_p \sin\varphi - \varphi R_p \cos\alpha\cos(\alpha+\varphi) \\ R_p\cos\varphi + \varphi R_p \cos\alpha\sin(\alpha+\varphi) \\ Z \\ 1 \end{bmatrix} = \begin{bmatrix} 48[\sin\varphi - \varphi\cos\alpha\cos(\alpha+\varphi)] \\ 48[\cos\varphi + \varphi\cos\alpha\sin(\alpha+\varphi)] \\ Z \\ 1 \end{bmatrix} \quad (6.244)$$

$\varphi_b \le \varphi \le \varphi_{add}$ and $0 \le Z \le Z_{gear}$

(4) *Boundary angles* φ_b and φ_{add}, Eqs. (6.242) and (6.243):

$$\sqrt{1+\varphi_b^2\frac{(1+\cos 2\alpha)}{2}+2\varphi_b\sin 2\alpha} = \cos\alpha$$

$$\sqrt{1+\varphi_{add}^2\frac{(1+\cos 2\alpha)}{2}+2\varphi_{add}\sin 2\alpha} = 1 + 1/12$$

Solutions of these equations are

$\varphi_b = -\tan\alpha$

$\varphi_{add} = 0.2096886$ (the second solution $\varphi_{add} = -0.9376291$ is not suitable).

(5) For $\varphi = 0$, the instantaneous position of the rack tooth profile, Eq. (6.233), becomes

$$\mathbf{r}_0 = \begin{bmatrix} -\varphi R_p\cos\varphi + u\sin(\alpha+\varphi) + R_p\sin\varphi \\ \varphi R_p\sin\varphi + u\cos(\alpha+\varphi) + R_p\cos\varphi \\ Z \\ 1 \end{bmatrix} = \begin{bmatrix} u\sin\alpha \\ 48 + u\cos\alpha \\ Z \\ 1 \end{bmatrix}$$

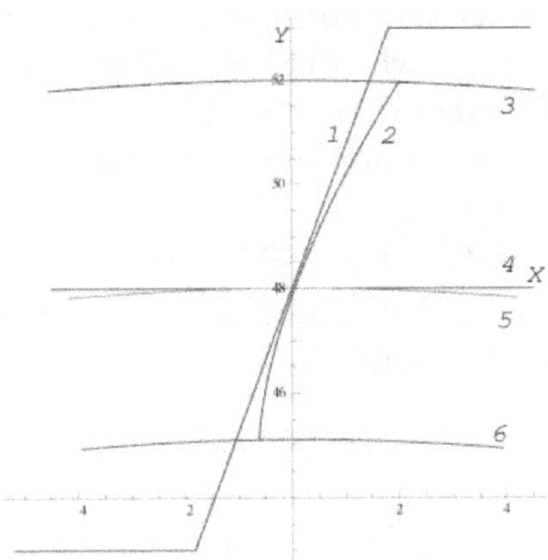

Figure 6-41 Involute profile (m = 4 mm, α = 20°) machined by the rack cutter: *1* – rack-cutter tooth profile; 2 – involute profile of the gear with teeth number N = 24; 3 – addendum circle; 4 – rack-cutter centrode; 5 – pitch circle (gear centrode); 6 – base circle.

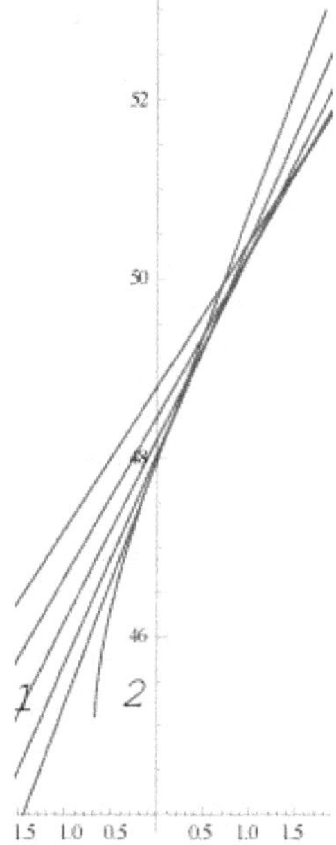

Figure 6-42 Five relative positions of the rack tooth *1* and machined profile 2.

(6) The involute profile 2 of the tooth side to be machined by rack tooth *1* (for $\varphi = 0$) is shown in the X_0Y_0 coordinate plane (Fig. 6.41). The machined gear tooth profile is confined between addendum circle 3 and base circle 6. The rack-cutter centrode 4 (straight line) and gear centrode 5 (pitch circle) are tangent at point $[x_0 = 0, y_0 = 48\text{ mm}]$.

(7) For illustration of the enveloping process, five positions of the rack tooth *1* relative to the machined profile 2 are shown in Figure 6.42. Remember that actually both kinematic motions – translation along the *X*-axis and rotation around the *Z*-axis – are performed by the rack cutter. Five positions of the rack tooth are shown for $\varphi = 0$, $\varphi_{add}/4$, $\varphi_{add}/2$, $3\varphi_{add}/4$, and φ_{add}, where $\varphi_{add} = 0.2096886$ rad.

(8) Length of the cutting pass entering into the FSF, Eq. (6.229), is $z = Z_{gear} + a + b$, where *a* and *b* are distances for entrance into and egress from the machined tooth.

6.5.2 Grinding a helical surface by tapered wheel

A layout of the machine for grinding helical surface is shown in Figure 6.43, and the links of the FSS are listed in Table 6.7. The coordinate code and the positional FSF are

$$k = \mathbf{63146} \tag{6.245}$$

$$\mathbf{r}_0 = \mathbf{A}^6(\varphi_1)\mathbf{A}^3(z)\,\mathbf{A}^1(a)\,\mathbf{A}^4(\Sigma)\mathbf{A}^6(\varphi_2)\,\mathbf{r}_5 \tag{6.246}$$

where $k_1 = 6$ corresponds to spindle rotation (feed motion) around the Z_0-axis associated with current angle φ_1; $k_2 = 3$ corresponds to the longitudinal feed along the Z_1-axis associated

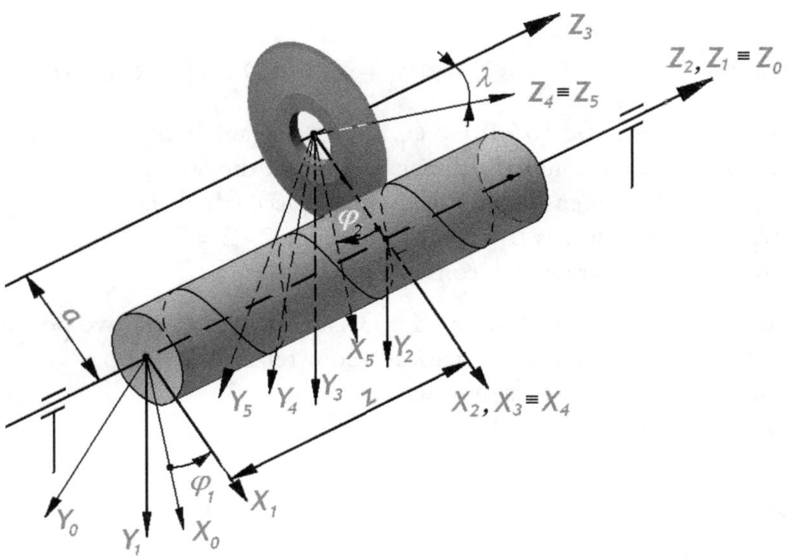

Figure 6-43 Coordinate system of thread-grinding machine with thin conical wheel.

Table 6-7 FSS of the thread grinding machine, k = 63146, K = 630146.

S_i	Link	Motion relative to previous link	Matrix ${}^{i-1}\mathbf{A}_i$
S_0	Machined screw	Conventionally fixed link	-
S_1	Workpiece table	Rotation along the Z_0-axis at angle φ_1	${}^0\mathbf{A}_1 = \mathbf{A}^6(-\varphi_1)$
S_2	Bed	Translation along the Z_1-axis at distance z	${}^1\mathbf{A}_2 = \mathbf{A}^3(-z)$
S_3	Grinding head	Translation along the X_2-axis at distance a	${}^2\mathbf{A}_3 = \mathbf{A}^1(a)$
S_4	Head drum	Swiveling around axis X_3-axis at angle Σ	${}^3\mathbf{A}_4 = \mathbf{A}^4(\Sigma)$
S_5	Wheel spindle	Rotation around the Z_4-axis at angle φ_2	${}^4\mathbf{A}_5 = \mathbf{A}^6(\varphi_2)$

with current distance z; $k_3 = 1$ corresponds to a displacement along the X_2-axis by distance a presenting the shortest distance between axes of the grinding wheel and the workpiece; $k_4 = 4$ corresponds to rotation around the X_3-axis through the thread helix angle Σ; $k_5 = 6$ corresponds to the grinding wheel rotation (primary motion)associated with current angle φ_2; and $\mathbf{r}_5$ is the position vector of the working points of the grinding wheel in coordinate frame S_5,

$$\mathbf{r}_5 = [R_w \cos\theta, R_w \sin\theta, H - R_w \tan\alpha, 1]^{\mathrm{T}} \tag{6.247}$$

R_w and θ is the polar radius and polar angle of the working surface of the wheel, respectively; H is the height of the cone; and α is the half angle of the cone.

Substitution vector $\mathbf{r}_5$, Eq. (6.247), into Eq. (6.246) yields the FSF

$$\mathbf{r}_0 = \begin{bmatrix} B\cos\varphi_1 - (C - D\sin\Sigma)\sin\varphi_1 \\ -B\sin\varphi_1 + (C - D\sin\Sigma)\cos\varphi_1 \\ -z + R_w \sin\Sigma\,\sin(\varphi_2 + \theta) + D\cos\Sigma \\ 1 \end{bmatrix} \tag{6.248}$$

with

$B = a + R_w \cos(\varphi_2 + \theta)$; $C = R_w \cos\Sigma \sin(\varphi_2 + \theta)$; and $D = H - R_w \tan\alpha$

There are seven variables in Eq. (6.248): $z, \varphi_1, a, \Sigma, \varphi_2, R_w$, and θ. In this case, the mobility number is $l = 5$ (five active motions), the dimensionality of the processing tool (the conical surface) is $M = 2$, and the dimensionality of the workpiece (the helical surface) is $m = 2$. Hence, the number of constraints is $L = n + M - m = 5 + 2 - 2 = 5$. Let us consider a setup for machining the cylindrical screw with constant pitch:

- There are three functional constraints ($L_f = 3$), among which are two positional ($a =$ Const and $\Sigma =$ Const) and one kinematical–screw relation $z = \pm p\varphi_1$;
- One implicit constraint ($L_i = 1$) $\varphi_2 + \theta = \varphi$;
- One enveloping constraint ($L_e = 1$)

$$\left| \frac{\partial r_0}{\partial \phi_1} \frac{\partial r_0}{\partial \phi} \frac{\partial r_0}{\partial R_w} \right| = 0 \tag{6.249}$$

After simplifications, equation (6.249) becomes

$$\cos\varphi\,(R_w \sec^2\alpha - H\tan\alpha) - \sin\varphi\,\tan\alpha\,(p + a\tan\Sigma) = p\tan\Sigma - a \tag{6.250}$$

There are two solutions of Eq. (6.250)

$$\varphi = \pm\left(\arccos\frac{D_3}{\sqrt{D_1^2 + D_2^2}} - \arctan\frac{D_2}{D_1}\right) \tag{6.251}$$

with

$D_1 = R_w \sec^2\alpha - H\tan\alpha$; $D_2 = \tan\alpha\,(p + a\tan\Sigma)$; and $D_3 = p\tan\Sigma - a$

Numerical Example 6.13 [8]. Grinding of the left side of the right hand thread of the K-worm [5]. **Given:** *Workpiece* of module $m = 4$ mm; tip and root cylinder diameters $D_c = 84$ mm and $D_r = 66$ mm, respectively; profile angle of the basic rack $\alpha = 12°$; the thread is a single-start, $z_1 = 1$; factor of tooth addendum depth $h_a = 1.0$; factor of total thread depth $h = 2.25$; and the worm threaded length grinding wheel $l = 70$ mm; *the grinding wheel* of external diameter $D_w = 300$ mm (Fig. 6.45). **Required:** Setup parameters.

Solution:

- The nearest distance between axes of the workpiece and the grinding wheel

$a = (D_w + D_r)/2 = (300 + 66)/2 = 183$ mm

- The factor of the pitch cylinder

$q = (D_c - 2h_a m)/(m z_1) = (84 - 2\times1\times4)/4 = 19$

- The helix angle for the pitch cylinder

$\Sigma = \arctan(z_1/q) = \arctan(1/19) = 0.052583$

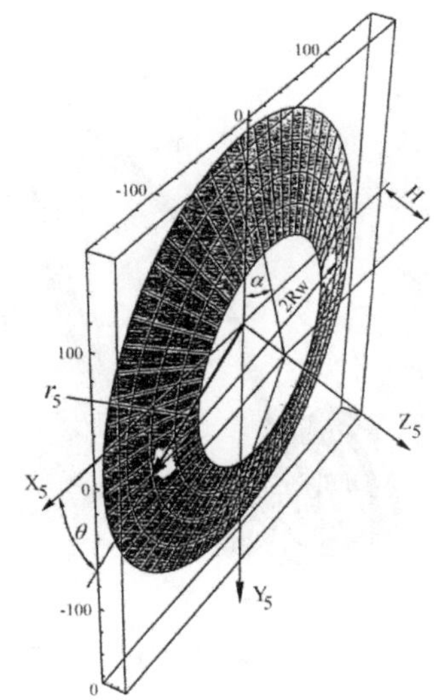

Figure 6-44 Conical grinding wheel with the associated coordinate frame S_5: $\mathbf{r}_5$ is the position vector of the current working point with the polar angle θ and radius *Rw*.

- The screw parameter $p = m\, z_1/(2 \cos \Sigma) = 2.0027$ mm

The FSF, Eq. (6.246), becomes

$$\mathbf{r}_0 = \mathbf{r}_0(\varphi_1, \varphi, R_w) = \mathbf{A}^6(\varphi_1)\mathbf{A}^3(\text{-}2.0027\ \varphi_1)\ \mathbf{A}^1(a)\ \mathbf{A}^4(\Sigma)\ \mathbf{A}^6(\varphi_2)\ \mathbf{r}_5(\varphi, R_w), \qquad (6.252)$$

$$141 \le R_w \le 150$$

Substituting enveloping condition $\varphi = \varphi(R_w)$, Eq. (6.251), into Eq. (6.252) gives the vector equation of the machined helical surface as $\mathbf{r}_0 = \mathbf{r}_0(\varphi_1, R_w)$, while fixing a value $\varphi_1 = \varphi_{1'fix}$ gives the current position of the grinding wheel $\mathbf{r}_w = \mathbf{r}_w(\varphi_{1'fix}, \varphi, R_w)$, see Figure 6.45. The relative position of the grinding tool and the left-hand side of the machined surface are shown in Figure 6.46 for $\varphi_{1'fix} = 6\ \varphi_1$.

6.6 Conclusion

The FSF theory-based approach is developed for modeling the machine setup problems. The model includes: (a) the machine hardware setup presenting an optional choice of the active

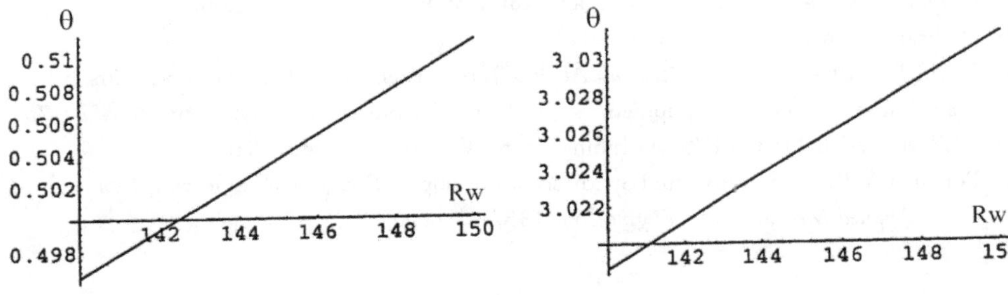

Figure 6-45 Two solutions of the tool-workpice contact conditions.

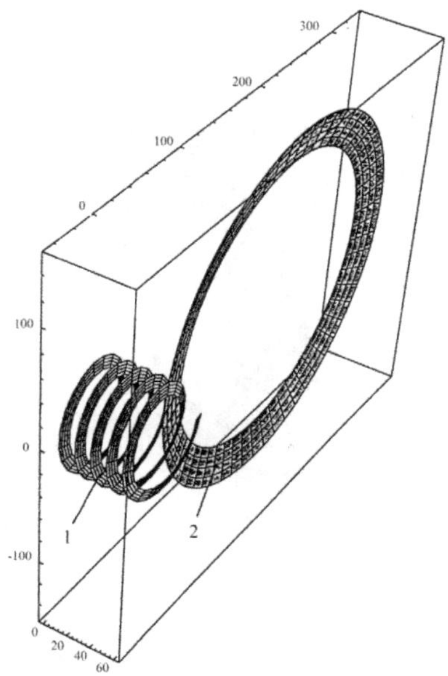

Figure 6-46 A current contact of the wheel *1* with the left side of the grinded thread *2*.

element (tool, probe, electrode, etc.) model in the form-shaping function; (b) the software setup fixing a set of the functional, enveloping and implicit constraints connecting parameters of the FSF; and (c) equaling the FSF to an equation of the target element (workpiece, assembly, welded connection, etc.) with following solution of the obtained simultaneous equations. In some cases, anyone step can be ignored. The number of constraints connecting parameters of the FSF and their type are defined through setup condition. Setup of the multi-axis machines and different machine tools are considered as application examples in this chapter. Setup of robots is considered in the next chapter.

References

[1] Freudenstein, F., 1973, Kinematics: Past, Presence, and Future, *Mechanism and Machine Theory*, **8**(2), pp. 151-161.

[2] Korn, G.A., Korn, T.M., 1968, *Mathematical Handbook for Scientists and Engineers*, 2nd ed., McGraw-Hill Book Co., New York, NY, pp. 101-102.

[3] Barles, R.H., Beary, J.C., and Barsky, B.A., 1987, *An Introduction to Splines for Use in Computer Graphics & Geometric Modeling*, Morgan Kaufmann Publishers, Burlington, MA.

[4] Wolfram Mathematica® 10 Online, 2017.

[5] Litvin, F.L.,1994, *Gear Geometry and Applied Theory*, Prentice-Hall, Englewood Cliffs, NJ.

[6] Hans Lapidary Tool and Equipment Co., Ltd, http://lapidarytec.en.ec21.com/, May, 2017.

[7] ALZMETALL, GS 600E/5 Machining Center, Catalog, Germany, 2016.

[8] Portman, V.T., 1996, Automated Synthesis of Enveloping Equations for Gearing, *Computer Methods in Applied Mechanics and Engineering*, **135**/1-2, pp. 63-83.

7. Kinematic setup of robots

7.1 Serial-type kinematics robots

7.1.1 Setup of the planar 3-DOF RRR manipulator

The FSF of the planar 3-DOF **RRR** robot-manipulator (Fig. 7.1) consists of six links ($n = 6$), among them three kinetmatic links rotating around the parallel axes and three geometric links fixing the distances between parallel axes. According to Eqs. (4.16-20), the FSF of the robot is

$$[\mathbf{r}_0\mathbf{c}_0] = {}^0\mathbf{A}_6\,[\mathbf{r}_6\mathbf{c}_6] \tag{7.1}$$

$${}^0\mathbf{A}_6 = \begin{bmatrix} \cos(\varphi_1+\varphi_2+\varphi_3) & -\sin(\varphi_1+\varphi_2+\varphi_3) & a_{14} \\ \sin(\varphi_1+\varphi_2+\varphi_3) & \cos(\varphi_1+\varphi_2+\varphi_3) & a_{24} \\ 0 & 0 & 1 \end{bmatrix} \tag{7.2}$$

$$a_{14} = a_1 \cos\varphi_1 + a_2 \cos(\varphi_1 + \varphi_2) + a_3 \cos(\varphi_1 + \varphi_2 + \varphi_3)$$

$$a_{24} = a_1 \sin\varphi_1 + a_2 \sin(\varphi_1 + \varphi_2) + a_3 \sin(\varphi_1 + \varphi_2 + \varphi_3) \tag{7.3}$$

$$\mathbf{r}_0 = [x_0, y_0, 1]^{\mathrm{T}} \tag{7.4}$$

$$\mathbf{r}_6 = [x_6, y_6, 1]^{\mathrm{T}} \tag{7.5}$$

$$\mathbf{c}_0 = [c_{0x}, c_{0y}, 0]^{\mathrm{T}} \tag{7.6}$$

$$\mathbf{c}_6 = [c_{6x}, c_{6y}, 0]^{\mathrm{T}} \tag{7.7}$$

where $\mathbf{r}_0$ and $\mathbf{r}_6$ are the position vectors of a point of the target element (line or point) in the coordinate frames S_0 and S_6, respectively; $\mathbf{c}_0$ and $\mathbf{c}_6$ are the unit vectors ($|\mathbf{c}_0| = |\mathbf{c}_6| = 1$) of the last link orientation in the same frames.

We consider the case, in which the robot executes the required operation by means of a single-point tool –the origin of the EE-associated with coordinate frameS_6 (Fig. 7.1) –and oriented along the link,

$$\mathbf{r}_6 = [x_6, y_6, 1]^{\mathrm{T}} = [0, 0, 1]^{\mathrm{T}} \tag{7.8}$$

$$\mathbf{c}_6 = [\cos\varphi_6, \sin\varphi_6, 0]^{\mathrm{T}} = [1, 0, 0]^{\mathrm{T}} \tag{7.9}$$

$$[\mathbf{r}_0\ \mathbf{c}_0] \equiv \begin{bmatrix} x_0 & c_{0x} \\ y_0 & c_{0y} \\ 1 & 0 \end{bmatrix} = \begin{bmatrix} \cos(\varphi_1+\varphi_2+\varphi_3) & -\sin(\varphi_1+\varphi_2+\varphi_3) & a_{14} \\ \sin(\varphi_1+\varphi_2+\varphi_3) & \cos(\varphi_1+\varphi_2+\varphi_3) & a_{24} \\ 0 & 0 & 1 \end{bmatrix} \begin{bmatrix} 0 & 1 \\ 0 & 0 \\ 1 & 0 \end{bmatrix} = \begin{bmatrix} a_{14} & \cos(\varphi_1+\varphi_2+\varphi_3) \\ a_{24} & \sin(\varphi_1+\varphi_2+\varphi_3) \\ 1 & 0 \end{bmatrix} \quad (7.10)$$

where a_{14} and a_{24} are defined through Eqs. (7.3).

The *TE* (processed element) is given in the form

$$\mathbf{r}_w = [x_w, y_w, 1]^{\mathrm{T}} \quad (7.11)$$

$$\mathbf{c}_w = [\cos\varphi_w, \sin\varphi_w, 0]^{\mathrm{T}} \quad (7.12)$$

The dimensionality of the *TE* is $m = 1$ (line-type link), dimensionality of the *AE* is $M = 0$ (point-type *AE*), and the mobility number is $l = 3$. Hence, the constraints number is $L = l + M - m + 1 = 3 + 0 - 1 + 1 = 3$. Two constraints are formulated by equating the right-hand sides of Eqs. (7.10) and (7.11):

$$a_{14} \equiv a_1 \cos\varphi_1 + a_2 \cos(\varphi_1 + \varphi_2) + a_3 \cos(\varphi_1 + \varphi_2 + \varphi_3) = x_w \quad (7.13)$$

$$a_{24} \equiv a_1 \sin\varphi_1 + a_2 \sin(\varphi_1 + \varphi_2) + a_3 \sin(\varphi_1 + \varphi_2 + \varphi_3) = y_w \quad (7.14)$$

The third constraint is obtained by equating the right-hand sides of Eqs. (7.10) and (7.12)

$$\cos(\varphi_1 + \varphi_2 + \varphi_3) = \cos\varphi_w \quad (7.15)$$

$$\sin(\varphi_1 + \varphi_2 + \varphi_3) = \sin\varphi_w \quad (7.16)$$

These two equations present one constraint whose solution is

$$\varphi_1 + \varphi_2 + \varphi_3 = \varphi_w \quad (7.17)$$

Substituting Eqs. (7.17) in Eqs. (7.13) and (7.14) yields

$$a_1 \cos\varphi_1 + a_2 \cos(\varphi_1 + \varphi_2) = x_w - a_3 \cos\varphi_w \quad (7.18)$$

$$a_1 \sin\varphi_1 + a_2 \sin(\varphi_1 + \varphi_2) = y_w - a_3 \sin\varphi_w \quad (7.19)$$

Summing the squares of Eqs. (7.18) and (7.19) yields

$$a_1^2 + a_2^2 + 2a_1a_2\cos\varphi_2 = (x_w - a_3\cos\varphi_w)^2 + (y_w - a_3\sin\varphi_w)^2 \quad (7.20)$$

Equation (7.20) results in two possible values φ_2

$$\varphi_2 = \pm\arccos u \quad (7.21)$$

$$u = \frac{x_w^2 + y_w^2 - a_1^2 - a_2^2 + a_3^2 - 2a_3(x_w\cos\varphi_w + y_w\sin\varphi_w)}{2a_1a_2}$$

The solution is possible if $-1 \le u \le 1$, i.e.,

$$(a_1 - a_2)^2 \le x_w^2 + y_w^2 + a_3^2 - 2a_3(x_w \cos\phi_w + y_w \sin\phi_w) \le (a_1 + a_2)^2 \tag{7.22}$$

To solve Eqs. (7.18) and (7.19) with respect to φ_1, one can represent them in the form

$$a_1 \cos\varphi_1 + a_2(\cos\varphi_1 \cos\varphi_2 - \sin\varphi_1 \sin\varphi_2) = x_w - a_3 \cos\varphi_w \tag{7.23}$$

$$a_1 \sin\varphi_1 + a_2(\sin\varphi_1 \cos\varphi_2 + \cos\varphi_1 \sin\varphi_2) = y_w - a_3 \sin\varphi_w \tag{7.24}$$

Hence,

$$(a_1 + a_2\cos\varphi_2)\cos\varphi_1 - (a_2\sin\varphi_2)\sin\varphi_1 = x_w - a_3\cos\varphi_w \tag{7.25}$$

$$(a_1 + a_2\cos\varphi_2)\sin\varphi_1 + (a_2\sin\varphi_2)\cos\varphi_1 = y_w - a_3\sin\varphi_w \tag{7.26}$$

Simultaneous Eqs. (7.25) and (7.26) can be solved with respect to $\sin\varphi_1$ and $\cos\varphi_1$ that yields the solution for angle φ_1 through atan2 function

$$\sin\varphi_1 = \frac{y_w(a_1 + a_2\cos\varphi_2) - a_1 a_3 \sin\varphi_w - a_2 a_3 \sin(\varphi_w - \varphi_2) - a_2 x_w \sin\varphi_2}{a_1^2 + a_2^2 + 2a_1 a_2 \cos\varphi_2} \tag{7.27}$$

$$\cos\varphi_1 = \frac{x_w(a_1 + a_2\cos\varphi_2) - a_1 a_3 \cos\varphi_w - a_2 a_3 \cos(\varphi_w - \varphi_2) + a_2 y_w \sin\varphi_2}{a_1^2 + a_2^2 + 2a_1 a_2 \cos\varphi_2} \tag{7.28}$$

$$\varphi_1 = \text{atan2}(\sin\varphi_1, \cos\varphi_1) \tag{7.29}$$

Finally, the angle φ_3 is calculated from Eq. (7.17)

$$\varphi_3 = \varphi_w - \varphi_1 - \varphi_2 \tag{7.30}$$

where φ_w is the given angle; and φ_1 and φ_2 are defined through Eqs. (7.29) and (7.21), respectively.

Obviously, there are two possible values φ_1 and φ_3 corresponding to two values φ_2, Eq. (7.21). Hence there are two possible configurations of the manipulator, which are chosen using the robot assembly requirements.

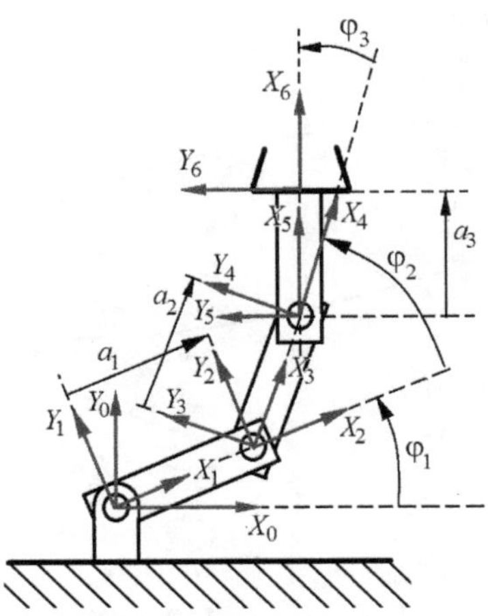

Figure 7-1 Planar 3-DOF (RRR) manipulator.

Table 7.1. Angles: Solution of *Numerical Example* 7.1.

x_0, m	φ_1	φ_2	φ_3
-0.2	1.3047 rad = 74.755°	1.4595 rad = 83.621°	-1.1934 rad = -68.375°
0	0.72973 rad = 41.810°	1.6821 rad = 96.379°	-0.84107 rad = -48.190°
0.2	0.37742 rad = 21.625°	1.45946 rad = 83.621°	-0.26608 rad = -15.245°
0.4	0.445561 rad = 25.528°	0.67967 rad = 38.942°	0.445561 rad = 25.529°

Numerical Example 7.1.

Given: $a_1 = a_2 = 0.3$ m, $a_3 = 0.2$ m. The working point at the end effector moves along the straight line parallel to the X_0-axis, $y_w = 0.6$, with $-0.2 \le x_w \le 0.4$ m. The gripper with an attached point-type tool is normally oriented relative to the given trajectory along the positive direction of the Y_0-axis as is shown in Figure 7.1, $\varphi_w = \pi/2$.

Required: Calculation of the angles φ_1, φ_2, and φ_3.

Solution:

(1) According to initial data, Eqs. (7.11) and (7.12) become

$$\mathbf{r}_w = [x_w, 0.6, 0, 1]^T, \text{ with } -0.2 \le x_w \le 0.4 \tag{7.31}$$

(2) Equations (7.15) and (7.16) yield

$$\varphi_1 + \varphi_2 + \varphi_3 = \varphi_w = \pi/2 \tag{7.32}$$

(3) Equation (7.21) yields

$$\varphi_2 = \pm\arccos u \tag{7.33}$$

$$u = \frac{x_w^2 + y_w^2 - a_1^2 - a_2^2 + a_3^2 - 2a_3(x_w \cos\phi_w + y_w \sin\phi_w)}{2a_1a_2}$$

$$= \frac{x_w^2 + 0.6^2 - 0.3^2 - 0.3^2 + 0.2^2 - 2\times 0.3\times 0.6}{2\times 0.3\times 0.3} = 5.55556x_w^2 - 0.11111 \tag{7.34}$$

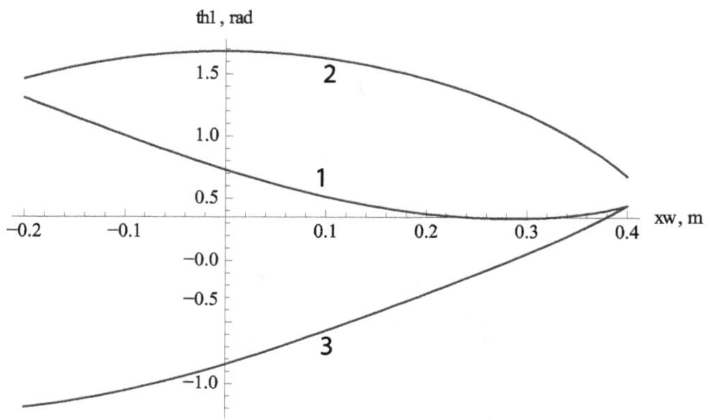

Figure 7-2 Angles φ_1, φ_2, and φ_3 (curves 1, 2, and 3, respectively) vs. x_w during motion along the line $y_0 = 0.6$ m with the normally oriented last link.

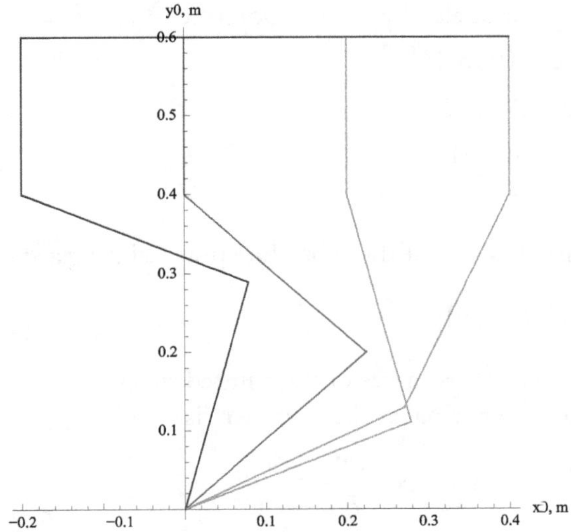

Figure 7-3 Four configurations of 3-RRR planar robot during motion along the straight line $y_w = 0.6$ m with the normally oriented last link.

As examples, angle φ_2 values for two boundaries are given in the following table.

$x_w = -0.2$ m $\quad u = 0.11111 \quad \varphi_2 = \pm 1.45946$ rad

$x_w = 0.4$ m $\quad u = 0.77778 \quad \varphi_2 = \pm 0.679674$ rad

Obviously, both u value boundaries meet Eq. (7.22), and then Eq. (7.33) has real solutions. Furthermore, Eqs. (7.27–29) yield the angle φ_1, and Eq. (7.30) yields the angle φ_3. As an example, the robot assembly defined by positive value $\varphi_2 = \arccos u$, Eq. (7.21), is considered. For this assembly, the values of angles φ_1, φ_2, and φ_3 are shown in Figure 7.2. Four robot configurations corresponding to x_w values listed in Table 7.1 are shown in Figure 7.3.

7.1.2 Spherical robot

The FSF of the spherical robot, Eq. (4.36), is

$$\mathbf{r}_0 = {}^0\mathbf{A}_4 \mathbf{r}_4 \tag{7.35}$$

$${}^0\mathbf{A}_4 = \begin{bmatrix} \cos\varphi & -\cos\theta\sin\varphi & \sin\theta\sin\varphi & -\rho\cos\theta\sin\varphi \\ \sin\varphi & \cos\theta\cos\varphi & -\sin\theta\cos\varphi & \rho\cos\theta\cos\varphi \\ 0 & \sin\theta & \cos\theta & h+\rho\sin\theta \\ 0 & 0 & 0 & 1 \end{bmatrix} \tag{7.36}$$

$$\mathbf{r}_0 = \begin{bmatrix} x_0 \\ y_0 \\ z_0 \\ 1 \end{bmatrix} \text{ and } \mathbf{r}_4 = \begin{bmatrix} x_4 \\ y_4 \\ z_4 \\ 1 \end{bmatrix} \tag{7.37}$$

If the active element is modeled by a single point ($x_4 = y_4 = z_4 = 0$), the FSF becomes

$$\mathbf{r}_0 = {}^0\mathbf{A}_4\,\mathbf{r}_4 = \begin{bmatrix} -\rho\cos\theta\sin\Phi \\ \rho\cos\theta\cos\Phi \\ h+\rho\sin\theta \\ 1 \end{bmatrix} \tag{7.38}$$

As an example, consider setup of the robot for spray painting a surface given by equation

$$\mathbf{r}_w = [x_w(u, v), y_w(u, v), z_w(u, v)]^T, \tag{7.39}$$

where u, v are the Gaussian coordinates of the painted surface.

The unit vector $\mathbf{n}_w$ of the normal at the surface $\mathbf{r}_w$ is

$$\mathbf{n}_w = \mathbf{N}_w / (\mathbf{N}_w \cdot \mathbf{N}_w)^{1/2}, \text{ with } \mathbf{N} = \frac{\partial \mathbf{r}_w}{\partial u} \times \frac{\partial \mathbf{r}_w}{\partial v} \tag{7.40}$$

The painting is worked out by scanning the surface to be painted under the following conditions:

- The central point of the nozzle face is considered as a point-type active element;
- The nozzle is offset from the painted point by face $L_{off} = 150 \div 250$ mm;.
- The paint sprayer axis is attached to the arm-mounted gripper and oriented along the positive direction of the Y_4-axis (Fig. 7.4);
- The sprayer axis is oriented to a current painted point, and the paint stream axis holds the straight form.

Under these conditions, the aiming point located at the center of the nozzle face lies on a beam originated from the sphere center O_1 (the origin of the frame $X_1Y_1Z_1$) to the painted point $\mathbf{r}_w$, Eq. (7.39). The distance (radius) d_w of the painted point from the sphere center and the unit vector $\mathbf{e}_b$ of the beam direction are

$$d_w = \sqrt{x_w^2 + y_w^2 + (z_w - h)^2} \tag{7.41}$$

$$\mathbf{e}_b = \frac{1}{d_w}\begin{bmatrix} x_w \\ y_w \\ z_w - h \end{bmatrix} = \frac{1}{\sqrt{x_w^2 + y_w^2 + (z_w - h)^2}}\begin{bmatrix} x_w \\ y_w \\ z_w - h \end{bmatrix} \tag{7.42}$$

The beam equation is as follows:

$$\mathbf{r}_b = \mathbf{r}_{O1} + t_b\mathbf{e}_b = \begin{bmatrix} 0 \\ 0 \\ h \end{bmatrix} + \frac{t_b}{d_w}\begin{bmatrix} x_w \\ y_w \\ z_w - h \end{bmatrix} = \begin{bmatrix} x_w t_b / d_w \\ y_w t_b / d_w \\ h + (z_w - h)t_b / d_w \end{bmatrix} \tag{7.43}$$

where t_b is the line parameter, $0 \le t_b \le 1$.

Hence, the setup equations become

$$\rho = d_w - L_{off} = \sqrt{x_w^2 + y_w^2 + (z_w - h)^2} - L_{off} \tag{7.44}$$

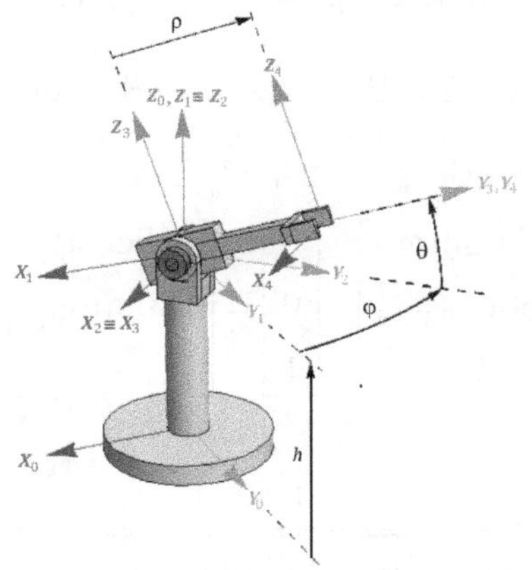

Figure 7-4 Spherical robot: links and associated coordinate frames; h, φ, θ, and ρ are the FSF parameters.

$$\theta = \arcsin \frac{z_w - h}{\sqrt{x_w^2 + y_w^2 + (z_w - h)^2}} \tag{7.45}$$

$$\Phi = -\arctan \frac{x_w}{y_w} \tag{7.46}$$

where ρ, θ, and φ are shown in Figure 7.4.

Obviously, the spherical robot is best suited to painting of the internal spherical surfaces if its radius is more than the minimal overhang ρ and the center can be brought into coincidence with the robot center. In this case, the arm axis direction is identical with the normal to the painted surface, and the paint stream is permanently coincide with the arm axis. Angle Λ between the arm (and paint stream) axis direction and the normal of the painted surface at the current painted point has to satisfy the painting requirement limits. The angle is defined through the unit vector $\mathbf{n}_w$, (Eq. 7.6), of the normal and the unit vector $\mathbf{e}_b$, Eq. (7.8), of the arm direction,

$$\Lambda = \arccos\left(\mathbf{n}_w \cdot \mathbf{e}_b\right) \tag{7.47}$$

Numerical Example 7.2.

Given: Sphere center of the spherical robot is at height h = 400 mm (Fig. 7.4); distance from the current painted point is L_{off}= 200 mm. The robot is used for spray painting of a fragment of the internal surface of an elliptical cylinder

$$\mathbf{r}_w = [x_c + R_{\max} \cos \varphi_w, y_c + R_{\min} \sin \varphi_w, z_c + H_w]^T \tag{7.48}$$

where R_{max} and R_{min} are the major and minor half-axes of the ellipse, R_{max} = 180 and R_{min} = 160 mm; x_c, y_c, and z_c are the coordinates of the ellipse center in the cross-section plane, x_c = 0, y_c= 440, and z_c=h= 400 mm; φ_w is the polar angle; and H_w is the z-distance from z_c.

The painted fragment is bounded by

$$\pi/3 \le \varphi_w \le 2\pi/3 \text{ and } -100 \le H_w \le 100 \text{ mm} \tag{7.49}$$

Required: Calculation of the overhang ρ and angles θ, Φ, and Λ, Eq. (7.47).
Solution:
Equations of the painted surface, Eq. (7.48), and the normal, Eq. (7.40), are

$$\mathbf{r}_w = [180\cos\varphi_w, 440 + 160\sin\varphi_w, 400 + H_w]^T \tag{7.50}$$

$$\mathbf{N}_w = \frac{\partial \mathbf{r}_w}{\partial \varphi_w} \times \frac{\partial \mathbf{r}_w}{\partial z_w} = \begin{bmatrix} -180\sin\varphi_w \\ 160\cos\varphi_w \\ 0 \end{bmatrix} \times \begin{bmatrix} 0 \\ 0 \\ 1 \end{bmatrix} = \begin{bmatrix} 160\cos\varphi_w \\ 180\sin\varphi_w \\ 0 \end{bmatrix}$$

$$\mathbf{n}_w = \mathbf{N}_w / \left(\mathbf{N}_w \cdot \mathbf{N}_w\right)^{1/2} = \frac{1}{\sqrt{25600 + 6800\sin^2\varphi_w}} \begin{bmatrix} 160\cos\varphi_w \\ 180\sin\varphi_w \\ 0 \end{bmatrix} \tag{7.51}$$

(2) The distance of the painted point from the sphere center, Eq. (7.41), and the unit vector of arm orientation, Eq. (7.42), are

$$d = \sqrt{x_w^2 + y_w^2 + (z_w - h)^2} = \sqrt{x_w^2 + y_w^2 + H_w^2} \tag{7.52}$$

$$\mathbf{e}_b = \frac{1}{d}\begin{bmatrix} x_w \\ y_w \\ z_w - h \end{bmatrix} = \frac{1}{d}\begin{bmatrix} x_w \\ y_w \\ H_w \end{bmatrix} \tag{7.53}$$

(3) The setup equations are

$$\rho = d_w - L_{off} = \sqrt{x_w^2 + y_w^2 + H_w^2} - L_{off} = -200 + \sqrt{x_w^2 + y_w^2 + H_w^2} \tag{7.54}$$

$$\theta = \arcsin\frac{z_w - h}{\sqrt{x_w^2 + y_w^2 + (z_w - h)^2}} = \arcsin\frac{H_w}{\sqrt{x_w^2 + y_w^2 + H_w^2}} \tag{7.55}$$

$$\Phi = -\arctan\frac{x_w}{y_w} \tag{7.56}$$

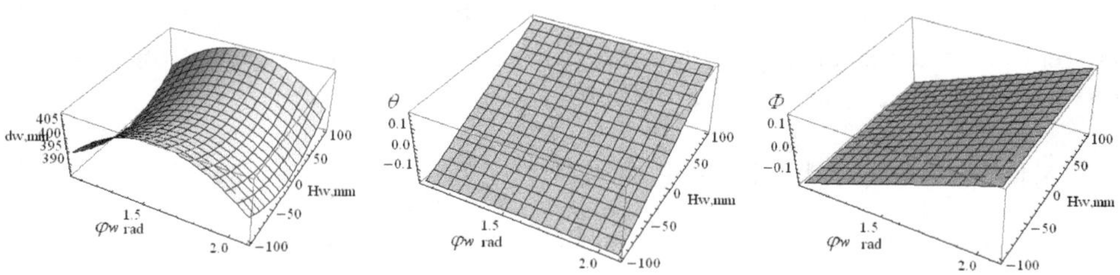

(a) (b) (c)

Figure 7-5 Setup for painting vs. angle φ_w and height H_w: (a) arm overhang ρ; (b) and (c) orientation angles θ and Φ, respectively.

In Figure 7.5, the setup parameters – the arm overhang ρ, Eq. (7.54), and the angles θ and Φ are shown vs. angle φ_w and height H_w.

(4) Consider the robot setup when the paint stream processes two points of the surface $\mathbf{r}_w$, Eq. (7.50): the central point O_1 and one of the points farthest from the center.

(4a) Point O_1: $H_w = 0$ and $\varphi_w = \pi/2$. Thus, coordinates of this point are

$$x_{wc} = x_c + R_{max} \cos \varphi_w = 0 + 180 \cos \pi/2 = 0$$
$$y_{wc} = y_c + R_{min} \sin \varphi_w = 440 + 160 \sin \pi/2 = 600;$$
$$z_{wc} = z_c + H_w = 400 \text{ mm} + 0 = 400 \text{ mm}$$

The corresponding setup parameters, Eqs. (7.54-56), are

$$d_w = \sqrt{x_w^2 + y_w^2 + H_w^2} = 600 \text{ mm};$$

$$\rho = d_w - L_{off} = 600 - 200 = 400;$$

$$\theta = \arcsin \frac{H_w}{\sqrt{x_w^2 + y_w^2 + H_w^2}} = 0;$$

$$\Phi = -\arctan \frac{x_w}{y_w} = 0$$

In the central point, setup parameters are

$$d = \sqrt{x_w^2 + y_w^2 + H_w^2} = \sqrt{0^2 + 600^2 + (400 - 400)^2} = 600 \text{ mm}$$

$$\rho = d - L_{off} = 600 - 200 = 400$$

$$\theta = \arcsin \frac{H_w}{\sqrt{x_w^2 + y_w^2 + H_w^2}} = 0$$

$$\Phi = -\arctan \frac{x_w}{y_w} = 0$$

$$\mathbf{e}_b = \frac{1}{d} \begin{bmatrix} x_w \\ y_w \\ H_w \end{bmatrix} = \frac{1}{600} \begin{bmatrix} 0 \\ 600 \\ 0 \end{bmatrix} = \begin{bmatrix} 0 \\ 1 \\ 0 \end{bmatrix}$$

$$\mathbf{n}_w = \frac{1}{\sqrt{25600 + 6800 \sin^2 \varphi_w}} \begin{bmatrix} 160 \cos \varphi_w \\ 180 \sin \varphi_w \\ 0 \end{bmatrix} = \frac{1}{\sqrt{25600 + 6800}} \begin{bmatrix} 0 \\ 180 \\ 0 \end{bmatrix} = \begin{bmatrix} 0 \\ 1 \\ 0 \end{bmatrix}$$

$$\Lambda = \arccos (\mathrm{n}_w \cdot \mathrm{e}_b) = \arccos 1 = 0$$

Hence, in the central point, the vectors $\mathbf{n}_w$ and $\mathbf{e}_b$ are coincide, and the best painting conditions are there.

(4b) Coordinates of one of the points farthest from the centre ($H_{wf} = 100$ and $\varphi_{wf} = \pi/3$) are

$x_{wf} = x_c + R_{max} \cos \varphi_{wf} = 0 + 180 \cos \pi/3 = 90;$

$y_{wf} = y_c + R_{min} \sin \varphi_{wf} = 440 + 160 \sin \pi/3 = 578.564$

$z_{wf} = z_c + H_{wf} = 400 + 100 = 500$ mm

The setup parameters:

$$d_{wf} = \sqrt{x_w^2 + y_w^2 + H_w^2} = \sqrt{90^2 + 578.564^2 + 100^2} = 594 \text{ mm}$$

$$\rho = d_w - L_{off} = 594 - 200 = 394 \text{ mm}$$

$$\theta_f = \arcsin\frac{H_w}{d} = \arcsin\frac{100}{594} = 0.169656 = 9.692^\circ$$

$$\Phi_f = -\arctan\frac{x_w}{y_w} = -\arctan\frac{90}{578.564} = -0.15432 = -8.842^\circ$$

$$\mathbf{e}_{bf} = \frac{1}{d_w}\begin{bmatrix} x_w \\ y_w \\ H_w \end{bmatrix} = \frac{1}{594}\begin{bmatrix} 90 \\ 578.564 \\ 100 \end{bmatrix} = \begin{bmatrix} 0.151515 \\ 0.974013 \\ 0.16835 \end{bmatrix}$$

$$\mathbf{n}_{wf} = \frac{1}{\sqrt{25600 + 6800\sin^2\frac{\pi}{3}}}\begin{bmatrix} 160\cos\frac{\pi}{3} \\ 180\sin\frac{\pi}{3} \\ 0 \end{bmatrix} = \begin{bmatrix} 0.456584 \\ 0.88968 \\ 0 \end{bmatrix}$$

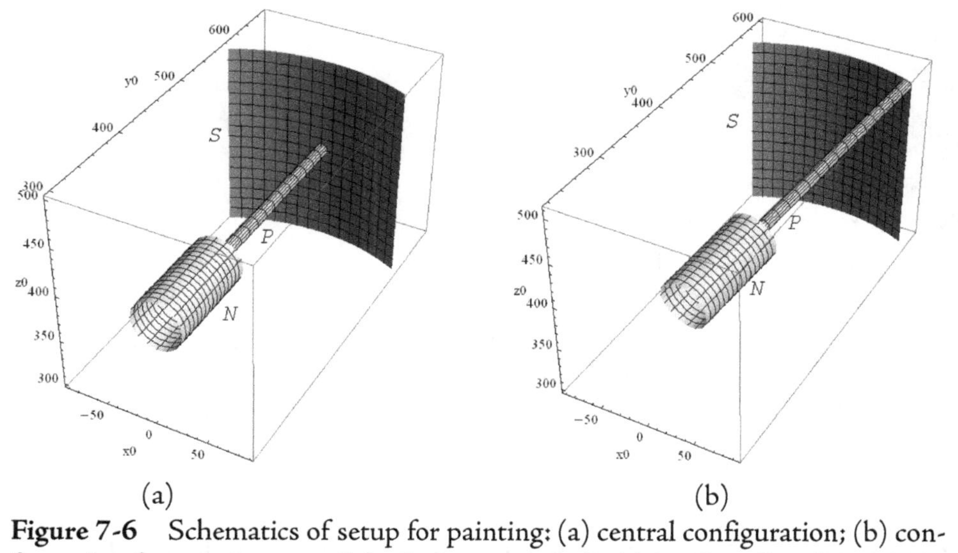

(a) (b)

Figure 7-6 Schematics of setup for painting: (a) central configuration; (b) configuration for painting one of the farthest points. S – painted surface; P – painting stream; and N – nozzle.

$$\Lambda_f = \arccos\left(\mathbf{n}_{wf} \cdot \mathbf{e}_{bf}\right) = \arccos 0.93574 = 0.36045 \text{ rad} = 20.65^\circ$$

Hence, in the farthest point, the angle between vectors $\mathbf{n}_{wf}$ and $\mathbf{e}_{bf}$ is about 20 degs, and it is necessary to check whether varying the angle from 0 to 20° meets painting conditions. The general view of the setup (Fig. 7.6) schematizes the relative positions of the painted surface *S*, nozzle *N*, and paint stream *P* during painting process for two current configurations: painting in the area of the central (Fig. 7.6a) and farthest (Fig. 7.6b) points.

7.1.3 Setup of the Scorbot robot

The structure of the 5-DOF Scorbot robot combines a fixed column and mobile links connected together in series by five revolute joints (Fig. 7.7). The setup problem includes calculations of current values of five angles: angle φ of body rotation relative to the base; angle ψ_1 of shoulder swivel; angle ψ_2 of elbow rotation; angle ψ_3 of the wrist pitch; and angle θ of the wrist roll. Since there is no yaw joint in the wrist, the gripper axis is constantly fixed in the plane parallel to the plane of arm motion. Similar to other 5-DOF robots, the Scorbot robot is useful for spray painting, resistance and arc welding, sealant applications, and so on.

As an example, a setup for resistance spot welding is considered (Fig. 7.8). For this joining operation, a spot-welding gun is held in the gripper. The welded sheets of thickness 0.5÷3

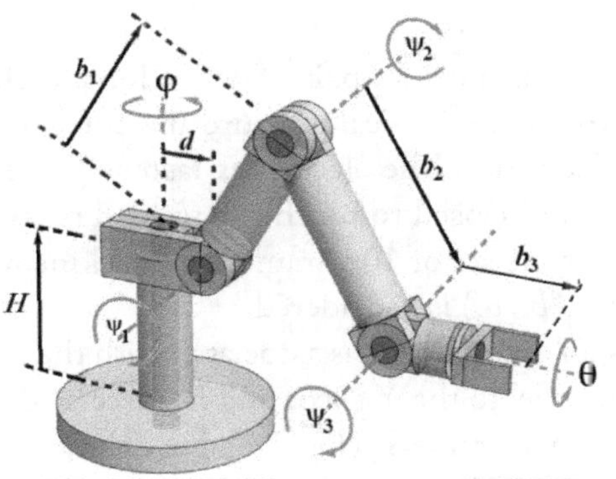

(a)

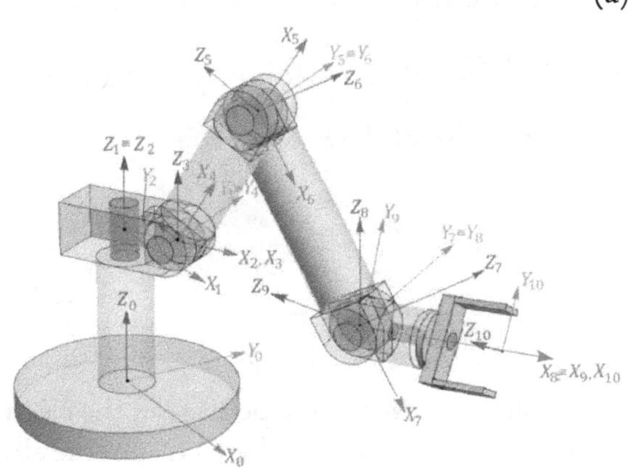

(b)

Figure 7-7 Scorbot arm robot: (a) kinematic parameters; (b) coordinate frames.

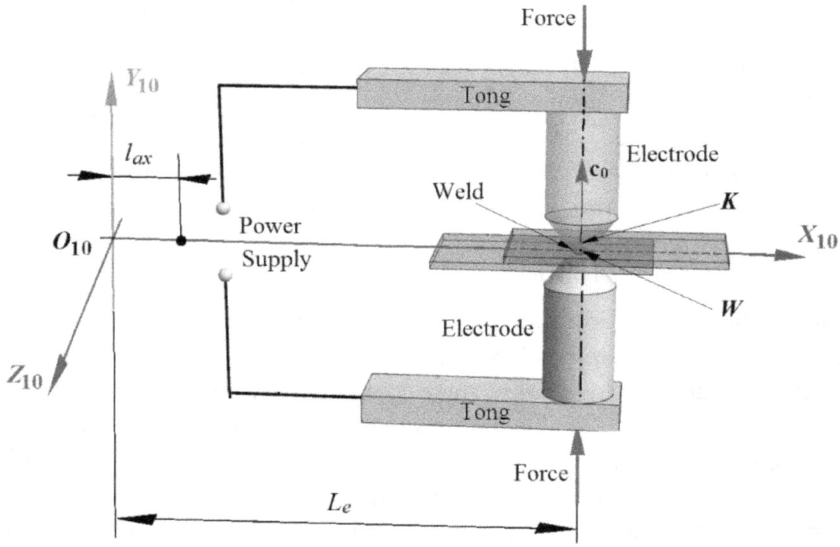

Figure 7-8 The resistance spot welding scheme: W is the welding center; K is the point on the working end of the upper electrode axis; and $\mathbf{c}_0$ is the unit vector of the electrode axis.

mm are pressed together by means of a pair of active forces, which are equal in value and opposite in direction. Both the mechanical pressure and current are supplied through a pair of shaped copper alloy electrodes. The electrodes fastened with the tongs can be pushed apart a necessary distance and closed to pinch the welded parts with required (reasonably high) force. Remember that the set of 10 parameters (five kinematic angles φ, ψ_1, ψ_2, ψ_3, θ, and five dimensions H, d, b_1, b_2, b_3) is considered.

The gripper with the spot-welding gun is associated with the coordinate frame S_{10}, and the electrode axis is pointed parallel to the Y_{10}-axis and is offset by distance L_e along the X_{10}-axis from the origin of the coordinate frame S_{10} (Fig. 7.8). For definiteness, only parameters describing one electrode of the pair are calculated. The robot manipulation matrix,Eq. (4.32), is

$$^0\mathbf{A}_{10} = \mathbf{A}^3(H)\,\mathbf{A}^6(\varphi)\,\mathbf{A}^1(d)\,\mathbf{A}^5(\psi_1)\,\mathbf{A}^1(a_1)\,\mathbf{A}^5(\psi_2)\mathbf{A}^1(a_2)\,\mathbf{A}^5(\psi_3)\mathbf{A}^4(\theta)\,\mathbf{A}^1(a_3) \tag{7.57}$$

For the considered setup, the FSF of the Scorbot robot is

$$[\mathbf{r}_{0K}\mathbf{c}_0] = {}^0\mathbf{A}_{10}\,[\mathbf{r}_{10K}\mathbf{c}_{10}] \tag{7.58}$$

$$\mathbf{r}_{10K} = [L_e, 0, 0, 1]^{\mathrm{T}} \tag{7.59}$$

$$\mathbf{c}_{10} = \pm[0,\ 1,\ 0,\ 0]^{\mathrm{T}} \tag{7.60}$$

where $\mathbf{r}_{0K}$ and $\mathbf{r}_{10K}$ are the position vectors of the point K of the electrode axis in coordinate frames S_0 and S_{10}, respectively; and $\mathbf{c}_0$ and $\mathbf{c}_{10}$ are the vectors of the electrode axis orientation in coordinate frames S_0 and S_{10}, respectively.

Plus or minus sign for $\mathbf{c}_{10}$, Eq. (7.60), describe the upper and lower electrodes, respectively. Only the electrode associated with the plus sign is considered.

Subsitution of the vector $\mathbf{r}_{10K}$, Eq. (7.59), and vectors $\mathbf{c}_{10K}$, Eq. (7.60), in the FSF function, Eq. (7.58), yields

$$\mathbf{r}_{0K} = {}^{0}\mathbf{A}_{10}\,\mathbf{r}_{10K} = \begin{bmatrix} B_1 \cos\varphi \\ B_1 \sin\varphi \\ B_2 \\ 1 \end{bmatrix} \tag{7.61}$$

$$\mathbf{c}_0 = {}^{0}\mathbf{A}_{10}\,\mathbf{c}_{10K} = \begin{bmatrix} -\sin\varphi\cos\theta + \cos\varphi\ \sin\psi_{123}\sin\theta \\ \cos\varphi\cos\theta + \sin\varphi\ \sin\psi_{123}\sin\theta \\ \cos\psi_{123}\sin\theta \\ 0 \end{bmatrix} \tag{7.62}$$

with

$$\psi_{12} = \psi_1 + \psi_2;\ \psi_{123} = \psi_1 + \psi_2 + \psi_3 \tag{7.63}$$

$$B_1 = d + b_1 \cos\psi_1 + b_2 \cos\psi_{12} + (b_3 + L_e)\cos\psi_{123} \tag{7.64}$$

$$B_2 = H - b_1 \sin\psi_1 - b_2 \sin\psi_{12} - (b_3 + L_e)\sin\psi_{123} \tag{7.65}$$

where φ is the current angle of rotation of the body relative to the column; ψ_1, ψ_2, ψ_3 are the current angles of rotation of the arm, forearm, and gripper around the Y-axes of the shoulder, elbow, and wrist pitch, respectively; and b_1, b_2, and b_3 are the axis-to-axis distances along the X-axes of the arm, forearm, and gripper, respectively.

Solution of the inverse kinematics problem with respect to five unknown angles (φ, ψ_1, ψ_2, ψ_3, and θ) is obtained by equating the computed vectors $\mathbf{r}_{0K}$, Eqs. (7.61), and $\mathbf{c}_0$, Eq. (7.62), respectively, to the given vectors $\mathbf{r}_w$ and $\mathbf{c}_w$, respectively,

$$\mathbf{r}_{0K} = \mathbf{r}_w \tag{7.66}$$

$$\mathbf{c}_0 = \mathbf{c}_w \tag{7.67}$$

Both $\mathbf{r}_w$ and $\mathbf{c}_w$ are given in the coordinate frame S_0. The position vector $\mathbf{r}_W$ of the point W is associated with the current center of the weld and $\mathbf{c}_w$ defines the electrode orientation

$$\begin{bmatrix} B_1 \cos\varphi \\ B_1 \sin\varphi \\ B_2 \\ 1 \end{bmatrix} = \begin{bmatrix} x_w \\ y_w \\ z_w \\ 1 \end{bmatrix} \tag{7.68}$$

$$\begin{bmatrix} -\sin\varphi\cos\theta + \cos\varphi\ \sin\psi_{123}\sin\theta \\ \cos\varphi\cos\theta + \sin\varphi\ \sin\psi_{123}\sin\theta \\ \cos\psi_{123}\sin\theta \\ 0 \end{bmatrix} = \begin{bmatrix} c_{wx} \\ c_{wy} \\ c_{wz} \\ 0 \end{bmatrix} \tag{7.69}$$

To simplify the solution procedure, it is worth transferring the vectors, Eqs. (7.68) and (7.69), in the coordinate frame of the arm. For this purpose, both sides of Eqs. (7.68-69) have to be multiplied by the matrix

$$({}^{1}\mathbf{A}_2)^{-1} = \mathbf{A}^6(-\varphi) = \begin{bmatrix} \cos\varphi & \sin\varphi & 0 & 0 \\ -\sin\varphi & \cos\varphi & 0 & 0 \\ 0 & 0 & 1 & 0 \\ 0 & 0 & 0 & 1 \end{bmatrix}$$

Hence,

$\mathbf{A}^6(-\varphi)\,[\mathbf{r}_{0K}] = \mathbf{A}^6(-\varphi)[\mathbf{r}_K]$
$\mathbf{A}^6(-\varphi)\,\mathbf{c}_0 = \mathbf{A}^6(-\varphi)\mathbf{c}_w$
i.e.,

$$\begin{bmatrix} B_1 \\ 0 \\ B_2 \\ 1 \end{bmatrix} = \begin{bmatrix} x_w\cos\varphi + y_w\sin\varphi \\ y_w\cos\varphi - x_w\sin\varphi \\ z_w \\ 1 \end{bmatrix} \tag{7.70}$$

$$\begin{bmatrix} \sin\psi_{123}\sin\theta \\ \cos\theta \\ \cos\psi_{123}\sin\theta \\ 0 \end{bmatrix} = \begin{bmatrix} c_{wx}\cos\varphi + c_{wy}\sin\varphi \\ c_{wy}\cos\varphi - c_{wx}\sin\varphi \\ c_{wz} \\ 0 \end{bmatrix} \tag{7.71}$$

Vector equations (7.70) and (7.71) represent six non-trivial scalar simultaneous equations. Solution with respect to angles φ, ψ_1, ψ_2, ψ_3, θ is carried out as follows:
Step 1: The second elements of vector Eq. (7.70) yields
$\tan\varphi = y_w/x_w$
i.e.,

$$\varphi = \begin{cases} \arctan(y_w/x_w) \\ \pi + \arctan(y_w/x_w) \end{cases} \tag{7.72}$$

Step 2: The second row of the vector equation (7.71) yields

$$\cos\theta = c_{wy}\cos\varphi - c_{wx}\sin\varphi$$
$$\theta = \pm\arccos(c_{wy}\cos\varphi - c_{wx}\sin\varphi) \tag{7.73}$$

Step 3: For each of four combinations of the values φ and θ, the first and third elements of the vector equation (7.71) give

$$\sin\psi_{123} = (c_{wx}\cos\varphi + c_{wy}\sin\varphi)/\sin\theta;$$
$$\cos\psi_{123} = c_{wz}/\sin\theta \tag{7.74}$$

Hence,

$\psi_{123} = \text{atan2}(y, x) = \text{atan2}(\sin \psi_{123}, \cos \psi_{123})$

where atan2(y, x) is the function of two arguments (Appendix 1).
Step 4: Calculation of angle ψ_2. Taking into account expressions (7.64-65) for B_1 and B_2, we can represent the first and third elements of the vector equation (7.70) as two equations with respect to angles ψ_1 and ψ_{12}:

$$B_1 = d + b_1 \cos \psi_1 + b_2 \cos \psi_{12} + (b_3 + L_e) \cos \psi_{123} = x_w \cos \varphi + y_w \sin \varphi$$

$$B_2 = H - b_1 \sin \psi_1 - b_2 \sin \psi_{12} - (b_3 + L_e) \sin \psi_{123} = z_w \tag{7.75}$$

i.e.,

$$b_1 \cos \psi_1 + b_2 \cos \psi_{12} = F \tag{7.76}$$

$$-b_1 \sin \psi_1 - b_2 \sin \psi_{12} = G \tag{7.77}$$

with

$$F = x_w \cos \varphi + y_w \sin \varphi - d - (b_3 + L_e) \cos \psi_{123}$$

$$G = z_w - H + (b_3 + L_e) \sin \psi_{123} \tag{7.78}$$

Since $\psi_{12} = \psi_1 + \psi_2$, Eqs. (7.76) and (7.77) become

$$b_1 \cos \psi_1 + b_2 (\cos \psi_1 \cos \psi_2 - \sin \psi_1 \sin \psi_2) = F \tag{7.79}$$

$$-b_1 \sin \psi_1 - b_2 (\sin \psi_1 \cos \psi_2 + \cos \psi_1 \sin \psi_2) = G \tag{7.80}$$

This system can be solved with respect to angle ψ_2 by summing the squares of their parts

$$b_1^2 + b_2^2 + 2b_1b_2 \cos\psi_2 = F^2 + G^2$$

$$\psi_2 = \pm\arccos\frac{F^2 + G^2 - b_1^2 - b_2^2}{2b_1b_2} \tag{7.81}$$

Step 5: Calculation of ψ_1. Since angle ψ_2 is obtained, Eqs. (7.79) and (7.80) can be represented as two simultaneous equations and solved with respect to $\sin \psi_1$ and $\cos \psi_1$:

$$(b_1 + b_2 \cos \psi_2) \cos \psi_1 - (b_2 \sin \psi_2) \sin \psi_1 = F$$

$$-(b_2 \sin \psi_2) \cos \psi_1 - (b_1 + b_2 \cos \psi_2) \sin \psi_1 = G$$

$$\cos\psi_1 = \frac{F(b_1 + b_2 \cos\psi_2) - Gb_2\sin\psi_2}{b_1^2 + b_2^2 + 2b_1b_2 \cos\psi_2} \tag{7.82}$$

$$\sin\psi_1 = -\frac{G(b_1 + b_2 \cos\psi_2) + Fb_2\sin\psi_2}{b_1^2 + b_2^2 + 2b_1b_2 \cos\psi_2} \tag{7.83}$$

Equations (7.82) and (7.83) uniquely determine angle ψ_1

$$\psi_1 = \text{atan2}(\sin\psi_1, \cos\psi_1) \tag{7.84}$$

Step 6: Calculation of angle

$$\psi_3 = \psi_{123} - \psi_1 - \psi_2, \tag{7.85}$$

where the angles ψ_{123}, ψ_1, and ψ_2 are defined through Eqs. (7.74), (7.81), and (7.84), respectively.

Since there are two solutions for each of angles φ, θ, and ψ_2, the inverse kinematics problem can have eight possible solutions. A suitable set of the angles φ, θ, ψ_1, ψ_2, and ψ_3 is chosen from additional geometrical requirements relating to the performed operation.
Step 7: An opening angle of the tong jaws with the electrodes is

$$\mu = 2\arctan\left[t/(L_e - l_{ax})\right] \approx 2t/(L_e - l_{ax}) \tag{7.86}$$

where t is the welded sheet thickness, and l_{ax} is the X-coordinate of the tong axis in the gripper associated frame S_{10}.

Numerical Example 7.3.
Setup for point-to-point spot welding of horizontal planar sheets (Fig. 7.9).
Given: Dimensions of the robot are H = 410 mm, b_1 = 220 mm, b_2= 220 mm, b_3 = 90 mm, d = 100 mm, L_e = 80 mm, and l_{ax} = 20 mm. Coordinates of the weld points are x_w = 460 mm, $-200 \le y_w \le 200$ mm, z_w = 420 mm. The thickness of the welded sheets is t = 2 mm. The electrodes are parallel to the Z_0-axis, $c_{wx} = 0$, $c_{wy} = 0$, $c_{wz} = \pm 1$, where plus and minus correspond to the upper and lower electrodes, respectively (for definiteness, only $c_{wz} = 1$ is considered below). **Required:** Calculation of the angles φ, θ, ψ_1, ψ_2, ψ_3, and μ.

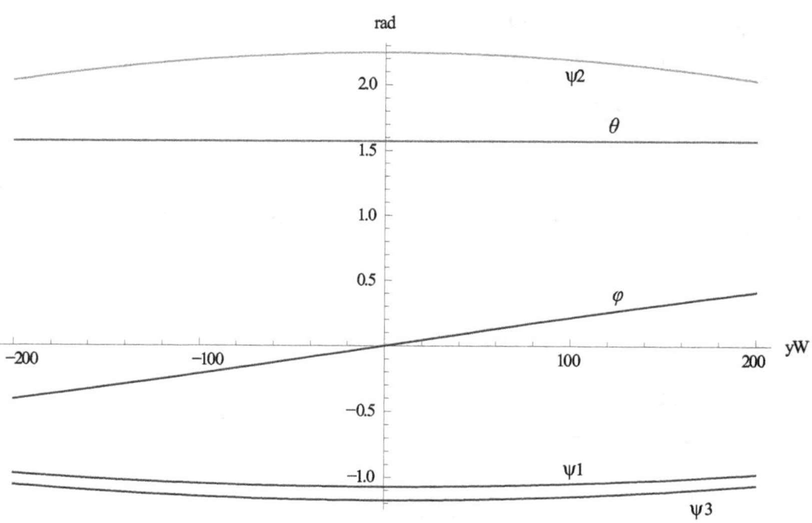

Figure 7-9 Current angles φ, θ, ψ_1, ψ_2, and ψ_3 according to the Numerical example data.

Solution:

(1) The position and orientation vectors as applied to the upper electrode become

$\mathbf{r}_w = [460, y_w, 420, 1]^T$ and $\mathbf{c}_w = [0,0, 1, 0]^T$

(2) Equation (7.80) yields

$\varphi = \arctan(y_w/x_w) = \arctan(y_w/460)$

The second solution $\varphi = \pi + \arctan(y_w/460)$ is *inconsistentwith the given weld location.*

(3) Equation (7.81) yields

$\theta = \pm\arccos(c_{wz}\sin\varphi - c_{wy}\cos\varphi) = \pm\arccos 0 = \pm\pi/2$

(4) Equations (7.74) yield

$\sin\psi_{123} = (c_{wx}\cos\varphi + c_{wy}\sin\varphi)/\sin\theta = 0$

$$\cos\psi_{123} = c_{wz}/\sin\theta = 1/\sin\theta = \begin{cases} 1, \text{ for } \theta = \pi/2 \\ -1, \text{ for } \theta = -\pi/2 \end{cases}$$

Hence, there are two possible values

$$\psi_{123} = \begin{cases} 0, \text{ for } \theta = \pi/2 \\ \pi, \text{ for } \theta = -\pi/2 \end{cases}$$

(5) Substitution of angles φ, θ, and ψ_{123} in Eq. (7.81) yields four possible values

$$\psi_2 = \pm\arccos\frac{F^2 + G^2 - b_1^2 - b_2^2}{2b_1b_2} =$$

$$\begin{cases} (\psi_2)_{1,2} = \pm\arccos\left[\dfrac{y_w^2 + 60(3130 - 9\sqrt{211600 + y_w^2}}{96800}\right], \text{ for } \theta = \dfrac{\pi}{2} \text{ and } \psi_{123} = 0 \\ (\psi_2)_{3,4} = \pm\arccos\left[\dfrac{y_w^2 + 20(5990 + 7\sqrt{211600 + y_w^2}}{96800}\right], \text{ for } \theta = -\dfrac{\pi}{2} \text{ and } \psi_{123} = \pi \end{cases} \tag{7.87}$$

The two last solutions are impossible since there are no real values of $(\psi_2)_{3,4}$ for y_w in the range between −200 and 200 mm, and only values $(\psi_2)_{1,2}$ are considered. Equations (7.82-84) lead to rather cumbersome numerical expressions of angles ψ_1 and ψ_3, Eq. (7.85). In Figure 7.10, angles φ, θ, ψ_1, ψ_2, and ψ_3 vs. y_w are shown for the positive value of angle ψ_2, Eq. (7.87):

Table 7.2. Configuration for $y_w = 0$.

Unit	φ	θ	ψ_1	ψ_2	ψ_3
Radians	0	$\pi/2$	-1.07104	2.24725	-1.17621
Degrees	0	90°	-61.37°	128.76°	-67.39°

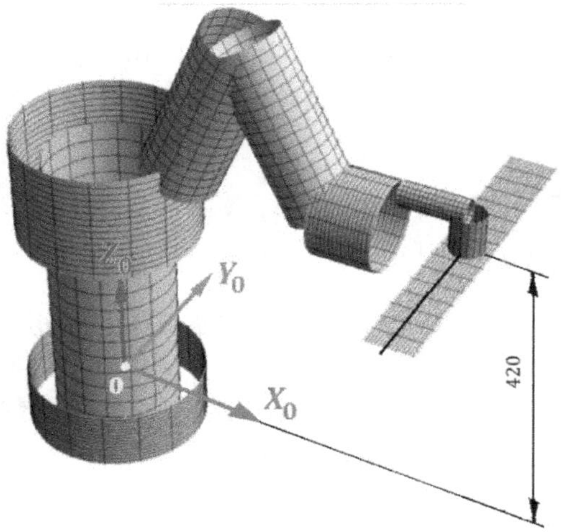

Figure 7-10 Configuration of the Scorbot robot according to Numerical Example 7.3.

$$(\psi_2)_1 = \arccos\left[\frac{y_w^2 + 60(3130 - 9\sqrt{211600 + y_w^2}}{96800}\right] \tag{7.88}$$

(6) A current configuration with $y_w = 0$ corresponding to angle ψ_2, Eq. (7.88), is shown in Figure 7.10. Only the upper electrode and the contact plane between two welded sheets are illustrated. The configuration angles are listed in Table 7.2.

(7) According to Eq. (7.86), the tong jaws with the electrodes are opened at angle

$\mu = \arctan\,[2t/(L_e - l_{ax})] = 2 \times 2/(80 - 20) = 3.8$ degrees.

7.2 Setup of parallel- and hybrid-kinematics machines

7.2.1 Setup procedure

The setup process for parallel- and hybrid-kinematics machines is reduced to solution of the inverse kinematics problem. At first, the FSF of the machine, Eq. (4.16–20), and the regular setup conditions, Eqs. (6.1) and (6.2), for the kinematic chain are synthesized:

$$[\mathbf{r}_0\mathbf{c}_0] = {}^0\mathbf{A}_n[\mathbf{r}_n\mathbf{c}_n]$$

$$\mathbf{r}_0 = \mathbf{r}_w$$

$$\mathbf{c}_0 = \mathbf{c}_w$$

Remember that in this case, there is a multi-parametric matrix, describing from 2- to 6-DOF moving platform motion) among n co-factor matrices comprising the manipulation matrix ${}^0\mathbf{A}_n$.

Then, the inverse-kinematics problem for *the parallel-kinematics link*is solved with respect to the actuator parameters based on information about the link pose:

$$q_{i_1} = q_{i_1}(x_i, y_i, z_i, \theta_i, \psi_i, \varphi_i)$$
$$q_{i_2} = q_{i_2}(x_i, y_i, z_i, \theta_i, \psi_i, \varphi_i)$$
....
$$q_{i_6} = q_{i_6}(x_i, y_i, z_i, \theta_i, \psi_i, \varphi_i)$$

7.2.2 Setup of the hexapod-based PKM

As is obtained in Section 5.3.1, the FSF of the hexapod-based PKM, Eqs. (5.55–58), is

$$[\mathbf{r}_0 \mathbf{c}_0] = {}^0\mathbf{A}_4 [\mathbf{r}_4 \mathbf{c}_4] \tag{7.89}$$

with

$\mathbf{r}_0 = [x_0, y_0, z_0, 1]^T$ and $\mathbf{r}_4 = [x_4, y_4, z_4, 1]^T$

$\mathbf{c}_0 = [c_{0x}, c_{0y}, c_{0z}, 0]^T$ and $\mathbf{c}_4 = [c_{4x}, c_{4y}, c_{4z}, 0]^T$

${}^0\mathbf{A}_4 = \mathbf{A}^1(q_1)\,\mathbf{A}^2(q_2)\,\mathbf{A}^3(q_3)\,\mathbf{A}^4(\theta_2)\,\mathbf{A}^5(\psi_2)\,\mathbf{A}^6(\varphi_2 + \varphi_3) =$

$$\begin{bmatrix} c\varphi_{23}\ c\psi_2 & -s\varphi_{23}\ c\psi_2 & s\psi_2 & z_p - x_h + h\ s\psi_2 \\ c\varphi_{23}\ s\psi_2\ s\theta_2 + s\varphi_{23}\ c\theta_2 & c\varphi_{23}\ c\theta_2 - s\varphi_{23}\ s\psi_2\ s\theta_2 & -c\psi_2\ s\theta_2 & y_p - y_h - h\ c\psi_2\ s\theta_2 \\ -c\varphi_{23}\ s\psi_2\ c\theta_2 + s\varphi_{23}\ s\theta_2 & s\varphi_{23}\ s\psi_2\ c\theta_2 + c\varphi_{23}\ s\theta_2 & c\psi_2\ c\theta_2 & z_p - z_h + h\ c\psi_2\ c\theta_2 \\ 0 & 0 & 0 & 1 \end{bmatrix} \tag{7.90}$$

where $c\varphi_{23} \equiv \cos(\varphi_2 + \varphi_3)$ and $s\varphi_{23} \equiv \sin(\varphi_2 + \varphi_3)$; x_h, y_h, z_h are the coordinates of the origin

Table 7.3. Form-shaping system of the hexapod-based machine.

S_i	Link	Motion relative to previous link	Matrix ${}^{i-1}\mathbf{A}_i$
S_0	Home-point on the base	Fixed link	
S_1	Workpiece (WP)	Transfer from the home-point to the WP	${}^0A_1(-x_1, -y_1, -z_1)$
S_2	Moving platform	3-DOF translation + 3-DOF rotation	${}^1A_2(x, y, z, \theta, \psi, \varphi)$
S_3	Main spindle	Rotation around the Z_2 axes at angle φ_{sp}	${}^2A_3 = A^6(\varphi_{sp})$
S_4	Milling cutter	Translation along the Z_3 axes at distance h	${}^3A_4 = A^3(h)$

Parameters: x_1, y_1, z_1 are the coordinates of the origin of the workpiece coordinate system as referred to of the home position of the base-associated coordinate frame; x, y, z are the platform centroid position; θ, ψ, φ are Euler's angles of the platform rotation; φ_{sp} is the angle of the spindle rotation; and h is the offset of the cutting tool relative to the platform.

of the workpiece coordinate system referring to the home position of the base-associated coordinate frame; x_p, y_p, z_p are the platform centroid position; $\theta_2, \psi_2, \varphi_2$ are the angles of the platform rotation; φ_3 is the angle of the spindle rotation; and h is the offset of the cutting tool relative to the platform; q_1, q_2, and q_3 are parameters of the spindle point translations,

$$q_1 = x_p - x_h + h \sin \psi_2$$

$$q_2 = y_p - y_h - h \cos \psi_2 \sin \theta_2$$

$$q_3 = z_p - z_h + h \cos \psi_2 \cos \theta_2$$

Performing drilling, milling, and grinding operations requires 5–DOF in motion of the moving platform relative to the workpiece. Reducing the platform motion from six to five DOF is executed by condition

$$\varphi_2 = 0 \tag{7.91}$$

In this case, the FSF manipulation matrix becomes:

$$ {}^0\mathbf{A}_{4(\text{with } \varphi 2 = 0)} = \begin{bmatrix} c\varphi_3\ c\psi_2 & -s\varphi_3\ c\psi_2 & s\psi_2 & x_p - x_h + h\ s\psi_2 \\ c\varphi_3\ s\psi_2\ s\theta_2 + s\varphi_3\ c\theta_2 & c\varphi_3\ c\theta_2 - s\varphi_3\ s\psi_2\ s\theta_2 & -c\psi_2\ s\theta_2 & y_p - y_h - h\ c\psi_2\ s\theta_2 \\ -c\varphi_3\ s\psi_2\ c\theta_2 + s\varphi_3\ s\theta_2 & s\varphi_3\ s\psi_2\ c\theta_2 + c\varphi_3\ s\theta_2 & c\psi_2\ c\theta_2 & z_p - z_h + h\ c\psi_2\ c\theta_2 \\ 0 & 0 & 0 & 1 \end{bmatrix} \tag{7.92}$$

To define the moving platform pose, five parameters must be established: linear displacements x_p, y_p, z_p and angles θ_2 and ψ_2. In the general case (Fig. 7.11), the FSS of the machine is described through the coordinate systems listed in Table 7.3: S_0– the home-point located on the base, S_1– the *TE* (workpiece), S_2– the moving platform in the initial pose, S_3– the moving platform in the working pose (during or after translational and rotational motions), and S_4– the *AE* (cutting tool).

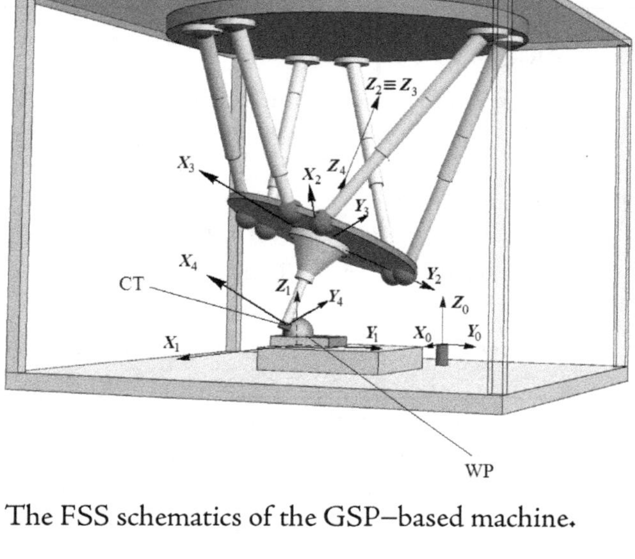

Figure 7-11 The FSS schematics of the GSP–based machine.

Numerical Example 7.4.

Given: Dimensions of the 3–3 GSP (Fig. 7.12): $r_1 = 0.7$ mm, $r_2 = 0.25$ m; polar angles of points M_i: $\alpha_1 = \alpha_6 = -\pi/3$; $\alpha_2 = \alpha_3 = \pi/3$; $\alpha_4 = \alpha_5 = \pi$; and polar angles of points Q_i: $\beta_i = \beta_{i+1} = (i-1)\pi/3$, with $i = 1, 3, 5$. The mobile platform is horizontally oriented, i.e., the Euler angles are $\theta = \varphi = \psi = 0$, and platform height is $z_p = 0.75$ m. The central point O_p of the mobile platform moves along a circle of radius $R_c = 0.1$ m, the circle center is located at point $x_c = 0.05$ m, $y_c = 0.1$ m.

Required: Calculation of the limb lengths during platform motion along the circle.
Solution: Substitution of the initial data in Eq. (5.39) yields

$$\mathbf{r}_i = \begin{bmatrix} x_p + r_2\cos\psi\cos(\varphi+\beta_i) - r_1\cos\alpha_i \\ y_p + r_2[\cos\theta\sin(\varphi+\beta_i) + \sin\theta\sin\psi\cos(\varphi+\beta_i)] - r_1\sin\alpha_i \\ z_p + r_2[\sin\theta\sin(\varphi+\beta_i) - \cos\theta\sin\psi\cos(\varphi+\beta_i)] \\ 0 \end{bmatrix} = \begin{bmatrix} x_p + 0.25\cos\beta_i - 0.7\cos\alpha_i \\ y_p + 0.25\sin\beta_i - 0.7\sin\alpha_i \\ 0.75 \\ 0 \end{bmatrix};$$

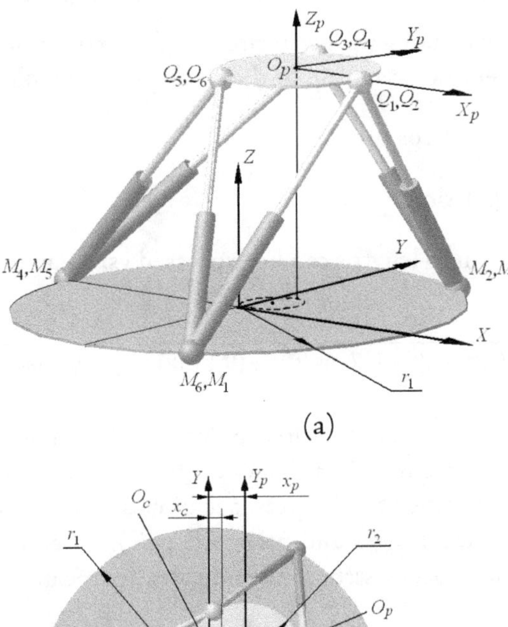

(a)

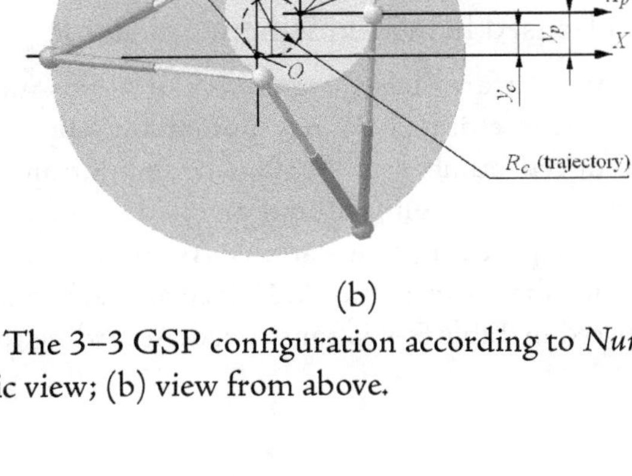

(b)

Figure 7-12 The 3–3 GSP configuration according to *Numerical Example 7.4*: (a) axonometric view; (b) view from above.

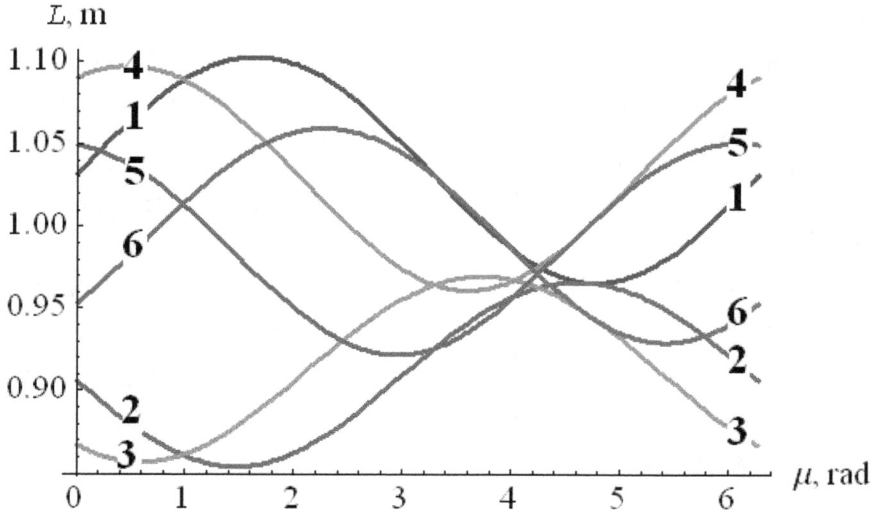

Figure 7-13 Lengths of the limbs: Solution of *Numerical Example 7.4.*

The moving platform moves along the given circle, and the coordinates of the point O_p are expressed through the current polar angle μ of the circle points as follows:

$$x_p = x_c + R_c \cos\mu = 0.05 + 0.1\cos\mu$$

$$y_p = y_c + R_c \sin\mu = 0.1 + 0.1\sin\mu$$

Hence, Eqs. (5.40) of the limb lengths can be expressed vs. polar angle μ of the machined object, where $0 \le \mu \le 2\pi$,

$$L(\mu) = \sqrt{(0.05 - 0.1\cos\mu + 0.25\cos\beta_i - 0.7\cos\alpha_i)^2 + (0.1 + 0.1\sin\mu + 0.25\sin\beta_i - 0.7\sin\alpha_i)^2 + 0.75^2}$$

The limb lengths vs. polar angle μ of the circle points are shown in Figure 7.12. As is clear from the graphics, the process requires 12 reverses of the actuators controlling limb lengths. Each of the reverses is accompanied by the presence of clearances in all limb joints (in this case, the universal and spherical hinges and prismatic pair). To eliminate or minimize the effect on the process accuracy, special structural and software means are used (see examples in Chapter 9).

7.2.3 Setup of the tripod-based hybrid-kinematics machine

In comparison with Stewart-platform-based structures of 5-axis machine tools, the tripod-based machine tool architecture [1-3] has potentially superior stiffness, greater working volume, and a reduced number of singular configurations. In particular, these advantages result from direct application (without any additional structural transformations) of the guideways for reciprocating translations. As an example, the kinematic setup of the tripod-based machine shown in Figure 5.31 (see also schematics in Figure 7.14), whose the FSF links are listed in Table 5.4, is considered.According to Eq. (5.64), the FSF manipulation matrix is

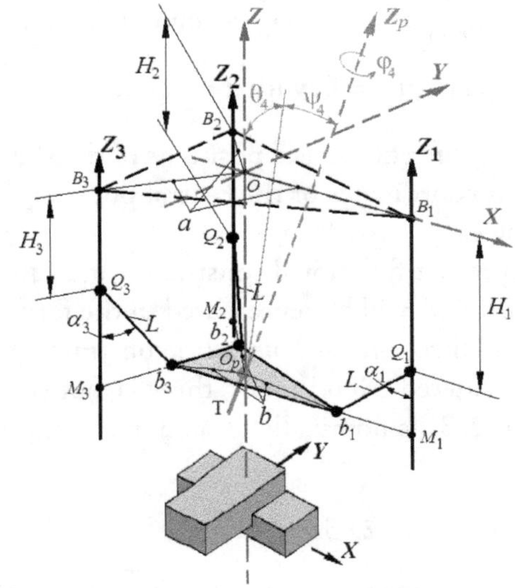

Figure 7-14 Schematics of the tripod-based machine.

$$^0\mathbf{A}_6 = {}^0\mathbf{A}_1(-x_h, -y_h, -z_h)\ \mathbf{A}^2(-y_c)\ \mathbf{A}^1(-x_t)\ {}^2\mathbf{A}_3\ (z_p, \theta_4, \psi_4)\ \mathbf{A}^6(\varphi_{sp})\ \mathbf{A}^3(h); \qquad (7.93)$$

$$^0\mathbf{A}_1(-x_h,-y_h,-z_h) = \begin{bmatrix} 1 & 0 & 0 & -x_h \\ 0 & 1 & 0 & -y_h \\ 0 & 0 & 1 & -z_h \\ 0 & 0 & 0 & 1 \end{bmatrix}; {}^1\mathbf{A}_2 = \mathbf{A}^2(-y_c) = \begin{bmatrix} 1 & 0 & 0 & 0 \\ 0 & 1 & 0 & -y_c \\ 0 & 0 & 1 & 0 \\ 0 & 0 & 0 & 1 \end{bmatrix}; {}^2\mathbf{A}_3 = \mathbf{A}^1(-x_t) = \begin{bmatrix} 1 & 0 & 0 & -x_t \\ 0 & 1 & 0 & 0 \\ 0 & 0 & 1 & 0 \\ 0 & 0 & 0 & 1 \end{bmatrix}$$

$$^3\mathbf{A}_4 = \mathbf{A}_{Eu}(x_p, y_p, z_p, \theta_4, \psi_4, \varphi_4) =$$

$$\begin{bmatrix} c\varphi_4\ c\psi_4 & -s\varphi_4\ c\psi_4 & s\psi_4 & x_p \\ c\varphi_4\ s\psi_4\ s\theta_4 + s\varphi_4\ c\theta_4 & c\varphi_4\ c\theta_4 - s\varphi_4\ s\psi_4\ s\theta_4 & -c\psi_4\ s\theta_4 & y_p \\ -c\varphi_4\ s\psi_4\ c\theta_4 + s\varphi_4\ s\theta_4 & s\varphi_4\ s\psi_4\ c\theta_4 + c\varphi_4\ s\theta_4 & c\psi_4\ c\theta_4 & z_p \\ 0 & 0 & 0 & 1 \end{bmatrix}$$

$$^4\mathbf{A}_5 = \mathbf{A}^6(\varphi_{sp}) = \begin{bmatrix} \cos\varphi_{sp} & -\sin\varphi_{sp} & 0 & 0 \\ \sin\varphi_{sp} & \cos\varphi_{sp} & 0 & 0 \\ 0 & 0 & 1 & 0 \\ 0 & 0 & 0 & 1 \end{bmatrix}; \text{ and, } {}^5\mathbf{A}_6 = \mathbf{A}^3(h) = \begin{bmatrix} 1 & 0 & 0 & 0 \\ 0 & 1 & 0 & 0 \\ 0 & 0 & 1 & h \\ 0 & 0 & 0 & 1 \end{bmatrix} \qquad (7.94)$$

where y_c and x_t are the parameters of the feed motions executed by the table and carriage; z_4 is the height position of the mobile platform; θ_4 and ψ_4 are the tilting angles of the platform orientation; φ_{sp} is the angle of the spindle rotation (the main motion); $^3\mathbf{A}_4$ is the 4×4 matrix describing 3-DOF motion of the mobile platform expressed through extended Euler matrix $\mathbf{A}_{Eu}$; x_p, y_p, z_p are the current coordinates of the central point O_p of the mobile platform; and θ_4, ψ_4, and φ_4 are the Euler angles of rotation around the X_3–, Y_3–, and Z_3–axes, respectively.

Configuration parameters $x_p, y_p, z_p, \theta_4, \psi_4, \varphi_4$ are connected through six kinematic constraints

$$F_k(x_p, y_p, z_p, \theta_4, \psi_4, \varphi_4, H_1, H_2, H_3) = 0, \text{ with } k = 1, 2,\ldots,6 \tag{7.95}$$

where H_1, H_2, and H_3 are the current vertical positions of the sliders 2, 3, and 4, respectively and x_4, y_4, z_4 are the current coordinates of the central point O_p of the mobile platform; θ_4, ψ_4, and φ_4 are the Euler angles.

Six equations (7.95) represent functional constraints and can be divided to two groups. Three *layout functional constraints* will be represented in the form $x_p = x_p(\theta_4, \psi_4)$, $y_p = y_p(\theta_4, \psi_4)$, and $\varphi_4 = \varphi_4(\theta_4, \psi_4)$; and three *inverse kinematics* constraints become $H_i = H_i(z_p, \theta_4, \psi_4)$, with i= 1, 2, 3. For the considered tripod robot, three layout constraints are formulated as follows: the ith limb (i = 1, 2, 3) is nominally located in a vertical plane defined by triangle OB_iQ_i (Fig. 7.13),

$$F_i \equiv (\mathbf{r}_i \cdot \mathbf{n}_i) = 0, \text{ with } i = 1, 2, 3 \tag{7.96}$$

where $\mathbf{r}_i$ is the vector of the ith connecting rod, $\mathbf{r}_i = \overline{b_iQ_i}$; and $\mathbf{n}_i$ is the normal of the plane wherein the ith rod moves,

$$\mathbf{r}_i \equiv \overline{b_iQ_i} = \left[x_p, y_p, z_p, 0\right]^{\mathrm{T}} = \mathbf{A}_{Eu}\,\mathbf{b}_i - \mathbf{r}_{Qi} \tag{7.97}$$

$$\mathbf{n}_i = [-\sin \mu_i, \cos \mu_i, 0, 0]^{\mathrm{T}}, \text{ with } i = 1, 2, 3 \tag{7.98}$$

$$\mathbf{b}_i = [b \cos \nu_i, b \sin \nu_i, 0, 1]^{\mathrm{T}} \tag{7.99}$$

$$\mathbf{r}_{Qi} = [a \cos \mu_i, a \sin \mu_i, H_i, 1]^{\mathrm{T}} \tag{7.100}$$

$\mathbf{A}_{Eu}$ is the 4 × 4 extended Euler matrix, Eq. (7.94); b is the radius of the resident circle of points b_i at the platform plane (Fig. 7.13); ν_i (i = 1, 2, 3) is the polar angle of points b_i; a is radius of the resident circle of points M_i at the base; μ_i (i = 1, 2, 3) is the polar angle of the position of point B_i and H_i is the current height position of the ith slider – active controlled parameter.

Substituting Eqs. (7.97-117) into Eqs. (7.96) yields

$$F_i \equiv x_p \sin \mu_i - y_p \cos \mu_i + b \sin (\mu_i - \varphi_4) (\cos \nu_i \cos \psi_4 + \sin \nu_i \sin \theta_4 \sin \psi_4) -$$
$$b \cos (\mu_i - \varphi_4) \sin \nu_i \cos \theta_4 = 0 \tag{7.101}$$

For symmetrical positions of the six platform joints, i.e., $\mu_1 = \nu_1 = 0$; $\mu_2 = \nu_2 = 2\pi/3$; and $\mu_3 = \nu_3 = 4\pi/3$, we obtain four solutions of simultaneous equations [4]:

$$(x_p)_{1,2} = -Q/P;\ (y_p)_{1,2} = -R/P;\ (\varphi_3)_{1,2} = \pm\arccos (S/P) \tag{7.102}$$

$$(x_p)_{3,4} = Q/P;\ (y_p)_{3,4} = R/P;\ (\varphi_4)_{1,2} = \pm\arccos (-S/P) \tag{7.103}$$

with

$$Q = (b/2) (\cos^2\theta_4 - \cos^2\psi_4 + \sin^2\theta_4 \sin^2\psi_4);\ R = b\cos \psi_4 \sin \theta_4 \sin \psi_4$$

$$S = \cos \theta_4 + \cos \psi_4;\ \text{and } P = 1 + \cos \theta_4 \cos \psi_4$$

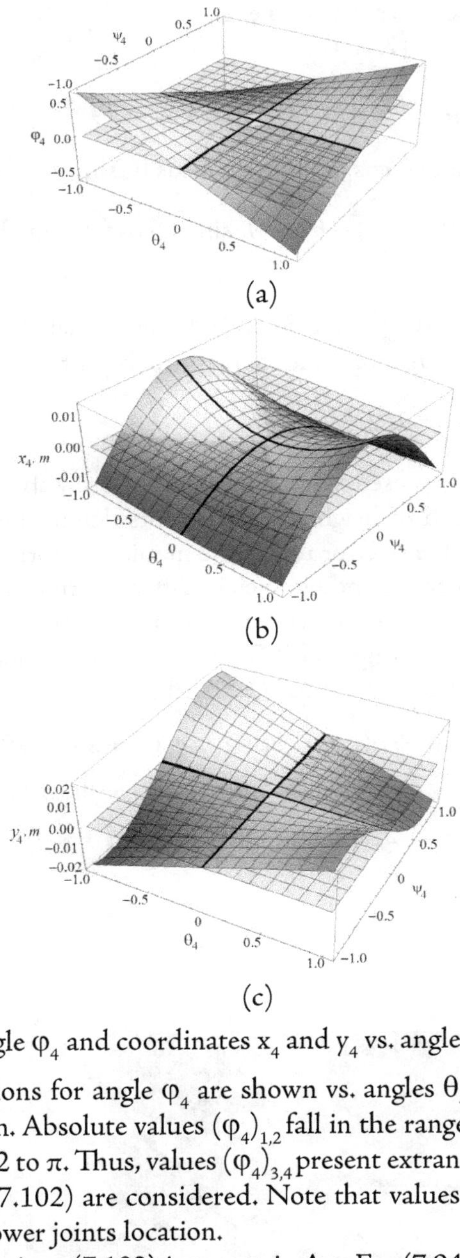

Figure 7-15 Angle φ_4 and coordinates x_4 and y_4 vs. angles θ_4 and ψ_4.

In Figure 7.15, solutions for angle φ_4 are shown vs. angles θ_4 and ψ_4 lying in the range $\pm\pi/2$, with $r_2 = 100$ mm. Absolute values $(\varphi_4)_{1,2}$ fall in the range between 0 and $\pi/2$, while $(\varphi_4)_{3,4}$ lie in the range $\pi/2$ to π. Thus, values $(\varphi_4)_{3,4}$ present extraneous solutions and, in what follows, only solutions (7.102) are considered. Note that values x_p, y_p, and φ_4 are independent of radius a of the lower joints location.

Substitution of expressions (7.102) into matrix $\mathbf{A}_{Eu}$, Eq. (7.94), results in cofactor-matrix ${}^3\mathbf{A}_4 = {}^3\mathbf{A}_4(z_p, \theta_4, \psi_3)$. Correspondingly, the FSF matrix becomes ${}^0\mathbf{A}_6 = {}^0\mathbf{A}_6(x_t, y_c, z_p, \theta_4, \psi_4, \varphi_{sp}, h)$. The magnitude of vector $\overline{b_iQ_i}$ associated with the ith connecting rod has the nominally permanent value of length L_i resulting from equation

$$F_{i+3} \equiv L^2 - (\mathbf{r}_i)^2 = 0, \text{ with } i = 1, 2, 3 \qquad (7.104)$$

To realize the orientation angles θ_4 and ψ_4 of the mobile platform and its height position z_p, three sliders have to execute corresponding control motions. With the assumption that the connecting rods are of identical length, i.e., $L_1 = L_2 = L_3 = L$, we obtain after substituting vectors $\mathbf{r}_i$, Eq. (7.97), into Eqs. (7.104):

$$F_{i+3} \equiv -L^2 + [x_p - a \cos \mu_i + b \cos \psi_4 \cos (\nu_i + \varphi_4)]^2 +$$

$$[y_p - a \sin \mu_i + b \sin \theta_4 \sin \psi_4 \cos (\nu_i + \varphi_4) + b \cos \theta_4 \sin (\nu_i + \varphi_4)]^2 +$$

$$[-H_i + z_p + b \sin \theta_4 \sin (\nu_i + \varphi_3) - b \cos \theta_4 \sin \psi_3 \cos (\nu_i + \varphi_4)]^2 = 0 \qquad (7.105)$$

with $i = 1, 2, 3$

Substituting solutions (7.105) into Eqs. (7.104) yields analytical solutions for heights of the sliders H_1, H_2, and H_3 vs. current values of platform height z_4 and tilting angles θ_4 and ψ_4:

$$H_i = H_i(z_p, \theta_4, \psi_4), \text{ with } i = 1, 2, 3 \qquad (7.106)$$

The *i*th equation (7.104) presents a quadratic equation with respect to H_i. Thus, there are, if any, two roots for each of heights H_i. The roots define alternate assembly versions: the slider can be located either over or under the mobile platform, i.e., the total number of assembly versions is eight, corresponding to eight different triplets H_1, H_2, and H_3.

The setup for drilling of radial-oriented holes in the half-sphere (Fig. 7.16) is considered. Since the active point of the cutting tool (CT) and the CT orientation are given by vectors $\mathbf{r}_6 = (0, 0, 0, 1)^T$ and $\mathbf{c}_6 = (0, 0, 1, 0)^T$, Eq. (7.93) yields

$$\mathbf{r}_0 = {}^0\mathbf{A}_6\,\mathbf{r}_6 = \begin{pmatrix} x_h - x_t + x_p + h(\cos\theta_4 \sin\psi_4 \cos\varphi_4 + \sin\theta_4 \sin\varphi_4) \\ y_h - y_c + y_p + h(\cos\theta_4 \sin\psi_4 \sin\varphi_4 - \sin\theta_4 \cos\varphi_4) \\ z_h + z_p + h\cos\theta_4 \cos\psi_4 \\ 1 \end{pmatrix} \qquad (7.107)$$

Figure 7-16 A half–sphere workpiece.

$$\mathbf{c}_0 = {}^0\mathbf{A}_6\,\mathbf{c}_6 = \begin{pmatrix} \cos\theta_4 \sin\psi_4 \cos\varphi_4 + \sin\theta_4 \sin\varphi_4 \\ \cos\theta_4 \sin\psi_4 \sin\varphi_4 - \sin\theta_4 \cos\varphi_4 \\ \cos\theta_4 \cos\psi_4 \\ 0 \end{pmatrix} \tag{7.108}$$

Substitution of Eqs. (7.96) and (7.97) into Eq. (7.107) and (7.108) yields

$$\mathbf{r}_0 = \begin{pmatrix} x_h - x_t + x_p + h\sin\psi_4 \\ y_h - y_c + y_p - h\sin\theta_4\cos\psi_4 \\ z_h + z_p + h\cos\theta_4\cos\psi_4 \\ 1 \end{pmatrix} \text{and } \mathbf{c}_0 = \begin{pmatrix} \sin\psi_4 \\ -\sin\theta_4\cos\psi_4 \\ \cos\theta_4\cos\psi_4 \\ 0 \end{pmatrix} \tag{7.109}$$

Let us consider the setup for drilling the radially-oriented bores in a half-sphere workpiece (Fig. 7.16). Equations of the bore axis and its orientation are

$$(\mathbf{r}_w, \mathbf{c}_w) = \begin{pmatrix} x_h + c_x s & c_x \\ y_h + c_y s & c_y \\ z_h + c_z s & c_z \\ 1 & 0 \end{pmatrix} \tag{7.110}$$

where s is the parameter defined along the bore axis.

Equating the right-hand sides of Eqs. (7.109) and (7.110) yields

$$\begin{pmatrix} x_h + c_x s & c_x \\ y_h + c_y s & c_y \\ z_h + c_z s & c_z \\ 1 & 0 \end{pmatrix} = \begin{pmatrix} x_h - x_t + x_p + h\sin\psi_4 & \sin\psi_4 \\ y_h - y_c + y_p - h\sin\theta_4\cos\psi_4 & -\sin\theta_4\cos\psi_4 \\ z_h + z_p + h\cos\theta_4\cos\psi_4 & \cos\theta_4\cos\psi_4 \\ 1 & 0 \end{pmatrix} \tag{7.111}$$

Thus, five setup parameters as functions of line parameter s are, with $0 \le s \le l_b$,

$$x_t(s) = x_p - c_x s + h\sin\psi_4$$

$$y_c(s) = y_p - c_y s - h\sin\theta_4\cos\psi_4$$

$$z_p(s) = c_z s - h\cos\theta_4\cos\psi_4$$

$$\theta_4 = \operatorname{atan2}(c_z/\cos\psi_4, -c_y/\cos\psi_4)$$

$$\psi_4 = \arcsin c_x \tag{7.112}$$

where atan2(y, x) is the single-valued two-argument function.

The inverse kinematics problem yields the values of the current height position of the slider – active controlled parameter H_1, H_2, and H_3– as solutions of three separate square equations

$$[-H_i + z_4 + b \sin \theta_4 \sin (\nu_i + \varphi_3) - b \cos \theta_4 \sin \psi_3 \cos (\nu_i + \varphi_4)]^2$$

$$=L^2 - [x_4 - a \cos \mu_i + b \cos \psi_4 \cos (\nu_i + \varphi_4)]^2 - [y_4 - a \sin \mu_i +$$

$$b \sin \theta_4 \sin \psi_4 \cos (\nu_i + \varphi_4) + b \cos \theta_4 \sin (\nu_i + \varphi_4)]^2, \text{with } i = 1,2,3 \tag{7.113}$$

where ν_i (i = 1, 2, 3) is the polar angle of points b_i; a is radius of the resident circle of points M_i at the base; μ_i (i = 1, 2, 3) is the polar angle of the position of point B_i at the base.

Substitution of Eqs. (7.112) in Eqs. (7.113) becomes the current heights H_i as functions of parameter s:

$$H_i = H_i(s), i = 1, 2, 3 \tag{7.114}$$

7.3 Conclusion

The FSF theory-based approach described in the previous chapter for modeling kinematic setup problems of serial-kinematics machine tools can also be successfully employed for serial-, parallel-, and hybrid-kinematics robots.

References

[1] Weck, M. and Staimer, D., 2002, Parallel Kinematic Machine Tools – Current State and Future Potentials, Annals of the CIRP, Vol. 51/2, pp. 671–684.

[2] Xi, F., Zhang, D., Xu, Z., Mechefske, C.M., 2003, A comparative study of tripod units for machine tools, International Journal of Machine Tools & Manufacture, 43, 721–730.

[3] Xi, F., Zhang, D., Mechefske, C.M. and Lang, Y.T., 2004, Global kinetostatic modeling of tripod-based parallel kinematic machine, Mechanism and Machine Theory, 39, pp. 357–377.

[4] Portman, V.T., Shneor, Y., Chapsky, V.S., Shapiro, A., 2015, Form-shaping function theory expansion: Stiffness model of multi-axis machines, International Journal of Advanced Manufacturing Technology, Volume 76(5), pp. 1063–1078.

Part II: Geometric and Kinematic Accuracy

8. Error budget

8.1 Mathematical technique

8.1.1 Infinitesimal displacements and infinitesimal matrices

8.1.1.1 Error smallness and Abbe errors

Actual machine motions are inevitably accompanied by small non-designed deviations of position and orientation of all the links, relative to each other. The sum of the elementary errors together with their transfer coefficients presents the *error budget* [1]. The number of the errors associated with all interacting links is several dozens, and an effective way to analyze their contribution to the machine accuracy and precision is formulated as *total variation* of the FSF [2]. In this chapter, a linearized mathematical model for a budget of geometric and kinematic errors suitable for the FSS of serial-, parallel-, and hybrid-kinematics machines is described.

From the mathematical viewpoint, it is worth to bring into sharp focus an important partial case of the body motion parameters when they are very small, i.e., the moving body carries out an infinitesimal displacement. In this case, there exists a possibility of drastic simplification of complicated mathematical expressions by means of linearization technique. A precise mathematical definition of the "infinitesimal" value is a variable having zero as a limit. However, in accuracy-related theories —approximate calculations, numerical methods, engineering metrology, and so on—the term "infinitesimal" is usually applied to value δ, which is so small that we can neglect the values δ^n, where $n \geq 2$, without sacrificing required precision, and take to be that $\delta^2 \approx 0$; $\delta_i\delta_j \approx 0$; $\delta^3 \approx 0$, etc.

If an angle presenting a rotation parameter is small, the following *first order* approximations for trigonometric functions take place:

$$\sin\delta \approx \delta, \text{ with deviation from } \sin\delta \text{ about } \delta^3/6 \qquad (8.1)$$

$$\cos\delta \approx 1, \text{ with deviation from } \cos\delta \text{ about } \delta^2/2 \qquad (8.2)$$

Since approximation (8.1) is more precise than (8.2), just the latter defines the calculation accuracy when the approximations are used. The regular linearization condition relating to the product of two infinitesimal values (it makes no difference whether they are linear or angular values) has to be added to Eqs. (8.1) and (8.2):

$$\delta_i\delta_j \approx 0, \text{ for } i, j = x, y, z, \theta, \psi, \varphi \qquad (8.3)$$

Numerical Example 8.1. The angle of rotation is $\delta = 0.05$ rad (i.e., 2.865°). The approximate values and their deviations from the exact values are as follows.

(a)Exact value of the sinus is sin 0.05 = 0.0499792.

- Approximation $\sin\delta \approx \delta = 0.05$ results in the deviation $\delta - \sin\delta = 0.05 - 0.0499792 = 0.0000208307$ from the exact value $\sin\delta$.
- A relative error of the approximation is $(\delta - \sin\delta)/\sin\delta = 0.00042 = 0.042\%$.
- The relative error with a high degree of accuracy is evaluated through Eq. (8.1): $0.05^3/6 = 0.0000208333$.

(b)Exact value of the cosine is $\cos 0.05 = 0.9987502$.

- Approximation $\cos\delta = 1$ yields deviation from the exact value $\cos 0.05 - 1 = -0.00124974$.
- The relative error of the approximation is $(1 - \cos 0.05)/\cos 0.05 = -0.001251 = -0.1251\%$.
- The relative error with a high degree of accuracy evaluated through Eq. (8.2) is $0.05^2/2 = -0.0012500$.

Since all deviations from the exact values are small enough, approximations (8.1) and (8.2) give acceptable accuracy for the majority of calculations. The considered approach is applied below for linearization of quite complicated matrix expressions describing *an actual* (non-ideal) motion of the machines. As a simple example of a real-life application, calculation of *Abbe errors* was considered in Chapter 1 (Fig. 1.9). The Abbe errors are small *linear displacements* caused by *small angular errors* during body motion when there is an *offset* between the body point of interest and the line of control and measurement of a body's travel.

Similarly to modelling regular (non-small) motions, the small displacement of the rigid body can be interpreted through a pair of coordinate frames (Fig. 8.1). In the basic (fixed) coordinate frame $OXYZ$, there is the point of interest K defined by position vector $\mathbf{r} = \overrightarrow{OK}$,

$$\mathbf{r} = [x, y, z, 1]^T \tag{8.4}$$

If the moving coordinate frame $O_pX_pY_pZ_p$ is displaced at small distances $\delta_x, \delta_y, \delta_z$ composing small vector OO_p relative to the frame $OXYZ$ and rotated at small angles $\delta_\theta, \delta_\psi$, and δ_φ, the point of interest is displaced together with the moving frame and goes into point Q. In the moving frame $O_pX_pY_pZ_p$, the point Q has the same coordinates x, y, z as the point K in the frame $OXYZ$,

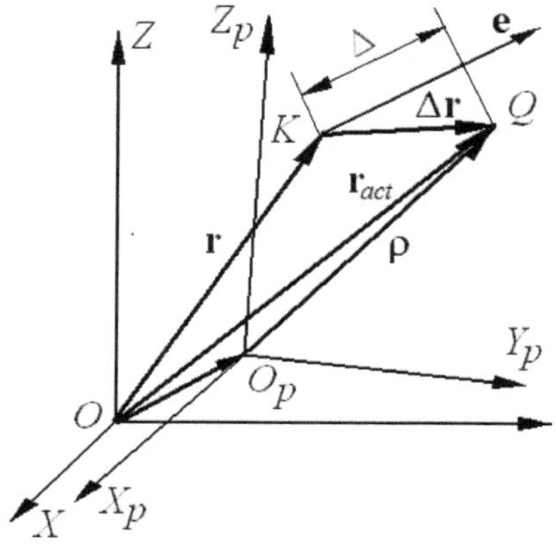

Figure 8-1 Infinitesimal displacement of a point of the body.

$$\overrightarrow{O_pQ} = \boldsymbol{\rho} = [x, y, z, 1]^T$$

In turn, the vector $\mathbf{r}_{act}$ defining the position of the point Q as referred to the frame $OXYZ$ can be represented as a sum of nominal vector $\overrightarrow{OK} = \mathbf{r}$ and error vector $\overrightarrow{KQ} = \Delta\mathbf{r}$ presenting the deviation of the actual position of the point from the nominal one

$$\mathbf{r}_{act} = \mathbf{r} + \Delta\mathbf{r} \tag{8.5}$$

8.1.1.2 Error matrix: standard form

In the general case of the body motion (i.e., the 6-DOF motion), the transformation matrix is the six-parametric extended Euler matrix

$$^{i-1}\mathbf{A}_i = (\mathbf{A}_{Eu})_i = \begin{bmatrix} \cos\varphi\ \cos\psi & -\sin\varphi\ \cos\psi & \sin\psi & p_x \\ \cos\varphi\ \sin\psi\ \sin\theta + \sin\varphi\ \cos\theta & \cos\varphi\ \cos\theta - \sin\varphi\ \sin\psi\ \sin\theta & -\cos\psi\ \sin\theta & p_y \\ -\cos\varphi\ \sin\psi\ \cos\theta + \sin\varphi\ \sin\theta & \sin\varphi\ \sin\psi\ \cos\theta + \cos o\ \sin\theta & \cos\psi\ \cos\theta & p_z \\ 0 & 0 & 0 & 1 \end{bmatrix}_i \tag{8.6}$$

where, as before, p_{xi}, p_{yi}, p_{zi} are the parameters of the ith body translation along the X_{i-1}-, Y_{i-1}-, and Z_{i-1}-axis, respectively; θ_i, ψ_i, φ_i are the Euler angles of successive rotations around the X_{i-1}-, Y_{i-1}-, and Z_{i-1}-axis, respectively (in the given order!).

If the six displacement parameters of matrix $\mathbf{A}_{Eu}$, Eq. (8.6), are infinitesimal, the matrix can be simplified using linearization of the trigonometric functions, Eqs (8.1-2). After substitutions of the small values δ_x, δ_y, δ_z, δ_θ, δ_ψ, and δ_φ for the regular values p_x, p_y, p_z, θ, ψ, and φ, respectively, in matrix $\mathbf{A}_{Eu}$, its entries obey linearization conditions

$a_{11} = \cos\delta_\varphi \cos\delta_\psi \approx 1$; $a_{12} = -\sin\delta_\varphi \cos\delta_\psi \approx -\delta_\varphi$; $a_{13} = \sin\delta_\psi \approx \delta_\psi$; $a_{14} = \delta_x$

$a_{21} = \cos\delta_\varphi \sin\delta_\psi \sin\delta_\theta + \sin\delta_\varphi \cos\delta_\theta \approx \delta_\psi\delta_\theta + \delta_\varphi \approx \delta_\varphi = -a_{12}$, etc.

Hence, a matrix of small displacements of the link caused by six infinitesimal position and orientation displacements of the link becomes

$$\mathbf{W} = \mathbf{A}_{Eu}(\delta_x, \delta_y, \delta_z, \delta_\theta, \delta_\psi, \delta_\varphi) \approx \begin{bmatrix} 1 & -\delta_\varphi & \delta_\psi & \delta_x \\ \delta_\varphi & 1 & -\delta_\theta & \delta_y \\ -\delta_\psi & \delta_\theta & 1 & \delta_z \\ 0 & 0 & 0 & 1 \end{bmatrix} \tag{8.7}$$

where δ_x, δ_y, δ_z, are the translational errors (small linear displacements along the coordinate axes); and δ_θ, δ_ψ, δ_φ are the rotational errors (small rotations around the coordinate axes).

Obviously, matrix $\mathbf{W}$ can be represented as a sum of two 4×4 matrices

$$\mathbf{W} = \mathbf{I}_{4\times4} + \mathbf{E} \tag{8.8}$$

where $\mathbf{I}_{4\times4}$ is the 4×4 identity matrix; and **E** is the infinitesimal matrix called *the standard error matrix* (in what follows "the error matrix"),

$$\mathbf{E} = \begin{bmatrix} 0 & -\delta_\varphi & \delta_\psi & \delta_x \\ \delta_\varphi & 0 & -\delta_\theta & \delta_y \\ -\delta_\psi & \delta_\theta & 0 & \delta_z \\ 0 & 0 & 0 & 0 \end{bmatrix} \tag{8.9}$$

Matrix **E** combines (see also Appendix *A2.1.2*):

(a) The entries δ_x, δ_y, and δ_z of the fourth column modelling the small translations of the link along the coordinate axes;

(b) The 3 × 3 upper-left anti-symmetric (skew) block-matrix composing the small angular displacements δ_θ, δ_ψ, and δ_φ around the *X*-, *Y*-, *Z*-coordinate axes;

(c) The zeros in the fourth row.

Because of the significant importance of the error matrix in the following material, its specific features are formulated below.

8.1.1.3 Features of the error matrix

Feature 1. Decomposition through matrices of rotation and translation: Matrix **E** can be represented as a sum of the skew symmetric matrix $\mathbf{E}_{rot}$ of angular errors (δ_θ, δ_ψ, and δ_φ) and matrix $\mathbf{E}_{tr}$ of translational errors (δ_{xi}, δ_y, and δ_z)

$$\mathbf{E} = \mathbf{E}_{rot} + \mathbf{E}_{tr} = \begin{bmatrix} 0 & -\delta_\varphi & \delta_\psi & 0 \\ \delta_\varphi & 0 & -\delta_\theta & 0 \\ -\delta_\psi & \delta_\theta & 0 & 0 \\ 0 & 0 & 0 & 0 \end{bmatrix} + \begin{bmatrix} 0 & 0 & 0 & \delta_x \\ 0 & 0 & 0 & \delta_y \\ 0 & 0 & 0 & \delta_z \\ 0 & 0 & 0 & 0 \end{bmatrix} \tag{8.10}$$

Feature 2. Decomposition through six elementary matrices: Matrix **E** can be represented as a sum of six matrices of elementary errors $\mathbf{E}^1, \mathbf{E}^2, \ldots, \mathbf{E}^6$, each of which depends on a single error

$$\mathbf{E} = \sum_{j=1}^{6} \mathbf{E}^j \tag{8.11}$$

with

$$\mathbf{E}^1 = \begin{bmatrix} 0 & 0 & 0 & \delta_x \\ 0 & 0 & 0 & 0 \\ 0 & 0 & 0 & 0 \\ 0 & 0 & 0 & 0 \end{bmatrix}; \mathbf{E}^2 = \begin{bmatrix} 0 & 0 & 0 & 0 \\ 0 & 0 & 0 & \delta_y \\ 0 & 0 & 0 & 0 \\ 0 & 0 & 0 & 0 \end{bmatrix}; \mathbf{E}^3 = \begin{bmatrix} 0 & 0 & 0 & 0 \\ 0 & 0 & 0 & 0 \\ 0 & 0 & 0 & \delta_z \\ 0 & 0 & 0 & 0 \end{bmatrix};$$

$$\mathbf{E}^4 = \begin{bmatrix} 0 & 0 & 0 & 0 \\ 0 & 0 & -\delta_\theta & 0 \\ 0 & \delta_\theta & 0 & 0 \\ 0 & 0 & 0 & 0 \end{bmatrix}; \mathbf{E}^5 = \begin{bmatrix} 0 & 0 & \delta_\psi & 0 \\ 0 & 0 & 0 & 0 \\ -\delta_\psi & 0 & 0 & 0 \\ 0 & 0 & 0 & 0 \end{bmatrix}; \mathbf{E}^6 = \begin{bmatrix} 0 & -\delta_\phi & 0 & 0 \\ \delta_\phi & 0 & 0 & 0 \\ 0 & 0 & 0 & 0 \\ 0 & 0 & 0 & 0 \end{bmatrix}$$

In turn, the sum (8.11) can be decomposed into a linear form of 4 × 4 constant matrices with infinitesimal variable factors:

$$\mathbf{E} = \sum_{j=1}^{6} \delta_j \mathbf{D}^j \tag{8.12}$$

where δ_j are the elementary errors; and $\mathbf{D}^j$ are the 4 × 4 constant matrices (called also "projection matrices"), $\mathbf{D}^j = \partial\mathbf{E}/\partial\delta_j$, whose elements are only 0, 1, and -1,

$$\delta_1 = \delta_x;\ \delta_2 = \delta_y;\ \delta_3 = \delta_z;\ \delta_4 = \delta_\theta;\ \delta_5 = \delta_\psi;\ \text{and}\ \delta_6 = \delta_\varphi \tag{8.13}$$

$$\mathbf{D}^1 = \frac{\partial\mathbf{E}}{\partial\delta_1} = \begin{bmatrix} 0 & 0 & 0 & 1 \\ 0 & 0 & 0 & 0 \\ 0 & 0 & 0 & 0 \\ 0 & 0 & 0 & 0 \end{bmatrix}; \mathbf{D}^2 = \frac{\partial\mathbf{E}}{\partial\delta_2} = \begin{bmatrix} 0 & 0 & 0 & 0 \\ 0 & 0 & 0 & 1 \\ 0 & 0 & 0 & 0 \\ 0 & 0 & 0 & 0 \end{bmatrix}; \mathbf{D}^3 = \frac{\partial\mathbf{E}}{\partial\delta_3} = \begin{bmatrix} 0 & 0 & 0 & 0 \\ 0 & 0 & 0 & 0 \\ 0 & 0 & 0 & 1 \\ 0 & 0 & 0 & 0 \end{bmatrix};$$

$$\mathbf{D}^4 = \frac{\partial\mathbf{E}}{\partial\delta_4} = \begin{bmatrix} 0 & 0 & 0 & 0 \\ 0 & 0 & -1 & 0 \\ 0 & 1 & 0 & 0 \\ 0 & 0 & 0 & 0 \end{bmatrix}; \mathbf{D}^5 = \frac{\partial\mathbf{E}}{\partial\delta_5} = \begin{bmatrix} 0 & 0 & 1 & 0 \\ 0 & 0 & 0 & 0 \\ 1 & 0 & 0 & 0 \\ 0 & 0 & 0 & 0 \end{bmatrix}; \mathbf{D}^6 = \frac{\partial\mathbf{E}}{\partial\delta_6} = \begin{bmatrix} 0 & -1 & 0 & 0 \\ 1 & 0 & 0 & 0 \\ 0 & 0 & 0 & 0 \\ 0 & 0 & 0 & 0 \end{bmatrix} \tag{8.14}$$

Feature 3. Partitioning form: The error matrix **E** can be represented through sub-matrices

$$\mathbf{E} = \begin{bmatrix} \Phi & \boldsymbol{\delta}_{tr} \\ \mathbf{0}^T & 0 \end{bmatrix} \tag{8.15}$$

where $\mathbf{0}^T = [0, 0, 0]$; Φ is the 3 × 3 anti-symmetric sub-matrix of angular errors; and $\boldsymbol{\delta}_{tr}$ is the vector of small linear displacements $\delta_x, \delta_y, \delta_z$,

$$\Phi = \begin{bmatrix} 0 & -\delta_\varphi & \delta_\psi \\ \delta_\varphi & 0 & \delta_\theta \\ -\delta_\psi & -\delta_\varphi & 0 \end{bmatrix} \text{and}\ \boldsymbol{\delta}_{tr} = \begin{bmatrix} \delta_x \\ \delta_y \\ \delta_z \end{bmatrix} \tag{8.16}$$

Feature 4. Addition of the error matrices: If several perturbations act on the body and each of these perturbations results in small displacements of the body, the corresponding infinitesimal matrices and their similar elements are summed,

$$\mathbf{E} = \sum_{k=1}^{m} \mathbf{E}_k \tag{8.17}$$

$$\mathbf{E} = \begin{bmatrix} 0 & -\delta_\varphi & \delta_\psi & \delta_x \\ \delta_\varphi & 0 & -\delta_\theta & \delta_y \\ -\delta_\psi & \delta_\theta & 0 & \delta_z \\ 0 & 0 & 0 & 0 \end{bmatrix} \text{and } \mathbf{E}_k = \begin{bmatrix} 0 & -\delta_{\varphi k} & \delta_{\psi k} & \delta_{xk} \\ \delta_{\varphi k} & 0 & -\delta_{\theta k} & \delta_{yk} \\ -\delta_{\psi k} & \delta_{\theta k} & 0 & \delta_{zk} \\ 0 & 0 & 0 & 0 \end{bmatrix}$$

with

$$\delta_x = \sum_{k=1}^{m} \delta_{xk}; \delta_y = \sum_{k=1}^{m} \delta_{yk}; \delta_z = \sum_{k=1}^{m} \delta_{zk}; \delta_\theta = \sum_{k=1}^{m} \delta_{\theta k}; \delta_\psi = \sum_{k=1}^{m} \delta_{\psi k}; \delta_\varphi = \sum_{k=1}^{m} \delta_{\varphi k} \tag{8.18}$$

where m is the number of the sources of perturbations relating to *the same* rigid body.

Feature 5. Similarity transformation [3] of the error matrix **E**, Eq. (8.9), with coordinate transformation matrix **A** preserves the standard form of the error matrix:

$$\tilde{\mathbf{E}} = \mathbf{A}\,\mathbf{E}\,\mathbf{A}^{-1} \tag{8.19}$$

where **E** and $\tilde{\mathbf{E}}$ are the error matrices before and after coordinate transformation, respectively.

To demonstrate the feature, consider the general case of transformation (8.19), in which matrix **A** presents the extended Euler matrix $\mathbf{A}_{Eu}$, Eq. (3.147):

$$\mathbf{A} = \mathbf{A}_{Eu} = \mathbf{A}^1(x)\,\mathbf{A}^2(y)\,\mathbf{A}^3(z)\,\mathbf{A}^4(\theta)\,\mathbf{A}^5(\psi)\,\mathbf{A}^6(\varphi) \tag{8.20}$$

According to *Property 4* of the HTMs, Eq. (3.145), the inverse matrix of the product of HTMs is equal to the product of the same matrices, of the argument with an opposite sign, taken in the inverse order,

$$\mathbf{A}^{-1} = (\mathbf{A}_{Eu})^{-1} = \mathbf{A}^6(-\varphi)\mathbf{A}^5(-\psi)\,\mathbf{A}^4(-\theta)\,\mathbf{A}^3(-z)\,\mathbf{A}^2(-y)\,\mathbf{A}^1(-x) \tag{8.21}$$

Substitution of Eqs. (8.20) and (8.21) in Eq. (8.19)yields the infinitesimal matrix

$$\tilde{\mathbf{E}} = \mathbf{A}_{Eu}\,\mathbf{E}\left(\mathbf{A}_{Eu}\right)^{-1} = \begin{bmatrix} 0 & \delta_{12} & \delta_{13} & \delta_{14} \\ \delta_{21} & 0 & \delta_{23} & \delta_{24} \\ \delta_{31} & \delta_{32} & 0 & \delta_{34} \\ 0 & 0 & 0 & 0 \end{bmatrix} \tag{8.22}$$

$$\delta_{21} = -\delta_{12} = \delta_\varphi \cos\theta\cos\psi + \cos\varphi\,(\delta_\psi \sin\theta - \delta_\theta \cos\theta\sin\psi) + \sin\varphi(\delta_\psi \cos\theta\sin\psi + \delta_\theta \sin\theta);$$

$$\delta_{31} = -\delta_{13} = -\delta_\varphi \sin\theta\cos\psi + \cos\varphi(\delta_\theta \sin\theta\sin\psi + \delta_\psi \cos\theta) + \sin\varphi(\delta_\theta \cos\theta - \delta_\psi \sin\theta\sin\psi);$$

$$\delta_{32} = -\delta_{23} = \delta_\theta \cos\varphi\cos\psi - \delta_\psi \cos\psi\sin\varphi + \delta_\varphi \sin\psi \qquad (8.23)$$

Hence, the matrix $\tilde{\mathbf{E}}$ combines the infinitesimal anti-symmetric 3×3 upper-left block and the 3×1 infinitesimal upper-right vector, i.e., the matrix obtained after general similarity transformation of the error matrix $\mathbf{E}$ is the error matrix in its standard form, Eq. (8.9).

Feature 6. Relationships with velocity matrix: Since the standard error matrix combines the set of infinitesimal displacements of the rigid body, there are close relationships of the matrix $\mathbf{E}$ with the velocity matrix $\mathbf{V}$ consisting of the 3×3 rotation block $\boldsymbol{\Omega}$ and the 3×1 translation vector $\mathbf{v}$:

$$\mathbf{E} = \mathbf{V}dt \qquad (8.24)$$

$$\mathbf{V} = \begin{bmatrix} \boldsymbol{\Omega} & \mathrm{v} \\ 0^{\mathrm{T}} & 0 \end{bmatrix} = \begin{bmatrix} 0 & -\omega_z & \omega_y & v_x \\ \omega_z & 0 & -\omega_x & v_y \\ -\omega_y & \omega_x & 0 & v_z \\ 0 & 0 & 0 & 0 \end{bmatrix}$$

$$\Omega = \begin{bmatrix} 0 & -\omega_z & \omega_y \\ \omega_z & 0 & -\omega_x \\ -\omega_y & \omega_x & 0 \end{bmatrix} \text{and } \mathbf{v} = \begin{bmatrix} v_x \\ v_y \\ v_z \end{bmatrix} \qquad (8.25)$$

where dt is the differential of time; ω_x, ω_y, and ω_z are the angular velocities of the rotation of the rigid body around the X-, Y-, and Z-axis, respectively ($\omega_x \approx \delta_\theta/dt$; $\omega_y \approx \delta_\psi/dt$; $\omega_z \approx \delta_\varphi/dt$); and v_x, v_y, and v_z are the linear velocities along the same axes ($v_x \approx \delta_x/dt$; $v_y \approx \delta_y/dt$; $v_z \approx \delta_z/dt$).

8.1.1.4 Axis and angle of small rotation

A finite rotation of the body can be described through establishing the unit vector $\mathbf{e}_{rot}$ of a rotation axis and the angle α of rotation around the axis, Eqs. (3.167) and (3.168),

$$\alpha = \arccos\frac{a_{11} + a_{22} + a_{33} - 1}{2} \text{and } \mathbf{e}_{rot} = \frac{1}{2\sin\alpha}\begin{bmatrix} a_{32} - a_{23} \\ a_{13} - a_{31} \\ a_{21} - a_{12} \\ 0 \end{bmatrix} \qquad (8.26)$$

If the three Euler angles are small, the *first order approximation*, Eqs. (8.1) and (8.2), cannot be used for calculation of the small angle δ_α since the *first order approximation* results in $a_{11} = a_{22} = a_{33} = 1$ and hence $\alpha = 0$. Thus, the *second order approximation* has to be examined

$$\cos\delta \approx 1 - \frac{\delta^2}{2} \text{ and, as before, } \sin\delta \approx \delta \tag{8.27}$$

Substitution of the diagonal elements a_{11}, a_{22}, and a_{33} from matrix $\mathbf{A}_{Eu}$, Eq. (8.6), in Eq. (8.26) yields

$$\cos\alpha = \frac{a_{11} + a_{22} + a_{33} - 1}{2} = \frac{\cos\varphi\ \cos\psi + \cos\varphi\ \cos\theta - \sin\varphi\ \sin\psi\ \sin\theta + \cos\psi\ \cos\theta - 1}{2} \tag{8.28}$$

Substituting approximations (8.27) for small angles α, θ, ψ, and φ in both parts of Eq. (8.28) yields:

$$1 - \frac{\delta_{rot}^2}{2} = \frac{1}{2}[(1 - \frac{\delta_\varphi^2}{2})(1 - \frac{\delta_\psi^2}{2}) + (1 - \frac{\delta_\varphi^2}{2})(1 - \frac{\delta_\theta^2}{2}) + (1 - \frac{\delta_\psi^2}{2})(1 - \frac{\delta_\theta^2}{2}) - 1]$$
$$= 1 - \frac{\delta_\theta^2 + \delta_\psi^2 + \delta_\varphi^2}{2}$$

Then,

$$\delta_{rot} = \sqrt{\delta_\theta^2 + \delta_\psi^2 + \delta_\varphi^2} \tag{8.29}$$

Substitution of the off-diagonal entries of matrix **E**, Eq. (8.9), into Eq. (8.26) taking into account the second order approximation, Eq. (8.27) results in the unit vector of an axis of the body rotation

$$\mathbf{e}_{rot} = \frac{1}{2\sin\delta_{rot}} \begin{bmatrix} a_{32} - a_{23} \\ a_{13} - a_{31} \\ a_{21} - a_{12} \end{bmatrix} = \frac{1}{\delta_{rot}} \begin{bmatrix} \delta_\theta \\ \delta_\psi \\ \delta_\varphi \end{bmatrix} = \boldsymbol{\delta} / \delta_{rot} \tag{8.30}$$

Hence,

$$\mathbf{e}_{rot} = \begin{bmatrix} e_{rot,x} \\ e_{rot,y} \\ e_{rot,z} \end{bmatrix} = \boldsymbol{\delta} / \delta_{rot} \tag{8.31}$$

where $\boldsymbol{\delta}$ is the angular error vector; and $e_{rot,x}$, $e_{rot,y}$, and $e_{rot,z}$ are components of the unit vector of an axis of the body rotation

$$\boldsymbol{\delta} = [\delta_\theta, \delta_\psi, \delta_\varphi]^T \tag{8.32}$$

$$e_{rot,x} = \delta_\theta/\delta_{rot};\ e_{rot,y} = \delta_\psi/\delta_{rot};\ e_{rot,z} = \delta_\varphi/\delta_{rot}; \tag{8.33}$$

Angle δ_c of rotation around an axis given by unit vector $\mathbf{c} = [c_x, c_y, c_z]^T$, $||\mathbf{c}|| = 1$, presents projection of vector $\boldsymbol{\delta}$ on vector $\mathbf{e}$ and calculated by dot product

$$\delta_c = \mathbf{c} \cdot \boldsymbol{\delta} = c_x\delta_\theta + c_y\delta_\psi + c_z\delta_\varphi \tag{8.34}$$

Example: Angle of rotation around the X-axis is δ_θ is $\delta_c = \begin{bmatrix} 1 \\ 0 \\ 0 \end{bmatrix} \cdot \begin{bmatrix} \delta_\theta \\ \delta_\psi \\ \delta_\varphi \end{bmatrix} = \delta_\theta$.

8.1.2 Position and orientation errors

As a result of the small displacement and rotation of the rigid body given by infinitesimal matrix **E**, Eq. (8.9), a body point obtains displacements (coordinate errors)

$$\Delta\mathbf{r} \equiv \begin{bmatrix} \Delta x \\ \Delta y \\ \Delta z \\ 0 \end{bmatrix} = \mathbf{E}\,\mathbf{r} = \begin{bmatrix} 0 & -\delta_\varphi & \delta_\psi & \delta_x \\ \delta_\varphi & 0 & -\delta_\theta & \delta_y \\ -\delta_\psi & \delta_\theta & 0 & \delta_z \\ 0 & 0 & 0 & 0 \end{bmatrix} \begin{bmatrix} x \\ y \\ z \\ 1 \end{bmatrix} = \begin{bmatrix} \delta_x - \delta_\varphi y - \delta_\psi z \\ \delta_y + \delta_\varphi x - \delta_\theta z \\ \delta_z - \delta_\psi x + \delta_\theta y \\ 0 \end{bmatrix} \tag{8.35}$$

Correspondingly, a direction of the unit vector **c** ($||\mathbf{c}|| = 1$) associated with the body orientation obtains the following non-designed distortions (orientation errors)

$$\Delta\mathbf{c} \equiv \begin{bmatrix} \Delta c_x \\ \Delta c_y \\ \Delta c_z \\ 0 \end{bmatrix} = \mathbf{E}\,\mathbf{c} = \begin{bmatrix} 0 & -\delta_\varphi & \delta_\psi & \delta_x \\ \delta_\varphi & 0 & -\delta_\theta & \delta_y \\ -\delta_\psi & \delta_\theta & 0 & \delta_z \\ 0 & 0 & 0 & 0 \end{bmatrix} \begin{bmatrix} c_x \\ c_y \\ c_z \\ 0 \end{bmatrix} = \begin{bmatrix} -\delta_\varphi c_y + \delta_\psi c_z \\ \delta_\varphi c_x - \delta_\theta c_z \\ -\delta_\psi c_x - \delta_\theta c_y \\ 0 \end{bmatrix} \tag{8.36}$$

In the planar case, the error matrix **E**, position error vector $\Delta\mathbf{r}$, and orientation error Δc become

$$\mathbf{E} = \begin{bmatrix} 0 & -\delta_\varphi & \delta_x \\ \delta_\varphi & 0 & \delta_y \\ 0 & 0 & 0 \end{bmatrix}; \tag{8.37}$$

$$\Delta\mathbf{r} \equiv \begin{bmatrix} \Delta x \\ \Delta y \\ 0 \end{bmatrix} = \mathbf{E}\mathbf{r} = \begin{bmatrix} 0 & -\delta_\varphi & \delta_x \\ \delta_\varphi & 0 & \delta_y \\ 0 & 0 & 0 \end{bmatrix} \begin{bmatrix} x \\ y \\ 1 \end{bmatrix} = \begin{bmatrix} \delta_x - \delta_\varphi y \\ \delta_y + \delta_\varphi x \\ 0 \end{bmatrix}; \text{ and } \Delta c = \delta\varphi \tag{8.38}$$

8.1.3 Displacements of the body point and the Plücker coordinates of the line

8.1.3.1 Displacement along the given direction

A small relative displacements of the body point from its nominal position K (Fig. 8.1) in the actual position Q is defined through error vector $\Delta\mathbf{r}$, Eq. (8.35). A projection of the small displacement vector $\Delta\mathbf{r}$ onto a line, whose direction is specified by unit vector $\mathbf{e}$ ($\|\mathbf{e}\| = 1$), can be expressed through position and angular deviations of the rigid body defined by six small values δ_x, δ_y, δ_z, δ_θ, δ_ψ, and δ_φ. The projection is a scalar value calculated through the dot product

$$\Delta = \Delta\mathbf{r} \cdot \mathbf{e} = \begin{bmatrix} \delta_x - \delta_\varphi y + \delta_\psi z \\ \delta_y + \delta_\varphi x - \delta_\theta z \\ \delta_z - \delta_\psi x + \delta_\theta y \\ 0 \end{bmatrix} \cdot \begin{bmatrix} e_x \\ e_y \\ e_z \\ 0 \end{bmatrix} =$$

$$e_x\delta_x + e_y\delta_y + e_z\delta_z + \left(ye_z - ze_y\right)\delta_\theta + \left(ze_x - xe_z\right)\delta_\psi + \left(xe_y - ye_x\right)\delta_\varphi \tag{8.39}$$

where e_x, e_y and e_z are the direction cosines forming the unit vector $\mathbf{e}$.

The bracketed coefficients at small angles δ_θ, δ_ψ, δ_φ are the moments of the unit vector $\mathbf{e}$ relative to the X-, Y-, Z-axis, respectively, composing the moment vector

$$\mathbf{n} = [n_x, n_y, n_z]^{\mathrm{T}} = \mathbf{r} \times \mathbf{e}$$

with

$$n_x = \begin{vmatrix} y & z \\ e_y & e_z \end{vmatrix} = e_z y - e_y z; n_y = \begin{vmatrix} z & x \\ e_z & e_x \end{vmatrix} = e_x z - e_z x; n_z = \begin{vmatrix} x & y \\ e_x & e_y \end{vmatrix} = e_y x - e_x y \tag{8.40}$$

Hence, displacement Δ, Eq. (8.39), along the direction given through unit vector $\mathbf{e}$ can be represented as

$$\Delta = e_x\delta_x + e_y\delta_y + e_z\delta_z + n_x\delta_\theta + n_y\delta_\psi + n_z\delta_\varphi \tag{8.41}$$

Six errors δ_x, δ_y, δ_z, δ_θ, δ_ψ, and δ_φ entering into the error matrix $\mathbf{E}$, Eq. (8.9), can be combined into the 6×1 vector called *twist*

$$\boldsymbol{\delta} = [\delta_x, \delta_y, \delta_z, \delta_\theta, \delta_\psi, \delta_\varphi]^{\mathrm{T}} \tag{8.42}$$

Six real numbers e_x, e_y, e_z, n_x, n_y, and n_z entering into Eq. (8.41) are called the Plücker coordinates of the line. The line goes through the point with coordinates x, y, z and its direction cosines combines the unit vector $\mathbf{e}$. The vector of the Plücker coordinates is

$$\mathbf{u} = [e_x, e_y, e_z, n_x, n_y, n_z]^{\mathrm{T}} \tag{8.43}$$

In these terms, Eq. (8.41) can be rewritten through the matrix product of the above implemented vectors of the sixth order

$$\Delta = \mathbf{u}^{T}\boldsymbol{\delta} \tag{8.44}$$

It is important to emphasize that both Eq. (8.44) and Eq. (8.41) define the identical values and will be used in one or another form depending on the situation. Note that only four of six Plücker coordinates are independent since two following connections take place

$$||\mathbf{e}|| = 1 \tag{8.45}$$

$$\mathbf{e} \cdot \mathbf{n} = \mathbf{e} \cdot [\mathbf{r} \times \mathbf{e}] = 0 \tag{8.46}$$

Vector $\Delta\mathbf{c}$ of the orientation error is defined through Eq. (8.36). Applying equation (8.34) for small angle δ_c of rotation around the given axis **e** we can define the vector **u** of the Plücker coordinates for the angular parameter

$$\mathbf{u} = [0, 0, 0, e_x, e_y, e_z]^{T} \tag{8.47}$$

$$\delta_c = \mathbf{u}^{T}\mathbf{e} = e_x\,\delta_\theta + e_y\,\delta_\psi + e_z\,\delta_\varphi \tag{8.48}$$

8.1.3.2 *Jacobian definition and application*

If the six errors δ_x, δ_y, δ_z, δ_θ, δ_ψ, and δ_φ defining the body position and orientation are unknown, they can be found through $N \geq 6$ linear simultaneous equations of type (8.41) and/or (8.48). For example, if there are only equations (8.41), the system becomes

$$e_{x1}\,\delta_x + e_{y1}\,\delta_y + e_{z1}\,\delta_z + n_{x1}\,\delta_\theta + n_{y1}\,\delta_\psi + n_{z1}\,\delta_\varphi = \Delta_1$$

$$e_{x2}\,\delta_x + e_{y2}\,\delta_y + e_{z2}\,\delta_z + n_{x2}\,\delta_\theta + n_{y2}\,\delta_\psi + n_{z2}\,\delta_\varphi = \Delta_2$$

....

$$e_{xN}\,\delta_x + e_{yN}\,\delta_y + e_{zN}\,\delta_z + n_{xN}\,\delta_\theta + n_{yN}\,\delta_\psi + n_{zN}\,\delta_\varphi = \Delta_N \tag{8.49}$$

where Δ_i ($i = 1,\ldots,N$) are known displacements, e.g., results of measurements. The system can be rewritten in the matrix form

$$\mathbf{J}\,\boldsymbol{\delta} = \boldsymbol{\Delta} \tag{8.50}$$

$$\mathbf{J} = \begin{bmatrix} e_{x1} & e_{y1} & e_{z1} & n_{x1} & n_{y1} & n_{z1} \\ e_{x2} & e_{y2} & e_{z2} & n_{x2} & n_{y2} & n_{z2} \\ \cdots & \cdots & \cdots & \cdots & \cdots & \cdots \\ e_{xN} & e_{yN} & e_{zN} & n_{xN} & n_{yN} & n_{zN} \end{bmatrix} \tag{8.51}$$

$$\boldsymbol{\delta} = [\delta_x, \delta_y, \delta_z, \delta_\theta, \delta_\psi, \delta_\varphi]^T$$

$$\boldsymbol{\Delta} = [\Delta_1, \Delta_2, \ldots, \Delta_N]^T \tag{8.52}$$

where **J** is the $N \times 6$ structural matrix of system (8.49) composed N rows of the Plücker coordinates, Eq.(8.41), of the body supports; $\boldsymbol{\delta}$ is the twist (error vector), Eq. (8.42); and $\boldsymbol{\Delta}$ is the vector of the right-hand sides of Eqs. (8.49).

Obviously, Eqs. (8.49) or (what is the same) matrix equation (8.51) allow calculation of small displacements of the N body points given by its coordinates along the given directions. The matrix **J** presents the Jacobian matrix of the system of simultaneous equations (8.49). In what follows, matrix **J** and its determinant Det[**J**] will be called the support *Jacobian matrix* and the *Jacobian*, respectively.

The simultaneous equations (8.49) have to be solved with respect to the six unknowns δ_x, δ_y, δ_z, δ_θ, δ_ψ, and δ_φ with the purpose of selecting the error magnitudes that ensures the compensation of the known output error. The solution procedure is considered under the conditions that $\boldsymbol{\Delta} \neq 0$ and the matrix **J** has full rank,

$$\text{rank}[\mathbf{J}] = 6 \tag{8.53}$$

It is evident that this requirement means that points $\mathbf{r}_i$ and directions $\mathbf{e}_i$ cannot be chosen arbitrarily. First, to satisfy requirement (8.52), among the simultaneous equations must be no fewer than three equations for linear displacements, Eq. (8.44). Second, the solution has different forms depending on whether $N = 6$ or $N > 6$. In the case $N = 6$, Eq. (8.53) becomes the requirement for the determinant of the Jacobian matrix

$$\text{Det}[\mathbf{J}] \neq 0 \tag{8.54}$$

Hence, the inverse Jacobian matrix exists, and the system twist $\boldsymbol{\delta}$, Eq. (8.42), becomes

$$\boldsymbol{\delta} = \mathbf{J}^{-1}\boldsymbol{\Delta} \tag{8.55}$$

If there are two sets of measurements Δ_1 and Δ_2 associated with two Jacobian matrices $\mathbf{J}_1$ and $\mathbf{J}_2$(with Det[$\mathbf{J}_1$] $\neq 0$ and Det[$\mathbf{J}_2$] $\neq 0$), respectively, both relating to *the same rigid body*, one can use Eq. (8.55) to obtain an interrelationship between Δ_1 and Δ_2:

$$\boldsymbol{\delta} = (\mathbf{J}_1)^{-1}\Delta_1 = (\mathbf{J}_2)^{-1}\Delta_2$$

$$\Delta_1 = \mathbf{J}_1(\mathbf{J}_2)^{-1}\Delta_2 \text{ or } \Delta_2 = \mathbf{J}_2(\mathbf{J}_1)^{-1}\Delta_1 \tag{8.56}$$

If $N < 6$ and condition (8.53)takes place, solution of simultaneous equations (8.49) presents, in sense, the best estimate of the unknowns. Estimate $\hat{\boldsymbol{\delta}}_i$ is calculated using the least mean square method. In the assumption that the values $\Delta_1, \Delta_2, \ldots, \Delta_N$ are known with identical precision (for example, they are obtained from the similar measurements), the estimation of the unknowns entering in system (8.48) is obtained [4] using the $N\times 6$ Jacobian matrix **J**, Eq. (8.51),

$$\hat{\boldsymbol{\delta}} = (\mathbf{J}^T\mathbf{J})^{-1}\mathbf{J}^T\boldsymbol{\Delta} \tag{8.57}$$

8.1.3.3 Planar case

The derived formulae are valid also for the planar case with the following modifications:

(a) The *twist* combines three errors – small linear displacements δ_x and δ_y, and small rotation angle δ_φ, i.e.,

$$\boldsymbol{\delta} = [\delta_x, \delta_y, \delta_\varphi]^T \tag{8.58}$$

(b) The error matrix **E**, Eq. (8.9), becomes the 3 × 3 matrix

$$\mathbf{E} = \begin{bmatrix} 0 & -\delta_\varphi & \delta_x \\ \delta_\varphi & 0 & \delta_y \\ 0 & 0 & 0 \end{bmatrix} \tag{8.59}$$

Equivalent representations of the error matrix are

$$\mathbf{E} = \mathbf{E}_{rot} + \mathbf{E}_{tr} = \begin{bmatrix} 0 & -\delta_\varphi & 0 \\ \delta_\varphi & 0 & 0 \\ 0 & 0 & 0 \end{bmatrix} + \begin{bmatrix} 0 & 0 & \delta_x \\ 0 & 0 & \delta_y \\ 0 & 0 & 0 \end{bmatrix} = \delta_x \begin{bmatrix} 0 & 0 & 1 \\ 0 & 0 & 0 \\ 0 & 0 & 0 \end{bmatrix} + \delta_y \begin{bmatrix} 0 & 0 & 1 \\ 0 & 0 & 0 \\ 0 & 0 & 0 \end{bmatrix} + \delta_\varphi \begin{bmatrix} 0 & -1 & 0 \\ 1 & 0 & 0 \\ 0 & 0 & 0 \end{bmatrix}$$

(c) The vector of the Plücker coordinates, Eq. (8.43), becomes

$$\mathbf{u} = [e_x, e_y, n_z]^T \tag{8.60}$$

where e_x and e_y are the direction cosines ($e_x^2 + e_y^2 = 1$) of the unit vector **e** defined the line orientation, $\mathbf{e} = [e_x, e_y]^T$; and n_z is the moment of vector **e** with respect to the origin,

$$n_z = \begin{vmatrix} x & y \\ e_x & e_y \end{vmatrix} = x e_y - y e_x \tag{8.61}$$

(d) The 3 × 3 Jacobian matrix is

$$\mathbf{J} = \begin{bmatrix} e_{x1} & e_{y1} & n_{z1} \\ e_{x2} & e_{y2} & n_{z2} \\ e_{x3} & e_{y3} & n_{z3} \end{bmatrix}, \tag{8.62}$$

(e) Projection Δ of the small displacement vector (twist) $\Delta\mathbf{r}$ onto the line specified by unit vector $\mathbf{e} = [e_x, e_y, 0]^T$, Eq. (8.60), becomes:

$$\Delta = \mathbf{u}^T\boldsymbol{\delta} = e_x\delta_x + e_y\delta_y + n_z\delta_\varphi \tag{8.63}$$

8.2 Error Budget

8.2.1 Manipulation Errors

Definition 1: The 4 × 4 matrix ${}^{0}\mathbf{A}_n$, Eq. (3.153), entering in the FSF will be further called the *nominal* manipulation matrix

$$ {}^{0}\mathbf{A}_n = \prod_{i=1}^{n} {}^{i-1}\mathbf{A}_i \tag{8.64} $$

where n is the number of the FSS links; and ${}^{i-1}\mathbf{A}_i (i = 1, 2,\ldots,n)$ is the 4 × 4 HTM, Eq. (3.131), associated with the ith link.

Definition 2: An *actual* manipulation matrix ${}^{0}\hat{\mathbf{A}}_n$ describing motion of the FSS links accompanied by their position-orientation (geometric and kinematic) errors is

$$ {}^{0}\hat{\mathbf{A}}_n = \prod_{i=1}^{n} {}^{i-1}\hat{\mathbf{A}}_i \tag{8.65} $$

where ${}^{i-1}\hat{\mathbf{A}}_i$ is the 4 × 4 HTM associated with the real motion of the ith link, expressing through nominal matrix ${}^{i-1}\mathbf{A}_i$ and matrix of small motion $\mathbf{V}_i$, Eq. (8.7),

$$ {}^{i-1}\hat{\mathbf{A}}_i = {}^{i-1}\mathbf{A}_i\mathbf{V}_i \tag{8.66} $$

Equation (8.66) means that the real motion of the ith link is interpreted as two successive motions of the link: the nominal motion and a small motion caused by six position-orientation errors, in that order.

Definition 3: The matrix of manipulation errors ${}^{0}\Delta\mathbf{A}_n$ resulting from small position-orientation deviations of the FSS links from their nominal motion presents the difference

$$ {}^{0}\Delta\mathbf{A}_n = {}^{0}\hat{\mathbf{A}}_n - {}^{0}\mathbf{A}_n \tag{8.67} $$

Statement: The manipulation error matrix ${}^{0}\Delta\mathbf{A}_n$, comprises 6 errors of the initial link with respect to the base coordinate frame and 6n elementary errors of n FSS links, and is expressed through the HTMs, Eq. (3.131), and the standard error matrices $\mathbf{E}_i$, Eq. (8.9),

$$ {}^{0}\Delta\mathbf{A}_n = \sum_{i=1}^{n} {}^{0}\mathbf{A}_i\mathbf{E}_i\,{}^{i}\mathbf{A}_n \tag{8.68} $$

$$ {}^{0}\mathbf{A}_i = \begin{cases} \mathbf{I}_{4\times 4}, \text{ for } j = 0 \\ \prod_{j=1}^{i} {}^{j-1}\mathbf{A}_j, \text{ for } j > 0 \end{cases} \tag{8.69} $$

$$ {}^{i}\mathbf{A}_n = \begin{cases} \prod_{j=i}^{n} {}^{j}\mathbf{A}_n, & \text{for } j < n \\ \mathbf{I}_{4\times 4}, & \text{for } j = n \end{cases} \tag{8.70} $$

$$ \mathbf{E}_i = \begin{bmatrix} 0 & -\delta_{\varphi i} & \delta_{\psi i} & \delta_{xi} \\ \delta_{\varphi i} & 0 & -\delta_{\theta i} & \delta_{yi} \\ -\delta_{\psi i} & \delta_{\theta i} & 0 & \delta_{zi} \\ 0 & 0 & 0 & 0 \end{bmatrix}, i = 0,\ldots,n \tag{8.71} $$

where $\mathbf{E}_0$ is the error matrix comprised the location and orientation errors of the initial link with respect to the base coordinate frame S_0; and $\mathbf{E}_i$ $(i = 1,\ldots,n)$ is the error matrix of the location and orientation errors of the ith link as referring to its own coordinate frame S_i.

Proof: Substituting Eq. (8.8) for matrix $\mathbf{V}_i$ in Eq. (8.66) yields

$$ {}^{i-1}\hat{\mathbf{A}}_i = {}^{i-1}\mathbf{A}_i\mathbf{V}_i = {}^{i-1}\mathbf{A}_i(\mathbf{I} + \mathbf{E}_i) = {}^{i-1}\mathbf{A}_i + {}^{i-1}\mathbf{A}_i\mathbf{E}_i \tag{8.72} $$

where $\mathbf{E}_i$ is the standard error matrix associated with the ith link, Eq. (8.9).

After substitution this expression into Eq. (8.65), we obtain

$$ {}^{0}\hat{\mathbf{A}}_n = \prod_{i=1}^{n} ({}^{i-1}\mathbf{A}_i + {}^{i-1}\mathbf{A}_i\mathbf{E}_i) \tag{8.73} $$

The scalar linearization conditions, Eq. (8.6), can be extended as applied to the error matrices

$$ \mathbf{E}_i\mathbf{E}_k \approx \mathbf{0} \text{ and } {}^{0}\mathbf{A}_i\mathbf{E}_i\,{}^{i}\mathbf{A}_k\mathbf{E}_k\,{}^{k}\mathbf{A}_n \approx \mathbf{0} \tag{8.74} $$

where $\mathbf{0}$ is the 4×4 zero matrix; $\mathbf{E}_i$ and $\mathbf{E}_k$ are the error matrices for the ith and kth links; ${}^{0}\mathbf{A}_i$, ${}^{i}\mathbf{A}_k$, and ${}^{k}\mathbf{A}_n$ are the 4×4 matrices defined similar to Eqs. (8.69) and (8.70).

Multiplication of the matrices in the right-hand side of Eq. (8.73) taking into account Eqs. (8.74) results in a linearized expression

$$ {}^{0}\hat{\mathbf{A}}_n = {}^{0}\mathbf{A}_n + {}^{0}\mathbf{A}_1\mathbf{E}_1\,{}^{1}\mathbf{A}_n + {}^{0}\mathbf{A}_2\mathbf{E}_2\,{}^{2}\mathbf{A}_n + \ldots + {}^{0}\mathbf{A}_{n-1}\mathbf{E}_{n-1}\,{}^{n-1}\mathbf{A}_n + {}^{0}\mathbf{A}_n\mathbf{E}_n \tag{8.75} $$

Hence, the manipulation error matrix ${}^{0}\Delta\mathbf{A}_n$, Eq. (8.67), becomes

$$ {}^{0}\Delta\mathbf{A}_n = {}^{0}\hat{\mathbf{A}}_n - {}^{0}\mathbf{A}_n = $$

$$ {}^{0}\mathbf{A}_1\mathbf{E}_1\,{}^{1}\mathbf{A}_n + {}^{0}\mathbf{A}_2\mathbf{E}_2\,{}^{2}\mathbf{A}_n + \ldots + {}^{0}\mathbf{A}_{n-1}\mathbf{E}_{n-1}\,{}^{n-1}\mathbf{A}_n + {}^{0}\mathbf{A}_n\mathbf{E}_n = \sum_{i=1}^{n} {}^{0}\mathbf{A}_i\mathbf{E}_i\,{}^{i}\mathbf{A}_n \tag{8.76} $$

8.2.2 Error budget of the body point position: Vector representation

8.2.2.1 *Direct form of error budget*

The nominal position and orientation of the ending link associated with the *AE* (working tool) of the FSS with respect to the initial link associated with the *TE* (element to be processed) is defined by the FSF, Eq. (3.164),

$$[\mathbf{r}_0\ \mathbf{c}_0] = {}^0\mathbf{A}_n\ [\mathbf{r}_n\ \mathbf{c}_n] \tag{8.77}$$

or

$$\begin{bmatrix} x_0 & c_{0x} \\ y_0 & c_{0y} \\ z_0 & c_{0z} \\ 1 & 0 \end{bmatrix} = {}^0\mathbf{A}_n \begin{bmatrix} x_n & c_{nx} \\ y_n & c_{ny} \\ z_n & c_{nz} \\ 1 & 0 \end{bmatrix}$$

Vector $\Delta\mathbf{r}_0$ of the position error and vector $\Delta\mathbf{c}_0$ of the orientation error of the active element relative to the target element are defined as follows:

$$\Delta\mathbf{r}_0 = \begin{bmatrix} \Delta_x \\ \Delta_y \\ \Delta_z \\ 0 \end{bmatrix} \text{and}\ \Delta\mathbf{c}_0 = \begin{bmatrix} \Delta_\theta \\ \Delta_\psi \\ \Delta_\varphi \\ 0 \end{bmatrix} \tag{8.78}$$

Their values are calculated by substitution of the manipulation error matrix ${}^0\Delta\mathbf{A}_n$, Eq. (8.68), for the nominal manipulation matrix ${}^0\mathbf{A}_n$, Eq. (8.77):

$$[\Delta\mathbf{r}_0\ \Delta\mathbf{c}_0] = {}^0\Delta\mathbf{A}_n\ [\mathbf{r}_n\ \mathbf{c}_n] \tag{8.79}$$

i.e.,

$$\begin{bmatrix} \Delta_x & \Delta_\theta \\ \Delta_y & \Delta_\psi \\ \Delta_z & \Delta_\varphi \\ 0 & 0 \end{bmatrix} = {}^0\Delta\mathbf{A}_n \begin{bmatrix} x_n & c_{nx} \\ y_n & c_{ny} \\ z_n & c_{nz} \\ 1 & 0 \end{bmatrix} \tag{8.80}$$

Taking into account Eq. (8.68) we obtain

$$[\Delta\mathbf{r}_0\ \Delta\mathbf{c}_0] = \sum_{i=0}^{n} {}^0\mathbf{A}_i\mathbf{E}_i\,{}^i\mathbf{A}_n[\mathbf{r}_n\ \mathbf{c}_n] \tag{8.81}$$

The *i*th addendum in sum (8.81) presents contribution of the *i*th link errors into the total error

$$[\Delta\mathbf{r}_{0i}\ \Delta\mathbf{c}_{0i}] = {}^0\mathbf{A}_i\mathbf{E}_i\,{}^i\mathbf{A}_n[\mathbf{r}_n\ \mathbf{c}_n],\ i = 0, 1, \ldots, n \tag{8.82}$$

The error vector, Eq. (8.81), can be rewritten in the extended form as a sum of the contributions of all links including the locating error

$$[\Delta\mathbf{r}_0\ \Delta\mathbf{c}_0] = (\mathbf{E}_0{}^0\mathbf{A}_1\ {}^1\mathbf{A}_2 \ldots {}^{n-1}\mathbf{A}_n + {}^0\mathbf{A}_1\mathbf{E}_1\ {}^1\mathbf{A}_2 \ldots {}^{n-1}\mathbf{A}_n + {}^0\mathbf{A}_1{}^1\mathbf{A}_2\mathbf{E}_2 \ldots {}^{n-1}\mathbf{A}_n + \ldots +$$

$${}^0\mathbf{A}_1{}^1\mathbf{A}_2 \ldots \mathbf{E}_{n-1}{}^{n-1}\mathbf{A}_n + {}^0\mathbf{A}_1{}^1\mathbf{A}_2 \ldots {}^{n-1}\mathbf{A}_n\mathbf{E}_n)\ [\mathbf{r}_n\ \mathbf{c}_n]$$

For illustration, consider two simple examples related to two-step chains.
Numerical Example 8.2: The *XY-translational stage* (Fig. 8.2). Since the number of links is $n = 2$, manipulation matrix ${}^0\mathbf{A}_n = {}^0\mathbf{A}_2$ and error matrix ${}^0\Delta\mathbf{A}_n = {}^0\Delta\mathbf{A}_2$ are

$$ {}^0\mathbf{A}_2 = \mathbf{A}^1(x)\mathbf{A}^2(y) = \begin{pmatrix} 1 & 0 & 0 & x \\ 0 & 1 & 0 & 0 \\ 0 & 0 & 1 & 0 \\ 0 & 0 & 0 & 1 \end{pmatrix} \begin{pmatrix} 1 & 0 & 0 & 0 \\ 0 & 1 & 0 & y \\ 0 & 0 & 1 & 0 \\ 0 & 0 & 0 & 1 \end{pmatrix} = \begin{pmatrix} 1 & 0 & 0 & x \\ 0 & 1 & 0 & y \\ 0 & 0 & 1 & 0 \\ 0 & 0 & 0 & 1 \end{pmatrix} \quad (8.83)$$

$$ {}^0\Delta\mathbf{A}_2 = \mathbf{E}_0\mathbf{A}^1(x)\mathbf{A}^2(y) + \mathbf{A}^1(x)\mathbf{E}_1\mathbf{A}^2(y) + \mathbf{A}^1(x)\mathbf{A}^2(y)\mathbf{E}_2 = $$

$$ = \begin{bmatrix} 0 & -\delta_{\varphi 0,1,2} & \delta_{\psi 0,1,2} & \delta_{x0,1,2} - \delta_{\psi 0,1}y \\ \delta_{\varphi 0,1,2} & 0 & -\delta_{\theta 0,1,2} & \delta_{y0,1,2} + \delta_{\psi 0}x \\ -\delta_{\psi 0,1,2} & \delta_{\theta 0,1,2} & 0 & \delta_{z0,1,2} - \delta_{\psi 0}x + \delta_{\theta 0,1}y \\ 0 & 0 & 0 & 0 \end{bmatrix} \quad (8.84)$$

where $\delta_{x0,1} = \delta_{x0} + \delta_{x1}$; $\delta_{x0,1,2} = \delta_{x0} + \delta_{x1} + \delta_{x2}$; $\delta_{y0,1,2} = \delta_{y0} + \delta_{y1} + \delta_{y2}$; and so on.
The position error $\Delta\mathbf{r}_0$, Eq. (8.35), and orientation error $\Delta\mathbf{c}_0$, Eq. (8.36), become

$$[\Delta\mathbf{r}_0\ \Delta\mathbf{c}_0] = {}^0\Delta\mathbf{A}_2[\mathbf{r}_2\ \mathbf{c}_2] = \begin{bmatrix} 0 & -\delta_{\varphi 0,1,2} & \delta_{\psi 0,1,2} & \delta_{x0,1,2} - \delta_{\psi 0,1}y \\ \delta_{\varphi 0,1,2} & 0 & -\delta_{\theta 0,1,2} & \delta_{y0,1,2} + \delta_{\psi 0}x \\ -\delta_{\psi 0,1,2} & \delta_{\theta 0,1,2} & 0 & \delta_{z0,1,2} - \delta_{\psi 0}x + \delta_{\theta 0,1}y \\ 0 & 0 & 0 & 0 \end{bmatrix} \begin{bmatrix} x_2 & c_{2x} \\ y_2 & c_{2y} \\ z_2 & c_{2z} \\ 1 & 0 \end{bmatrix} =$$

$$\begin{bmatrix} \delta_{x0,1,2} - \delta_{\varphi 0,1}y - \delta_{\varphi 0,1,2}y_2 + \delta_{\psi 0,1,2}z_2 & -\delta_{\varphi 0,1}c_{y2} + \delta_{\psi 0,1,2}c_{z2} \\ \delta_{y0,1,2} + \delta_{\varphi 0}x + \delta_{\varphi 0,1,2}x_2 - \delta_{\theta 0,1,2}z_2 & \delta_{\varphi 0,1}c_{x2} - \delta_{\theta 0,1,2}c_{z2} \\ \delta_{z0,1,2} - \delta_{\psi 0}x - \delta_{\psi 0,1,2}x_2 + \delta_{\theta 0,1}y + \delta_{\theta 0,1,2}y_2 & -\delta_{\psi 0,1,2}c_{x2} + \delta_{\theta 0,1,2}c_{y2} \\ 0 & 0 \end{bmatrix} \quad (8.85)$$

Numerical Example 8.3. Facing operation on the lathe (Fig. 8.3). The manipulation matrix ${}^0\mathbf{A}_n$ and associated error matrix ${}^0\Delta\mathbf{A}_n$, Eq. (8.81), are

$$ {}^0\mathbf{A}_2 = \mathbf{A}^6(\varphi)\mathbf{A}^1(x) = \begin{bmatrix} \cos\varphi & -\sin\varphi & 0 & 0 \\ \sin\varphi & \cos\varphi & 0 & 0 \\ 0 & 0 & 1 & 0 \\ 0 & 0 & 0 & 1 \end{bmatrix} \begin{bmatrix} 1 & 0 & 0 & x \\ 0 & 1 & 0 & 0 \\ 0 & 0 & 1 & 0 \\ 0 & 0 & 0 & 1 \end{bmatrix} = \begin{bmatrix} \cos\varphi & -\sin\varphi & 0 & x\cos\varphi \\ \sin\varphi & \cos\varphi & 0 & x\sin\varphi \\ 0 & 0 & 1 & 0 \\ 0 & 0 & 0 & 1 \end{bmatrix}$$

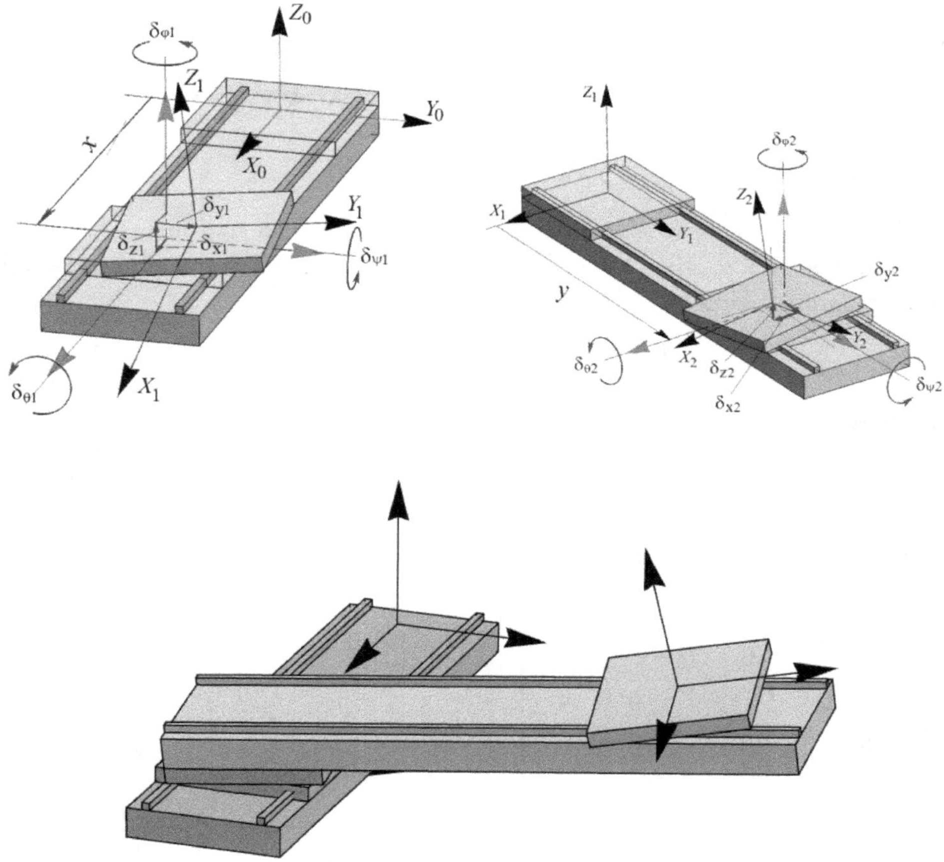

Figure 8-2 The XY translational stage with errors of the carriage positions and motions.

$$^{0}\mathbf{\Delta A}_2 = \mathbf{E}_0\, \mathbf{A}^6(\varphi)\mathbf{A}^1(x) + \mathbf{A}^6(\varphi)\mathbf{E}_1\, \mathbf{A}^1(x) + \mathbf{A}^6(\varphi)\,\mathbf{A}^1(x)\mathbf{E}_2 =$$

$$\begin{bmatrix} -\delta_{\varphi 0,1,2}\sin\varphi & -\delta_{\varphi 0,1,2}\cos\varphi & \delta_{\psi 0}+\delta_{\psi 1,2}\cos\varphi+\delta_{\theta 1,2}\sin\varphi & \Delta_{14} \\ \delta_{\varphi 0,1,2}\cos\varphi & -\delta_{\varphi 0,1,2}\sin\varphi & -\delta_{\theta 0}-\delta_{\theta 1,2}\cos\varphi+\delta_{\psi 1,2}\sin\varphi & \Delta_{24} \\ -\delta_{\psi 0}\cos\varphi+\delta_{\theta 0}\sin\varphi+\delta_{\psi 1,2} & \delta_{\theta 0}\cos\varphi+\delta_{\psi 0}\sin\varphi+\delta_{\theta 1,2} & 0 & \Delta_{34} \\ 0 & 0 & 0 & 0 \end{bmatrix}$$

with

$$\Delta_{14} = \delta_{x0} + \delta_{x1,2}\cos\varphi - (\delta_{y1,2} + \delta_{\varphi 0,1}\, x)\sin\varphi$$

$$\Delta_{24} = \delta_{y0} + \delta_{x1,2}\sin\varphi + (\delta_{y1,2} + \delta_{\varphi 0,1}\, x)\cos\varphi$$

$$\Delta_{34} = \delta_{z0,1,2} - \delta_{\psi 0}\, x\cos\varphi + \delta_{\theta 0}\, x\sin\varphi - \delta_{\psi 1}\, x \tag{8.86}$$

Then, position errors are calculated $\Delta\mathbf{r} = {}^{0}\Delta\mathbf{A}_2\mathbf{r}_2$

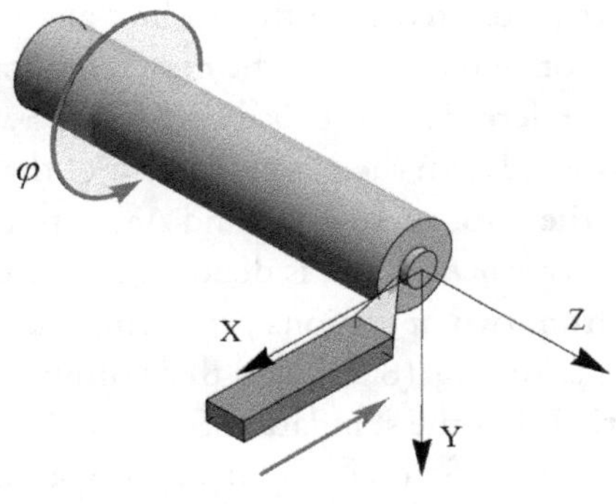

Figure 8-3 Facing operation motions.

8.2.2.2 *Standard form of the error budget*

In the general case, the error budget matrix ${}^0\Delta\mathbf{A}_n$, Eq. (8.79), is not of the standard form **E**, Eq. (8.9), since its 3 × 3 upper-left block is not an anti-symmetrical matrix. As an example, see Eq. (8.86). However, the error budget matrix can be represented in the standard form

$$[\Delta\mathbf{r}_0\ \Delta\mathbf{e}_0]= \mathbf{E}[\mathbf{r}_0\ \mathbf{e}_0] \tag{8.87}$$

with

$$\mathbf{E} = \sum_{i=0}^{n} {}^0\mathbf{A}_i\mathbf{E}_i({}^0\mathbf{A}_i)^{-1} \tag{8.88}$$

$$\mathbf{E} = \begin{bmatrix} 0 & -\delta_\varphi & \delta_\psi & \delta_x \\ \delta_\varphi & 0 & -\delta_\theta & \delta_y \\ -\delta_\psi & \delta_\theta & 0 & \delta_z \\ 0 & 0 & 0 & 0 \end{bmatrix} \tag{8.89}$$

where δ_x, δ_y, δ_z are the linear deviations of the origin of the frame S_0; and δ_θ, δ_ψ, and δ_φ are the small rotations (angular deviations) around the axes of the frame S_0.

Actually, since the FSF manipulation matrix ${}^0\mathbf{A}_n$, Eq. (3.153), is non-singular, one can invers the FSF and express vectors $\mathbf{r}_n$ and $\mathbf{c}_n$ through $\mathbf{r}_0$ and $\mathbf{c}_0$:

$$[\mathbf{r}_n\,\mathbf{c}_n] = ({}^0\mathbf{A}_n)^{-1}[\mathbf{r}_0\,\mathbf{c}_0]$$

Matrix ${}^0\mathbf{A}_n$ can be represented as a product of two matrices ${}^0\mathbf{A}_n = {}^0\mathbf{A}_i\,{}^i\mathbf{A}_n$, where $0 \le i \le n$, and ${}^0\mathbf{A}_i$ and ${}^i\mathbf{A}_n$ are defined by Eqs. (8.69) and (8.70). Since both ${}^0\mathbf{A}_i$ and ${}^i\mathbf{A}_n$ are non-singular,the inverse matrix $({}^0\mathbf{A}_n)^{-1}$can be represented as $({}^0\mathbf{A}_n)^{-1}= ({}^i\mathbf{A}_n)^{-1}\,({}^0\mathbf{A}_i)^{-1}$. Hence,

$${}^i\mathbf{A}_n[\mathbf{r}_n\,\mathbf{c}_n] = {}^i\mathbf{A}_n\,({}^0\mathbf{A}_n)^{-1}[\mathbf{r}_0\,\mathbf{c}_0] = {}^i\mathbf{A}_n({}^i\mathbf{A}_n)^{-1}({}^0\mathbf{A}_i)^{-1}[\mathbf{r}_0\,\mathbf{c}_0] = ({}^0\mathbf{A}_i)^{-1}[\mathbf{r}_0\,\mathbf{c}_0] \tag{8.90}$$

Substitution of Eq. (8.90) in Eq. (8.81) yields

$$[\Delta\mathbf{r}_0\,\Delta\mathbf{c}_0]= \sum_{i=0}^{n} {}^0\mathbf{A}_i\mathbf{E}_i({}^0\mathbf{A}_i)^{-1}\ [\mathbf{r}_0\,\mathbf{c}_0] \tag{8.91}$$

The *i*th addendum-matrix in the right-hand side of Eq. (8.91) is denoted as

$$\mathbf{E}_{i0} = {}^0\mathbf{A}_i\mathbf{E}_i({}^0\mathbf{A}_i)^{-1} \tag{8.92}$$

Error matrices $\mathbf{E}_{i0}$ and $\mathbf{E}_i$ are associated with the *i*th link and connected by the similarity transformation–pre-multiplication of matrix $\mathbf{E}_i$ by the matrix of motion ${}^0\mathbf{A}_i$ and post-multiplication by its inverse. Matrix $\mathbf{E}_i$ is referred to the coordinate frame S_i, while matrix $\mathbf{E}_{i0}$ is referred to the coordinate frame S_0 associated with the target element of the FSS. A sum of *n* matrices connecting through Eq. (8.87) the error vectors $\Delta\mathbf{r}_0$ and $\Delta\mathbf{e}_0$ with their nominal position vectors $\mathbf{r}_0$ and $\mathbf{e}_0$ presents the *total error matrix* and is denoted as **E**, Eq. (8.89). The total manipulation error matrix **E** results from two operations preserving the standard form of the error matrix: the similarity transformation, Eq. (8.92), and their summation, Eq. (8.88). According to Eqs. (8.17) and (8.19), matrix **E** has the standard form of the error matrix, Eq. (8.9).

Remember that the six deviations δ_x, δ_y, δ_z, δ_θ, δ_ψ, and δ_φ in the right-hand side of Eq. (8.89) depend on (6*n* + 6) elementary errors δ_{xi}, δ_{yi}, δ_{zi}, $\delta_{\theta i}$, $\delta_{\psi i}$, and $\delta_{\varphi i}$ entering into matrices $\mathbf{E}_i$ ($i = 1, 2,\dots,n$). In what follows, we will also use the designation $({}^0\mathbf{A}_i)^{-1} = {}^i\mathbf{A}_0$. In these terms, Eq. (8.88) can be rewritten in the more compact form:

$$\mathbf{E} = \sum_{i=0}^{n} {}^0\mathbf{A}_i \mathbf{E}_i {}^i\mathbf{A}_0 \tag{8.93}$$

Position and orientation errors are calculated through Eq. (8.87)

$$[\Delta\mathbf{r}_0\ \Delta\mathbf{c}_0] = \mathbf{E}[\mathbf{r}_0\ \mathbf{c}_0] =$$

$$\begin{bmatrix} 0 & -\delta_\varphi & \delta_\psi & \delta_x \\ \delta_\varphi & 0 & -\delta_\theta & \delta_y \\ -\delta_\psi & \delta_\theta & 0 & \delta_z \\ 0 & 0 & 0 & 0 \end{bmatrix} \begin{bmatrix} x_0 & c_{0x} \\ y_0 & c_{0y} \\ z_0 & c_{0z} \\ 1 & 0 \end{bmatrix} = \begin{bmatrix} \delta_x - \delta_\varphi y_0 + \delta_\psi z_0 & -\delta_\varphi c_{0y} + \delta_\psi c_{0z} \\ \delta_y + \delta_\varphi x_0 - \delta_\theta z_0 & \delta_\varphi c_{0x} - \delta_\theta c_{0z} \\ \delta_z - \delta_\psi x_0 + \delta_\theta y_0 & -\delta_\psi c_{0x} + \delta_\theta c_{0y} \\ 0 & 0 \end{bmatrix} \tag{8.94}$$

8.2.2.3 Linear expansion of the budget

The position matrix **E** can be represented through Eq. (8.14) as a sum of six addenda, consisting of only one elementary error each. This allows representation of the position error vector $\Delta\mathbf{r}_0$ and orientation error vector $\Delta\mathbf{c}_0$ as a sum of (6*n* + 6) addenda as follows:

$$\Delta\mathbf{r}_0 \equiv \begin{bmatrix} \Delta_x \\ \Delta_y \\ \Delta_z \\ 0 \end{bmatrix} = \sum_{i=0}^{n} \sum_{j=1}^{6} \mathbf{b}_{ij} \delta_{ij} \tag{8.95}$$

$$\Delta\mathbf{c}_0 \equiv \begin{bmatrix} \Delta_\theta \\ \Delta_\psi \\ \Delta_\varphi \\ 0 \end{bmatrix} = \sum_{i=0}^{n} \sum_{j=1}^{6} \mathbf{d}_{ij} \delta_{ij} \tag{8.96}$$

where δ_{ij} ($i = 0, 1,\dots,n$; $j = 1, 2,\dots,6$) are the elementary errors ordered in the lexicographic form,

$$\delta_{i1} = \delta_{ix};\ \delta_{i2} = \delta_{iy};\ \delta_{i3} = \delta_{iz};\ \delta_{i4} = \delta_{i\theta};\ \delta_{i5} = \delta_{i\psi};\ \delta_{i6} = \delta_{i\varphi}; \tag{8.97}$$

and **b***ij* and **d***ij* are the vector transfer factors for elementary input error δ_{ij}.
Remember that the transfer factors can be expressed through either direct matrix form, Eq. (8.68), or reflective, Eq. (8.92), matrix form, and both of them are equivalent:

$$\mathbf{b}_{ij} \equiv \begin{bmatrix} b_x \\ b_y \\ b_z \\ 0 \end{bmatrix}_{ij} = {}^0\mathbf{A}_i \mathbf{D}^{j\,i}\mathbf{A}_n \mathbf{r}_n = {}^0\mathbf{A}_i \mathbf{D}^{j\,i}\mathbf{A}_0 \mathbf{r}_0 \tag{8.98}$$

$$\mathbf{d}_{ij} \equiv \begin{bmatrix} d_x \\ d_y \\ d_z \\ 0 \end{bmatrix}_{ij} = {}^0\mathbf{A}_i \mathbf{D}^{j\,i}\mathbf{A}_n \mathbf{c}_n = {}^0\mathbf{A}_i \mathbf{D}^{j\,i}\mathbf{A}_0 \mathbf{c}_0 \tag{8.99}$$

8.2.3 Error budget of the body point position: scalar presentation

An error along a given direction defines a scalar error budget. It is obtained by projecting the vector error budget onto unit vector $\mathbf{c}_0$ defining the chosen direction and derived through the dot-product of the small vector $\Delta\mathbf{r}_0$ and $\Delta\mathbf{c}_0$, Eqs. (8.79) and (8.87), by unit vector $\mathbf{e}_0$:

$$\Delta_0 = \Delta\mathbf{r}_0 \cdot \mathbf{e}_0 \tag{8.100}$$

$$\Delta c_0 = \Delta\mathbf{e}_0 \cdot \mathbf{e}_0 \tag{8.101}$$

where $\mathbf{e}_0 = [e_{x0}, e_{y0}, e_{z0}, 0]^T$

These definitions can be represented in scalar terms:

$$\Delta_0 = \sum_{i=0}^{n}\sum_{j=1}^{6} b_{ij}\delta_{ij} \tag{8.102}$$

$$\Delta c_0 = \sum_{i=0}^{n}\sum_{j=4}^{6} d_{ij}\delta_{ij} \tag{8.103}$$

where b_{ij} and d_{ij} are defined through dot products of the unit vector $\mathbf{e}_0$ by vector $\mathbf{b}_{ij}$, Eq. (8.98), or $\mathbf{d}_{ij}$, Eq. (8.99), respectively,

$$b_{ij} = \mathbf{b}_{ij} \cdot \mathbf{e}_0 \text{ and } d_{ij} = \mathbf{d}_{ij} \cdot \mathbf{e}_0 \tag{8.104}$$

Note that the scalar budgets Δ_0 and Δc_0 differ in the maximal number of their non-zero terms: $(6n + 6)$ and $(3n + 3)$, respectively. The influence of the error δ_{ij} on the output accuracy Δ_0 and Δc_0 is characterized by b_{ij} and d_{ij} values.

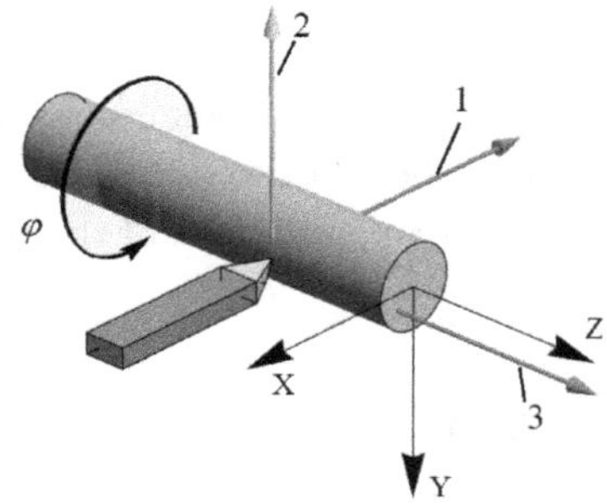

Figure 8-4 Sensitive (1) and non-sensitive (2 and 3) directions during cylinder turning by a single point edge.

If the error vector $\mathbf{b}_{ij}$ (or $\mathbf{d}_{ij}$) is located at right angle to the line axis, Eq. (8.104) yields $b_{ij} = \mathbf{b}_{ij} \cdot \mathbf{e}_0 = 0$ and $d_{ij} = \mathbf{b}_{ij} \cdot \mathbf{e}_0 = 0$. In this case, vector $\mathbf{e}_0$ defines a *non-sensitive direction* (in the first order approximation), along which the error δ_{ij} makes no impact on the Δ_0 (or Δc_0) value. Otherwise, vector $\mathbf{e}_0$ defines a *sensitive direction*.

Turning the cylinder by a single-point edge is considered as a simple example (Fig. 8.4). Displacement of the working point at small distance δ_x along the X-axis (arrow 1) changes the diameter of the machined cylinder by $\Delta D = 2\delta_x$. Displacement at small distance δ_z along the Z-axis (arrow 3) changes a position of the machined face by $\Delta_l = \delta_z$. Otherwise, unit vectors of the X-axis $\mathbf{e}_0 = (1, 0, 0)^T$ and the Z-axis $\mathbf{e}_0 = (0, 0, 1)^T$ define two *sensitive directions*.

By contrast, the small displacement δ_y along the Y-axis (arrow 2) does not change the diameter in the first order approximation, and hence, defines the *non-sensitive direction*. Actually, as is clear from the triangle OPQ (Fig. 8.5), the radius is changed by a value of the second order, $\Delta R = OQ - R = \sqrt{R^2 + \delta_y^2} - R \approx (R + \delta_y^2/2) - R = \delta_y^2/2$.

There are other ways to derive the scalar error budget depending on presentation of vector $\Delta \mathbf{r}_0$. In particular, substitution of Eq. (8.87) in definition (8.100) yields:

$$\Delta_0 = (\mathbf{E}\mathbf{r}_0) \cdot \mathbf{e}_0 = \begin{bmatrix} \delta_x - \delta_\varphi y_0 + \delta_\psi z_0 \\ \delta_y + \delta_\varphi x_0 - \delta_\theta z_0 \\ \delta_z - \delta_\psi x_0 + \delta_\theta y_0 \\ 0 \end{bmatrix} \begin{bmatrix} e_{x0} \\ e_{y0} \\ e_{z0} \\ 0 \end{bmatrix} =$$

$$e_{x0}\,\delta_x + e_{y0}\,\delta_y + e_{z0}\,\delta_z + n_{x0}\,\delta_\theta + n_{y0}\,\delta_\psi + n_{z0}\,\delta_\varphi \tag{8.105}$$

with

$$n_{x0} = y_0 e_{z0} - z_0 e_{y0};\; n_{y0} = z_0 e_{x0} - x_0 e_{z0};\; n_{z0} = x_0 e_{y0} - y_0 e_{x0} \tag{8.106}$$

where x_0, y_0, and z_0 are the nominal coordinates of the body point as referred to coordinate frame S_0; n_x, n_y and n_z are the moments of the unit vector $\mathbf{e}_0$ with respect to the S_0 axes; and $\delta_x, \delta_y, \delta_z, \delta_\theta, \delta_\psi$, and δ_φ are defined through Eq. (8.89).

Equations (8.105) and (8.106) allow representation in the following most compact form through two vectors of the sixth order – the twist $\boldsymbol{\delta}$ and the vector of the Plücker coordinates $\mathbf{u}_0$:

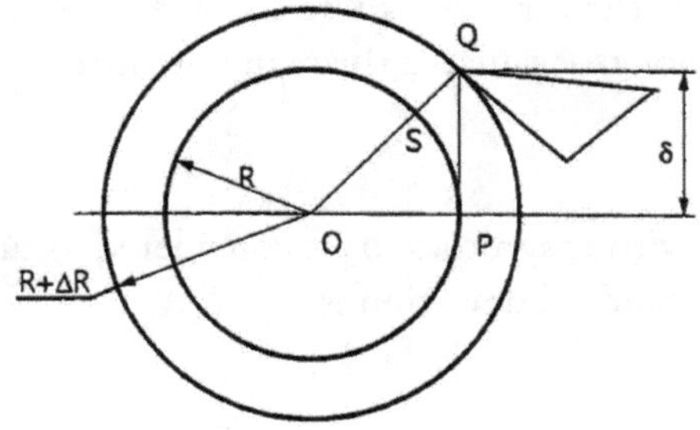

Figure 8-5 Forming the second order error of the machined cylinder with vertical displacement of the tool tip.

$$\Delta_0 = \mathbf{u}_0^{\mathrm{T}} \boldsymbol{\delta} \tag{8.107}$$

8.2.4 Constraint variation

To perform the target motion of the machine, parameters $q_1, q_1, \ldots, q_m$ of the FSF are encompassed by functional constraint(s)

$$f(q_1, q_2, \ldots q_m) = 0$$

Correspondingly, there are constraints between the elementary errors depending on constraint parameters. In the first order approximation, these constraints have the form

$$\frac{\partial f}{\partial q_1}\delta q_1 + \frac{\partial f}{\partial q_2}\delta q_2 + \ldots + \frac{\partial f}{\partial q_1}\delta q_m = \delta f \tag{8.108}$$

where $\delta q_1, \delta q_2, \ldots, \delta q_m$ are some of the elementary errors δ_{ij}.

For example, in thread machining, the kinematic constraint links the axial displacement z of the cutting tool with the angle of rotation φ of the machined thread $z - P\varphi = 0$, where P is the thread parameter expressed through helix lead L, $P = L/(2\pi)$. If the angle φ is the independent variable ($\delta_\varphi = 0$), Eq. (8.108) becomes

$$\delta_z - \delta_p \varphi = \delta_f \tag{8.109}$$

where δ_p is the kinematic error of the tread-cutting chain, which can present a mechanical chain in the traditional machine tools or computer-based electronic connection in the CNC machines; and δ_f is the error of the initial machine setup.

8.2.5 Statistical evaluation

The elementary errors δ_{ij} ($i = 0, 1, \ldots, n$; $j = 1, 2, \ldots, 6$) entering in Eqs. (8.102) and (8.103) are considered as random values in nature, while the coefficients b_{ij} and d_{ij} are non-random scalars or scalar functions depending only on machine geometry and kinematics. In what follows, some simplified assumptions will be used:

- All the errors δ_{ij} are independent of one another;
- The errors are distributed with mean value $M[\delta_{ij}] = \Delta_{ij}$
- The standard deviation σ_{ij} of the error δ_{ij} is defined depending on its physical sense. In particular,

- In the majority of cases, the error δ_{ij} can be considered as a normal-distributed value and its standard deviation σ_{ij} evaluated through tolerance T_{ij} using the "six-sigma rule":

$$\sigma_{ij} = T_{ij}/6 \tag{8.110}$$

- A runout component with magnitude δ_{ij} is considered as an arcsine-distributed random value. Hence, its standard deviation is

$$\sigma_{ij} = \delta_{ij} / \sqrt{2} \tag{8.111}$$

- A kinematic error is a composite of errors of power transmissions and supports combining the kinematic chain. In this case, the angular error can be expanded in Fourier series

$$\delta_{ij} = \left[\frac{1}{2}\mu_0 + \sum_{k=1}^{n} \mu_k \cos(k\varphi + \lambda_k) \right]_{ij}$$

where k is the ordered number of harmonics; φ is the current angle; μ_k and λ_k are the magnitude and phase angle of the kth harmonics.

In these assumptions,

(a) The average value of the total position deviation Δ, Eq. (8.102), and the total orientation deviation Δc, Eq. (8.103), is the algebraic sum of the average deviations of the components

$$M[\Delta] = \sum_{i=0}^{n} \sum_{j=1}^{6} \frac{b_{ij}\delta_{ij}}{6(n+1)} \tag{8.112}$$

$$M[\Delta c] = \sum_{i=0}^{n} \sum_{j=4}^{6} \frac{d_{ij}\delta_{ij}}{3(n+1)} \tag{8.113}$$

(b) The standard deviations σ_Δ of the total position error Δ and the standard deviations $\sigma_{\Delta c}$ of the total orientation error Δc_0 are defined through the square summation rule

$$\sigma_\Delta = \sqrt{\sum_{i=0}^{n} \sum_{j=1}^{6} (b_{ij}\sigma_{ij})^2} \tag{8.114}$$

$$\sigma_{\Delta c} = \sqrt{\sum_{i=0}^{n} \sum_{j=1}^{6} (d_{ij}\sigma_{ij})^2} \tag{8.115}$$

where b_{ij} and d_{ij} are defined through Eq. (8.104).

(c) If the number of elementary errors is reasonably large, they have only a weak relative dependence, and their relative values look much the same scale, i.e., among them there are no dominant components, the distribution of the total error Δ_0 is of near-normal. Since all these conditions usually take place for the random value Δ_0, its limit values can be evaluated using the "six sigma area" rule: the random value Δ with 99.73% of confidence lies in the interval

$$M[\Delta] - 3\sigma_\Delta < \Delta < M[\Delta] + 3\sigma_\Delta \tag{8.116}$$

where $M[\Delta]$ and σ_Δ are defined through Eqs. (8.112) and (8.115), respectively.

A similar equation takes place for $\sigma_{\Delta c}$.

8.2.6 Diagnostic procedures

Equations (8.110-116) describe the case, in which there is an explicit expression for the evaluated random value. In the general case, the error budget equations has the form

$$\Delta(u,v) = \sum_{i=1}^{m} a_i(u,v)\delta_{qi} \tag{8.117}$$

where δ_{qi} ($i = 1, 2,\ldots,m$) is the elementary error of parameter q_i; and a_i is the known transmission coefficient (or functions of u, v) associated with δ_{qi}.

It is required to estimate parameters $\delta_{q1}, \delta_{q2},\ldots,\delta_{qm}$ if the results of n-fold measurements of quantity $\Delta_i(u, v)$ at no less than m different points are given:

$$\Delta_1(u, v) = a_{11}\,\delta_{q1} + a_{12}\,\delta_{q2} +\ldots+ a_{1m}\,\delta_{qm}$$

$$\Delta_2(u, v) = a_{21}\,\delta_{q1} + a_{22}\,\delta_{q2} +\ldots+ a_{2m}\,\delta_{qm}$$

$$\ldots$$

$$\Delta_n(u, v) = a_{n1}\,\delta_{q1} + a_{n2}\,\delta_{q2} +\ldots+ a_{nm}\,\delta_{qm} \tag{8.118}$$

The similar problem was considered for the special case $m = 6$, Eqs. (8.49), with respect to six unknowns $\delta_x, \delta_y, \delta_z, \delta_\theta, \delta_\psi$, and δ_φ. The set of coefficients a_{ij} ($i = 1, 2,\ldots, n; j = 1, 2,\ldots, m$) entering in simultaneous Eqs. (8.118) forms the $n \times m$ *structural* matrix [4] presenting the Jacobian matrix of the system (8.118)

$$\mathbf{J}_s = \begin{bmatrix} a_{11} & a_{12} & \cdots & a_{1m} \\ a_{21} & a_{22} & \cdots & a_{2m} \\ \cdots & \cdots & \cdots & \cdots \\ a_{n1} & a_{n2} & \cdots & a_{nm} \end{bmatrix} \tag{8.119}$$

The Jacobian matrix $\mathbf{J}_s$ connects the $n\times 1$ vector of results of measurements $\boldsymbol{\Delta}$ with the $m\times 1$ vector of parameters δ,

$$\boldsymbol{\Delta} = [\Delta_1, \Delta_2,\ldots,\Delta_n]^T \tag{8.120}$$

$$\boldsymbol{\delta} = [\delta_{q1}, \delta_{q2},\ldots,\delta_{qn}]^T \tag{8.121}$$

$$\boldsymbol{\Delta} = \mathbf{J}_s\boldsymbol{\delta} \tag{8.122}$$

As before, the solution procedure is considered under the conditions that $n > m$, $\mathbf{\Delta} \neq 0$ and the matrix $\mathbf{J}_s$ has the full rank,

$$\text{rank}[\mathbf{J}_s] = m \tag{8.123}$$

Under these conditions, solution of simultaneous equations (8.118) can be obtained in the form of the best estimate $\hat{\delta}_{qi}$ $(i = 1, 2,\dots,m)$ of the unknowns δ_{qi} using the least mean square (LMS) method,

$$\hat{\boldsymbol{\delta}} = \mathbf{B}^{-1}\mathbf{J}_s^{\mathrm{T}}\mathbf{D}^{-1}\mathbf{\Delta} \tag{8.124}$$

$$\mathbf{D} = \begin{bmatrix} \sigma_1^2 & \sigma_1\sigma_2\rho_{12} & \cdots & \sigma_1\sigma_n\rho_{1n} \\ & \sigma_2^2 & \cdots & \sigma_2\sigma_2\rho_{2n} \\ & & \cdots & \cdots \\ sym & & & \sigma_n^2 \end{bmatrix}$$

where $\hat{\boldsymbol{\delta}} = \left[\hat{\delta}_{q1}, \hat{\delta}_{q2}, \dots, \hat{\delta}_{qm}\right]^{\mathrm{T}}$ is the vector of the estimated parameters; $\mathbf{D}$ is the non-degenerated $n \times n$ covariant matrix of the set of the results of n measurements Δ_i $(i = 1, 2,\dots,n)$ comprising the dispersions σ_i of values δ_{qi}; and the correlation coefficients ρ_{ij} $(i,j = 1, 2,\dots,m)$ of values δ_{qi} and δ_{qj}; and $\mathbf{B}$ is the $m \times n$ matrix compiled by the LMS method,

$$\mathbf{B} = \mathbf{J}_s^{\mathrm{T}}\mathbf{D}^{-1}\mathbf{J}_s \tag{8.125}$$

The wide spectrum of the actual accuracy problems can be solved in the assumption that the values δ_{q1}, $\delta_{q2},\dots,\delta_{qm}$ are non-correlated and known with similar precision (for example, they are obtained from the similar measurements). In this case, Eq. (124) is simplified and becomes

$$\hat{\boldsymbol{\delta}} = (\mathbf{J}_s^{\mathrm{T}}\mathbf{J}_s)^{-1}\mathbf{J}_s^{\mathrm{T}}\mathbf{\Delta} \tag{8.126}$$

The main applications of this equation are diagnostics of the machine accuracy and the based on this error compensation using the measurements of the output accuracy of the machines, e.g., compensation of geometric and kinematic errors of the machine tool using results of measurements of the machined parts. An example of calculations by Eq. (8.126) can be found in Section 10.2.1.

8.3 Summary

8.3.1 Main equations

Generally, the elementary error δ_{ij} entering into vector δ_s of input errors can be alternatively classified as follows:

- *According to type of motions:* linear errors $(j = 1, 2, 3)$ and angular errors $(j = 4, 5, 6)$;
- *According to impact on the output error:* elementary error δ_{ij} acts in a sensitive

direction when its transfer factor has non-zero value; otherwise it acts along a non-sensitive direction;

- *According to discrimination possibility*: two errors can be discriminative from each another when their transfer factors are different; otherwise they are indiscriminative.

The standard form of the error matrix **E**, Eqs. (8.9), and its equivalent representations:

$$\mathbf{E} = \begin{cases} \begin{bmatrix} 0 & -\delta_\varphi & \delta_\psi & \delta_x \\ \delta_\varphi & 0 & -\delta_\theta & \delta_y \\ -\delta_\psi & \delta_\theta & 0 & \delta_z \\ 0 & 0 & 0 & 0 \end{bmatrix}, & \text{for spatial case} \\ \begin{bmatrix} 0 & -\delta_\phi & \delta_x \\ \delta_\varphi & 0 & \delta_y \\ 0 & 0 & 0 \end{bmatrix}, & \text{for planar case} \end{cases}$$

Different presentations of the error matrix are

$$\mathbf{E} = \mathbf{E}_{rot} + \mathbf{E}_{tr} = \sum_{j=1}^{n} \mathbf{E}^j = \sum_{j=1}^{n} \delta_j \mathbf{D}_j$$

The position and orientation error vectors are

$$[\Delta\mathbf{r}\ \Delta\mathbf{c}] = \mathbf{E}\,[\mathbf{r}\ \mathbf{c}]$$

where matrix **E** and vectors **r** and **c** are defined through Eqs. (8.9), (8.35), and (8.36).

8.3.2 Algorithm of accuracy evaluation

Step 1. The manipulation matrix in the extended form is used:

$${}^0\mathbf{A}_n = \prod_{i=1}^{n} {}^{i-1}\mathbf{A}_i(q_i) = {}^0\mathbf{A}_1 {}^1\mathbf{A}_2 \ldots {}^{n-1}\mathbf{A}_n$$

Step 2. The total manipulation error matrix **E**, Eq. (8.89), is calculated:

$$\mathbf{E} = \sum_{i=0}^{n} {}^0\mathbf{A}_i \mathbf{E}_i ({}^0\mathbf{A}_i)^{-1} = \begin{bmatrix} 0 & -\delta_\varphi & \delta_\psi & \delta_x \\ \delta_\varphi & 0 & -\delta_\theta & \delta_y \\ -\delta_\psi & \delta_\theta & 0 & \delta_z \\ 0 & 0 & 0 & 0 \end{bmatrix}$$

Step 3. The point of interest and a critical direction, along which the scalar error is considered, are defined with respect to the basic coordinate frame S_0:

$$\mathbf{r}_0 = [x_0, y_0, z_0]^T$$

$$\mathbf{c}_0 = [c_{x0}, c_{y0}, c_{z0}]^T$$

where x_0, y_0, and z_0 are the nominal coordinates of the body point; and c_{x0}, c_{y0}, and e_{z0} are the direction cosines composing the unit vector $\mathbf{c}_0(||\mathbf{c}_0|| = 1)$.
Step 4. The scalar budget at the point of interest along a critical direction is formulated:

$$\Delta_0 = \mathbf{u}_0{}^T\boldsymbol{\delta},$$

where $\boldsymbol{\delta}$ is the twist, $\boldsymbol{\delta} = [\delta_x, \delta_y, \delta_z, \delta_\theta, \delta_\psi, \delta_\varphi]^T$; and $\mathbf{u}_0$ is the vector of Plücker coordinates as referring to the basic coordinate frame S_0, $\mathbf{u}_0 = [e_{x0}, e_{y0}, e_{z0}, n_{x0}, n_{y0}, n_{z0}]^T$.
Step 5. Deviations of the FSF parameters interrelated by means of constraints F_k $(k = 1, 2,\ldots,L)$ are connected to one another

$$\sum_{i=1}^{L}\frac{\partial F_k}{\partial q_i}dq_i + \sum_{j=1}^{N_a}\frac{\partial F_k}{\partial a_j}da_j = 0, k = 1, 2,\ldots,L$$

where L is the number of constraints, $L = l + m - M$; q_i $(i = 1,2,\ldots, L)$ is a parameter of the FSF; and a_j $(j = 1,2,\ldots,N_a)$ is the jth parameter entering in the kth constraint but the parameter presents none of the FSF parameter q_i.

8.4 Conclusion

A linearized mathematical model for a budget of geometric and kinematic errors suitable for the FSS of serial-, parallel-, and hybrid-kinematics machines is described. The mathematical tool providing solution of a wide spectrum of geometric accuracy problems related to an individual link of the FSS and the whole FSS is entered. The matrix technique is extended to cover error definition and propagation. For this purpose, the standard form of the error matrix is entered. The developed approach is illustrated by some practical examples: error budgets of robots, machine tools, and CMMs.

References

[1] Reshetov, D.N., Portman, V.T., 1988, *Accuracy of Machine Tools*, ASME Press, New York, NY.
[2] Donaldson, R., 1980, in: Hocken, R., Machine Tool Accuracy, *Technology of Machine Tools*, Vol. 5, Lawrence Livermore Lab., University of California, UCRL-52960-5, Livermore, CA.
[3] Korn, G., Korn, T., 1968, *Mathematical Handbook for Scientists and Engineers*, McGraw-Hill, NY, p. 412.
[4] Hudson, D.J., 1964, Statistics, European Organization for Nuclear Research, Geneva.

9. Accuracy of an axis

9.1 Linear motion stage

9.1.1 Structure

9.1.1.1 Guideways

A 1-DOF translational link, also called a linear joint or reciprocating joint, realizes a function of a prismatic kinematic pair and presents one of the basic units of different machines and robots. The link provides the 1-DOF translation by means of an actuator and constraints nominally preventing the relative movement along and about five DOFs. A triplet of linear stages performing the *XYZ* motions arranged in series presents a regular part of the 3-D Cartesian robots, coordinate measuring machines (CMM), milling machines, CNC milling and machining centers, etc. To provide high stiffness, accuracy and precision, and loading capacity, the linear guideways, also called linear motion bearings, are widely applied. In feed units of machine tools, the most widespread structure of the linear motion stage is a slider actuated by a motor-driven ball screw and moved along prismatic guideways of different forms (Fig. 9.1). Here, three types of prismatic guideways are shown: flat guideways, combined guideways (a set of flat- and V-form), and double V-form guideways.

The guideways of direct mechanical (metal-to-metal or metal-to-plastic) contact are classified as *sliding and rolling* linear motion bearings. The *sliding* guideways are the oldest, simplest

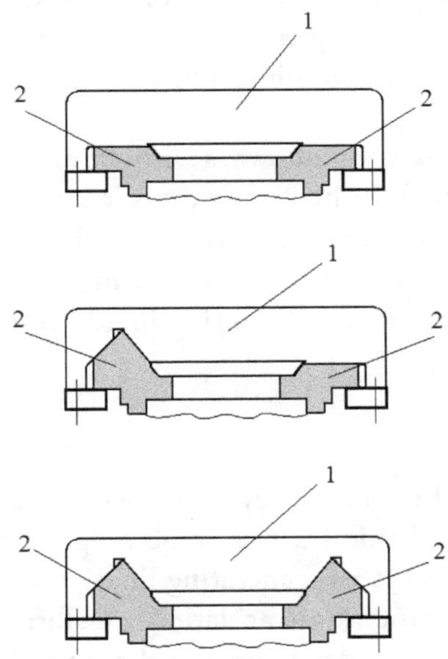

Figure 9-1 Prismatic guideways of machine tools: 1 – a moving unit; 2 – a bed or a saddle

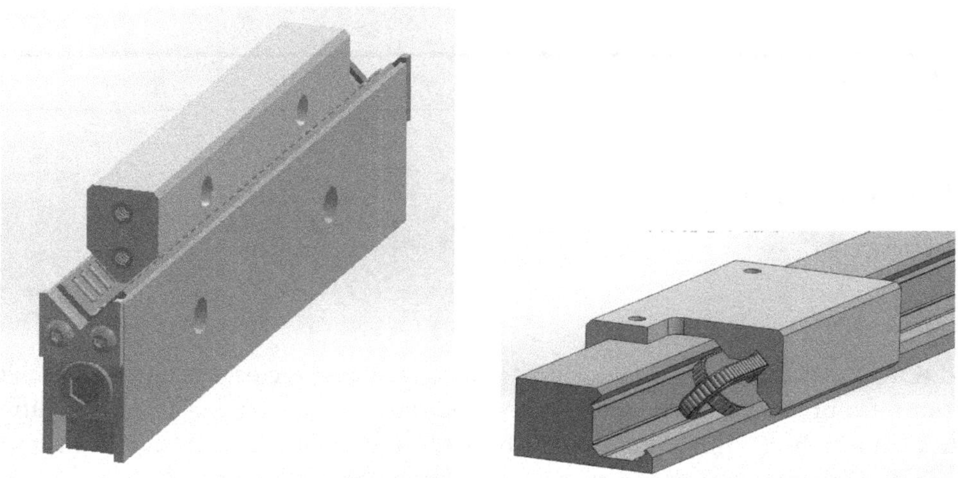

(a) (b)

Figure 9-2 Linear rolling guideways:(a) the base-arranged guideways; (b) the saddle-arranged recirculating guideways.

and least expensive ones. They have the high damping responsibility, but the relatively high friction coefficients in the mating pair, the stick-slip phenomenon during very low velocities (some mm/min or lower) and limited durability result in general transfer to *rolling* guideways.

Rolling contact between two relatively moving objects of the kinematic pair via balls, rollers, etc., significantly decreases friction and is very suitable for high precision operations. Two versions of rolling guideways are shown in Figure 9.2: (a) the base-arranged rolling guideways and (b) the linear recirculating rolling guideways mounted on the slider (saddle, table and so on) and performing a translation motion together with the slider. The recent trend is application of the complete linear drive units with central location of the ball leadscrews (Fig. 9.3).

For ultra-precision machines, the guideways without metal-to-metal contact between moving elements can be applied. In hydrostatic and aerostatic linear motion bearings, the elements of the friction pair are separated by full fluid- or gas-film lubrication, respectively. Native drawbacks of them are relatively high cost and the need for an additional hydraulic (or pneumatic) system resulting in additional difficulties in manufacturing and exploitation.

9.1.1.2 Linear drive

For a long time, hydraulic and pneumatic cylinders were used as the main producers of linear motion. Now electromechanical drives combining a *rotating* electric motor and a mechanical power transmission system generating linear motion are usually applied as a power supply. The most popular rotation-translation transformer is a screw-nut pair mechanism called leadscrew or ballscrew depending on the type of friction. In the CNC-machines, the ballscrew drives, wherein the balls or rollers separate helicoidal surfaces of the screw-nut pair, are regularly used instead of conventional leadscrews.

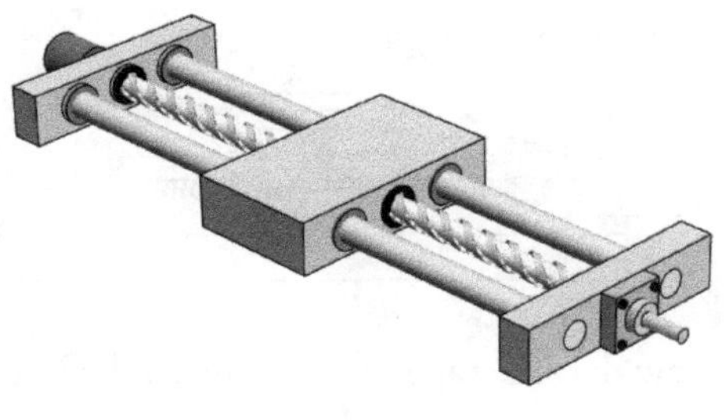

Figure 9-3 Linear drive units with central location of the ball lead screw.

In the industrial robots, there is a solution wherein the slider has a built-in nut (Fig. 9.3), i.e., the part simultaneously acts as the ballscrew's nut and the guideway's block. The solution features by compact and light weight, high accuracy, and high stiffness. A complete set of the linear stage for machine tools is shown in Figure 9.4. They include an engine decoupled with the lead screw mounted in two double bearing supports. The nut is rigidly connected with the carriage working out the reciprocating movement along the guideways. In this open-loop control system, step motors are usually applied. In Figure 9.5, a fragment of the ball screw with the external recirculation tube is shown.

The electro-mechanical drive system shown in Figure 9.4 presents an open-loop positioning system with a stepping motor that is of considerable current use owing to its comparatively simple and cheap structure, control simplicity, sufficiently high accuracy and repeatability. The stepping motor is a brushless DC electrical motor whose output shaft is driven by *electrical pulses* generated by the machine (or robot) control system. Each pulse rotates the rotor a fraction of one revolution

$$\alpha = 360/n_s \tag{9.1}$$

where α is the step angle, degrees; and n_s is the integer (e.g., $n_s = 100, 150$, and so on) presenting the number of pulses per revolution of the motor shaft.

The angle A [degrees] of the rotor shaft rotation resulting from n_p pulses of the motor shaft received by the motor is given by

$$\mathrm{A} = \alpha n_p \tag{9.2}$$

The angle A results in the table displacement on length

$$l = \frac{\mathrm{A}p}{360u} = \frac{\alpha n_p L}{360u}$$

where L is the lead of the screw, i.e., the screw pitch multiplied by the number of the thread starts; and u is the transfer ratio defined through the motor shaft rotational speed N_m and the leadscrew rotation speed N_{sc},

$$u = N_m/N_{sc} \tag{9.3}$$

Thus, the number of pulses n_p required for displacement l of the table is

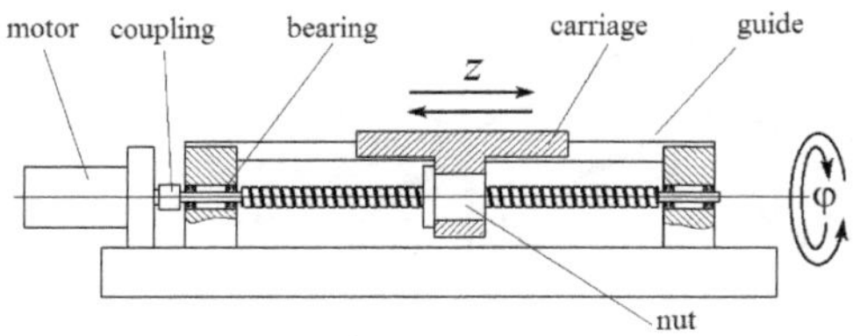

Figure 9-4 A leadscrew driven stage of machine tool. Open-loop positioning system.

$$n_p = 360lu/(\alpha L) \tag{9.4}$$

The rotational speed of the motor shaft N_{sc}, rpm, and leadscrew N_{sc}, rpm, depend on the controlled pulse frequency

$$N_m = 60f_p/n_s \text{ and } N_{sc} = 60f_p/un_s \tag{9.5}$$

where f_p is the pulse frequency, Hz (pulses/sec); and n_s is defined through Eq. (9.1).

The feed rate of the driven unit f_r, mm/min, is defined through the rotational speeds

$$f_r = N_{sc}L \text{ and } f_r = N_mL/u \tag{9.6}$$

When the feed rate f_r is established the required pulse frequency is defined

$$f_p = \frac{N_m n_s}{60} = \frac{f_r n_s u}{60L} \tag{9.7}$$

For NC machines and robots and other applications, the resolution, i.e., the smallest magnitude Δ_l to be distinguished from the controlled displacement, is a very important parameter of the drive system. There are two consistent methods for decreasing the displacement Δ_l: mechanical and electronical. Generally, the Δ_l value depends on the step angle α, the screw lead L, mechanical transfer ratio u, and micro positioning factor k, which enhance the resolution using *electronical subdivision* of each full step into k parts (e.g., $k = 10$ or 50):

$$\Delta_l = L/(n_s uk) \tag{9.8}$$

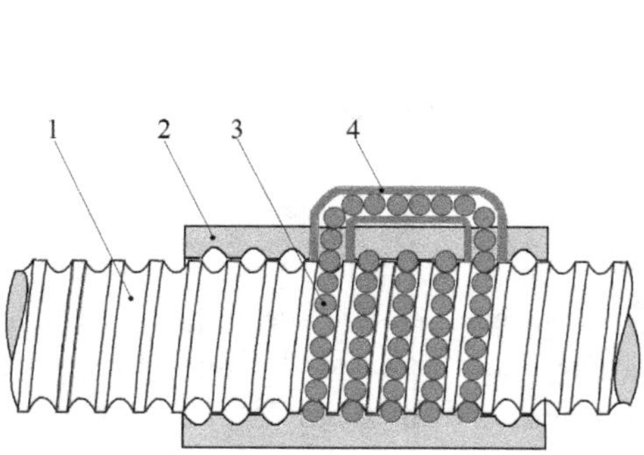

Figure 9-5 A ball screw-nut pair: *1* – the screw; 2 – the nut; 3 – the ball; *4* – the ball return tube.

Numerical Example 9.1. **Given:** The stepping motor has $n_s = 200$ step angles. The motor shaft is directly coupled with the leadscrew, $u = 1$; lead $L = 10$ mm, micropositioning factor $k = 10$. The table moves a distance $l = 270$ mm, the required feed rate is $f_p = 300$ mm/min. **Required:** The set of the drive through definition (a) pulse frequency f_p; the required angle of the rotor shaft rotation A; the required number of pulses n_p; the motor shaft rotational speed N_m and the leadscrew rotation speed N_{sc}; and system resolution Δ_l.

Solution

(1) The step angle, Eq. (9.1): $\alpha = 360/n_s = 360/200 = 1.8°$;

(2) The number of required pulses n_p, Eq. (9.4), and the total angle of the rotor shaft rotation, Eq. (9.2):

$$n_p = 360lu/(\alpha L) = 360\times 270\times 1/(1.8\times 10) = 5400 \text{ pulses}$$

$$A = n_p\alpha = 5400\times 1.8° = 9720° = 27 \text{ revolutions}$$

(3) The required pulse frequency, Eq. (9.7):

$$f_p = \frac{f_r n_s u}{60L} = \frac{300\times 200\times 1}{60\times 10} = 100\ Hz$$

(4) The rotational speeds, Eq. (9.3):

$$N_{sc} = N_m = 60f_p/n_s = 60\times 100/200 = 30 \text{ rpm}$$

(5) The system resolution, Eq. (9.8):

$$\Delta_l = L/(n_s uk) = 10/(200\times 1\times 10) = 0.005 \text{ mm.}$$ ■

Another type of linear positioning systems is the closed-loop one, presenting an electromechanical drive(Fig. 9.6) combining servo motors, a ball screw-nut pair, and a feedback chain. The chain includes a rotary encoder used in both machine tools (Fig. 9.6a) and robots (Fig. 9.6b). The encoder usually presents an optical sensor transforming the mechanical signal (the current angular position of the leadscrew motion) to an analog or digital code for control system. This configuration is featured by simple and cheap structure, high accuracy and repeatability, good static and dynamic properties. Another common type of the rotary angle measurement sensors is the resolver presenting the small brushless electrical motor having a fixed stator and rotor mounted on the shaft.

To increase the motion accuracy, linear encoders providing accurate and high-speed movement are used (Fig. 9.6c). Typically, the linear encoder combines a scale paired with a readout head that encodes a current position of the moving unit in order to convert the encoded position into the signal for the control system. The feedback includes not only the rotary screw, but also the actual translation of the moving unit. The system is widespread in robotics, machine tools, pick-and-place assembly equipment, printers, etc.

Currently, the linear induction motors are also increasingly used as linear actuators. This electrical motor has a stator and unrolled rotor producing a linear force along its length. For the time being, accuracy and load capacity properties of these motors rank below the ballscrew driven stages.

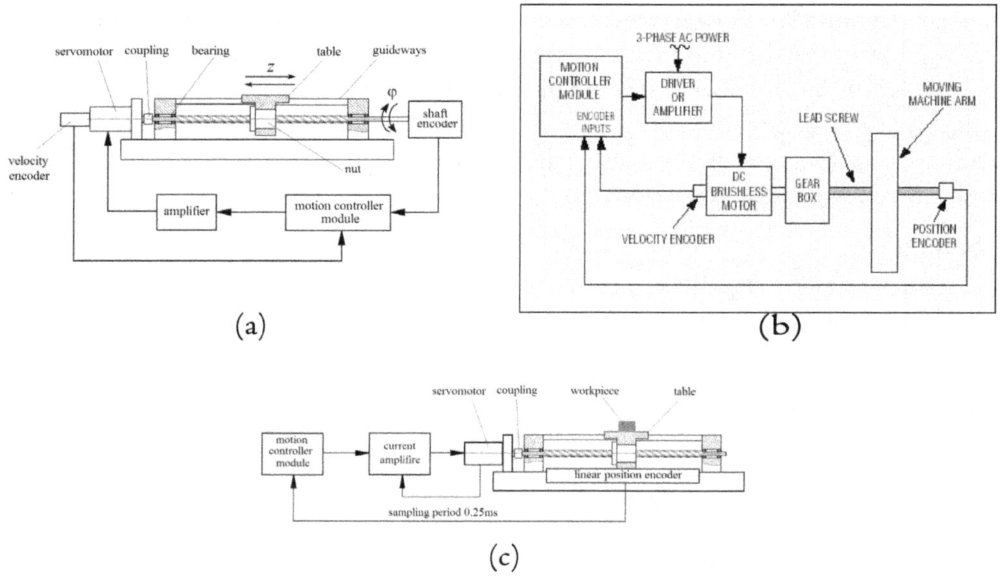

Figure 9-6 Closed-loop positioning systems with different feedback output: (a) output from the lead- or ball-screw for a machine tool table; (b) output from the lead- or ball-screw for a robot arm; (c) output from the table linear position.

To extend resolution and accuracy as applied to micro-, submicro-, and nano-motion, the piezo motors are increasingly used as linear actuators for linear and angular motions and blocking at rest. Technical parameters lie in a relatively wide range: stroke from some mm up to 10^3 mm, velocities $5 \div 100$ mm/sec, resolution from less than 10^{-9} up to 10^{-7} m, etc. This is an intensively developed area with applications in optics, medical instruments and devices, microscopes, astronomy devices, etc. For illustration, a micro-cutting device using the piezo-electric actuator for the ultra-precision diamond turning provides accuracy of ± 0.1 μm with stiffness 80 N/μm and resolution 5 nm [1].

9.1.2 Linear stage accuracy

9.1.2.1 *Linear stage errors: definitions*

A set of errors of the link having 1-DOF in its *nominal motion* (Fig. 9.7) includes a single kinematic error and five geometrical errors. Error definitions by ISO 230-1 [2] associated with a link carrying out the linear motion along the *Z*-axis are shown in Figure 9.7a. The error components are symbolized through three letters. Initial letter E symbolizes an error, the second and third letters are the DOF symbols: the former is the direction of the error, and the latter is the motion direction. For example, EXZ is the error-associated small displacement along the *X*-axis during motion along the *Z*-axis. Correspondence between the entries of the error matrix **E**, Eq. (8.9), where six errors are defined (translation errors δ_x, δ_y, δ_z and for rotation errors δ_θ, δ_ψ, δ_φ), and the six error symbols by ISO 230-1 are established in Table 9.1. Note that the explanation is associated with the case when the

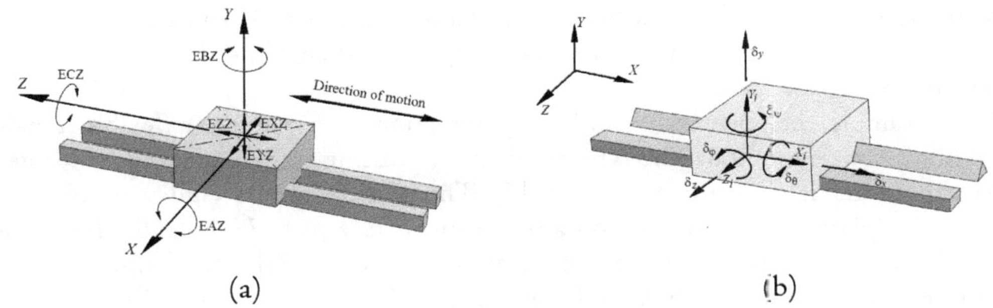

Figure 9-7 Definitions of six components of the geometric accuracy for: (a) the Z-linear stage according to ISO 230-1 [2]; (b) the X-linear stage in terms of the error matrix **E**, Eq. (9.9).

XZ plane is horizontal and the unit moves along the Z-axis. Obviously, the six error components are functions of the current nominal position z during the unit motion, $\delta_k = \delta_k(z)$, where index k takes the values x, y, z, θ, ψ, φ, and the error matrix is

$$\mathbf{E}(z) = \begin{bmatrix} 0 & -\delta_\varphi(z) & \delta_\psi(z) & \delta_x(z) \\ \delta_\varphi(z) & 0 & -\delta_\theta(z) & \delta_y(z) \\ -\delta_\psi(z) & \delta_\theta(z) & 0 & \delta_z(z) \\ 0 & 0 & 0 & 0 \end{bmatrix} \tag{9.9}$$

The error components δ_{xi}, δ_{yi}, δ_{zi}, $\delta_{\theta i}$, $\delta_{\psi i}$, and $\delta_{\varphi i}$ for X-linear stage associated with coordinate frame S_i are shown in Figure 9.7b.

There are different methods of measurements of geometric and kinematic errors. Control of three linear displacements can be carried out at one, two, or three points of the linear stage in three mutually perpendicular directions. They allow evaluation of linear kinematic error (positioning error) and two straightness deviations along the axes perpendicular to the axis of motion. Different devices for measurements of linear displacements can be used: the dial indicators with graduations from 0.01 to 0.001 mm, the linear variable differential transformers (LVDT) with graduations from 1 to 0.01 μm (10 nm), and so on.

Control of three small angles of rotation of the body relative to three mutually perpendicular axes includes evaluation of parallelism of the direction of the body displacements in two planes related to the given axis of motion (pitch and yaw error motion), and roll error of motion (rocking around the axis of motion). Angles are measured using two systems: (a) direct measurement by means of levels, collimators, autocollimators, etc.; (b) indirect measurement by means of devices for linear measurement. In both systems, a relative movement of the measured object and the measuring apparatus is necessary.

9.1.2.2 *Error measurement and evaluation*

From the structural viewpoint, there are two types of accuracy-defined elements of a linear stage: (a) *a prismatic joint* capable of five constraints considered as a source of five geometric errors; and (b) a *linear actuator* accompanied by a single kinematic error. During translation, any point of the moving body nominally moves along a line, that is parallel to the motion axis of the fixed body.

As an example, the power table for boring operations entering into a flexible transfer line is shown in Figure 9.8. The typical scheme of testing translational units for geometrical accuracy is shown in Figure 9.9 [3]. The table combines guideways plate *Gd* with moving platform *Pf* mounted on a precision surface plate *Pl*. Five dial indicators 1–5 with scale graduation 0.001 mm are rigidly connected to the platform. Their finger probes are in contact with the reference straight-edge rulers 6 and 7 mounted on supports 8 and 9 parallel to the table guideways. In Figure 9.10, the results of measurements are shown vs. coordinate z of the platform nominal motion.

The results of measurement allow calculation of the five geometrical errors of the platform in its current position during translation. Five points A_1, A_2, A_3, A_4, and A_5 are the points of measurements of geometrical errors located on the moving platform. In another experiment, a kinematic error is measured at point A_6 separately from the geometrical errors. The point A_6 is located on the axis of the drive screw-nut pair. The unit vector $\mathbf{e}_i$ ($i = 1, 2,\ldots,6$) combining the direction cosines e_{xi}, e_{yi}, e_{zi} shows the direction of the ith measurement (see arrows in Figure 9.9),

$$\mathbf{e}_i = [e_{xi}, e_{yi}, e_{zi}]^{\mathrm{T}} \tag{9.10}$$

In Table 9.2, the unit vectors $\mathbf{e}_i$ and their moments $\mathbf{n}_i$ relative to the coordinate axes are shown, where

$$\mathbf{n}_i = [n_{xi}, n_{yi}, n_{zi}]^{\mathrm{T}} \tag{9.11}$$

Table 9.1. Translation along the Z-axis: ISO 230-1 error definitions.

Error matrix entry	ISO 230-1 symbol	Error type	Comprehensive description of motion of Z in the horizontal XZ plane
$\delta_x = \delta_x(z)$	EXZ	Geometric	Horizontal straightness error motion of Z (straightness error motion of Z in X direction)
$\delta_y = \delta_y(z)$	EYZ	Geometric	Vertical straightness error motion of Z (straightness error motion of Z in Y direction)
$\delta_z = \delta_z(z)$	EZZ	Kinematic	Positioning error
$\delta_\theta = \delta_\theta(z)$	EAZ	Geometric	Yaw error motion of Z (tilt error motion of Z around the X-axis)
$\delta_\psi = \delta_\psi(z)$	EBZ	Geometric	Pitch error motion of Z (tilt error motion of Z around the Y-axis)
$\delta_\varphi = \delta_\varphi(z)$	ECZ	Geometric	Roll error motion of Z

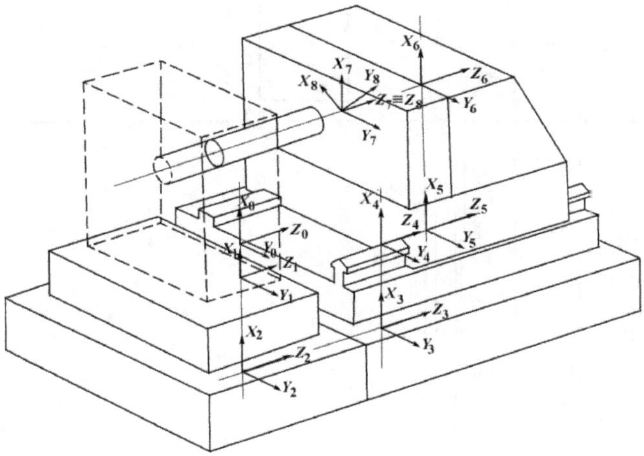

Figure 9-8 Power table for boring operations [3].

$$n_{xi} = e_{zi}y_i - e_{yi}z_i;\; n_{yi} = e_{xi}z_i - e_{zi}x_i;\; n_z = e_{yi}x_i - e_{xi}y_i \tag{9.12}$$

Coordinates of the ith unit vector $\mathbf{e}_i$, Eq. (9.10), and moments $\mathbf{n}_i$, Eq. (9.11), compose the ith row ($i = 1, 2,\ldots, 6$) of the 6×6 Jacobian matrix describing the measurement schematics:

$$\mathbf{J} = \begin{bmatrix} e_{x1} & e_{y1} & e_{z1} & n_{x1} & n_{y1} & n_{z1} \\ e_{x2} & e_{y2} & e_{z2} & n_{x2} & n_{y2} & n_{z2} \\ \cdots & \cdots & \cdots & \cdots & \cdots & \cdots \\ e_{x6} & e_{y6} & e_{z6} & n_{x6} & n_{y6} & n_{z6} \end{bmatrix} = \begin{bmatrix} -1 & 0 & 0 & 0 & -z_1 & y_1 \\ -1 & 0 & 0 & 0 & -z_1 - L_1/2 & y_1 + B \\ 0 & 1 & 0 & -z_3 & 0 & x_3 \\ 0 & 1 & 0 & -z_3 + L_2 & 0 & x_3 \\ -1 & 0 & 0 & 0 & -z_1 + L_1/2 & y_1 + B \\ 0 & 0 & 1 & y_6 & -x_6 & 0 \end{bmatrix} \tag{9.13}$$

Determinant of the Jacobian $\mathbf{J}$ has nonzero value, $\mathrm{Det}[\mathbf{J}] = -L_1L_2B \neq 0$. Hence, the Jacobian matrix is regular and the inverse matrix $\mathbf{J}^{-1}$ exists. In engineering design terms, the points and directions of measurements are chosen well, and position and orientation errors in current table positions is

$$\boldsymbol{\delta} = \mathbf{J}^{-1}\boldsymbol{\Delta} \tag{9.14}$$

$$\boldsymbol{\delta} = [\delta_x, \delta_y, \delta_z, \delta_\theta, \delta_\psi, \delta_\varphi]^T; \tag{9.15}$$

$$\boldsymbol{\Delta} = [\Delta_1, \Delta_2,\ldots,\Delta_6]^T; \tag{9.16}$$

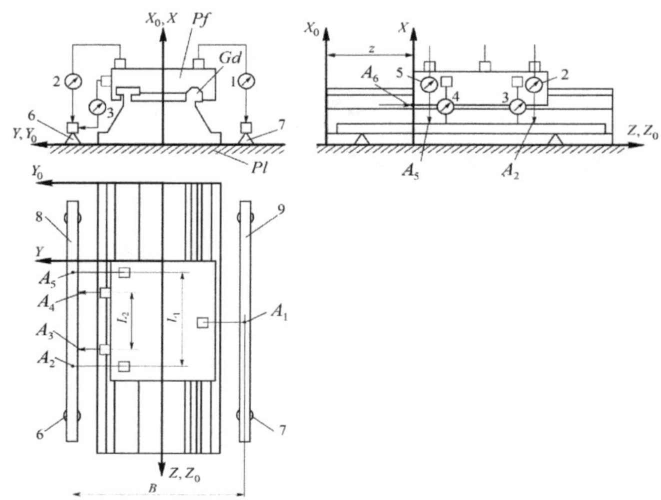

Figure 9-9 Testing the power table for geometrical accuracy: 1, 2, 3, 4, 5 —the dial indicators; 6, 7 – the rules' supports; 8, 9 – the reference rules.

$$\mathbf{J}^{-1} = \begin{bmatrix} -1-\dfrac{y_1}{B} & \dfrac{y_1}{2B}+\dfrac{z_1}{L_1} & 0 & 0 & \dfrac{y_1}{2B}-\dfrac{z_1}{L_1} & 0 \\ \dfrac{x_3}{B} & -\dfrac{x_3}{2B} & 1-\dfrac{z_3}{L_2} & \dfrac{z_3}{L_2} & -\dfrac{x_3}{2B} & 0 \\ 0 & -\dfrac{x_6}{L_1} & \dfrac{y_6}{L_2} & -\dfrac{y_6}{L_2} & \dfrac{x_6}{L_1} & 1 \\ 0 & 0 & -\dfrac{1}{L_2} & \dfrac{1}{L_2} & 0 & 0 \\ 0 & -\dfrac{1}{L_1} & 0 & 0 & \dfrac{1}{L_1} & 0 \\ -\dfrac{1}{B} & \dfrac{1}{2B} & 0 & 0 & \dfrac{1}{2B} & 0 \end{bmatrix} \tag{9.17}$$

Table 9.2. Initial data for Jacobian matrix J.

Point		Cartesian coordinates			Plücker coordinates					
A_i	Δ_i	x_i	y_i	z_i	e_{xi}	e_{yi}	e_{zi}	n_{xi}	n_{yi}	n_{zi}
A_1	Δ_1	x_1	y_1	z_1	-1	0	0	0	$-z_1$	y_1
A_2	Δ_2	x_1	y_1+B	$z_1+L_1/2$	-1	0	0	0	$-z_1-L_1/2$	y_1+B
A_3	Δ_3	x_3	y_3	z_3	0	1	0	$-z_3$	0	x_3
A_4	Δ_4	x_3	y_3	z_3-L_2	0	1	0	$-z_3+L_2$	0	x_3
A_5	Δ_5	x_1	y_1+B	$z_1-L_1/2$	-1	0	0	0	$-z_1+L_1/2$	y_1+B
A_6	Δ_6	x_6	y_6	z_6	0	0	1	y_6	$-x_6$	0

where $\boldsymbol{\delta}$ is the twist composing position and orientation errors; $\boldsymbol{\Delta}$ is the vector of results of measurements; and $\mathbf{J}^{-1}$ is the inverse Jacobian matrix.

Substitution of Eqs. (9.17) and (9.16) in Eq. (9.14) yields

$$\begin{bmatrix} \delta_x \\ \delta_y \\ \delta_z \\ \delta_\theta \\ \delta_\psi \\ \delta_\varphi \end{bmatrix} = \begin{bmatrix} -1-\frac{y_1}{B} & \frac{y_1}{2B}+\frac{z_1}{L_1} & 0 & 0 & \frac{y_1}{2B}-\frac{z_1}{L_1} & 0 \\ \frac{x_3}{B} & -\frac{x_3}{2B} & 1-\frac{z_3}{L_2} & \frac{z_3}{L_2} & -\frac{x_3}{2B} & 0 \\ 0 & -\frac{x_6}{L_1} & \frac{y_6}{L_2} & -\frac{y_6}{L_2} & \frac{x_6}{L_1} & 1 \\ 0 & 0 & -\frac{1}{L_2} & \frac{1}{L_2} & 0 & 0 \\ 0 & -\frac{1}{L_1} & 0 & 0 & \frac{1}{L_1} & 0 \\ -\frac{1}{B} & \frac{1}{2B} & 0 & 0 & \frac{1}{2B} & 0 \end{bmatrix}$$

$$\begin{bmatrix} \Delta_1 \\ \Delta_2 \\ \Delta_3 \\ \Delta_4 \\ \Delta_5 \\ \Delta_6 \end{bmatrix} = \begin{bmatrix} -\Delta_1-(2\Delta_1-\Delta_2-\Delta_5)\frac{y_1}{2B}+(\Delta_2-\Delta_5)\frac{z_1}{L_1} \\ \Delta_3+(2\Delta_1-\Delta_2-\Delta_5)\frac{x_3}{2B}-(\Delta_3-\Delta_4)\frac{z_3}{L_2} \\ \Delta_6-(\Delta_2-\Delta_5)\frac{x_6}{L_1}+(\Delta_3-\Delta_4)\frac{y_6}{L_2} \\ \frac{-\Delta_3+\Delta_4}{L_2} \\ \frac{-\Delta_2+\Delta_5}{L_1} \\ \frac{-2\Delta_1+\Delta_2+\Delta_5}{2B} \end{bmatrix} \tag{9.18}$$

In compact form, this result is

$$\delta_\theta = \frac{-\Delta_3+\Delta_4}{L_2};\ \delta_\psi = \frac{-\Delta_2+\Delta_5}{L_1};\ \delta_\varphi = \frac{-2\Delta_1+\Delta_2+\Delta_5}{2B};$$

$$\delta_x = -\Delta_1+y_1\delta_\varphi-z_1\delta_\psi;\ \delta_y = \Delta_3-x_3\delta_\varphi+z_3\delta_\theta;\ \delta_z = \Delta_6+x_6\delta_\psi-y_6\delta_\theta \tag{9.19}$$

As is clear from Eqs. (9.18), the five geometric errors (δ_x, δ_y, δ_θ, δ_ψ, δ_φ) are independent of the kinematic deviation Δ_6 and coordinates x_6, y_6 and z_6 of the drive application point. This fact

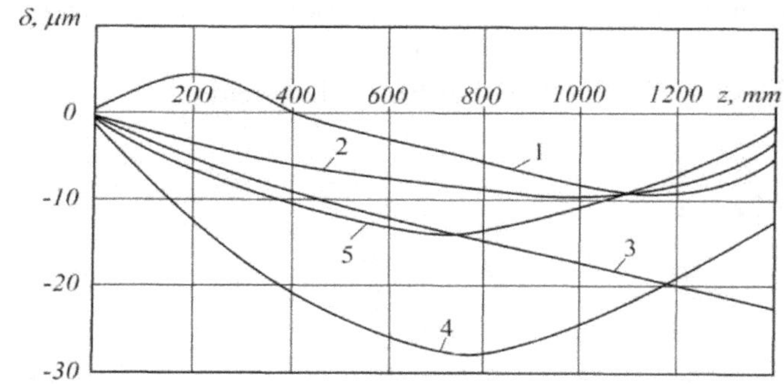

Figure 9-10 Results of measurements of geometrical accuracy: 1–5 are the numbers of the indicators shown in Figure 9.9.

allows evaluation of geometrical errors using only five measuring devices $A_1,\ldots,A_5$ as is done in the experiment shown in Figure 9.9.

Numerical Example 9.2. The schematics in Figure 9.9 and results of measurements in Figure 9.10 are considered. **Given:** L_1 = 1100 mm; L_2 = 780 mm; B = 900 mm; coordinate of the points of measurements are given in Table 9.3 as referred to the platform-associated coordinate frame. **Required:** (1) Geometric errors $\delta_y, \delta_z, \delta_\theta, \delta_\psi, \delta_\varphi$ for the current position of the moving platform z = 800 mm. (2) Position accuracy of point Q of a part arranged on the moving platform, x_Q = 1100; y_Q = 60; z_Q = 1080.

Solution: (1) According to Eqs. (9.19) we obtain

$$\delta_\theta = \frac{-\Delta_3 + \Delta_4}{L_2} = \frac{0.0145 - 0.027}{780} = -1.7857\times10^{-5}\,\text{rad};$$

$$\delta_\psi = \frac{-\Delta_2 + \Delta_5}{L_1} = \frac{0.009 - 0.013}{1100} = 7.0\times10^{-5}\,\text{rad};$$

$$\delta_\varphi = \frac{-2\Delta_1 + \Delta_2 + \Delta_5}{2B} = \frac{2\times0.0055 - 0.009 - 0.013}{2\times700} = 5.0\times10^{-7}\,\text{rad};$$

$$\delta_x = -\Delta_1 + y_1\delta - z_1\delta_\psi = 0.0055 - 350 \times 5.0 \times 10^{-7} - 700 \times 7.0 \times 10^{-5} = -0.043675\ \text{mm};$$

$$\delta_y = \Delta_3 - x_3\delta_\varphi + z_3\delta_\theta = -0.0145 - 95\times 5.0 \times 10^{-7} + 1090\times(-1.7657 \times 10^{-5}) =$$
-0.034012 mm

$$\delta_z = \Delta_6 + x_6\delta_\psi - y_6\delta_\theta = 0.012 + 720 \times 7.0 \times 10^{-5} + 40\times(-1.7657 \times 10^{-5}) = 0.037686\ \text{m}$$

(2) Position errors of point Q are defined through Eq.(8.35):

Table 9.3. Coordinates of the points and the results of measurements.

Point	$x_{i0,}$ mm	$y_{i0,}$ mm	$z_{i0,}$ mm	$\Delta_{i,}$ µm
A_1	140	−350	700	−5.5
A_2	120	350	1250	−9.0
A_3	95	290	1090	−14.5
A_4	95	290	310	−27.0
A_5	140	350	150	−13.0
A_6	720	−40	600	12.0

$$\Delta \mathbf{r}_Q \equiv \begin{bmatrix} \Delta x_Q \\ \Delta y_Q \\ \Delta z_Q \\ 0 \end{bmatrix} = \begin{bmatrix} \delta_x - \delta_\varphi y_Q + \delta_\psi z_Q \\ \delta_y + \delta_\varphi x_Q - \delta_\theta z_Q \\ \delta_z - \delta_\psi x_Q + \delta_\theta y_Q \\ 0 \end{bmatrix}, \text{i.e.,}$$

$$\Delta x_Q = -0.043675 - 60 \times 5.0 \times 10^{-7} + 1080 \times 7.0 \times 10^{-5} = 0.03190 \text{ mm}$$

$$\Delta y_Q = -0.034012 + 1100 \times 5.0 \times 10^{-7} + 1080 \times 1.7857 \times 10^{-5} = 0.01418 \text{ mm}$$

$$\Delta z_Q = 0.037686 - 1100 \times 7.0 \times 10^{-5} - 60 \times 1.7857 \times 10^{-5} = 0.04039 \text{ mm}$$

9.1.2.3 Guideways' accuracy

A guideways accuracy model is considered on the assumptions that:

- Both bed and carriage are the rigid bodies;
- There are exactly five contact points between the guideways planes of the carriage and base. In these points, the contact is constant;
- There is a single point of contact with the last link of the drive (the nut of the lead-screw, the tooth of a rack gear, etc.);
- During the motion, the position of the contact points remains unchanged relative to the moving unit (the carriage), and the number of contact points remains equal to six;
- Form deviations along the guideways faces and kinematic error are given.

To determine the points of contact, the three-dimensional contact problem will be solved under consideration of the stiffness-compliance problems. In Figure 9.11, one of the possible variants of such a problem is depicted. Five points (*A*, *B*, *C*, *D*, and *E*) are located on the guides, and not more than two of them can be on the one face. The sixth point (*K*) is the central point of the nut. Points and directions of measurements of geometrical accuracy:

- Δ_A, Δ_B, Δ_C, Δ_D, and Δ_E are the deviations from linearity of the guideways edges;
- A_N, B_N, C_N, D_N, and E_N are the nominal position of the contact points;

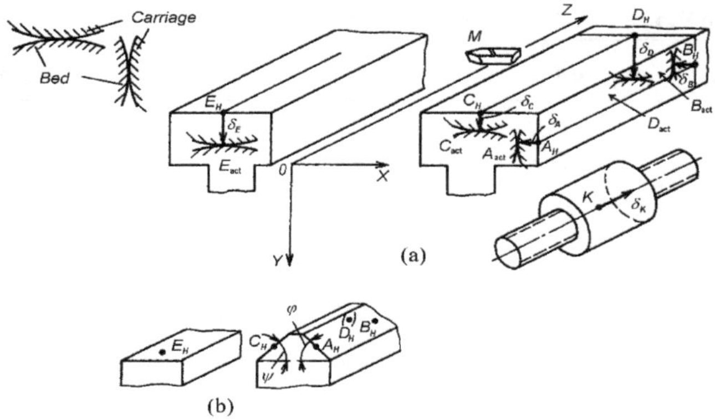

Figure 9-11 Schematic diagram for calculation of accuracy of carriage displacement along the Z-axis: (a) rectangular guides; (b) combined guides.

- A_{act}, B_{act}, C_{act}, D_{act}, and E_{act} are the actual position of the same points.

The coordinate of the points, directions of displacements and their Plücker coordinates as referred to the frame associated with the base are given in Table 9.4. Calculations are identical to those described by Eqs. (9.10-12). For example, the Plücker coordinates of a line, along which the deviation Δ_A is directed, are as follows: (a) since the deviation Δ_A is directed along the negative X–axis, the direction cosines are $e_{xA} = -1$; $e_{yA} = 0$; and $e_{zA} = 0$; (b) since the line of deviation passes through point A, the moments of vector $\mathbf{e}_A$ are (see the first line in Table 9.4)

$n_{xA} = y_A\, e_{zA} - z_A\, e_{yA} = 0$; $n_{yA} = z_A\, e_{xA} - x_A\, e_{zA} = -\, z_A$; and $n_{zA} = x_A\, e_{yA} - y_A\, e_{xA} = y_A$

The Jacobian matrix and its determinant become

$$\mathbf{J}_s = \begin{bmatrix} -1 & 0 & 0 & 0 & -z_A & y_A \\ -1 & 0 & 0 & 0 & -z_B & y_B \\ 0 & 1 & 0 & -z_C & 0 & x_C \\ 0 & 1 & 0 & -z_D & 0 & x_D \\ 0 & 1 & 0 & -z_E & 0 & x_E \\ 0 & 0 & 1 & y_K & -x_K & 0 \end{bmatrix} \qquad (9.20)$$

$$\mathrm{Det}[\mathbf{J}_s] = (z_A - z_B)[x_E\,(z_C - z_D) - x_C\,(z_E - z_D) + x_D\,(z_E - z_C)].$$

If the points of measurements are located as is shown in Figure 9.11, we have $x_A = x_B$; $y_A = y_B$; $x_C = x_D$; $y_C = y_D$; and $z_C = z_E$. Since $x_E \neq x_C$, $z_B \neq z_A$; $z_D \neq z_C$, the determinant is

$$\mathrm{Det}[\mathbf{J}_s] = (x_E - x_C)\,(z_B - z_A)\,(z_D - z_C) \neq 0 \qquad (9.21)$$

Since the matrix $\mathbf{J}_s$ is non-singular, the inverse matrix $\mathbf{J}_s^{-1}$ exists, and the system can be solved with respect to unknown twist $\delta = [\delta_x, \delta_y, \delta_z, \delta_\theta, \delta_\psi, \delta_\varphi]^T$ using the given errors of the guideways $\Delta = [\Delta_A, \Delta_B, \Delta_C, \Delta_D, \Delta_E, \Delta_K]^T$

Table 9.4. Position errors of the linear stage: Initial data.

Point	Cartesian coordinates				Plücker coordinates					
Symbol	Δ_j	x_j	y_j	z_j	e_{xj}	e_{yj}	e_{zj}	n_{xj}	n_{yj}	n_{zj}
A	Δ_A	x_A	y_A	z_A	-1	0	0	0	$-z_A$	y_A
B	Δ_B	x_B	y_B	z_B	-1	0	0	0	$-z_B$	y_B
C	Δ_C	x_C	y_C	z_C	0	1	0	$-z_C$	0	x_C
D	Δ_D	x_D	y_D	z_D	0	1	0	$-z_D$	0	x_D
E	Δ_E	x_E	y_E	z_E	0	1	0	$-z_E$	0	x_E
K	Δ_K	x_K	y_K	z_K	0	0	1	y_K	$-x_K$	0

$$\boldsymbol{\delta} = \mathbf{J}_s^{-1}\boldsymbol{\Delta} = \begin{bmatrix} -1 & 0 & 0 & 0 & -z_A & y_A \\ -1 & 0 & 0 & 0 & -z_B & y_B \\ 0 & 1 & 0 & -z_C & 0 & x_C \\ 0 & 1 & 0 & -z_D & 0 & x_D \\ 0 & 1 & 0 & -z_E & 0 & x_E \\ 0 & 0 & 1 & y_K & -x_K & 0 \end{bmatrix}^{-1} \begin{bmatrix} \Delta_A \\ \Delta_B \\ \Delta_C \\ \Delta_D \\ \Delta_E \\ \Delta_K \end{bmatrix} \tag{9.22}$$

The solution can be represented in the form

$$\delta_\theta = \frac{\Delta_C - \Delta_D}{z_D - z_C}; \delta_\psi = \frac{\Delta_A - \Delta_B}{z_B - z_A}; \delta_\varphi = \frac{\Delta_C - \Delta_E}{x_C - z_E};$$

$$\delta_x = -\Delta_A - z_A\delta_\psi + y_A\delta_\varphi;\ \delta_y = \Delta_C + z_C\delta_\theta - x_C\delta_\varphi;\ \text{and}\ \delta_z = \Delta_K - y_K\delta_\theta + x_K\delta_\psi \quad (9$$

Physically, the displacement of the points occurs due to errors of manufacturing, mations, and wear of the guides. The geometric and kinematic errors depend on the position z of the linear stage: $\delta_x = \delta_x(z)$, $\delta_y = \delta_y(z)$, $\delta_z = \delta_z(z)$, $\delta_\theta = \delta_\theta(z)$, $\delta_\psi = \delta_\psi(z$ $= \delta_\varphi(z)$ since all measured deviations present functions of the current value of the ment z. Correspondingly, the error of motion is defined as the difference of tw errors:

$$\boldsymbol{\delta}_{ij} = \boldsymbol{\delta}(z_j) - \boldsymbol{\delta}(z_i)$$

$$\mathbf{E}_{ij} = \mathbf{E}(z_j) - \mathbf{E}(z_i)$$

where $\boldsymbol{\delta}(z)$ and $\mathbf{E}(z)$ are the twist, Eq. (9.22), and the error matrix, Eq. (' associated with the current position z of the linear stage.

9.1.2.4 *Analysis of the guideways form deviations*

Some sources of quasistatic errors can be diagnosed in terms of low of the guideways form deviations. For example, the lead error of the lead or ball screws results in the lead errors of the produced screws. In turn, this affects the profile deviation (waviness) of the parts machined on the machine tool, wherein these lead-screws are used as feed screws. For diagnostic goals, it is worth representing guideways form deviation Δ by means of the Fourier series expansion

$$\Delta = \frac{1}{2}(a_0 + b_0 z) + \sum_{k=1}^{n} a_k \sin[\frac{k\pi}{L}(z - c_k)] \tag{9.26}$$

where a_0 and b_0 are the small numbers displaying the initial deviation and angular trend of the linear component of the accumulated error; n is the number of harmonics of interest in results of measurements; z is the current distance of stage motion; L is the length of the working stroke of the unit; a_k and c_k are the amplitude of the kth component and location of its maximum.

Components entering into the deviation Δ, Eq. (9.26), are illustrated in Figure 9.12, where the three first components and the total deviations of the geometric error are shown. The kinematic error Δ_K combines the cumulative positioning error $\Delta_{cum}(z)$ and the cyclic error $\Delta_t(z)$ of the leadscrew

$$\Delta_K(z) = \Delta_{cum}(z) + \Delta_t(z) \tag{9.27}$$

$$\Delta_{cum}(z) = \frac{1}{2}(u_0 + v_0 z) + \sum_{k=1}^{n} u_k \sin[\frac{k\pi}{L}(z - c_k)] \tag{9.28}$$

$$= \sum_{i=1}^{N} u_k \sin[\frac{2k\pi}{t_s}(z - d_k)] \tag{9.29}$$

...nts of the cumulative errors $\Delta_{cum}(z)$ are obtained for working length of the ... error $\Delta_t(z)$ presents the Fourier expansion of the cycle error for the lead-

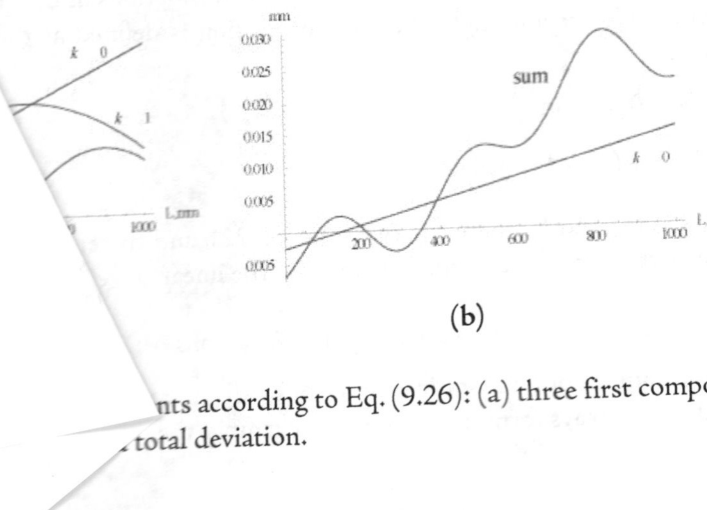

(b)

...nts according to Eq. (9.26): (a) three first compo-... total deviation.

9.1.3 Analysis of axis accuracy with experimental validation

9.1.3.1 *Accuracy components*

There are some standard procedures for definition of the position accuracy of the coordinate systems. In modern machinery, the commonly used translational stage of feed drives combines a separate electrical motor and a ball screw-nut pair (Figs. 9.4 and 9.5). During unidirectional motion, the leadscrew-caused errors can be expressed through the current coordinate x of the nominal machine table position

$$\delta_x = \tfrac{1}{2}A(x) + \sum_{k=1}^{N} a_k \sin[\frac{2\pi k}{t_s}(x - \xi_k)] + \delta_{back} \tag{9.30}$$

where a_k is the amplitude of the periodic components (cyclic errors) caused by pitch deviations; k is the harmonic number; t_s is the pitch of the lead screw; ξ_k is the linear displacements characterizing the phases of the kth harmonic, $0 \le \xi_k \le t_s/k$; and δ_{back} is the component appearing while reversing the machine table.

The term $A(x)$ can be presented as a sum of accumulated errors $A_0 + A_1x$ (aperiodic components) and some low-frequency components representing expansion with respect to trigonometric functions whose frequency is proportional to double-length $2L$ of the working part of the leadscrew

$$A(x) = A_0 + A_1x + A_2\cos\left[\frac{\pi}{L}(x - c_2)\right] + A_3\cos\left[\frac{2\pi}{L}(x - c_3)\right] + \ldots \tag{9.31}$$

In Figure 9.13 the terms entering in Eqs. (9.30) and (9.31) are separately illustrated (the errors are increased ×10000 compared with the nominal size): (a) the linear component; (b) the semi-wave deviation; (c) the full-wave deviation; and (d) the cyclic error.

9.1.3.2 *Issues of axial errors*

In the experiments shown below, interrelationships between the linear component of a position error of the machine table (component $A_0 + A_1x$ in Figure 9.13) and a cumulative error of the equipped lead screw is explicitly demonstrated. Experimental data shown in Figure

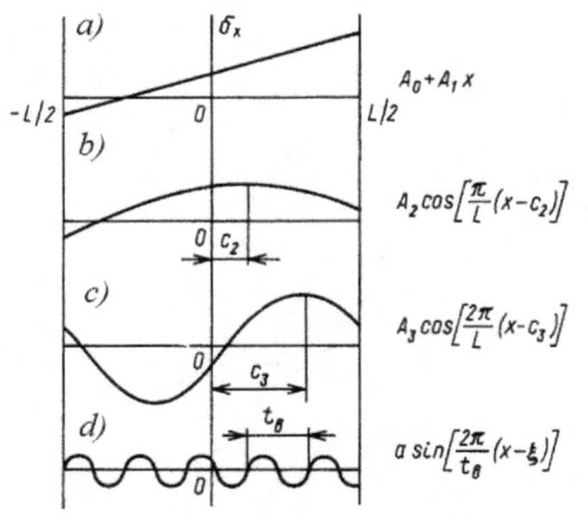

Figure 9-13 Position error components.

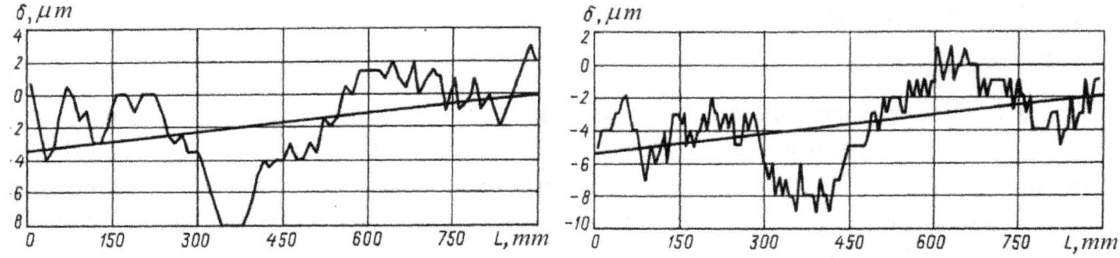

(a) (b)

Figure 9-14 (a) Lead error variation along the screw length; (b) a table-positioning error along the table travel length.

9.14 are obtained during measurements of the NC contour-grinding machine MA396F3, ENIMS production. The machine structure is shown in Figure 4.19.

A lead error variation along the screw working length (Fig. 9.14a) and table-positioning error along the machine table travel length (Fig. 9.14b) are compared. In both cases, the straight line presents a linear regression estimating the accumulated error. The machine table travel errors are measured with an interval of 6 mm by microscope and measuring wire. The measurements were taken during shop machine-tool certifying procedure following its complete assembly. The feed drive of the table is equipped with two leadscrews having a lead of 12 mm, which were certified on the measuring machine with increment of 6 mm. As is evident from Figure 9.14, the character of variation of the functions and straight line trends are similar for both realizations. This means that the linear component of the accumulated errors of the lead screw totally enters into the positional error and determines the main part of the error systematic components.

The main accuracy parameters of the screws depend on the lead precision. As an example, the SKF data [4] for tolerance values e_p [μm] over specified travel l [mm] for three accuracy grades, that comply with the ISO 286-2 [5] standard tolerances G5, G7, and G9 are shown in Figure 9.15.

Following subtraction of the linear trend, we obtain a centered realization that can be assumed stationary. The autocorrelation functions for these realizations (Fig. 9.16) allow

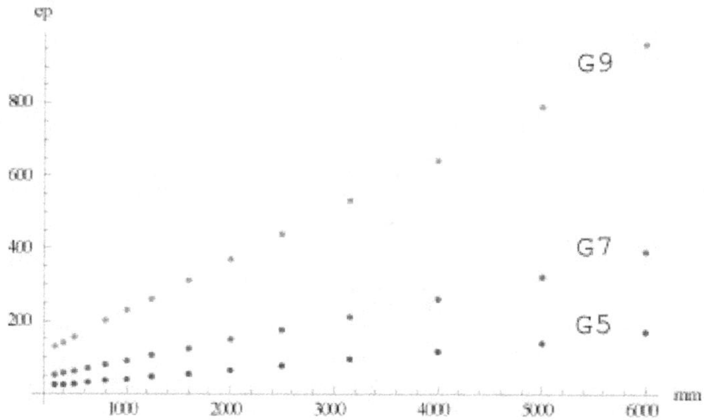

Figure 9-15 Tolerance e_p [μm] over the specified travel [4].

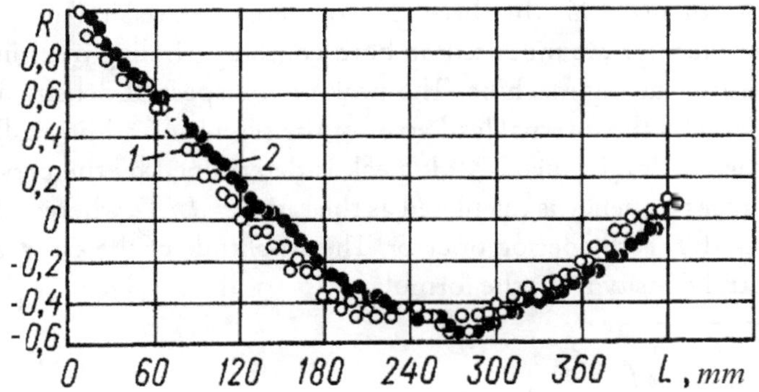

Figure 9-16 Autocorrelation function of the errors: (*1*) of the lead screw mounted on the machine tool; (2) of the table-positioning vs. longitudinal displacement.

one to take into account the effect of the distance between two points on the accuracy of the distance between the centers when machining holes. If L_{ij} is the distance between the points L_i and L_j, i.e., $L_{ij} = L_i - L_j$, then the limit error of the distance

$$\delta L_{ij} = bL_{ij} + 6\sqrt{\sigma_i^2 + \sigma_j^2 - 2r_{ij}\sigma_i\sigma_j} \tag{9.32}$$

where b is the slope of the straight-line trend; σ_i and σ_j are the standard deviations of the centered realization; r_{ij} is the coefficient of correlation between the *i*th and *j*th sections that is determined from the graph of the autocorrelation function.

Analysis of the frequency domain is carried out by computing the spectral density of the random event that characterizes the distribution of the values of the process at a fixed point of the frequency domain. Spectral density of poisoning accuracy is built along the longitudinal coordinates of the machine table for two machine tools of the same model. In Figure 9.17, curve *1* presents the error density of the lead screw mounted on one of the contour-grinding machines, and curves 2 and 3 are the density of the table-positioning errors under longitudinal displacement for two machines. Note that the curves *1* and 2 characterize the same machine tool.

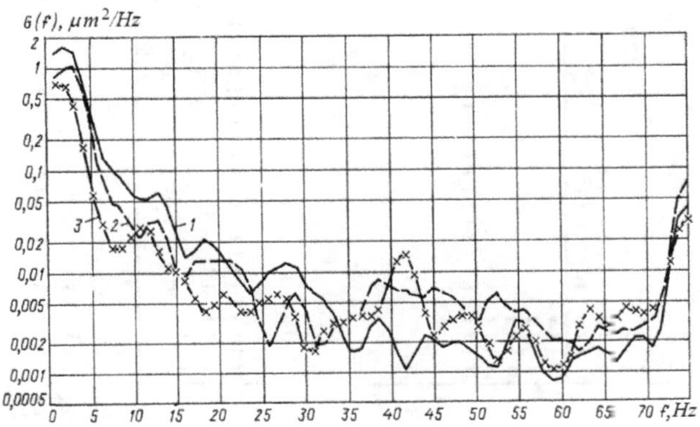

Figure 9-17 Error spectral densities: (*1*) of the lead screw mounted on a machine; (2 and 3) of the table-positioning for two machines vs. longitudinal displacement.

As is clear from Figure 9.17, the three spectral densities are similar, which is explained by the fact that the leadscrews mounted on both contour-grinding machines are machined on a common thread-cutting machine. The peak of the spectral density at a frequency of 75 Hz is determined by the internal lead error of the screw and additionally by other cyclic components of the system. In turn, a 2Hz peak is determined by the presence of accumulated error. The error frequency is calculated as the ratio $f = L/T_f$, where L is the total length of measurement and T_f is the period of error. The amplitude of the error is found from the graph of the spectral density from the formula

$$A_i = \sqrt{2\int_{f_i - B_i/2}^{f_i + B_i/2} G_x(f)df}\,, \tag{9.33}$$

where $G_x(f)$ is the estimate of the spectral density of the error in point f; B_i is the width of the spectral band used in calculation; for example, the amplitude of the internal lead error is $A \approx 1.2$ μm, which corresponds to the magnitude of the actual internal lead error of the leadscrew.

The accumulated error comprises roughly 70% of the total error. The histograms of error distribution (with exception of the linear part) were determined with the aid of filters selecting the entire analyzed quantity in the desired frequency range. The histograms in the low- and high-frequency ranges are represented in Figure 9.18. Both distribution densities have a form close to the arcsinus, from which it can be concluded that the representation of the error by a sinusoid is sufficiently close to reality.

9.1.3.3 *Backlash components*

Component δ_{back} results from the screw axial float and guideway play. There are two main types of backlash issues: the air gap δ_{ag} and elastic axial deformation δ_{eld} (elastic hysteresis) in the nut-screw pair of the translational drives. These errors appear after reversing the forward (unidirectional) motion of the machine table

$$\delta_{back} = \delta_{ag} + \delta_{eld} = \delta_{ag} + F_{ax}/k_d \tag{9.34}$$

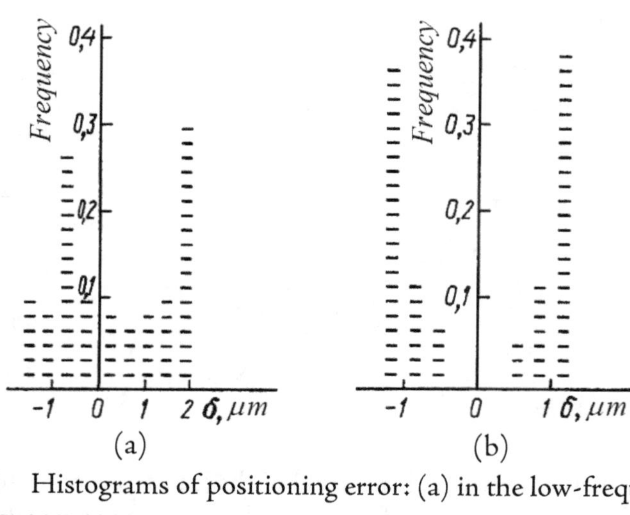

Figure 9-18 Histograms of positioning error: (a) in the low-frequency range; (b) in the high-frequency range.

where F_{ax} is the axial feed force; and k_d is the stiffness value of the drive considered as a linear spring of rate k_d.

The backlash results in so-called "dead zone" whose duration t_{dead} combines the time for loosening the preloading of the axial transmission elements and the time necessary for their preloading after reverse. Throughout time t_{dead}, the control command is not be able to be turned to a predesigned reversing feed motion. Two examples of the backlash issues are shown in Figure 9.19, where typical kinematic pairs participating in linear motion – gearing and screw-nut – are displayed. When the gear drive operates in the given direction the airgap comes into existence between out-of-work profiles of teeth. In an analogous way the backlash between the helicoid profiles in the screw-nut pairs is manifested. Another type of physical phenomena is realized in guideways where the backlash affects the accuracy under changes of the active force direction rather than direction of motion.

There are two basic physical (structural) concepts of the backlash elimination or minimization using preloading in the contact zone: (a) including special structural elements producing an additional force by means of relative displacement of contact elements on the backlash values or more, and (b) application of the intermediate elements (e.g., balls and rolls) of oversize dimensions.

The most commonly used examples of the former concept are application of tensioning idler pulley for belt drives and power transmission chains as well as different techniques for preloading rolling friction bearings [6]. In Figure 9.20, fixed and flexible schemes providing preloading of the teeth contact in gearing are shown. The fixed preloading is obtained through preliminary small displacement of either gear base with its following fixation by bolts (Fig. 9.20a), and the flexible preloading is organized by means of the extension spring (Fig. 9.20b). Another way is dividing the gear of width B into two half-gears of width $B/2$. The halves are displaced relative to each other at a small angle under spring force (Fig. 9.20c). This results in increasing of the actual tooth thickness while the tooth pitch is unchanged.

In Figure 9.21, the backlash in leadscrew-nut pair is eliminated by the small relative axial approach of two separated half-nuts up to sufficient preloading in the screw-nut contact (Fig. 9.21a). In Figure 9.21b, this is reached using a spring built-into the anti-backlash assembly. An additional load is about 7–8.5% of the dynamic basic load value [4].

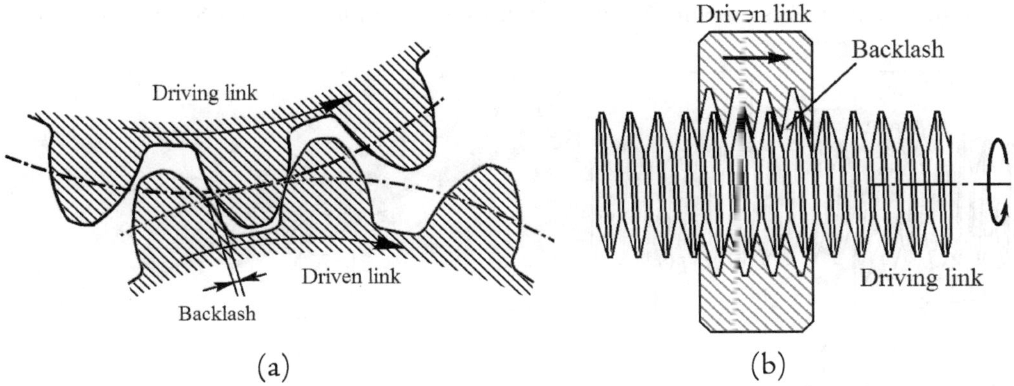

Figure 9-19 Backlash in power transmissions: (a) the gearing; (b) the screw-nut pair.

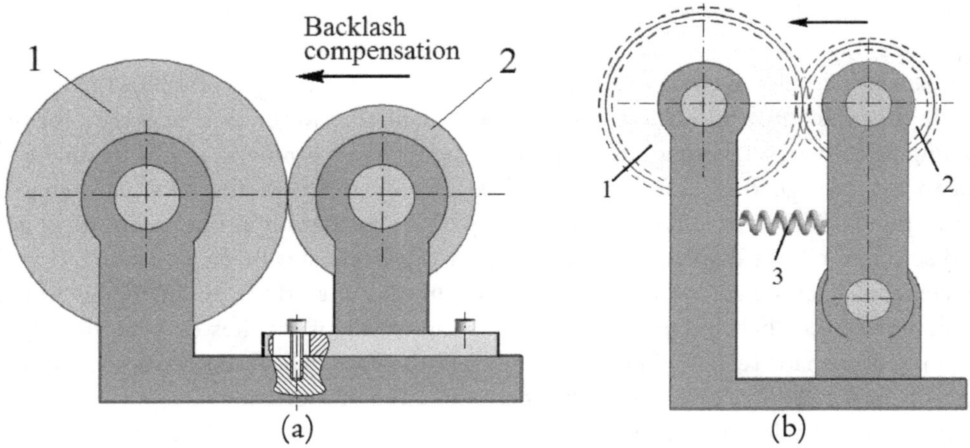

Figure 9-20 Preloading in gearing: (a) backlash-free gearing using setup regulation of the axis-to-axis distance; (b) tight mesh by spring pull. 1 – the fixed gear; 2 – the regulated gear; 3 – the extension spring.

The concept (b) is illustrated by the ballscrew-nut pairs, in which balls of enlarged diameter are used (Fig. 9.22a) to accomplish the backlash-free structure. This structure is optionally used by SKF for rolled thread ball screws with recirculation through an insert on user request. Contrary to the spring-based anti-backlash assembly resulting in two-point contact (Fig. 9.22b), the structure in Figure 22a provides four-point contact of the ball with the screw and nut. In both cases, the balls – bodies of revolution between screw and nut helicoid surfaces – obtain additional stresses providing both "balls-screw" and "balls-nut" tight butts for the given force range, but simultaneously allowing sufficiently free relative rotation of the pair elements.

Remember that in Figures 5.15 and 5.16 another type of backlash-free structures was shown and briefly described — elastic hinges, which are permanently preload by internal deformation-caused stresses.

9.1.3.4 *Tests of position accuracy*

In Figure 9.23, the position error during forward and subsequent reverse motions of the table along the *X*-axis is shown. A curve describing the unidirectional forward motion is

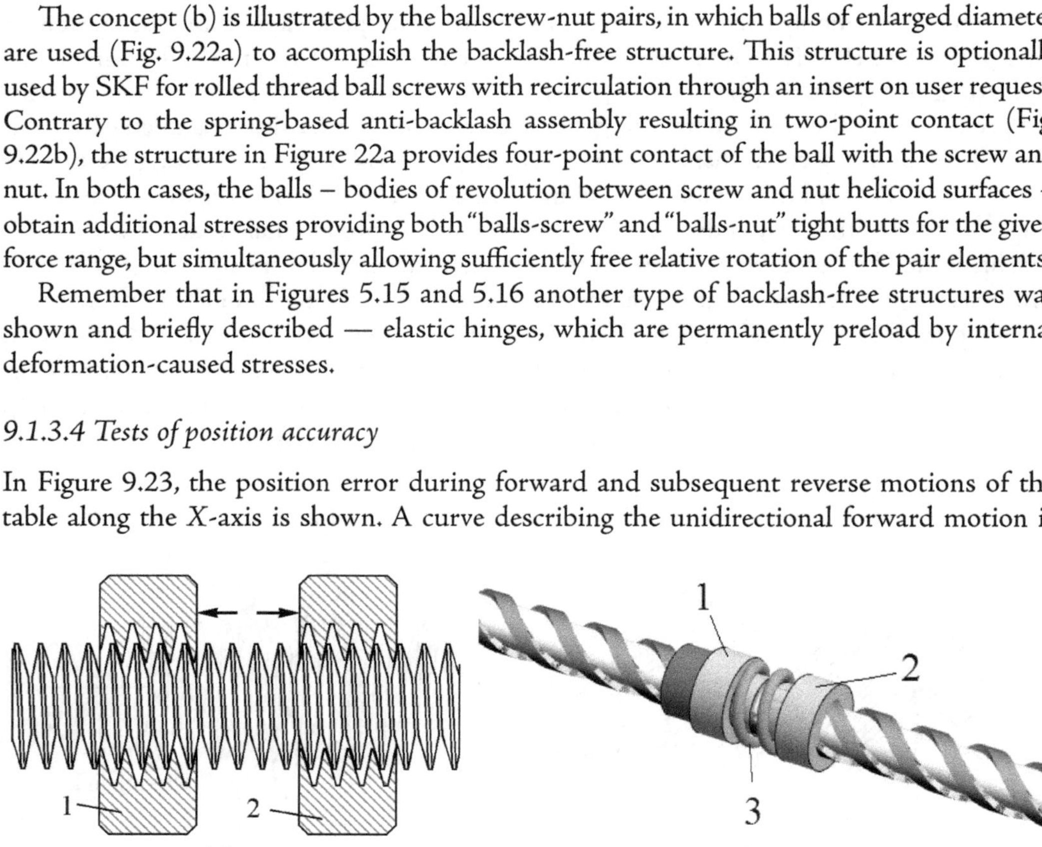

Figure 9-21 Preloading in the nut-screw pair: (a) a small relative axial displacement of half-nuts; (b) a spring-based anti-backlash assembly for leadscrew nut [4]. 1 and 2 –half-nuts of the split nut; 3 – compression spring for force-separation.

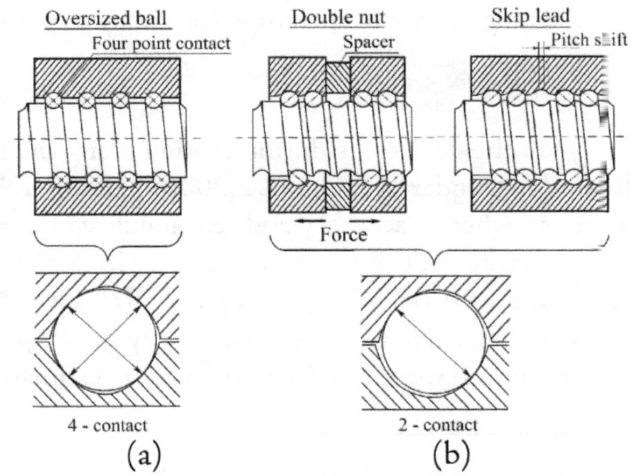

Figure 9-22 Structural elimination of the backlash in the ballscrew-nut pair using: (a) four-point contact; (b) two-point contact [4].

finished at point *1*. The curve portion the between points *1* and *2* describes a process of loosening the preloading of the axial transmission elements. A jump observed at point 2 is equal to the axial airgap; and then the preloading of the drive mechanism elements for reversal motion, also with lack of the motion, is started and lasts for a short time.

Note that the graph in Figure 9.23 is built on the assumption that magnitudes A_k, a_k, and phase parameters ξ_k and c_k figured in Eqs. (9.30) have fixed values, i.e., the shown curves can be considered as average values of the position errors. Their limit values $\delta_{x,lim}$ are estimated using standard deviation σ as follows:

$$\delta_{lim}(x) = \delta_x \pm 3\sigma \tag{9.35}$$

$$\sigma = \sqrt{\sum a_k^2 + \sum_{k>1} A_k^2} \tag{9.36}$$

where a_k and A_k are magnitudes of harmonics defined in Eq. (9.30).

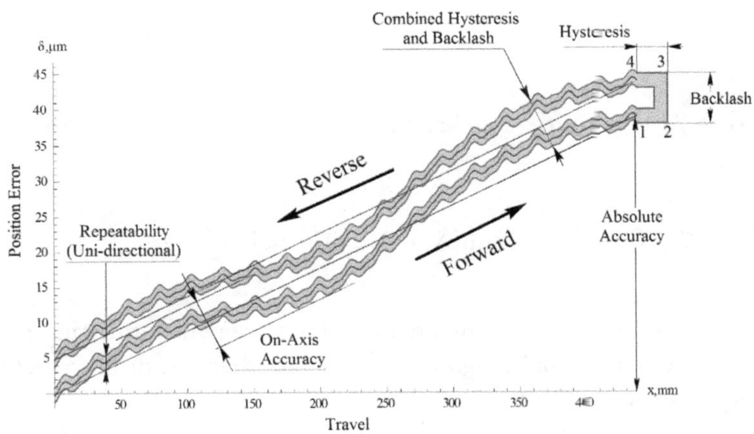

Figure 9-23 Position error during forward/reverse motion along the X-axis: *1-2* and *3-4* are the forward/reverse elastic backlash segments (×5000); *2-3* is the air gap segment.

For each given $x = x_i$, the range of the bi-directional accuracy of positioning of an axis is

$$\delta_{lim}(x_{if}) - 3\sigma \le \delta_{lim}(x_i) \le \delta_{lim}(x_{ir}) + 3\sigma \tag{9.37}$$

where $\delta_{lim}(x_{if})$ and $\delta_{lim}(x_{ir})$ are the deviations during motion along the forward and reverse directions, respectively, used in standards ISO 230-2:2006 [7] and VDI/DQG 3441.

Acomputerized procedure, whereat accuracy and repeatability of positioning of numerically-controlled axes is statistically evaluated is established by ISO 230-2:2006 [7]. The test provides $m = 5$ repeat measurements ($m = 7$ according to VDI/DQG 3441) for N positions x_i (e.g., $N = 10 \div 20$) along the machine workspace signing by subscript f and r as applied to *forward* and *reverse* axial motions, respectively. In each position, estimates of the mean value and standard deviation are

$$\delta_{if} = \frac{1}{m}\sum_{j=1}^{m}\delta_{ijf}\ ;\ s_{if} = \sqrt{\frac{1}{m-1}\sum_{j=1}^{m}(\delta_{ijf} - \delta_{if})^2} \tag{9.38}$$

$$\delta_{ir} = \frac{1}{m}\sum_{j=1}^{m}\delta_{ijr}\ ;\ s_{ir} = \sqrt{\frac{1}{m-1}\sum_{j=1}^{m}(\delta_{ijr} - \delta_{ir})^2} \tag{9.39}$$

These values are used for estimation of the *axis positioning*:

- *Systematic deviation:*

Unidirectional $E_{if} = \max[\delta_{if}] - \min[\delta_{if}]$ and $E_{ir} = \max[\delta_{ir}] - \min[\delta_{ir}]$

Bi-directional $E_i = \max[\delta_{if}; \delta_{ir}] - \min[\delta_{if}; \delta_{ir}]$

- *Repeatability:*

Unidirectional $R_{if} = \max[4s_{if}]$ and $R_{ir} = \max[4s_{ir}]$

Bi-directional $R_i = \max[R_{if}; R_{ir}]$

- *Accuracy:*

Unidirectional

$A_{if} = \max[\delta_{if} + 2s_{if}] - \min[\delta_{if} - 2s_{if}]$ and $A_{ir} = \max[\delta_{ir} + 2s_{ir}] - \min[\delta_{ir} - 2s_{ir}]$;

Bi-directional

$A_i = \max[\delta_{if} + 2s_{if}; \delta_{ir} + 2s_{ir}] - \min[\delta_{if} - 2s_{if}; \delta_{ir} - 2s_{ir}]$, etc.

Note that:

(a) Number of positions of measurements provides no way of estimating the cyclic errors since the leadscrew pitch is usually significantly smaller than the distance between two successive positions of measurements;

(b) The probability of the recommended estimates of repeatability and accuracy are well below the commonly accepted requirements for accuracy of measurements (99.73% or $\pm 3\sigma$ interval) since both the sample size $m = 5$ and the tolerance interval $\pm 2\ s$ are small.

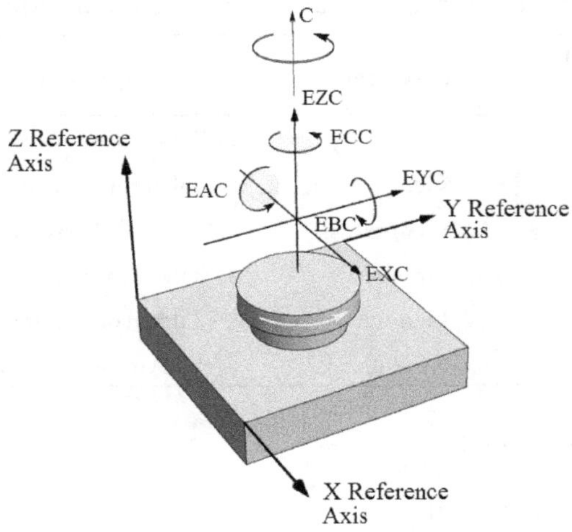

Figure 9-24 Error definitions for rotation around the Z-axis (C-motion). Error definitions according to ISO 230-1 [1].

9.2 Geometric accuracy of spindle unit

9.2.1 Structure and model

Error definitions associated with a link carrying out the rotary motion (in this case, around the Z-axis) according to ISO 230-1 [1] are shown in Figure 9.24 and listed in Table 9.5. The six error components are functions of the current rotary angle φ, i.e., $\delta_k = \delta_k(\varphi)$.

As an example of the revolute pair, the spindle unit is considered (Fig. 9.25). The geometrical accuracy of the spindle unit is usually controlled by measuring radial runout Δ_A, Δ_B, Δ_C, and Δ_D at points A and B on the fitting ring of the spindle and points C and D on the mandrel mounted in the opening of the spindle. Furthermore, axial runout Δ_Q of the spindle at point Q and the kinematic error Δ_{kin} of the angle of rotation are measured. In this case, the measuring devices are mounted on the fixed base (column, bed, headstock, cross-rail, etc.). The initial data for the system are presented in Table 9.6.

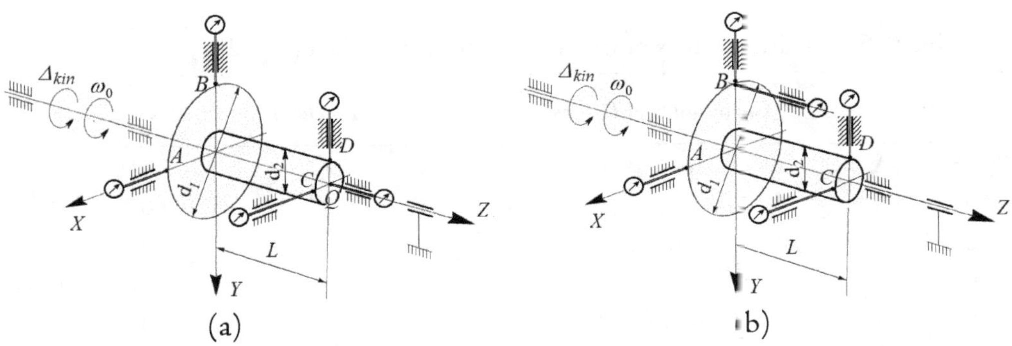

Figure 9-25 Two schematics of the spindle accuracy measurement: (a) the axial runout is measured at point Q; (b) the face runout is measured at point B.

Table 9-5 Rotation around the vertical Z-axis: Error definitions.

Error matrix entry	ISO 230–1 symbol	Error type	Comprehensive description
$\delta_x = \delta x(\varphi)$	EXC	Geometry	Radial error motion of C in X direction
$\delta_y = \delta y(\varphi)$	EYC	Geometry	Radial error motion of C in Y direction
$\delta_z = \delta z(\varphi)$	EZC	Geometry	Axial error motion of C
$\delta_\theta = \delta_\theta(\varphi)$	EAC	Geometry	Tilt error motion of C around X
$\delta_\psi = \delta_\psi(\varphi)$	EBC	Geometry	Tilt error motion of C around Y
$\delta_\varphi = \delta_\varphi(\varphi)$	ECC	Kinematic	Angular positioning error

The support Jacobian matrix $\mathbf{J}_s$ is

$$\mathbf{J}_s = \begin{bmatrix} -1 & 0 & 0 & 0 & 0 & 0 \\ 0 & 1 & 0 & 0 & 0 & 0 \\ -1 & 0 & 0 & 0 & -L & 0 \\ 0 & 1 & 0 & -L & 0 & 0 \\ 0 & 0 & -1 & 0 & 0 & 0 \\ 0 & 0 & 0 & 0 & 0 & 1 \end{bmatrix} \tag{9.40}$$

Since $\mathrm{Det}[\mathbf{J}_s] = -L^2 \neq 0$, position errors of the spindle unit is defined

$$\boldsymbol{\delta} \equiv \begin{bmatrix} \delta_x \\ \delta_y \\ \delta_z \\ \delta_\theta \\ \delta_\psi \\ \delta_\varphi \end{bmatrix} = \mathbf{J}_s^{-1}\boldsymbol{\Delta} = \begin{bmatrix} -1 & 0 & 0 & 0 & 0 & 0 \\ 0 & 1 & 0 & 0 & 0 & 0 \\ -1 & 0 & 0 & 0 & -1 & 0 \\ 0 & 1/L & 0 & -1/L & 0 & 0 \\ 1/L & 0 & -1/L & 0 & 0 & 0 \\ 0 & 0 & 0 & 0 & 0 & 1 \end{bmatrix} \begin{bmatrix} \Delta_A \\ \Delta_B \\ \Delta_C \\ \Delta_D \\ \Delta_Q \\ \Delta_{kin} \end{bmatrix} = \begin{bmatrix} -\Delta_A \\ \Delta_B \\ -\Delta_Q \\ (\Delta_B - \Delta_D)/L \\ (\Delta_A - \Delta_C)/L \\ \Delta_{kin} \end{bmatrix} \tag{9.41}$$

In Figure 9.25a, the axial runout is measured at point Q on the datum axis. Instead of this, the runout in the axial direction is sometimes tested on the large diameter. Let, for

Table 9-6 Position errors of the rotary unit: Initial data (Fig. 25a).

Point		***Cartesian coordinates***			***Plücker's coordinates***					
Symbol	Δ_i	x_i	y_i	z_i	e_{xi}	e_{yi}	e_{zi}	n_{xi}	n_{yi}	n_{zi}
A	Δ_A	$d_1/2$	0	0	−1	0	0	0	0	0
B	Δ_B	0	$-d_1/2$	0	0	1	0	0	0	0
C	Δ_C	$d_2/2$	0	L	−1	0	0	0	$-L$	0
D	Δ_D	0	$-d_2/2$	L	0	1	0	$-L$	0	0
Q	Δ_Q	0	0	L	0	0	−1	0	0	0
Rotary axis	Δ_{kin}	0	0	0	0	0	0	0	0	1

Table 9-7 Position errors of the rotary unit: Initial data (Fig 25b).

Point		*Cartesian coordinates*			*Plücker's coordinates*					
Symbol	Δ_i	x_i	y_i	z_i	e_{xi}	e_{yi}	e_{zi}	n_{xi}	n_{yi}	n_{zi}
A	Δ_A	$d_1/2$	0	0	−1	0	0	0	0	0
B	Δ_B	0	$-d_1/2$	0	0	1	0	0	0	0
C	Δ_C	$d_2/2$	0	L	−1	0	0	0	$-L$	0
D	Δ_D	0	$-d_2/2$	L	0	1	0	$-L$	0	0
B	Δ_{Ba}	0	$-d_1/2$	0	0	0	−1	$-d_1/2$	0	0
Rotary axis	Δ_{kin}	0	0	0	0	0	0	0	0	1

example, the measuring device, oriented in the axial direction opposite to the Z-axis, contacts with the spindle unit at point B on diameter d_1 (Fig. 9.25b). In the Table 9.7, the initial data are shown which differ from those in Table 9.6 by substitution of coordinates of point B for Q. In this case, the face runout at point B is measured instead of axial runout at point Q, and results of measurements in the axial direction δ_z depend also on runouts in the radial directions Δ_B and Δ_D. According to Eq. 9.14, the position errors of the spindle unit are

$$\boldsymbol{\delta} \equiv \begin{bmatrix} \delta_x \\ \delta_y \\ \delta_z \\ \delta_\theta \\ \delta_\psi \\ \delta_\varphi \end{bmatrix} = \mathbf{J}^{-1}\boldsymbol{\Delta} = \begin{bmatrix} -\Delta_A \\ \Delta_B \\ -\Delta_{Ba} + (\Delta_B - \Delta_D)\dfrac{d_1}{2L} \\ (\Delta_B - \Delta_D)/L \\ (\Delta_A - \Delta_C)/L \\ \Delta_{kin} \end{bmatrix}$$

9.2.2 Error analysis

Results of measurements of the rotary unit present the functions of angle φ of spindle rotation and can be represented by Fourier series expansion in terms of frequency ω of the spindle rotation and harmonics ω_k. For example,

$$\Delta_A = \frac{1}{2}a_0 + \sum_{k=1}^{n} a_k \sin(\frac{\omega_k}{\omega_0}\varphi + \theta_k) \tag{9.42}$$

$$\Delta_C = \frac{1}{2}c_0 + \sum_{k=1}^{n} c_k \sin(\frac{\omega_k}{\omega}\varphi + \psi_k) \tag{9.43}$$

where n is the number of harmonics of interest in the results of measurements; φ is the curren tangle of spindle rotation; ω_k is the frequency of the kth harmonic multiple by frequency ω, $\omega_k = k\omega$; a_k and c_k are the amplitudes at frequency ω_k; θ_k and ψ_k are the phase angles.

Substitution of Δ_A and Δ_C, Eqs. (9.42) and (9.43), in expression $\delta_\psi = (\Delta_A - \Delta_C)/L$ obtained from Eq. (9.41) yields

$$\delta_y = \frac{1}{L}\left[\frac{1}{2}(a_0 - c_0) + \sum_{k=1}^{n} e_k \sin\left(\frac{\omega_k}{\omega_0}\varphi + \lambda_k\right)\right] \tag{9.44}$$

with

$$e_k = \sqrt{a_k^2 + c_k^2 - 2a_k c_k \cos(\theta_k - \psi_k)} \text{ and } \tan\lambda_k = \frac{a_k \sin\theta_k - c_k \sin\psi_k}{a_k \cos\theta_k - c_k \cos\psi_k} \tag{9.45}$$

A similar expression can be derived for the small angle $\delta_\theta = (\Delta_B - \Delta_D)/L$, Eq. (9.41). Equations (9.42-44) can be used for diagnostics of the rotary units. In particular,

- The free members a_0, c_0, and e_0 describe the preliminary installation of the measuring probes;
- The first member $c_1\sin(\varphi + \theta_1)$ describes an eccentricity of the mandrel; and,
- The members $c_2\sin(2\varphi + \theta_2)$, $c_3\sin(3\varphi + \theta_3)$,..., $c_n\sin(n\varphi + \theta_n)$, i.e., without the eccentricity member, describe the *runout of the axis of the spindle rotation.*

As an example, the results of measurements of the runout of the spindle rotation are shown in Figure 9.26a. In Figure 9.26b, the first harmonic caused by the mandrel eccentricity and the runout of *the axis* of the spindle rotation, i.e., the results of measurements without the eccentricity component, are shown.

The runout affects the accuracy of the element to be processed. For example, in machine tools, the runout of the main spindle is usually directly transferred to a workpiece. In particular, the form accuracy of the workpiece depends on the harmonics of the second and higher orders.

9.3 Workpiece locating

9.3.1 The 3–2–1 locating principle

The geometric scheme "rigid body on six rigid supports" can be demonstrated in locating a workpiece with respect to a cutting tool on machine tools. In these structures, a jig or fixture consists of six fixed or adjustable pins representing six points of the "workpiece-jig" contact.

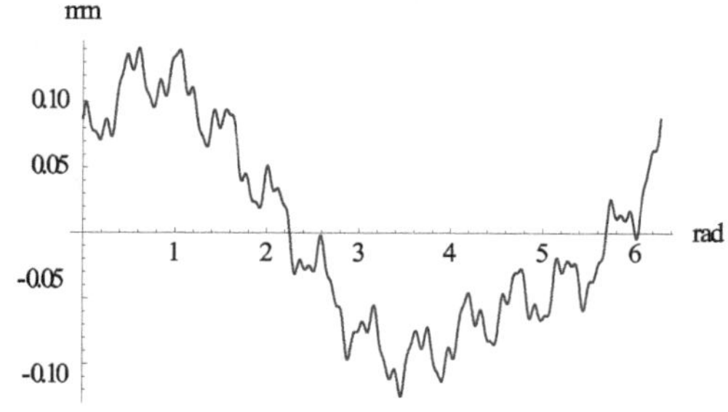

(a)

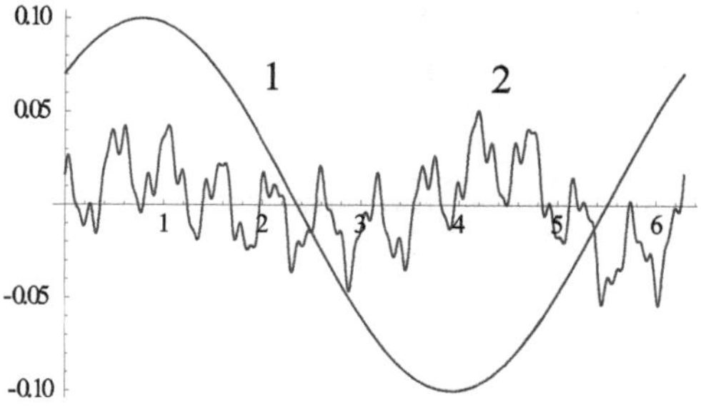

(b)

Figure 9-26 (a) Results of measurements of the spindle runout; (b) 1- the first harmonic with amplitude 0.1 mm; 2 – the runout of the axis of the spindle rotation (without the first harmonic).

The popular scheme, the so-called 3-2-1 locating principle, is shown in Figure 9.27. Three pins 1, 2, and 3 form the primary locating datum (a plane), which establishes three DOFs of the based part – two translations and one rotation. Two pins 4 and 5 form the secondary locating datum (a line), which establishes two DOFs; and pin 6 presents the tertiary locating datum (a point) fixing the sixth DOF. This scheme allows careful adjustment of the workpiece by small displacement of the pins. Equations connecting these displacements and controlled displacements of the workpiece have the abovementioned form, Eqs. (9.10), (9.15), etc.

$$\mathbf{J}_s\boldsymbol{\delta} = \boldsymbol{\Delta} \tag{9.46}$$

where $\boldsymbol{\delta}$ is the twist describing location/orientation of the workpiece; $\boldsymbol{\Delta}$ is the vector of the pins' heights Δ_k ($k = 1,2,\ldots,6$); and $\mathbf{J}_s$ is the 6×6 Jacobian matrix of the supports,

$$\boldsymbol{\delta} = [\delta_x, \delta_y, \delta_z, \delta_\theta, \delta_\psi, \delta_\varphi]^{\mathrm{T}} \tag{9.47}$$

$$\boldsymbol{\Delta} = [\Delta_1, \Delta_2, \Delta_3, \Delta_4, \Delta_5, \Delta_6]^{\mathrm{T}} \tag{9.48}$$

$$\mathbf{J}_s = \begin{bmatrix} 0 & 0 & 1 & y_1 & -x_1 & 0 \\ 0 & 0 & 1 & y_2 & -x_2 & 0 \\ 0 & 0 & 1 & y_3 & -x_3 & 0 \\ 0 & 1 & 0 & -z_4 & 0 & x_4 \\ 0 & 1 & 0 & -z_5 & 0 & x_5 \\ 1 & 0 & 0 & 0 & z_6 & -y_6 \end{bmatrix} \tag{9.49}$$

$$\mathrm{Det}[\mathbf{J}_s] = (x_4 - x_5)[x_3(y_2 - y_1) + x_2(y_3 - y_1) + x_1(y_3 - y_2)] \tag{9.50}$$

If the pins are located according to Figure 9.27, i.e., $x_1 = x_2 = x_4$; $x_3 = x_5$; $y_3 = y_6$; $y_4 = y_5$; $z_1 = z_2 = z_3$; and $z_4 = z_5$, but $x_1 \neq x_3$ and $y_1 \neq y_2$, the determinant, Eq. (9.50), becomes

$$\mathrm{Det}[\mathbf{J}_s] = (x_1 - x_3)^2\,(y_1 - y_2) \neq 0 \tag{9.51}$$

Since the matrix $\mathbf{J}_s$ is non-singular, the inverse matrix $\mathbf{J}_s^{-1}$ exists, and the located workpiece twist is

$$\boldsymbol{\delta} = \mathbf{J}_s^{-1}\boldsymbol{\Delta} \tag{9.52}$$

Numerical Example 9.3. Coordinate of the pins [mm] are given in the Table. Let us consider two interrelated problems.

k	x_k	y_k	z_k
1	350	50	0
2	350	300	0
3	50	175	0
4	350	0	150
5	50	0	150
6	0	175	150

(a) **Given:** The small variation of the workpiece position and orientation, vector **δ**: $\delta_x = 0.1$ mm; $\delta_y = -0.03$ mm; $\delta_z = 0.05$ mm; $\delta_\theta = 1.2 \times 10^{-5}$ rad; $\delta_\psi = 1.5 \times 10^{-5}$ rad; and $\delta_\varphi = 3 \times 10^{-5}$ rad. **Required:** Displacements of the pins (adjustable rests) resulting in the given variation.

Solution: According to Eq. (9.46):

$$\begin{bmatrix}\Delta_1\\\Delta_2\\\Delta_3\\\Delta_4\\\Delta_5\\\Delta_6\end{bmatrix}=\begin{bmatrix}0&0&1&y_1&-x_1&0\\0&0&1&y_2&-x_2&0\\0&0&1&y_3&-x_3&0\\0&1&0&-z_4&0&x_4\\0&1&0&-z_5&0&x_5\\1&0&0&0&z_6&-y_6\end{bmatrix}\begin{bmatrix}\delta_x\\\delta_y\\\delta_z\\\delta_\theta\\\delta_\psi\\\delta_\varphi\end{bmatrix}=\begin{bmatrix}\delta_z+y_1\delta_\theta-x_1\delta_\psi\\\delta_z+y_2\delta_\theta-x_2\delta_\psi\\\delta_z+y_3\delta_\theta-x_3\delta_\psi\\\delta_y-z_4\delta_\theta+x_4\delta_\varphi\\\delta_y-z_5\delta_\theta+x_5\delta_\varphi\\\delta_x+z_6\delta_\psi-y_6\delta_\varphi\end{bmatrix}=$$

$$\begin{bmatrix}0.05+50\times1.2\times10^{-5}-350\times1.5\times10^{-5}\\0.05+300\times1.2\times10^{-5}-350\times1.5\times10^{-5}\\0.05+175\times1.2\times10^{-5}-50\times1.5\times10^{-5}\\-0.03-150\times1.2\times10^{-5}+350\times3\times10^{-5}\\-0.03-150\times1.2\times10^{-5}+50\times3\times10^{-5}\\0.1+150\times1.5\times10^{-5}-175\times3\times10^{-5}\end{bmatrix}=\begin{bmatrix}0.04535\\0.04835\\0.05135\\-0.2913\\-0.3003\\0.097\end{bmatrix}\text{mm}$$

(b) **Given:** Displacements of the adjustable rests, vector **Δ**: $\Delta_1 = 0.05$ mm; $\Delta_2 = 0.03$ mm; $\Delta_3 = 0.04$ mm; $\Delta_4 = 0.01$ mm; $\Delta_5 = 0.08$ mm; and $\Delta_6 = 0.15$ mm. **Required:** The small variation of the workpiece position and orientation, vector **δ**, resulting from the rests displacements.

Solution: According to Eq. (9.52):

$$\begin{bmatrix}\delta_x\\\delta_y\\\delta_z\\\delta_\theta\\\delta_\psi\\\delta_\varphi\end{bmatrix}=\mathbf{J}_s^{-1}\begin{bmatrix}\Delta_1\\\Delta_2\\\Delta_3\\\Delta_4\\\Delta_5\\\Delta_6\end{bmatrix}=\begin{bmatrix}0&0&1&50&-350&0\\0&0&1&300&-350&0\\0&0&1&175&-50&0\\0&1&0&-150&0&350\\0&1&0&-150&0&50\\1&0&0&0&150&-175\end{bmatrix}^{-1}\begin{bmatrix}\Delta_1\\\Delta_2\\\Delta_3\\\Delta_4\\\Delta_5\\\Delta_6\end{bmatrix}=$$

$$\begin{bmatrix}0.25&0.25&-0.5&0.58333&-0.58333&1\\-0.6&0.6&0&-0.16667&0.16667&0\\0.61667&-0.78333&1.16667&0&0&0\\-0.004&0.004&0&0&0&0\\-0.0016667&-0.0016667&0.0033333&0&0&0\\0&0&0&0.0033333&-0.0033333&0\end{bmatrix}\begin{bmatrix}0.05\\0.03\\0.04\\0.01\\0.08\\0.15\end{bmatrix}=\begin{bmatrix}0.109167\text{ mm}\\0.079667\text{ mm}\\0.054\text{ mm}\\-0.00008\text{ rad}\\0\text{ rad}\\-0.0002333\text{ rad}\end{bmatrix}$$

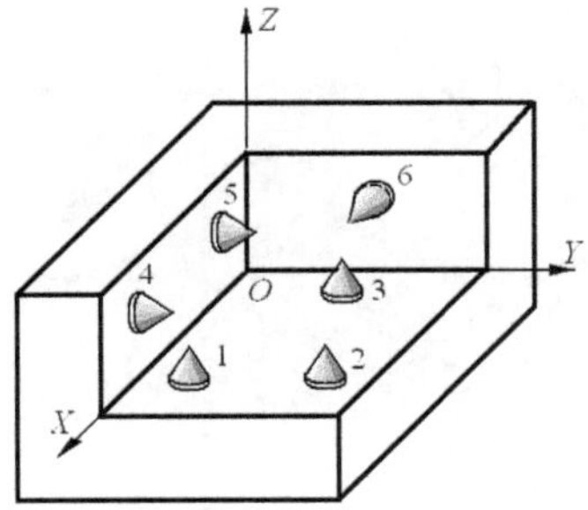

Figure 9-27 The 3–2–1 locating scheme.

As a real-life example, consider locating a car unit (Fig. 9.28), where the three primary locating points, A1, A2, and A3, control three degrees of freedom: translation in Z, rotation around X, and rotation around Y. The two secondary locating points, B1 and B2, control two degrees of freedom: translation in X and rotation around Z. The last, tertiary locating point C1 controls one degree of freedom. translation along Y.

The Jacobian of the locating system combines the six last column of the Table 9.8

$$\mathbf{J}_s = \begin{bmatrix} 0 & 0 & 1 & -y_A/2 & -a & 0 \\ 0 & 0 & 1 & y_A/2 & a & 0 \\ 0 & 0 & 1 & 0 & -a-x_A & 0 \\ 1 & 0 & 0 & 0 & h & y_B/2 \\ 1 & 0 & 0 & 0 & h & -y_B/2 \\ 0 & -1 & 0 & H & 0 & -a/2 \end{bmatrix} \tag{9.53}$$

Equation (8.50) yields

$\mathbf{\Delta} = \mathbf{J}_s\,\boldsymbol{\delta}$, i.e.,

$$\begin{bmatrix} \Delta_{A1} \\ \Delta_{A2} \\ \Delta_{A3} \\ \Delta_{B1} \\ \Delta_{B2} \\ \Delta_{C1} \end{bmatrix} = \begin{bmatrix} 0 & 0 & 1 & -y_A/2 & -a & 0 \\ 0 & 0 & 1 & y_A/2 & a & 0 \\ 0 & 0 & 1 & 0 & -a-x_A & 0 \\ 1 & 0 & 0 & 0 & h & y_B/2 \\ 1 & 0 & 0 & 0 & h & -y_B/2 \\ 0 & -1 & 0 & H & 0 & -a/2 \end{bmatrix} \begin{bmatrix} \delta_x \\ \delta_y \\ \delta_z \\ \delta_\theta \\ \delta_\psi \\ \delta_\varphi \end{bmatrix} = \begin{bmatrix} \delta_z - \delta_\psi a - \delta_\theta y_A/2 \\ \delta_z - \delta_\psi a + \delta_\theta y_A/2 \\ \delta_z - \delta_\psi (a + x_A) \\ \delta_x + \delta_\psi h + \delta_\varphi y_B/2 \\ \delta_x + \delta_\psi h - \delta_\varphi y_B/2 \\ -\delta_y + \delta_\theta H - \delta_\varphi a/2 \end{bmatrix} \tag{9.54}$$

Table 9-8 Cartesian and the Plücker coordinates of the pins.

Point	x_i	y_i	z_i	e_{xi}	e_{yi}	e_{zi}	n_{xi}	n_{yi}	n_{zi}
A_1	$x_1 = a$	$y_1 = -y_A/2$	0	0	0	1	$-y_A/2$	$-a$	0
A_2	$x_2 = a$	$y_2 = y_A/2$	0	0	0	1	$y_A/2$	a	0
A_3	$x_3 = a + x_A$	$y_3 = 0$	0	0	0	1	0	$-a - x_A$	0
B_1	$x_4 = b$	$y_4 = -y_B/2$	h	1	0	0	0	h	$y_B/2$
B_2	$x_5 = b$	$y_5 = y_B/2$	h	1	0	0	0	h	$-y_B/2$
C_1	$x_6 = a/2$	y_6	H	0	−1	0	H	0	$-a/2$

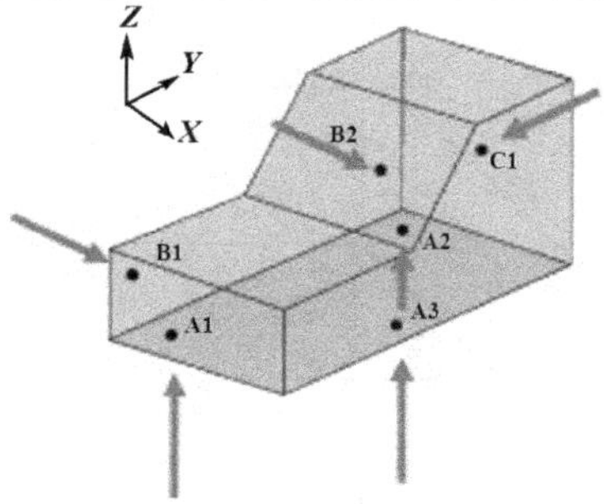

Figure 9-28 The 3–2–1 locating principle as applied to locating of a car unit [8].

Further calculations are similar to those considered in the *Numerical example* 9.3. Another version of the workpiece locating 3–1–1–1 is shown in Figure 9.29.

9.3.2 Workpiece locating during centerless grinding

Finish machining of the bearing rollers are performed by means of centerless grinding (Fig. 9.30). The workpiece 1 of diameter $2r$ is supported on blade 3 and regulating (drive) wheel 2. Apart from its supporting function, wheel 2 actuates rotation of the workpiece in the direction defined by vector δ_C'. Note that the workpiece translation in the axial direction is not considered. Hence, the current location of the workpiece is fixed by three directions given by vectors δ_B, δ_C, and δ_C'.

In this layout, displacement δ_A of point A of contact of workpiece 1 with grinding wheel 3 is determined as a function of the known displacements δ_B and δ_C of points B and C of the workpiece that are in contact with drive wheel 2 and blade 3, respectively. Displacements δ_B and δ_C are normally oriented with respect to drive wheel 2 and blade 3 and depend on the roundness of the workpiece surface before machining. Deviations δ_A caused by δ_B and δ_C result in planar motion of the workpiece and define variations of the allowance for machining (grinding). In Table 9.9, the unit vectors of the lines given vectors δ_A, δ_B, δ_C, and δ_C' as referred to coordinate frame OXY associated with the rotated workpiece are listed.

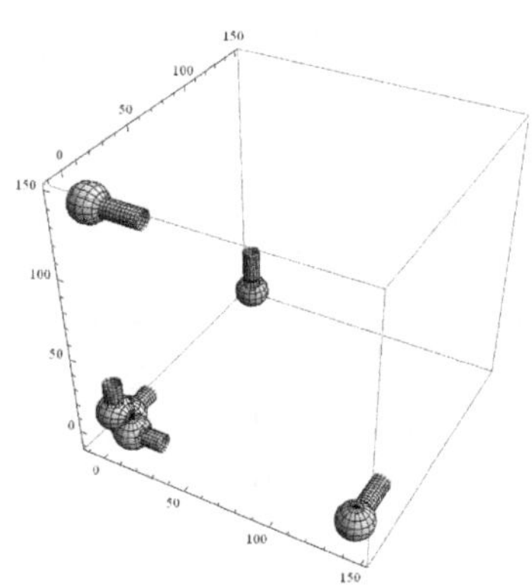

Figure 9-29 The 3–1–1 locating schematics.

Table 9-9 Plücker coordinates of the workpiece locating (Fig. 9.30).

Point	Vector	Displacement	e_{xi}	e_{zi}	n_{zi}
A	δ_A	Δ_A	$-\sin(\alpha + \beta)$	$\cos(\alpha + \beta)$	0
B	δ_B	Δ_B	$-\cos(\lambda - \beta)$	$\sin(\lambda - \beta)$	0
C	δ_C	Δ_C	0	–1	0
C	$\delta_{C'}$	0	1	0	r

Since the planar problem is considered, the geometric accuracy of the workpiece is calculated using the 3 × 3 Jacobian, Eq. (8.59), composing the Plücker coordinates of lines given by vectors δ_B, δ_C, and δ_C',

$$\mathbf{J} = \begin{bmatrix} -\cos(\lambda-\beta) & \sin(\lambda-\beta) & 0 \\ 0 & -1 & 0 \\ 1 & 0 & r \end{bmatrix} \tag{9.55}$$

The displacements δ_x, δ_y, δ_φ of the workpiece are

$$\boldsymbol{\delta} = \begin{bmatrix} \delta_x \\ \delta_y \\ \delta_\varphi \end{bmatrix} = \mathbf{J}^{-1}\boldsymbol{\Delta} = \begin{bmatrix} -\cos(\lambda-\beta) & \sin(\lambda-\beta) & 0 \\ 0 & -1 & 0 \\ 1 & 0 & r \end{bmatrix}^{-1} \begin{bmatrix} \Delta_B \\ \Delta_C \\ \Delta_{C'} \end{bmatrix} =$$

$$\begin{bmatrix} -\dfrac{\Delta_B}{\cos(\lambda-\beta)} - \Delta_C \tan(\lambda-\beta) \\ -\Delta_C \\ \dfrac{\Delta_B}{r\cos(\lambda-\beta)} + \dfrac{\Delta_C}{r}\tan(\lambda-\beta) + \dfrac{\Delta_{C'}}{r} \end{bmatrix} \tag{9.56}$$

The displacements Δ_A of point A is calculated through Eq. (8.63)

$$\Delta_A = (\mathbf{u}_A)^{\mathrm{T}}\boldsymbol{\delta} \tag{9.57}$$

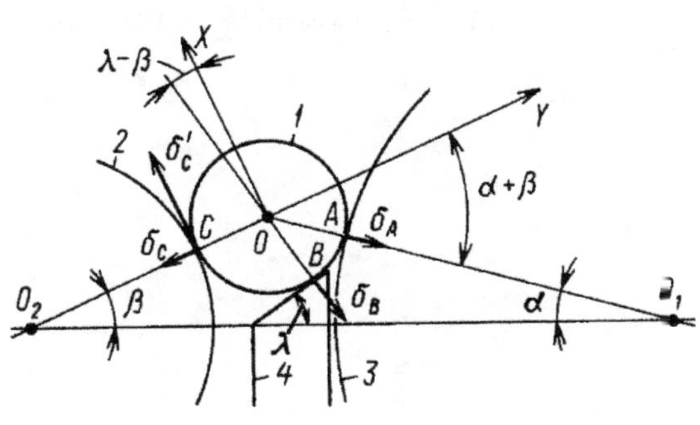

Figure 9-30 Geometrical error during centerless grinding *1* – workpiece; 2 – moving grinding wheel; 3 – stationary grinding wheel; 4 – work rest.

$$\mathbf{u}_A = [-\sin(\alpha + \beta), \cos(\alpha + \beta), 0]^T \tag{9.58}$$

where $\mathbf{u}_A$ is the vector of Plücker coordinates (see the first row of Table 9.9).

Substitutions of the expression for $\boldsymbol{\delta}$, Eq. (9.56), and $\mathbf{u}_A$ from the first raw of Table 9.9 yield

$$\Delta_A = [-\sin(\alpha+\beta),\ \cos(\alpha+\beta),\ 0]\begin{bmatrix} -\dfrac{\Delta_B}{\cos(\lambda-\beta)} - \Delta_C \tan(\lambda-\beta) \\ -\Delta_C \\ \dfrac{\Delta_B}{r\cos(\lambda-\beta)} + \dfrac{\Delta_C}{r}\tan(\lambda-\beta) + \dfrac{\Delta_{C'}}{r} \end{bmatrix} =$$

$$[\Delta_B \sin(\alpha + \beta) - \Delta_C \cos(\alpha + \beta)]/\cos(\lambda - \beta) \tag{9.59}$$

This expression is well-known from the investigation of the kinematics and dynamics of the centerless grinding and serves as a basis for real-life setups providing high roundness of the machined rollers.

References:

[1] Hara, Y., Motonishi, S., Yoshida, K., 1990, A New Micro-Cutting Device with High Stiffness and Resolution, submitted by Ikawa, *CIRP Annals – Manufacturing Technology*, Vol. 39/1, 375-378.

[2] ISO 230-1-2012. Test code for machine tools. Part 1: Geometric accuracy of machines operating under no-load or quasi-static conditions.

[3] Cherpakov, B.I., Portman, V.T., Feofanov, A.I., Genin, D.V., 1988, Accuracy research of power tables of flexible transfer lines, 1988, *Stanki i instrument*, No. 6, pp. 3-4 (*in Russian*).

[4] SKF.com, 2017, Precision rolled ball screws, Google, http://www.skf.com/group/products/, Catalog on line.

[5] ISO 286-2:1988. ISO system of limits and fits – Part 2: Table of standard tolerance grades and limit deviations for holes and shafts.

[6] Rivin, E.I., 2010, *Handbook on Stiffness & Damping in Mechanical Design*, ASME Press, New York, NY.

[7] ISO 230-2:2006. Test code for machine toos – Part 2: Determination of accuracy and repeatability of positioning numerically controlled axes.

[8] Söderberg, R., Wickman, C., Lindkvist, L., 2008, Improving decision making by simulating and visualizing geometrical variation in non-rigid assemblies, *CIRP Annals - Manufacturing Technology*, 57/1, pp. 175-178.

10. Infinitesimal kinematics of serial-kinematics machines

10.1 Error propagation: serial-type robots

10.1.1 Planar 2-DOF (RR) manipulator

To clear up an interrelationship between the nominal displacements and their deviations in the accuracy-related formulae, the links of the 2-DOF manipulator are renumbered as is shown in Figure 10.1. In these terms, the manipulation matrix, Eqs. (4.7),becomes

$$^{0}\mathbf{A}_4 = \mathbf{A}^6(\varphi_1)\mathbf{A}^1(a_2)\mathbf{A}^6(\varphi_3)\mathbf{A}^1(a_4) = \begin{bmatrix} \cos(\varphi_1+\varphi_3) & -\sin(\varphi_1+\varphi_3) & a_2\cos\varphi_1 + a_4\cos(\varphi_1+\varphi_3) \\ \sin(\varphi_1+\varphi_3) & \cos(\varphi_1+\varphi_3) & a_2\sin\varphi_1 + a_4\sin(\varphi_1+\varphi_3) \\ 0 & 0 & 1 \end{bmatrix} \tag{10.1}$$

$$\mathbf{A}^6(\varphi_i) = \begin{bmatrix} \cos\varphi_i & -\sin\varphi_i & 0 \\ \sin\varphi_i & \cos\varphi_i & 0 \\ 0 & 0 & 1 \end{bmatrix}; i = 1,3; \text{ and } \mathbf{A}^1(a_i) = \begin{bmatrix} 1 & 0 & a_i \\ 0 & 1 & 0 \\ 0 & 0 & 1 \end{bmatrix}; i = 2,4 \tag{10.2}$$

The 3 × 3 total manipulation error matrix **E** and error vector $\Delta\mathbf{r}_0$ are

$$\mathbf{E} = \sum_{i=0}^{4} {}^{0}\mathbf{A}_i \mathbf{E}_i ({}^{0}\mathbf{A}_i)^{-1} \tag{10.3}$$

$$\mathbf{E}_i = \begin{bmatrix} 0 & -\delta_{\varphi i} & \delta_{xi} \\ \delta_{\varphi i} & 0 & \delta_{yi} \\ 0 & 0 & 0 \end{bmatrix}, (i = 0,1,2,3,4) \quad (1$$

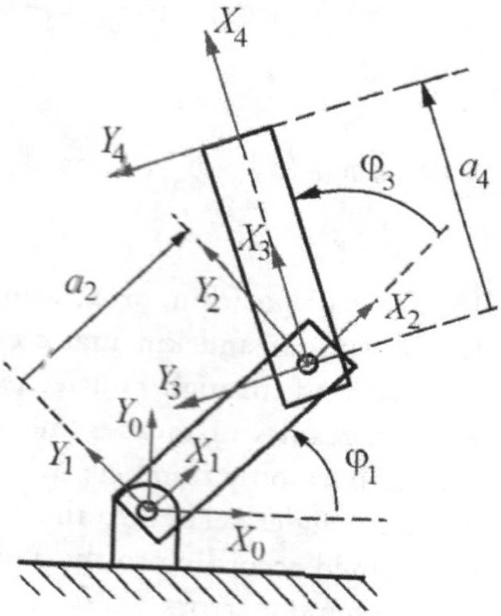

Figure 10-1 Planar 2-DOF (**RR**) manipulator.

$$\Delta\mathbf{r}_0 \equiv \begin{bmatrix} \Delta_x \\ \Delta_y \\ 0 \end{bmatrix} = \mathbf{E}\mathbf{r}_0 = \begin{bmatrix} 0 & -\delta_\varphi & \delta_x \\ \delta_\varphi & 0 & \delta_y \\ 0 & 0 & 0 \end{bmatrix} \begin{bmatrix} x_0 \\ y_0 \\ 0 \end{bmatrix} = \begin{bmatrix} \delta_x - \delta_\varphi y_0 \\ \delta_y + \delta_\varphi x_0 \\ 0 \end{bmatrix} \tag{10.5}$$

where δ_x are δ_y are the linear deviations of the origin of the end-effector (EE) as referred to coordinate frame S_0 along the X_0 and Y_0-axis; δ_φ is the small rotation (angular deviation) of the EE around the Z_0-axis; $\mathbf{E}_0$ is the error matrix associated with the base; and $\mathbf{E}_i$ ($i = 1, 2, 3, 4$) is the error matrix associated with the ith link.

The expanded form of matrix **E**, Eq. (10.2), is

$$\mathbf{E} \equiv \begin{bmatrix} 0 & -\delta_\varphi & \delta_x \\ \delta_\varphi & 0 & \delta_y \\ 0 & 0 & 0 \end{bmatrix} = \mathbf{E}_0 + \mathbf{A}^6(\varphi_1)\mathbf{E}_1\mathbf{A}^6(-\varphi_1) + \mathbf{A}^6(\varphi_1)\mathbf{A}^1(a_2)\mathbf{E}_2\mathbf{A}^1(-a_2)\mathbf{A}^6(-\varphi_1)$$

$$+\mathbf{A}^6(\varphi_1)\,\mathbf{A}^1(a_2)\mathbf{A}^6(\varphi_3)\,\mathbf{E}_3\mathbf{A}^6(-\varphi_3)\,\mathbf{A}^1(-a_2)\,\mathbf{A}^6(-\varphi_1)\; + {}^0\mathbf{A}_4\mathbf{E}_4({}^0\mathbf{A}_4)^{-1} \tag{10.6}$$

where ${}^0\mathbf{A}_4$ is manipulation matrix, Eq. (10.1). See also Appendix *A2.2.4*.

Some parameters entering into the error matrices have no effect on the processed object. ·his case, $\delta_{\varphi 0} = \delta_{\varphi 2} = \delta_{\varphi 4} = 0$, and calculations according to Eq. (10.6) yield

$$\delta_{x0} + \delta_{x1,2}\cos\varphi_1 + \delta_{x3,4}\cos(\varphi_1+\varphi_3) + (\delta_{\varphi 2,3,4}a_2 - \delta_{y1,2})\sin\varphi_1 + (\delta_{\varphi 4}a_4 - \delta_{y3,4})\sin(\varphi_1+\varphi_3);$$

$$(\delta_{y1,2} - \delta_{\varphi 2,3,4}a_2)\cos\varphi_1 + (\delta_{y3,4} - \delta_{\varphi 4}a_4)\cos(\varphi_1+\varphi_3) + \delta_{x1,2}\sin\varphi_1$$
$$\ldots(\varphi_1+\varphi_3);\ \delta_\varphi = \delta_{\varphi 0,1,2,3,4} \tag{10.7}$$

$\ldots\delta_{x2}$; $\delta_{x3,4} = \delta_{x3} + \delta_{x4}$;...; and $\delta_{\varphi 0,1,2,3,4} = \delta_{\varphi 0} + \delta_{\varphi 1} + \delta_{\varphi 2} + \delta_{\varphi 3} + \delta_{\varphi 4}$.

$\ldots\Delta_y$ of the position error vector $\Delta\mathbf{r}_0$ can be expressed through x_0 and y_0:

$$\ldots + \delta_{x1,2}\cos\varphi_1 + \delta_{x3,4}\cos(\varphi_1+\varphi_3) + (\delta_{\varphi 2,3,4}a_2 - \delta_{y1,2})\sin\varphi_1$$
$$\ldots\sin(\varphi_1+\varphi_3) - \delta_\varphi y_0 \tag{10.8}$$

$$\ldots_{1,2} - \delta_{\varphi 2,3,4}a_2)\cos\varphi_1 + (\delta_{y3,4} - \delta_{\varphi 4}a_4)\cos(\varphi_1+\varphi_3)$$
$$\ldots\sin(\varphi_1+\varphi_3) + \delta_\varphi x_0 \tag{10.9}$$

...he 15 errors of position, orientation, and movements of ...).1. The runout and kinematic errors associated with ... represented through Fourier expansions, where n is ...test in results of measurements; φ_1 and φ_3 are the ... the harmonic number; δ_{xik}, δ_{yik}, $\delta_{\varphi ik}$ ($i = 1, 3$) are ...ase angles related to the rotation given by angle ...fined according to the axis-to-axis dimensional ...sembly errors δ_{y2}, δ_{y4} are defined according to

$$\Delta\mathbf{r}_0 \equiv \begin{bmatrix} \Delta_x \\ \Delta_y \\ 0 \end{bmatrix} = \mathbf{E}\mathbf{r}_0 = \begin{bmatrix} 0 & -\delta_\varphi & \delta_x \\ \delta_\varphi & 0 & \delta_y \\ 0 & 0 & 0 \end{bmatrix} \begin{bmatrix} x_0 \\ y_0 \\ 0 \end{bmatrix} = \begin{bmatrix} \delta_x - \delta_\varphi y_0 \\ \delta_y + \delta_\varphi x_0 \\ 0 \end{bmatrix} \tag{10.5}$$

where δ_x are δ_y are the linear deviations of the origin of the end-effector (EE) as referred to coordinate frame S_0 along the X_0 and Y_0-axis; δ_φ is the small rotation (angular deviation) of the EE around the Z_0-axis; $\mathbf{E}_0$ is the error matrix associated with the base; and $\mathbf{E}_i$ $(i = 1, 2, 3, 4)$ is the error matrix associated with the ith link.

The expanded form of matrix **E**, Eq. (10.2), is

$$\mathbf{E} \equiv \begin{bmatrix} 0 & -\delta_\varphi & \delta_x \\ \delta_\varphi & 0 & \delta_y \\ 0 & 0 & 0 \end{bmatrix} = \mathbf{E}_0 + \mathbf{A}^6(\varphi_1)\mathbf{E}_1\mathbf{A}^6(-\varphi_1) + \mathbf{A}^6(\varphi_1)\mathbf{A}^1(a_2)\mathbf{E}_2\mathbf{A}^1(-a_2)\mathbf{A}^6(-\varphi_1)$$

$$+\mathbf{A}^6(\varphi_1)\,\mathbf{A}^1(a_2)\mathbf{A}^6(\varphi_3)\,\mathbf{E}_3\mathbf{A}^6(-\varphi_3)\,\mathbf{A}^1(-a_2)\,\mathbf{A}^6(-\varphi_1)\; + {}^0\mathbf{A}_4\mathbf{E}_4({}^0\mathbf{A}_4)^{-1} \tag{10.6}$$

where ${}^0\mathbf{A}_4$ is manipulation matrix, Eq. (10.1). See also Appendix *A2.2.4*.

Some parameters entering into the error matrices have no effect on the processed object. In this case, $\delta_{\varphi 0} = \delta_{\varphi 2} = \delta_{\varphi 4} = 0$, and calculations according to Eq. (10.6) yield

$$\delta_x = \delta_{x0} + \delta_{x1,2}\cos\varphi_1 + \delta_{x3,4}\cos(\varphi_1+\varphi_3) + (\delta_{\varphi 2,3,4}a_2 - \delta_{y1,2})\sin\varphi_1 + (\delta_{\varphi 4}\,a_4 - \delta_{y3,4})\sin(\varphi_1+\varphi_3);$$

$$\delta_y = \delta_{y0} + (\delta_{y1,2} - \delta_{\varphi 2,3,4}a_2)\cos\varphi_1 + (\delta_{y3,4} - \delta_{\varphi 4}\,a_4)\cos(\varphi_1+\varphi_3) + \delta_{x1,2}\sin\varphi_1$$
$$+ \delta_{x3,4}\sin(\varphi_1+\varphi_3);\; \delta_\varphi = \delta_{\varphi 0,1,2,3,4} \tag{10.7}$$

where $\delta_{x1,2} = \delta_{x1} + \delta_{x2}$; $\delta_{x3,4} = \delta_{x3} + \delta_{x4}$;...; and $\delta_{\varphi 0,1,2,3,4} = \delta_{\varphi 0} + \delta_{\varphi 1} + \delta_{\varphi 2} + \delta_{\varphi 3} + \delta_{\varphi 4}$.

Entities Δ_x and Δ_y of the position error vector $\Delta\mathbf{r}_0$ can be expressed through x_0 and y_0:

$$\Delta_x = \delta_x - \delta_\varphi y_0 = \delta_{x0} + \delta_{x1,2}\cos\varphi_1 + \delta_{x3,4}\cos(\varphi_1+\varphi_3) + (\delta_{\varphi 2,3,4}\,a_2 - \delta_{y1,2})\sin\varphi_1$$
$$+ (\delta_{\varphi 4}\,a_4 - \delta_{y3,4})\sin(\varphi_1+\varphi_3) - \delta_\varphi y_0 \tag{10.8}$$

$$\Delta_y = \delta_y + \delta_{\varphi x0} = \delta_{y0} + (\delta_{y1,2} - \delta_{\varphi 2,3,4}\,a_2)\cos\varphi_1 + (\delta_{y3,4} - \delta_{\varphi 4}\,a_4)\cos(\varphi_1+\varphi_3)$$
$$+ \delta_{x1,2}\sin\varphi_1 + \delta_{x3,4}\sin(\varphi_1+\varphi_3) + \delta_\varphi x_0 \tag{10.9}$$

The geometrical sense of the 15 errors of position, orientation, and movements of the links are listed in Table 10.1. The runout and kinematic errors associated with the frames of links S_1 and S_3 are represented through Fourier expansions, where n is the number of harmonics of interest in results of measurements; φ_1 and φ_3 are the current angles of link rotation; k is the harmonic number; δ_{xik}, δ_{yik}, $\delta_{\varphi ik}$ $(i = 1, 3)$ are the amplitudes; and λ_{ik}, μ_{ik} are the phase angles related to the rotation given by angle φ_i. Geometric errors δ_{x2} and δ_{x4} are defined according to the axis-to-axis dimensional tolerances, locating errors δ_{x0}, δ_{y0} and assembly errors δ_{y2}, δ_{y4} are defined according to setup conditions.

10. Infinitesimal kinematics of serial-kinematics machines

10.1 Error propagation: serial-type robots

10.1.1 Planar 2-DOF (RR) manipulator

To clear up an interrelationship between the nominal displacements and their deviations in the accuracy-related formulae, the links of the 2-DOF manipulator are renumbered as is shown in Figure 10.1. In these terms, the manipulation matrix, Eqs. (4.7),becomes

$$ {}^{0}\mathbf{A}_4 = \mathbf{A}^6(\varphi_1)\mathbf{A}^1(a_2)\mathbf{A}^6(\varphi_3)\mathbf{A}^1(a_4) = \begin{bmatrix} \cos(\varphi_1+\varphi_3) & -\sin(\varphi_1+\varphi_3) & a_2\cos\varphi_1 + a_4\cos(\varphi_1+\varphi_3) \\ \sin(\varphi_1+\varphi_3) & \cos(\varphi_1+\varphi_3) & a_2\sin\varphi_1 + a_4\sin(\varphi_1+\varphi_3) \\ 0 & 0 & 1 \end{bmatrix} \tag{10.1} $$

$$ \mathbf{A}^6(\varphi_i) = \begin{bmatrix} \cos\varphi_i & -\sin\varphi_i & 0 \\ \sin\varphi_i & \cos\varphi_i & 0 \\ 0 & 0 & 1 \end{bmatrix}; i=1,3; \text{ and } \mathbf{A}^1(a_i) = \begin{bmatrix} 1 & 0 & a_i \\ 0 & 1 & 0 \\ 0 & 0 & 1 \end{bmatrix}; i = 2,4 \tag{10.2} $$

The 3×3 total manipulation error matrix $\mathbf{E}$ and error vector $\Delta\mathbf{r}_0$ are

$$ \mathbf{E} = \sum_{i=0}^{4} {}^{0}\mathbf{A}_i \mathbf{E}_i ({}^{0}\mathbf{A}_i)^{-1} \tag{10.3} $$

$$ \mathbf{E}_i = \begin{bmatrix} 0 & -\delta_{\varphi i} & \delta_{xi} \\ \delta_{\varphi i} & 0 & \delta_{yi} \\ 0 & 0 & 0 \end{bmatrix}, (i = 0,1,2,3,4) \tag{10.4} $$

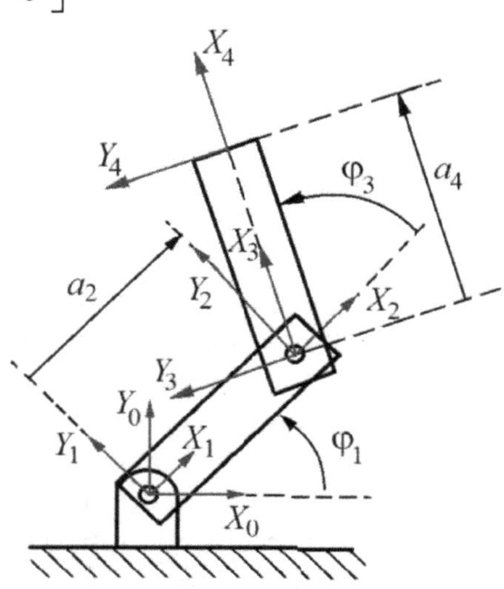

Figure 10-1 Planar 2-DOF (**RR**) manipulator.

lead or ball screws results in the lead errors of the produced screws. In turn, this affects the profile deviation (waviness) of the parts machined on the machine tool, wherein these lead-screws are used as feed screws. For diagnostic goals, it is worth representing guideways form deviation Δ by means of the Fourier series expansion

$$\Delta = \frac{1}{2}(a_0 + b_0 z) + \sum_{k=1}^{n} a_k \sin[\frac{k\pi}{L}(z - c_k)] \tag{9.26}$$

where a_0 and b_0 are the small numbers displaying the initial deviation and angular trend of the linear component of the accumulated error; n is the number of harmonics of interest in results of measurements; z is the current distance of stage motion; L is the length of the working stroke of the unit; a_k and c_k are the amplitude of the kth component and location of its maximum.

Components entering into the deviation Δ, Eq. (9.26), are illustrated in Figure 9.12, where the three first components and the total deviations of the geometric error are shown. The kinematic error Δ_K combines the cumulative positioning error $\Delta_{cum}(z)$ and the cyclic error $\Delta_t(z)$ of the leadscrew

$$\Delta_K(z) = \Delta_{cum}(z) + \Delta_t(z) \tag{9.27}$$

$$\Delta_{cum}(z) = \frac{1}{2}(u_0 + v_0 z) + \sum_{k=1}^{n} u_k \sin[\frac{k\pi}{L}(z - c_k)] \tag{9.28}$$

$$\Delta_t(z) = \sum_{j=1}^{N} u_k \sin[\frac{2k\pi}{t_s}(z - d_k)] \tag{9.29}$$

where components of the cumulative errors $\Delta_{cum}(z)$ are obtained for working length of the screw L; and cyclic error $\Delta_t(z)$ presents the Fourier expansion of the cycle error for the lead-screw with lead t_s.

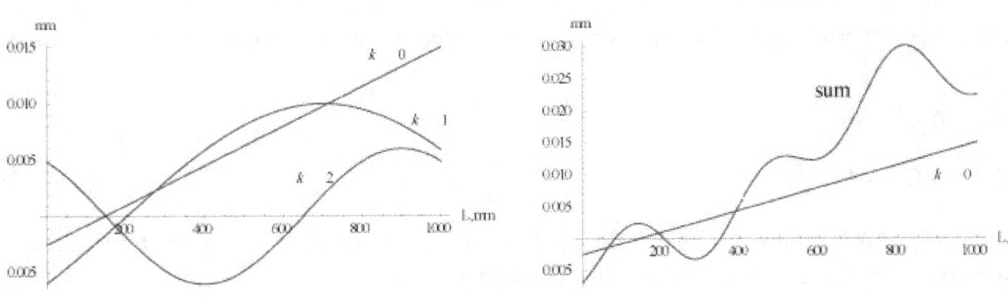

(a) (b)

Figure 9-12 Deviation components according to Eq. (9.26): (a) three first components; (b) linear component and total deviation.

Table 9.4. Position errors of the linear stage: Initial data.

Point	Cartesian coordinates				Plücker coordinates					
Symbol	Δ_j	x_j	y_j	z_j	e_{xj}	e_{yj}	e_{zj}	n_{xj}	n_{yj}	n_{zj}
A	Δ_A	x_A	y_A	z_A	−1	0	0	0	$-z_A$	y_A
B	Δ_B	x_B	y_B	z_B	−1	0	0	0	$-z_B$	y_B
C	Δ_C	x_C	y_C	z_C	0	1	0	$-z_C$	0	x_C
D	Δ_D	x_D	y_D	z_D	0	1	0	$-z_D$	0	x_D
E	Δ_E	x_E	y_E	z_E	0	1	0	$-z_E$	0	x_E
K	Δ_K	x_K	y_K	z_K	0	0	1	y_K	$-x_K$	0

$$\boldsymbol{\delta} = \mathbf{J}_s^{-1}\boldsymbol{\Delta} = \begin{bmatrix} -1 & 0 & 0 & 0 & -z_A & y_A \\ -1 & 0 & 0 & 0 & -z_B & y_B \\ 0 & 1 & 0 & -z_C & 0 & x_C \\ 0 & 1 & 0 & -z_D & 0 & x_D \\ 0 & 1 & 0 & -z_E & 0 & x_E \\ 0 & 0 & 1 & y_K & -x_K & 0 \end{bmatrix}^{-1} \begin{bmatrix} \Delta_A \\ \Delta_B \\ \Delta_C \\ \Delta_D \\ \Delta_E \\ \Delta_K \end{bmatrix} \tag{9.22}$$

The solution can be represented in the form

$$\delta_\theta = \frac{\Delta_C - \Delta_D}{z_D - z_C}; \delta_\psi = \frac{\Delta_A - \Delta_B}{z_B - z_A}; \delta_\varphi = \frac{\Delta_C - \Delta_E}{x_C - z_E};$$

$$\delta_x = -\Delta_A - z_A\delta_\psi + y_A\delta_\varphi;\ \delta_y = \Delta_C + z_C\delta_\theta - x_C\delta_\varphi;\ \text{and}\ \delta_z = \Delta_K - y_K\delta_\theta + x_K\delta_\psi \tag{9.23}$$

Physically, the displacement of the points occurs due to errors of manufacturing, deformations, and wear of the guides. The geometric and kinematic errors depend on the current position z of the linear stage: $\delta_x = \delta_x(z)$, $\delta_y = \delta_y(z)$, $\delta_z = \delta_z(z)$, $\delta_\theta = \delta_\theta(z)$, $\delta_\psi = \delta_\psi(z)$, and $\delta_\varphi = \delta_\varphi(z)$ since all measured deviations present functions of the current value of the displacement z. Correspondingly, the error of motion is defined as the difference of two position errors:

$$\boldsymbol{\delta}_{ij} = \boldsymbol{\delta}(z_j) - \boldsymbol{\delta}(z_i) \tag{9.24}$$

$$\mathbf{E}_{ij} = \mathbf{E}(z_j) - \mathbf{E}(z_i) \tag{9.25}$$

where $\boldsymbol{\delta}(z)$ and $\mathbf{E}(z)$ are the twist, Eq. (9.22), and the error matrix, Eq. (8.9), respectively, associated with the current position z of the linear stage.

9.1.2.4 Analysis of the guideways form deviations

Some sources of quasistatic errors can be diagnosed in terms of low-frequency harmonics of the guideways form deviations. For example, the lead error of the machine producing the

Table 10-1 List of deviations.

Body	Frame	Symbol	Mechanical sense	Description
Base	S_0	δ_{x0}	Locating error	
		δ_{y0}	Locating error	
		$\delta_{\varphi 0}$	Locating error	
	S_1	δ_{x1}	X_1 runout	$\frac{1}{2}\delta_{x10}\cos\lambda_{10} + \sum_{k=1}^{n}\delta_{x1k}\cos(k\varphi_1 + \lambda_{1k})$
		δ_{y1}	Y_1 runout	$\frac{1}{2}\delta_{y10}\sin\lambda_{10} + \sum_{k=1}^{n}\delta_{y1k}\sin(k\varphi_1 + \lambda_{1k})$
		$\delta_{\varphi 1}$	Kinematic error	$\frac{1}{2}\delta_{\varphi 10} + \sum_{k=1}^{n}\delta_{\varphi 1k}\cos(k\varphi_1 + \mu_{1k})$
Link 1	S_2	δ_{x2}	Dimensional error	Axis-to-axis tolerance Δ_{x2}
		δ_{y2}	Assembly error and deformations*	
		$\delta_{\varphi 2}$	Assembly error and deformations	
	S_3	δ_{x3}	X_3 runout	$\frac{1}{2}\delta_{x30}\cos\lambda_{30} + \sum_{k=1}^{n}\delta_{x3k}\cos(k\varphi_3 + \lambda_{3k})$
		δ_{y3}	Y_3 runout	$\frac{1}{2}\delta_{y30}\sin\lambda_{30} + \sum_{k=1}^{n}\delta_{y3k}\sin(k\varphi_3 + \lambda_{3k})$
		$\delta_{\varphi 3}$	Kinematic error	$\frac{1}{2}\delta_{\varphi 30} + \sum_{k=1}^{n}\delta_{\varphi 3k}\cos(k\varphi_3 + \mu_{3k})$
Link 2	S_4	δ_{x4}	Dimensional error	Axis-to-axis tolerance Δ_{x4}
		δ_{y4}	Assembly error and deformations*	
		$\delta_{\varphi 4}$	Assembly error and deformations	

* Deformation problem is considered in Part III of the book

A specified trajectory performed by the robot end-effector is

$$\mathbf{r}_w = \mathbf{r}_w(t) = \begin{bmatrix} x_w(t) \\ y_w(t) \\ 1 \end{bmatrix} \tag{10.10}$$

where t dis the trajectory parameter.

According to Eq. (3.30), the curve (10.10) has a normal at every regular point. Its unit vector has the form $\mathbf{n} = [n_x, n_y, 0]$, Eq. (3.30), and the normal error Δ_n at every regular point of the trajectory is defined through dot-product of the error vector $\Delta\mathbf{r}_0$ and the normal

$$\Delta_n = \mathbf{n}\cdot \Delta\mathbf{r}_0 \tag{10.11}$$

or

$$\Delta_n = \begin{bmatrix} n_x \\ n_y \\ 0 \end{bmatrix} \cdot \begin{bmatrix} \delta_x - \delta_\varphi y_w \\ \delta_y + \delta_\varphi x_w \\ 0 \end{bmatrix} = n_x\delta_x + n_y\delta_y + (x_w n_y - y_w n_x)\delta_\varphi \tag{10.12}$$

where x_w and y_w are the nominal coordinates of the trajectory point as referred to coordinate frame S_0; n_x and n_y are the components of the unit vector $\mathbf{n}$ with respect to the S_0 axes; and δ_x, δ_y, and δ_φ are defined through Eqs. (10.7).

10.1.2 Planar 3-DOF (**RRR**) manipulator

According to schematics in Figure 10.2, the manipulation matrix ${}^0\mathbf{A}_6$, Eq. (4.14), becomes

$${}^0\mathbf{A}_6 = \prod_{i=1}^{6} {}^{i-1}\mathbf{A}_i(q_i) = \mathbf{A}^6(\varphi_1)\,\mathbf{A}^1(a_2)\,\mathbf{A}^6(\varphi_3)\,\mathbf{A}^1(a_4)\,\mathbf{A}^6(\varphi_5)\,\mathbf{A}^1(a_6) \tag{10.13}$$

$$\mathbf{A}^6(\varphi_i) = \begin{bmatrix} \cos\varphi_i & -\sin\varphi_i & 0 \\ \sin\varphi_i & \cos\varphi_i & 0 \\ 0 & 0 & 1 \end{bmatrix} (i = 1,\ 3,\ 5) \text{ and } \mathbf{A}^1(a_i) = \begin{bmatrix} 1 & 0 & a_i \\ 0 & 1 & 0 \\ 0 & 0 & 1 \end{bmatrix} (i = 2,\ 4,\ 6) \tag{10.14}$$

Thus, the total error matrix is

$$\mathbf{E} = \sum_{i=0}^{6} {}^0\mathbf{A}_i \mathbf{E}_i ({}^0\mathbf{A}_i)^{-1} \tag{10.15}$$

$$\mathbf{E}_i = \begin{bmatrix} 0 & -\delta_{\varphi i} & \delta_{xi} \\ \delta_{\varphi i} & 0 & \delta_{yi} \\ 0 & 0 & 0 \end{bmatrix}, i = 0,1,\ldots,6 \tag{10.16}$$

$$\mathbf{E} = \begin{bmatrix} 0 & -\delta_\varphi & \delta_x \\ \delta_\varphi & 0 & \delta_y \\ 0 & 0 & 0 \end{bmatrix} \tag{10.17}$$

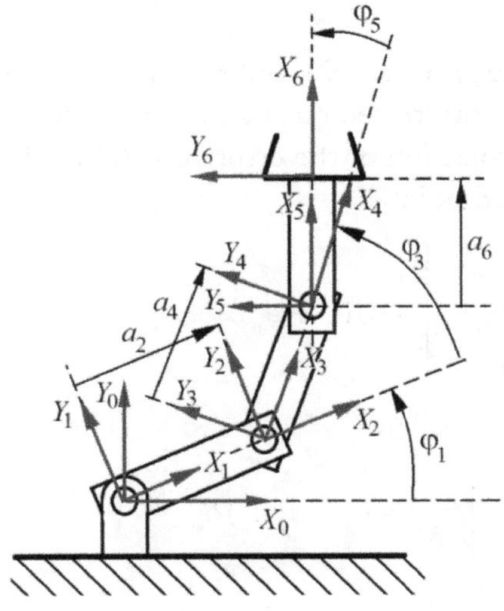

Figure 10-2 Planar 3-DOF (**RRR**) manipulator.

where $\mathbf{E}_0$ is the error matrix associated with the base coordinate frame S_0; and $\mathbf{E}_i$ $(i = 1,2,...,6)$ is the error matrix associated with the *i*th link; δ_{xi}, δ_{yi}, and $\delta_{\varphi i}(i = 0, 1,...,6)$ are the position and orientation errors associated with the *i*th link; δ_{xi} and δ_{yi} are the linear errors along the X_i- and Y_i-axis, respectively; and $\delta_{\varphi i}$ is the *i*th angular error.

Calculations according to Eqs. (10.15)–(10.17) yield

$$\delta_x = \delta_{x0} + \delta_{x1,2} \cos \varphi_1 + \delta_{x3,4} \cos (\varphi_1 + \varphi_2) + \delta_{x5,6} \cos (\varphi_1 + \varphi_2 + \varphi_3) + (\delta_{\varphi 2,3,4,5,6}\, a_2 - \delta_{y1,2}) \\ \times \sin \varphi_1 + (\delta_{\varphi 4.5.6}\, a_4 - \delta_{y3,4}) \sin (\varphi_1 + \varphi_2) - \delta_{y5,6} \sin (\varphi_1 + \varphi_2 + \varphi_3) \tag{10.18}$$

$$\delta_y = \delta_{y0} + (\delta_{y1,2} - \delta_{\varphi 2,3,4,5,6}\, a_2) \cos \varphi_1 + (\delta_{y3,4} - \delta_{\varphi 4,5,6}\, a_4) \cos (\varphi_1 + \varphi_2) + (\delta_{y5,6} - \delta_{\varphi 6}\, a_6) \\ \times \cos (\varphi_1 + \varphi_2 + \varphi_3) + \delta_{x1,2} \sin \varphi_1 + \delta_{x3,4} \sin (\varphi_1 + \varphi_2) + \delta_{x5,6} \sin (\varphi_1 + \varphi_2 + \varphi_3) \tag{10.19}$$

$$\delta_\varphi = \delta_{\varphi 0,1,2,3,4,5,6} \tag{10.20}$$

where, as before, $\delta_{x1,2} = \delta_{x1} + \delta_{x2}$; $\delta_{y1,2} = \delta_{y1} + \delta_{y2}$;...; $\delta_{\varphi 0,1,2,3,4,5,6} = \delta_{\varphi 1} + \delta_{\varphi 2} + ... + \delta_{\varphi 6}$.

Position error $\Delta\mathbf{r}_0$ of the point given by the position vector $\mathbf{r}_0$ and orientation errors $\Delta\mathbf{c}_0$ are

$$[\Delta\mathbf{r}_0\, \Delta\mathbf{c}_0] = \mathbf{E}[\mathbf{r}_0\, \mathbf{c}_0] = \begin{bmatrix} 0 & -\delta_\varphi & \delta_x \\ \delta_\varphi & 0 & \delta_y \\ 0 & 0 & 0 \end{bmatrix} \begin{bmatrix} x_0 & c_{0x} \\ y_0 & c_{0y} \\ 1 & 0 \end{bmatrix} = \begin{bmatrix} \delta_x - \delta_\varphi y_0 & -\delta_\varphi c_y \\ \delta_y + \delta_\varphi x_0 & \delta_\varphi c_x \\ 0 & 0 \end{bmatrix} \tag{10.21}$$

i.e.,

$$\Delta_x = \delta_x - \delta_\varphi\, y_0 \text{ and } \Delta_y = \delta_y + \delta_\varphi x_0 \tag{10.22}$$

$$\Delta c_x = -\,\delta_\varphi\, c_y \text{ and } \Delta c_x = \delta_\varphi c_x \tag{10.23}$$

where **E** is the total error matrix, Eq. (10.17), and $\mathbf{r}_0$ is the FSF of the manipulator.

Accuracy of the plane trajectory traced out by the end-effector (EE) and the EE orientation are obtained through projection of the error vector, Eq. (10.21), on the normal of a plane curve describing by the EE point,

$$\Delta_e = \mathbf{e}\,.\,\Delta\mathbf{r}_0 = \begin{bmatrix} e_x \\ e_y \\ 0 \end{bmatrix} . \begin{bmatrix} \delta_x - \delta_\varphi y_0 \\ \delta_x + \delta_\varphi x_0 \\ 0 \end{bmatrix} = e_x\,\delta_x + e_y\,\delta_y + \,(x_w\, e_y - y_w\, e_x)\,\delta_\varphi \tag{10.24}$$

$$\Delta_c = \mathbf{e}\,.\,\Delta\mathbf{c}_0 = \begin{bmatrix} e_x \\ e_y \\ 0 \end{bmatrix} . \begin{bmatrix} -\delta_\varphi c_y \\ \delta_\varphi c_x \\ 0 \end{bmatrix} = \,(c_x\, e_y - c_y\, e_x)\,\delta_\varphi \tag{10.25}$$

The geometrical and mechanical sense of the 21 errors of position, orientation and motions of the links are listed in Table 10.2, presenting an extension of Table 10.1:

- Runout and kinematic errors associated with the frames of links S_1, S_3, and S_5 are represented as Fourier expansions, where n is the number of harmonics of interest in results of measurements; φ_1, φ_3, and φ_5 are the current angles of link rotation; k is the harmonic number; δ_{xik}, δ_{yik}, $\delta_{\varphi ik}$ $(i = 1, 3, 5)$ are the amplitudes and λ_{ik}, μ_{ik} are the phase angles related to the rotation given by angle φ_i.
- Geometric errors δ_{x2}, δ_{x4}, and δ_{x6} are defined according to the axis-to-axis dimensional tolerances and include the errors of part machining and assembly.
- Locating errors are defined according to setup conditions.

Numerical Example 10.1 (extension of Numerical Example 7.1).

Given:

(1) *Dimensions:* $a_1 = a_2 = 0.3$, $a_3 = 0.2$ m.

(2) *The processed subject:* The working point moves along the straight line parallel to the X_0-axis. The gripper with an attached point-type tool is normally oriented relative to the given trajectory along the positive direction of the Y_0-axis:

$$y_0 = 0.6, \text{ with } -0.2 \le x_0 \le 0.4 \text{ m} \tag{10.26}$$

$$(\varphi_1 + \varphi_3 + \varphi_5) = \pi/2$$

(3) *Errors values:*

(a) The locating errors are equal to zero, $\delta_{x0} = \delta_{y0} = \delta_{\varphi 0} = \delta_{y2} = \delta_{\varphi 2} = \delta_{y4} = \delta_{\varphi 4} = \delta_{y6} = \delta_{\varphi 6} = 0$.

(b) Three dimensional errors are chosen according to tolerance ±0.10 mm as follows:

$\delta_{x2} = 0.1$ mm; $\delta_{x4} = -0.05$ mm; and $\delta_{x6} = 0.08$ mm.

Table 10-2 Error definitions.

Body	Frame	Error	Mechanical sense	Formula
Base	S_0	δ_{x0}	Locating error	
		δ_{y0}	Locating error	
		$\delta_{\varphi 0}$	Locating error	
	S_1	δ_{x1}	X_1 runout	$\frac{1}{2}\delta_{x10}\cos\lambda_{10}+\sum_{k=1}^{n}\delta_{x1k}\cos(k\varphi_1+\lambda_{1k})$
		δ_{y1}	Y_1 runout	$\frac{1}{2}\delta_{y10}\sin\lambda_{10}+\sum_{k=1}^{n}\delta_{y1k}\sin(k\varphi_1+\lambda_{1k})$
		$\delta_{\varphi 1}$	Kinematic error	$\frac{1}{2}\delta_{\varphi 10}+\sum_{k=1}^{n}\delta_{\varphi 1k}\cos(k\varphi_1+\mu_{1k})$
Link 1	S_2	δ_{x2}	Dimensional error	Axis-to-axis tolerance Δ_{x2}
		δ_{y2}	Assembly error	
		$\delta_{\varphi 2}$	Assembly error	
	S_3	δ_{x3}	X_3 runout	$\frac{1}{2}\delta_{x30}\cos\lambda_{30}+\sum_{k=1}^{n}\delta_{x3k}\cos(k\varphi_3+\lambda_{3k})$
		δ_{y3}	Y_3 runout	$\frac{1}{2}\delta_{y30}\sin\lambda_{30}+\sum_{k=1}^{n}\delta_{y3k}\sin(k\varphi_3+\lambda_{3k})$
		$\delta_{\varphi 3}$	Kinematic error	$\frac{1}{2}\delta_{\varphi 30}+\sum_{k=1}^{n}\delta_{\varphi 3k}\cos(k\varphi_3+\mu_{3k})$
Link 2	S_4	δ_{x4}	Dimensional error	Axis-to-axis tolerance Δ_{x4}
		δ_{y4}	Assembly error	
		$\delta_{\varphi 4}$	Assembly error	
	S_5	δ_{x5}	X_5 runout	$\frac{1}{2}\delta_{x50}\cos\lambda_{50}+\sum_{k=1}^{n}\delta_{x5k}\cos(k\varphi_5+\lambda_{5k})$

(Continued)

Table 10-2 Error definitions (*Continued*)

		δ_{y5}	Y_5 runout	$\frac{1}{2}\delta_{y50}\sin\lambda_{50}+\sum_{k=1}^{n}\delta_{y5k}\sin(k\varphi_5+\lambda_{5k})$
		$\delta_{\varphi5}$	Kinematic error	$\frac{1}{2}\delta_{\varphi50}+\sum_{k=1}^{n}\delta_{\varphi5k}\cos(k\varphi_5+\mu_{5k})$
Link 3				
	S_6	δ_{x6}	Dimensional error	Axis-to-axis tolerance Δ_{x6}
		δ_{y6}	Assembly error	
		$\delta_{\varphi6}$	Assembly error	

(c) The runout within three pairs (*X*- and *Y*-components) of the revolute joints are as follows:

$\delta_{x1} = 0.1\cos(\varphi_1 + \pi/3)$; $\delta_{y1} = 0.1\sin(\varphi_1 + \pi/3)$;

$\delta_{x3} = 0.1\cos(\varphi_3 + \pi/6)$; $\delta_{y3} = 0.1\sin(\varphi_3 + \pi/6)$;

$\delta_{x5} = 0.1\cos(\varphi_5 + \pi/4)$; $\delta_{y5} = 0.1\sin(\varphi_5 + \pi/4)$;

(d) Three kinematic errors are as follows:

$\delta_{\varphi1} = 0.03\cos(\varphi_1 + \pi/3) + 0.01\cos(2\varphi_1 + \pi/12)$;

$\delta_{\varphi3} = 0.03\cos\varphi_3 + 0.01\cos(2\varphi_3 + \pi/4)$; and,

$\delta_{\varphi5} = 0.03\cos(\varphi_5 + \pi/4) + 0.01\cos 2\varphi_5$.

Required: Accuracy of the drawn line, Eq. (10.26).
Solution:

(1) According to results obtained in Eqs. (7.27–29), nominal values of the angles of the links rotations for configurations in Figure 10.2 are as follows:

$$\varphi_2 = \arccos(5.5556\,x_0^2 - 0.1111) \tag{10.27}$$

$$\varphi_1 = \arcsin\left[\frac{2}{3} - \frac{5x_0\sin\varphi_2}{3(1+\cos\varphi_2)}\right] = \arcsin\left[\frac{2}{3} - \frac{5x_0\sin[\arccos(5,5556x_0^2 - 0.1111)]}{2.6667 + 16,6667x_0^2}\right] \tag{10.28}$$

$$\varphi_3 = \pi/2 - \varphi_1 - \varphi_2 \tag{10.29}$$

Remember that functions $\varphi_1(x_0)$, $\varphi_2(x_0)$, and $\varphi_3(x_0)$ for initial data according to the *Numerical Example* are shown in Figure 7.2.

(2) Deviations Δ_0 from the given end-effector trajectory, $y_0 = 0.6$, is obtained by projection of the position error $\mathbf{\Delta r}_0$, Eq. (10.21), on the unit vector $\mathbf{n}$ of the normal at the end-effector trajectory

$$\Delta_0 = \mathbf{\Delta r}_0 \cdot \mathbf{n} = \begin{bmatrix} \delta_x - \delta_\varphi y_0 \\ \delta_y + \delta_\varphi x_0 \\ 0 \end{bmatrix} \cdot \begin{bmatrix} 0 \\ 1 \\ 0 \end{bmatrix} = \delta_y + \delta_\varphi x_0 =$$

$$\delta_{y0} + \delta_{x1,2} \sin\varphi_1 + \delta_{y1,2} \cos\varphi_1 + \delta_{x3,4} \sin\varphi_{12} + \delta_{y3,4} \cos\varphi_{12} + \delta_{x5,6} \sin\varphi_{123} + \delta_{y5,6} \cos\varphi_{123}$$
$$- \delta_{\varphi 2,3,4,5,6}\, a_1 \cos\varphi_1 - \delta_{\varphi 4,5,6}\, a_2 \cos\varphi_{12} - \delta_{\varphi 6}\, a_3 \cos\varphi_{123} + \delta_{\varphi 0,1,2,3,4,5,6}\, x_0 \qquad (10.30)$$

where δ_y and δ_φ are defined through Eqs. (10.19) and (10.20), respectively.

(3) Substitution of the errors, Eqs. (10.18–20) and angles φ_1, φ_2, and φ_3, Eqs. (10.27–29), in expression Δ_0, Eq. (10.30), yields the results shown in Figure 10.2. As is seen in Figure 10.3, the mean deviation (curve *1*) falls within the broad range between +7.802 mm (for $x_0 = -0.2$ m) and +1.8357 mm (for $x_0 = 0.07067$ m). Curves 2 and 3 illustrate the maximal and minimal deviations ($\pm 3\sigma_\Delta$ interval), where σ_Δ is the standard deviation of Δ_0 value.

10.1.3 SCARA arm robot

The 4-DOF SCARA arm (Fig. 4.3) has three articulated links as in the **RRR** planar manipulator and an actuated translational link (guide rod) sliding along the axis of rotation of the last link. Since the arm has only four DOF, direction of the end-effector translation is always oriented at the right angle to the planes of the rotating links. Nevertheless, presence of the translational link allows useful applications of the robot, such as assembling plain PC boards.

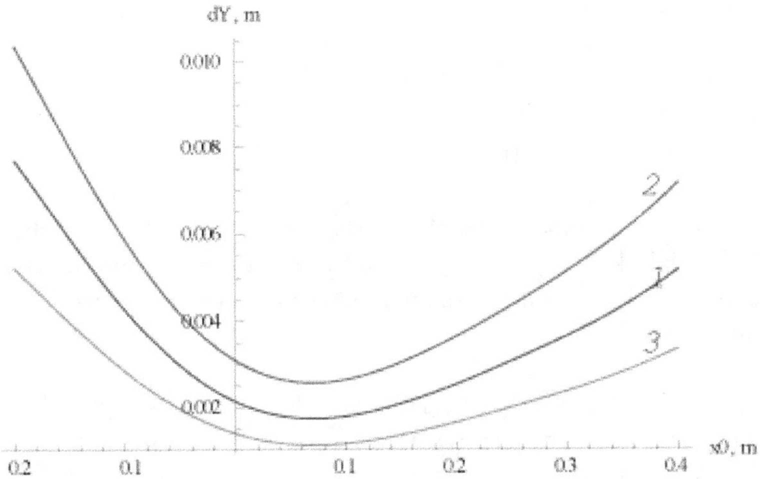

Figure 10-3 *Numerical Example 10.1*: Linear deviations of the drawn straight line $y_0 = 0.6$ m. *1* – mean value, 2 – maximal deviation, 3 – minimal deviation.

The manipulation matrix ${}^0\mathbf{A}_8$, Eq. (4.22), of the robot is

$${}^0\mathbf{A}_7 = \mathbf{A}^6(\varphi_1)\,\mathbf{A}^2(a_1)\,\mathbf{A}^6(\varphi_2)\,\mathbf{A}^2(a_2)\,\mathbf{A}^3(z)\,\mathbf{A}^6(\varphi_3)\,\mathbf{A}^2(a_3)$$

$$= \begin{bmatrix} \cos(\varphi_1+\varphi_2+\varphi_3) & -\sin(\varphi_1+\varphi_2+\varphi_3) & 0 & a_{14} \\ \sin(\varphi_1+\varphi_2+\varphi_3) & \cos(\varphi_1+\varphi_2+\varphi_3) & 0 & a_{24} \\ 0 & 0 & 1 & z \\ 0 & 0 & 0 & 1 \end{bmatrix}, \qquad (10.31)$$

with

$$a_{14} = a_1 \cos\varphi_1 + a_2 \cos(\varphi_1+\varphi_2) + a_3 \cos(\varphi_1+\varphi_2+\varphi_3)$$

$$a_{24} = a_1 \sin\varphi_2 + a_2 \sin(\varphi_1+\varphi_2) + a_3 \sin(\varphi_1+\varphi_2+\varphi_3) \qquad (10.32)$$

The total error matrix is

$$\mathbf{E} = \sum_{i=0}^{7} {}^0\mathbf{A}_i \mathbf{E}_i ({}^0\mathbf{A}_i)^{-1} \qquad (10.33)$$

$$\mathbf{E} = \begin{bmatrix} 0 & -\delta_\varphi & \delta_\psi & \delta_x \\ \delta_\varphi & 0 & -\delta_\theta & \delta_y \\ -\delta_\psi & \delta_\theta & 0 & \delta_z \\ 0 & 0 & 0 & 0 \end{bmatrix} \qquad (10.35)$$

$$\mathbf{E}_i = \begin{bmatrix} 0 & -\delta_{\varphi i} & \delta_{\psi i} & \delta_{xi} \\ \delta_{\varphi i} & 0 & -\delta_{\theta i} & \delta_{yi} \\ -\delta_{\psi i} & \delta_{\theta i} & 0 & \delta_{zi} \\ 0 & 0 & 0 & 0 \end{bmatrix}, \; i = 0, 1, \ldots, 7 \qquad (10.34)$$

where $\mathbf{E}_0$ is the error matrix associated with the base; and $\mathbf{E}_i$ ($i = 1,2,\ldots,7$) is the error matrix associated with the ith link; δ_{xi}, δ_{yi}, and δ_{zi} ($i = 0, 1,\ldots,7$) are the position (linear) errors associated with the ith link; $\delta_{\theta i}$, $\delta_{\psi i}$ and $\delta_{\varphi i}$ are the orientation (angular) errors.

After calculation, we obtain

$$\begin{aligned}\delta_x = {} & \delta_{x0} + \delta_{x1,2}\cos\varphi_1 - \delta_{y1,2}\sin\varphi_1 + \delta_{x3,4}\cos(\varphi_1+\varphi_2) - \delta_{y3,4}\sin(\varphi_1+\varphi_2) + \\ & \delta_{x5,6,7}\cos(\varphi_1+\varphi_2+\varphi_3) - \delta_{y5,6,7}\sin(\varphi_1+\varphi_2+\varphi_3) + \delta_{\varphi 2,3,4,5,6,7}\, a_1 \sin\varphi_1 + \\ & \delta_{\varphi 4,5,6,7}\, a_2 \sin(\varphi_1+\varphi_2) + (\delta_{\varphi 6,7}\, a_3 - \delta_{\theta 7}\, z)\sin(\varphi_1+\varphi_2+\varphi_3) - \\ & \delta_{\psi 7}\, z\cos(\varphi_1+\varphi_2+\varphi_3);\end{aligned} \qquad (10.36)$$

$$\delta_y = \delta_{y0} + \delta_{x1,2} \sin \varphi_1 + \delta_{y1,2} \cos \varphi_1 + \delta_{x3,4} \sin (\varphi_1 + \varphi_2) + \delta_{y3,4} \cos (\varphi_1 + \varphi_2) + \delta_{x5,6,7} \sin (\varphi_1 + \varphi_2 + \varphi_3) + \delta_{y5,6,7} \cos (\varphi_1 + \varphi_2 + \varphi_3) - \delta_{\varphi 2,3,4,5,6,7}\, a_1 \cos \varphi_1 - \delta_{\varphi 4,5,6,7}\, a_2 \cos (\varphi_1 + \varphi_2) - (\delta_{\varphi 6,7}\, a_3 - \delta_{\theta 7}\, z) \cos (\varphi_1 + \varphi_2 + \varphi_3) - \delta_{\psi 7}\, z \sin (\varphi_1 + \varphi_2 + \varphi_3); \tag{10.37}$$

$$\delta_z = \delta_{z0,1,2,3,4,5,6,7} + \delta_{\psi 2}\, a_1 + \delta_{\psi 4}\, a_2 + \delta_{\psi 6,7}\, a_3 + \delta_{\psi 3,4}\, a_1 \cos \varphi_2 + \delta_{\psi 5,6,7}\, a_2 \cos \varphi_3 + \delta_{\psi 5,6,7}\, a_1 \cos (\varphi_2 + \varphi_3) + \delta_{\theta 3,4}\, a_1 \sin \varphi_2 + \delta_{\theta 5,6,7} [a_2 \sin \varphi_3 + a_1 \sin (\varphi_2 + \varphi_3)]; \tag{10.38}$$

$$\delta_\theta = \delta_{\theta 0} + \delta_{\theta 1,2} \cos \varphi_1 + \delta_{\theta 3,4} \cos (\varphi_1 + \varphi_2) + \delta_{\theta 5,6,7} \cos (\varphi_1 + \varphi_2 + \varphi_3) - \delta_{\psi 1,2} \sin \varphi_1 - \delta_{\psi 3,4} \sin (\varphi_1 + \varphi_2) - \delta_{\psi 5,6,7} \sin (\varphi_1 + \varphi_2 + \varphi_3); \tag{10.39}$$

$$\delta_\psi = \delta_{\psi 0} + \delta_{\psi 1,2} \cos \varphi_1 + \delta_{\psi 3,4} \cos (\varphi_1 + \varphi_2) + \delta_{\psi 5,6,7} \cos (\varphi_1 + \varphi_2 + \varphi_3) + \delta_{\theta 1,2} \sin \varphi_1 + \delta_{\theta 3,4} \sin (\varphi_1 + \varphi_2) + \delta_{\theta 5,6,7} \sin (\varphi_1 + \varphi_2 + \varphi_3); \tag{10.40}$$

$$\delta_\varphi = \delta_{\varphi 0,1,2,3,4,5,6,7} \tag{10.41}$$

10.1.4 Spherical robot

The manipulation matrix of the spherical robot, Eq. (4.35), is

$$^{0}\mathbf{A}_4 = \prod_{j=1}^{4} {}^{j-1}\mathbf{A}_j(q_j) = \mathbf{A}^3(h)\mathbf{A}^6(\varphi)\mathbf{A}^4(\theta)\mathbf{A}^2(\rho) \tag{10.42}$$

where h is the distance between the column pedestal and base for rotation of the elevated pivot; φ is the angle of rotation of the elevated pivot around the Z_1-axis; θ is the angle of rotation of the telescoping arm around the X_2-axis; and ρ is the current extension of the sliding sleeve.

The total error matrix **E**, Eq. (8.9), is

$$\mathbf{E} = \sum_{i=0}^{4} {}^{0}\mathbf{A}_i \mathbf{E}_i ({}^{0}\mathbf{A}_i)^{-1} = \begin{bmatrix} 0 & -\delta_\varphi & \delta_\psi & \delta_x \\ \delta_\varphi & 0 & -\delta_\theta & \delta_y \\ -\delta_\psi & \delta_\theta & 0 & \delta_z \\ 0 & 0 & 0 & 0 \end{bmatrix} \tag{10.43}$$

with

$$\delta_x = \delta_{x0,1} - h\,\delta_{\psi 1} + (\delta_{x2,3,4} - h\,\delta_{\psi 2} - h \cos \theta\, \delta_{\psi 3,4} + h \sin \theta\, \delta_{\varphi 3,4} + \rho\, \delta_{\varphi 4}) \cos \varphi - (\delta_{y2} + \cos \theta\, \delta_{y3,4} - \sin \theta\, \delta_{z3,4} + h\, \delta_{\theta 2,3,4} + \rho \sin \theta\, \delta_{\theta 4}) \sin \varphi \tag{10.44}$$

$$\delta_y = \delta_{y0,1} + h\,\delta_{\theta 1} + (\delta_{y2,3,4} - h\,\delta_{\psi 2} - h\cos\theta\,\delta_{\psi 3,4} + h\sin\theta\,\delta_{\varphi 3,4} + \rho\,\delta_{\varphi 4})\sin\varphi + (\delta_{y2} + \cos\theta\,\delta_{y3,4} - \sin\theta\,\delta_{z3,4} + h\,\delta_{\theta 2,3,4} + \rho\sin\theta\,\delta_{\theta 4})\cos\varphi \tag{10.45}$$

$$\delta_z = \delta_{z0,1,2} + \cos\theta\,\delta_{z3,4} + \sin\theta\,\delta_{y3,4} - \rho\cos\theta\,\delta_{\theta 4} \tag{10.46}$$

$$\delta_\theta = \delta_{\theta 0,1} + \delta_{\theta 2,3,4}\cos\varphi - (\delta_{\psi 2} + \delta_{\psi 3,4}\cos\theta - \delta_{\varphi 3,4}\sin\theta)\sin\varphi; \tag{10.47}$$

$$\delta_\psi = \delta_{\psi 0,1} + \delta_{\theta 2,3,4}\sin\varphi + (\delta_{\psi 2} + \delta_{\psi 3,4}\cos\theta - \delta\varphi_{3,4}\sin\theta)\cos\varphi \tag{10.48}$$

$$\delta_\varphi = \delta_{\varphi 0,1,2} + \cos\theta\,\delta_{\varphi 3,4} + \sin\theta\,\delta_{\psi 3,4} \tag{10.49}$$

10.1.5 Scorbot robot

The FSS links and associated coordinate frames are shown in Figure 4.4. Their motions are listed in Table 4.4 according to the structural code k_{st} = **3615151541**, Eq. (4.31). The set of kinematic parameters and 10 coordinate frames are shown in Figure 7.7. The 4 × 4 manipulation matrix $^0\mathbf{A}_{10}$, Eqs. (4.32-33), of the robot is

$$^0\mathbf{A}_{10} = \mathbf{A}^3(H)\,\mathbf{A}^6(\varphi)\,\mathbf{A}^1(d)\,\mathbf{A}^5(\psi_1)\,\mathbf{A}^1(a_1)\,\mathbf{A}^5(\psi_2)\mathbf{A}^1(a_2)\,\mathbf{A}^5(\psi_3)\mathbf{A}^4(\theta)\,\mathbf{A}^1(a_3) \tag{10.50}$$

The corresponding total error matrix is as follows:

$$\mathbf{E} = \sum_{i=0}^{4} {}^0\mathbf{A}_i\mathbf{E}_i({}^0\mathbf{A}_i)^{-1} = \begin{bmatrix} 0 & -\delta_\varphi & \delta_\psi & \delta_x \\ \delta_\varphi & 0 & -\delta_\theta & \delta_y \\ -\delta_\psi & \delta_\theta & 0 & \delta_z \\ 0 & 0 & 0 & 0 \end{bmatrix} \tag{10.51}$$

with

$$\delta_x = \delta_{x0.1} + [\delta_{x2.3} + \delta_{x4,5}\cos\psi_1 + \delta_{x6,7}\cos\psi_{1,2} + \delta_{x8,9,10}\cos\psi_{1,2,3}]\cos\varphi - [\delta_{y2.3,4,5,6,7,8} + \delta_{y9,10}\cos\theta]\sin\varphi + \delta_{y9,10}\sin\theta\sin\psi_{1,2,3}\cos\varphi + [\delta_{z4,5}\sin\psi_1 + \delta_{z6,7}\sin\psi_{1,2} + \delta_{x8}\sin\psi_{1,2,3} + \delta_{z9,10}\cos\theta\sin\psi_{1,2,3}]\cos\varphi + \delta_{z9,10}\sin\theta\sin\varphi -$$

$$[\delta_{\theta 2.3}H + \delta_{\theta 4,5}(d\sin\psi_1 + H\cos\psi_1) + \delta_{\theta 6,7}(d\sin\psi_{1,2} + H\cos\psi_{1,2} + a_1\sin\psi_2) + \delta_{\theta 8,9,10} \times (d\sin\psi_{1,2,3} + H\cos\psi_{1,2,3} + a_1\sin\psi_2\cos\psi_3 + a_1\cos\psi_2\sin\psi_3 + a_2\sin\psi_3)]\sin\varphi - \delta_{\psi 1}H + [-\delta_{\psi 2,3,4,5,6,7,8}H + \delta_{\psi 5,6,7,8}a_1\sin\psi_1 + \delta_{\psi 7,8}a_2 s\text{ in }\psi_{1,2} + \delta_{\psi 9,10}(H + a_1\sin\psi)\cos\theta + \delta_{\psi 10}a_3\cos\theta\sin\psi_{1,2,3}]\cos\varphi + \delta_{\psi 9,10}[(-H\sin\psi_{1,2,3} + d\cos\psi_{1,2,3} + a_1\cos\psi_{2,3} + a_2\cos\psi_3) + \delta_{\psi 10}a_3]\sin\theta\sin\varphi + \delta_{\psi 10}a_3 +$$

$$[\delta_{\varphi 3}\, d + \delta_{\varphi 4,5}\,(d \cos \psi_1 - H \sin \psi_1) + \delta_{\varphi 5}\, a_1 + \delta_{\varphi 6,7}\,(d \cos \psi_{1,2} - H \sin \psi_{1,2} + a_1 \cos \psi_2) +$$
$$\delta_{\varphi 7}\, a_2 + \delta_{\varphi 8}\,(d \cos \psi_{1,2,3} - H \sin \psi_{1,2,3} + a_1 \cos \psi_{2,3} + a_2 \cos \psi_3) +$$
$$\delta_{\varphi 9}\,(d \cos \psi_{1,2,3} - H \sin \psi_{1,2,3} + a_1 \cos \psi_{2,3} + a_2 \cos \psi_3) \cos \theta] \sin \varphi +$$
$$\delta_{\varphi 9}\,(H - a_1 \sin \psi_1 - a_2 \sin \psi_{1,2}) \sin \theta \cos \varphi +$$
$$\delta_{\varphi 10}\,[(d \cos \psi_{1,2,3} - H \sin \psi_{1,2,3} + a_1 \cos \psi_{2,3} + a_2 \cos \psi_3 + a_3) \cos \theta] \sin \varphi$$
$$+ (H - a_1 \sin \psi_1 - a_2 \sin \psi_{1,2} - a_3 \sin \psi_{1,2,3}) \sin \theta \cos \varphi; \qquad (10.52)$$

....

$$\delta_\theta = \delta_{\theta 0,1} + [\delta_{\theta 2,3} + \delta_{\theta 4,5} \cos \psi_1 + \delta_{\theta 6,7} \cos (\psi_1+\psi_2) + \delta_{\theta 8,9,10} \cos (\psi_1+\psi_2+\psi_3)] \cos \varphi$$
$$- [\delta_{\psi 2,3,4,5,6,7,8} + \delta_{\psi 9,10} \cos \theta] \sin \varphi + \delta_{\psi 9,10} \sin \theta \sin (\psi_1+\psi_2+\psi_3) \cos \varphi + [\delta_{\varphi 4,5} \sin \psi_1$$
$$+ \delta_{\varphi 6,7} \sin (\psi_1+\psi_2) + \delta_{\varphi 8} \sin (\psi_1+\psi_2+\psi_3) + \delta_{\varphi 9,10} \cos \theta \sin (\psi_1+\psi_2+\psi_3)] \cos \varphi +$$
$$\delta_{\varphi 9,10} \sin \theta \sin \varphi; \qquad (10.53)$$

$$\delta_\psi = [\delta_{\theta 2,3} + \delta_{\theta 4,5} \cos \psi_1 + \delta_{\theta 6,7} \cos (\psi_1+\psi_2) + \delta_{\theta 8,9,10} \cos (\psi_1+\psi_2+\psi_3)] \sin \varphi + \delta_{\psi 0,1}$$
$$+ [\delta_{\psi 2,3,4,5,6,7,8} + \delta_{\psi 9,10} \cos \theta + \delta_{\psi 9,10} \cos \theta] \cos \varphi + \delta_{\psi 9,10} \sin \theta \sin (\psi_1+\psi_2+\psi_3)$$
$$\times \sin \varphi + [\delta_{\varphi 4,5} \sin \psi_1 + \delta\varphi_{6,7} \sin (\psi_1+\psi_2) + \delta_{\varphi 8} \sin (\psi_1+\psi_2+\psi_3) +$$
$$\delta_{\varphi 9,10} \cos \theta \sin (\psi_1+\psi_2+\psi_3)] \sin \varphi - \delta_{\varphi 9,10} \sin \theta \cos \varphi; \qquad (10.54)$$

$$\delta_\varphi = \delta_{\theta 4,5} \sin \psi_1 - \delta_{\theta 6,7} \sin \psi_{1,2} - \delta_{\theta 8,9,10} \sin_{\psi 1,2,3} + \delta_{\psi 9,10} \sin \theta \cos (\psi_1+\psi_2+\psi_3)$$
$$+ \delta_{\varphi 0,1,2,3} + \delta_{\varphi 4,5} \cos \psi_1 + \delta_{\varphi 6,7} \cos \psi_{1,2} + (\delta_{\varphi 8} + \delta_{\varphi 9,10} \cos \theta) \cos \psi_{1,2,3} \qquad (10.55)$$

Consider as an example, the setup accuracy for spot welding (Fig. 7.8) described in Section 7.1.3. In this case, the position error Δ_0 of the weld point is

$$\Delta_0 = e_{x0}\, \delta_x + e_{y0}\, \delta_y + e_{z0}\, \delta_z + n_{x0}\, \delta_\theta + n_{y0}\, \delta_\psi + n_{z0}\, \delta_\varphi \qquad (10.56)$$

with

$$n_{x0} = y_0\, e_{z0} - z_0\, e_{y0};\; n_{y0} = z_0\, e_{x0} - x_0\, e_{z0};\; n_{z0} = x_0\, e_{y0} - y_0\, e_{x0} \qquad (10.57)$$

where x_0, y_0, and z_0 are the nominal coordinates of the body point as referred to coordinate frame S_0; n_{xt}, n_{yt}, and n_{zt} are the moments of the unit vector $\mathbf{e}_0$ with respect to the S_0 axes; and δ_{xt}, δ_{yt}, δ_{zt}, $\delta_{\theta t}$, $\delta_{\psi t}$, and $\delta_{\varphi t}$ are defined through Eq. (10.52-55).

Using initial data from *Numerical Example 7.3,* the position vector of the welding points is

$$\mathbf{r}_0 = \mathbf{r}_w = [460, y_w, 420, 1]^T \tag{10.58}$$

where y_w is the point coordinate along the weld.

Displacement of the points along the perpendicular to the straight line, Eq. (10.58), in the horizontal plane is of interest: $\mathbf{e}_0=\mathbf{e}_w= [1, 0, 0, 0]^T$. The moments, Eq. (10.57), become:

$$n_{x0} = y_0 e_{z0} - z_0 e_{y0} = 0; n_{y0} = z_0 e_{x0} - x_0 e_{z0} = z_0 = 420; \text{and } n_{z0} = -y_w$$

Hence,

$$\Delta_0 = \delta_x + 420\delta_\psi - y_w\delta_\varphi \tag{10.59}$$

where δ_x, δ_ψ and δ_φ are defined through Eqs. (10.52), (10.53), and (10.55), respectively.

10.2 Error Propagation: Lathe

10.2.1 Error budget

The FSF of the lathe is obtained above in two versions: the kinematic chain, code k = **631**, Eq. (3.91), and the structure-kinematic chain, code K_{st} = **62321**, Eq. (3.93). For methodological goals, the error budget will be considered here for both versions.

10.2.1.1 Kinematic chain version

The manipulation matrix and error matrix for the FSS shown in Figure 10.4 are

$$^0\mathbf{A}_3 = \mathbf{A}^6(\varphi)\mathbf{A}^3(z)\,\mathbf{A}^1(x) \tag{10.60}$$

$$\mathbf{E} = \sum_{i=0}^{3} {}^0\mathbf{A}_i\mathbf{E}_i\,{}^i\mathbf{A}_3 = \mathbf{E}_0 + \mathbf{A}^6(\varphi)\,\mathbf{E}_1\mathbf{A}^6(-\varphi) + \mathbf{A}^6(\varphi)\,\mathbf{A}^3(z)\mathbf{E}_2\,\mathbf{A}^3(-z)\mathbf{A}^6(-\varphi)$$
$$+\mathbf{A}^6(\varphi)\,\mathbf{A}^3(z)\,\mathbf{A}^1(x)\mathbf{E}_3\,\mathbf{A}^1(-x)\,\mathbf{A}^3(-z)\,\mathbf{A}^6(-\varphi) \tag{10.61}$$

where $\mathbf{E}_0$ is the error matrix associated with the base; and $\mathbf{E}_i$ (i = 1,...,5) is the error matrix associated with the ith link,

$$\mathbf{E}_i = \begin{bmatrix} 0 & -\delta_{\varphi i} & \delta_{\psi i} & \delta_{xi} \\ \delta_{\varphi i} & 0 & -\delta_{\theta i} & \delta_{yi} \\ -\delta_{\psi i} & \delta_{\theta i} & 0 & \delta_{zi} \\ 0 & 0 & 0 & 0 \end{bmatrix}, \; i = 0,\ldots,3 \tag{10.62}$$

$$\mathbf{E} = \begin{bmatrix} 0 & -\delta_\varphi & \delta_\psi & \delta_x \\ \delta_\varphi & 0 & -\delta_\theta & \delta_y \\ -\delta_\psi & \delta_\theta & 0 & \delta_z \\ 0 & 0 & 0 & 0 \end{bmatrix} \tag{10.63}$$

with

$$\delta_x = \delta_{x0} + (\delta_{x1,2,3} - \delta\psi_{2,3}\, z)\cos\varphi - (\delta_{y1,2,3} + \delta_{\theta 2,3}\, z - \delta_{\varphi 3}\, x)\sin\varphi \tag{10.64}$$

$$\delta_y = \delta_{y0} + (\delta_{x1,2,3} - \delta_{\psi 2,3}\, z)\sin\varphi + (\delta_{y1,2,3} + \delta_{\theta 2,3}\, z - \delta_{\varphi 3}\, x)\cos\varphi \tag{10.65}$$

$$\delta_z = \delta_{z0,1,2,3} + \delta_{\psi 3}\, x \tag{10.66}$$

$$\delta_\theta = \delta_{\theta 0} + \delta_{\theta 1,2,3}\cos\varphi - \delta_{\psi 1,2,3}\sin\varphi \tag{10.67}$$

$$\delta_\psi = \delta_{\psi 0} + \delta_{\psi 1,2,3}\cos\varphi + \delta_{\theta 1,2,3}\sin\varphi \tag{10.68}$$

$$\delta_\varphi = \delta_{\varphi 0,1,2,3} \tag{10.69}$$

10.2.1.2 Structural chain version

To assure automatic inclusion of the Abbe-offset-caused errors in the error budget, the structural version of the FSF is used. For this purpose, two additional geometric links are incorporated in the FSF (Fig. 4.7): (a) distance y_2 represented through transfer from the spindle axis to the longitudinal-guideways-associated plane, and (b) distance y_4 modelling transfer from the longitudinal to transverse guideways-associated plane. The manipulation matrix ${}^0\mathbf{A}_5$ and position FSF of the quasi-kinematics chain are given by Eqs. (4.44) and (4.45), respectively. The corresponding total error matrix **E** is

$$\mathbf{E} \equiv \begin{bmatrix} 0 & -\delta_\varphi & \delta_\psi & \delta_x \\ \delta_\varphi & 0 & -\delta_\theta & \delta_y \\ -\delta_\psi & \delta_\theta & 0 & \delta_z \\ 0 & 0 & 0 & 0 \end{bmatrix} = \sum_{i=0}^{5} {}^0\mathbf{A}_i \mathbf{E}_i ({}^0\mathbf{A}_i)^{-1} =$$

$$\mathbf{E}_0 + \mathbf{A}^6(\varphi)\,\mathbf{E}_1\mathbf{A}^6(-\varphi) + \mathbf{A}^6(\varphi)\mathbf{A}^2(y_2)\mathbf{E}_2\mathbf{A}^2(-y_2)\mathbf{A}^6(-\varphi) +$$

$$\mathbf{A}^6(\varphi)\mathbf{A}^2(y_2)\mathbf{A}^3(z)\,\mathbf{E}_3\mathbf{A}^3(-z)\,\mathbf{A}^2(-y_2)\mathbf{A}^6(-\varphi) +$$

$$\mathbf{A}^6(\varphi)\mathbf{A}^2(y_2)\mathbf{A}^3(z)\mathbf{A}^2(-y_4)\,\mathbf{E}_4\,\mathbf{A}^2(y_4)\mathbf{A}^3(-z)\,\mathbf{A}^2(-y_2)\mathbf{A}^6(-\varphi)$$

$$+\,\mathbf{A}^6(\varphi)\mathbf{A}^2(y_2)\mathbf{A}^3(z)\mathbf{A}^2(-y_4)\mathbf{A}^1(x)\,\mathbf{E}_5\,\mathbf{A}^1(-x)\mathbf{A}^2(y_4)\mathbf{A}^3(-z)\,\mathbf{A}^2(-y_2)\mathbf{A}^6(-\varphi) \tag{10.70}$$

where φ is the angle of the spindle rotation; x and z are the displacements of the tip of the cutter across and along the axis of spindle rotation, respectively;y_2 is the distance between the spindle axis and longitudinal guideways-associated horizontal plane; and y_4 is the distance between the transverse and longitudinal guideways planes; $\mathbf{E}_0$ is the error matrix associated with the base; and $\mathbf{E}_i$ ($i = 1,\ldots,5$) is the error matrix associated with the ith link,

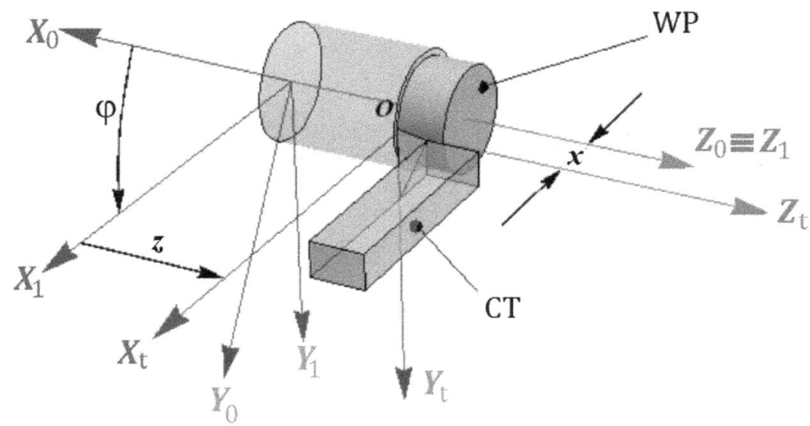

Figure 10-4 The set of coordinate frames associated with lathe form-shaping system.

$$\mathbf{E}_i = \begin{bmatrix} 0 & -\delta_{\varphi i} & \delta_{\psi i} & \delta_{xi} \\ \delta_{\varphi i} & 0 & -\delta_{\theta i} & \delta_{yi} \\ -\delta_{\psi i} & \delta_{\theta i} & 0 & \delta_{zi} \\ 0 & 0 & 0 & 0 \end{bmatrix}, i = 0,\ldots,5 \tag{10.71}$$

$$\delta_x = \delta_{x0} + (\delta_{x1,2,3,4,5} - \delta_{\psi 3,4,5}\, z + \delta_{\varphi 2,3,4,5}\, y_2 + \delta_{\varphi 4,5}\, y_4)\cos\varphi - (\delta_{y1,2,3,4,5} + \delta_{\theta 3,4,5}\, z - \delta_{\varphi 5}\, x)\sin\varphi \tag{10.72}$$

$$\delta_y = \delta_{y0} + (\delta_{x1,2,3,4,5} - \delta_{\psi 3,4,5}\, z + \delta_{\varphi 2,3,4,5}\, y_2 - \delta_{\varphi 4,5}\, y_4)\sin\varphi - (\delta_{y1,2,3,4,5} + \delta_{\theta 3,4,5}\, z - \delta_{\varphi 5}\, x)\cos\varphi \tag{10.73}$$

$$\delta_z = \delta_{z0,1,2,3,4,5} - \delta_{\theta 2,3}\, y_2 - \delta_{\theta 4,5}\,(y_2 - y_4) + \delta_{\psi 3}\, x \tag{10.74}$$

$$\delta_\theta = \delta_{\theta 0} + \delta_{\theta 1,2,3,4,5}\cos\varphi - \delta_{\psi 1,2,3,4,5}\sin\varphi \tag{10.75}$$

$$\delta_\psi = \delta_{\psi 0} + \delta_{\theta 1,2,3,4,5}\sin\varphi + \delta_{\psi 1,2,3,4,5}\cos\varphi \tag{10.76}$$

$$\delta_\varphi = \delta_{\varphi 0,1,2,3,4,5} \tag{10.77}$$

where, as before, $\delta_{\varphi 01} = \delta_{\varphi 0} + \delta_{\varphi 1}$; $\delta_{x12345} = \sum_{k=1}^{5} \delta_{xk}$; $\delta_{\varphi 2345} = \sum_{k=2}^{5} \delta_{\varphi k}$; etc.

The 36 elementary errors composing matrices $\mathbf{E}_0, \mathbf{E}_1, \ldots, \mathbf{E}_5$ and their geometric and kinematic interpretations are listed in Table 10.3. For illustration, the cross carriage (CC) model is shown in Figure 10.5. The CC is the linear stage moved along the X_5-axis and presents the fifth link ($i = 5$) of the FSS associated with the coordinate frame S_5. A correspondence between the geometrical interpretations of the errors (position error, straightness, roll, pitch, and yaw), on the one hand, and the six deviations of this link (δ_{x5}, δ_{y5}, δ_{z5}, $\delta_{\theta 5}$, $\delta_{\psi 5}$, and $\delta_{\varphi 5}$), on the other hand, are listed after sub-title *matrix* $\mathbf{E}_5$.

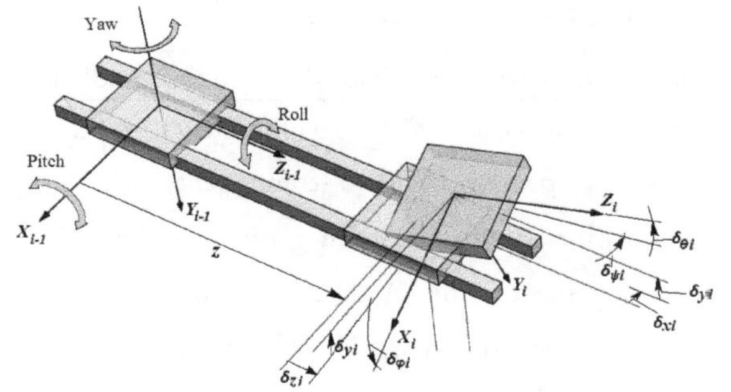

Figure 10-5 Linear stage moving along the *X*- axis. Error definitions according to Table 9.1.

10.2.2 Error budgets for different setups

10.2.2.1 Formulation

The structural form of the FSF of the lathe for the single-point machining, Eq. (4.45), is

$$\mathbf{r}_0 = {}^0\mathbf{A}_5\mathbf{r}_5 = \mathbf{A}^6(\varphi)\mathbf{A}^2(y_2)\mathbf{A}^3(z)\mathbf{A}^2(-y_4)\mathbf{A}^1(x)\mathbf{r}_5 = \begin{bmatrix} (x+x_5)\cos\varphi - (y_2 - y_4 + y_5)\sin\varphi \\ (x+x_5)\sin\varphi + (y_2 - y_4 + y_5)\cos\varphi \\ z + z_5 \\ 1 \end{bmatrix} \quad (10.78)$$

To setup the machine, eight parameters entering in Eq. (10.78)–φ, x, x_5, y_2, y_4, y_5, z, and z_5 – have to be covered by six constraints. Five of them are geometric and the sixth is the kinematic constraint. Two of the geometric constraints are structure-fixed dimensions independent of setup requirements,

$$y_2 = H;\ y_4 = H_g, \quad (10.79)$$

where H is the centre height (basic machine size); and H_g is the distance between two planes fixed with respect to the longitudinal and transverse guideways.

Three geometrical constraints fix initial work points fixed *during machine setup*

$$x_5 = x_{in} \quad (10.80)$$

$$y_2 - y_4 + y_5 = h \quad (10.81)$$

$$z_5 = z_{in} \quad (10.82)$$

where x_{in} is the initial work point for the workpiece diameter; h is the height of the cutting tool with respect to the line of machine centres; and z_{in} is the initial work point for workpiece length.

The kinematic constraint in its general form is

$$f(\varphi, x, z) = 0 \quad (10.83)$$

Table 10-3 Geometrical and kinematic errors of the lathe.

Matrix $\mathbf{E}_0$: Accuracy of mutual location of the workpiece (*WP*) and the spindle axis (*SA*)	
δ_{x0}	– Eccentricity of the *WP* with respect to the *SA* (*X*-component)
δ_{y0}	– Eccentricity of the *WP* with respect to the *SA* (*Y*-component)
δ_{z0}	– Displacement of the *WP* with respect to the home-position of the *SA*
$\delta_{\theta 0}$	– Parallelism of the *WP* with respect to the *SA* (*X*-component)
$\delta_{\psi 0}$	– Parallelism of the *WP* with respect to the *SA* (*Y*-component)
$\delta_{\varphi 0}$	– Relative angular displacement of the *WP* with respect to the spindle
Matrix $\mathbf{E}_1$: Accuracy of spindle rotation[(1)]	
δ_{x1}	– Radial runout (*X*-component) of the *SA*([2])
δ_{y1}	– Radial runout (*Y*-component) of the *SA*([2])
δ_{z1}	– Axial runout of the *SA*
$\delta_{\theta 1}$	– Angular runout (*X*-component) of the *SA*([2])
$\delta_{\psi 1}$	– Angular runout (*Y*-component) of the *SA*([2])
$\delta_{\varphi 1}$	– **Kinematic error** of the spindle rotation
Matrix $\mathbf{E}_2$: Accuracy of mutual location and orientation of the longitudinal guideways (*LG*) relative to the *SA* (both *LG* and *SA* are geometrical elements associated with the bed)	
δ_{x2}	– *X*-deviation of the distance between the *LG* and *SA*
δ_{y2}	– *Y*-deviation of the distance between the *LG* and *SA*
δ_{z2}	– *Z*-deviation of the distance between the *LG* and *SA*
$\delta_{\theta 2}$	– Parallelism of the horizontal plane of the *LG* relative to the SA in the *YZ*-plane
$\delta_{\psi 2}$	– Parallelism of the horizontal plane of the *LG* relative to the SA in the *XZ*-plane
$\delta_{\varphi 2}$	– Parallelism of the vertical plane of the *LG* relative to the SA in the *XY*-plane
Matrix $\mathbf{E}_3$: Accuracy of motion of the longitudinal carriage (*LC*)	
δ_{x3}	– Straightness of the *LC* trajectory in the horizontal (XZ) plane
δ_{y3}	– Straightness of the *LC* trajectory in the vertical (*YZ*) plane
δ_{z3}	– **Kinematic error** of the CG position
$\delta_{\theta 3}$	– Parallelism of the *LC* motion relative to the SA in the vertical plane (pitch error)
$\delta_{\psi 3}$	– Parallelism of the *LC* motion relative to the SA in the horizontal plane (yaw error)
$\delta_{\varphi 3}$	– Roll error motion
Matrix $\mathbf{E}_4$: Accuracy of mutual location and orientation of cross guideways (CG) relative to the *LG* (both CG and *LG* are geometrical elements associated with the longitudinal carriage)	
δ_{x4}	– *X*-deviation of the distance between the *LG* and CG
δ_{y4}	– *X*-deviation of the distance between the *LG* and CG
δ_{z4}	– *Z*-deviation of the distance between the *LG* and CG

(Continued)

Table 10-3 Geometrical and kinematic errors of the lathe. *(Continued)*

$\delta_{\theta 4}$ – Parallelism of the horizontal planes of the CG to *LG* in the *YZ*-plane
$\delta_{\psi 4}$ – Perpendicularity of the vertical plane of the CG related to the vertical plane of the *LG*
$\delta_{\varphi 4}$ – Parallelism of the horizontal plane of the CG to that of the LG in the *XY*-plane
Matrix $\mathbf{E}_5$: Accuracy of motion of the cross carriage (CC)
δ_{x5} – **Kinematic error** of the CC position
δ_{y5} – Straightness of the cross slide trajectory in the horizontal (*XZ*) plane
δ_{z5} – Straightness of the cross slide trajectory in the vertical (*XY*) plane
$\delta_{\theta 5}$ – Roll error
$\delta_{\psi 5}$ – Yaw error
$\delta_{\varphi 5}$ – Pitch error

(1)The errors combining matrices $\mathbf{E}_0$ and $\mathbf{E}_1$ enter in Eq. (10.70) and (10.71) with the opposite sign.
(2)With exclusion of the component depending on rotational speed.

Variation to constraints (10.79)-(10.83) yields

$$\delta_{y2} = \delta_{H};\ \delta_{y4} = \delta_{Hg};\ \delta_{x5} = \delta_{xin};\ \delta_{z5} = \delta_{zin};$$

$$\delta_{y2} - \delta_{y4} + \delta_{y5} = \delta_{h}$$

$$\frac{\partial f}{\partial \varphi}\delta_{\varphi 1} + \frac{\partial f}{\partial z}\delta_{z3} + \frac{\partial f}{\partial x}\delta_{x5} = 0$$

where $\delta_{\varphi 1}$, δ_{x5}, and δ_{z3} are the kinematic errors associated with nominal motions φ, x, and z, respectively (the kinematic errors are printed in bold in Table 10.3).

10.2.2.2 Cylinder turning

Equations(10.80–83) for machining the cylinder of diameter $D = 2R$ by single-point cutting tool become(Fig. 10.6a)

$$y_2 = H;\ y_4 = H_g;\ x_5 = 0;\ z_5 = 0 \tag{10.84}$$

$$y_2 - y_4 + y_5 = 0 \tag{10.85}$$

$$x = R \tag{10.86}$$

Equation of the machined cylinder $\mathbf{r}_0(\varphi, z)$, unit vector of the normal at the cylinder $\mathbf{e}_0\,(\varphi, z)$, and vector $\mathbf{u}_0(\varphi, z)$, Eq. (8.109), of the Plücker coordinates are

$$\mathbf{r}_0(\varphi, z) = \begin{bmatrix} R\cos\varphi \\ R\sin\varphi \\ z \\ 1 \end{bmatrix};\ 0 \le z \le z_{max} \tag{10.87}$$

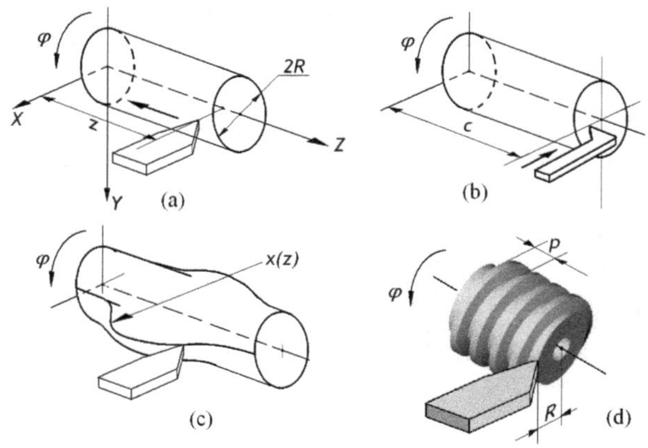

Figure 10-6 Setup for machining of different surfaces: (a), (b), and (c) machining the cylinder, face, and body of revolution, respectively, by single-point tool; (d) machining the helical surface by linear edge-type cutting tool.

$$\mathbf{e}_0(\varphi, z) = \begin{bmatrix} \cos\varphi \\ \sin\varphi \\ 0 \end{bmatrix} \tag{10.88}$$

$$\mathbf{u}_0(\varphi, z) = [\cos\varphi, \sin\varphi, 0, -z\sin\varphi, z\cos\varphi, 0]^{\mathrm{T}} \tag{10.89}$$

where φ, z are the Gaussian coordinates of the machined surface: the polar angle φ identical with the angle of spindle rotation and the coordinate z is identical with the distance of the longitudinal carriage from the origin of the basic coordinate frame.

The normal error Δ_0, Eq. (8.107), at each point of the cylinder composes six total errors:

$$\Delta_0 = \mathbf{u}_0^{\mathrm{T}}\delta = [\cos\varphi, \sin\varphi, 0, -z\sin\varphi, z\cos\varphi, 0]\cdot[\delta_x, \delta_y, \delta_z, \delta_\theta, \delta_\psi, \delta_\varphi]^{\mathrm{T}} =$$

$$= \delta_x\cos\varphi + \delta_y\sin\varphi - \delta_\theta z\sin\varphi + \delta_\psi z\cos\varphi \tag{10.90}$$

Substitution of Eqs. (10.72-77) in Eq. (10.90) with $y_2 - y_4 = -y_5$ yields

$$\Delta_0 = \delta_{x0}\cos\varphi + \delta_{x1.2.3,4,5} + \delta_{y0}\sin\varphi - \delta_{\theta 0}z\sin\varphi + \delta_{\psi 0}z\cos\varphi + \delta_{\psi 1,2}z + \delta_{\varphi 2,3}y_2 - \delta_{\varphi 4,5}y_5 \tag{10.91}$$

Taking into account constraints (10.84)-(10.86), we obtain

$$\delta_{y2} = \delta_H;\ \delta_{y4} = \delta_{Hg};\ \delta_{x5} = 0;\ \delta_{z5} = 0 \tag{10.92}$$

$$\delta_{y2} + \delta_{y4} + \delta_{y5} = 0 \tag{10.93}$$

$$\delta_{x5} = \delta_R \tag{10.94}$$

Simplification of Eq. (10.91) taking into account Eq. (10.86) yields

$$\Delta_0 = \sqrt{\delta_{x0}^2 + \delta_{y0}^2}\,\sin(\varphi+\lambda) + \delta_{x1,2,3,4} + \delta_R + \sqrt{\delta_{\theta 0}^2 + \delta_{\psi 0}^2}\,z\sin(\varphi-\mu) + \delta_{\psi 1,2}\,z + \delta_{\varphi 2,3}\,y_2 - \delta_{\varphi 4,5}\,y_5 \qquad (10.95)$$

where $\lambda = \arctan\dfrac{\delta_{y0}}{\delta_{x0}}$; and $\mu = \arctan\dfrac{\delta_{\theta 0}}{\delta_{\psi 0}}$

According to Eqs. (10.95), 15 of 36 elementary errors act along the sensitive direction, i.e., along the unit vector of the normal $\mathbf{e}_0(\varphi, z)$, Eq. (10.88), at the point $\mathbf{r}_0(\varphi, z)$, Eq. (10.22). Remember that all coefficients at the angular errors in Eq. (10.95) are the Abbe offsets. For example, coefficient y_2 is the Abbe offset for angles $\delta_{\varphi 2}$ and $\delta_{\varphi 3}$, coefficient $-y_5$ is the Abbe offset for angles $\delta_{\varphi 4}$ and $\delta_{\varphi 5}$, and so on. Let us consider some typical cases of the interrelationships between 15 elementary input errors δ_{xi} $(i = 0, 1,\ldots,5)$, δ_{y0}, $\delta_{\theta 0}$, $\delta_{\psi i}$ $(i = 0, 1, 2)$, $\delta_{\varphi 0}$ $(i = 2, 3, 4, 5)$, δ_{xi} $(i = 0, 1,\ldots,5)$, δ_{y0}, $\delta_{\theta 0}$, $\delta_{\psi i}$ $(i = 0, 1, 2)$, and $\delta_{\varphi i}$ $(i = 2, 3, 4, 5)$, on the one hand, and the form-and-position error Δ_0, on the other. A possible contribution of input errors in total radial deviation Δ_0 varied with current values of the curvilinear coordinates φ and z is analyzed below.

- *Eccentric initial position of the workpiece* δ_{x0} and δ_{y0} results in eccentricity $\delta_{ecc} = \sqrt{\delta_{x0}^2 + \delta_{y0}^2}$ of the workpiece axis.
- *The spindle runout* $\delta_{x2} = \delta_{x2}(\varphi)$ results in lobing in the workpiece cross-section. The profile spectrum of the section can be represented through a partial sum consisting of N harmonics of the Fourier series

$$\delta_{x2} = a_0/2 + \sum_{k=1}^{N} a_k \cos(k\varphi + \lambda_k) \qquad (10.96)$$

where a_0 is a systematic setup displacement; a_k and λ_k are the magnitude and phase of the kth harmonic associated with the profile concentricity deviation $(k = 1)$ and form deviations $(k > 1)$: ovality for $k = 2$ and k-lobing with $k > 2$.

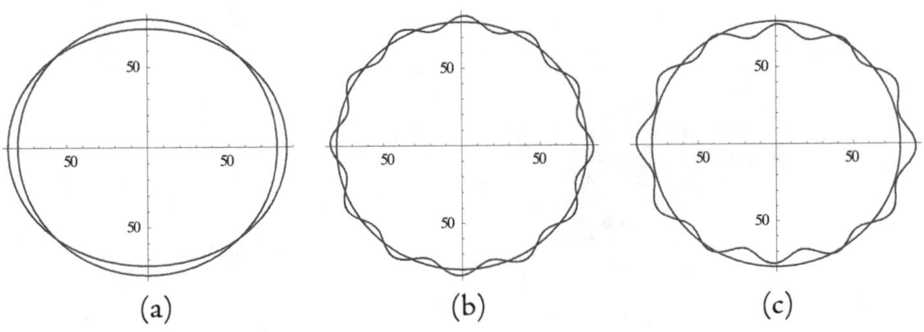

Figure 10-7 Profile examples: (a) 2-lobe (oval) profile, $k = 2$; (b) the 12-lobe profile, $k = 12$; and (c) the sum of profiles (a) and (b) with phase $\lambda_{12} = \pi/10$ (the deviations are shown on an enlarged scale).

For example, 2- and 12-lobe profiles caused by the 2nd and 12th harmonics and their sum are shown in Figures 10.7(a), (b), and (c), respectively. When the harmonics phase is variable, $\lambda_k = \lambda_k(z)$, waviness of the axial section takes place

$$\delta_{x2} = \delta_{x2}(\varphi, z) \tag{10.97}$$

If the kth harmonics has the monotonic-variable phase $\lambda_k = \lambda_k(z)$ with pitch h, the lobes become helical in form, and Eq. (10.97) becomes

$$\delta_{x2} = \delta_{x2}(\varphi, z) = a_k \cos(k\varphi + 2\pi z/p) \tag{10.98}$$

where p is the pitch of the helicoid and k_p is its lead.

An example with $k = 12$ and $p = 160$ is shown in Figure 10.8.

- *Radial deviation of the longitudinal motion* $\delta_{x2} = \delta_{x2}(z)$ results in form deviations of the workpiece axial section. The deviation can be represented through a sum of an aperiodic part considered as a z-polynomial and a partial sum combining N harmonics of the Fourier series,

$$\delta_{x2}(z) = b_0(z) + \sum_{k=1}^{N} b_k \cos\left[\frac{\pi k}{L}(z + z_k)\right] \tag{10.99}$$

where $b_0(z)$ is the z-polynomial of small displacements; b_k and z_k are the magnitude and phase of the kth harmonic; and L is a distance of the longitudinal motion, for example, the length of the portion of the machined surface.

In Figure 10.9, the case in which an aperiodic part takes into account a linear trend of the displacements and the periodic part includes two components with lead $k = 1$ (i.e., the lead equals the double length $2L$) and $k = 2$ (i.e., the lead equals L). In this case, Eq. (10.99) becomes

$$\delta_{x2}(z) = B_0 + B_1 z + B_2 \cos\frac{\pi}{L}(z + z_2) + B_3\frac{2\pi}{L}(z + z_3) \tag{10.100}$$

where B_2 and B_3 are the magnitudes of the harmonics with leads equal to the double length $2L$ and L, respectively.

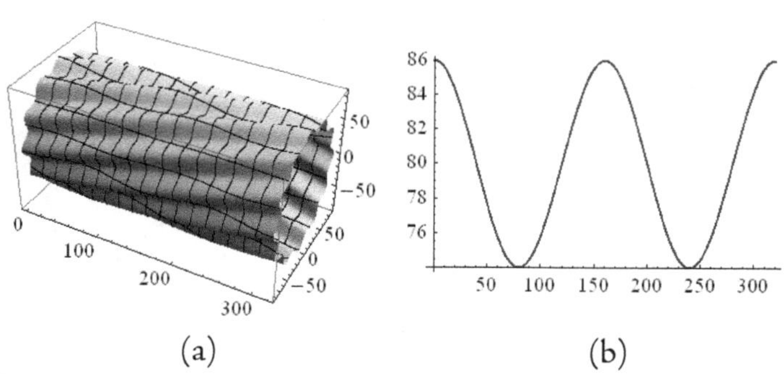

Figure 10-8 (a) The 12-lobe cylindrical part and (b) its axial section (the deviations are shown on an enlarged scale).

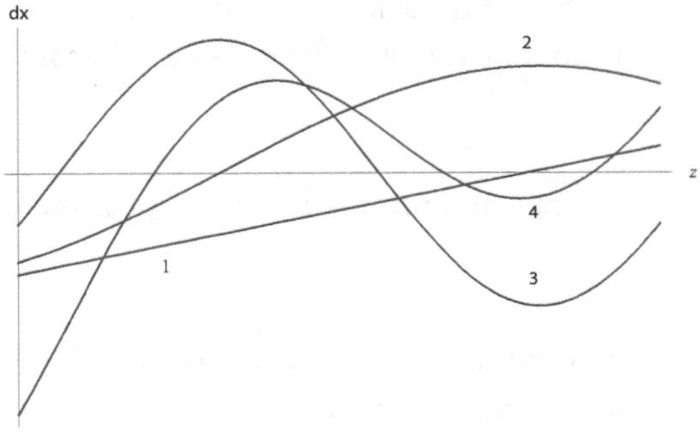

Figure 10-9 Curves 1, 2, 3 are the separate terms in Eq. (10.100) and 4 is the total error.

In Figure 10.9, the terms entering into Eq. (10.100) are displayed, where *1* is the linear term $B_0 + B_1 z$; *2* is the half-wave term $B_2 \cos\frac{\pi}{L}(z+z_2)$; *3* is the single-wave term $B_3 \frac{2\pi}{L}(z+z_3)$; and *4* is the total error $\delta_{x2}(z)$. In Figure 10.10, the actual form of the cylinder associated with the error sum according to Eq. (10.100) is shown.

- *Roll deviations of the longitudinal motion* $\delta_{\varphi 2,3} \equiv \delta_{\varphi 2} + \delta_{\varphi 3}$ affects workpiece form as a consequence of the Abbe offset y_2 – the distance between the longitudinal motion and the axis of the workpiece rotation. This deviation can be represented in a manner like Eq. (10.100).

$$\delta_\varphi(z) = B_0 + B_1 z + B_2 \cos\frac{\pi}{L}(z+z_2) + B_3\frac{2\pi}{L}(z+z_3) \tag{10.101}$$

where $\delta_{\varphi 0}(z)$ is the z-polynomial of small displacements; $\delta_{\varphi k}$ and z_k are the magnitude and phase of the kth harmonic; and L is a distance of the longitudinal motion, for example the length of the machined surface.

Let us consider the case in which only the linear term takes place

$$\delta_\varphi(z) = B_0 + B_1 z \tag{10.102}$$

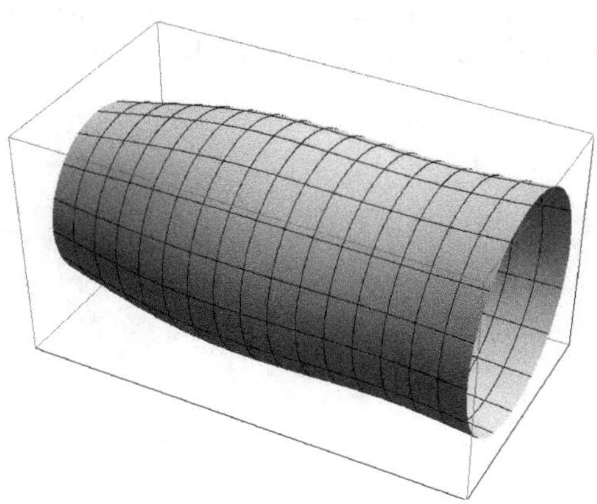

Figure 10-10 The cylinder with form errors according to Eq. (10.100).

This error results in a conical form deviation of the machined cylinder where the difference of the maximal and minimal diameters of the cylinder becomes

$$\Delta_D = 2\, y_2\, B_1\, |z_{max} - z_{min}|$$

where z_{max} and z_{min} are the maximal and minimal nominal position of the tool working point.

10.2.2.3 Face turning

The setup equations for single-point machining of the cylinder face of diameter $D = 2R$ positioned at interval c from the origin (Fig. 10.11) are

$$x_5 = 0 \tag{10.103}$$

$$y_2 - y_4 + y_5 = 0 \tag{10.104}$$

$$z_5 = 0 \tag{10.105}$$

$$z = c \tag{10.106}$$

Hence, the vector equations of the machined face $\mathbf{r}_0(\varphi, x)$, unit vector of the normal $\mathbf{e}(\varphi, x)$, and vector $\mathbf{u}_0(\varphi, x)$ of the Plücker coordinates are

$$\mathbf{r}_0(\varphi, x) = \begin{bmatrix} x\cos\varphi \\ x\sin\varphi \\ c \\ 1 \end{bmatrix}; 0 < x \leq R \tag{10.107}$$

$$\mathbf{e}(\varphi, x) = [0, 0, 1]^{\mathrm{T}} \tag{10.108}$$

$$\mathbf{u}_0(\varphi, x) = [0, 0, 1, x \sin\varphi, -x\cos\varphi, 0]^{\mathrm{T}} \tag{10.109}$$

where φ, x are the Gaussian coordinates of the machined surface: polar angle φ is identical with the angle spindle rotation, and the current radius coordinate x is identical with the distance of the working point of the cutting edge from the Z_0-axis.

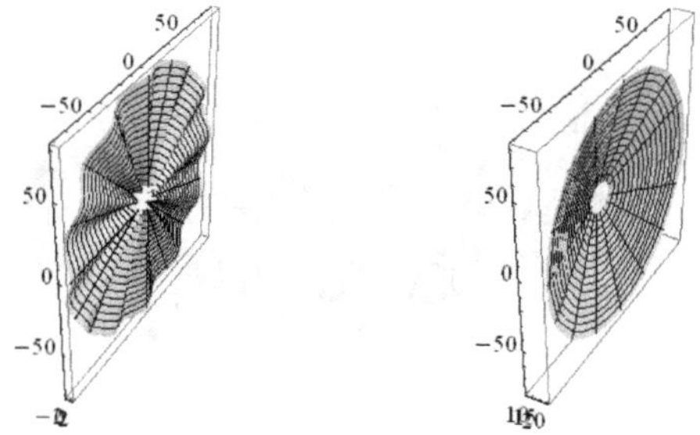

(a) (b)

Figure 10-11 Face form deviations: (a) the face runout; (b) the conical deviation of flatness.

Substituting Eq. (10.103) in Eq. (10.89) yields

$$\Delta_0 = \delta_{z0,1,2,3,4,5} - (\delta_{\psi 0}\cos\varphi - \delta_{\theta 0}\sin\varphi + \delta_{\psi 1,2,3,4})x - \delta_{\theta 2,3}\,y_2 - \delta_{\theta 4,5}\,(y_2 - y_4) \tag{10.110}$$

Taking into account setup Eq. (10.104) we can rewrite Eq. (10.110) as follows:

$$\Delta_0 = \delta_{z0,1,2,3,4,5} - (\delta_{\psi 0}\cos\varphi - \delta_{\theta 0}\sin\varphi + \delta_{\psi 1,2,3,4})\,x - \delta_{\theta 2,3}\,y_2 + \delta_{\theta 4,5}\,y_5 \tag{10.111}$$

where $\delta_{z3} + \delta_{z5} = \delta_C$ presents setup error as this follows from Eqs. (10.105) and (10.106).

In Figure 10.11, contributions in the total axial deviation Δ_0 from the sinusoidal axial runout $\delta_{z5} = b\cos 12\varphi$ (Fig. 10.11a) and the conical deviation $\delta_{\theta 4,5} = \delta_{\theta 4} + \delta_{\theta 5} = B_1 + B_2 x$ (Fig. 10.11b) are shown.

10.2.2.4 Sphere machining

The setup equations for single-point machining a body of revolution (Fig. 10.6c) are given through Eq. (10.83). As an example, setup constraints for machining a sphere are

$$x = R\cos\theta \tag{10.112}$$

$$z = a + R\sin\theta \tag{10.113}$$

$$y_2 - y_4 + y_5 = 0 \tag{10.114}$$

$$x_5 = z_5 = 0 \tag{10.115}$$

where R is the sphere radius; θ is the angle in the axial section of the sphere; φ is the angle in the cross-section identical to the angle of spindle rotation; and a is the sphere centre coordinate along the Z-axis.

The FSF $\mathbf{r}_0(\varphi, \theta)$, unit vector $\mathbf{e}_0(\varphi, \theta)$ of the normal, the moment of the unit vector $\mathbf{n}_0$ with respect to the origin, and vector of the Plücker coordinates $\mathbf{u}_0(\varphi, \theta)$ are

$$\mathbf{r}_0(\varphi,\theta) = \begin{bmatrix} R\cos\theta\cos\varphi \\ R\cos\theta\sin\varphi \\ a + R\sin\theta \\ 1 \end{bmatrix} (0 \le \varphi \le 2\pi; -\pi/2 < \theta_{min} \le \theta \le \theta_{max} < \pi/2) \tag{10.116}$$

$$\mathbf{e}_0(\varphi,\theta) = \begin{bmatrix} \cos\theta\cos\varphi \\ \cos\theta\sin\varphi \\ \sin\theta \\ 0 \end{bmatrix} \tag{10.117}$$

$$\mathbf{n}_0(\varphi, \theta) = \mathbf{r}_0 \times \mathbf{e}_0 = [-a \cos\theta \sin\varphi, a \cos\theta \cos\varphi, 0]^T$$

$$\mathbf{u}_0(\varphi, \theta) = [\cos\theta \cos\varphi, \cos\theta \sin\varphi, \sin\theta, -a \cos\theta \sin\varphi, a \cos\theta \cos\varphi, 0]^T \qquad (10.118)$$

where θ_{min} and θ_{max} are the angular boundaries of the spherical segment.

In Figure 10.12a, the segment machined without reversing the radial motion of the transverse carriage is shown. To machine the segment shown in Figure 10.12b, this motion has to be reversed.

Substituting Eq. (10.118) in Eq. (10.90) yields expression of normal error Δ_0 and actual surface $\mathbf{r}_{act}$ in the form

$$\Delta_0 = [(\delta_{x0} + a\,\delta_{\psi 0}) \cos\varphi + (\delta_{y0} - a\,\delta_{\theta 0}) \sin\varphi + \delta_{x1,2,3,4,5} + a\,\delta_{\psi 1,2} - \delta_{\psi 3,4} R \sin\theta + \delta_{\varphi 2,3,4,5}\, y_2 + \delta_{\varphi 4,5}\, y_4] \cos\theta + (\delta_{z0,1,2,3,4,5} - \delta_{\theta 2,3,4,5}\, y_2 - \delta_{\theta 4,5}\, y_4) \sin\theta \qquad (10.119)$$

$$\mathbf{r}_{act} = \mathbf{r}_0 + \Delta_0\,\mathbf{e}_0 = \begin{bmatrix} R\cos\varphi\cos\theta \\ R\sin\varphi\cos\theta \\ a + R\sin\theta \\ 1 \end{bmatrix} + \Delta_0 \begin{bmatrix} \cos\varphi\cos\theta \\ \sin\varphi\cos\theta \\ \sin\theta \\ 0 \end{bmatrix} = \begin{bmatrix} (R+\Delta_0)\cos\varphi\cos\theta \\ (R+\Delta_0)\sin\varphi\cos\theta \\ a + (R+\Delta_0)\sin\theta \\ 1 \end{bmatrix} \qquad (10.120)$$

This is the well-known condition, that the normal at a sphere point is aligned with the radial line from the centre to the point. As an example, a profile deviation of the sphere caused by pitch errors $\delta_{x4} = \delta_{x4}(x)$ of the X-leadscrew (the screw of the transverse feed) is

$$\delta_{x4} = b \cos(2\pi x/p) \qquad (10.121)$$

where b is the magnitude of the pitch error; and p is the X-leadscrew pitch.

Substitution of Eq. (10.112) in Eq. (10.121) and Eq. (10.121) in Eq. (10.120) yields

$$\Delta_0 = b \cos(2\pi R \cos\theta / p) \cos\theta \qquad (10.122)$$

Actual profile of the sphere consisting of this error is shown in Figure 10.13.

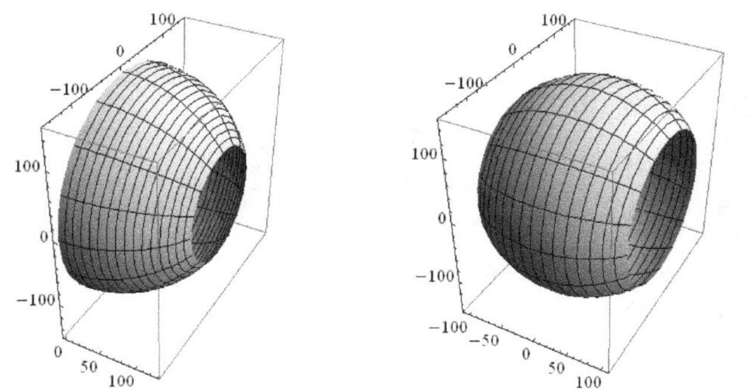

(a) (b)

Figure 10-12 Spherical segments machined (a) without and (b) with reversing the transversal motion.

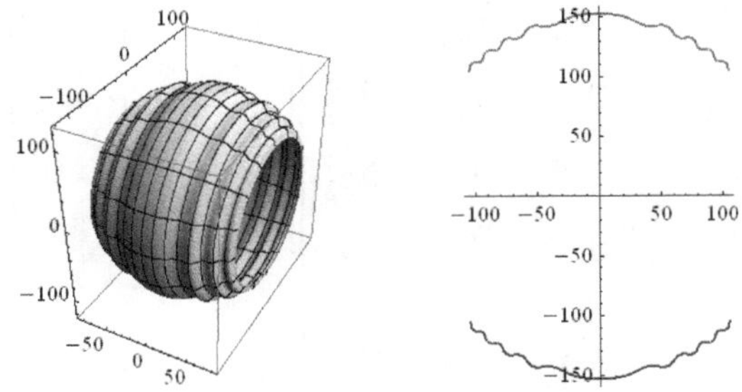

Figure 10-13 (a) Actual spherical surface; (b) axial section of the spherical segment.

10.3 Error Propagation: Multi-Axis Machine Tools

10.3.1 The *XYZ* translational triplet

Three orthogonal translational motions mounted in series along the Cartesian axes present a fundamental part of modern multi-axis machine tools. In this section, the accuracy problem for the *XYZ*-triplet is considered at first, and then the accuracy for multi-axis machines is analyzed as its extension. A pioneering matrix technique-based investigation of the *XYZ*-scheme accuracy was carried out by Schultschik [1] as applied to the coordinate measuring machines. The manipulation matrix of the Cartesian **PPP** set is obtained through multiplication of three matrices of translational motions:

$$
{}^{0}\mathbf{A}_3 = \mathbf{A}^1(x)\,\mathbf{A}^2(y)\,\mathbf{A}^3(z) =
$$

$$
\begin{bmatrix} 1 & 0 & 0 & x \\ 0 & 1 & 0 & 0 \\ 0 & 0 & 1 & 0 \\ 0 & 0 & 0 & 1 \end{bmatrix}
\begin{bmatrix} 1 & 0 & 0 & 0 \\ 0 & 1 & 0 & y \\ 0 & 0 & 1 & 0 \\ 0 & 0 & 0 & 1 \end{bmatrix}
\begin{bmatrix} 1 & 0 & 0 & 0 \\ 0 & 1 & 0 & 0 \\ 0 & 0 & 1 & z \\ 0 & 0 & 0 & 1 \end{bmatrix}
=
\begin{bmatrix} 1 & 0 & 0 & x \\ 0 & 1 & 0 & y \\ 0 & 0 & 1 & z \\ 0 & 0 & 0 & 1 \end{bmatrix}
\tag{10.123}
$$

The total manipulation error matrix **E**, Eq. (8.88), is

$$
\mathbf{E} \equiv \begin{bmatrix} 0 & -\delta_\varphi & \delta_\psi & \delta_x \\ \delta_\varphi & 0 & -\delta_\theta & \delta_y \\ -\delta_\psi & \delta_\theta & 0 & \delta_z \\ 0 & 0 & 0 & 0 \end{bmatrix}
= \sum_{i=0}^{3} {}^{0}\mathbf{A}_i \mathbf{E}_i ({}^{0}\mathbf{A}_i)^{-1} =
\begin{bmatrix} 0 & -\delta_{\varphi 0} & \delta_{\psi 0} & \delta_{x0} \\ \delta_{\varphi 0} & 0 & -\delta_{\theta 0} & \delta_{y0} \\ -\delta_{\psi 0} & \delta_{\theta 0} & 0 & \delta_{z0} \\ 0 & 0 & 0 & 0 \end{bmatrix}
$$

$$
+ \begin{bmatrix} 1 & 0 & 0 & x \\ 0 & 1 & 0 & 0 \\ 0 & 0 & 1 & 0 \\ 0 & 0 & 0 & 1 \end{bmatrix}
\begin{bmatrix} 0 & -\delta_{\varphi 1} & \delta_{\psi 1} & \delta_{x1} \\ \delta_{\varphi 1} & 0 & -\delta_{\theta 1} & \delta_{y1} \\ -\delta_{\psi 1} & \delta_{\theta 1} & 0 & \delta_{z1} \\ 0 & 0 & 0 & 0 \end{bmatrix}
\begin{bmatrix} 1 & 0 & 0 & -x \\ 0 & 1 & 0 & 0 \\ 0 & 0 & 1 & 0 \\ 0 & 0 & 0 & 1 \end{bmatrix}
$$

$$+\begin{bmatrix}1&0&0&x\\0&1&0&y\\0&0&1&0\\0&0&0&1\end{bmatrix}\begin{bmatrix}0&-\delta_{\varphi 2}&\delta_{\psi 2}&\delta_{x2}\\\delta_{\varphi 2}&0&-\delta_{\theta 2}&\delta_{y2}\\-\delta_{\psi 2}&\delta_{\theta 2}&0&\delta_{z2}\\0&0&0&0\end{bmatrix}\begin{bmatrix}1&0&0&-x\\0&1&0&-y\\0&0&1&0\\0&0&0&1\end{bmatrix}$$

$$+\begin{bmatrix}1&0&0&x\\0&1&0&y\\0&0&1&z\\0&0&0&1\end{bmatrix}\begin{bmatrix}0&-\delta_{\varphi 3}&\delta_{\psi 3}&\delta_{x3}\\\delta_{\varphi 3}&0&-\delta_{\theta 3}&\delta_{y3}\\-\delta_{\psi 3}&\delta_{\theta 3}&0&\delta_{z3}\\0&0&0&0\end{bmatrix}\begin{bmatrix}1&0&0&-x\\0&1&0&-y\\0&0&1&-z\\0&0&0&1\end{bmatrix} \tag{10.124}$$

Hence,

$$\delta_x = \delta_{x0,1,2,3} - \delta_{\psi 3}\, z + \delta_{\varphi 2,3}\, y$$

$$\delta_y = \delta_{y0,1,2,3} + \delta_{\theta 3}\, z - \delta_{\varphi 1,2,3}\, x$$

$$\delta_z = \delta_{z0,1,2,3} - \delta_{\theta 2,3}\, y + \delta_{\psi 1,2,3}\, x$$

$$\delta_\theta = \delta_{\theta 0,1,2,3}$$

$$\delta_\psi = \delta_{\psi 0,1,2,3}$$

$$\delta_\varphi = \delta_{\varphi 0,1,2,3} \tag{10.125}$$

where, as before, $\delta_{x0,1,2,3} = \sum_{k=0}^{3} \delta_{xk}$; $\delta_{\varphi 1,2,3} = \delta_{\varphi 1} + \delta_{\varphi 2} + \delta_{\varphi 3}$; $\delta_{\varphi 2,3} = \delta_{\varphi 2} + \delta_{\varphi 3}$; etc.

10.3.2 Five-axis machines

10.3.2.1 Machine with the tilting-swivel spindle head

A structure of a machine with the tilt-swivel spindle head is illustrated in Figures 4.21a and 4.22, and the FSS links are listed in Table 4.10. According to Eq. (4.66), the manipulation matrix of the machine is

$$^{0}\mathbf{A}_6 = \mathbf{A}^1(-x)\, \mathbf{A}^2(-y)\, \mathbf{A}^3(z)\, \mathbf{A}^6(\varphi_{sw})\, \mathbf{A}^4(\theta)\, \mathbf{A}^6(\varphi_s) \tag{10.126}$$

where x, y, z are translation movements of the machine table, saddle and spindle head, respectively; φ_{sw} and θ are the swivel and tilt angles, respectively, of rotation of the spindle head; and φ_s is the angle of the spindle rotation.

The total manipulation error matrix **E** can be represented as follows:

$$\mathbf{E} \equiv \begin{bmatrix} 0 & -\delta_{\varphi} & \delta_{\psi} & \delta_x \\ \delta_{\varphi} & 0 & -\delta_{\theta} & \delta_y \\ -\delta_{\psi} & \delta_{\theta} & 0 & \delta_z \\ 0 & 0 & 0 & 0 \end{bmatrix} = \mathbf{E}_0 + \mathbf{E}_{XYZ}(-x, -y, z) + \sum_{i=4}^{6} {}^0\mathbf{A}_i \mathbf{E}_i ({}^0\mathbf{A}_i)^{-1} \qquad (10.127)$$

where $\mathbf{E}_0$ and $\mathbf{E}_{XYZ}(-x, -y, z)$ are the error matrices defined through Eqs. (10.124) and (10.125); $\mathbf{E}_4$ and $\mathbf{E}_5$ are error matrices of the tilting and swivelling links, respectively; $\mathbf{E}_6$ is the error matrix of the spindle rotation; and ${}^0\mathbf{A}_4$, ${}^0\mathbf{A}_5$, and ${}^0\mathbf{A}_6$ are the manipulation matrices,

$${}^0\mathbf{A}_4 = \mathbf{A}^1(-x)\,\mathbf{A}^2(-y)\,\mathbf{A}^3(z)\,\mathbf{A}^6(\varphi_{sw})$$

$${}^0\mathbf{A}_5 = \mathbf{A}^1(-x)\,\mathbf{A}^2(-y)\,\mathbf{A}^3(z)\,\mathbf{A}^6(\varphi_{sw})\,\mathbf{A}^6(\theta)$$

$${}^0\mathbf{A}_6 = \mathbf{A}^1(-x)\,\mathbf{A}^2(-y)\,\mathbf{A}^3(z)\,\mathbf{A}^6(\varphi_{sw})\,\mathbf{A}^4(\theta)\,\mathbf{A}^6(\varphi_s).$$

Formal transformations result in six elementary errors composing the total error matrix **E**:

$$\begin{aligned}\delta_x = {} & \delta_{x0,1,2.3} - \delta_{\psi 3}\, z + \delta_{\varphi 2,3}\, y + \delta_{x4,5} \cos\varphi_{sw} + \delta_{x6} (\cos\varphi_{sw} \cos\varphi_s - \cos\theta \sin\varphi_{sw} \sin\varphi_s) - \\ & \delta_{y4} \sin\varphi_{sw} - \delta_{y5} \cos\theta \sin\varphi_{sw} - \delta_{y6} (\cos\varphi_{sw} \sin\varphi_s + \cos\theta \sin\varphi_{sw} \cos\varphi_s) + \\ & \delta_{z5,6} \sin\theta \sin\varphi_{sw} - \delta_{\theta 4,5}\, z \sin\varphi_{sw} - \delta_{\theta 6} (z \cos\theta \cos\varphi_{sw} \sin\varphi_s + z \sin\varphi_{sw} \cos\varphi_s + \\ & y \sin\theta \sin\varphi_{sw}) - \delta_{\psi 4}\, z \cos\varphi_{sw} - \delta_{\psi 5} (z \cos\theta \cos\varphi_{sw} + y \sin\theta) - \\ & \delta_{\psi 6} (z \cos\theta \cos\varphi_{sw} \cos\varphi_s - z \sin\varphi_{sw} \sin\varphi_s + y \sin\theta \cos\varphi_s) - \delta_{\varphi 4}\, y + \\ & \delta_{\varphi 5,6} (z \sin\theta \cos\varphi_{sw} - y \cos\theta)\end{aligned}$$

$$\begin{aligned}\delta_y = {} & \delta_{y0,1,2.3} + \delta_{\theta 3}\, z - \delta_{\varphi 1,2,3}\, x + \delta_{x4,5} \sin\varphi_{sw} + \delta_{x6} \sin\varphi_{sw} (\sin\varphi_{sw} \cos\varphi_s + \cos\theta \cos\varphi_{sw} \sin\varphi_s) \\ & + \delta_{y4} \cos\varphi_{sw} + \delta_{y5} \cos\theta \cos\varphi_{sw} - \delta_{y6} (\sin\varphi_{sw} \sin\varphi_s - \cos\theta \cos\varphi_{sw} \cos\varphi_s) - \\ & \delta_{z5,6} \sin\theta \cos\varphi_{sw} + \delta_{\theta 4,5}\, z \cos\varphi_{sw} - \delta_{\theta 6} (z \cos\theta \sin\varphi_{sw} \sin\varphi_s - z \cos\varphi_{sw} \cos\varphi_s - \\ & x \sin\theta \sin\varphi_{sw}) - \delta_{\psi 4}\, z \sin\varphi_{sw} - \delta_{\psi 5} (z \cos\theta \sin\varphi_{sw} - x \sin\theta) - \\ & \delta_{\psi 6} (z \cos\theta \sin\varphi_{sw} \cos\varphi_s + z \cos\varphi_{sw} \sin\varphi_s + x \sin\theta \cos\varphi_s) + \delta_{\varphi 4}\, x + \\ & \delta_{\varphi 5,6} (z \sin\theta \sin\varphi_{sw} + x \cos\theta)\end{aligned}$$

$$\begin{aligned}\delta_z = {} & \delta_{z0,1,2,3} - \delta_{\theta 2,3}\, y + \delta_{\psi 1,2,3}\, x + \delta_{x6} \sin\theta \sin\varphi_{sw} + \delta_{y5} \sin\theta + \delta_{y6} \sin\theta \cos\varphi_s + \delta_{z4} + \delta_{z5,6} \cos\theta \\ & + \delta_{\theta 4,5} (y \cos\varphi_{sw} - x \sin\varphi_{sw}) + \delta_{\theta 6} [\cos\varphi_{sw} (y \cos\varphi_s - x \cos\theta \sin\varphi_s) - x \sin\varphi_{sw} \cos\varphi_s\end{aligned}$$

$$- y \cos\theta \sin\varphi_{sw} \sin\varphi_s] - \delta_{\psi4} (x \cos\varphi_{sw} + y \sin\varphi_{sw}) - \delta_{\psi5} (x \cos\theta \cos\varphi_{sw} + y \cos\theta \sin\varphi_{sw}) - \delta_{\psi6}[\cos\varphi_{sw} (y \sin\varphi_s + x \cos\theta \cos\varphi_s) - x \sin\varphi_{sw} \sin\varphi_s + y \cos\theta \sin\varphi_{sw} \sin\varphi_s] + \delta_{\varphi5,6} (x \sin\theta \cos\varphi_{sw} + y \sin\theta \sin\varphi_{sw})$$

$$\delta_\theta = \delta_{\theta0,1,2,3} + \delta_{\theta4,5} \cos\varphi_{sw} + \delta_{\theta6} (\cos\varphi_{sw} \cos\varphi_s - \cos\theta \sin\varphi_{sw} \sin\varphi_s) - \delta_{\psi4} \sin\varphi_{sw} - \delta_{\psi5} \cos\theta \sin\varphi_{sw} - \delta_{\psi6} (\cos\varphi_{sw} \sin\varphi_s + \cos\theta \sin\varphi_{sw} \cos\varphi_s) + \delta_{\varphi5,6} \sin\varphi_{sw} \sin\varphi_s$$

$$\delta_\psi = \delta_{\psi0,1,2,3} + \delta_{\theta4,5} \sin\varphi_{sw} + \delta_{\theta6}(\cos\varphi_s \sin\varphi_{sw} + \cos\theta \cos\varphi_{sw} \sin\varphi_s) + \delta_{\psi4} \cos\varphi_{sw} + \delta_{\psi5} \cos\theta \cos\varphi_{sw} + \delta_{\psi6} (\cos\theta \cos\varphi_{sw} \cos\varphi_s - \sin\varphi_{sw} \sin\varphi_s) - \delta_{\varphi5,6} \cos\varphi_{sw} \sin\varphi_s$$

$$\delta_\varphi = \delta_{\varphi0,1,2,3,4} + \delta_{\theta6} \sin\theta \sin\varphi_s + \delta_{\psi5} \sin\theta + \delta_{\psi6} \sin\theta \sin\varphi_s + \delta_{\varphi5,6} \cos\theta \qquad (10.128)$$

10.3.2.2 Five-axis machine with trunnion rotary table

The error budget of the other 5-axis machines is formed in perfect analogy. For example, the manipulation matrix of a machine with an integrated trunnion rotary table (Figs. 4.21b and 4.25) is given through Eq. (4.68):

$${}^0\mathbf{A}_6 = \mathbf{A}^6(-\varphi_{rot})\, \mathbf{A}^4(-\theta)\, \mathbf{A}^1(-x)\, \mathbf{A}^2(-y)\, \mathbf{A}^3(z)\, \mathbf{A}^6(\varphi_{sp}) \qquad (10.129)$$

where x and y are translation movements of the machine table and saddle, respectively; z is the spindle head translation; φ_{rot} are the rotation angle of the rotary table and θ is its tilting angle; and φ_{sp} is the angle of the spindle rotation.

Hence, the total manipulation error matrix is

$$\mathbf{E} \equiv \begin{bmatrix} 0 & -\delta_\varphi & \delta_\psi & \delta_x \\ \delta_\varphi & 0 & -\delta_\theta & \delta_y \\ -\delta_\psi & \delta_\theta & 0 & \delta_z \\ 0 & 0 & 0 & 0 \end{bmatrix} = \sum_{i=0}^{6} {}^0\mathbf{A}_i \mathbf{E}_i ({}^0\mathbf{A}_i)^{-1} = \mathbf{E}_0 + \mathbf{A}^6(-\varphi_{rot})\, \mathbf{E}_1 \mathbf{A}^6(\varphi_{rot})$$

$$+ \mathbf{A}^6(-\varphi_{rot})\, \mathbf{A}^4(-\theta)\, [\mathbf{E}_2 + \mathbf{E}_{XYZ}(-x, -y, z)]\, \mathbf{A}^4(\theta)\, \mathbf{A}^6(\varphi_{rot}) + {}^0\mathbf{A}_6\, \mathbf{E}_6 ({}^0\mathbf{A}_6)^{-1} \qquad (10.130)$$

where $\mathbf{E}_{XYZ}(-x, -y, z)$ is the translation error matrix, Eq. (10.124), with substitution of subscripts 1, 2, 3 for 3, 4, 5, respectively; and ${}^0\mathbf{A}_6$ is defined through Eq. (10.129).

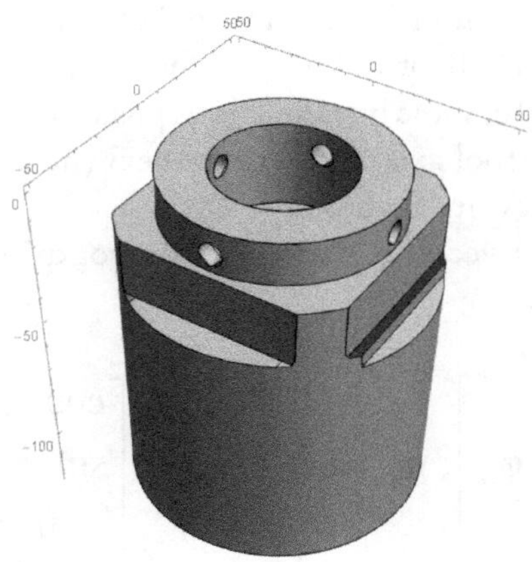

Figure 10-14 The standard testing part [2].

10.3.3 Case study: machining of cylindrical surface

10.3.3.1 Cylinder profiling on contour grinding machine

The multi-axis machines are designed to operate using joint action of the coordinate axes. Here the profile machining of a cylindrical surface on the contour grinding machine is illustrated as an example. Remember that the contour grinding machine operates in the 2.5-axis machine mode.

As is known, the cylinders as well as planes are the regular surfaces applied in machine parts. In particular, they present a part of the standard workpiece (Fig. 10.14) used for inspection of the kinematic accuracy of the NC and CNC machine tools according to NAS 979 piece cutting test [2]. The circular pass tracing called *the circle test* provides the basis of the Renishaw ball-bar test for CNC machines [3].The wireless ball-bar system (Fig. 10.15) combines the high-accuracy telescoping linear sensor *1* with two precision balls *2* and *3* mounted on the sensor ends. There are two precision magnetic mounts attached to the machine table (mount *4*) and to the machine spindle (mount *5*). The sensor measures small deviations of the ball-to-ball distance from its nominal value. The deviations are caused when the machine table makes the programmed circular trajectory whose radius equals the nominal ball-to-ball distance and the center is located on the spindle axis.

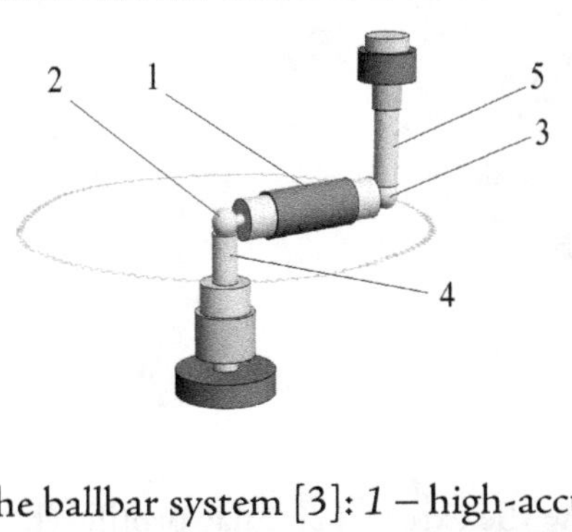

Figure 10-15 The ballbar system [3]: *1* – high-accuracy telescoping linear sensor; *2* and *3* – precision balls; *4* and *5* – precision magnetic mounts attached to the machine table and to the machine spindle, respectively.

According to Eq. (6.94), the cylindrical cutting tool of radius R – milling cutter or grinding wheel – moved along the circle of radius r performs profiling of the cylinder of outer radius $R_w = r - R$. In Figure 10.16, some instantaneous positions of the rotating cutting tool *1*, the trajectory *2* of the cutting tool axis, and workpiece *3* (machined cylinder) fitted to the machine table, are shown in view from above.

According to Eqs. (6.77-78), vector equations of the workpiece $\mathbf{r}_w$, unit normal vector $\mathbf{n}_w$, and Plücker coordinates $\mathbf{u}_w$ are

$$\mathbf{r}_w(z_w, \varphi_w) = \begin{bmatrix} a + R_w \cos\varphi_w \\ b + R_w \sin\varphi_w \\ z_w \end{bmatrix}; \mathbf{n}_w = \mathbf{n}(z_w, \varphi_w) = \begin{bmatrix} \cos\varphi_w \\ \sin\varphi_w \\ 0 \end{bmatrix} \tag{10.131}$$

$$\mathbf{n}_w = \mathbf{n}(z_w, \varphi_w) = [\cos\varphi_w, \sin\varphi_w, 0]^{\mathrm{T}} \tag{10.132}$$

$$\mathbf{u}_w = [\cos\varphi_w, \sin\varphi_w, 0, -Z\sin\varphi_w, Z\cos\varphi_w, b\cos\varphi_w - a\sin\varphi_w]^{\mathrm{T}} \tag{10.133}$$

where a and b are the X- and Y-coordinates of the workpiece and tool centroid; $Z = z_t + z_w$ is instantaneous Z-coordinate of points of the tool; z_w is the coordinate of the working point of the tool; and z_t is the tool working motion along the Z-axis.

A normal error Δ_0, Eq. (10.90), in each point of the machined cylinder is

$$\Delta_0 = \mathbf{u}_w^{\mathrm{T}}\boldsymbol{\delta} = [\cos\varphi_w, \sin\varphi_w, 0, -Z\sin\varphi_w, Z\cos\varphi_w, b\cos\varphi_w$$

$$- a\sin\varphi_w][\delta_x, \delta_y, \delta_z, \delta_\theta, \delta_\psi, \delta_\varphi]^{\mathrm{T}} = (\delta_x + \delta_\psi Z + \delta_\varphi b)\cos\varphi_w$$

$$+(\delta_y - \delta_\theta Z - \delta_\varphi a)\sin\varphi_w \tag{10.134}$$

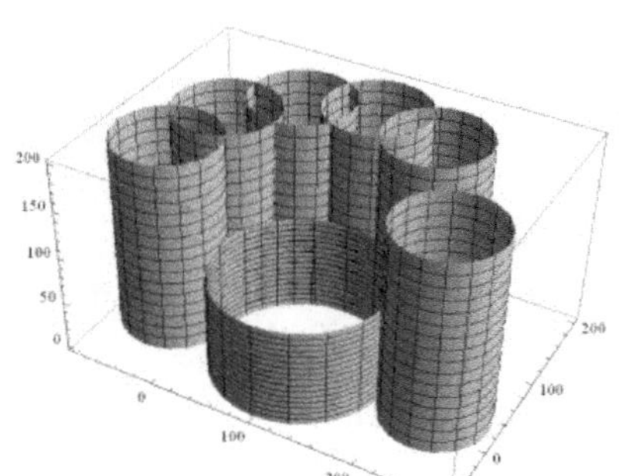

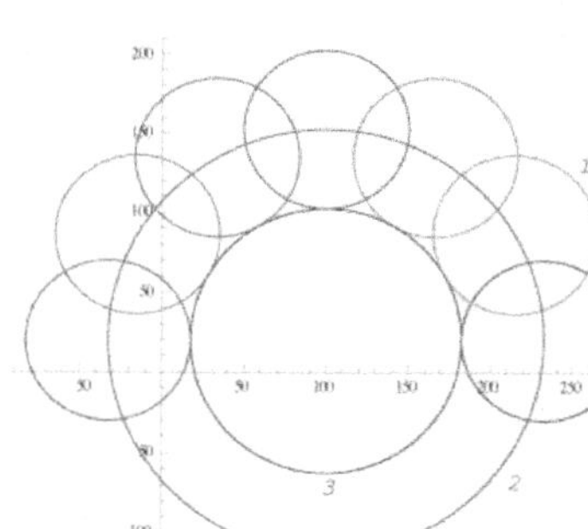

(a) (b)

Figure 10-16 (a) Grinding the external circular profile; (b) view from above: *1* – instantaneous positions of the cutting tool; *2* – the tool axis trajectory; *3* – the machined profile.

On the other hand, the real profile can obtained through Eq. (10.120) in the form

$$\mathbf{r}_a = \mathbf{r}_w + \Delta_0\, \mathbf{n}_w = \begin{bmatrix} a + R_w \cos\, \varphi_w \\ b + R_w \sin\ \varphi_w \\ z_w \end{bmatrix} + \Delta_0 \begin{bmatrix} \cos\varphi_w \\ \sin\, \varphi_w \\ 0 \end{bmatrix} = \begin{bmatrix} a + (R_w + \Delta_0)\cos\, \varphi_w \\ b + (R_w + \Delta_0)\sin\, \varphi_w \\ z_w \end{bmatrix} \tag{10.135}$$

Hence, the normal error Δ_0 presents a deviation of the nominal radius of the machined cylinder from its nominal value $R_w = r - R$. This is evident from the fact that the normal at the cylinder at a given point is aligned with the radius at the same point. The radial error is

$$\delta R_w = (R_w + \Delta_0) - R_w = \Delta_0$$

Hence,

$$\delta R_w = (\delta_x + \delta_\psi Z + \delta_\varphi b) \cos \varphi_w + (\delta_y - \delta_\theta Z - \delta_\varphi a) \sin \varphi_w = d_0 \sin (\varphi_w + \lambda) \tag{10.136}$$

with

$$d_0 = \sqrt{(\delta_x + \delta_\psi Z + \delta_\varphi b)^2 + (\delta_y - \delta_\theta Z - \delta_\varphi a)^2} \tag{10.137}$$

$$\lambda = \arctan\, [(\delta_y - \delta_\theta Z - \delta_\varphi a)/(\delta_x + \delta_\psi Z + \delta_\varphi b)] \tag{10.138}$$

According to Eqs. (10.133)–(10.138), the radial error δR_w combines 24 elementary components. Let us consider the effect of positioning errors δ_{x1} and δ_{y2} of the feed drives along the X- and Y-axis

$$\delta_{feed} = \delta_{x1} \cos \varphi_w + \delta_{y2} \sin \varphi_w \tag{10.139}$$

where δ_{x1} and δ_{y2} are the errors that are expressed through current coordinate x and y of the nominal machine table position,

$$\delta_x = ½\, A(x) + \sum_{k=1}^{N} a_k \sin[\frac{2\pi k}{L_s}(x - \xi_k)] + \delta_{xb} \tag{10.140}$$

$$\delta_y = ½\, B(y) + \sum_{k=1}^{N} b_k \sin[\frac{2\pi k}{L_s}(y - \zeta_k)] + \delta_{yb} \tag{10.141}$$

where a_k and b_k are amplitudes of the periodic components (cyclic errors) caused by pitch deviations; k is the number of the harmonic; L_s is the lead of the leadscrew; ξ_k and ζ_k are linear displacements characterizing the phases of the kth harmonic, $0 \le (\xi_k, \zeta_k) \le L_s/k$; and δ_{xb} and δ_{yb} are components appearing during reversing the machine table.

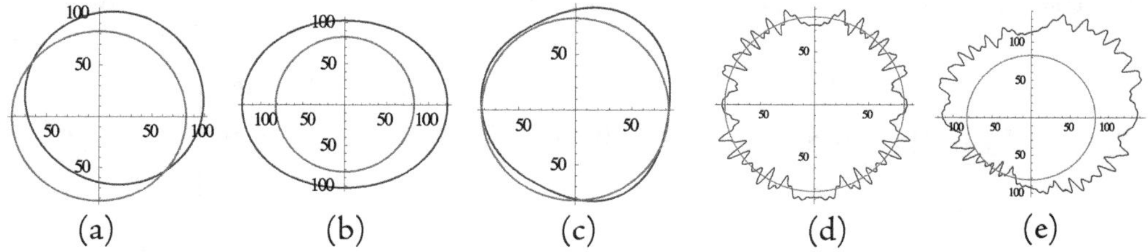

Figure 10-17 Different types of errors (the error are increased compared with the nominal circle): (a) constant terms $A_0 \neq 0, B_0 \neq 0$; (b) linear terms $A_1 \neq 0, B_1 \neq 0$; (c) half- and full-wave terms $A_2 \neq 0, B_2 \neq 0, A_3 \neq 0, B_3 \neq 0$; (d) cyclic terms $a \neq 0$, $b \neq 0$; (e) all types of errors take place simultaneously.

The aperiodic terms $A(x)$ and $B(y)$ can be presented as follows:

$$A(x) = A_0 + A_1 x + A_2 \cos[\frac{\pi}{L_x}(x - c_2)] + A_3 \cos[\frac{2\pi}{L_x}(x - c_3)] + \ldots \quad (10.142)$$

$$B(y) = B_0 + B_1 y + B_2 \cos[\frac{\pi}{L_y}(y - c_2)] + B_3 \cos[\frac{2\pi}{L_y}(y - c_3)] + \ldots \quad (10.143)$$

where L_x and L_y are the lengths of the working part of the leadscrews used for the X- and Y-axis motion, respectively.

Deviations of size and form of the circular profile caused by some types of errors are shown in Figure 10.17. The constant terms A_0 and B_0 results in deviations of the centre (Fig. 10.17a), the linear and low-frequency terms deviate both the position and the form (Fig. 10.17b, c), and the cyclic error produces the waviness of the profile with a pitch equal or proportional to the leadscrew lead (Fig. 10.17d). In Figure 10.17e, the join action of all listed errors is shown.

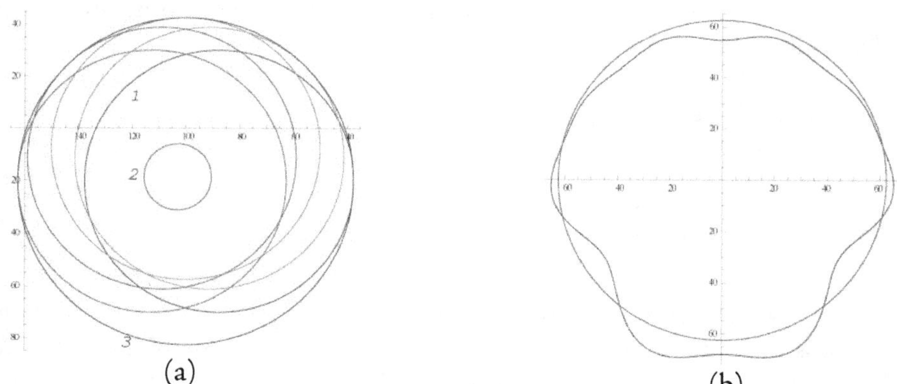

Figure 10-18 (a) Profiling the inner cylindrical surface (cross-section): *1* – an instantaneous position of the cutting tool; *2* – the tool axis trajectory; *3* – the machined profile. (b) Waviness resulting from the cyclic errors.

Similar results are obtained for profile machining the inner diameter of the cylinder of radius $R_w = r + R$, where, as before, R and r are radii or the tool and the tool centre trajectory, respectively. The unit normal vector $\mathbf{e}_0$ and Plücker coordinates $\mathbf{u}_0$ are identical to those given by Eqs. (10.132) and (10.133), and the vector equation of the workpiece $\mathbf{r}_w$ becomes

$$\mathbf{r}_w = \begin{bmatrix} -a+(R+r)\cos u \\ -b+(R+r)\sin u \\ Z \\ 1 \end{bmatrix} \tag{10.144}$$

The mutual arrangement of the cutting tool and workpiece during the internal grinding is shown in Figure 10.18a, where some instantaneous position of the tool *1* relative to the workpiece 3 and the tool axis trajectory 2 are shown. In Figure 10.18b, the roundness in the cross-sections is shown in the case, in which only the cyclic error takes place. This error is caused by natural deviations in the leadscrew production and/or leadscrew mounting. Cyclic error is manifested through profile waviness with variable amplitude and reaches the maximal amplitude at axis reversal. The number of waves on the circular profile caused by cyclic errors of two axial lead screws is

$$N = \begin{cases} 4(R_w + R)/L_s, & \text{for machining of external diameter} \\ 4(R_w - R)/L_s, & \text{for machining of internal diameter} \end{cases} \tag{10.145}$$

where, as before, L_s is the leadscrew lead; R is the radius of the cutting tool; R_w is the radius of the machined cylinder; machining of external and internal cylinder diameters is shown in Figures10.16 and 10.18a, respectively.

Since the roundness measurement presents a standard inspection procedure, and the number of waves is calculated very simply, Eq. (10.145) can serve as a useful tool for diagnostics of the accuracy of translational motions in multi-axis machines.

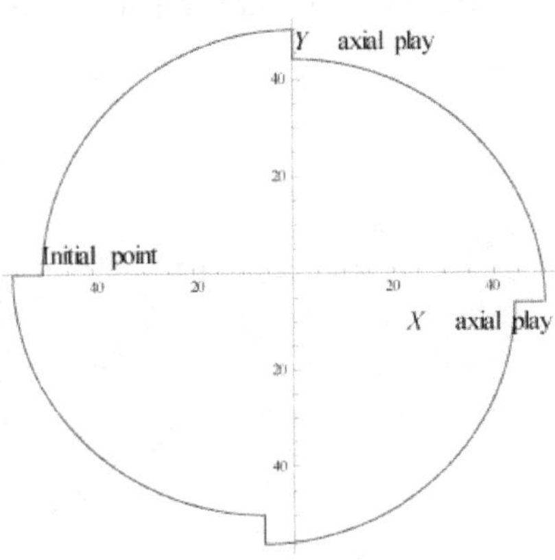

Figure 10-19 Form deviation of the circle caused by end play of the drive.

To machine the cylindrical surface in the profiling mode, each leadscrew has to make two reverses. If the backlash presents the air gap, this results in four sharp form deviations appearing at the reverse points $\varphi = 0, \pi/2, \pi$, and $3\pi/2$ (Fig. 10.19).

10.3.3.2 Experimental investigation of the positioning system

As an object of investigation, an NC contour-grinding machine MA396F3 [4] was used (Fig. 4.19). The machine is fitted with a 2.5-axis NC-system, in which the tool axis position is controlled in the horizontal (XY) plane, while the fixed cycle of the grinding head presents a reciprocal motion along the Z-axis. For grinding complex faces, where very accurate shape and dimensions are required, the machine is provided with various devices for accurately determining the positions of the assemblies. The table-positioning error of the machine is shown in Figure 9.14b.

The Talyrond charts [5] of the machined cylinder after external and internal grinding are shown in Figure 10.20. The waviness of the machined profiles testifies that the non-compensated cyclic error takes place. For the experiment shown in Figure 10.20a, the grinding wheel of diameter $D = 100$ mm was used for machining the external cylinder of diameter $D_{w,ex} = 165$ mm. The screw lead was $L_s = 12$ mm. According to Eq. (10.145), the number of waves on the external profile is

$$N_{ex} = 4(R_w + R)/L_s = 4(165/2 + 100/2)/12 = 44.17 \text{ peaks} \tag{10.146}$$

The corresponding 44 peaks on the circumference of the external diameter are shown and numbered in Figure 10.20a.

The chart of the machined hole after internal grinding is shown in Figure 10.20b. In this case, the same grinding wheel of diameter $D = 100$ mm was used for machining of the cylindrical hole of diameter $D_w = 125$ mm. The number of waves on the internal profile is

$$N_{in} = 4(R_w + R)/L_s = 4\,(125/2 - 100/2)/12 = 4.2 \tag{10.147}$$

The four peaks on the circumference of the internal diameter are clearly distinguished and numbered in Figure 10.20b. Note that the waviness shown in Figure 10.20a and b is similar to that in Figure 10.17. Hence, the experimental results stated above show that knowing

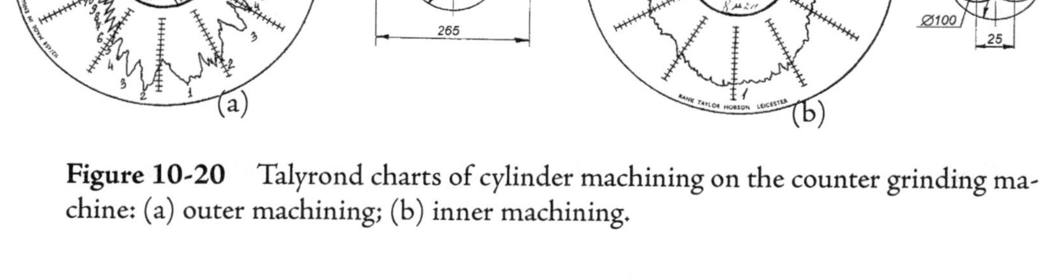

Figure 10-20 Talyrond charts of cylinder machining on the counter grinding machine: (a) outer machining; (b) inner machining.

the errors of the feed leadscrews from its certification allows prediction of the errors of the machined workpiece with fairly high degree of detailization.

10.4 Structural Jacobian matrix of serial-kinematics systems

According to Eqs. (8.35) and (8.36), there are the following connections between nominal position and orientation vectors and corresponding position and orientation errors:

$$[\Delta\mathbf{r}_0\ \Delta\mathbf{c}_0] = \mathbf{E}\ [\mathbf{r}_0\ \mathbf{c}_0] \tag{10.148}$$

$$\mathbf{E} = \sum_{i=0}^{n} {}^0\mathbf{A}_i \mathbf{E}_i\, {}^i\mathbf{A}_0 = \begin{bmatrix} 0 & -\delta_\varphi & \delta_\psi & \delta_x \\ \delta_\varphi & 0 & -\delta_\theta & \delta_y \\ -\delta_\psi & \delta_\theta & 0 & \delta_z \\ 0 & 0 & 0 & 0 \end{bmatrix} \tag{10.149}$$

where the linear deviations δ_x, δ_y, δ_z are the linear deviations of the origin of the frame S_0; and δ_θ, δ_ψ, δ_φ are the small rotations around the axes of the frame S_0.

The six deviations entering in matrix **E** can be combined in a sixth order vector called *twist*

$$\boldsymbol{\delta} = [\delta_x, \delta_y, \delta_z, \delta_\theta, \delta_\psi, \delta_\varphi]^{\mathrm{T}} \tag{10.150}$$

The twist can be computed through $(6n + 6)$ elementary errors δ_{xi}, δ_{yi}, δ_{zi}, $\delta_{\theta i}$, $\delta_{\psi i}$, and $\delta_{\varphi i}$ $(i = 1, 2,\ldots,n)$

$$\boldsymbol{\delta} = \mathbf{J}_t \boldsymbol{\delta}_t \tag{10.151}$$

where $\boldsymbol{\delta}_t$ is the total error vector composing $(6n + 6)$ elementary errors δ_{ij} $(i = 0, 1, 2,\ldots,n$ and $j = 1, 2\ldots,6)$ of $(n + 1)$ links of the FSS listed in the lexicographic order

$$\boldsymbol{\delta}_t = [\delta_1, \delta_2,\ldots,\delta_{6n+6}]^{\mathrm{T}} \tag{10.152}$$

with

$$\delta_1 = \delta_{0x};\ \delta_2 = \delta_{0y};\ \delta_3 = \delta_{0z};\ \delta_4 = \delta_{0\theta};\ \delta_5 = \delta_{0\psi};\ \delta_6 = \delta_{0\varphi};\ \delta_7 = \delta_{1x};\ldots;\ \delta_{6n+1} = \delta_{nx};\ \delta_{6n+2} = \delta_{ny};$$

$$\delta_{6n+3} = \delta_{nz};\ \delta_{6n+4} = \delta_{n\theta};\ \delta_{6n+5} = \delta_{n\psi};\ \text{and}\ \delta_{6n+6} = \delta_{n\varphi};$$

and $\mathbf{J}_t$ is the $6 \times (6n + 6)$ total Jacobian matrix composing of transfer factors of the $(6n + 6)$ elementary errors δ_{ij},

$$\mathbf{J}_t = \frac{\partial\boldsymbol{\delta}}{\partial\boldsymbol{\delta}_t} = \frac{\partial(\delta_x, \delta_y, \ldots, \delta_\varphi)}{\partial(\delta_1, \delta_2, \ldots, \delta_{6n+6})} = \begin{bmatrix} d_{11} & d_{12} & d_{13} & \ldots\ldots & d_{1,6n+5} & d_{1,6n+6} \\ d_{21} & d_{22} & d_{23} & \ldots\ldots & d_{2,6n+5} & d_{2,6n+6} \\ \ldots\ldots & \ldots\ldots & \ldots\ldots & \ldots\ldots & \ldots\ldots & \ldots\ldots \\ d_{61} & d_{11} & d_{11} & \ldots\ldots & d_{6,6n+5} & d_{6,6n+6} \end{bmatrix} \tag{10.153}$$

with

$$d_{11} = \frac{\partial \delta_x}{\partial \delta_1};\ d_{12} = \frac{\partial \delta_x}{\partial \delta_2};\ldots;\ d_{1,6n+6} = \frac{\partial \delta_x}{\partial \delta_{6n+6}};\ldots;d_{61} = \frac{\partial \delta_\varphi}{\partial \delta_1};\ d_{62} = \frac{\partial \delta_\varphi}{\partial \delta_2};\ldots;\ \text{and}\, d_{6,6n+6} = \frac{\partial \delta_\varphi}{\partial \delta_{6n+6}}.$$

Obviously, the Jacobian matrix $\mathbf{J}_t$ is independent of errors and its entities present functions of the FSF components. If some of the columns of the Jacobian are identical, corresponding elements δ_{ij} of the vector δ are linearly dependent. In this case, these errors can be represented as a single error equalling their sum, and the corresponding columns of the Jacobian matrix $\mathbf{J}_t$ may unite into one column. Note also that the linearly-dependent errors cannot be discriminated one from the other during diagnostic procedures when the element δ_{ij} is verified using a predesigned variation of the twist δ_t and can be compensated using a single compensated value. Two simple examples are considered here.

(a) Jacobian matrix of the 2-DOF manipulator

- The twist $\boldsymbol{\delta}$ composing the three total errors δ_{xt}, δ_{yt}, and $\delta_{\varphi t}$ describing deviations of the end-effector relative to the base coordinate frame,

$$\boldsymbol{\delta}_t = [\delta_{xt}, \delta_{yt}, \delta_{\varphi t}]^{\mathrm{T}} \tag{10.154}$$

- The vector δ_s composing 15 elementary errors δ_{xi}, δ_{yi}, and $\delta_{\varphi i}(i = 0, 1, 2, 3, 4)$ presenting the deviations of the links with respect to their own coordinate frames,

$$\boldsymbol{\delta}_s = [\delta_{x0}, \delta_{y0}, \delta_{\varphi 0}, \delta_{x1}, \delta_{y1}, \delta_{\varphi 1}, \delta_{x2}, \delta_{y2}, \delta_{\varphi 2}, \delta_{x3}, \delta_{y3}, \delta_{\varphi 3}, \delta_{x4}, \delta_{y4}, \delta_{\varphi 4}]^{\mathrm{T}} \tag{10.155}$$

The error vectors $\boldsymbol{\delta}_t$ and $\boldsymbol{\delta}_s$ are connected by the 3 × 15 total Jacobian matrix $\mathbf{J}_t$,

$$\boldsymbol{\delta}_t = \mathbf{J}\boldsymbol{\delta}_s \tag{10.156}$$

$$\mathbf{J}_t = \frac{\partial \boldsymbol{\delta}_t}{\partial \boldsymbol{\delta}_s} = \begin{bmatrix} 1 & 0 & 0 & c\varphi_1 & -s\varphi_1 & 0 & c\varphi_1 & -s\varphi_1 & a_2 s\varphi_1 & c\varphi_{1,3} & -s\varphi_{1,3} & a_2 s\varphi_1 & c\varphi_{1,3} & -s\varphi_{1,3} & a_2 s\varphi_1 + a_4 s\varphi_{1,3} \\ 0 & 1 & 0 & s\varphi_1 & c\varphi_1 & 0 & s\varphi_1 & c\varphi_1 & -a_2 c\varphi_1 & s\varphi_{1,3} & c\varphi_{1,3} & -a_2 c\varphi_1 & s\varphi_{1,3} & c\varphi_{1,3} & -a_2 c\varphi_1 - a_4 c\varphi_{1,3} \\ 0 & 0 & 1 & 0 & 0 & 1 & 0 & 0 & 1 & 0 & 0 & 1 & 0 & 0 & 1 \end{bmatrix} \tag{10.157}$$

where symbols $c\varphi$ and $s\varphi$ are used as instead of cos φ and sin φ, respectively,

$$c\varphi_1 = \cos\varphi_1;\ c\varphi_{1,3} = \cos(\varphi_1 + \varphi_2);\ s\varphi_1 = \sin\varphi_1;\ \text{and}\ s\varphi_{1,3} = \sin(\varphi_1 + \varphi_3) \tag{10.158}$$

Since some of the transfer factors in Jacobian $\mathbf{J}_t$, Eq. (10.157), are identical, the order of the vector $\boldsymbol{\delta}_s$, Eq. (10.155), and the Jacobian can be reduced

$$\mathbf{J}_t = \begin{bmatrix} 1 & 0 & 0 & c\varphi_1 & -s\varphi_1 & a_2 s\varphi_1 & c\varphi_{1,3} & -s\varphi_{1,3} & a_2 s\varphi_1 + a_4 s\varphi_{1,3} \\ 0 & 1 & 0 & s\varphi_1 & c\varphi_1 & -a_2 c\varphi_1 & s\varphi_{1,3} & c\varphi_{1,3} & -a_2 c\varphi_1 - a_4 c\varphi_{1,3} \\ 0 & 0 & 1 & 0 & 0 & 1 & 0 & 0 & 1 \end{bmatrix} \quad (10.159)$$

$$\boldsymbol{\delta}_t = [\delta_{x0}, \delta_{y0}, \delta_{\varphi 0,1}, \delta_{x1,2}, \delta_{y1,2}, \delta_{\varphi 2,3}, \delta_{x3,4}, \delta_{y3,4}, \delta_{\varphi 4}]^{\mathrm{T}} \quad (10.160)$$

where $\delta_{\varphi 0,1} = \delta_{\varphi 0} + \delta_{\varphi 1}$; $\delta_{x1,2} = \delta_{x1} + \delta_{x2}$, etc.

The Mathematica® code for synthesis of Jacobian, Eq. (10.157), is given in Appendix *A2.2.6.*

(b) Jacobian of the lathe

The error budget of the lathe, Eqs. (10.64–69), can be represented in the Jacobian form. Since the number of the FSS links is $n = 3$, the twist and total error vector compose 6 and 24 components, respectively, connected through the 6×24 total Jacobian matrix

$$\boldsymbol{\delta} = \mathbf{J}_t \boldsymbol{\delta}_s \quad (10.161)$$

with

$$\boldsymbol{\delta} = [\delta_x, \delta_y, \delta_z, \delta_\theta, \delta_\psi, \delta_\varphi]^{\mathrm{T}} \quad (10.162)$$

$$\boldsymbol{\delta}_s = [\delta_{x0}, \delta_{y0}, \delta_{z0}, \delta_{\theta 0}, \delta_{\psi 0}, \delta_{\varphi 0}, \delta_{x1}, \delta_{y1}, \ldots, \delta_{\theta 4}, \delta_{\psi 4}, \delta_{\varphi 4}]^{\mathrm{T}} \quad (10.163)$$

$$\mathbf{J}_t = \frac{\partial \boldsymbol{\delta}}{\partial \boldsymbol{\delta}_s} = \frac{\partial(\delta_x, \delta_y, \ldots, \delta_\varphi)}{\partial(\delta_1, \delta_2, \ldots, \delta_{24})} = \begin{bmatrix} d_{11} & d_{12} & d_{13} & \ldots\ldots & d_{1,23} & d_{1,24} \\ d_{21} & d_{22} & d_{23} & \ldots\ldots & d_{2,23} & d_{2,24} \\ \ldots\ldots & \ldots\ldots & \ldots\ldots & \ldots\ldots & \ldots\ldots & \ldots\ldots \\ d_{61} & d_{62} & d_{63} & \ldots\ldots & d_{6,23} & d_{6,24} \end{bmatrix} =$$

$$\left[\mathbf{I}_{6\times 6} \quad \begin{matrix} c\varphi & -s\varphi & 0 & 0 & 0 & 0 & c\varphi & -s\varphi & 0 & -z\,s\varphi & -z\,c\varphi & 0 & c\varphi & -s\varphi & 0 & -z\,s\varphi & -z\,c\varphi & x\,s\varphi \\ s\varphi & c\varphi & 0 & 0 & 0 & 0 & s\varphi & c\varphi & 0 & z\,c\varphi & -z\,s\varphi & 0 & s\varphi & c\varphi & 0 & z\,c\varphi & -z\,s\varphi & -x\,c\varphi \\ 0 & 0 & 1 & 0 & 0 & 0 & 0 & 0 & 1 & 0 & 0 & 0 & 0 & 0 & 1 & 0 & x & 0 \\ 0 & 0 & 0 & c\varphi & -s\varphi & 0 & 0 & 0 & 0 & c\varphi & -s\varphi & 0 & 0 & 0 & 0 & c\varphi & -s\varphi & 0 \\ 0 & 0 & 0 & s\varphi & c\varphi & 0 & 0 & 0 & 0 & s\varphi & c\varphi & 0 & 0 & 0 & 0 & s\varphi & c\varphi & 0 \\ 0 & 0 & 0 & 0 & 0 & 1 & 0 & 0 & 0 & 0 & 0 & 1 & 0 & 0 & 0 & 0 & 0 & 1 \end{matrix} \right] \quad (10.164)$$

Table 10-4 Results of measurements.

Value	Point number i											
	1	2	3	4	5	6	7	8	9	10	11	12
φ_i°	0	60	120	180	240	300	0	60	120	180	240	300
z_i, mm	0	0	0	0	0	0	100	100	100	100	100	100
Δ_i, mm	0.010	0.020	0.020	0.030	0.020	0.010	0.020	0.030	0.030	0.030	0.020	0.020

where $\mathbf{I}_{6\times6}$ is the six columns composing the 6×6 identity matrix; and $c\varphi$ and $s\varphi$ are defined through Eqs. (10.158).

10.5 Diagnostics procedure: methodological example

Since the error budget directly links elementary (input) errors of machines and robots with accuracy parameters (output errors) of the FSS target (processed) element (*TA*), it can be applied for diagnostics of the sources of machine errors based on results of the *TA* measurements. As an example, consider diagnostics procedure, in which the input errors of the lathe are evaluated using the results of measurements of a cylindrical mandrel (Fig. 9.25) fixed in the main spindle. The mandrel is measured in two cross-sections at six points in each section in 60 degree interval (Table 10.4).

The normal error Δ_0, Eq. (10.90), at each point of the cylinder composes six total errors:

$$\Delta_0 = \delta_x \cos\varphi + \delta_y \sin\varphi - \delta_\theta z \sin\varphi + \delta_\psi z \cos\varphi \tag{10.165}$$

where φ, z are the Gaussian coordinates of the machined surface: the polar angle φ identical with the angle of spindle rotation and the coordinate z is identical with the distance of the longitudinal carriage from the origin of the basic coordinate frame.

Substitution of expressions for δ_x, δ_y, δ_θ, and δ_ψ from Eqs. (10.64), (10.65) (10.67), and (10.68), respectively, yields

$$\Delta_0 = \cos\varphi\, \delta_{x0} + \sin\varphi\, \delta_{y0} - (z \sin\varphi)\, \delta_{\theta0} + (z\cos\varphi)\, \delta_{\psi0} + z\, \delta_{\psi1} + \delta_{x1} + \delta_{x2} + \delta_{x3} \tag{10.166}$$

For simplicity assume that all errors are independent of z and φ. Under this assumption, coefficients for values δ_{x1}, δ_{x2}, and δ_{x3} are identical, and the number of linear independent errors in system of simultaneous equations (8.118) is reduced from eight to $m = 6$:

$$\delta_{q1} = \delta_{x0};\ \delta_{q2} = \delta_{y0};\ \delta_{q3} = \delta_{\theta0};\ \delta_{q4} = \delta_{\psi0};\ \delta_{q5} = \delta_{\psi1};\ \delta_{q6} = \delta_{x1} + \delta_{x2} + \delta_{x3} \tag{10.167}$$

$$\Delta_0(\varphi_i, z_i) = \cos\varphi_i\, \delta_{q1} + \sin\varphi_i \delta_{q2} - (z_i \sin\varphi_i)\, \delta_{q3} + (z_i \cos\varphi_i)\, \delta_{q4} + z_i\, \delta_{q5} + \delta_{q6} \tag{10.168}$$

$i = 1, 2, \ldots, 12$

Taking into account the numerical data from Table 10.4, the 12 × 6 structural matrix $\mathbf{J}_s$, Eq. (8.119), becomes

$$\mathbf{J}_s = \begin{bmatrix} \cos\varphi_1 & \sin\varphi_1 & -z_1 \sin\varphi_1 & z_1 \cos\varphi_1 & z_1 & 1 \\ \cos\varphi_2 & \sin\varphi_2 & -z_2 \sin\varphi_2 & z_2 \cos\varphi_2 & z_2 & 1 \\ \ldots & \ldots & \ldots & \ldots & \ldots & \ldots \\ \cos\varphi_{12} & \sin\varphi_{12} & -z_{12} \sin\varphi_{12} & z_{12} \cos\varphi_{12} & z_{12} & 1 \end{bmatrix} =$$

$$\begin{bmatrix} \cos 0 & \sin 0 & 0 & 0 & 0 & 1 \\ \cos(\pi/3) & \sin(\pi/3) & 0 & 0 & 0 & 1 \\ \dots & \dots & \dots & \dots & \dots & \dots \\ \cos(5\pi/3) & \sin(5\pi/3) & -100\sin(5\pi/3) & 100\cos(5\pi/3) & 100 & 1 \end{bmatrix} \tag{10.169}$$

The matrix $\mathbf{J}_s$ has full rank,

$$\text{rank}\,[\mathbf{J}_s] = 6 \tag{10.170}$$

For example, the determinant of the 6 × 6 minor comprising 1st, 2nd, 3rd, 7th, 8th, and 9th rows of matrix $\mathbf{J}_s$ is

$$\text{Det}\begin{bmatrix} 1 & 0 & 0 & 0 & 0 & 1 \\ 1/2 & \sqrt{3}/2 & 0 & 0 & 0 & 1 \\ -1/2 & \sqrt{3}/2 & 0 & 0 & 0 & 1 \\ 1 & 0 & 0 & 100 & 100 & 1 \\ 1/2 & \sqrt{3}/2 & -50\sqrt{3} & 50 & 100 & 1 \\ -1/2 & \sqrt{3}/2 & -50\sqrt{3} & 50\sqrt{3} & 100 & 1 \end{bmatrix} = 375000(\sqrt{3}-1) \neq 0$$

Matrix form of simultaneous equations (10.168) becomes

$$\mathbf{\Delta} = \mathbf{J}_s\boldsymbol{\delta} \tag{10.171}$$

$$\mathbf{\Delta} = [\Delta_1, \Delta_2, \dots, \Delta_{12}]^T = [0.01, 0.02, 0.02, 0.03, 0.02, 0.01, 0.02, 0.03, 0.03, 0.03, 0.02, 0.02]^T \tag{10.172}$$

$$\boldsymbol{\delta} = [\delta_{q1}, \delta_{q2}, \dots, \delta_{q6}]^T \tag{10.173}$$

Since according to Eq. (10.170) rank$[\mathbf{J}_s]$= 6, the system (10.170) can be solved with respect to six unknowns δ_{qi} (i = 1, 2,...,6) in the form of the best estimate $\hat{\delta}_{qi}$, Eq. (8.124). In the assumption that the values $\delta_{q1}, \delta_{q2}, \dots, \delta_{q6}$ are non-correlated and known with similar precision, we can apply Eq. (8.126)

$$\hat{\boldsymbol{\delta}} = (\mathbf{J}_s^T\mathbf{J}_s)^{-1}\mathbf{J}_s^T\mathbf{\Delta} \tag{10.174}$$

Solution of this equation is (see Mathematica-based numerical procedure, Appendix *A2.2.7*)

$\hat{\delta}_{q1} = \delta_{x0} = -0.00576$ mm; $\hat{\delta}_{q2} = \delta_{y0} = 0.00289$ mm; $\hat{\delta}_{q3} = \delta_{\theta 0} = -0.0000311$ rad.; $\hat{\delta}_{q4} = \delta_{\psi 0} = -0.00000277$ rad.; $\hat{\delta}_{q5} = \delta_{\psi 1} = 0.0000667$; $\hat{\delta}_{q6} = \delta_{x1} + \delta_{x2} + \delta_{x3} = 0.0183$ mm.

Introducing the machine corrections whose values are opposite to the obtained estimates allows compensation of the error of the machine tool. The compensation problem is not often solved uniquely. For example, compensation for the error $\hat{\delta}_{q6} = 0.021$ mm can be realized either by moving any of the systems S_1, S_2, S_3 along the X-axis by the value −0.021 mm, or can be realized by any combination of these displacements that give in summing the value −0.021 mm. The simplest way is, understandably, to shift the cutter, i.e., to shift system S_3.

References

[1] Schultschik, R., 1977, The components of the volumetric accuracy, CIRP Annals - Manufacturing Technology, Vol. 25/1, pp. 223-228.

[2] Burdekin, M.S., 1980, Cutting tests for accuracy assessment, in: Hocken, R.J., Technology of Machine Tools, Volume 5: Machine Tool Accuracy, Lawrence Livermore Laboratory, Livermore, CA, pp. 9.10-13.

[3] Renishaw QC20-W ballbar testing for CNC machines, International websites, Renishaw plc., 2017.

[4] Portman, V.T., Sakharova, O.P., and Raybach, Y.S., 1979, Investigation of Position Accuracy of Contour-Grinding NC-Machine, *Stanki i Instrument*, No. 6, pp. 4-6 (in Russian).

[5] Taylor Hobson® official site, 2017, http://www.taylor-hobson.com/products/roundness-form.html

11. Infinitesimal mechanics of the PKMs

11.1 Position accuracy

11.1.1 Position error formulation: Jacobian matrix

Consideration of the position error problem uses the models entered for solution of inverse kinematics for parallel-kinematics machines. Obviously, an actual motion of the moving platform of the l_i-DOF mechanism ($l_i = 2 \div 6$) is accompanied with the same sextet of position-orientation errors δ_{xi}, δ_{yi}, δ_{zi}, $\delta_{\theta i}$, $\delta_{\psi i}$, $\delta_{\varphi i}$, composing the twist $\boldsymbol{\delta}$ for 1-DOF body. Since the set of drives of the moving platform is simultaneously served as a movement issue and as a suspension system, the moving platform can be modelled as a rigid body on six rigid supports whose supports are the limbs. Hence, the solution obtained for the guideways accuracy (Subsection *9.1.2.3*) can be applied to the position accuracy problem of the moving platform of the parallel mechanism taking into account that:

- The position and orientation of the supports is various and defined by instantaneous position and orientation of the limbs.
- The separation between kinematic and geometric errors of the l_i-DOF mechanism differs in comparison with the 1-DOF unit, i.e., there are l_i kinematic and $(6 - l_i)$ geometric errors. Both kinematic and geometric errors are expressed through small motion deviations of the limbs.

As an example, in the Gough-Stewart mechanism small displacements of the moving platform depend on displacements of the six limbs

$$\boldsymbol{\delta} = \mathbf{J}^{-1}\boldsymbol{\Delta} \tag{11.1}$$

$$\boldsymbol{\delta} = [\delta_x, \delta_y, \delta_z, \delta_\theta, \delta_\psi, \delta_\varphi]^{\mathrm{T}} \tag{11.2}$$

$$\boldsymbol{\Delta} = [\Delta_1, \Delta_2, \ldots, \Delta_6]^{\mathrm{T}} \tag{11.3}$$

where δ is the twist combining three small linear displacements δ_x, δ_y, δ_z and three small angular displacements δ_θ, δ_ψ, δ_φ of the moving platform; $\boldsymbol{\Delta}$ is the vector of errors $\Delta_1, \Delta_2, \ldots, \Delta_6$ of the limbs; and $\mathbf{J}$ is the 6×6 Jacobian matrix

$$\mathbf{J} = \begin{bmatrix} e_{x1} & e_{y1} & e_{z1} & n_{x1} & n_{y1} & n_{z1} \\ e_{x2} & e_{y2} & e_{z2} & n_{x2} & n_{y2} & n_{z2} \\ \cdots & \cdots & \cdots & \cdots & \cdots & \cdots \\ e_{x6} & e_{y6} & e_{z6} & n_{x6} & n_{y6} & n_{z6} \end{bmatrix} \tag{11.4}$$

$$\mathrm{Det}[\mathbf{J}] \neq 0 \tag{11.5}$$

Geometrical sense of the Jacobian matrix entities is illustrated in Figure 11.1:

- Elements e_{xi}, e_{yi}, e_{zi} ($i = 1, 2, \ldots, 6$) are the direction cosines entering into the ith unit vector $\mathbf{e}_i$ describing an instantaneous orientation of the ith limb M_iN_i,

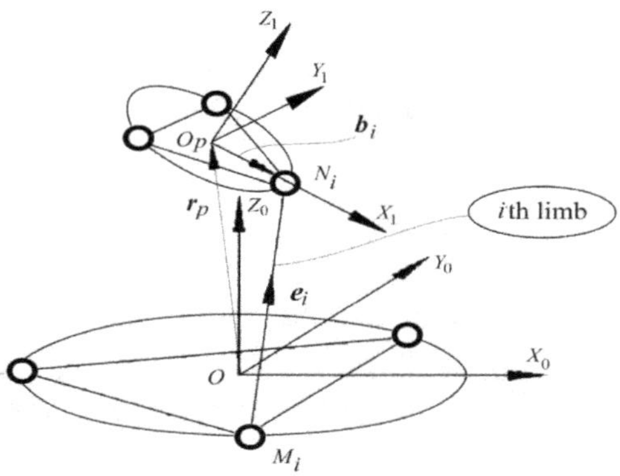

Figure 11-1 The GSP limb formalization.

$$\mathbf{e}_i = \begin{bmatrix} e_{xi} \\ e_{yi} \\ e_{zi} \end{bmatrix} = (\mathbf{r}_{Ni} - \mathbf{r}_{Mi}) / \|\mathbf{r}_{Ni} - \mathbf{r}_{Mi}\| \tag{11.6}$$

$$\|\mathbf{e}_i\| = (e_{xi})^2 + (e_{yi})^2 + (e_{zi})^2 = 1 \tag{11.7}$$

- Elements n_{xi}, n_{yi}, n_{zi} make up vector $\mathbf{n}_i$ presenting the moment of the unit vector $\mathbf{e}_i$ with respect to the centre of the moving platform O_p,

$$\mathbf{n}_i = \begin{bmatrix} n_{xi} \\ n_{yi} \\ n_{zi} \end{bmatrix} = \mathbf{b}_i \times \mathbf{e}_i = \begin{bmatrix} x_{pi} \\ y_{pi} \\ z_{pi} \end{bmatrix} \times \begin{bmatrix} e_{xi} \\ e_{yi} \\ e_{zi} \end{bmatrix} = \begin{bmatrix} y_{pi}e_{zi} - z_{pi}e_{yi} \\ z_{pi}e_{xi} - x_{pi}e_{zi} \\ x_{pi}e_{yi} - y_{pi}e_{xi} \end{bmatrix}, \tag{11.8}$$

i.e.,

$$n_{xi} = y_{pi}e_{zi} - z_{pi}e_{yi};\ n_{yi} = z_{pi}e_{xi} - x_{pi}e_{zi};\ \text{and}\ n_{zi} = x_{pi}e_{yi} - y_{pi}e_{xi} \tag{11.9}$$

where $\mathbf{b}_i$ is the position vector O_pN_i shown in Figure 11.1, $\mathbf{b}_i = \mathbf{r}_{Ni} - \mathbf{r}_{Op}$.

Note that a failure of the non-singularity condition, Eq. (11.5), results in singular configurations, which will be considered below (see Section 11.2).

11.1.2 Lower-mobility parallel structures

In modern micro- and nano-positioning stages with different DOF number, multi-DOF accelerometers and some other devices, the moving parts of the mechanical system can perform only very small displacements from an initial pose. Hence, the mechanism is associated with a nearly constant configuration (platform pose), and its Jacobian matrix is nearly constant. For example, the 6-DOF micro- and nano-positioning stage [1] presents a cube with edge 70 mm suspended on six supports (Fig. 11.2) served as the end-effector (moving platform) of a 6-DOF fine motion mechanism. The suspension system is orthogonal and fully symmetric. Each of six limbs presents the PZT actuator connecting the cube with the fixed

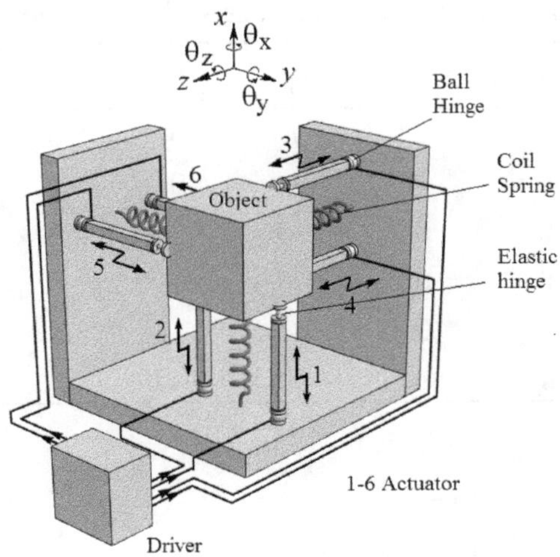

Figure 11-2 Orthogonal suspension system of the superfine motion mechanism [1].

frame through one spherical-type ball hinge and one universal-type elastic hinge. A pair of parallel limbs perpendicular to a cube face connects the face with the frame, and hence, the set of six limbs forms the 2-2-2 isotropic suspension system.

A schema of this suspension system is shown in Figure 11.3. The Cartesian coordinates x_i, y_i, z_i $(i = 1, 2,...,6)$ of the actuators' application points, unit vectors of the support directions, and moments of the unit vectors relative to central point O are listed in Table 11.1. The Jacobian matrix of the system, Eq. (11.4), is

$$\mathbf{J} = \begin{bmatrix} -1 & 0 & 0 & 0 & -a & 0 \\ -1 & 0 & 0 & 0 & a & 0 \\ 0 & -1 & 0 & 0 & 0 & -a \\ 0 & -1 & 0 & 0 & 0 & a \\ 0 & 0 & -1 & -a & 0 & 0 \\ 0 & 0 & -1 & a & 0 & 0 \end{bmatrix} \tag{11.10}$$

Vector $\mathbf{\Delta}$, Eq. (11.3), of the PZT-actuators displacements and twist δ of the end-effector pose errors, Eq. (11.2), are interrelated as $\mathbf{\Delta} = \mathbf{J}\delta$, i.e.,

$$\begin{bmatrix} \Delta_{A1} \\ \Delta_{A2} \\ \Delta_{A3} \\ \Delta_{A4} \\ \Delta_{A5} \\ \Delta_{A6} \end{bmatrix} = \begin{bmatrix} -1 & 0 & 0 & 0 & -a & 0 \\ -1 & 0 & 0 & 0 & a & 0 \\ 0 & -1 & 0 & 0 & -0 & -a \\ 0 & -1 & 0 & 0 & 0 & a \\ 0 & 0 & -1 & -a & 0 & 0 \\ 0 & 0 & -1 & a & 0 & 0 \end{bmatrix} \begin{bmatrix} \delta_x \\ \delta_y \\ \delta_z \\ \delta_\theta \\ \delta_\psi \\ \delta_\varphi \end{bmatrix} = \begin{bmatrix} -\delta_x - \delta_\psi a \\ -\delta_x + \delta_\psi a \\ -\delta_y - \delta_\psi a \\ -\delta_y + \delta_\psi a \\ -\delta_z - \delta_\theta a \\ -\delta_z + \delta_\theta a \end{bmatrix} \tag{11.11}$$

Table 11-1. Cartesian and the Plücker coordinates of the supports.

Point	x_i	y_i	z_i	e_{xi}	e_{yi}	e_{zi}	n_{xi}	n_{yi}	n_{zi}
A_1	a	0	a	-1	0	0	0	$-a$	0
A_2	a	0	$-a$	-1	0	0	0	a	0
A_3	a	a	0	0	-1	0	0	0	$-a$
A_4	$-a$	a	0	0	-1	0	0	0	a
A_5	0	a	a	0	0	-1	$-a$	0	0
A_6	0	$-a$	a	0	0	-1	a	0	0

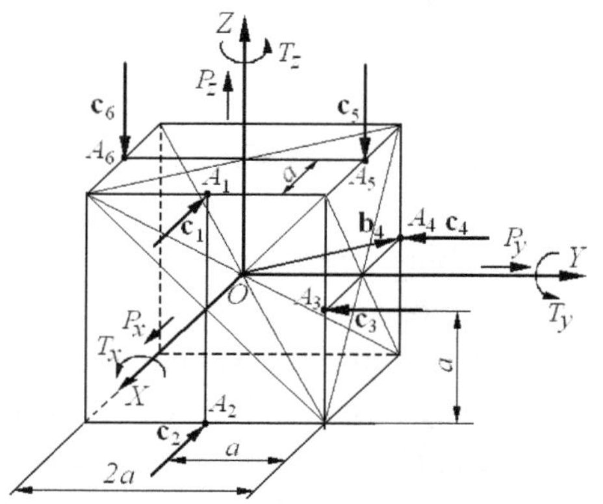

Figure 11-3 Free-body diagram of the cube on the orthogonal suspended system.

Since the determinant of matrix **J**, Eq. (11.10), is non-zero, Det[**J**] $= 8a^3 \neq 0$, the matrix is non-degenerated and must be inversed, $\delta = \mathbf{J}^{-1}\boldsymbol{\Delta}$,

$$\begin{bmatrix} \delta_x \\ \delta_y \\ \delta_z \\ \delta_\theta \\ \delta_\psi \\ \delta_\varphi \end{bmatrix} = \frac{1}{2} \begin{bmatrix} -1 & -1 & 0 & 0 & 0 & 0 \\ 0 & 0 & -1 & -1 & 0 & 0 \\ 0 & 0 & 0 & 0 & -1 & -1 \\ 0 & -1 & 0 & 0 & -1/a & 1/a \\ -1/a & 1/a & 0 & 0 & 0 & 0 \\ 0 & 0 & -1/a & 1/a & 0 & 0 \end{bmatrix} \begin{bmatrix} \Delta_{A1} \\ \Delta_{A2} \\ \Delta_{A3} \\ \Delta_{A4} \\ \Delta_{A5} \\ \Delta_{A6} \end{bmatrix} = \frac{1}{2} \begin{bmatrix} -\Delta_{A1} - \Delta_{A2} \\ -\Delta_{A3} - \Delta_{A4} \\ -\Delta_{A5} - \Delta_{A6} \\ (-\Delta_{A5} + \Delta_{A6})/a \\ (-\Delta_{A1} + \Delta_{A2})/a \\ (-\Delta_{A3} + \Delta_{A4})/a \end{bmatrix} \tag{11.12}$$

11.1.3 Gough-Stewart platform

Among non-orthogonal locating schemes, the Gough-Stewart-platform based system is popular. Here the 3-3 Gough-Stewart platform mechanism (Fig. 11.4) is considered, in which both upper and lower joints are coupled in three symmetrically located points. Universal joints M_1, M_2, M_3 and spherical joints Q_1, Q_2, Q_3 are located on the circles of radii r_1 and r_2, respectively. The joint centres make up the equilateral triangles $M_1M_2M_3$ and $Q_1Q_2Q_3$ lying in the base and platform plane, respectively. As a simple example, the positioning accuracy of the moving platform in its *central parallel-to-base pose* is analysed. As referred to the base-associated coordinate frame $OXYZ$, the centre O_p of the moving platform lies on the Z-axis, $x_p = y_p = 0$, it is oriented parallel to the base ($\theta = \psi = \varphi = 0$), and only motion along the Z-axis is available. Position vectors of points M_i and Q_i ($i = 1, 2,\ldots,6$) are established in the coordinate frame $OXYZ$ associated with the fixed frame:

$$\mathbf{r}_{M1} = \mathbf{r}_{M6} = r_1 \begin{pmatrix} 1 \\ 0 \\ 0 \end{pmatrix}; \mathbf{r}_{M2} = \mathbf{r}_{M3} = \frac{r_1}{2} \begin{pmatrix} -1 \\ \sqrt{3} \\ 0 \end{pmatrix}; \text{ and } \mathbf{r}_{M4} = \mathbf{r}_{M5} = -\frac{r_1}{2} \begin{pmatrix} 1 \\ \sqrt{3} \\ 0 \end{pmatrix} \tag{11.13}$$

$$\mathbf{r}_{Q1} = \mathbf{r}_{Q2} = \begin{pmatrix} r_2/2 \\ r_2\sqrt{3}/2 \\ z_p \end{pmatrix}; \mathbf{r}_{Q3} = \mathbf{r}_{Q4} = \begin{pmatrix} -r_2 \\ 0 \\ z_p \end{pmatrix}; \text{ and } \mathbf{r}_{Q5} = \mathbf{r}_{Q6} = \begin{pmatrix} r_2/2 \\ -r_2\sqrt{3}/2 \\ z_p \end{pmatrix} \tag{11.14}$$

where r_1 and r_2 are the radii of the joint locations on the base and the platform plane, respectively; and z_p is the height of the platform location.

The vector $\mathbf{r}_i$ of the limb M_iQ_i ($i = 1,\ldots,6$) presents a difference:

$$\mathbf{r}_i = \begin{pmatrix} x_i \\ y_i \\ z_i \end{pmatrix} = \mathbf{r}_{Qi} - \mathbf{r}_{Mi} = \begin{pmatrix} x_{Qi} - x_{Mi} \\ y_{Qi} - y_{Mi} \\ z_{Qi} - z_{Mi} \end{pmatrix}, i = 1,\ldots,6 \tag{11.15}$$

i.e.,

$$\mathbf{r}_1 = \begin{pmatrix} x_1 \\ y_1 \\ z_1 \end{pmatrix} = \begin{pmatrix} x_{Q1} - x_{M1} \\ y_{Q1} - y_{M1} \\ z_{Q1} - z_{M1} \end{pmatrix} = \begin{pmatrix} r_2/2 - r_1 \\ r_2\sqrt{3}/2 \\ z_p \end{pmatrix}, \text{ etc.} \tag{11.16}$$

Length L_i of the limb M_iQ_i ($i = 1,\ldots,6$) is the distance between the ends of the limb

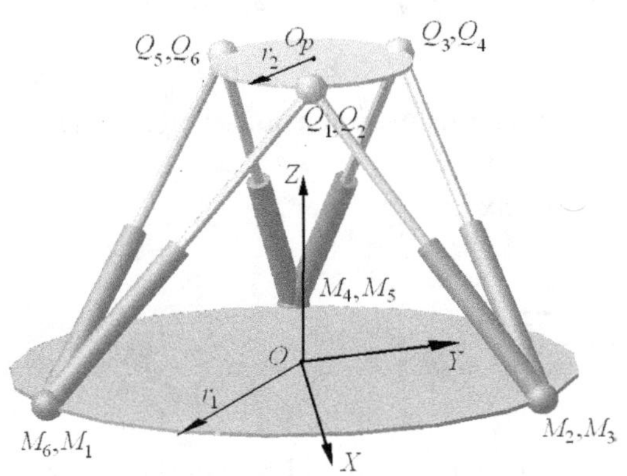

Figure 11-4 The 3-3 Gough-Stewart platform: central position of the horizontally oriented moving platform.

$$L_i = |M_iQ_i| = \|\mathbf{r}_i\| = \sqrt{x_i^2 + y_i^2 + z_i^2} \tag{11.17}$$

Because of symmetry of the configuration all the lengths are identical $L_1 = L_2 = \ldots = L_6 = L$. Thus, the length can be calculated through coordinates of the limb M_1Q_1, Eq. (11.16),

$$L = L_1 = \sqrt{x_1^2 + y_1^2 + z_1^2} = \sqrt{(r_2/2 - r_1)^2 + (-r_2\sqrt{3}/2)^2 + z_p^2} = \sqrt{r_1^2 - r_1 r_2 + r_2^2 + z_p^2} \tag{11.18}$$

The unit vector $\mathbf{c}_i$ of the orientation of the limb M_iQ_i (i = 1,...,6) and its moment (row-vector) $\mathbf{n}_i$ relative to the platform centre O_p (not origin O of the coordinate frame!) are

$$\mathbf{e}_i = \mathbf{r}_i / L_i \tag{11.19}$$

$$\mathbf{n}_i = \mathbf{b}_i \times \mathbf{e}_i \tag{11.20}$$

with

$$\mathbf{b}_i = \overline{O_pQi} = \mathbf{r}_{Qi} - \begin{pmatrix} 0 \\ 0 \\ z_p \end{pmatrix}$$

The Jacobian matrix for the configuration shown in Figure 11.4 is

$$\mathbf{J} = \begin{bmatrix} \mathbf{e}_1^T & \mathbf{n}_1^T \\ \mathbf{e}_2^T & \mathbf{n}_2^T \\ \cdots\cdots & \cdots\cdots \\ \mathbf{e}_6^T & \mathbf{n}_6^T \end{bmatrix} = \frac{r_2}{2L} \begin{bmatrix} 1-2\mu & \sqrt{3} & 2z_p/r_2 & \sqrt{3}z_p & -z_p & \sqrt{3}r_1 \\ 1+\mu & \sqrt{3}(1-\mu) & 2z_p/r_2 & \sqrt{3}z_p & -z_p & -\sqrt{3}r_1 \\ -2+\mu & -\sqrt{3}\mu & 2z_p/r_2 & 0 & 2z_p & \sqrt{3}r_1 \\ -2+\mu & \sqrt{3}\mu & 2z_p/r_2 & 0 & 2z_p & -\sqrt{3}r_1 \\ 1+\mu & -\sqrt{3}(1-\mu) & 2z_p/r_2 & -\sqrt{3}z_p & -z_p & \sqrt{3}r_1 \\ 1-2\mu & -\sqrt{3} & 2z_p/r_2 & -\sqrt{3}z_p & -z_p & -\sqrt{3}r_1 \end{bmatrix} \tag{11.21}$$

where z_p, r_1, and r_2 are the mechanism dimensions (Fig. 11.4); L is the limb length, Eq. (11.18); and $\mu = r_1/r_2$.

The determinant of this matrix is

$$\text{Det}[\mathbf{J}] = \frac{81\sqrt{3}}{4}\left(\frac{r_1 r_2 z_p}{r_1^2 + r_2^2 - r_1 r_2 + z_p^2}\right)^3 \tag{11.22}$$

The condition of the non-singularity, Eq. (11.5), is valid if $z_p \neq 0$ and z_p does not tend to infinity. In this case, matrix $\mathbf{J}$ is inversed and the twist $\boldsymbol{\delta}$, Eq. (11.2), can be calculated through the limb displacements $\mathbf{\Delta}$, Eq. (11.3), according to equation $\boldsymbol{\delta} = \mathbf{J}^{-1}\mathbf{\Delta}$. Substitution of Jacobian matrix, Eq. (11.21), yields

$$\boldsymbol{\delta} \equiv \begin{bmatrix} \delta_x \\ \delta_y \\ \delta_z \\ \delta_\theta \\ \delta_\psi \\ \delta_\varphi \end{bmatrix} = \mathbf{J}^{-1} \begin{bmatrix} \Delta_1 \\ \Delta_2 \\ \Delta_3 \\ \Delta_4 \\ \Delta_5 \\ \Delta_6 \end{bmatrix} = \frac{L}{3r_1} \begin{bmatrix} -\Delta_1 + \Delta_2 + \Delta_5 - \Delta_6 \\ \dfrac{\Delta_1 - \Delta_2 - 2(\Delta_3 - \Delta_4) + \Delta_5 - \Delta_6}{\sqrt{3}} \\ \dfrac{(\Delta_1 + \Delta_2 + \Delta_3 + \Delta_4 + \Delta_5 + \Delta_6) r_1}{2z_p} \\ \dfrac{(\Delta_1 - \Delta_6)(2r_1 - r_2) + (\Delta_2 - \Delta_5)(r_1 + r_2) + (\Delta_3 - \Delta_4)(-r_1 + 2r_2)}{\sqrt{3} z_p r_2} \\ \dfrac{-(\Delta_1 + \Delta_6) r_2 + (\Delta_2 + \Delta_5)(-r_1 + r_2) + (\Delta_3 + \Delta_4) r_1}{z_p r_2} \\ \dfrac{\Delta_1 - \Delta_2 + \Delta_3 - \Delta_4 + \Delta_5 - \Delta_6}{\sqrt{3} r_2} \end{bmatrix} \tag{11.23}$$

where $\Delta_1, \Delta_2, \ldots, \Delta_6$ are the deviations of the limbs; in the partial case, in which all displacements of the limbs are identical, $\Delta_1 = \Delta_2 = \ldots = \Delta_6 = \Delta$, we obtain

$$\boldsymbol{\delta} = \mathbf{J}^{-1} \begin{bmatrix} \Delta \\ \Delta \\ \Delta \\ \Delta \\ \Delta \\ \Delta \end{bmatrix} = \frac{L}{z_p} \begin{bmatrix} 0 \\ 0 \\ \Delta \\ 0 \\ 0 \\ 0 \end{bmatrix} \tag{11.24}$$

This result illustrates the contention that the parallel kinematics systems in some configurations can be more precise than serial ones: the non-zero deviation $\delta_z = (L/z_p)\,\Delta$ is of the order of one Δ, while the serial system can give the deviation about 6Δ in the similar case.

11.2 Singularity problem

11.2.1 General description

All configurations of the PKMs can be alternatively classified into regular and singular. In regular configurations, the DOF number of the PKM equals that calculated by the Grübler-Kutzbach equation, Eq. (5.1). In regular configurations, the Gough-Stewart platform mechanism has just six DOF, the 3RRR mechanism has three DOF, and so on. However, in singular configurations the mechanical system changes its fundamental property: a serial-type mechanism *loses* one or more DOF, while a parallel mechanism *gains* one or more DOF [2]. Here two simple cases relating to the mechanisms with minimum DOF are considered: a single-DOF serial mechanism and a two-DOF parallel mechanism.

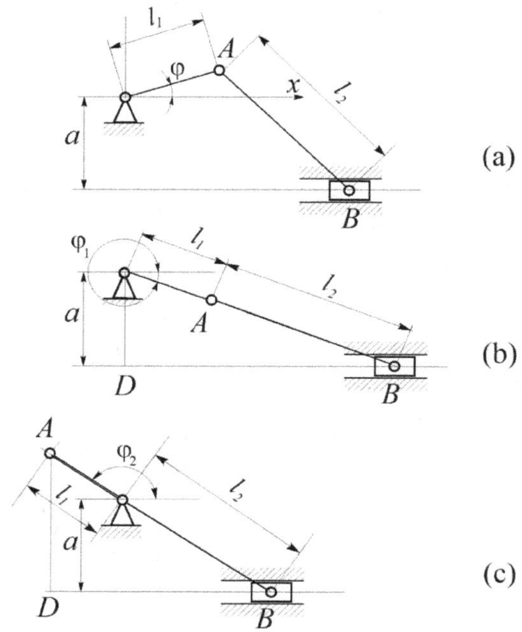

Figure 11-5 Configurations of the crank-slider mechanism: (a) regular configuration; (b) and (c) singular configurations.

An example of singular configurations of serial-type mechanisms is dead points (or toggle points) of the crank-slide mechanism (Fig. 11.5). The dead point is defined as the configuration of the mechanism in which the input link cannot be moved without assistance from the motion of other links. In regular configurations (Fig. 11.5a), the mechanism has one DOF, i.e., rotation of crank is transformed by means of connecting-rod into translational motion of slide and conversely, translation of slide causes rotation of crank However, there are two specific configurations (Fig. 11.5b and c) in which a motion of the slide (the piston) cannot induce rotation of the crank, i.e., the mechanism loses its unique DOF when the motion is actuated by piston. As is clear from Figure 11.5b, and angle φ defining the angular positions of the crank in two different singular configurations is

$$\varphi_1 = 2\pi - \arcsin\left[a/(l_1 + l_2)\right] \tag{11.25}$$

$$\varphi_2 = \pi - \arcsin\left[a/(l_1 - l_2)\right] \tag{11.26}$$

In both singular configurations, the force applied to the piston cannot move the crank and results only in tension or compression of crank and connecting rod. In the assumption that the links of the mechanism are absolutely rigid bodies, an active force applied to the piston results in zero displacement of the crank. In the stiffness-related terms, this means that the stiffness value of the mechanism in the singular configuration becomes infinite (or the compliance value becomes zero), while (Fig. 11.5a) the mechanism has finite non-zero stiffness values in all regular configurations.

In all the cases, it is necessary at the design stage to provide for the exit from all possible singular configurations or to avoid the entering into them. An example of overcoming the

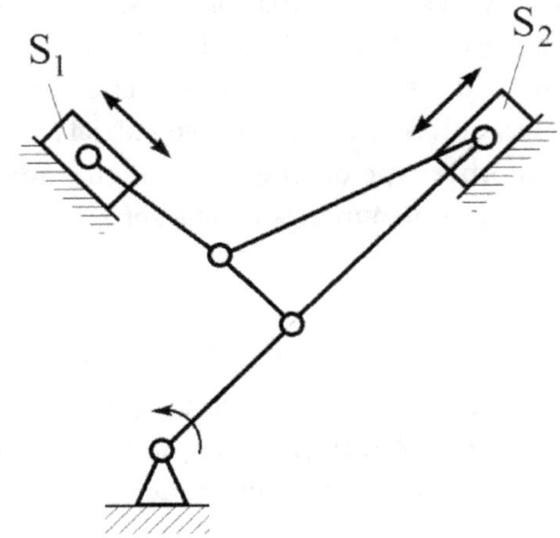

Figure 11-6 The free-dead-point two-slider mechanism.

dead point of the crank mechanism of the V-engine is shown in Figure 11.6: a pair of sliders S_1 and S_2 are located and oriented so that they cannot simultaneously fall into the singular configuration (note that slider S_2 is shown in the singular configuration).

In contrast to serial-type mechanisms, parallel kinematics mechanisms in singular configurations gain extra degree(s) of freedom, i.e., obtain the possibility to perform a local non-designed motion even if all actuated joints (all the drives) are locked [2]. Hence, singular configurations of the parallel mechanisms cannot be tolerated during their functioning, and overcoming these configurations presents one of top priorities for the mechanism design.

Generally, the structures of the parallel mechanisms in their singular configurations cannot resist or balance external forces and/or moments applied to the output link since their stiffness values in a corresponding direction vanish. For the 6-DOF mechanism, it means that there is (no less than one) sextet of infinitesimal displacements (three small translations + three small rotations), which results in no reactions from the mechanism links, and the mechanism has zero stiffness value. This can result in a general failure or permanent damage to the mechanism and surrounding equipment. Thus, the analysis of the singular configurations is very important and must be considered at the earliest step of the design procedure for parallel-kinematics machines and robots [3]. A designer must avoid entering into and even coming close to singular configurations. For comparison, in serial-type mechanisms it is sufficient to specify the way to exit from singular configurations.

There is a general quantitative indicator of singular configurations: the determinant Det(**J**) of the Jacobian matrix **J** associated with the considered configuration has zero value,

$$\mathrm{Det}(\mathbf{J}) = 0 \tag{11.27}$$

In other terms, the Jacobian matrix associated with the singular configuration is also singular. And vice versa, the configuration is *regular* (non-degenerated) if $\mathrm{Det}(\mathbf{J}) \neq 0$. Usually, using the singular configurations is unacceptable and they must be eliminated from the robot workspace, because motion of the end-effector in this configuration or in its immediate vicinity is disrupted. In other terms, non-singularity is the regular functional requirement for any configuration reached in the workspace. Search of the singular configurations by means of solution of Eq. (11.27) is a reasonably complicated problem since the determinant depends on six parameters of position and orientation of the end-effector, and the equation is high-order non-linear. Here consideration is restricted to two partial cases.

Note also that there is another type of singular configurations of the PKMs, for which determinant of the inverse-to-Jacobian matrix equals zero,

$$\mathrm{Det}(\mathbf{J}^{-1}) = 0 \tag{11.28}$$

The singularity of type (11.28) is usually less dangerous than that of type (11.27) since the stiffness value in configurations associated with Eq. (11.28) tends to infinity, and it is sufficient to design the way to come out from these configurations, while the configurations associated with Eq. (11.27) must be, as a rule, precluded.

11.2.2 Singular configurations of the GSP

11.2.2.1 Central configurations of the 3-3 GSP

The pose of the Gough-Stewart platform, in which the moving platform is located in the central position ($x_p = y_p = 0$) and oriented parallel to the base ($\theta = \psi = \varphi = 0$), characterizes the central configuration shown in Figure11.4. For the considered configuration, the Jacobian matrix, Eq. (11.21), becomes

$$\mathbf{J} = \frac{r_2}{2\sqrt{r_1^2 + r_2^2 - r_1 r_2 + z_p^2}} \begin{bmatrix} 1-2\mu & \sqrt{3} & 2z_p/r_2 & \sqrt{3}z_p & -z_p & \sqrt{3}r_1 \\ 1+\mu & \sqrt{3}(1-\mu) & 2z_p/r_2 & \sqrt{3}z_p & -z_p & -\sqrt{3}r_1 \\ -2+\mu & -\sqrt{3} & 2z_p/r_2 & 0 & 2z_p & \sqrt{3}r_1 \\ -2+\mu & \sqrt{3}\mu & 2z_p/r_2 & 0 & 2z_p & -\sqrt{3}r_1 \\ 1+\mu & -\sqrt{3}(1-\mu) & 2z_p/r_2 & -\sqrt{3}z_p & -z_p & \sqrt{3}r_1 \\ 1-2\mu & -\sqrt{3} & 2z_p/r_2 & -\sqrt{3}z_p & -z_p & -\sqrt{3}r_1 \end{bmatrix} \tag{11.29}$$

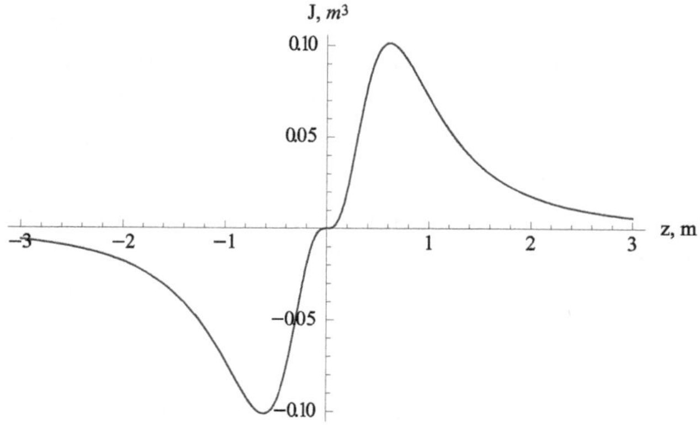

Figure 11-7 The determinant of the Jacobian matrix (in m³) associated with the central configuration of the 3-3 GSP vs. distance z_p (in m).

where $z_p = h$ is the distance between the centre O_p of the moving platform and base centre; r_1 and r_2 are the radii of the base and moving platform, respectively (Fig. 11.4); and $\mu = r_1/r_2$.

The determinant value Eq. (11.22), for this configuration with $r_1 = 0.7$ m $r_2 = 0.25$ m is shown in Figure 11.7.

As has been found, two values of variable z_p define singular configurations. In Figure 11.8, a singular configuration with $z_p = 0$ is shown and a regular configuration (in red) is added for comparison. Substitution of $z_p = 0$ in Eq. (11.29) yields:

$$\mathbf{J} = \frac{1}{2L}\begin{pmatrix} -r_1+2r_2 & \sqrt{3}r_1 & 0 & 0 & 0 & \sqrt{3}r_1r_2 \\ -r_1+2r_2 & \sqrt{3}r_1 & 0 & 0 & 0 & -\sqrt{3}r_1r_2 \\ -r_1-r_2 & \sqrt{3}(-r_1+r_2) & 0 & 0 & 0 & \sqrt{3}r_1r_2 \\ 2r_1-r_2 & \sqrt{3}r_2 & 0 & 0 & 0 & -\sqrt{3}r_1r_2 \\ 2r_1-r_2 & -\sqrt{3}r_2 & 0 & 0 & 0 & \sqrt{3}r_1r_2 \\ -r_1-r_2 & \sqrt{3}(r_1-r_2) & 0 & 0 & 0 & -\sqrt{3}r_1r_2 \end{pmatrix} \quad (11.30)$$

The three zero columns in the Jacobian matrix correspond to three infinitesimal motions relating to three non-designed DOF, and locations of these columns correspond to the axes of the small additional DOF: the third column corresponds to translation δ_z along the Z-axis, and the fourth and fifth columns correspond to rotations δ_θ and δ_ψ around the X- and Y-axis, respectively. Obviously, this configuration has only methodological interest since the working area of the upper platform in regular machine and robot structures usually corresponds to $z_p >> 0$ or $z_p << 0$. However, this singularity case explains the non-trivial fact that in the majority of actual structures the workspace of the platform-type mechanism lies on one side of the basic plane to eliminate the transfer through the singular configuration.

In configuration with $z_p \to \infty$, the six limbs approach to become parallel with one another (Fig. 11.9). In this case, the first, second, and sixth columns of the Jacobian matrix, Eq. (11.29), become zero columns, i.e., the mechanism gains three non-designed DOFs: translations along the X- and Y-axis and rotation around the Z-axis.

Equation(11.23) allows consideration of the variation of the moving platform position depending on different errors of limb lengths as the platform approaches the singular

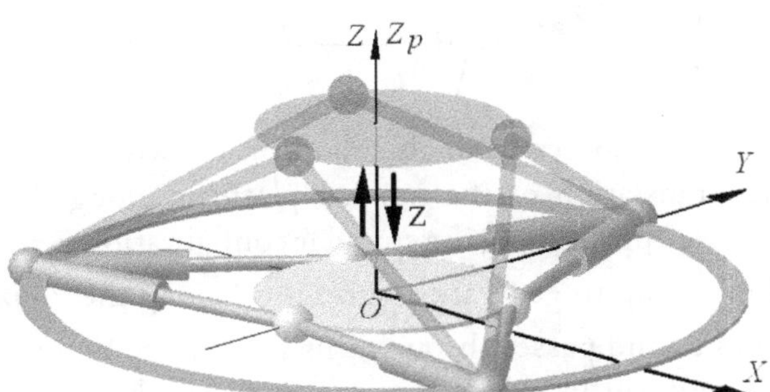

Figure 11-8 The 3-3 GSP in its central pose with moving platform parallel to the base: a regular configuration (in red) and the singular configuration with $z_p = 0$.

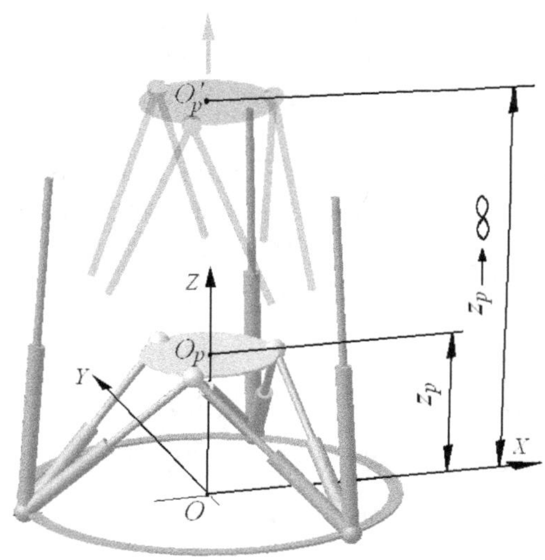

Figure 11-9 The 3-3 GSP in its central configuration with moving platform parallel to the base: the singular configuration with $z_p \to \infty$.

configuration. The numerical example below shows this fact as applied to the GSP in the central position of the moving platform.

Numerical Example 11.1.

Given: The 3-3 Gough-Stewart pose: $x_p = y_p = 0$ and $\theta = \psi = \varphi = 0$ (Fig. 11.4). Base radius $r_1 = 0.7$ m and moving platform radius $r_2 = 0.25$ m. **Required:** Calculation of the platform position while approaching the singular configurations when the limb length errors are as follows: (a) all limb errors are uniform in size, $\Delta_1 = \Delta_2 = \ldots = \Delta_6 = 1$; and (b) only error Δ_1 of the first limb takes place, $\Delta_1 = 1$ and $\Delta_2 = \ldots = \Delta_6 = 0$.

Solution:

(a) $\Delta_1 = \Delta_2 = \ldots = \Delta_6 = 1$. Substitution of expression for limb length L, Eq. (11.18), in Eq. (11.24) gives $\delta_x = \delta_y = \delta_\theta = \delta_\psi = \delta_\varphi = 0$ and

$$\delta_z = \frac{L}{z_p}\Delta = \frac{\sqrt{r_1^2 - r_1 r_2 + r_2^2 + z_p^2}}{z_p}\Delta = \sqrt{1 + \frac{r_1^2 - r_1 r_2 + r_2^2}{z_p^2}}\,\Delta \tag{11.31}$$

In Figure 11.10a, the position error of the platform caused by deviations of the limb length Δ is shown vs. translation of the moving platform along the Z-axis: the error increases indefinitely while approaching the singular configuration $z_p = 0$. This is explained by the fact that the limbs stop to generate any reaction along the Z-axis when the moving platform approaches to and passes through this singular configuration. Note that the error δ_z caused by six uniform in side deviations of the limb length Δ goes down and approaches $\Delta = 1$ with increase in the moving platform height (up to the singular configuration $zp \to \infty$).

(b) If $\Delta_1 = 1$ and $\Delta_2 = \ldots = \Delta_6 = 0$, we obtain (Fig. 11.10b)

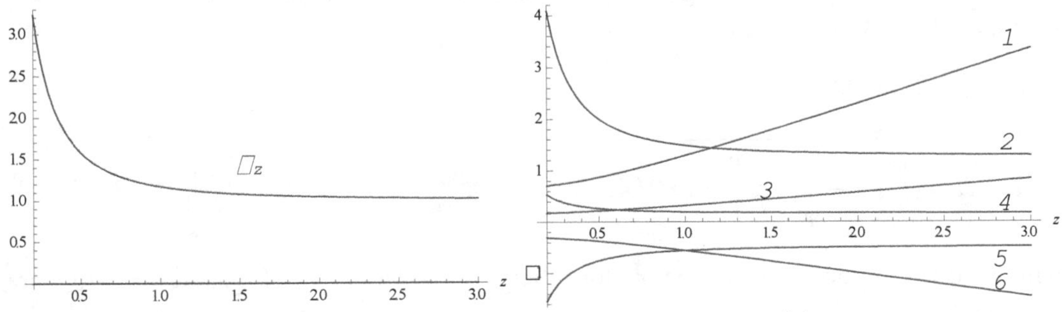

(a) (b)

Figure 11-10 The position error δ_z caused by uniform in side deviations of the limb length vs. translation of the moving platform along the Z-axis: (a) the case $\Delta_1 = \Delta_2 = \ldots = \Delta_6 = 1$; (b) the case $\Delta_1 = 1$ and $\Delta_2 = \ldots = \Delta_6 = 0$, where *1* – deviation δ_φ; *2* – deviation δ_θ; *3* – deviation δ_y; *4* – deviation δ_z; *5* – deviation δ_ψ; and *6* – deviation δ_x.

$$\boldsymbol{\delta} = \mathbf{J}^{-1}\begin{bmatrix}1\\0\\0\\0\\0\\0\end{bmatrix} = \frac{\sqrt{0.3757 + z_p^2}}{z_p}\begin{bmatrix}-0.4762z_p\\0.2749z_p\\1/6\\1.2645\\-0.4762\\1.0997z_p\end{bmatrix} \tag{11.32}$$

Hence, the errors δ_y, δ_φ, and δ_y increase indefinitely in magnitude while approaching to the singular configuration . It is clear that all considered singular configurations are associated with indefinite increasing of one or more of six components of twist $\boldsymbol{\delta}$.

11.2.2.2 Hunt-type singular configuration of the 3-3 hexapod

A singular configuration investigated by Hunt [4] takes place when two limbs and the moving platform lie in the same plane. Such a case is shown in Figure 11.11: limbs 1 and 2 lie in plane $Q_1Q_3Q_5$ of the moving platform. In this configuration, the six limbs intersect straight line Q_3Q_5: limbs 3 and 4 pass through point Q_2, limbs 5 and 6 pass through point Q_3, and continuations of limbs 1 and 2 intersect the line Q_3Q_5. Thus, a sum of the moments relative to line Q_3Q_5 of six reactions produced in the limbs equals zero, and there is no possibility to counter balance the active torque acting around the axis Q_3Q_5. Obviously the mechanism has an additional degree of freedom: an infinitesimal rotation around axis Q_3Q_5 can be carried out without any resistance whatsoever. Obviously, the same is true for a force applied at point Q_1 (Q_2) acting along the normal at the moving platform plane.

For definiteness, consider the case when the moving platform centroid O_p is located at the Z-axis, i.e., $x_p = y_p = 0$, and the angles of rotation of the platform around the X- and Z-axis are equal to zero (Fig. 11.11). In this configuration, lines O_pQ and M_1M_2 lie in the same plane. Hence, they are intersected, and their intersection point is marked Q. Angle O_pQO

equals the angle ψ of the moving platform rotation around the Y-axis. One can obtain from triangles OO_pQ and M_1Q_1Q:

$$z_p = (r_1/2)\tan\psi$$

$$l_1 = l_2 = \sqrt{\left(\frac{r_1}{2\cos\psi} - r_2\right)^2 + 3r_1^2/4}, \tag{11.33}$$

where z_p is the Z-coordinate of the platform height, $z_p = OO_p$; $l_1 = M_1Q_1$ and $l_2 = M_2Q_2$ are the limb lengths.

To prove that the Hunt-type configuration (Fig.11.11) is singular, we have to build the Jacobian matrix. Position vectors $\mathbf{r}_{M1},\ldots,\mathbf{r}_{M6}$ of the lower limb ends and vectors $\mathbf{r}_{Q1},\ldots,\mathbf{r}_{Q6}$ of the upper limb ends are obtained by Eqs. (11.29) and (11.30). Position vectors of the upper limb ends $Q_1, Q_2,\ldots,Q_6$ in the base associated coordinate frame are

$$\mathbf{r}_{Q1} = \mathbf{r}_{Q2} = \mathbf{r}_p + \mathbf{A}_{rot}\,\mathbf{b}_{Q1} = \begin{bmatrix} r_2\cos\psi \\ 0 \\ z_p - r_2\sin\psi \end{bmatrix};$$

$$\mathbf{r}_{Q3} = \mathbf{r}_{Q4} = \mathbf{r}_p + \mathbf{A}_{rot}\,\mathbf{b}_{Q3} = \begin{bmatrix} -(1/2)r_2\cos\psi \\ r_2\sqrt{3}/2 \\ z_p + (1/2)r_2\sin\psi \end{bmatrix};$$

$$\mathbf{r}_{Q5} = \mathbf{r}_{Q6} = \mathbf{r}_p + \mathbf{A}_{rot}\,\mathbf{b}_{Q5} = \begin{bmatrix} -(1/2)r_2\cos\psi \\ -r_2\sqrt{3}/2 \\ z_p + (1/2)r_2\sin\psi \end{bmatrix}, \tag{11.34}$$

where $\mathbf{r}_p$ is position vector of point O_p, $\mathbf{r}_p = [0, 0, z_p]^{\mathrm{T}}$; and $\mathbf{A}_{rot}$ is the matrix of rotation around the Y-axis,

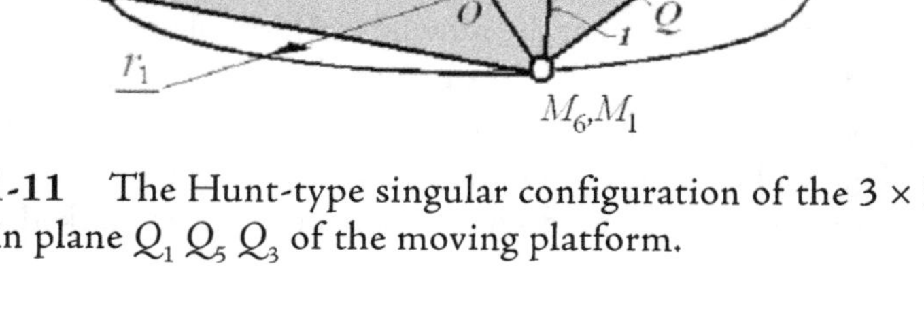

Figure 11-11 The Hunt-type singular configuration of the 3 × 3 GSP: limbs *1* and 2 lie in plane $Q_1\,Q_5\,Q_3$ of the moving platform.

$$\mathbf{A}_{rot} = \mathbf{A}^4(\psi) = \begin{bmatrix} \cos\psi & 0 & \sin\psi \\ 0 & 1 & 0 \\ -\sin\psi & 0 & \cos\psi \end{bmatrix} \tag{11.35}$$

Vector $\mathbf{r}_i$ of the *i*th limb ($i = 1,\ldots,6$) is $\mathbf{r}_i = \mathbf{r}_{Qi} - \mathbf{r}_{Mi}$, where $\mathbf{r}_{Qi}$ and $\mathbf{r}_{Mi}$ are given through Eqs. (11.34) and (11.29), respectively:

$$\mathbf{r}_1 = \mathbf{r}_{Q1} - \mathbf{r}_{M1} = \begin{bmatrix} r_2\cos\psi \\ 0 \\ z_p - r_2\sin\psi \end{bmatrix} - \frac{r_1}{2}\begin{bmatrix} 1 \\ -\sqrt{3} \\ 0 \end{bmatrix} = \begin{bmatrix} r_2\cos\psi - r_1/2 \\ r_1\sqrt{3}/2 \\ z_p - r_2\sin\psi \end{bmatrix} \tag{11.36}$$

The limb lengths l_i are defined through equation $l_i = \sqrt{x_i^2 + y_i^2 + z_i^2}$. The considered configuration has specular symmetry, and hence the limbs have identical lengths in pairs:

$$l_1 = l_2 = \sqrt{r_1^2 + r_2^2 + z_p^2 - r_1 r_2\cos\psi - 2r_2 z_p\sin\psi}$$

$$l_3 = l_6 = \sqrt{r_1^2 - 3r_1 r_2/2 + r_2^2 + z_p^2 + (1/2)r_1 r_2\cos\psi + r_2 z_p\sin\psi}$$

$$l_4 = l_5 = \sqrt{r_1^2 + r_2^2 + z_p^2 - r_1 r_2\cos\psi + 2r_2 z_p\sin\psi} \tag{11.37}$$

The unit vector of the current orientation of the limbs is

$$\mathbf{e}_i = \mathbf{r}_i/l \tag{11.38}$$

For example, unit vector $\mathbf{e}_1$ of the orientation of the limb M_1Q_1 is

$$\mathbf{e}_1 = \begin{pmatrix} e_{x1} \\ e_{y1} \\ e_{z1} \end{pmatrix} = \mathbf{r}_1/l_1 = \frac{1}{\sqrt{r_1^2 + r_2^2 + z_p^2 - r_1 r_2\cos\psi - 2r_2 z_p\sin\psi}}\begin{bmatrix} r_2\cos\psi - r_1/2 \\ r_1\sqrt{3}/2 \\ z_p - r_2\sin\psi \end{bmatrix}$$

where $\mathbf{r}_1$ is given by Eq. (11.36); and l_1 is the length of the limb No. 1, Eq. (11.37).

The moments $\mathbf{n}_i$ of the vector $\mathbf{e}_i$ with respect to the centroid O_p of the moving platform are calculated as the vector product of the arm $\mathbf{b}_i$ by the unit vector $\mathbf{e}_i$ referring to the base associated coordinate frame:

$$\mathbf{n}_i = \begin{pmatrix} n_{x1} \\ n_{y1} \\ n_{z1} \end{pmatrix} = \mathbf{b}_i \times \mathbf{e}_i = \left(\mathbf{r}_{Qi} - \mathbf{r}_p\right) \times \mathbf{e}_i \left(i = 1, 2, \ldots, 6\right) \tag{11.39}$$

For example, moment $\mathbf{n}_1$ of the unit vector $\mathbf{e}_1$ (the unit vector oriented along limb M_1Q_1) is:

$$\mathbf{n}_1 = \mathbf{b}_1 \times \mathbf{e}_1 = \begin{bmatrix} r_2 \cos\psi \\ 0 \\ r_2 \sin\psi \end{bmatrix} \times \frac{1}{l_1} \begin{bmatrix} r_2 \cos\psi - r_1/2 \\ r_1\sqrt{3}/2 \\ z_p - r_2 \sin\psi \end{bmatrix} = \frac{r_2}{2l_1} \begin{bmatrix} \sqrt{3}r_1 \sin\psi \\ r_1 \sin\psi - 2z_p \cos\psi \\ \sqrt{3}r_1 \cos\psi \end{bmatrix}$$

$$= \frac{1}{2\sqrt{r_1^2 + r_2^2 + z_p^2 - r_1 r_2 \cos\psi - 2r_2 z_p \sin\psi}} \begin{bmatrix} \sqrt{3}r_1 r_2 \sin\psi \\ r_1 r_2 \sin\psi - 2r_2 z_p \cos\psi \\ \sqrt{3}r_1 r_2 \cos\psi \end{bmatrix} \tag{11.40}$$

According to Eq. (11.4), the Jacobian matrix is

$$\mathbf{J} = \mathbf{J}\left(z_p\right) = \begin{bmatrix} e_{x1}(z_p) & e_{y1}(z_p) & e_{z1}(z_p) & n_{x1}(z_p) & n_{y1}(z_p) & n_{z1}(z_p) \\ e_{x2}(z_p) & e_{y2}(z_p) & e_{z2}(z_p) & n_{x2}(z_p) & n_{y2}(z_p) & n_{z2}(z_p) \\ \cdots & \cdots & \cdots & \cdots & \cdots & \cdots \\ e_{x6}(z_p) & e_{y6}(z_p) & e_{z6}(z_p) & n_{x6}(z_p) & n_{y6}(z_p) & n_{z6}(z_p) \end{bmatrix}, \tag{11.41}$$

where $e_{xi} = e_{xi}(z_p)$ and $n_{xi} = n_{xi}(z_p)$ $(i = 1, 2, \ldots, 6)$ are defined through Eqs. (11.38) and (11.39), respectively.

The determinant of the Jacobian matrix, Eq. (11.41), is

$$\mathrm{Det}[\mathbf{J}] = -\frac{81\sqrt{3}r_1^3 r_2^3 \cos\frac{\psi}{2}[z_p \cos\frac{\psi}{2} + (r_1 - r_2)\sin\frac{\psi}{2}](-2z_p \cos\psi + r_1 \sin\psi)(2z_p + r_2 \sin\psi)}{16 l_1^2 l_3^2 l_5^2} \tag{11.42}$$

where l_1, l_3, and l_5 are the limb lengths, Eqs. (11.37).

The singular configurations are obtained using equation Det[J] = 0 as applied to the determinant by Eq. (11.42). The solution fixes three singular configurations:

$$z_{p1} = -(r_2/2)\sin\psi \tag{11.43}$$

$$z_{p2} = (r_2 - r_1)\tan(\psi/2) \tag{11.44}$$

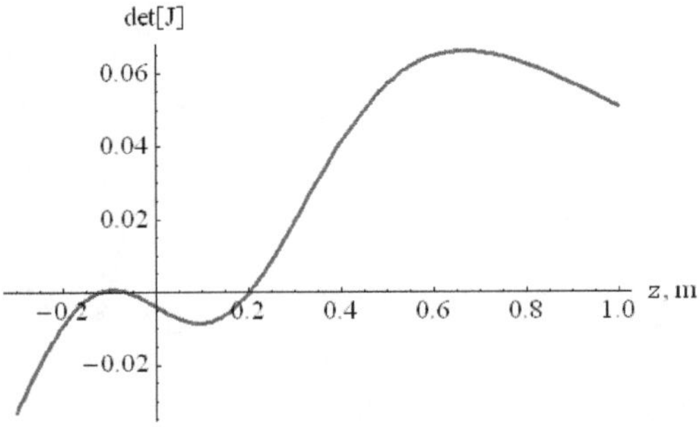

Figure 11-12 Determinant of the Jacobian matrix, Eq. (11.41), with $\psi = \pi/6$, $r_1 = 0.7$, and $r_2 = 0.25$ m.

$$z_{p3} = (r_1/2)\tan\psi \tag{11.45}$$

Solution z_{p3} is identical to that by Eq. (11.33). Solutions z_{p1} and z_{p2} (with $r_2 > r_1$) describe the singular configurations taking place when the centre O_p of the moving platform is located in the negative half-plane, $z_{p1,2} < 0$. In the *Numerical Example 11.2* below, the solution obtained from the geometrical consideration is verified by formal application of Eq. (11.27) as applied to the determinant, Eq. (11.42).

Numerical Example 11.2:
Given: Configuration of the 3 × 3 GSP is given by $x_p = y_p = 0$; $\theta = 0$; $\psi = \pi/6$; $\varphi = 0$; $r_1 = 0.7$ m; and $r_2 = 0.25$ m. **Required:** Calculation of the Jacobian and fixing the singular configurations during motion along the Z-axis.
Solution: A graph of the determinant vs. distance z_p is shown in Figure 11.12. The three values z_p, Eqs. (11.43–45) are $z_{p1} = -0.120577$; $z_{p2} = -0.0625$; and $z_{p3} = 0.202073$ m. In particular, the configuration shown in Figure 11.11 is associated with $z_p = 0.202073$ m. The corresponding Jacobian matrix is

$$\mathbf{J}(\text{for } z_p = z_{p3}) = \begin{bmatrix} -0.213416 & 0.96916 & 0.123216 & 0.121145 & 0 & 0.209829 \\ -0.213416 & -0.96916 & 0.123216 & -0.121145 & 0 & -0.209829 \\ -0.697315 & -0.593017 & 0.402595 & 0.124228 & 0 & 0.215169 \\ 0.865884 & 0.316807 & 0.387141 & 0.064018 & 0.096027 & -0.221765 \\ 0.865884 & -0.316807 & 0.387141 & -0.064018 & 0.096027 & 0.221765 \\ -0.697315 & 0.593017 & 0.402595 & -0.124228 & 0 & -0.215169 \end{bmatrix} \tag{11.46}$$

Det[**J**] = 0

■

As mentioned above, in each singular configuration, there is at least one direction along which the mechanism can obtain a spontaneous infinitesimal displacement. To establish this displacement, it is necessary to solve the eigenvalue problem for the Jacobian matrix associated with the singular configuration. For the singular conditions of the 3×3 GSP (Figure 11.11), the direction is given through the unit vector perpendicular to the plane of the moving platform, $\mathbf{p} = [\sin 30°, 0, \cos 30°]^T$. This vector lies in the XOZ plane and line Q_3Q_5 presents an instantaneous axis of the infinitesimal rotation.

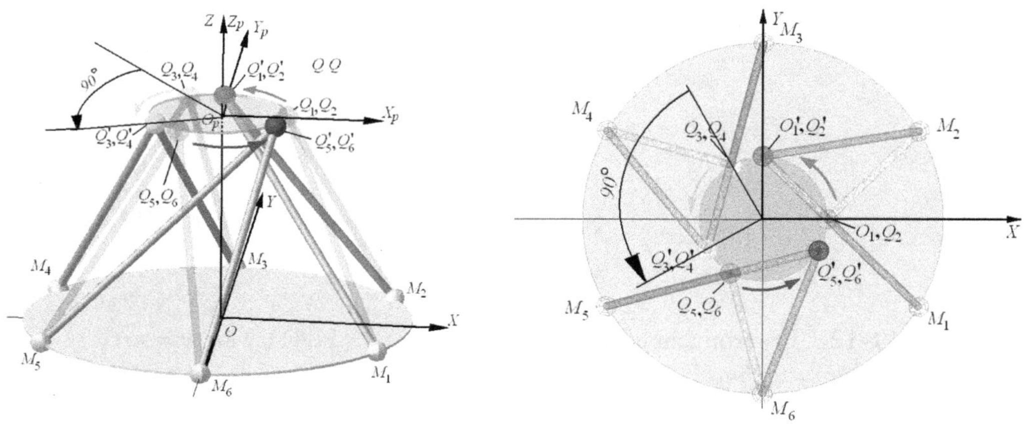

Figure 11-13 The Fichter's singular configuration of the 3-3 GSP: (a) frontal view; (b) view from above.

11.2.2.3 Fichter's singular configuration of the GSP

In Figure 11.13, the 3 × 3 GSP is shown after rotation of the moving platform at the angle $\pi/2$ around the vertical axis. In this case, the moving platform pose is $x_p = y_p = 0$; $\theta = \psi = 0$; and $\varphi = \pi/2$. After synthesis of the Jacobian matrix, we can prove that this configuration is singular as is shown by Fichter [5]. Substitution of the data inEq. (5.39) yields:

$$\mathbf{r}_i = \begin{bmatrix} x_p + r_2\cos\psi\cos(\varphi+\beta_i) - r_1\cos\alpha_i \\ y_p + r_2[\cos\theta\sin(\varphi+\beta_i) + \sin\theta\sin\psi\cos(\varphi+\beta_i)] - r_1\sin\alpha_i \\ z_p + r_2[\sin\theta\sin(\varphi+\beta_i) - \cos\theta\sin\psi\cos(\varphi+\beta_i)] \\ 0 \end{bmatrix}$$

$$= \begin{bmatrix} r_2\cos 0\cos(\pi/2+\beta_i) - r_1\cos\alpha_i \\ r_2[\cos 0\sin(\pi/2+\beta_i) + \sin 0\sin 0\cos(\pi/2+\beta_i)] - r_1\sin\alpha_i \\ z_p + r_2[\sin 0\sin(\pi/2+\beta_i) - \cos 0\sin 0\cos(\pi/2+\beta_i)] \\ 0 \end{bmatrix} = \begin{bmatrix} -r_2\sin\beta_i - r_1\cos\alpha_i \\ r_2\cos\beta_i - r_1\sin\alpha_i \\ z_p \\ 0 \end{bmatrix}$$

The current length L_i of the ith limb and the limb orts are

$$L_i = \sqrt{x_i^2 + y_i^2 + z_i^2} = \sqrt{(-r_2\sin\beta_i - r_1\cos\alpha_i)^2 + (r_2\cos\beta_i - r_1\sin\alpha_i)^2 + z_p^2}$$

$$= [r_1^2 + r_2^2 - 2r_1 r_2\sin(\beta_i - \alpha_i) + z_p^2]^{1/2} \; (i = 1,\ldots,6),$$

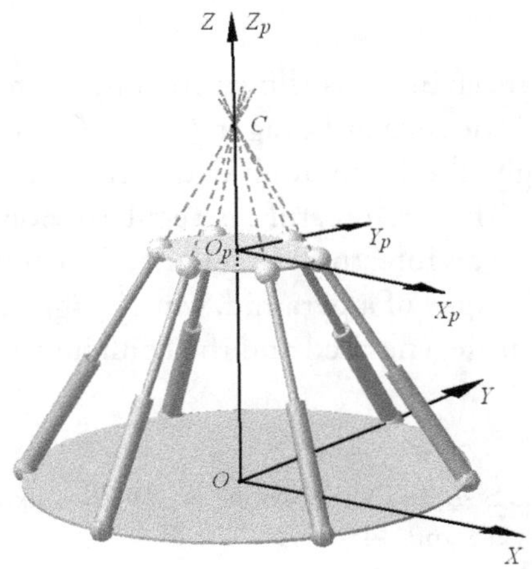

Figure 11-14 Architectural singularity

$$\mathbf{e}_i = \mathbf{r}_i / L_i = \frac{1}{L_i}\begin{bmatrix} -r_2\sin\beta_i - r_1\cos\alpha_i \\ r_2\cos\beta_i - r_1\sin\alpha_i \\ z_p \\ 0 \end{bmatrix}$$

The arm of the limb orts and moments of orts are

$$\mathbf{b}_i = r_2\begin{bmatrix} \cos\psi\cos(\varphi+\beta_i) \\ \cos\theta\sin(\varphi+\beta_i)+\sin\theta\sin\psi\cos(\varphi+\beta_i) \\ \sin\theta\sin(\varphi+\beta_i)-\cos\theta\sin\psi\cos(\varphi+\beta_i) \\ 0 \end{bmatrix} = r_2\begin{bmatrix} -\sin\beta_i \\ \cos\beta_i \\ 0 \\ 0 \end{bmatrix}$$

$$\mathbf{n}_i = \mathbf{b}_i \times \mathbf{e}_i = r_2\begin{bmatrix} -\sin\beta_i \\ \cos\beta_i \\ 0 \\ 0 \end{bmatrix} \times \frac{1}{L_i}\begin{bmatrix} -r_2\sin\beta_i - r_1\cos\alpha_i \\ r_2\cos\beta_i - r_1\sin\alpha_i \\ z_p \\ 0 \end{bmatrix} = \frac{r_2}{L_i}\begin{bmatrix} z_p\cos\beta_i \\ z_p\sin\beta_i \\ r_1\cos(\alpha_i-\beta_i) \\ 0 \end{bmatrix}$$

The Jacobian matrix for the configuration shown in Figure 11.13 is

$$\mathbf{J} = \begin{bmatrix} \mathbf{e}_1^{\mathrm{T}} & \mathbf{n}_1^{\mathrm{T}} \\ \mathbf{e}_2^{\mathrm{T}} & \mathbf{n}_2^{\mathrm{T}} \\ \cdots\cdots & \cdots\cdots \\ \mathbf{e}_6^{\mathrm{T}} & \mathbf{n}_6^{\mathrm{T}} \end{bmatrix}$$

Det[J] = 0, for all z_p, r_1, and r_2.

Mathematica® code for the Fichter's singular configuration of the 3-3 GSP is given in Appendix A2.2.8.

11.2.2.4 Architectural singularity

In Figure 11.14, a 6-6 GSP mechanism is shown the upper limb ends of the mechanism are arranged in the vertices of the regular hexagon (i.e., in 60 degree interval) on the moving platform. Correspondingly, the lower limb ends are arranged in the vertices of the regular hexagon on the base. This architecture is regularly non-used since in the central horizontal pose it has so-called architectural singularity. The limb axes extensions meet at a single point C presenting the apex of a pyramid, whose edges are limb axes.The Jacobian matrix for this configuration is degenerated and the configuration is singular.

References

[1] Henmi, N., Sato, K., Wada, S.-I., and Shimokohbe, A., 1992, A six-degree of freedom fine motion mechanism, Mechatronics, 2(5), pp. 445-457.
[2] Tsai, L.-W., 1999, Robot Analysis, John Wiley & Sons, Inc., New York, NY.
[3] Merlet, J.-P., 2006, Parallel Robots, 2nd ed., Springer, Heidelberg, 394 pp.
[4] Hunt, K.H., 1990, Kinematic Geometry of Mechanisms, Oxford Engineering Science Series, Paperback Edition, Oxford University, Oxford, UK.
[5] Fichter, E.F., 1986, A Stewart platform based manipulator: general theory and practical application, International Journal of Robotics Research, 5(2), 157-182.

12. Non-linear problems of machine accuracy

12.1 Statement of the problem

The error budgeting technique presented in Chapter 8 and developed in Chapters 9-11 uses linearization of trigonometric functions realized through the standard error matrices **E**, Eq. (8.9). However, in studies of accuracy problems, there are specific cases in which the linearization technique is unsuitable for accuracy calculations, and a non-linear problem has to be solved. There are two main groups of such situations:

Group 1: The linear component *exists* but does not meet the accuracy requirements since the terms of the higher order are non-small in comparison with the terms of the first order.

Group 2: The first approximation results in *zero error*, and the higher-order approximation is necessary to arrive at a useful answer [1-3], i.e., only higher order components give the information about an output value of the small displacements.

There is a fundamental difference between these groups. With computations in Group 1, the higher order approximation terms have to quantitatively refine the output value since the error smallness hypothesis breaks down: the deviations are not small compared with nominal dimension. This is the typical case for gear teeth and helicoids machining. In conventional cases, the refinement is very small. For example, a relative dimensional error δ for the tolerances according to the 1T7—1T9 classes of the ISO tolerancing system is $\delta \leq 0.00058 \div 0.00144$ (the ratio between the tolerance and the nominal dimension). For these classes, the second order error is a relatively small value, $\Delta^{(2)} = 0.5\ \delta^2 \leq 1.7 \cdot 10^{-7} \div 1.0 \cdot 10^{-6}$, i.e., adds about (10 ÷ 15) percent to the first approximation of the value δ. Taking into account or ignoring this addition depends on the accuracy requirements.

The situation within Group 2 differs fundamentally from the former. In this case, the first order terms are rigorously equal to zero, and a non-linear problem has to be solved. These cases are considered in this chapter. Real-world applications are largely connected with computing a form error, a form tolerancing, and a form modification.

An important form of the machining errors of this type are the so-called "layout errors" of machine tools [2]. A machine can have these errors even when it is free of geometrical and kinematic errors, i.e., the layout errors deliberately results in deviations of the form of the machined surface from the nominal form. Usually, machines and setups with application of the layout errors are designed to simplify machine kinematics, actual form of the complicated cutting tool, constraints between the parameters of motions, or to replace one type of parameter relations by another. The layout errors manifest themselves as systematic components of the total error and have non-linear dependence on their sources. Typical examples include the following.

(a) By turning with a single-point round-nosed cutting tool, the machined surface is scalloped. The magnitude and frequency depend on the feed rate and radius of the nose. The error manifests itself in the kinematic component of the surface integrity.

(b) To mill spur gears, one uses formed cutters [4]. These cutters are generally made in sets. Each such cutter is used for cutting gears with a specified range of number of teeth. For example, the No. 3 cutter from the set of the Brown and Sharpe Involute Geartooth Cutters is intended for cutting the gears with 20 different teeth numbers, from 35 to 54. Therefore, only a gear with a single teeth number is machined without systematic profile errors.

(c) Helical groove machining by conical grinding wheel results in the ZK-worm surface [5]. It may be considered as an approximate Archimedes (ZA) worm surface with systematic deviations from the nominal form. This deviation can be eliminated by using a specific profile for the grinding wheel. A similar problem was considered for grinding helical drill flutes [6].

Since the linear error budget is invalid, the non-linear problems that come into existence require a solution of transcendental equations either in a closed form or by means of higher order approximation.

As an example, an error of the tooth direction of the gear tooth profile grinded by the periphery of a flat grinding wheel is considered below. The error results from the fact that the generation of the profile in the mean cross section is a normal involute and in the other cross-sections is a shortened involute. The error value depends on the value $\delta_y^{1.5}$ where δ_y is the deflection sag of arc of the grinding wheel on the width of the gear wheel. Naturally, the error seems to be equal to zero by a first order approximation calculation, but it can be determined correctly either by an exact method or by a higher order approximation method.

Another typical example of a situation from Group 2 considered below is the grinding of surfaces of revolution of sophisticated forms (crowning conic surfaces, hyperboloids, etc.). The method of grinding, when the axes of the grinding wheel and the surface to be machined are crossed with a relatively small crossing angle λ, is known. Since small displacements of both surfaces take place in their common tangent plane, the first order approximation for calculation of the form deviation of the grounded surface, as a function of the crossing angle, results in zero and this form is a function of λ^2. A practical application of this method of grinding will be described in detail in section 12.3.3.

Similar geometry is considered in an example of turning hyperboloid on the lathe when the workpiece axis is crossed with center line of the lathe.

In the previous chapters, the actual surface has resulted from variation (small disturbances) of the nominal surface, and only linear disturbances were considered. Here, two different surfaces are given, and it is suggested that both of them are obtained from a single form-shaping system through application of two different sets of parametrical constraints. The proposed approach, named the "double-FSF method", has the goal to evaluate deviations between two FSFs. In this approach, two geometrically closely-related FSFs are synthesized using the regular procedures described in Part I, and then an equation point-to-point correspondence between two FSFs is formulated and solved in an explicit or a higher-order approximation form.

Some real-life examples are considered below. Generally, the problem is reduced to solution of a vector equation with respect to unknown distances between two surfaces defined along the normal to either surface. In the considered examples, the linearization gives zero value since one of the form-shaping motions is carried out in a plane tangent to the nominal surface, and hence, the solution of the non-linear problem is necessary.

If the non-linear equation allows only an approximate solution, there are two different ways to find the roots of the transcendental equations, and both ways significantly use specific FSF features:

- Symbolic solution is based on the fact that the linear approximation of the FSF errors results in zero value of the form deviation. This is used for formulation of the second order approximation in the explicit form allowing an effective solution procedure.
- Numerical solution using iteration procedures requires an appropriate trial solution for the unknown [7, p. 719]. A distinctive property of this method lies in the fact that the known coordinates of the nominal surface can be used as the trial solution for the vector equation mentioned above [2].

As before, cumbersome analytical transformations, numerical calculations, and their graphical presentations are worked out by means of an application of the Mathematica® software system [8]. In the considered real-life examples, the problems of the accuracy assessment and a specific setup are solved and experimentally tested.

12.2 Double-FSF approach

12.2.1 Distance between nominal and actual surfaces

Two types of surfaces are considered: the nominal surface $\mathbf{r}_0$, and the actual surface $\mathbf{r}_a$:

$$\mathbf{r}_0 = \mathbf{r}_0(u, v), \text{ with } \mathbf{r}_u \times \mathbf{r}_v \neq 0 \tag{12.1}$$

$$\mathbf{r}_a = \mathbf{r}_a(U, V), \text{ with } \mathbf{r}_U \times \mathbf{r}_V \neq 0 \tag{12.2}$$

where u, v and U, V are the Gaussian coordinates of the nominal and actual surfaces, respectively; and $\mathbf{r}_u$ and $\mathbf{r}_v$ ($\mathbf{r}_U$ and $\mathbf{r}_V$) are the partial derivatives of the position vector $\mathbf{r}_0$ ($\mathbf{r}_a$) with respect to parameters u and v (U and V), respectively.

These surfaces are closely located to one another if there is a possibility to establish a correspondence between two surfaces so that the difference

$$\Delta\mathbf{r}_0 = \mathbf{r}_a - \mathbf{r}_0 \tag{12.3}$$

be comes a small vector.

Schematics of a correspondence of this type is shown in Figure 12.1: the point of the actual surface $\mathbf{r}_a(U, V)$ and the point of the nominal surface $\mathbf{r}_0(u, v)$ present the point of intersections the surfaces with the normal to the nominal surface

$$\mathbf{N} = \mathbf{r}_u \times \mathbf{r}_v \tag{12.4}$$

The normal passes through both surfaces, and the distance between the intersection points is defined as the normal deviation $\Delta_n(u, v)$ (Δ_n is a scalar value) at a current point of the actual surface from the nominal one. The deviation is calculated as a solution of a vector equation

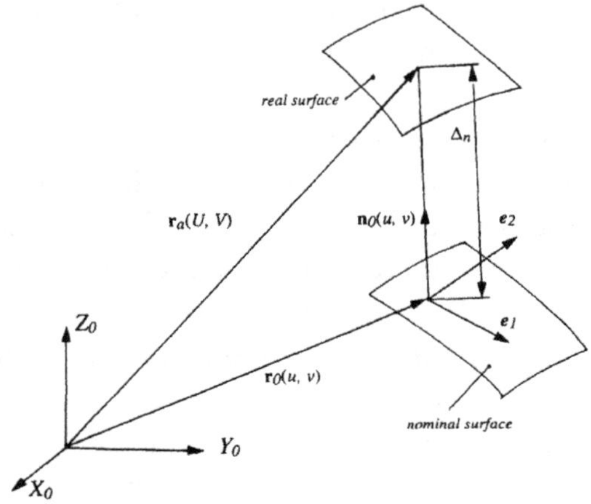

Figure 12-1 A normal distance between two surfaces.

$$\mathbf{r}_a(U, V) - \mathbf{r}_0(u, v) = \Delta_n(u, v)\, \mathbf{n}_0(u, v) \tag{12.5}$$

$$\mathbf{n} = \mathbf{N}/||\mathbf{N}|| = [\mathbf{r}_u \times \mathbf{r}_v]/||\mathbf{r}_u \times \mathbf{r}_v||$$

where $\mathbf{n}(u, v)$ is the unit vector of the normal $\mathbf{N}$ (Fig. 12.1).

Thus, the problem is to solve transcendental vector equation (12.5) in three unknowns U, V and Δ_n as functions of u, v. The *first order* approximation for Δ_n is obtained assuming that. $U = u$ and $V = v$, i.e.,

$$\Delta_n = [\mathbf{r}_a (U, V) - \mathbf{r}_0(u, v)]\, \mathbf{n}_0(u, v) \tag{12.6}$$

To solve the non-linear Eq. (12.5) in the general case, the three vectors in Eq. (12.5) are projected at three unit vectors forming a moving trihedron of the nominal surface at the point specified by the Gaussian coordinates u, v. The trihedron (Fig. 12.1) is composed of the unit vector of normal $\mathbf{n}_0(u, v)$ and two tangent unit vectors: $\mathbf{e}_1 = \mathbf{r}_u/||\mathbf{r}_u||$ and $\mathbf{e}_2 = \mathbf{r}_v/||\mathbf{r}_v||$.

Since $\mathbf{n}_0 \cdot \mathbf{n}_0 = 1$, a multiplication of both sides of Eq. (12.5) on $\mathbf{n}_0$ yields an expression for $\Delta_n(u, v)$; and since $\mathbf{n}_0 \cdot \mathbf{e}_1 = \mathbf{n}_0 \cdot \mathbf{e}_2 = 0$, multiplications on $\mathbf{e}_1$ and $\mathbf{e}_2$ yield two equations f_1 and f_2 in four Gaussian coordinates U, V, u, and v:

$$f_1(U, V, u, v) \equiv [\mathbf{r}_a(U, V) - \mathbf{r}_0(u, v)] \cdot \mathbf{e}_1(u, v) = 0 \tag{12.7}$$

$$f_2(U, V, u, v) \equiv [\mathbf{r}_a(U, V) - \mathbf{r}_0(u, v)] \cdot \mathbf{e}_2(u, v) = 0 \tag{12.8}$$

$$[\mathbf{r}_a(U, V) - \mathbf{r}_0(u, v)] \cdot \mathbf{n}_0(u, v) = \Delta_n(u, v) \tag{12.9}$$

A convenient way of solving the simultaneous Eqs. (12.7-12.9) is as follows:

(a) Since Eqs. (12.7) and (12.8) are free of unknown Δ_n, parameters U and V can be expressed through another pair of parameters u and v. Preliminarily, it is necessary to examine the fulfilment of the condition that the Jacobian of functions f_1 and f_2 with respect to u and v does not equal zero in the neighborhood of the point defined by u and v, i.e.,

$$\left|\frac{\partial(f_1, f_2)}{\partial(u,v)}\right| \neq 0$$

If this is true, according to theorem of implicit function [7, p. 101], a solution of Eqs. (12.8) and (12.9) may be represented in the form

$$U = U(u, v);\ V = V(u, v) \qquad (12.10)$$

Correspondingly, if $\left|\frac{\partial(f_1, f_2)}{\partial(U,V)}\right| \neq 0$, we can obtain an alternative expression

$$u = u(U, V);\ v = v(U, V) \qquad (12.10a)$$

(b) Then the deviations $\Delta_n(u, v)$ or $\Delta_n(U, V)$ are calculated by substitution of Eqs. (12.10) or (12.10a), respectively, in Eq. (12.9). Obviously, application of either of the alternative ways depends on the considered problem.

(c) A numerical solution of Eq. (12.10) can be obtained using the Mathematica® software system [8] using a function for solving equations

FindRoot $[\{f_1 == 0, f_2 == 0\}, \{U,u_0\}, \{V,v_0\}]$ (12.11)

or

FindRoot $[\{f_1 == 0, f_2 == 0\}, \{u,U_0\}, \{v,V_0\}]$ (12.11a)

with

$$f_1 = (\mathbf{r}_a - \mathbf{r}_0)\cdot\mathbf{e}_1$$

$$f_2 = (\mathbf{r}_a - \mathbf{r}_0)\cdot\mathbf{e}_2$$

where two first expressions determine f_1 and f_2 as functions of u, v and U, V according to Eqs. (12.7) and (12.8), respecively; **FindRoot** is the operator for solution of the set of transcendental equations; u_0 and v_0 (U_0 and V_0) are the trial solution to start the numerical solution procedure.

The key for this solution lies in the fact that the initial values u_0 and v_0 used in Eq. (12.11) are already known, since the surfaces $\mathbf{r}_a$ and $\mathbf{r}_0$ are little different from each other. Note that the preliminary preparation of Eq. (12.11), i.e., synthesis of Eqs. (12.1), (12.2), and (12.7)–(12.9) can also be carried out in the symbolic form via the Mathematica® software system. Equation (12.11) is the first step in the solution procedure when we have to apply a numerical operation. The real-life problem is considered below.

12.2.2 Layout errors of gear grinding with flat wheel

12.2.2.1 Description of the machine

To illustrate how the proposed method is employed, grinding a tooth flank on a gear grinding machine is considered. A typical setting of these machines (Fig. 12.2) is characterized by an arrangement of the wheel at a pressure angle $\alpha = 0^{\circ}$. With this setup, the working plane of the grinding wheel simulates a flank of the rectangular tooth of the gear rack, and the pitch diameter of the ground gear coincides (during machining) with the base-circle diameter $2r_b$. The goal of this example is to generate tooth flank equations for two settings of the machine and to calculate the layout errors that characterize one of these setups.

Form-shaping system links are shown in Figure 12.2 and listed in Table 12.1, where, as before, the relative motions of the links relative to the preceding ones are described.

The manipulation matrix and the FSF are

$$^{0}\mathbf{A}_4 = {}^{0}\mathbf{A}_1{}^{1}\mathbf{A}_2{}^{2}\mathbf{A}_3{}^{3}\mathbf{A}_4 = \mathbf{A}^6(\varphi_n + \varphi)\mathbf{A}^1(-x)\mathbf{A}^2(a)\mathbf{A}^4(\theta)$$

$$\mathbf{r}_0 = {}^{0}\mathbf{A}_4\mathbf{r}_4(\theta_w, R) \tag{12.12}$$

where φ_n is the dividing angle; φ is the angle of gear rotation within the gear-wheel meshing; x is the translation of the grinding head in the gear-wheel meshing; a is the axis-to-axis distance between the workpiece and grinding wheel; θ is the current angle of the wheel rotation in the cutting motion; and $\mathbf{A}^6(\varphi_n + \varphi)$, $\mathbf{A}^1(-x)$, $\mathbf{A}^2(a)$, and $\mathbf{A}^4(\theta)$ are the 4 x 4 homogeneous transformation matrices (Table 3.5),

$$^{0}\mathbf{A}_1 = \mathbf{A}^6(\varphi_n) = \begin{bmatrix} \cos(\varphi_n+\varphi) & -\sin(\varphi_n+\varphi) & 0 & 0 \\ \sin(\varphi_n+\varphi) & \cos(\varphi_n+\varphi) & 0 & 0 \\ 0 & 0 & 1 & 0 \\ 0 & 0 & 0 & 1 \end{bmatrix}; \mathbf{A}_2 = \mathbf{A}^1(-x) = \begin{bmatrix} 1 & 0 & 0 & -x \\ 0 & 1 & 0 & 0 \\ 0 & 0 & 1 & 0 \\ 0 & 0 & 0 & 1 \end{bmatrix}$$

$$^{2}\mathbf{A}_3 = \mathbf{A}^2(a) = \begin{bmatrix} 1 & 0 & 0 & 0 \\ 0 & 1 & 0 & a \\ 0 & 0 & 1 & 0 \\ 0 & 0 & 0 & 1 \end{bmatrix}; \text{ and } {}^{3}\mathbf{A}_4 = \mathbf{A}^4(\theta) = \begin{bmatrix} 1 & 0 & 0 & 0 \\ 0 & \cos\theta & -\sin\theta & 0 \\ 0 & \sin\theta & \cos\theta & 0 \\ 0 & 0 & 0 & 1 \end{bmatrix}$$

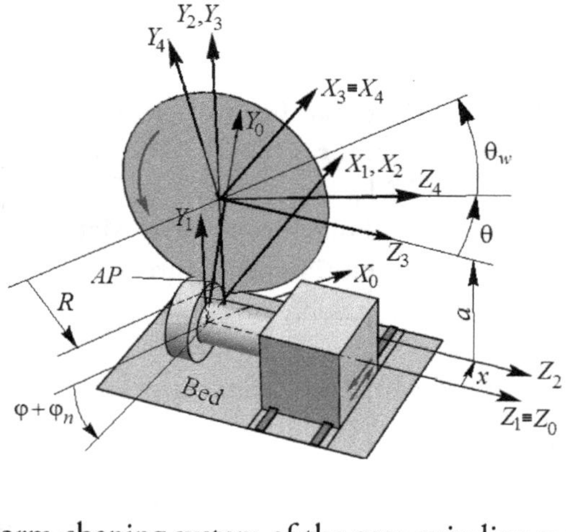

Figure 12-2 Form-shaping system of the gear-grinding machine.

Table 12-1 FSS of the gear grinding machine.

S_i	Link	Motion relative to previous link	Matrix ${}^{i\text{-}1}\mathbf{A}_i$
S_0	Workpiece spindle	Conventionally fixed link	–
S_1	Headstock	Rotation around the Z_0-axis at angle $\varphi_n + \varphi$	${}^0\mathbf{A}_1 = \mathbf{A}^6(\varphi_n + \varphi)$
S_2	Bed	Translation along the X_1-axes at distance x	${}^1\mathbf{A}_2 = \mathbf{A}^1(-x)$
S_3	Grinding head	Translation along the Y_2-axes at distance a	${}^2\mathbf{A}_3 = \mathbf{A}^2(a)$
S_4	Main spindle	Rotation around the X_3-axes at angle θ	${}^3\mathbf{A}^4 = \mathbf{A}^4(\theta)$

The position vector $\mathbf{r}_4$ entering in the FSF, Eq. (12.12), defines the current point of the grinding wheel in its own coordinate frame S_4

$$\mathbf{r}_4 = [0, -R\cos\theta_w, -R\sin\theta_w, 1]^T \tag{12.13}$$

where θ_w and R are the current polar angle and radius, respectively, of the active point (AP in Figure 12.2) of the wheel.

Substitution of Eq. (12.13) in Eq. (12.12) gives

$$\mathbf{r}_0 = \mathbf{A}^6(\varphi_n + \varphi)\,\mathbf{A}^1(-x)\,\mathbf{A}^2(a)\,\mathbf{A}^4(\theta)\,\mathbf{r}_5(\theta_w, R)$$

$$= \begin{bmatrix} -x\cos(\varphi_n+\varphi)+[R\cos(\theta+\theta_w)-a]\sin(\varphi_n+\varphi) \\ -x\sin(\varphi_n+\varphi)-[R\cos(\theta+\theta_w)-a]\cos(\varphi_n+\varphi) \\ -R\sin(\theta+\theta_w) \\ 1 \end{bmatrix} \tag{12.14}$$

In Figure12.3, the relative arrangement of grinding wheel *1* and the ground gear (the latter is represented by base cylinder *4* and tooth flanks *2, 3*) is shown for the stage of the machining operation when the portion of the involute profile of flank *2*, located near base cylinder *4*, is ground. Note that the tooth dimensions are shown on an enlarged scale compared with the dimensions of the grinding wheel and the base cylinder. The geometrical objects described through Eq. (12.14) and shown in Figure 12.3 are drawn via the Mathematica® software system [8] using commands **ParametricPlot3D** for a current relative position of the machined gear and the grinding wheel when they form the"gear-wheel" meshing.

12.2.2.2 FSF of the nominal machined surface

Six variable parameters (φ_n, φ, x, R, θ, and θ_w) entering in the FSF, Eq. (12.12), are connected by four constraints:

- the implicit constraint $\theta + \theta_w = \Theta$;
- the geometric constraint $\varphi_n = 2\pi(n-1)/z_t$, where n is the tooth number ($n = 1, 2, \ldots z_t$); z_t is the teeth number;
- the kinematic constraint (meshing relation) $x + r_b\,\varphi = 0$, where r_b is the base circle radius;
- the one-parametric enveloping constraint.

To calculate the enveloping constraint, it is necessary preliminarily to substitute the implicit, geometric, and kinematic constraints in Eq. (12.14)

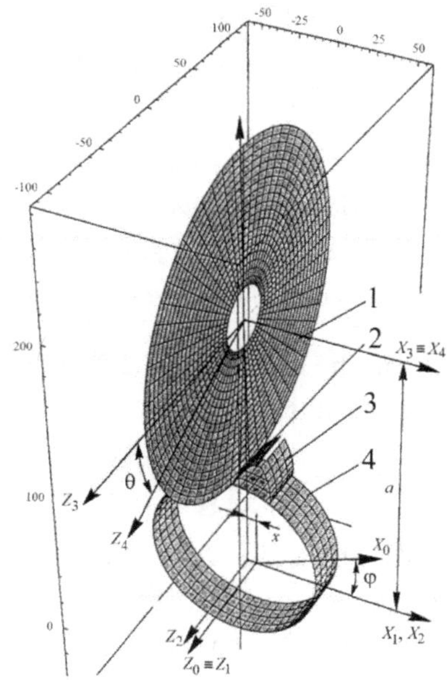

Figure 12-3 The "gear–wheel" meshing.

$$\mathbf{r}_0 = \begin{bmatrix} r_b\varphi\cos(\varphi_n+\varphi)+(R\cos\Theta-a)\sin(\varphi_n+\varphi) \\ r_b\varphi\sin(\varphi_n+\varphi)-(R\cos\Theta-a)\cos(\varphi_n+\varphi) \\ -R\sin\Theta \\ 1 \end{bmatrix}$$

The one-parametric enveloping constraint becomes

$$\left[\frac{\partial \mathbf{r}_0}{\partial \varphi}\times\frac{\partial \mathbf{r}_0}{\partial R}\right]\cdot\frac{\partial \mathbf{r}_0}{\partial \Theta} = R(-a+r_b+R\cos\Theta) = 0$$

Solution of this equation results in the enveloping constraint

$$R = (a-r_b)/\cos\Theta \tag{12.15}$$

Substitution of all constraints in Eq. (12.14) yields the equation of the machined surface:

$$\mathbf{r}_0 = \mathbf{r}_0(\varphi_n,\varphi,\Theta) = \begin{bmatrix} r_b\varphi\cos(\varphi_n+\varphi)-r_b\sin(\varphi_n+\varphi) \\ r_b\varphi\sin(\varphi_n+\varphi)+r_b\cos(\varphi_n+\varphi) \\ -(a-r_b)\tan\Theta \\ 1 \end{bmatrix} \tag{12.16}$$

Equation (12.16) presents the conventional involute cylinder forming the tooth flank [5]. Hence, it can be considered as a *nominal machined surface* of the *n*th tooth, where $u = \varphi$ and

$v = \Theta$. For simplicity, the first tooth of the gear ($n = 1$ and $\varphi_n = 0$) will be considered. The unit vectors of the moving trihedron for the nominal surface are as follows:

$$\mathbf{e}_1 = \frac{\partial \mathbf{r}_0}{\partial \varphi} \Big/ \left\| \frac{\partial \mathbf{r}_0}{\partial \varphi} \right\| = \begin{bmatrix} -\sin\varphi \\ \cos\varphi \\ 0 \\ 0 \end{bmatrix}; \mathbf{e}_2 = \frac{\partial \mathbf{r}_0}{\partial \theta} \Big/ \left\| \frac{\partial \mathbf{r}_0}{\partial \theta} \right\| = \begin{bmatrix} 0 \\ 0 \\ 1 \\ 0 \end{bmatrix}; \mathbf{n}_0 = \mathbf{e}_1 \times \mathbf{e}_2 = \begin{bmatrix} \cos\varphi \\ \sin\varphi \\ 0 \\ 0 \end{bmatrix} \tag{12.17}$$

The limit values of the parameters are: $\varphi_{min} \leq \varphi \leq \varphi_{add}$, where φ_{add} corresponds to the addendum circle of the tooth, $\varphi_{min} \geq 0$ ($\varphi_{min} = 0$, if and only if the involute of the tooth profile starts from the base circle); $-\Theta_{lim} \leq \Theta \leq \Theta_{lim}$, where Θ_{lim} corresponds to the lateral side of the machined gear,

$$\varphi_{add} = \sqrt{\left(r_a^2 - r_b^2\right)} / r_b; \Theta_{lim} = \arcsin\left(b / 2R_w\right) \tag{12.18}$$

where r_a is the radius of the addendum circle of the tooth; b is the tooth width; and R_w is the external radius of the grinding wheel.

A schematic diagram of the setup for machining the nominal surface, Eq. (12.16), for the first tooth ($\varphi_n = 0$) is shown in Figure12.4a: the edge of the grinding wheel is lowered by values $c = 1.2 \div 1.8$ mm below the pressure line NN, which is the tangent to the base circle and the normal to the tooth profile. The axis-to-axis distance between the grinding wheel and ground gear is

$$a = r_b + R_w - c \tag{12.19}$$

Substitution of Eq. (12.19) in Eq. (12.16) yields

$$\mathbf{r}_0 = \mathbf{r}_0(\varphi_n, \varphi, \Theta) = \begin{bmatrix} r_b \varphi \cos(\varphi_n + \varphi) - r_b \sin(\varphi_n + \varphi) \\ r_b \varphi \sin(\varphi_n + \varphi) + r_b \cos(\varphi_n + \varphi) \\ -(R_w - c)\tan\Theta \\ 1 \end{bmatrix}$$

In this setting, the cutting force acts on the grinding wheel in such a way that the large component F_n is directed along the normal to the wheel face. The wheel, which is very thin ($t = 4 \div 6$ mm), has a low level of stiffness in this direction, and this limits the depth of cut.

12.2.2.3 FSF of the actual machined surface

An alternative setup of the machine is shown in Figure 12.4b. In this setup, the wheel edge is not lowered below line NN, i.e., $c = 0$. The larger force component in this method is F_t, which acts in the direction of the maximum stiffness of the cutting system. This enables us to remove the allowance in one pass, but this method has a layout error, which must be estimated. According to Figure 12.4b, the grinding wheel forms the ground surface only by means of the points located on the external diameter $D_w = 2R_w$ of the wheel. Thus, instead of meshing condition Eq. (12.15) and axis-to-axis condition Eq. (12.19), we have for the setup in Figure 12.4b

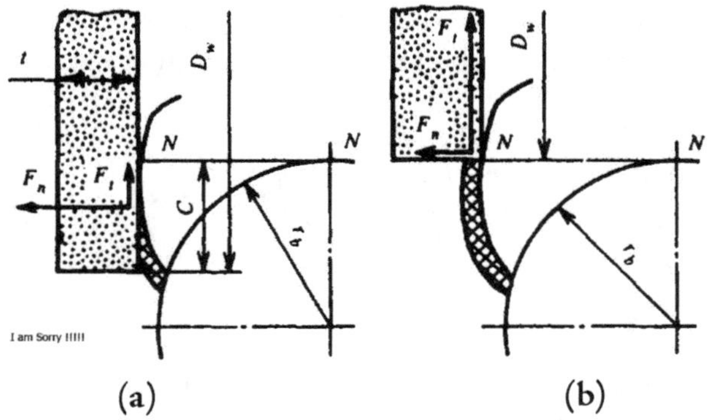

Figure 12-4 Two setups of the gear–grinding machine; the feed is directed: (a) along the pressure line NN; (b) perpendicular to the pressure line NN.

$$R = R_w \text{ and } a = r_b + R_w \tag{12.20}$$

Substitution of these conditions in the FSF, Eq. (12.14), results in the FSF for the *actual surface* ground according schematics in Figure 12.4b

$$\mathbf{r}_0 = \begin{bmatrix} r_b\varphi\cos(\varphi_n+\varphi) - g\sin(\varphi_n+\varphi) \\ r_b\varphi\sin(\varphi_n+\varphi) + g\cos(\varphi_n+\varphi) \\ -R_w\sin\Theta \\ 1 \end{bmatrix} \tag{12.21}$$

with $g = r_b + R_w(1 - \cos\Theta)$.

Equation (12.21) gives the ordinary involute with $\Theta = 0$, and the shortened involute in the other sections, $\Theta \neq 0$. As in Eq. (12.16), we will consider the first tooth of the gear (the dividing angle $\varphi_n = 0$). In this case, a distance δ_y between the external diameter of the grinding wheel and line NN (Fig. 12.4b) increases from the mean section of the ground tooth to its lateral sides according to the formula

$$\delta_y = g - r_b = R_w(1 - \cos\Theta) \approx b^2/(8R_w) \tag{12.22}$$

The limit values of the angle φ are $\varphi_0 \leq \varphi \leq \varphi_{add}$, where φ_{add} is the angle φ for the profile point on addendum circle of radius r_a; and φ_0 is shown in Figure 12.5c

$$\varphi_{add} = \sqrt{(r_a^2 - r_b^2)} / r_b;$$

$$\varphi_0 = \arccos[r_b/(r_b + \delta_y)] \tag{12.23}$$

The limit values of the angle Θ (Fig. 12.5b) are

$$-\arctan(b/2R_w) = -\Theta_{lim} \leq \Theta \leq \Theta_{lim} = \arctan(b/2R_w)$$

12.2.2.4 Accuracy problem

Two surfaces defined through Eqs. (12.16) and (12.21) can be considered as the pair given by Eqs. (12.1) and (12.2), where an ordinary involute cylinder, Eq. (12.16), is the nominal

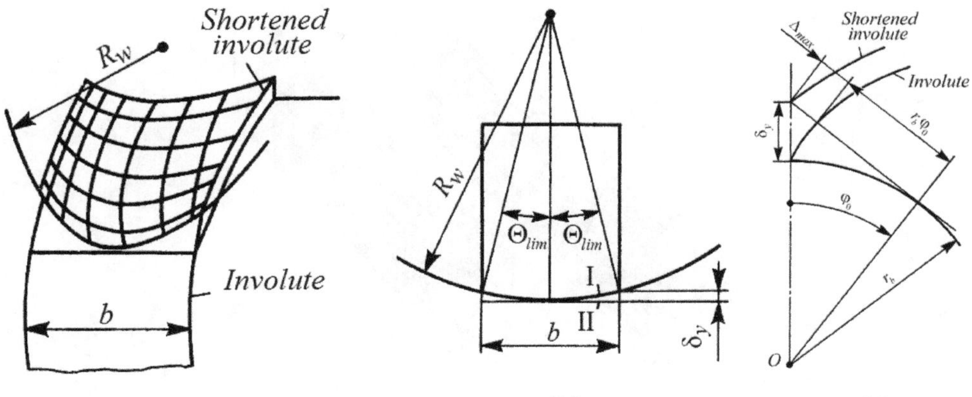

(a) (b) (c)

Figure 12-5 Surface pattern machined by setup in Figure 12.4b: (a) shortened and regular in volutes; (b) limits of the lateral tooth surface; (c) limits of the angle.

surface, and the surface given by Eq. (12.21) is the actual one. Both surfaces are shown in Figure12.6.

Substitution of vectors (12.16), (12.18), and (12.21) into Eqs. (12.7)–(12.9) gives

$$f_1 \equiv r_b + r_b\,\varphi_2 \sin(\varphi_1 - \varphi_2) - g\cos(\varphi_1 - \varphi_2) = 0 \tag{12.24}$$

$$f_2 \equiv (a - r_b)\tan\Theta_1 - R_w \sin\Theta_2 = 0 \tag{12.25}$$

$$f_3 \equiv r_b[\varphi_2 \cos(\varphi_1 - \varphi_2) - \varphi_1] + g\sin(\varphi_1 - \varphi_2) = \Delta_n \tag{12.26}$$

where $g = r_b + R_w(1 - \cos\Theta_2)$; and the subscripts 1 and 2 are labeled the nominal surface, Eq. (12.16), and actual surface, Eq.(12.21), respectively.

Using Eq. (12.25) we can express parameter Θ_1 through parameter Θ_2 in the symbolic form, and then, using Eq. (12.24) allows expression of parameter φ_1 through φ_2 and Θ_2:

$$\Theta_1 = \arctan\frac{R_w \sin\Theta_2}{R_w - c} \text{ or } \Theta_1 = \pi + \arctan\frac{R_w \sin\Theta_2}{R_w - c}$$

$$\varphi_1 = \varphi_2 - \arctan\frac{\varphi_2 r_b}{g} \pm \arccos\frac{r_b}{\sqrt{g^2 + (\varphi_2 r_b)^2}} \tag{12.27}$$

Two values of φ_1 correspond to the left or right involute, and two values of $\Theta_1$correspond to two diametrically opposite positions of the machined gear relative to the grinding wheel in space. The substitution of Eqs. (12.27) in Eq. (12.26) yields the value $\Delta_n = \Delta_n(\Theta_2, \varphi_2)$.

Numerical calculations can be worked out through Eq. (12.11a), which becomes in this case:

FindRoot $[\{(f_1 == 0, f_2 == 0\}, \{\varphi_1, (\varphi_2)_0\}, \{\Theta_1, (\Theta_2)_0\}]$

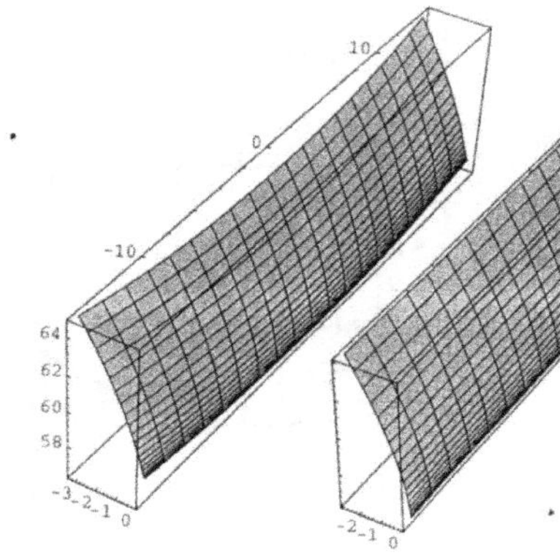

Figure 12-6 The tooth profile for setup according to: (a) Figure 12.4b (actual surface); (b) Figure 12.4a (nominal surface).

where $(\varphi_2)_0$ and $(\Theta_2)_0$ are the φ_2 and Θ_2 values in the *i*th points of the actual surface, Eq. (12.21); and f_1 and f_2 are the left sides of Eqs. (12.24) and (12.25), respectively.

We obtain numerical solutions for φ_1 and Θ_1 in the form $\varphi_1 = \varphi_1[(\varphi_2)_{0,} (\Theta_2)_0]$ and $\Theta_1 = \Theta_1[(\varphi_2)_{0,} (\Theta_2)_0]$. Then, substitution of $\varphi_1, \Theta_1, (\varphi_2)_{0,}$ and $(\Theta_2)_0$ in Eq. (12.26) results in the normal deviation (the layout error) associated with the chosen point of the actual surface, Eq. (12.21),

$$\Delta_n = \Delta_n(\Theta_2, \varphi_2)$$

The surface by Eq. (12.21) is seen in Fig. 12.4 to have the error f_{fr} of the tooth profile. In Figure 12.7, the difference between profiles defined through Eqs. (12.16) and (12.21) is evaluated by application of the standard metrological assessments. The error $(f_{fr})_{full}$ of the actual profile *2* is measured as a distance between nominal surface *1* and another ordinary involute 3, which passes through the initial point M_0 of the actual profile 2. In workshop practice, the other assessment is also used – the error $(f_{fr})_{lat}$ of the profile on the lateral side of flank 3. This error is measured as the distance between involute cylinders 3 and 4. Ordinary involute 4 passes through the last point M_1 of actual profile *2*; this point is located on the addendum cylinder on the lateral side of flank 2.

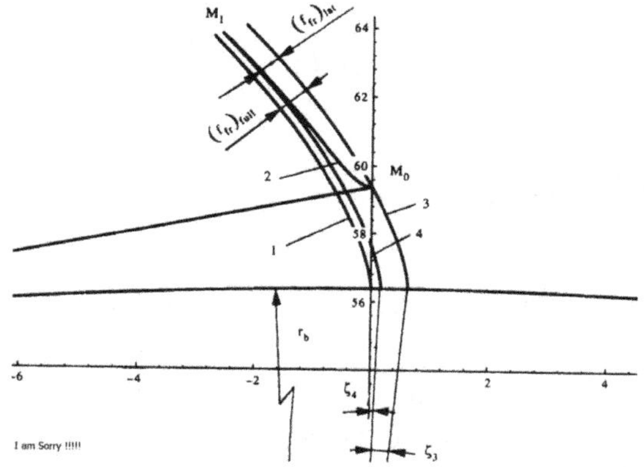

Figure 12-7 Metrological assessments of the form accuracy for the profile according to Figure 12.6a.

The profile angle on the lateral side of the gear (i.e., for $\Theta_2=\Theta_{2,lim}$) located between involutes 3 and 4 is $\zeta_{ex} = \varphi_{2,min}$ and $\zeta_{in} = \varphi_{2,add}$, where subscript 2 marks the limit angles through Eq. (12.23) for the surface, Eq. (12.21). The error is

$$(f_{fr})_{lat} = \Delta_n(\varphi_{2,add}, \Theta_{2,lim}) - \Delta_n(\varphi_{2,min}, \Theta_{2,lim}) \tag{12.28}$$

The error for the full profile $(f_{fr})_{full}$ of the gear located between involutes *1* and *3* is

$$(f_{fr})_{full} = \Delta_n(\varphi_{2,min}, \Theta_{2,lim}) - \Delta_n(\varphi_{1,min}, 0) \tag{12.29}$$

Two additional notes:

- As is seen from Eq. (12.26), the first order approximation results in $\Delta_n = 0$. Actually, substitution $\varphi_1 = \varphi_2$, which takes place in the first order approximation, in Eq. (12.26) gives $\Delta_n = r_b[\varphi_1 \cos 0 - \varphi_1] + g \sin 0 = 0$. Hence, the non–linear problem has to be solved.
- The maximum Δ_n value, Eq. (12.26), is reached in the lateral side of the machined tooth, where $\varphi_2 = 0$ and $g = r_b + R_w(1 - \cos\Theta_2) = r_b + \delta_y$. Hence,

$$\Delta_{max} = g \sin\varphi_1 - r_b\varphi_1 = [r_b + R_w(1 - \cos\Theta_2)]\sin\varphi_1 - r_b\varphi_1 = (r_b + \delta_y)\sin\varphi_0 - r_b\varphi_0$$

As is clear from Figure 12.5c, $\varphi_0 = \arccos[r_b/(r_b + \delta_y]$. Hence,

$$\Delta_{max} = \sqrt{2r_b\delta_y} - r_b \arccos[r_b/(r_b + \delta_y] \approx \sqrt{\frac{8\delta_y^3}{9r_b}}$$

12.2.2.5 Experimental investigation and verification

In Figure 12.8, the results of experimental machining by two methods are compared. The spur gears of $m = 6$ mm and $z = 21$ were machined under following one-pass cutting conditions:

- Method 1 (Fig. 12.4a): depth $t = 0.1$ mm and time of the generating process $\tau = 8$ s, and value $c = 1.0$ mm;
- Method 2 (Fig. 12.4b): depth $t = 0.2$ mm and time of the generating process $\tau = 12$ s.

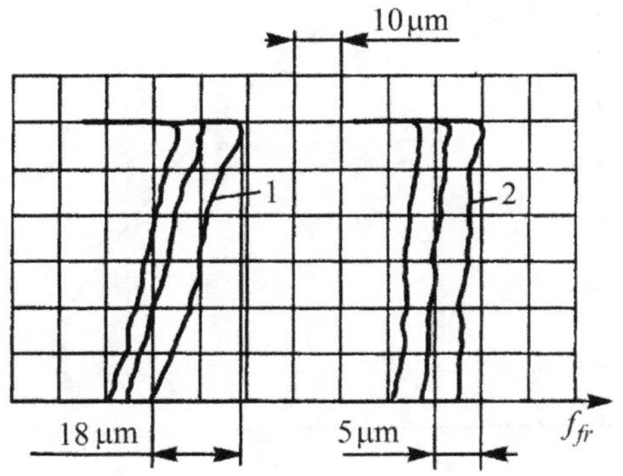

Figure 12-8 Comparison of the involute charts of tooth profiles machined by two methods.

The profile accuracy was measured along the direction of the involute curve by an involute measuring device. A deviation of the actual tooth profile from the ideal one along the normal to the involute surface was measured. In Figure 12.8, the results of measurements are shown on the involute charts: methods 1 and 2 result in 18 μm and 5 μm, respectively. Note that the similar accuracy upon grinding by method 1 can be reached only after four additional passes. Hence, the peripheral grinding considerably increases machining accuracy within specific range of parameters of the gearwheels compared with the traditional grinding method.

For testing the theoretical results, the experimental data [9] was used. The experiments were performed on the gear-grinding machine HSS 60/80, Maag. The geometrical data for the workpiece (gear) and the grinding wheel are: module $m = 4$ mm; teeth number $z_t = 30$; addendum modification shift $x = 0$; addendum $h_a = m = 4$ mm; gear face width $b = 30$ mm; base-circle radius $r_b = mz_t \cos 20° = 56.38$ mm; and grinding wheel external diameter $2R_w = 220$ mm. The machining method differs from the schema in Figure 12.2: the grinding wheel additionally moves along the axes of the machined gear. In this method, the profiles of the ground teeth are identical in all the cross-sections. The shortened involute profile may be machined if, and only if, the distance between the axes is $a > r_b + R_w$. In the experiment, this was $r_b + R_w + \delta_y$, where δ_y was set on the five levels: $\delta_y = -0.8$; -0.3; 0.2; 0.7; and 1.2 mm.

The teeth to be ground were preliminarily machined by the protuberance-type hob to remove the part of the effective tooth profile on the dedendum part. The effective tooth profile started from the angle $\varphi_{min} = (r_b\ \varphi_{min})/r_b = 12/56.38 = 0.2128414$. The gears were ground by two passes, and the wheels were dressed after each pass. The profile of ground tooth was measured in the mean cross-section by the involute-measuring gage (C. Zeiss, Jena).

In Figure 12.9, the theoretical curve according to Eq. (12.18) and the experimental results are shown. The values f_{fr} in the theoretical curve were corrected according to the systematic error δ_{sys} of the grinding machine. The value of δ_{sys} was adopted as an average of those experimental results, for which $\delta_y < 0$, i.e., $\delta_y = -0.8$ and -0.3 mm. The correction was $\delta_{sys} = -2$ μm. For example, when $\delta_y < 0$, we have $f_{fr} = 0 + \delta_{sys} = -2$ μm (Fig. 12.9). It is evident from a comparison of the results that they agree nicely with each other. The maximal relative error is 12.7% for $\delta_y = 1.2$ mm.

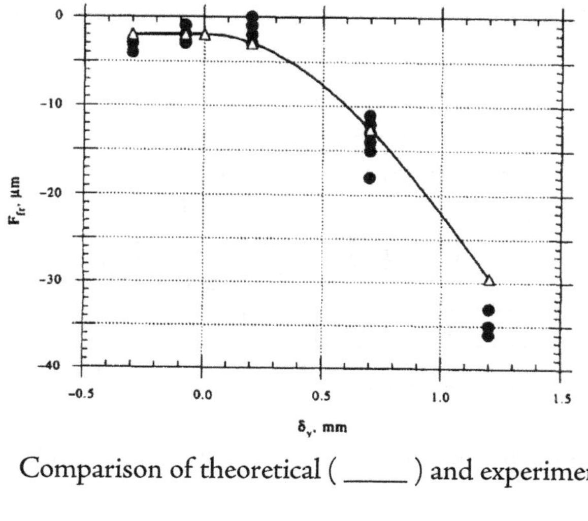

Figure 12-9 Comparison of theoretical (_____) and experimental (•••) results.

Numerical Example 12.1.

Let us consider the calculation of the error f_{fr} for $\delta_y = 0.7$ mm. The parameter g in Eq. (12.22) is $g = r_b + \delta_y = 56.38 + 0.7 = 57.08$ mm. The limit values, Eq. (12.23), of the angle $\Theta_{2,lim} = \arctan(30/220) = 0.135528$ and those of the angle Θ_2 are

$$\varphi_{2,add} = \sqrt{(0.5mz_t + 1)^2 - r_b^2} / r_b = \sqrt{(0.5 \times 4 \times 30 + 1)^2 - 56.38^2} / 56.38 = 0.537192$$

$\varphi_{2,\min} = 0.2128414$

For these limits values, Eqs. (12.11a) and (12.27) become:

$$\textbf{FindRoot}\ [\{(f_1)_0 == 0, (f_2)_0 == 0\}, \{\varphi_1, 0.537192\}, \{\Theta_1, 0.135528\}] \quad (12.30)$$

where $(f_1)_0$ and $(f_2)_0$ are the left sides of Eqs. (12.24) and (12.25) for $\varphi_2 = 0.537192$ and $\Theta_2 = 0.135528$, i.e.,

$$(f_1)_0 = 56.38 + (56.38)(0.537192) \sin(\varphi_1 - 0.537192) - 57.08 \cos(\varphi_1 - 0.537192)$$

$$(f_2)_0 = 220 \tan 0.135528 - 220 \sin \Theta_1$$

The numerical solution of this system is $\varphi_{1,add} = 0.559824$; $\Theta_1 = 0.13679$. Again, note that the values $\Theta_{2,\lim}$ and is $\Theta_{2,add}$ are used as initial values of Θ_1 and $\Theta_{1,add}$, respectively, when solving Eqs. (12.24), and (12.25) by means of Eq. (12.30). The substitution of the values $\varphi_{1,\mathrm{add}}$ and $\varphi_{2,\mathrm{add}}$ into Eq. (12.26) gives

$$\Delta_{n,\,add} = r_b[\varphi_{2,\,add} \cos(\varphi_{1,\,add} - \varphi_{2,\,add}) - \varphi_{1,\,\mathrm{add}}] + g \sin(\varphi_{1,\,add} - \varphi_{2,\,add}) =$$

$$56.38[0.537192 \cos(0.559824 - 0.538192) - 0.559824] + 57.08 \sin(0.559824 -$$

$$0.537192) = 0.007976 \text{ mm}$$

The similar calculations for $\varphi_{\min}$ give $\varphi_{1,\,\min} = 0.264783$, and

$$\Delta_{n,\min} = r_b[\varphi_{2,\min} \cos(\varphi_{1\min} - \varphi_{2,\min}) - \varphi_{1,\,\min}] + g \sin(\varphi_{1\min} - \varphi_{2,\min})$$

$$= 56.38[0.2128414 \cos(0.023078 - 0.2128414) - 0.023078] + 57.08 \sin(0.023078 - 0.2128414) = 0.018842 \text{ mm}$$

The profile error, Eq. (12.28), is

$$f_{fr} = (f_{fr})_{\mathrm{lat}} = \Delta_{n,add} - \Delta_{n,\,\min} = 0.007976 - 0.018842 = -0.010866 \text{ mm}$$

The corresponding point in Figure 12.9 has coordinates $\delta_y = 0.7$ mm and

$$f_{fr} = (f_{fr})_{lat} + \delta_{sys} = -10.866 + (-2) = -12.9\ \mu\text{m}.$$ ■

12.3 Higher order approximation approach

12.3.1 Mathematical formulation

Solution of the non-linear problem in Section 12.3 was carried out using direct symbolic definition of the error. Here, the calculation of the normal error, i.e., the projection of error onto the normal at the machined surface, is continued. For this purpose, the method using the high order approximation is considered in the general form, and the real-life examples are solved. In Figure 12.10, the schematics previously shown in Figure 12.1 is modernized: the position vectors of the nominal surface F_0 and actual surface F defined through position vectors $\mathbf{r}_0(u, v)$ and $\mathbf{r}(u, v)$, respectively, and unit-vector $\mathbf{n}$ of the normal to the surface F_0 passing through a point $\mathbf{r}(u, v)$ of the surface F are shown. As before, it is necessary to define the point M on the surface F_0, which is the foot of the normal. The situation is described by two equations:

$$\Delta_n \cdot \mathbf{n}_0(u + \Delta u, v + \Delta v) = \mathbf{r}(u, v) - \mathbf{r}_0(u + \Delta u, v + \Delta v) \tag{12.31}$$

$$\mathbf{r}(u, v) = \mathbf{r}_0(u, v) + \Delta\mathbf{r}_0(u, v) \tag{12.32}$$

where u and v are curvilinear coordinates; Δu and Δv are their increments; and $\Delta\mathbf{r}_0(u, v)$ is the vector of position error of a point of the surface F with respect to the points of the surface F_0.

Taking into account that Δu and Δv are small values, it is possible to use Taylor's formula for $\mathbf{r}_0(u + \Delta u, v + \Delta v)$ and $\mathbf{n}_0(u + \Delta u, v + \Delta v)$,

$$\Delta_n[\mathbf{n}_0(u,v) + \mathbf{n}_u \Delta u + \mathbf{n}_v \Delta v + O(\delta^2)]$$
$$= \Delta\mathbf{r}_0 - [\mathbf{r}_u \Delta u + \mathbf{r}_v \Delta v + \frac{1}{2}\mathbf{r}_{uu}\Delta u^2 + \mathbf{r}_{uv}\Delta u \Delta v + \frac{1}{2}\mathbf{r}_{vv}\Delta v^2 + O(\delta^3)] \tag{12.33}$$

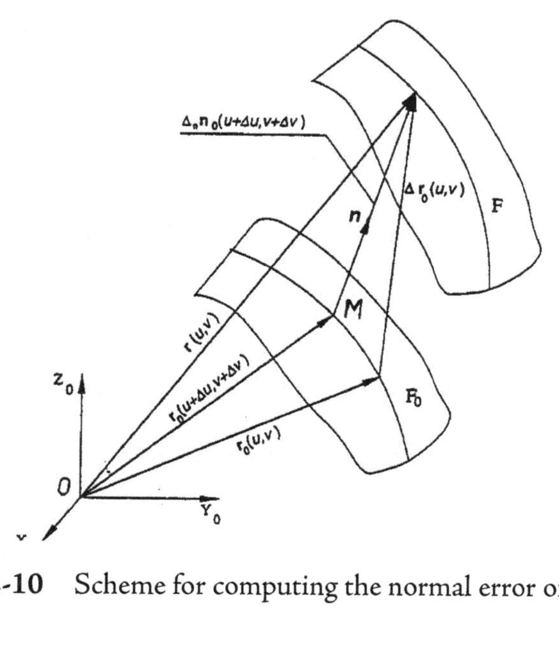

Figure 12-10 Scheme for computing the normal error of the machined surface.

where subscripts of vectors denote the partial derivatives with respect to u, v, i.e., $\mathbf{n}_u = \partial\mathbf{n}/\partial u$; $\mathbf{n}_v = \partial\mathbf{n}/\partial v$; $\mathbf{r}_u = \partial\mathbf{r}/\partial u$; $\mathbf{r}_v = \partial\mathbf{r}/\partial v$; $\mathbf{r}_{uu} = \partial^2\mathbf{r}_0/\partial u^2$; $\mathbf{r}_{uv} = \partial^2\mathbf{r}_0/\partial u\partial v$; $\mathbf{r}_{vv} = \partial^2\mathbf{r}_0/\partial v^2$, and δ^2 and δ^3 are the values of the second and third order of values Δu and Δv.

Since the error $\Delta\mathbf{r}_0$ is the function of some elementary errors (e.g., linear and angular errors associated with FSS links), it may be represented according to MacLaurin's series expansion

$$\Delta\mathbf{r}_0 = \mathbf{d}_1 + (1/2)\mathbf{d}_2 + O(\delta^3) \tag{12.34}$$

where $\mathbf{d}_1$ and $\mathbf{d}_2$ are vectors of the first and second order approximation, respectively, determined by the values of the partial derivatives of the position vector $\mathbf{r} = \mathbf{r}_0 + \Delta\mathbf{r}_0$ of the real surface with respect to the elementary errors

$$\mathbf{d}_1 = \sum_{i=1}^{n} \left.\frac{\partial\mathbf{r}}{\partial\lambda_i}\right|_{\lambda_i=0} \lambda_i\,;\ \mathbf{d}_2 = \sum_{i,j=1}^{n} \left.\frac{\partial^2\mathbf{r}}{\partial\lambda_i\partial\lambda_j}\right|_{\lambda_i=\lambda_j=0} \lambda_i\lambda_j \tag{12.35}$$

λ_i $(i = 1,\ldots,n)$ is the generalized designation of the elementary errors; and the partial derivatives of vector $\mathbf{r}$ with respect to λ_i are determined at point $\lambda_i = 0$.

Let us denote the sum of three members entering in Eq. (12.33) and relating only to the *first order* approximation by

$$\mathbf{d}^{(1)} = \Delta_n^{(1)}\mathbf{n}_0(u,v) + \mathbf{r}_u\Delta u^{(1)} + \mathbf{r}_v\Delta v^{(1)} \tag{12.36}$$

Since $\mathbf{d}^{(1)}$ is a part of the vector $\Delta\mathbf{r}_0$, Eq. (12.34), with respect to terms of the first order only, then $\mathbf{d}^{(1)} = \mathbf{d}_1$, Eq.(12.35).The solution of the vector equation (12.36) has the form [7, p. 151]

$$\Delta_n^{(1)} = \mathbf{d}^{(1)}\cdot\mathbf{n}_0;\ \Delta u^{(1)} = \frac{(\mathbf{d}^{(1)}\cdot\mathbf{r}_v\cdot\mathbf{n}_0)}{[\mathbf{r}_u\times\mathbf{r}_v]};\ \text{and}\ \Delta v^{(1)} = \frac{(\mathbf{r}_u\cdot\mathbf{d}^{(1)}\cdot\mathbf{n}_0)}{[\mathbf{r}_u\times\mathbf{r}_v]} \tag{12.37}$$

In a similar manner, the sum of the members entering in Eq. (12.33) and relating only to the *second order* approximation is denoted by

$$\mathbf{d}^{(2)} = \Delta_n^{(2)}\cdot\mathbf{n}_0 + \mathbf{r}_u\Delta u^{(2)} + \mathbf{r}_v\Delta v^{(2)} \tag{12.38}$$

On the other hand, the members of Eq. (12.33) relating to the second order approximation are

$$\mathbf{d}_2 = \Delta\mathbf{r}_0 - \frac{1}{2}\mathbf{r}_{uu}(\Delta u^{(1)})^2 - \mathbf{r}_{uv}\Delta u^{(1)}\Delta v^{(1)} + \frac{1}{2}\mathbf{r}_{vv}(\Delta v^{(1)})^2 - (\mathbf{n}_u\Delta u^{(1)} + \mathbf{n}_v\Delta v^{(1)})\Delta_n^{(1)} \tag{12.39}$$

where $\Delta u^{(1)}$, $\Delta v^{(1)}$ are the increments through formulae (12.37).

The solution of Eq. (12.38) is

$$\Delta_n^{(2)} = \mathbf{d}^{(2)}\cdot\mathbf{n}_0 \tag{12.40}$$

Expressions for $\Delta u^{(2)}$ and $\Delta v^{(2)}$ are obtained in the form of Eq. (12.37) with substitution $\mathbf{d}^{(2)}$ in place of $\mathbf{d}^{(1)}$. The approximations of the next orders may be obtained in a similar manner.

In an important particular case, in which $\mathbf{d}^{(1)} \neq 0$, but $\Delta^{(1)} = 0$ since $\mathbf{d}^{(1)}$ is located on the tangent plane of the surface F_0 (Fig. 12.10), Eq. (12.40) becomes

$$\Delta_n^{(2)} = \Delta\mathbf{r}_0^{(2)}\,\mathrm{n}_0 - \frac{1}{2}\,\mathbf{II}_0 \tag{12.41}$$

where $\mathbf{II}_0$ is the second fundamental form of the nominal surface [5, p. 572]

$$\mathbf{II}_0 = (\mathbf{r}_{uu}\cdot\mathbf{n})du^2 + 2(\mathbf{r}_{uv}\cdot\mathbf{n})dudv + (\mathbf{r}_{vv}\cdot\mathbf{n})dv^2, \tag{12.42}$$

where $du = \Delta u$; $dv = \Delta v$; and the partial derivatives are calculated for the nominal surface $\mathbf{r}_0$.

In particular, the case in which the vector $\Delta\mathbf{r}_0$ contains a single small deviation λ can be applied for the machine setups when surfaces of sophisticated forms (crowned, helical, etc.) are machined. Let the setting displacement Δ be small. Hence, putting in Eqs (12.35) $n = 1$, we obtain:

$$\Delta = (\mathbf{d}_1\cdot\mathbf{n}_0) + \frac{1}{2}[(\mathbf{d}_2\cdot\mathbf{n}_0) - \mathbf{II}_0] \tag{12.43}$$

$$\mathbf{d}_1 = \left.\frac{\partial\mathbf{r}}{\partial\lambda}\right|_{\lambda=0}\lambda;\ \text{and}\ \mathbf{d}_2 = \left.\frac{\partial^2\mathbf{r}}{\partial\lambda^2}\right|_{\lambda=0}\lambda^2 \tag{12.44}$$

In this case, the following two-step procedure can be used for setup calculations:

Step 1. The first approximation is determined

$$\Delta^{(1)} = (\mathbf{d}_1\cdot\mathbf{n}_0) \tag{12.45}$$

If $\Delta^{(1)} \neq 0$, the higher order approximation is unnecessary.

Step 2. If $\Delta^{(1)} = 0$ and $\lambda \neq 0$, the second order approximation is

$$\Delta^{(2)} = \frac{1}{2}[(\mathbf{d}_2\cdot\mathbf{n}_0) - \mathbf{II}_0] \tag{12.46}$$

where increments du and dv of curvilinear coordinates u, v of the second fundamental form $\mathbf{II}_0$ are determined using the value $\mathbf{d}_1$, which is obtained as the intermediate result of the first step

$$\Delta u_1 = \frac{(\mathbf{d}_1\cdot\mathbf{r}_{0v}\cdot\mathbf{n}_0)}{|\mathbf{r}_{0u}\times\mathbf{r}_{0v}|};\ \text{and}\ \Delta v_1 = \frac{(\mathbf{r}_{0u}\cdot\mathbf{d}_1\cdot\mathbf{n}_0)}{|\mathbf{r}_{0u}\times\mathbf{r}_{0v}|}; \tag{12.47}$$

where $\mathbf{r}_{0u}$ and $\mathbf{r}_{0v}$ are partial derivatives of vector $\mathbf{r}_0$ with respect to u and v.

12.3.2 Methodological example: hyperboloid turning

Consider the kinematic method of machining hyperbolic shafts with a small specified bow (i.e., with small deviation of the hyperboloid points from the points of the inscribed cylinder). This method implies that the cylindrical workpiece is machined on the lathe by the longitudinal feed with a small crossing of the workpiece axis and the cutting tool trajectory.

The principal layout of this method is shown in Figure 12.11, where 1 is the hyperboloid to be machined, 2 is the cutting tool, and 3 is the cylinder, which is coaxial with hyperboloid 1. Radius R of the cylinder coincides with the minimal radius of the hyperboloid (i.e., radius of the hyperboloid throat). The trajectory MS of the cutting tool is directed at angle Λ toward the center line OO. As a result, the active point S of the cutting tool (the tip of the cutting blade) has a vertical displacement δ, which varies in direct proportion with distance z of the cutting tool from the minimal section of the hyperboloid; namely, $\delta = z\tan(\Lambda)$.

The form-shaping function for considered setup becomes

$$\mathbf{r}_0 = \mathbf{A}^6(\varphi)\mathbf{A}^1(R)\mathbf{A}^4(-\Lambda)\mathbf{A}^3(z)\,\mathbf{r}_4 \tag{12.48}$$

with

$$\mathbf{A}^6(\varphi) = \begin{bmatrix} \cos\varphi & -\sin\varphi & 0 & 0 \\ \sin\varphi & \cos\varphi & 0 & 0 \\ 0 & 0 & 1 & 0 \\ 0 & 0 & 0 & 1 \end{bmatrix}; \mathbf{A}^1(R) = \begin{bmatrix} 1 & 0 & 0 & R \\ 0 & 1 & 0 & 0 \\ 0 & 0 & 1 & 0 \\ 0 & 0 & 0 & 1 \end{bmatrix}; \mathbf{A}^4(-\Lambda) = \begin{bmatrix} 1 & 0 & 0 & 0 \\ 0 & \cos\Lambda & \sin\Lambda & 0 \\ 0 & -\sin\Lambda & \cos\Lambda & 0 \\ 0 & 0 & 0 & 1 \end{bmatrix};$$

$$\mathbf{A}^3(z) = \begin{bmatrix} 1 & 0 & 0 & 0 \\ 0 & 1 & 0 & 0 \\ 0 & 0 & 1 & z \\ 0 & 0 & 0 & 1 \end{bmatrix}; \text{ and } \mathbf{r}_4 = \begin{bmatrix} 0 \\ 0 \\ 0 \\ 1 \end{bmatrix}$$

where φ is the angle of the workpiece rotation; R is the radius of the hyperboloid throat; Λ is the crossing angle of the longitudinal feed with respect to center line; z is the longitudinal feed current position; and $\mathbf{r}_4$ is the position vector of the cutting edge in its own coordinate frame.

We will consider the case when the angle Λ is small. Hence, $\delta \approx z\,\Lambda$ and formula for the hyperboloid, Eq. (12.48), becomes

$$\mathbf{r} = \begin{bmatrix} R\cos\varphi + \Lambda z\sin\varphi \\ -R\sin\varphi + \Lambda z\cos\varphi \\ z \\ 1 \end{bmatrix} \tag{12.49}$$

$$\mathbf{r}_0 = \begin{bmatrix} R\cos\varphi \\ -R\sin\varphi \\ z \\ 1 \end{bmatrix} \tag{12.50}$$

For manipulation with derivatives, it is more convenient to use the position vectors of the third order.

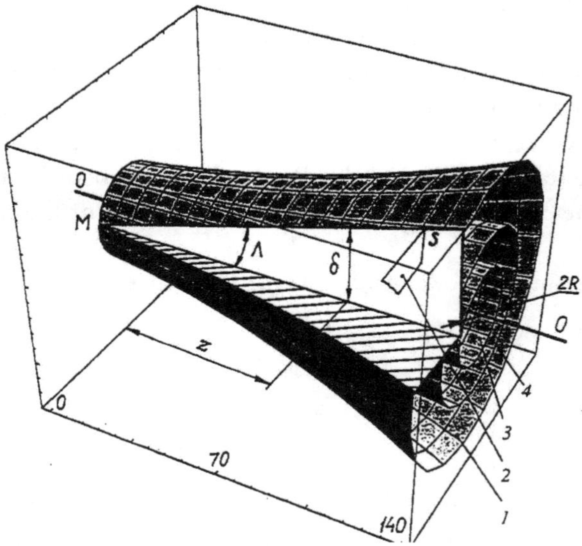

Figure 12-11 Machining a hyperboloid on the lathe.

$$\mathbf{r} = \mathbf{r}(\varphi, z) = \begin{bmatrix} R\cos\varphi + \Lambda z \sin\varphi \\ -R\sin\varphi + \Lambda z\cos\varphi \\ z \end{bmatrix} \text{and } \mathbf{r}_0 = \mathbf{r}_0(\varphi, z) = \begin{bmatrix} R\cos\varphi \\ -R\sin\varphi \\ z \end{bmatrix}$$

The partial derivatives, Eq. (12.44), of the hyperboloid, Eq. (12.49), with respect to small angular displacement Λ in the point $\Lambda = 0$ are

$$\mathbf{d}_1 = \left.\frac{\partial \mathbf{r}}{\partial \Lambda}\right|_{\Lambda=0} \Lambda = \begin{bmatrix} \sin\varphi \\ \cos\varphi \\ 0 \end{bmatrix} \Lambda z \text{ and } \mathbf{d}_2 = \left.\frac{\partial^2 \mathbf{r}}{\partial \Lambda^2}\right|_{\Lambda=0} \Lambda^2 = \begin{bmatrix} 0 \\ 0 \\ 0 \end{bmatrix}$$

The partial derivatives, Eq. (12.44), of the cylinder, Eq. (12.50), with respect to the Gaussian coordinates φ and z are

$$\mathbf{r}_{0\varphi} = \begin{bmatrix} -R\sin\varphi \\ -R\cos\varphi \\ 0 \end{bmatrix}; \mathbf{r}_{0z} = \begin{bmatrix} 0 \\ 0 \\ 1 \end{bmatrix}; \mathbf{r}_{0\varphi\varphi} = \begin{bmatrix} -R\cos\varphi \\ R\sin\varphi \\ 0 \end{bmatrix}; \mathbf{r}_{0\varphi z} = \mathbf{r}_{0zz} = \begin{bmatrix} 0 \\ 0 \\ 0 \end{bmatrix};$$

The normal to the cylinder is

$$\mathbf{n}_0 = \pm[\mathbf{r}_{0\varphi} \times \mathbf{r}_{0z}] / |\mathbf{r}_{0\varphi} \times \mathbf{r}_{0z}| = \pm \begin{bmatrix} -\cos\varphi \\ \sin\varphi \\ 0 \end{bmatrix}$$

As usual, the normal can be defined in two opposite directions. In this case, sign "plus" before the vector identifies the internal normal oriented from the surface to the axis. The external normal that in what follows will be used is

$$\mathbf{n}_0 = \begin{bmatrix} \cos\varphi \\ -\sin\varphi \\ 0 \end{bmatrix}$$

Step 1 of the setup procedure shows that the first approximation, Eq. (12.45), results in zero deviation value:

$$\Delta^{(1)} = (\mathbf{d}_1 \cdot \mathbf{n}_0) = \begin{bmatrix} \sin\varphi \\ \cos\varphi \\ 0 \end{bmatrix} \Lambda z \cdot \begin{bmatrix} \cos\varphi \\ -\sin\varphi \\ 0 \end{bmatrix} = 0$$

Since $\Delta^{(1)} = 0$, the higher order approximation is necessary.

Step 2. The second order approximation, Eq. (12.46), is calculated. Increments of the Gaussian coordinates, Eq. (12.47), are

$$d\varphi = \frac{(\mathbf{d}_1 \cdot \mathbf{r}_{0z} \cdot \mathbf{n}_0)}{|\mathbf{r}_{0\varphi} \times \mathbf{r}_{0z}|} = \frac{\Lambda z}{R} \begin{vmatrix} \sin\varphi & 0 & \cos\varphi \\ \cos\varphi & 0 & -\sin\varphi \\ 0 & 1 & 0 \end{vmatrix} = \frac{\Lambda z}{R} \tag{12.51}$$

$$dz = \frac{(\mathbf{r}_{0\varphi} \cdot \mathbf{d}_1 \cdot \mathbf{n}_0)}{|\mathbf{r}_{0\varphi} \times \mathbf{r}_{0z}|} = \frac{\Lambda z}{R} \begin{vmatrix} -R\sin\varphi & \cos\varphi & \cos\varphi \\ -R\cos\varphi & \sin\varphi & -\sin\varphi \\ 0 & 0 & 0 \end{vmatrix} = 0 \tag{12.52}$$

The second fundamental form of the nominal surface, Eq. (12.42), is

$$\mathbf{II}_0 = (\mathbf{r}_{0\varphi\varphi} \cdot \mathbf{n}_0)\, d\varphi^2 + 2(\mathbf{r}_{0\varphi z} \cdot \mathbf{n}_0)\, d\varphi dz + (\mathbf{r}_{0zz} \cdot \mathbf{n}_0)\, dz^2$$

$$= \begin{bmatrix} -R\cos\varphi \\ R\sin\varphi \\ 0 \end{bmatrix} \cdot \begin{bmatrix} \cos\varphi \\ -\sin\varphi \\ 0 \end{bmatrix} d\varphi^2 + 0 + 0 = -R\left(\frac{\Lambda z}{R}\right)^2 = -\frac{\Lambda^2 z^2}{R}$$

According to Eq. (12.46) we obtain deviations of the radius of the hyperboloid, Eq. (12.49), from the cylinder, Eq. (12.50),

$$\Delta R = \Delta^{(2)} = \frac{1}{2}[(\mathbf{d}_2 \cdot \mathbf{n}_0) - \mathbf{II}_0] = \frac{1}{2}[0 - \left(-\frac{\Lambda^2 z^2}{R}\right)] = \frac{\Lambda^2 z^2}{2R} \tag{12.53}$$

For an experimental verification of this result, a shaft of L_s = 140 mm in length was turned between centers (with a driver plate and a lathe dog) on the CAZENEUVE HB 500® lathe by a turning tool. The workpiece diameter was 32 mm. To obtain a layout according to Figure 12.11, the center of the tailstock was displaced (by means of steel washers on the guideways of the tailstock) upwards at the value $\delta = 1.40$ mm and in the horizontal plane at the value $\delta_{hor} = 0.70$ mm. The turned shaft was measured by a digital micrometer in 15 cross-sections (at intervals of $h = 10$ mm in length). In Figure 12.12, both experimental data 1 and calculated curve 2 present a difference between corresponding diameters of the hyperboloid 1 and inscribed cone 4 (both surfaces are shown in Figure 12.11). The calculated results are seen to agree satisfactorily with the experimental data.

Remark About the Special Case. When $\Lambda z = \delta$, we obtain $\Delta R = \delta^2/(2R)$. This is the well-known case in which the "non-sensitive" displacement of the tool with the longitudinal turning of the cylinder takes place (Fig. 8.5).

12.3.3 Real-life application: setup of internal grinding of bearing ring

12.3.3.1 Machining scheme

Grinding the crowned surface of an outer race (Fig. 12.13) of the tapered roller bearing is considered. As is known, this race is crowned for improvement of contact conditions between the bearing ring and rollers. The race is ground on an internal grinding machine (Fig. 12.14) using a cylindrical wheel (the preferable scheme) or a conical wheel. The setting of the machine includes two rotations of workpiece 1 relative to the grinding wheel (not shown in Figure 12.14). Headstock 1 with workpiece 2 rotates by an angle β (half of the cone angle) in the horizontal plane relative to plate 3 around axis *I*. In turn, plate 3 rotates by a small angle λ in a vertical plane relative to bed 4 around axis *II*.

In Figure 12.15, the results of setup rotations of the grinding wheel *1* are shown. After rotation around axis *I*, the conical surface 2 is obtained (Fig. 12.15a), and after the second rotation around axis *II*, the machined bore assumes the hyperbolic form 3 (Fig. 12.15b). To visually differ between two machined surfaces, one can compare the straight line generatrix of the conical surface 2 with hyperbolical generatrix of the hyperboloid 3.

The aim of the setup calculation consists of computing angle λ as a function of values Δ_1 and Δ_2 (Fig. 12.13) as well as distance z_m between the base section and the vertex point. As before,

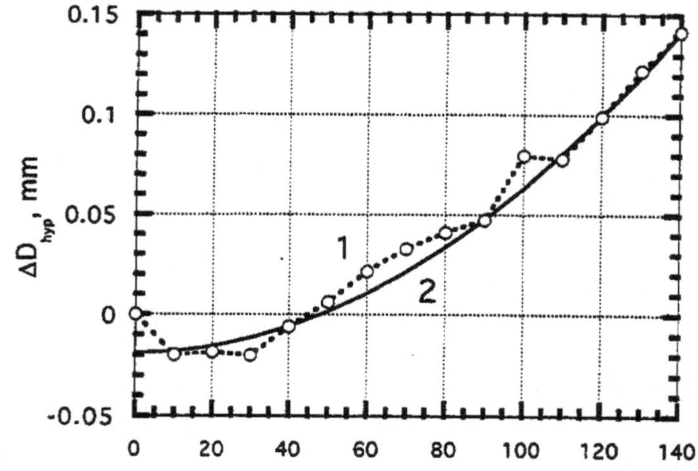

Figure 12-12 A correlation between the experimental (1) and calculated (2) data.

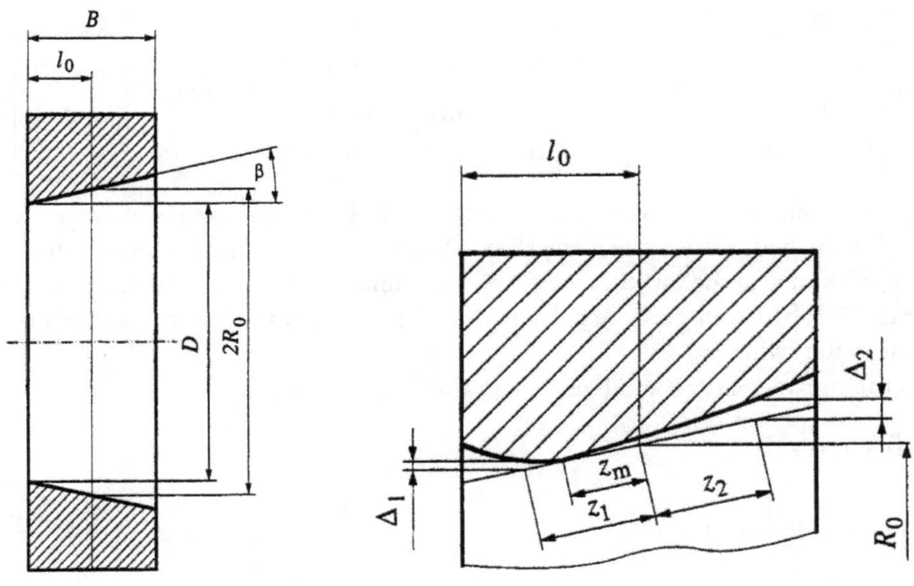

Figure 12-13 Outline of the outer ring of the tapered roller bearing: (a) the general view; (b) the crowned profile.

the calculation is performed with the Mathematica® software system. A set of coordinate frames S_i $(i = 0, 1,...,5)$ associated with the machine units is shown in Figure 12.16 (the case of machining by a cylindrical grinding wheel is considered) and listed in Table 12.2.

According to Table 12.2, the FSF of the machine is given by

$$\mathbf{r} = \mathbf{A}^6(\varphi)\mathbf{A}^3(-a)\mathbf{A}^5(\beta)\mathbf{A}^3(c)\mathbf{A}^4(-\lambda)\mathbf{r}_5 \tag{12.54}$$

$$\mathbf{A}^6(\varphi) = \begin{bmatrix} \cos\varphi & -\sin\varphi & 0 & 0 \\ \sin\varphi & \cos\varphi & 0 & 0 \\ 0 & 0 & 1 & 0 \\ 0 & 0 & 0 & 1 \end{bmatrix}; \mathbf{A}^3(-a) = \begin{bmatrix} 1 & 0 & 0 & 0 \\ 0 & 1 & 0 & 0 \\ 0 & 0 & 1 & -a \\ 0 & 0 & 0 & 1 \end{bmatrix}; \mathbf{A}^5(\beta) = \begin{bmatrix} \cos\beta & 0 & \sin\beta & 0 \\ 0 & 1 & 0 & 0 \\ -\sin\beta & 0 & \cos\beta & 0 \\ 0 & 0 & 0 & 1 \end{bmatrix};$$

Table 12-2. FSS of the bearing ring grinding machine.

S_i	Link	Motion relative to previous link	Matrix ${}^{i-1}\mathbf{A}_i$
S_0	Workpiece spindle	Conventionally fixed link	–
S_1	Headstock	Rotation around the Z_0–axis at angle φ	${}^0\mathbf{A}_1 = \mathbf{A}^6(\varphi)$
S_2	Bed	Translation along the X_1–axes at distance $-a$	${}^1\mathbf{A}_2 = \mathbf{A}^3(-a)$
S_3	Grinding head	Rotation around the Y_2-axes at angle β	${}^2\mathbf{A}_3 = \mathbf{A}^5(\beta)$
S_4	Tilt plate	Translation along the Z_3-axes at distance c	${}^3\mathbf{A}_4 = \mathbf{A}^3(c)$
S_5	Main spindle	Rotation around the X_4-axes at angle $-\lambda$	${}^3\mathbf{A}_4 = \mathbf{A}^4(-\lambda)$

$$\mathbf{A}^3(c)=\begin{bmatrix}1&0&0&0\\0&1&0&0\\0&0&1&c\\0&0&0&1\end{bmatrix};\mathbf{A}^4(-\lambda)=\begin{bmatrix}1&0&0&0\\0&\cos\lambda&\sin\lambda&0\\0&-\sin\lambda&\cos\lambda&0\\0&0&0&1\end{bmatrix};\text{ and } \mathbf{r}_5=\begin{bmatrix}R\cos\theta\\R\sin\theta\\z\\1\end{bmatrix}$$

where φ is the angle of the workpiece rotation; a and c are the setting parameters (Fig. 12.16b); β is the half of the cone angle (Fig. 12.13a); λ is the angle for hyperboloid setup (Fig. 12.13b); and $\mathbf{r}_5$ is the position vector of the cylindrical grinding wheel, R is the wheel radius (Fig. 12.16b), θ and z are the polar angle and the longitudinal coordinate, respectively, for a point of the cylinder.

The setup parameters a, c, and β are constrained

$$a\tan\beta + R\cos\beta - R_0 = 0$$

$$c\tan\beta + R - R_0/\cos\beta = 0$$

where R_0 is a circle radius within the given section of the machined surface (Fig. 12.13).

Parameters φ, θ, and z are related by a one-parameter enveloping constraint

$$[\mathbf{r}_\theta\, \mathbf{r}_\varphi\, \mathbf{r}_z] = 0 \tag{12.55}$$

where $\mathbf{r}_\theta$, $\mathbf{r}_\varphi$, $\mathbf{r}_z$ are the partial derivatives of vector $\mathbf{r}$, Eq. (12.48), with respect to θ, φ, and z, respectively.

By executing a formal manipulation according to formulae (12.54) and (12.55), the equation of the hyperboloid is obtained

$$\mathbf{r}=\begin{bmatrix}X\cos\varphi - Y\sin\varphi\\X\sin\varphi + Y\cos\varphi\\Z\end{bmatrix} \tag{12.56}$$

$$X = R\cos\theta\cos\beta + p\sin\beta$$

$$Y = R\sin\theta\cos\lambda + z\sin\lambda$$

$$Z = -R\cos\theta\sin\beta + p\cos\beta - a$$

$$p = z\cos\lambda + c - R\sin\theta\sin\lambda$$

$$\tan\theta = z\sin\lambda\cot\beta/(z + c\cos\lambda)$$

The equation of the cone surface obtained without rotation by angle λ is determined from Eq. (12.56) with λ = 0

$$\mathbf{r}_0 = \begin{bmatrix} (R_0 + z\sin\beta)\cos\varphi \\ (R_0 + z\sin\beta)\sin\varphi \\ z\cos\beta \end{bmatrix} \tag{12.57}$$

The partial derivatives $\mathbf{r}_{0\varphi}$, $\mathbf{r}_{0z}$, $\mathbf{r}_{0\varphi\varphi}$, and $\mathbf{r}_{0zz}$ for the cone, Eq. (12.57), are

$$\mathbf{r}_{0\varphi} = \begin{bmatrix} -(R_0 + z\sin\beta)\sin\varphi \\ (R_0 + z\sin\beta)\cos\varphi \\ 0 \end{bmatrix}; \mathbf{r}_{0z} = \begin{bmatrix} \sin\beta\cos\varphi \\ \sin\beta\sin\varphi \\ \cos\beta \end{bmatrix};$$

$$\mathbf{r}_{0\varphi\varphi} = \begin{bmatrix} -(R_0 + z\sin\beta)\cos\varphi \\ -(R_0 + z\sin\beta)\sin\varphi \\ 0 \end{bmatrix}; \mathbf{r}_{0\varphi z} = \begin{bmatrix} -\sin\beta\sin\varphi \\ \sin\beta\cos\varphi \\ 0 \end{bmatrix}; \mathbf{r}_{0zz} = \begin{bmatrix} 0 \\ 0 \\ 0 \end{bmatrix}$$

The unit vector $\mathbf{n}_0$ of the normal to the surface and the second fundamental form $\mathbf{II}_0$, Eq. (12.42), are

$$\mathbf{n}_0 = [\mathbf{r}_{0\varphi} \times \mathbf{r}_{0z}] / |\mathbf{r}_{0\varphi} \times \mathbf{r}_{0z}| = \begin{bmatrix} \cos\beta\cos\varphi \\ \cos\beta\sin\varphi \\ -\sin\beta \end{bmatrix}$$

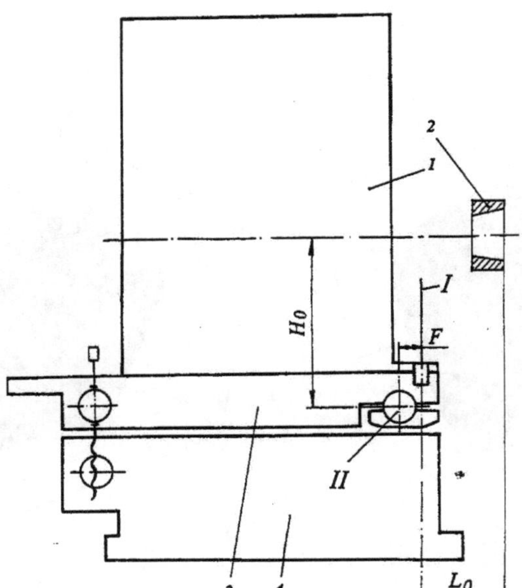

Figure 12-14 Outline of internal grinding machine: 1 - the workhead; 2 - the workpiece; *I* and *II*- the pivot axes.

$$\mathbf{II}_0 = (\mathbf{r}_{0\varphi\varphi}\cdot\mathbf{n}_0)d\varphi^2 + 2(\mathbf{r}_{0\varphi z}\cdot\mathbf{n}_0)d\varphi dz + (\mathbf{r}_{0zz}\cdot\mathbf{n}_0)dz^2 =$$

$$\begin{bmatrix}-(R_0+z\sin\beta)\cos\varphi\\-(R_0+z\sin\beta)\sin\varphi\\0\end{bmatrix}\cdot\begin{bmatrix}\cos\beta\cos\varphi\\\cos\beta\sin\varphi\\-\sin\beta\end{bmatrix}d\varphi^2 + 2\begin{bmatrix}-\sin\beta\sin\varphi\\\sin\beta\cos\varphi\\0\end{bmatrix}\cdot\begin{bmatrix}\cos\beta\cos\varphi\\\cos\beta\sin\varphi\\-\sin\beta\end{bmatrix}d\varphi dz+\begin{bmatrix}0\\0\\0\end{bmatrix}dz^2$$

$$= \cos\beta\,(R_0 + z\sin\beta)\,d\,\varphi^2 \qquad (12.58)$$

12.3.3.2 Setup problem

The purpose is to calculate the deviation Δ, Eq. (12.43), of the hyperboloid, Eq, (12.56), from the cone, Eq. (12.57). In terms of schematics shown in Figure 12.10, surface F is the hyperboloid model and surface F_0 is the cone model. In this case, the two-step procedure, Eqs.(12.45) and (12.46), is aimed at establishing the correlation between the setup angle λ and deviation Δ.

Step 1. According to Eqs. (12.44) and (12.45), the first order approximation yields

$$\mathbf{d}_1 = \left.\frac{\partial\mathbf{r}}{\partial\lambda}\right|_{\lambda=0}\lambda = z\lambda\frac{R_0+z\sin\beta}{R_0+z\sin\beta-R\cos\beta}\begin{bmatrix}-\sin\varphi\\\cos\varphi\\0\end{bmatrix}$$

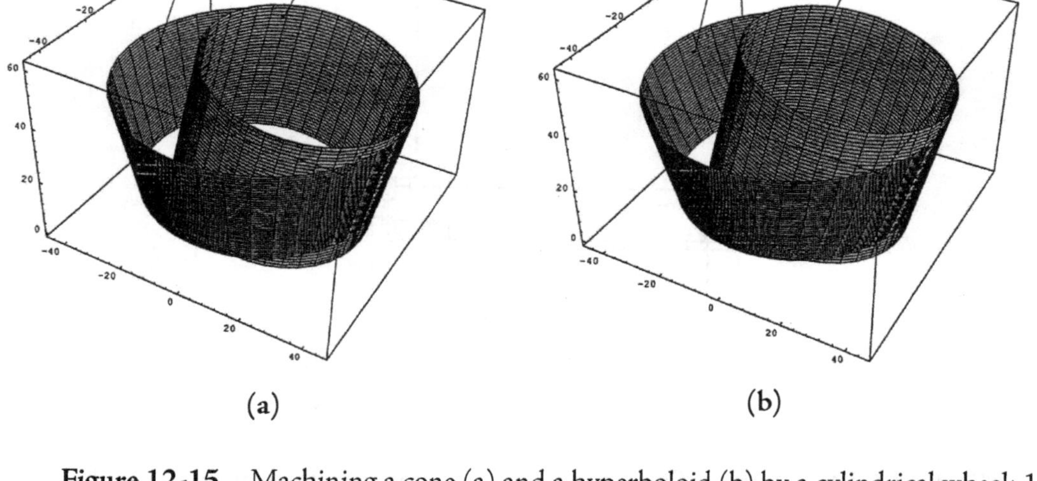

Figure 12-15 Machining a cone (a) and a hyperboloid (b) by a cylindrical wheel: 1 is a grinding wheel surface; 2 is the cone; and 3 is the hyperboloid.

$$\Delta^{(1)} = (\mathbf{d}_1 \cdot \mathbf{n}_0) = z\lambda \frac{R_0 + z\sin\beta}{R_0 + z\sin\beta - R\cos\beta} \begin{bmatrix} -\sin\varphi \\ \cos\varphi \\ 0 \end{bmatrix} \cdot \begin{bmatrix} \cos\beta\cos\varphi \\ \cos\beta\sin\varphi \\ -\sin\beta \end{bmatrix} = 0$$

Step 2. Since $\Delta^{(1)} = 0$, the second order approximation is determined. The increments of the curvilinear coordinate φ and z, Eqs. (12.47), are

$$d\varphi = \frac{[\mathbf{d}_1\, \mathbf{r}_{0z}\, \mathbf{n}_0]}{|\mathbf{r}_{0\varphi} \times \mathbf{r}_{0z}|} \text{ and } dz = \frac{[\mathbf{r}_{0\varphi}\, \mathbf{d}_1\, \mathbf{n}_0]}{|\mathbf{r}_{0\varphi} \times \mathbf{r}_{0z}|}$$

Substituting the increments d_φ and dz in the second fundamental form $\mathbf{II}_0$, Eq. (12.58), and then, substituting $\mathbf{II}_0$ and value $\mathbf{d}_2$, Eq. (12.44), in equation for the second order approximation,Eq. (12.46), we obtain the normal deviation Δ of the hyperboloid points form the cone (the

$$\Delta = \Delta^{(2)} = \frac{1}{2}[(\mathbf{d}_2 \cdot \mathbf{n}_0) - \mathbf{II}_0] = \frac{\lambda^2}{2} \frac{z^2\cos\beta}{R_0 + z\sin\beta - R\cos\beta}$$

The controlled deviations Δ_c measured at a section perpendicular to the cone axes become

$$\Delta_c = \frac{\Delta}{\cos\beta} = \frac{\lambda^2}{2} \frac{z^2}{R_0 + z\sin\beta - R\cos\beta} \quad (12.59)$$

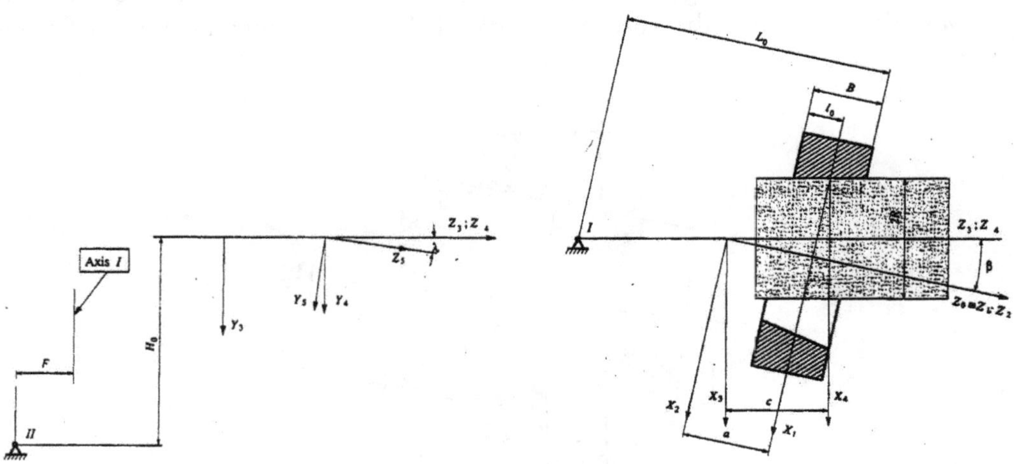

Figure 12-16 Coordinate systems associated with the units of the grinding machine: (a) frontal view; (b) view from above.

Using expression (12.58) allows determination of all the setup parameters. The angle λ for a given value Δ_{max} of the maximal crown is calculated on the sampling length $l_s = |z_1| + |z_2|$ (Fig. 12.13). If $z_1 = -z_2 = 0.5l_s$ and $z_m = 0$, Δ_{max} is the half-sum of Δ_1 and Δ_2, i.e.,

$$\Delta_{max} = (\Delta_1 + \Delta_2)/2 = (1/4)\,\lambda^2 f l_s/(f^2 - \sin^2\beta), \tag{12.60}$$

where $\Delta_1 = \Delta_c(z_1)$; $\Delta_2 = \Delta_c(-z_2)$; and $f = (R + l_0 \tan\beta - R\cos\beta)/l_s$.

The angle λ is calculated from Eq. (12.60). For this λ, the value Δ_{max} must also lie within the given tolerance after 20 percent wear of the grinding wheel. The actual angle β_a of the machined ring differs from the nominal angle β by the second order value depending on the angle λ,

$$\beta_a = \beta - (\Delta_2 - \Delta_1)/[(z_2 - z_1)\cos\beta] = \beta + (1/2)\,\lambda^2 \tan\beta/(f^2 - \sin^2\beta) \tag{12.61}$$

Equation (12.61) enables us to compute the linear deviation Δ_{ba} of the angle β on the base length l_s after 20 percent wear of the wheel:

$$(\Delta_{ba})_{20\%} = l_s \cos\beta[(\beta_a)_{new} - (\beta_a)_{20\%}], \tag{12.62}$$

where $(\beta_a)_{new}$ and $(\beta_a)_{20\%}$ are obtained through Eq. (12.61) for the radii R and $0.8\,R$, respectively.

In cases when the tolerances for Δ_{max} or Δ_{ba} become invalid after 20 percent wear of the grinding wheel, the machining layout is modified: the conical grinding wheel (Fig. 12.17) is used instead of the cylindrical wheel as shown in Figures 12.15 and 12.16. The cone angle γ of the grinding wheel in this machining method is used as an additional parameter of the setting. Modified equations (12.59–62) for this case are not considered here.

Numerical Example 12.2. The setup of the internal grinding machine ME326F2002.

Given:

- *Machine dimensions* (Figs. 12.14 and 12.16): the distance from the frontal face of the workpiece to axis I of the workhead rotation is $L_0 = 140.0$ mm; the distance from the

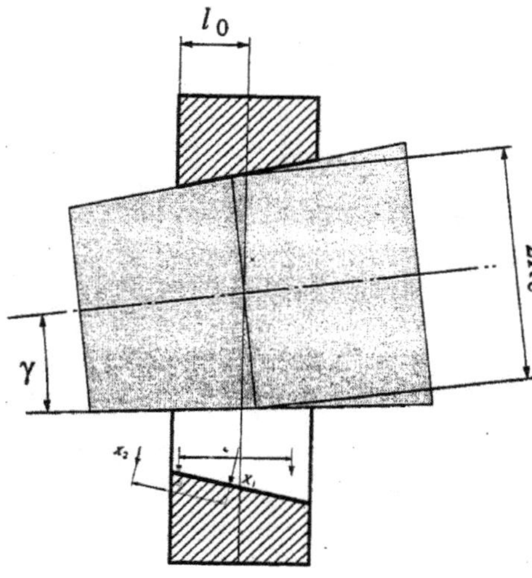

Figure 12-17 Schematics of grinding by conical wheel.

workpiece spindle axis to axis II is $H_0 = 278.0$ mm; distance between axes I and II is $F = 30.0$ mm;

- *Parameters of the machined ring* (Fig. 12.13): a small diameter of the conical bore on the ring face is $D = 76.21$ mm; the ring width is $B = 37.0$ mm; the half-angle of the cone is $\beta = 11°40'$; distance from the crown top to the base face is $l_0 = 18.3$ mm; the sampling length for the crown measurement is $l_s = 24$ mm; the measurement zone is symmetric, i.e., $z_1 = -z_2 = 12.0$ mm; $z_m = 0$; crown value is $\Delta_{max} = 0.014$ mm;
- *Grinding wheel:* Diameter of the new wheel is $2R = 63.0$ mm. Tolerances after 20 percent wear of the grinding wheel: the crown value must be $(\Delta_{max})_{tol} \geq 0.009$ mm; the deviation of the half-angle β at the sampling length l_s is $(\Delta_{ha})_{tol} \leq 0.002$ mm.

Required: Setup parameters.

Solution:

(1) Angle λ of the workhead rotation in vertical plane according to Eq. (12.60) is:

$$\lambda = \sqrt{2\Delta_{max}\left(f^2 - \sin^2\beta\right)/\left(0.5l_s f\right)} =$$

$$\sqrt{2\cdot 0.014(0.9195^2 - \sin^2 11°40')/(0.5\cdot 24\cdot 0.9195)} = 0.04519 = 2°35',$$

where

$f = (D/2 + l_0 \tan\beta - R\cos\beta)/(0.5\, l_s) = (76.21/2 + 18.3 \tan 11°40' - 31.5 \cos 11°40')/12$
$= 0.9195$ mm.

With this λ, the values Δ_1 and Δ_2 according to Eq. (12.59) are:

$\Delta_1 = \Delta_c(z_1 = 12) = 0.01094$ mm and $\Delta_2 = \Delta_c(z_2 = -12) = 0.01716$ mm.

(2) Angle β of the workhead rotation in the horizontal plane by Eq. (12.61) is

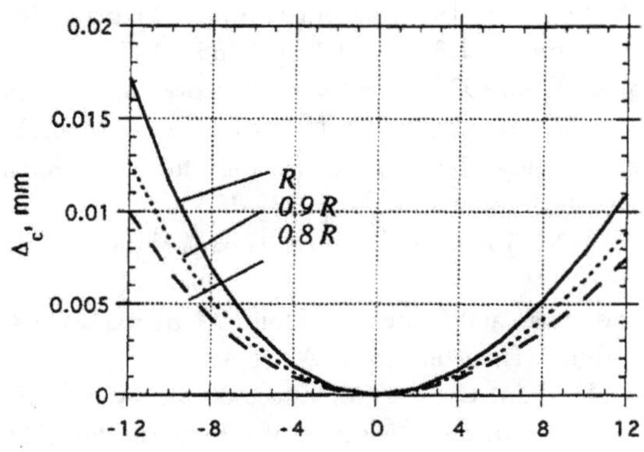

Figure 12-18 A dependence of the form of crowning profile on the radius R of the cylindrical grinding wheel.

$\beta_b = \beta_a = 11°40' + 0.5{\cdot}0.04519^2 \tan 11°40'/(0.9195^2 - \sin^2 11°40') = 11°39.67'$.

(3) A wheel wear leads to change in the form of the bore profile of the bearing ring. Three corresponding profiles obtained through Eq. (12.59) are shown in Figure 12.18: for new wheel and after 10% (radius 0.9 *R*) and 20% wear (radius 0.8 *R*). After 20% wear, with the calculated value $\lambda = 2°35'$, the crowning value $(\Delta_{max})_{20\%}$ and the deviation $(\Delta_{ba})_{20\%}$ of the angle β are as follows [the calculations through Eqs. (12.60) and (12.62) are omitted]: $(\Delta_{max})_{20\%} = 0.009$ mm; $(\Delta_{ba})_{20\%} = 0.004$ mm.

(4) Since $(\Delta_{ba})_{20\%} > (\Delta_{ba})_{tol} = 0.002$ mm, this setting-up is inoperative. Therefore, the setting with the conical wheel according to Figure 12.17 is used. For this case, the wheel has half of the cone angle $\gamma = 5°19'$, while the design diameter of the new wheel in the given section is $2R_c = 63$ mm (the corresponding calculations are omitted).

12.4 Conclusion

Although the linearization approach for the computation of the machine accuracy of machines is very effective in the majority of the practical problems, there are important actual cases, in which the first order approximation is vanishing and the corresponding approach is collapsed. Typical examples are the layout errors of machine tools, setup of the form-shaping system, in which one of the motions is located in the tangent plane to the processed element, and so on. Generally, the problem can be formulated as searching for a distance between two objects expressed through two form-shaping functions. In this chapter, two approaches to solution of the developed non-linear problem are considered using real-life examples: the direct solution and application of the higher order approximations.

References

[1] Reshetov, D. N. and Portman, V. T., 1988, *Accuracy of Machine Tools*, ASME Press, NY, New York.

[2] Portman, V.T., Weill, R., and Shuster, V.G., 1997, High-Order Approximation in Accuracy Computations of Machine Tools and Robots *Transactions of ASME: Journal of Manufacturing Science and Engineering* 119, pp. 732 - 742.

[3] Portman, V.T. and Shuster, V.G., 1997, Layout Errors of Machine Tools *International Journal of Machine Tools & Manufacture*, Vol. 37, No. 10, pp. 1485-1497.

[4] Colvin, F. H. and Stanley, F. A., 1937, *Gear Cutting Practice*, McGraw-Hill. New York, p. 57.

[5] Litvin, F. L., 1994, *Gear Geometry and Applied Theory*, Prentice Hall, Englewood Cliffs, NJ.

[6] Ehmann. K. F., 1990, Grinding wheel profile definition for the manufacture of drill flutes, *CIRP Annals – Manufacturing* Technology, 39(1), pp. 153-156.

[7] Korn, G. A., and Korn, T. M., 1968, *Mathematical Handbook for Scientists and Engineers*, McGraw-Hill, New York.

[8] Wolfram S., 1992, Mathematica: A System for Doing Mathematics by Computer, 2nd ed., *Addison –Wesley Publ. Co.*, Reading, Massachusetts, MA, USA.

[9] Trushakin, N. T. and Fradkin, E. I., 1978, Effect of the Wheel Arrangement on Machining Accuracy and Temperature with the 0^0 Method of Gear Grinding, *Machine Tools and Tooling*, 6, pp. 31-32 (in Russian).

Appendix 1: Function atan2

The ratio of two real numbers defines a single-valued function of two arguments:

$$\operatorname{atan2}(y,x) = \begin{cases} \arctan(y/x), \text{ for } x > 0 \\ \pi + \arctan(y/x), \text{ for } y \geq 0, x < 0 \\ -\pi + \arctan(y/x), \text{ for } y < 0, x < 0 \\ \pi/2, \text{ for } y > 0, x = 0 \\ -\pi/2, \text{ for } y < 0, x = 0 \\ \textit{undefined}, \text{ for } y = 0, x = 0 \end{cases}$$

Here, it is applied for calculation of a single value of angle φ given through sin φ and cos φ.

Example A1.1:

(a) Equation $\sin\varphi = 1/2$ has two solutions

$$\varphi = \begin{cases} \arcsin(1/2) \\ \pi - \arcsin(1/2) \end{cases} = \begin{cases} \pi/6 \\ 5\pi/6 \end{cases}$$

Equation $\cos\varphi = \sqrt{3}/2$ also has two solutions

$$\varphi = \begin{cases} \arccos(\sqrt{3}/2) \\ -\arccos(\sqrt{3}/2) \end{cases} = \begin{cases} \pi/6 \\ -\pi/6 \end{cases}$$

(b) Two simultaneous equation $y \equiv \sin\varphi = 1/2$ and $x \equiv \cos\varphi = \sqrt{3}/2$ results in a single value. Namely, since $x = \sqrt{3}/2 > 0$, the first line in the atan2 definition gives

$$\operatorname{atan2}(y, x) = \arctan(y/x) = \arctan\frac{1/2}{\sqrt{3}/2} = \pi/6.$$

Appendix 2: Synthesis of the FSF using Mathematica® software system

A2.1 Basic HTM matrices of the FSF theory

A2.1.1 Matrices of the elementary motions

In the following Mathematica® code the 4 × 4 HTM matrices of elementary motions (Table 3.5) are labelled m_1, m_2, m_3, m_4, m_5, and m_6,

```
  m1={{1,0,0,xx},{0,1,0,0},{0,0,1,0},{0,0,0,1}}
  m2={{1,0,0,0},{0,1,0,yy},{0,0,1,0},{0,0,0,1}}
  m3={{1,0,0,0},{0,1,0,0},{0,0,1,zz},{0,0,0,1}}
  m4={{1,0,0,0},{0,Cos[th],-Sin[th],0},{0,Sin[th],
Cos[th],0},{0,0,0,1}}
  m5={{Cos[ps],0,Sin[ps],0},{0,1,0,0},{-Sin[ps],0,
Cos[ps],0},{0,0,0,1}}
  m6={{Cos[ph],-Sin[ph],0,0},{Sin[ph],Cos[ph],0,0},
{0,0,1,0},{0,0,0,1}}
```
(A2.1)

A2.1.2 The error matrix

In the following Mathematica® code the 4 × 4 HTM error matrix, Eq. (8.9), are labelled **del**,

```
  del={{0,-dph,dps,dx},{dph,0,-dth,dy},
{-dps,dth,0,dz},{0,0,0,0}}
```
(A2.2)

For convenience, some error matrices associated with FSS links are prepared. If, for example, the number of the form-shaping system links is less than 11 (including the fixed link No. 0), the following 10 matrices are defined:

```
  del0={{0,-dph0,dps0,dx0},{dph0,0,-dth0,dy0},
{-dps0,dth0,0,dz0},{0,0,0,0}}
  del1={{0,-dph1,dps1,dx1},{dph1,0,-dth1,dy1},
{-dps1,dth1,0,dz1},{0,0,0,0}}
  del2={{0,-dph2,dps2,dx2},{dph2,0,-dth2,dy2},
{-dps2,dth2,0,dz2},{0,0,0,0}}
  ...
  del9={{0,-dph9,dps9,dx9},{dph9,0,-dth9,dy9},
{-dps9,dth9,0,dz9},{0,0,0,0}}
```
(A2.3)

A2.1.3 Saving main matrices

To append the listed matrix definitions to any developed file, the entered symbols are saved using the command

```
Save["matr", {m1, m2, m3, m4, m5, m6, del0, del1, del2, del3,
del4, del5, del6, del7, del8, del9}]
```
(A2.4)

where "**matr**" and "**del**" are the filenames.

To manipulate with the matrices using their symbols in any processed file, the name "**matr**" is entered as follows:

```
<<matr
```
(A2.5)

where command "<<" allows reading the symbols in Eq. (A4) and evaluation of all read files within the processed file.

A2.1.4 Basic HTM matrices for the planar (2-D) case

- *Three matrices of the elementary motions* associated with symbols *mx, my, mz* are defined:

```
mx={{1,0, xx},{0,1,0},{0,0,1}}
my={{1,0,0},{0,1, yy},{0,0,1}}
mph={{Cos[ph],Sin[ph],0},{-Sin[ph],Cos[ph],0},{0,0,1}}
```
(A2.6)

- *The error matrix*, Eq. (8.9), is defined:

```
eps ={{0,-dph,dx},{dph,0,dy},{0,0,0}}
```
(A2.7)

For convenience, four error matrices associated with FSS links are prepared,

```
eps0={{0,-dph0,dx0},{dph0,0,dy0},{0,0,0}}
eps1={{0,-dph1,dx1},{dph1,0,dy1},{0,0,0}}
eps2={{0,-dph2,dx2},{dph2,0,dy2},{0,0,0}}
eps3={{0,-dph3,dx3},{dph3,0,dy3},{0,0,0}}
```

To append the listed matrix definitions to a developed file, the entered symbols are saved:

```
Save["matr3", {mx, my, mph, eps0, eps1, eps2, eps3}]
```
(A2.8)

where "**matr3**" is the filename.

To manipulate with the matrices using only their symbols in any processed file, Eq. (A2.8), is entered as **<<matr3**.

A2.1.5 Transformation of the HTM matrices and vectors to the third order form

The 4×4 matrices **m4, m5, m6** (matrices of rotations) and the 4×1 vectors can be transformed into the regular matrices and vectors of the third order using the 3×4 matrix

$$T_{4to3} = \begin{bmatrix} 1 & 0 & 0 & 0 \\ 0 & 1 & 0 & 0 \\ 0 & 0 & 1 & 0 \end{bmatrix}$$

In terms of Mathematica® codes,

```
T4to3 ={{1,0,0,0},{0,1,0,0},{0,0,1,0}}
```
(A2.9)

Examples:

(a) The code for matrix transformation

```
In[1]:= m6 = {{Cos[ph],-Sin[ph],0,0},{Sin[ph],Cos
        [ph],0,0},{0,0,1,0}},{0,0,0,1}};
In[2]:= T4to3.m6.Transpose[T4to3]
Out[2]:= {{Cos[ph],-Sin[ph],0},{Sin[ph],Cos[ph],0},{0,0,1}}
```

(b) The code for vector transformation

```
In[1]:=  T4to3.{x,y,z,1}
Out[1]:= {x,y,z}
```

where the point "." stands for matrix multiplication; **In[n] :=** is the input information; and **Out[n] :=** is the *n*th output.

A2.2 Examples: Solutions of problems considered in this book

A2.2.1 Synthesis of manipulation matrix, Eq. (4.38)

Given: Coordinate code of the lathe is $k = 631$. **Required:** Synthesis of the manipulation matrix.
Solution:

- The code starts from the command **<<matr**, Eq. (A2.5), to add matrices of links motion m6, m3, and m1, Eq. (A2.1) to the processed file. The identifiers A01, A12, A23 are assigned to matrices m6, m3, and m1, respectively;
- Manipulation matrix A03 is obtained by the ordered multiplication A01.A12.A23;
- The result is shown in the matrix form.

```
In[1]:=   <<matr
          "Links motion matrices";
          A01=m6;
          A12=m3;
          A23=m1;
          "Manipulation matrix";
          A03=A01.A12.A23;
          "A03 ="MatrixForm[A03]
```

Out[n]:=

$$\mathbf{A03} = \begin{bmatrix} \cos ph & -\sin ph & 0 & xx\cos ph \\ \sin ph & \cos ph & 0 & -xx\sin ph \\ 0 & 0 & 1 & zz \\ 0 & 0 & 0 & 1 \end{bmatrix}$$

A2.2.2 Synthesis of the form-shaping function, Eq. (4.41)

Given: Coordinate code of the lathe is $k = 631$. **Required:** Synthesis of the form-shaping function (FSF). **Solution:**

- The code starts from seven commands of the *Example A2.1*;
- The single-point tool vector $r_3 = [0, 0, 0, 1]^T$ is entered in the form r3 = {0, 0, 0, 1} and multiplied by the manipulation matrix.
- The Output demonstrates the FSF in the matrix form.

```
In[1]:=   <<matr;
          "Links motion matrices";
          A01 = m6;
          A12 = m3;
          A23 = m1;
          "Manipulation matrix";
          A03 = A01.A12.A23;
          "Single point-tool vector";
          r3={0,0,0,1};
          r0=A03.r3;
          "r0 ="MatrixForm[r0]
```

Out[n]:= r0 =

$$\begin{bmatrix} xx\cos ph \\ -xx\sin ph \\ zz \\ 1 \end{bmatrix} \tag{A2.10}$$

A2.2.3 Setup (table 6.4)

Given: FSF of the lathe from the previous Example. On the lathe, the cylinder of radius R is machined using longitudinal turning. **Required:** (a) Equation of the machined cylinder; (b) equation of the unit vector of the normal to the machined cylinder; (c) graphical presentation of the cylinder of radius $R = 50$ mm and length 120 mm, the initial cross section is located on $z = 20$ mm. **Solution:**

(a) *Input description:*

- The code starts from vector r0, Eq. (A2.10);
- The single constraints $xx = R$ is applied using replacement operator /. xx –> R, and the HTM vector is rearranged to a Cartesian coordinate vector using matrix $T_{4to3,}$ Eq. (A2.9), with renaming r0 for rcyl;

- Partial derivatives dph with respect to ph and dz with respect to zz are defined using operator D[rcyl, ph] and D[rcyl, zz], respectively;
- The normal NN is defined through vector cross product Cross[dph, dz];
- Unit normal vector nn is defined and simplified through equation nn = NN/(NN.NN)$^{1/2}$;
- The cylinder rcyl50 of radius 50 is defined using replacement operator /. R –> 50;
- Using commands ParametricPlot3D[rcyl50, {ph, 0, 2 Pi}, {zz, 20, 140}], the cylinder plot is built. The command allows building the surfaces given in the two-parametric form, rcyl = rcyl(ph, zz). The parameter limits are $0 < ph \leq 2\pi$ and $20 < zz \leq 140$

(b) *Output description*:

Formula of the unit vector of the normal gives two possible values, nn = $\pm[R \cos \text{ph}, R \sin \text{ph}, 0]^T$;

- The plot shows the cylindrical surface between $z = 20$ and $z = 140$.

```
In[n]:=   <<matr;
          r0={xx Cos[ph], xx Sin[ph], zz, 1};
          rcyl=Tr4to3.r0 /. xx -> R
          dph=D[rcyl,ph];
          dz=D[rcyl, zz];
          NN=Cross[dph, dz];
          nn=Simplify[NN/(NN.NN)^(1/2)]
          rcyl50=rcyl /. R->50;
          ParametricPlot3D[rcyl50,{ph, 0, 2 Pi},{zz, 0, 100}]
```

Out[n]:= $\{-\frac{R\cos ph}{\sqrt{R^2}}, \frac{R\sin ph}{\sqrt{R^2}}, 0\}$

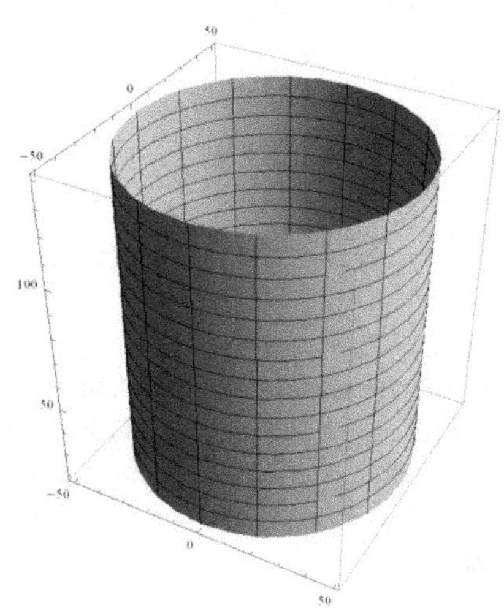

The conventional coordinate frame used in Mathematica differs from the frame accepted in the lathe (see system $X_1Y_1Z_1$ in Figure 4.7). To transform the coordinate systems, two successive rotations are entered in the previous code: around the Y-axis at 90° and around the Z-axis at −90°. See the following code, where the output is related to the lathe accepted frame:

```
In[n]:=   <<matr;
          r0={xx Cos[ph],-xx Sin[ph],zz,1};
          rL=(m5/.ps->Pi/2).(m6/.ph->-Pi/2).r0
          rcyl=Tr4to3.rL/.xx->R;
          dph=D[rcyl,ph];
          dc=D[rcyl,zz];
          NN=Cross[dph,dc];
          Nn=Simplify[NN/(NN.NN)^(1/2)]
          Rcyl=rcyl/.R->50;
          ParametricPlot3D[Rcyl,{ph,0,2 Pi},{zz,20,140}]
```

Out[n]:= $\{0, -\frac{R\cos ph}{\sqrt{R^2}}, \frac{R\sin ph}{\sqrt{R^2}}\}$

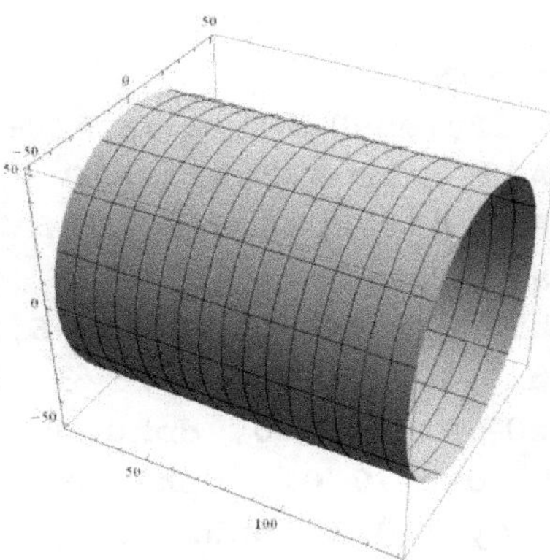

A2.2.4 Geometrical accuracy of the 2-DOF (RR) manipulator, Eq. (10.1-6)

Given: Dimensions of the 2-DOF manipulator (Fig. 10.1): the end-effector (EE) point is given through position vector $\mathbf{r}_0 = [0, 0, 1]^T$; $a_1 = a_2 = 250$ mm; geometrical errors are $\delta_{x0} = \delta_{y0} = 0$; $\delta_{\varphi 0} = 0$; $\delta_{x1} = \delta_{y1} = 0.01$ mm; $\delta_{\varphi 1} = 0.0005$; $\delta_{x2} = \delta_{y2} = 0.1$ mm; $\delta_{\varphi 2} = 0$; $\delta_{x3} = \delta_{y3} = 0.01$ mm; $\delta_{\varphi 3} = 0.0005$; $\delta_{x4} = \delta_{y4} = 0.1$ mm; and $\delta_{\varphi 4} = 0$. **Required:** The distance error δ_{AB} between the two position of the EE point: configuration A ($\varphi_{2A} = \pi/12$; and $\varphi_{4A} = \pi/6$) and configuration B ($\varphi_{2B} = \pi/6$; and $\varphi_{4A} = \pi/3$).
Solution: The In[n] commands are entered in the given order:

- The code fetches file matr3, Eq. (A2.9), with the list of 3 × 3 matrices;

The motion matrices A01 with new parameter pfi1, A12 with new parameter a2, A23 with new parameter pfi2, and A34 with new parameter a4;

- The manipulating matrix ${}^0A_4$, Eq. (10.1), called A04;
- The reciprocal matrices A10(−ph1), A21(−a2), A32(−ph3), and A43(−a4);
- The total error matrix E, Eq. (10.3), called "Et";
- The numerical data condA, condB, and condE;
- The position errors `erA` of point A and `erB` of point B;
- The distance error `dAB` is defined.

The Out[n] consists of a single number for verification of the code.

```
In[n]:=    <<matr3;
           "FSF manipulating matrix";
           A01=mph/.ph→phi1;
           A12=mx/.xx→a2;
           A23=mph/.ph→phi3;
           A34=mx/.xx→a4;
           A04=Simplify[A01.A12.A23.A34];eps0
           A10=mph/.ph→-phi1;
           A21=mx/.xx→-a2;
           A23=mph/.ph→-phi3;
           A34=mx/.xx→-a4;
           r4={0,0,1};
           r0=A04.r4;
           "Error matrix";
           Et = Simplify[eps0+A01.eps1.A10+A01.A12.eps2.A21.A10 +
           A01.A12.A23.eps3.A32.A21.A10 +
           A01.A12.A23.A34.eps4.A43.A32.A21.A10];
           "Numerical Example";
           condA={phi1→Pi/12, phi3→Pi/6, a2→250, a4→250};
           condB={phi1→Pi/6, phi3→Pi/3, a2→250, a4→250};
           condE={dx0→0, dy0→0, dph0→0,
           dx1→0.01, dy1→0.01, dph1→0.0005,
           dx2→0.1, dy2→0.1, dph2→0,
           dx3→0.01, dy3→0.01, dph3→0.0005,
           dx4→0.1, dy4→0.1, dph4→0};
           "Two points";
           erA=N[(Et.r0)/.condA/.condE];
           erB=N[(Et.r0)/.condB/.condE];
           "Distance AB";
           dAB=((erA-erB).(erA-erB))^(1/2)
Out[n]:= 0.333215 mm
```

A2.2.5 Inverse kinematics of the GSP platform, Eq. (5.38-40)

Given: The form-shaping system combines the base (coordinate frame S_0) and the moving platform (coordinate frame S_1); dimensions of the GSP platform (Fig. 5.26): $r_1 = 700$ mm; $r_2 = 250$ mm; angles of the limb ends location on the base plane are α_i ($i = 1,2,...,6$); angles of the limb ends location on the moving platform plane are β_i ($i = 1,2,...,6$). **Required:** (a) Lengths of the limbs in the general case; (b) Lengths of the limbs of the 3–3 platform in the configuration $x = y = 0$; $z = 400$ mm; $\theta = \psi = \varphi = 0$, with end limbs location $\alpha_1 = \alpha_6 = -\pi/3$; $\alpha_2 = \alpha_3 = \pi/3$; $\alpha_4 = \alpha_5 = \pi$; $\beta_1 = \beta_2 = 0$; $\beta_3 = \beta_4 = 2\pi/3$; $\beta_5 = \beta_6 = 4\pi/3$.

Solution:

```
In[n]:= <<matr;
      "FSF";
      A01 = m1.m2.m3.m4.m5.m6;
      "Lower ends of limbs in coordinate frame S0";
      rM1={r1 Cos[al1],r1 Sin[al1],0,1};
      rM2={r1 Cos[al2],r1 Sin[al2],0,1};
      rM3={r1 Cos[al3],r1 Sin[al3],0,1};
      rM4={r1 Cos[al4],r1 Sin[al4],0,1};
      rM5={r1 Cos[al5],r1 Sin[al5],0,1};
      rM6={r1 Cos[al6],r1 Sin[al6],0,1};
      "Upper ends of limbs in coordinate frame S1";
      rQ1={r2 Cos[bet1],r2 Sin[bet1],0,1};
      rQ2={r2 Cos[bet2],r2 Sin[bet2],0,1};
      rQ3={r2 Cos[bet3],r2 Sin[bet3],0,1};
      rQ4={r2 Cos[bet4],r2 Sin[bet4],0,1};
      rQ5={r2 Cos[bet5],r2 Sin[bet5],0,1};
      rQ6={r2 Cos[bet6],r2 Sin[bet6],0,1};
      "Upper ends of limbs in coordinate frame S0";
      rU1= A01.rQ1;
      rU2= A01.rQ2;
      rU3= A01.rQ3;
      rU4= A01.rQ4;
      rU5= A01.rQ5;
      rU6= A01.rQ6;
      "Limb lengths in the general case";
      L1=((rU1-rM1).(rU1-rM1))^(1/2);
      L2=((rU2-rM2).(rU2-rM2))^(1/2);
      L3=((rU3-rM3).(rU3-rM3))^(1/2);
      L4=((rU4-rM4).(rU4-rM4))^(1/2);
      L5=((rU5-rM5).(rU5-rM5))^(1/2);
      L6=((rU6-rM6).(rU6-rM6))^(1/2);
      "Numerical example: 3-3 GSP"
      cond={r1→700, r2→250,
        al1→ -Pi/3, al2→ Pi/3, al3→ Pi/3, al4→ Pi, al5→ Pi,
      al6→ -Pi/3,
        bet1→ 0, bet2→ 0, bet3→ 2Pi/3, bet4→ 2Pi/3, bet5→
      4Pi/3, bet6→ 4Pi/3, xx→0,yy→0,zz→400,th→0,ps→0,ph→0};
      "Limb lengths"
      N[L1/.cond]
      N[L2/.cond]
      N[L3/.cond]
      N[L4/.cond]
      N[L5/.cond]
      N[L6/.cond]
```

```
Out[n]:= Limb lengths
     733.144
     733.144
     733.144
     733.144
     733.144
     733.144
```

A2.2.6 Error matrix E and Jacobian J_t for 2RR robot

In the very beginning, the code fetches file "`matr3`" with the list of 3 × 3 matrices. Then, the following subjects are computed in the given order:

- The manipulating matrix ${}^0\mathbf{A}_4$, Eq. (10.1), called "A04";
- The total error matrix **E**, Eq. (10.149), called "Et";
- Twist δ, Eq. 10.154), called "tw"; and,
- The Jacobian matrix **J**, Eq. (10.157), called "Jt".

Remember that the command **`Simplify[`*`qqq`*`]`** simplifies of the bracketed expression *qqq*, the command **`Transpose[`*`mmm`*`]`** transposes the bracketed matrix *mmm*, and the command **`Inverse[`*`mmm`*`]`** carries out inversion of the square matrix *mmm*.

```
In[1] :=     "3RR robot";
             <<matr3;
             "FSF manipulating matrix";
             A01=mph /. phi→phi1;
             A12=mx/.xx→a2;
             A23=mph /. phi→phi3;
             A34=mx/.xx→a4;
             A04=Simplify[A01.A12.A23.A34];
             "FSF variation";
             Et= Simplify[eps0+A01.eps1.Inverse[A01]+
                 A01.A12.eps2.Inverse[A01.A12]+A01.A12.A23.eps3.
                 Inverse[A01.A12.A23]+A01.A12.A23.A34.eps4.
                 Inverse[A01.A12.A23.A34]];
             "Twist";
             tw={Et[[1,4]],Et[[2,4]],Et[[2,1]]};

             "Jacobian";

             Jt=Transpose[{D[tw,dx0],D[tw,dy0],D[tw,dph0],
             D[tw,dx1],D[tw,dy1],D[tw,dph1],D[tw,dx2],D[tw,dy2],
             D[tw,dph2],D[tw,dx3],D[tw,dy3],D[tw,dph3],
             D[tw,dx4],D[tw,dy4],D[tw,dph4]}];
```

A2.2.7 Diagnostic procedure: Computation of Eq. (10.174)

```
In[1]:=      "Structural matrix";
             js={{Cos[0], Sin[0],0,0,0,1},
```

```
{Cos[Pi/3], Sin[Pi/3],0,0,0,1},
{Cos[2Pi/3], Sin[2Pi/3],0,0,0,1},
{Cos[3Pi/3], Sin[3Pi/3],0,0,0,1},
{Cos[4Pi/3], Sin[4Pi/3],0,0,0,1},
{Cos[5Pi/3], Sin[5Pi/3],0,0,0,1},
{Cos[0], Sin[0],-100 Sin[0],100 Cos[0],100,1},
{Cos[Pi/3],Sin[Pi/3],100 Sin[Pi/3],100
   Cos[Pi/3],100,1},
{Cos[2Pi/3],Sin[2Pi/3],-100 Sin[2Pi/3],100
   Sin[2Pi/3],100,1},
{Cos[3Pi/3],Sin[3Pi/3],-100 Sin[3Pi/3],100
   Cos[3Pi/3], 100,1},
{Cos[4Pi/3],Sin[4Pi/3],-100 Sin[4Pi/3],100
   Cos[4Pi/3], 100,1},
{Cos[5Pi/3],Sin[5Pi/3],-100 Sin[5Pi/3],100
   Sin[5Pi/3],100,1}};

"Results of measurements, Eq. (10.172)";
meas={0.01,0.02,0.02,0.03,0.02,0.01,0.02,0.03,0.03,
0.03,0.02,0.02};

"Estimates Eq. (10.174)"
Inverse[N[Transpose[js]].js].N[Transpose[js]].meas

Out[n] := {-0.0057578, .00288675,-0.0000310551,-0.000002773,
        0.00006667,0.0183333}
```

A2.2.8 The Fichter's singular configuration of the 3-3 GSP matrix (Fig. 11.13)

The Jacobian matrix for Fichter's singular configuration is built. A part of the code is identical to that in Appendix A.2.2.5. The demonstrated results are the determinant graph of the Jacobian matrix vs. angle φ(see Out[65]) and zero value of the determinant for $\varphi = \pi/2$ (see Out[66]).

```
In[1]:= "Fichter singularity of the 3-3 GPS";

  <<matr;
    A01=m1.m2.m3.m4.m5.m6;

  "Lower ends of limbs";

  rM1={r1 Cos[al1],r1 Sin[al1],0,1};
  rM2={r1 Cos[al2],r1 Sin[al2],0,1};
  rM3={r1 Cos[al3],r1 Sin[al3],0,1};
  rM4={r1 Cos[al4],r1 Sin[al4],0,1};
  rM5={r1 Cos[al5],r1 Sin[al5],0,1};
```

```
rM6={r1 Cos[al6],r1 Sin[al6],0,1};

"Upper ends of limbs in S1";
rQ1={r2 Cos[bet1],r2 Sin[bet1],0,1};
rQ2={r2 Cos[bet2],r2 Sin[bet2],0,1};
rQ3={r2 Cos[bet3],r2 Sin[bet3],0,1};
rQ4={r2 Cos[bet4],r2 Sin[bet4],0,1};
rQ5={r2 Cos[bet5],r2 Sin[bet5],0,1};
rQ6={r2 Cos[bet6],r2 Sin[bet6],0,1};

"Upper ends of limbs in S0";
rU1=A01.rQ1;
rU2=A01.rQ2;
rU3=A01.rQ3;
rU4=A01.rQ4;
rU5=A01.rQ5;
rU6=A01.rQ6;

"Limb vectors";
l1=rU1-rM1;
l2=rU2-rM2;
l3=rU3-rM3;
l4=rU4-rM4;
l5=rU5-rM5;
l6=rU6-rM6;

"Limb lengths";
L1=(l1.l1)^(1/2);
L2=(l2.l2)^(1/2);
L3=(l3.l3)^(1/2);
L4=(l4.l4)^(1/2);
L5=(l5.l5)^(1/2);
L6=(l6.l6)^(1/2);

"Limb orts";
e1=l1/L1;
e2=l2/L2;
e3=l3/L3;
e4=l4/L4;
e5=l5/L5;
e6=l6/L6;

"Limb arms";
arm1=A01.rQ1/.zz-> 0;
arm2=A01.rQ2/.zz-> 0;
```

```
  arm3=A01.rQ3/.zz-> 0;
  arm4=A01.rQ4/.zz-> 0;
  arm5=A01.rQ5/.zz-> 0;
  arm6=A01.rQ6/.zz-> 0;

  "Ort moments";
  n1=Cross[T4to3.arm1, T4to3.e1];
  n2=Cross[T4to3.arm2, T4to3.e2];
  n3=Cross[T4to3.arm3, T4to3.e3];
  n4=Cross[T4to3.arm4, T4to3.e4];
  n5=Cross[T4to3.arm5, T4to3.e5];
  n6=Cross[T4to3.arm6, T4to3.e6];

  "Jacobian matrix synthesis"
  JJ={{e1[[1]],e1[[2]],e1[[3]],n1[[1]],n1[[2]],n1[[3]]},
     {e2[[1]],e2[[2]],e2[[3]],n2[[1]],n2[[2]],n2[[3]]},
     {e3[[1]],e3[[2]],e3[[3]],n3[[1]],n3[[2]],n3[[3]]},
     {e4[[1]],e4[[2]],e4[[3]],n4[[1]],n4[[2]],n4[[3]]},
     {e5[[1]],e5[[2]],e5[[3]],n5[[1]],n5[[2]],n5[[3]]},
     {e6[[1]],e6[[2]],e6[[3]],n6[[1]],n6[[2]],n6[[3]]}};

cond = {r1-> 0.7, r2-> 0.25, al1-> -Pi/3, al2-> Pi/3, al3->
Pi/3, al4-> Pi, al5-> Pi, al6-> -Pi/3, bet1-> 0, bet2-> 0,
bet3-> 2Pi/3, bet4-> 2Pi/3, bet5-> 4Pi/3, bet6-> 4Pi/3, xx->
0, yy-> 0, zz-> 0.4, th-> 0, ps-> 0};

  JJphi=JJ/.cond;
  detJ=Det[JJphi];
  Plot[detJ,{ph,0,Pi/2+Pi/24},AxesLabel->{phi,J}]

Out[65]:=
```

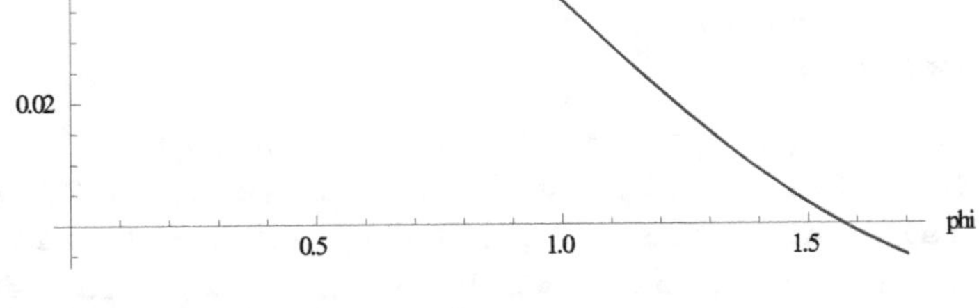


```
In[66]:= Chop[N[detJ/.ph-> Pi/2]]
Out[66]:= 0
```

Appendix 3: Some symbols and operations from the Mathematica® software system

A3.1 Global objects

A3.1.1 Input and output

`In[n]` is the *n*th input information entered by the user;
`Out[n]` is an output resulting from code processing generated by the software system.

A3.1.2 Algebraic symbols

Regular symbols "+, –, ×, /, and ()" are used for algebraic operations on scalar values and functions; "=" is the assignment symbol;

The power symbol is "^", i.e., x^y means x^y;

Dot " . " is the symbol of product of matrices, matrices by vectors, and the dot-product (scalar product) of two vectors.

Attention:

- instead of product symbol "×", the blank *is regularly used* for multiplication of scalars, i.e., $a \times b \equiv a\ b$;
- square brackets are used for giving parameter/s of commands and functions;
- braces are used a list of elements, vectors and matrices.

A3.2 Matrix and Vector Synthesis

- *Vectors of the third order*

```
In[n]:= r = {x, y, z}
Out[n]:= {x, y, z}
```

where "=" is the assignment symbol.

- *Position vector of the fourth order*

```
In[n]:= r = {x, y, z, 1}
Out[n]:= {x, y, z, 1}
```

- *Non-position vector of the forth order*

```
In[n]:= r = {x, y, z, 0}
Out[n]:= {x, y, z, 0}
```

- *Matrix of the 4×4 order (an expression in the braces is the matrix row)*

```
In[n]:=     A = {{a11, a12, a13, a14}, {a11, a12, a13, a14},
            {a11, a12, a13, a14}, {a11, a12, a13, a14}}
Out[n]:=    {{a11, a12, a13, a14}, {a11, a12, a13, a14}, {a11,
            a12, a13, a14}, {a11, a12, a13, a14}}
```

A3.3 Operations on Vectors and Matrices

- *Sum of the third order vectors*

```
In[n]:=     r1 = {x1, y1, z1};
            r2 = {x2, y2, z2};
            rs = r1 + r2
Out[n]:=    {x1 + x2, y1 + y2, z1 + z2}
```

- *Sum of position and non-position vectors*

```
In[n]:=     r1 = {x1, y1, z1, 1}
            r2 = {x2, y2, z2, 0}
            rs = r1 + r2
Out[n]:=    {x1 + x2, y1 + y2, z1 + z2, 1}
```

- Sum of two non-position vectors

```
In[n]:=     r1 = {x1, y1, z1, 0}
            r2 = {x2, y2, z2, 0}
            rs = r1 + r2
Out[n]:=    {x1 + x2, y1 + y2, z1 + z2, 0}
```

- *Sum of the matrices (e.g., of the 4×4 order)*

```
In[n]:=     A = {{a11, a12, a13, a14}, {a21, a22, a23, a24},
            {a31, a32, a33, a34},{a41, a42, a43,a44}};
            B = {{b11, b12, b13, b14}, {b21, b22, b23, b24},
            {b31, b32, b33, b34}, {b41, b42, b43, b44}};
            AS = A + B
Out[n]:=    {{a11 + b11, a12 + b12, a13 + b13, a14 + b14}, {a21
            + b21, a22 + b22, a23 + b23, a24 + b24}, {a31 +
            b31, a32 + b32, a33 + b33, a34 + b34}, {a41 + b41,
            a42 + b42, a43 + b43, a44 + b44}}
```

- *Product of two real or numbers*

```
In[n]:=     x = a;
            y = b;
            x y
Out[n]:=    a b
```

- *Dot product of vectors*

```
In[n]:=     r1 = {x1, y1, z1};
            r2 = {x2, y2, z2};
            rp = r1.r2
Out[n]:=    x1 x2 + y1 y2 + z1 z2
```

- *Vector (cross) product of two vectors of the third order*

```
In[n]:=     r1 = {x1, y1, z1};
            r2 = {x2, y2, z2};
            rp = Cross[r1,r2]
Out[n]:=    {-y2 z1 + y1 z2, x2 z1 - x1 z2, - x2 y1 + x1 y2}
```

- *Product of the matrix and vector*

```
In[n]:=     A = {{a11, a12, a13, a14}, {a21, a22, a23, a24},
            {a31, a32,a33, a34}, {a41, a42, a43, a44}};
            r1 = {x, y, z, v};
            r = A.r1
Out[n]:=    {a14 v + a11 x + a12 y + a13 z, a24 v+ a21 x + a22
            y + a23 z, a34 v + a31 x + a32 y + a33 z, a44 v+ a4
            1x + a42 y + a43 z}
```

- *Product of two matrices (e.g., the 3×4 matrix by 4×2 matrix)*

```
In[n]:=     A={{a11,a12,a13,a14},{a21,a22,a23,a24},{a31,a32,a33,
            a34}};
            B={{b11,b12},{b21,b22},{b31,b32},{b41,b42}};
            A.B
Out[n]:=
            {{a11 b11 + a12 b21 + a13 b31 + a14 b41,
            a11 b12 + a12 b22 + a13 b32 + a14 b42},
            {a21 b11 + a22 b21 + a23 b31 + a24 b41,
            a21 b12 + a12 b22 + a23 b32 + a24 b42},
            {a31 b11 + a32 b21 + a33 b31 + a34 b41,
            a31 b12 + a32 b22 + a33 b32 + a34 b42}}
```

- *Demonstration*
    - Symbol *expr;* evaluates expr, but ignores demonstration of the result in the Output.
    - Symbol "text" ignores evaluation of the text and demonstrates it as is.
    - Command MatrixForm[*matr*] demonstrates matrix *matr* in the regular matrix form:

```
In[n]:=     matr={{a11, a12}, {a21, a22}};
            MatrixForm[matr]
```

Out[n]:= $\begin{bmatrix} a11 & a12 \\ a21 & a22 \end{bmatrix}$.

Abbreviations

AE – active element of the FSS

CMM – coordinate measuring machine

CNC – computer numerical control

DC – dimensional chain

DOF – degree of freedom

EE – end-effector

FSF – form-shaping function

FSS – form-shaping system

GSP – Gough-Stewart platform

HTM – homogeneous transformation matrix

IR – industrial robot

IT – international tolerance

LMS – least-mean square

LVDT – linear variable differential transformer

MinZ – minimum zone

MT – machine tool

NC – numerical control

PWM – precision working machine

PKM – parallel-kinematics machine

SKM – serial-kinematics machine

TE – target element of the FSS

WM – working machine

Index

C

D

E

F

G

N

O

P

Q

R

T

U

V

W

X

Y

Z

About the Author

Prof. Vladimir T. Portman (Sc.D. 1988, PhD 1970, PE), Professor Emeritus, Department of Mechanical Engineering, Ben-Gurion University of the Negev (BGU), Beer-Sheva, Israel, Fellow (Em.) of the International Academy for Production Engineering (CIRP). After MSc graduation at Moscow State Technological University (STANKIN), Moscow, Russia, in 1958, he spent seven years as a designer in Production Technological Bureau in Moscow, and then joined the Experimental Research Institute for Metal Cutting Machine Tools (ENIMS), Moscow, Russia, where served as a researcher and in 1979 – 1990 as a head of the Laboratory for Computer Aided Calculations of Machine Tools and FMS.

After repatriation in Israel in 1990, he has been served as Professor of Machine Design and Mechanism and Machine Theory at BGU.

Professor Portman published over 200 papers, books, patents, and other academic works in the field of parallel kinematic machines, accuracy of machine tools and robots, precision machining, dimensional metrology, and FMS simulation. The most important work is the authored book in common with Prof. D. N. Reshetov – monograph *Accuracy of Machine Tools,* issued in 1986 in Russian by Mashinostorenie, Moscow, former USSR, and translated in English in 1988 (additional issues 1989, 1991, 1992) by ASME Press, New York, NY.

www.ingramcontent.com/pod-product-compliance
Lightning Source LLC
LaVergne TN
LVHW062244200926
848484LV00006B/1611